Physical Chemistry of Foods

ift Basic Symposium Series

Edited by
INSTITUTE OF FOOD TECHNOLOGISTS
221 N. LaSalle St.
Chicago, Illinois

1. Foodborne Microorganisms and Their Toxins: Developing Methodology, *edited by Merle D. Pierson and Norman J. Stern*
2. Water Activity: Theory and Applications to Food, *edited by Louis B. Rockland and Larry Beuchat*
3. Nutrient Interactions, *edited by C. E. Bodwell and John W. Erdman, Jr.*
4. Food Toxicology: A Perspective on the Relative Risks, *edited by Steven L. Taylor and Richard A. Scanlan*
5. Biotechnology and Food Process Engineering, *edited by Henry G. Schwartzberg and M. A. Rao*
6. Sensory Science Theory and Applications in Foods, *edited by Harry T. Lawless and Barbara P. Klein*
7. Physical Chemistry of Foods, *edited by Henry G. Schwartzberg and Richard W. Hartel*

Physical Chemistry of Foods

edited by

Henry G. Schwartzberg

Department of Food Science
University of Massachusetts
Amherst, Massachusetts

Richard W. Hartel

Department of Food Science
University of Wisconsin
Madison, Wisconsin

Marcel Dekker, Inc. New York • Basel • Hong Kong

Library of Congress Cataloging-in-Publication Data

Physical chemistry of foods / edited by Henry G. Schwartzberg, Richard W. Hartel.
p. cm. -- (IFT basic symposium series)
Includes bibliographical references and index.
ISBN 0-8247-8693-9
1. Food--Composition--Congresses. I. Schwartzberg, Henry . II. Hartel, Richard W. III. Series.
TP372.5.P495 1992
664--dc20 92-10185
CIP

This book is printed on acid-free paper.

MARCEL DEKKER, INC.
270 Madison Avenue, New York, New York 10016

Current printing (last digit):
10 9 8 7 6 5 4 3 2 1

PRINTED IN THE UNITED STATES OF AMERICA

Preface

Physical chemistry of foods is an important area of food science. In the United States it has been relatively neglected, but appreciation of its importance is growing. In Europe, the Food Chemistry Group of the (British) Royal Society of Chemistry organizes a two-day symposium on physicochemical aspects of foods every two years. Similar meetings have been held by the International Association of Colloid and Interface Scientists, and intensive short courses dealing with particular aspects of the subject are frequently given.

Each year the Institute of Food Technologists (IFT) and the International Union of Food Science and Technology sponsor a two-day Basic Symposium held in conjunction with the IFT annual meeting. The symposium is designed to cover new, basic developments in a major field in food science. Physical Chemistry of Foods, the 15th symposium in that series, took place May 31 and June 1, 1991, in Dallas before the IFT's 51st Annual Meeting and was designed to help meet demands for teaching and reference material dealing with the physical chemistry of foods. The chapters in this book deal with the major topics covered during that symposium. Four areas were covered: phase equilibria, reaction kinetics, physical chemistry of gels and disperse systems, and physicochemical effects occurring during consumption and use.

Physicochemical phenomena govern rates at which reactions occur in foods; phase equilibria (e.g., melting, freezing, crystallization, ingredient solubility); processes regulated by diffusion and flow in glassy and rubbery phases in foods; the release, availability, and recoverability of aromas and flavors; sensory processes; uptake and absorption of nutrients; rheological and textural behavior; functional properties of food ingredients; the behavior and stability of food dispersions; the formation and properties of gels; and the protective nature of packaging and food coatings.

A wide range of problems can be dealt with using physicochemical approaches, for example, selecting temperatures and pressures for supercritical extraction; controlling crystal size distributions; ensuring emulsion and foam stability; selecting emulsifying agents; choosing solutes that provide adequate levels of fluidity in semi-frozen slushes or that retard shrinkage during the cooling of baked goods; developing cellular structures in gels; manipulating and controlling puffing during high-intensity drying; providing delayed or regulated gas, flavor, or functional ingredient release; improving powder dispersibility; limiting the caking of powder; and controlling friability.

As the chapters in this book show, physical chemistry provides not only useful bases for understanding physical and chemical phenomena in foods, it also provides tools for manipulating and controlling those phenomena, and for creating and upgrading food products and processes.

Food science is an eclectic discipline. Therefore, it is fitting that the physical chemistry of foods should draw broadly from recent developments in polymer science; colloid, membrane, and surface chemistry and physics; molecular thermodynamics; neurology; and biochemistry. Our ability to measure and analyze physicochemical behavior in foods has also been greatly enhanced by recent advances in instrumentation and machine-based computation.

We acknowledge the help provided by present and former Basic Symposium Committee members: Barbara P. Klein (Chair), University of Illinois; Cavit Akin, Institute of Gas Technology; Allen V. Clark, the Coca Cola Company; Ron Richter, Texas A&M University; Peter Bluestein, Thomas J. Lipton, Inc.; Nevam Hettiararchchy, North Dakota State University; Merle Pierson, Virginia Polytechnic Institute and State University; Patricia Ann Kendall, Colorado State University; Les Smoot, ABC Research Corp.; Rakesh Singh, Purdue University; and Warren A. Widicus, Quaker Oats Co., all of whom contributed greatly to setting up the symposium.

The symposium organizers thank Dr. Daryl B. Lund, 1990–1991 IFT President, for his enthusiastic support and also thank Howard W. Mattson, IFT Executive Director; John B. Klis, Director of Publications; Anna May Schenck, Associate Scientific Editor; and other IFT staff members who provided support and coordination for the symposium.

Most of all, we want to thank the speakers for their contributions. Without their dedication, expertise, and hard work, timely publication of these proceedings would not have been possible.

HENRY G. SCHWARTZBERG
RICHARD W. HARTEL

Contributors

José Miguel Aguilera, Ph.D. Professor and Head, Department of Chemical Engineering, Catholic University, Santiago, Chile

Edward B. Bagley, Ph.D. Research Leader, Food Physical Chemistry Research, National Center for Agricultural Utilization Research, Agricultural Research Service, U.S. Department of Agriculture, Peoria, Illinois

Wendy M. Baisier, M.S. Research Assistant, Department of Food Science and Nutrition, University of Minnesota, St. Paul, Minnesota

Joseph G. Brand, Ph.D. Associate Director and Professor of Biochemistry, Monell Chemical Senses Center, Veterans Affairs Medical Center, and University of Pennsylvania, Philadelphia, Pennsylvania

Allan H. Clark, Ph.D. Senior Research Scientist, Department of Biomaterials Science, Unilever Research Laboratory, Sharnbrook, Bedford, England

E. Allen Foegeding, Ph.D. Associate Professor, Department of Food Science, North Carolina State University, Raleigh, North Carolina

Dennis T. Gordon, Ph.D. Professor, Department of Food Science and Human Nutrition, University of Missouri, Columbia, Missouri

Donald D. Hamann, Ph.D. Professor, Department of Food Science, North Carolina State University, Raleigh, North Carolina

Richard W. Hartel, Ph.D. Associate Professor of Food Engineering, Department of Food Science, University of Wisconsin, Madison, Wisconsin

R. Carl Hoseney, Ph.D. Professor, Department of Grain Science and Industry, Kansas State University, Manhattan, Kansas

Marcus Karel, Ph.D., Sc.D. h.c., State of New Jersey Professor of Food Science, Department of Food Science, Rutgers University, New Brunswick, New Jersey

Viwat Krisdhasima, Ph.D. Research Associate, Department of Bioresource Engineering, Oregon State University, Corvallis, Oregon

Theodore P. Labuza, Ph.D. Professor of Food Science and Technology, Department of Food Science and Nutrition, University of Minnesota, St. Paul, Minnesota

Marc Le Maguer, Ph.D. Professor and Chairman, Department of Food Science, University of Guelph, Guelph, Ontario, Canada

Harry Levine, Ph.D. Research Fellow, Fundamental Science Group, Nabisco Brands, Inc., East Hanover, New Jersey

Joseph McGuire, Ph.D., P.E. Assistant Professor, Department of Bioresource Engineering, Oregon State University, Corvallis, Oregon

Ganesan Narsimhan, Ph.D. Assistant Professor, Department of Agricultural Engineering, Purdue University, West Lafayette, Indiana

Micha Peleg, D.Sc. Professor of Food Engineering, Department of Food Science, University of Massachusetts, Amherst, Massachusetts

Lori Pellett, Ph.D. Research Assistant, Department of Food Science and Human Nutrition, University of Missouri, Columbia, Missouri

Pankaj S. Shah, Ph.D. Monell Chemical Senses Center, Philadelphia, Pennsylvania

Louise Slade, Ph.D. Research Fellow, Fundamental Science Group, Nabisco Brands, Inc., East Hanover, New Jersey

David W. Stanley, Ph.D. Professor, Department of Food Science, University of Guelph, Guelph, Ontario, Canada

Rickey Y. Yada, Ph.D. Associate Professor, Department of Food Science, University of Guelph, Guelph, Ontario, Canada

Contents

Physical Chemistry of Foods

1

Thermodynamics and Vapor-Liquid Equilibria

Marc Le Maguer

University of Guelph
Guelph, Ontario, Canada

INTRODUCTION

A major concern of food engineers is the design of equipment and processes for efficient conversion of large quantities of raw materials into finished products. In the design and operation of such processes, a large amount of reliable data on the equilibrium properties of materials is necessary. However, the limited availability of experimental data can hardly satisfy such an enormous demand. Engineers are thus required to rely primarily on a variety of methods for estimating physical properties of industrial interest.

Thermodynamics provides us with an efficient way to organize useful information from limited experimental measurements, and the prediction can be achieved by the use of a reliable and versatile thermodynamic model.

It is the purpose of this chapter to review and present the most recent developments in the area of vapor-liquid equilibrium for prediction of thermodynamic properties of liquid food solutions.

EXCESS GIBBS ENERGY AND THERMODYNAMIC MODELING OF SOLUTIONS

Introduction

Thermodynamics is an efficient tool which allows us to extract the maximum amount of information with a minimum amount of experimental effort. As Prausnitz (1977) once pointed out, "Thermodynamics is not magic but a tool; it tells us how the desired equilibrium property is related to some other equilibrium property." In other words, thermodynamics provides us with a time-saving bookkeeping system; only a few equilibrium properties need to be measured, the rest can be calculated. Thus, from an engineer's point of view, the main advantage of thermodynamics is that it reduces experimental effort.

Because of the basic thermodynamic relationships that exist between the excess Gibbs energy and other useful parameters, the Gibbs function serves as an accounting function which allows us to quantify and characterize the deviation from ideality of the system (Prausnitz, 1969), and through it all other thermodynamic and physical properties of the system can be derived (as shown in Fig. 1.1).

Thus, the best way to approach the prediction of the behavior of complex solutions is through the use of the excess Gibbs energy function to represent thermodynamic properties (Le Maguer, 1987). A model expressed in excess Gibbs energy can then be used to predict phase changes, water activity, activity coefficients of the solutes, and estimate partial equilibrium properties (Le Maguer, 1987, 1989) such as solubilities, relative volatilities and other much needed physicochemical properties for design purposes.

Review of Existing Models

Empirical representations of excess Gibbs energy: composition dependence. Generally, mathematical representation provides a convenient method to reduce extensive tables of experimental data into a few equations and a means to transmit data from the experimentalist to the engineer. As a result, many empirical expressions have been proposed. Some popular

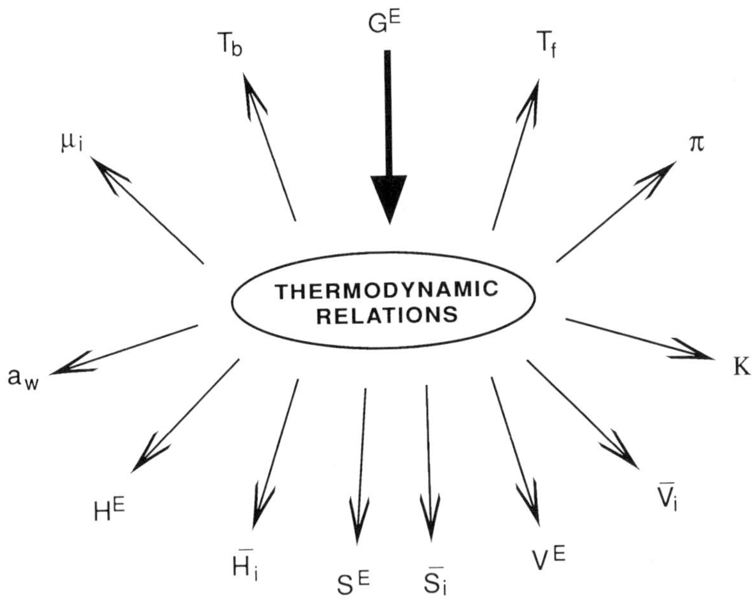

Fig. 1.1 Predictive power of the excess Gibbs energy.

expressions are those due to Margules, van Laar, Redlich-Kister, Scatchard-Hamer, and Wohl. These empirical expressions share a common purpose: to describe the nonideal behavior of the system investigated by fitting the experimental data through thermodynamically consistent correlations.

For a binary mixture, any expression for the excess Gibbs energy must obey the following two boundary conditions: $G^E = 0$ when $x_1 = 0$; $G^E = 0$ when $x_2 = 0$. The simplest polynomial expression satisfying these boundary conditions is $G^E = Ax_1x_2$, where A is an empirical parameter which is a complicated function of macroscopic and molecular properties, and is thus difficult to estimate a priori, but has to be determined experimentally.

The equation obtained is called the two-suffix Margules equation. This simple expression, however, can only represent liquid mixtures containing constituents of similar size, shape, and chemical nature (Sandler, 1977). More complex expressions are thus required for more complex mixtures.

One convenient extension would be:

$$G^E = x_1x_2\,[A + B\,(x_1 - x_2) + C\,(x_1 - x_2)^2 + \cdots]$$

where A, B, and C are empirical parameters. This equation is commonly known as the Redlich-Kister expansion (Redlich and Kister, 1948).

The Redlich-Kister expansion provides a flexible algebraic expression for representing the excess Gibbs energy of a complex liquid mixture and gives an indication of the apparent complexity of the mixture (Prausnitz et al., 1986) as accuracy of the experimental data and the fit desired. Other similar expressions are the n-suffix Margules equations.

Another general expression is Wohl's expansion. Wohl (1946) developed expressions for the excess Gibbs energy involving weighted mole fractions (called q-fractions) and interaction terms up to a four-suffix (a_{hijk}) term. One of the main advantages of this method is that some rough physical significance can be assigned to the parameters that appear in the equations; as a result, Wohl's expansion can be extended systematically to multicomponent solutions.

All these equations share the common feature of expressing the relative frequency of a particular grouping in terms of the products of the effective mole fractions of the molecular species involved (Chao and Greenkorn, 1975). Implied in this expression is that the molecular collisions are random and that no modes are preferred. Chao and Greenkorn (1975) thus described these empirical equations as the "random-solution" equations to distinguish them from equations that are based on the concept of nonrandomness.

Although the equations mentioned are frequently quite useful for correlating experimental data, or for making predictions (in a very limited sense) with a limited amount of experimental information, they are of little value in making predictions when no experimental data are available (Sandler, 1977). This is because their basis has been completely empirical, and little physical significance has been attributed to the parameters in these equations.

Furthermore, all random-solution equations are unable to predict multicomponent solution behavior from binary solution data alone. Multicomponent solution data are required for the evaluation of the higher interaction constants. This is rarely possible because of the scarcity of multicomponent solution data.

Theory of Van Laar. A large number of semi-empirical models have been developed to explain the thermodynamic properties of binary mixtures on a quantitative basis in the past decades. One of the first systematic attempts to quantitatively describe the properties of fluid mixtures was made by van der Waals and his coworkers early in the twentieth century; however, the significant contributions were made by van Laar, a student of van der Waals. Van Laar's work is still recognized today primarily because he chose good simplifying assumptions that made the problem solvable and yet did not greatly violate physical reality.

The theory of van Laar (1910, 1913) is based on the assumptions that (1) the binary mixture is composed of two species of similar size and energies of interaction, and (2) the van der Waals equation of state applies both to pure fluids and the binary mixture. The implication of assumption (1) is that the molecules of each species will be uniformly distributed throughout the mixture and the intermolecular spacing will be quite similar to that in the pure fluids. Consequently, at a given temperature and pressure, $V^E = 0$ and $S^E = 0$ so that $G^E = U^E$. These simplifying assumptions allow the computation of U^E along the convenient path chosen by van Laar.

Despite the simplicity and elegance of the derivation of the van Laar equations, they are almost inapplicable to most real solutions, particularly polar nonideal mixtures. Quantitative agreement between van Laar's equations and experimental data is not good. This poor agreement is mainly attributed to the use of the van der Waals equation of state and the mixing rules proposed by van der Waals (Acree, 1984), in addition to the simplifying assumptions.

Regular solution theory. Hildebrand (1929) first proposed the "regular solution" to describe solutions characterized by the absence of any specific interactions between molecules, such as association, strong dipole interactions, and chemical effects, and in which the distributions and orientations are random (Hildebrand and Scott, 1962; Hildebrand et al., 1970).

The theory of regular solutions is, in fact, an extension of the van Laar equation. The extension leads to an expression of the equation constants in terms of pure component properties, thus making it possible to make estimates of slight and moderate departures from ideal solution behavior from pure component properties.

Because of the simplicity and generality of the regular solution equations, they are widely used for hydrocarbon solutions. If regular solution calculations are compared with experimental vapor-liquid equilibrium data, the deviations are found to be generally within reasonable bounds for hydrocarbon solutions (Chao and Greenkorn, 1975). Despite good results for the Gibbs energy of hydrocarbon solutions, the equations give poor results for entropy and enthalpy. The results for the Gibbs energy are probably due to an apparent cancellation of errors in the entropy and enthalpy (Chao and Greenkorn, 1975). Furthermore, the extension of regular solution theory to multicomponent mixtures is an algebraically messy task (Sandler, 1977).

Lattice theory and related models. Most current work in the theory of solution utilizes the powerful methods of statistical mechanics which relate macroscopic (bulk) properties to microscopic (molecular) phenomena. Molecular considerations suggest that deviations from ideal behavior in liquid solutions are due primarily to the intermolecular forces (giving rise to a nonvanishing enthalpy of mixing), as well as the arrangements and orientations of unlike molecules in the mixture (giving rise to a nonideal entropy of mixing) (Prausnitz et al., 1986). As a result, recent models usually aim at describing the properties of solutions in terms of intermolecular forces and fundamental liquid structure.

One popular approach to solution theory considers a liquid mixture as quasi-crystalline (Guggenheim, 1952), i.e., it treats a liquid mixture as if the molecules were arranged on a lattice. In this the molecules do not translate fully in a chaotic manner as in a gas, but each molecule tends to stay in a small region, a more or less fixed position in space about which it vibrates back and forth (Prausnitz et al., 1986). Hildebrand has raised a number of arguments against the lattice model (Hildebrand and Scott, 1962). Nevertheless, a majority of investigators feel that the lattice model, though it may not truly represent the real liquid structure, gives a practical solution to the problem. Some facts tend to justify this "pro–lattice-theory" view: analysis by x-rays has shown that the mutual agreement of neighboring molecules in liquids is much like that in a solid (Guggenheim, 1952; Prigogine et al., 1957); the available volume per particle in the liquid phase near the melting point is only slightly (about 10%) larger than in the solid (Prigogine et al., 1957); at temperatures well below the critical temperature, the number of nearest neighbors has a fairly well-defined average value and, although there are fluctuations about this average, these fluctuations are not large and the geometrical relationship of each molecule to its immediate neighbors is on the average very similar to that in a crystal (Guggenheim, 1952). But x-ray data on liquids also indicate that there exists in liquids a local order, but no long-range order as in crystals (Eyring and Marchi, 1963).

Using the concept of a quasi-crystalline lattice as a model for liquid, Guggenheim (1935) devised an approximate treatment for regular solutions known as the quasi-chemical method. In this treatment, Guggenheim treated the molecules of the two components of a binary solution as spheres of approximately the same size with short-range fields of force.

When considering solutions containing molecules of differing sizes, it has been found advantageous to assume, at least as a first approximation, that the enthalpy of mixing is zero and to concentrate on the entropic contributions (Prausnitz et al., 1986). Solutions having zero enthalpy of mixing are referred to as athermal solutions (Prigogine and Defay, 1954). Flory (1942) and Huggins (1941, 1942) independently derived an expression for the entropy of mixing in an athermal solution containing molecules of different sizes by visualizing the molecules as occupying different numbers of contiguous sites in a liquid lattice. The derivation has been clearly presented in several references (e.g., Flory, 1953; Fast, 1970).

To apply the theoretical result of Flory and Huggins to real polymer solutions, it becomes necessary to add to the entropy computed a semiempirical term for the enthalpy of mixing. The form of this term is the same as that used in the Scatchard-Hildebrand theory of solutions (or the regular solution theory); the excess enthalpy is proportional to the volume of the solution and to the product of the volume fractions (Prausnitz et al., 1986).

The addition of an enthalpic term to the theoretical result is indeed an empirical modification required to obtain a reasonable expression for the Gibbs energy of mixing. Nevertheless, calorimetric studies indicate that the enthalpic term added does not properly describe the enthalpy of mixing of the solution (Prausnitz et al., 1986). Thus, though this relatively simple theory contains most of the essential features that distinguish solutions of large molecules from those containing molecules of approximately equal sizes, the Flory-Huggins equation for real polymer solutions is not an accurate description of the thermodynamic properties of such systems.

Local composition models. The concept of local composition has been widely applied in the study of phase equilibrium for mixtures and used in the formulation of excess free energy equations. Important examples are the Wilson equation (Wilson, 1964), the NRTL equation of Renon and Prausnitz (1968), and the UNIQUAC equation of Abrams and Prausnitz (1975).

The use of local composition is very appealing because it is an aspect of the structure of a liquid mixture that can be easily expressed in terms of intermolecular correlation functions and is thus amenable to statistical mechanical calculations (Seaton and Glandt, 1985). The following sections briefly review some successful local composition models and the underlying theories.

The Wilson Equation. Expecting a random arrangement of the molecules in solution is really a postulate. Indeed, it is more realistic to expect that the different sizes and forms of the molecules, as well as their interactions with each other or their segments, will interfere with their completely random arrangement. The quantitative description of such local inhomogeneity is achieved by the use of the local composition concept in which local mole fractions are defined. The concept is well illustrated by Cukor and Prausnitz (1969).

The introduction of the local composition concept has resulted in the development of many flexible and powerful models for excess Gibbs energy (Rasmussen, 1983). Wilson (1964) obtained an expression for representing G^E by considering the preferential orientation of molecules due to energy effects. He postulated that the local concentrations in the neighborhood of a central molecule differ from the bulk concentrations because of the competition among the various molecular species in their interactions with the central molecule.

In the Wilson equation, the free energy of mixing is assumed to be given by a relation similar to the athermal Flory-Huggins equation, except that the overall solution volume fractions are replaced by local volume fractions which are related to local molecule segregations caused by differing energies of interaction between pairs of molecules. And the distribution of molecules about a central molecule is assumed to be expressible in terms of Boltzmann factors. Based on these two assumptions, the equation is derived (Wilson, 1964).

An important aspect of the Wilson equation is that only binary interactions are considered, and the equation constants, including all those in the equations for multicomponent mixtures, can be determined by fitting binary-solution data only (Chao and Greenkorn, 1975). However, the usefulness of the Wilson equation is limited to completely miscible systems (Prausnitz, 1969).

Two-Liquid Theory and the NRTL Equation. The two-liquid theory (or the two-fluid model) provides a useful point of departure for deriving local composition equations to represent thermodynamic excess functions for highly nonideal systems (Prausnitz et al., 1986). The two-fluid model was devised independently by Scott (1956), Brown (1957) (who called it a semirandom approximation), and Prigogine (1957) (who called it the "second refined version of the average potential model"). It was introduced to allow for the departures of a real mixture from the randomness imposed by the one-fluid model.

Based on Scott's two-liquid theory, Renon and Prausnitz (1968) further developed Wilson's concept of local concentrations with the view of obtaining more universal expressions for liquid-phase activity coefficients. To define the local composition, Renon and Prausnitz made an assumption similar to that of the quasichemical theory of Guggenheim (1952). But unlike Wilson's model, the local composition is expressed in terms of mole fractions rather than volume fractions. Instead of using the Flory-

Huggins equation as in Wilson's model, Renon and Prausnitz used Scott's two-liquid theory of binary mixtures to obtain an expression for the excess Gibbs energy. The expression thus obtained is called the NRTL (non-random, two-liquid) equation.

One advantage that the NRTL equation offers over the Wilson equation is that the equation is applicable to partially miscible as well as completely miscible systems. However, the NRTL equation contains three adjustable parameters for each binary system. This complicates the calculations, particularly when the equation is extended to multicomponent system calculations.

The UNIQUAC Equation and Corresponding Models. In an attempt to develop a versatile model with fewer adjustable parameters, yet a more theoretical basis, Abrams and Prausnitz (1975) used statistical mechanics to derive an expression called UNIQUAC (universal quasi-chemical) for the excess Gibbs energy. The UNIQUAC model generalizes the quasi-chemical lattice theory developed by Guggenheim (1952) and extends it to mixtures of molecules that differ appreciably in size and shape. As in the Wilson and NRTL equations, local concentrations are used. However, rather than local volume fractions or local mole fractions, UNIQUAC uses the local area fractions as the primary concentration variable.

Because of its stronger theoretical base, the resulting equation, which uses two adjustable parameters per binary pair, is applicable to multi-component mixtures of nonpolar and polar liquids. Theoretically, reliable estimates can be made of both vapor-liquid and liquid-liquid equilibria using the same equation for the excess Gibbs energy (Abrams and Prausnitz, 1975).

As compared to the previous local composition models, UNIQUAC has three advantages: (1) it has only two (rather than three) adjustable parameters: (2) because of its better theoretical basis, UNIQUAC's parameters often have a smaller dependence on temperature; and (3) because the primary concentration variable is a surface fraction (rather than mole fraction), UNIQUAC is applicable to solutions containing small or large molecules, including polymers (Reid et al., 1977). All these are very important for engineering applications.

Applicability and Limitations of the Wilson, NRTL, and UNIQUAC Equations to Nonideal Systems. Since the development of the local composition models, the applicability and performance of the Wilson, NRTL, and UNIQUAC equations have been evaluated extensively. It can be demonstrated that, in weakly nonideal systems, the differences between the individual equations for the dependence of G^E on the composition are very small and approach zero as the system approaches ideality (Reid et al., 1977; Novak et al., 1987). Thus, there is little reason to choose one over another for these systems.

A study of the Wilson equation by Orye and Prausnitz (1965) showed that for approximately 100 miscible binary mixtures of various chemical types, activity coefficients could be adequately described by the Wilson equation. Its precision is approximately the same as the four or five constant Redlich-Kister equations (Acree, 1984). However, there are two major limitations of the Wilson equation. First, the equation is not applicable to a mixture which exhibits a miscibility gap; it is inherently unable, even qualitatively, to account for phase splitting (Reid et al., 1977). Hence, it cannot predict liquid-liquid equilibrium. Second, it is unable to predict maxima and minima in the activity coefficient–mole fraction relationship (Henley and Seader, 1981). Unlike Wilson's equation, the NRTL and UNIQUAC equations are applicable to both vapor-liquid and liquid-liquid equilibria (Reid et al., 1977).

The accuracy of the three models (Wilson, NRTL, and UNIQUAC) in phase equilibrium calculations has been studied. Monfort and Rojasr (1978) reported that vapor-liquid calculations for 34 binary systems and 9 ternary systems show a slight superiority for the UNIQUAC equation in the representation of experimental data.

As mentioned, of the three solution models, only the NRTL and UNIQUAC models predict liquid-liquid equilibrium. In a series of investigations, Sorensen and coworkers (1979a,b, 1980) evaluated the NRTL and UNIQUAC equations with respect to their ability to represent binary and ternary liquid-liquid equilibrium data. They concluded that the UNIQUAC equation appears to be more convenient to use and to correlate the data slightly better than does NRTL with the same number of parameters. Monfort and Rojasr (1978) reported similar results in liquid-liquid calculations.

The capability of the three models to represent the excess enthalpy function has been investigated. As reported by Vonka et al. (1975) and Hanks et al. (1978), the Wilson equation cannot exceed certain positive values of H^E/RT, so that the H^E-fit is poor for systems showing strong endothermic behavior, while the UNIQUAC and NRTL equations are able to represent the experimental H^E data satisfactorily for those systems (Gmehling and Kolbe, 1983). For systems showing strong exothermic behavior, the H^E-representation is satisfactory for all three equations (Gmehling and Kolbe, 1983). According to Gmehling and Kolbe, for all the three models studied, the NRTL gives a better fit of the experimental data because of the presence of three adjustable parameters. The three parameters, however, are strongly correlated, so that several solutions are obtained depending on the initial values taken (Gmehling and Kolbe,

1983). Both the NRTL and UNIQUAC equations are able to describe different signs for different compositions.

In general, the UNIQUAC model can mathematically represent thermodynamic excess properties with a reasonable degree of accuracy, and it has turned out to be a flexible and reliable model which allows satisfactory representation of vapor-liquid equilibria (VLE) and liquid-liquid equilibria (LLE). Nevertheless, previous studies (Anderson and Prausnitz, 1978; Nath and Bender, 1981; Brandani and Evangelista, 1984; Zou and Prausnitz, 1987) have shown that the original UNIQUAC model fails to give an accurate representation of the thermodynamic properties of highly nonideal polar mixtures, particularly water-containing systems. It has also been found that the model cannot in general represent the excess Gibbs energy and excess enthalpy simultaneously with satisfactory results (Skjold-Jorgensen et al., 1980; Renon, 1985). Moreover, a precise simultaneous representation of VLE and LLE is not always possible (Malanowski et al., 1981; Skjold-Jorgensen et al., 1982).

Modifications of the UNIQUAC Equation. The flexibility and reliability of the UNIQUAC equation in the representation of both the vapor-liquid and liquid-liquid equilibria of a large number of binary systems, along with its straightforward extension to multicomponent systems, have prompted a number of researchers to attempt: (1) to transform UNIQUAC into group contribution models—Fredenslund et al. (1975, 1977) developed the UNIFAC model, a group contribution model based on the UNIQUAC method; (2) to apply UNIQUAC to electrolyte solutions—Christensen and coworkers (1983) proposed a model for aqueous electrolyte systems based on an extension of the UNIQUAC equation; and (3) to apply the underlying concepts to derive similar expressions for G^E—Vera et al. (1977) proposed the Local Surface Guggenheim (LSG) and the Local Composition Guggenheim (LCG) equations which are quite similar to UNIQUAC and have no definite superiority over the UNIQUAC equation in terms of practical performance (Vera et al., 1977).

Application to Food Solutions

Empirical and semi-theoretical models. In food materials, water is the dominant component and may occur as an intracellular and/or extracellular component in vegetables and animal products (Le Maguer, 1987, 1989) and is found to be widely distributed throughout foods as a solvent system for sugars, salts, hydrophillic carbohydrates, gums, etc. (Zapsalis and Beck, 1985). Industrially, the presence of water as a liquid has at least two major advantages: a multitude of various processes can occur simultaneously within a relatively small volume (as compared to gas phase) of aqueous mixture, and water is able to retain and stabilize very complicated systems (e.g., living tissues) without limiting their mobility or interactive possibilities. In other words, the aqueous mixtures allow the creation of complex and multifunctional systems that are essential and useful for most food manufacturers.

Numerous studies on the physical and thermodynamic properties of aqueous carbohydrate solutions have been done. Nevertheless, the behavior of carbohydrates in solution is still not well understood. As there is at present no theory for predicting the properties of aqueous carbohydrate solutions with any degree of reliability, most effort has been devoted to developing empirical correlations between measurable quantities and the desired function, e.g., water activity.

Previous studies were focused mainly on the estimation or prediction of water activity. The importance of water activity in relation to food stability is well documented (Troller and Christian, 1978), and the effects of water activity on microbial growth, chemical reactions, and enzyme activity have been studied extensively (Scott, 1957; Karel, 1975; Labuza, 1975; Troller and Christian, 1978).

A number of correlating equations for the calculation of the water activity and activity coefficient of sugar solutions have been proposed in the literature. Those of Grover and Nicol (1940) and Money and Born (1951), which were probably best known in the confectionery industry (Norrish, 1966), and that of Dunning et al. (1951), which is valid only for pure sucrose solutions, are all entirely empirical correlations. These equations are in many ways unsatisfactory because no significance can be attached to the values of the adjustable constants that appear in them and no confidence can be placed in their use outside the range over which they have been fitted to the experimental results. In addition, neither Grover's equation nor that of Money and Born is thermodynamically consistent in the sense that the Gibbs-Duhem equation cannot be satisfied without postulating absurd functional relationships between the activity coefficients of the solutes and concentration (Norrish, 1966). As a result, semi-empirical correlations were developed.

Based on the identity of equilibrium relative humidity (erh) with the thermodynamic activity, Norrish (1966) derived an equation for the activity coefficient of water as a function of concentration in confectionery syrups. This thermodynamically based equation gives excellent accuracy with sugar solutions (Ross, 1975) and could be generalized to include multicomponent systems. However, it cannot be generalized to treat nonsolute components (Ross, 1975). Based on the Gibbs-Duhem equation, Ross (1975) presented an equation for the prediction of water activity of aqueous solutions. It treated effects produced by nonsolute materials, such as starches or proteins, which bind water and help to lower water activity.

Modifications of these two semi-empirical methods of Norrish (1966) and Ross (1975) have been reported. Chuang and Toledo (1976) presented a modification of Norrish's equation, in which water activities of binary solutions were fitted as linear functions of the solute mole fraction. Ferro Fontan and Chirife (1981) and Ruegg and Blanc (1981) proposed modified Ross's equations. The water activity values of aqueous solutions and many liquid food systems have also been correlated to the thermodynamic expressions in terms of freezing point depression by many researchers (Ferro Fontan and Chirife, 1981; Lerici et al., 1983; Chen, 1987a, b). However, reliable freezing point depression data for solutions at high concentrations are very limited.

All these shortcomings of the empirical or semi-empirical correlations indicate that there is no versatile model to give a satisfactory prediction of water activity in complex aqueous solutions. The attempt to apply thermodynamics in developing the correlations was crucial and critical but incomplete, as the focus of most studies was on the water activity only. Such thermodynamic representation of a single component, water, in a complex system is not sufficient for most food operations. For the application of process modeling techniques to food processes, a thermodynamic representation of each component in the system is necessary.

Several attempts at applying some popular G^E models to food solutions have been reported. Le Maguer (1981) adapted the UNIQUAC model (Maurer and Prausnitz, 1978) to the system terpene-water with successful results. Lebert and Richon (1984) cited the work of Allaneau (1979), who used a number of nonelectrolyte thermodynamic models to predict the infinite dilution activity coefficients of some aromatic compounds in solutions of olive oil from experimental data. Le Maguer (1985) applied equilibrium thermodynamics to describe the starch-water system. Sorrentino et al. (1986) used the ASOG (Derr and Deal, 1969) and the UNIFAC (Fredenslund et al., 1975) models to predict the infinite dilution activity coefficients of aroma compounds in water-carbohydrate and water-polyethylene glycol solutions at various concentrations. Choudhury and Le Maguer (1986) predicted the water activity in glucose solutions using the UNIFAC model (Fredenslund et al., 1975) with reasonable agreement between experimental and predicted values. Choudhury (1987) applied the UNIFAC model with free-volume correction term (Oishi and Prausnitz, 1978) to glucose and other sugar solutions; slight improvements in prediction were found. Lee (1987) used the extended UNIQUAC equation (Christensen et al., 1983) to represent an aqueous solution of ethanol, sucrose, and sodium chloride with good results. Loncin (1988) reports the use of UNIQUAC, Wilson and Margules equations to describe model solutions of wine and grapefruit juice. Saravacos (1990) also applied the UNIQUAC equation to describe the relative volatility of aroma compounds in sugar solutions. From all these attempts, it is clear that models expressed in excess Gibbs energy do give a reasonable and complete description of the systems investigated.

THERMODYNAMIC MODELING OF AQUEOUS CARBOHYDRATE SOLUTIONS

Brief Review of Some Properties of Carbohydrate Solutions

Contemporary perspectives of carbohydrates define them as polyhydroxyaldehydes or ketones (Zapsalis and Beck, 1985). Because of this polyhydroxy nature, carbohydrate molecules usually have a high affinity for water molecules. This is demonstrated by the solubility of the low molecular weight carbohydrate molecules and the sorption of water by some polysaccharides. The mixture of water and soluble carbohydrate molecules is generally highly structured by the presence of numerous hydrogen bonds (Rowlinson, 1969). As a result, some local orderliness is expected in the mixture, and the assumption of total randomness is unacceptable. This, indeed, justifies the use of local composition models to describe the thermodynamic behavior of aqueous carbohydrate solutions.

The thermodynamic behavior of aqueous solutions of some common sugars such as glucose and sucrose has been reported by a number of researchers (Scatchard et al., 1938; Gucker et al., 1939; Honig, 1953; Taylor and Rowlinson, 1955; Robinson and Stokes, 1959, 1961; Stokes and Robinson, 1966; Garrod and Herrington, 1970; Stroth and Schonert, 1977). Recent systematic studies on the excess thermodynamic properties of aqueous solutions of mono- and oligosaccharides and their derivatives were carried out by Barone and coworkers (1981a,b, 1982, 1983a,b,c, 1984, 1985, 1986). Based on these studies, some characteristic behavior of aqueous carbohydrate solutions can be outlined. All data reported for the binary solutions show a slight but general positive deviations from Raoult's and Henry's laws. The aqueous solutions of saccharides show a characteristic behavior in that the values of the excess functions (G^E, H^E, and TS^E) are all positive. This behavior is quite different from that of the aqueous solutions of other hydrophilic solutes, such as the urealike compounds (Barone et al., 1981c) or self-associating solutes, for which all values are negative. The positive values of the excess Gibbs energies are determined by the positive values of the excess enthalpies, which overwhelm the entropies. This fact suggests that the solute-solute interactions in aqueous solutions of monosaccharides are prevented or hidden by other more favorable interactions with the solvent. This is in agreement with spectroscopic and NMR data (Walrafen, 1966; Franks, 1977). On the other hand, the structure of liquid water is stabilized by a series of specific interactions with the polar groups of the solute molecules. During concentration processes, hydration partially reverses, the water molecules going back to the bulk state, richer in enthalpy and entropy. This reversal can account for the positive increasing values of H^E and TS^E. It is also found that, especially for the smaller saccharide molecules, the enthalpic and entropic terms are comparatively large with respect to the excess Gibbs energy. The relatively lower value of G^E is due to the phenomenon called the enthalpy-entropy compensation effect (Lumry and Rajander, 1970). This may indicate that water plays the main role in determining the properties of these solutions. In addition, the values of G^E were found to relate to the number of equatorial OH groups of the carbohydrate molecules.

These findings indicate the highly nonideal and complex behavior of aqueous carbohydrate solutions and consequently rule out use of some earlier solution theories and models such as van Laar's model, regular solution theory, the original lattice model, and Flory-Huggin's equation for polymer mixtures, to aqueous carbohydrate solutions due to the oversimplified assumptions used in their derivations.

Selecting a Model

Three models seem to offer considerable promise in their ability to represent such highly nonideal systems and to be readily generalized to multicomponent systems. These are the Wilson equation (Wilson, 1964), the NRTL equation of Renon and Prausnitz (1968), and the UNIQUAC equation of Abrams and Prausnitz (1975).

Although the low molecular weight carbohydrate molecules are soluble in water, the solubility is limited. As the molecular weight increases, the solubility decreases rapidly. The inability of the Wilson equation to represent heterogeneous systems severely restricts its application to aqueous carbohydrate systems.

Another major drawback of the Wilson equation is its incapability in fitting the higher values of heat of mixing (Vonka et al., 1975; Hanks et al., 1978). The aqueous solutions of saccharides have positive values of excess enthalpies, while the Wilson equation cannot exceed certain positive values of H^E. For homogeneous systems with large positive deviations from the Raoult's law, which are in the vicinity of the heterogeneous region, the Wilson equation leads to very low values of H^E. This fact cannot be eliminated even by introducing temperature-dependent parameters (Vonka et al., 1975). These limitations are not found in the NRTL and UNIQUAC equations.

Although the NRTL equation is not restricted to one liquid phase system as is Wilson's equation, according to the analysis by Hanks et al. (1978), the NRTL equation suffers from a more serious limitation in the case of highly nonideal solutions than does Wilson's equation, because the G^E expression for the NRTL equation degenerates to an ideal solution model. Another major disadvantage of the NRTL equation is that it contains three adjustable parameters for each binary system. The values of the third parameter (described as a nonrandomness factor) are somewhat arbitrary and have little physical significance (Wang and Chao, 1983). The introduction of a purely empirical factor much weakens the theoretical soundness of the model.

Among the three, the UNIQUAC model is a more refined local composition model. It takes into consideration the influences of molecular sizes, shapes, and energies of interaction on the distribution of species around a given molecules. Moreover, UNIQUAC has an advantage over NRTL, that is, it uses only two adjustable parameters per binary, instead of three.

The UNIQUAC Equation

Important features. Abrams and Prausnitz (1975) developed the UNIQUAC model from statistical mechanical arguments. A basic assumption of the model is that it is possible to define local mole fractions representative of concentrations on a microscopic scale. These concentrations are different from the directly measurable overall mole fractions. But it has also been pointed out by Flemr (1976) and McDermott and Ashton (1977) that the expression has an inherent weakness because certain conservation equations are not satisfied. In 1978, Maurer and Prausnitz rederived the model using phenomenological arguments based on the two-liquid theory, thereby eliminating the inconsistency pointed out by the previous authors. This settled the UNIQUAC model on a firmer theoretical foundation.

Some important features of the UNIQUAC model are as follows (Abrams and Prausnitz, 1975):

1. The excess Gibbs energy is considered to be the sum of a combinatorial term and a residual term.

$$G^E = G^E_{\text{(combinatorial)}} + G^E_{\text{(residual)}}$$

2. $G^E_{\text{(combinatorial)}}$ is derived from Staverman's potential (Staverman, 1950), an extension of the Flory-Huggins athermal theory (Flory, 1941, 1942), to molecules of different shapes as well as sizes.
3. $G^E_{\text{(residual)}}$ uses local area fraction as its concentration variable.

Conceptually, the combinatorial term describes the dominant entropic contribution (Prausnitz et al., 1986). It involves the contribution due to the arrangement of multisegment molecules in the mixture, in calculations in which all kinds of interactions between the molecules are left out of consideration (athermal concept). Under the assumption that each possible microstate is of identical energy, the combinatorial term can be evaluated by the use of statistical mechanics.

In a real mixture, the attractive-repulsive forces between the molecules or segments have to taken into account. These forces displace the molecules from their initial arrangement and distort the presumed arrangement. The residual term thus attempts to account for these energetic effects. It involves the contribution due primarily to intermolecular forces that are responsible for the enthalpy of mixing (Prausnitz et al., 1986).

In the UNIQUAC equation,

$$G^E_{\text{(comb.)}} = RT\left[\sum_i x_i \ln \frac{\phi_i}{x_i} + \frac{z}{2}\sum_i q_i x_i \ln \frac{\theta_i}{\phi_i}\right] \tag{1}$$

$$G^E_{\text{(res.)}} = -RT\left[\sum_i q_i x_i \ln\left(\sum_j \theta_j \tau_{ji}\right)\right] \tag{2}$$

where segment fraction ϕ_i, surface fraction θ_i, and the energetic parameters τ_{ji} are given by

$$\phi_i = \frac{r_i x_i}{\sum_j r_j x_j} \tag{3}$$

$$\theta_i = \frac{q_i x_i}{\sum_j q_j x_j} \tag{4}$$

$$\tau_{ji} + \exp\left[\frac{-(u_{ji} - u_{ii})}{RT}\right] = \exp\left(\frac{-a_{ji}}{T}\right) \tag{5}$$

x_i is the mole fraction, z is the coordination number, r_i and q_i are the two structural parameters, a_{ji} is the adjustable binary energy parameter, and $(u_{ji} - u_{ii})$ is the parameter representing the potential energy of interaction between molecules i and j.

By the Gibbs-Helmholtz equation,

$$\left[\frac{\partial(G^E/T)}{\partial T}\right]_{P,x} = \frac{-H^E}{T^2} \tag{6}$$

In the UNIQUAC equation, the parameters r_i and q_i are not temperature dependent. Assuming that the energy parameter a_{ji} is independent of temperature, we obtain

$$H^E = R\sum_i \left[\frac{q_i x_i}{\sum_j \theta_j \tau_{ji}}\left(\sum_j \theta_j \tau_{ji} a_{ji}\right)\right] \tag{7}$$

Comparing equations, we find

$$G^E_{(\text{residual})} \neq H^E$$

Since

$$-S^E = \left[\frac{\partial G^E}{\partial T}\right]_{p,x} \tag{8}$$

or

$$S^E = \frac{H^E - G^E}{T} \tag{9}$$

$$S^E = R\left[\frac{1}{T}\sum_i \frac{q_i x_i}{\sum_j \theta_j \tau_{ji}} \sum_j \theta_j \tau_{ji} a_{ji} - \sum_i x_i \ln \frac{\phi_i}{x_i} - \frac{z}{2}\sum_i q_i x_i \ln\left(\sum_j \theta_j \tau_{ji}\right)\right] \tag{10}$$

or

$$-TS^E = RT\left[\sum_i x_i \ln \frac{\phi_i}{x_i} + \frac{z}{2}\sum_i q_i x_i \ln \frac{\theta_i}{\phi_i} - \sum_i q_i x_i \ln\left(\sum_j \theta_j \tau_{ji}\right) - \frac{1}{T}\sum_i \frac{q_i x_i}{\sum_j \theta_j \tau_{ji}} \sum_j \theta_j \tau_{ji} a_{ji}\right] \tag{11}$$

Comparing equations, we find

$$G^E_{\text{(combinatorial)}} \neq -\ TS^E$$

It may be concluded that the two terms of the UNIQUAC equation are conceptually related to the H^E and TS^E functions, but they are not exactly the same, because the combinatorial term only accounts for the athermal entropic contribution (i.e., $-TS^E_{\text{(athermal)}}$) and all the effects not included in the combinatorial term are taken into consideration in the residual term. In other words, the residual term, in fact, consists of the nonathermal entropic contribution as well as the enthalpic contribution. That is,

$$G^E_{\text{(combinatorial)}} = -TS^E_{\text{(athermal)}}$$

The combinatorial part is determined only by the composition and by the sizes and shapes of the molecules; it requires only the structural parameters, r_i and q_i. The residual part, however, requires also the energy parameters, a_{ji}.

The structural parameters

Importance of the r_j and q_i Parameters. ϕ_i is the segment fraction; it measures the size of the molecule and thus directly relates to the volume fraction. If ϕ_i is replaced by the bulk volume fraction, the G^E equation will become the Flory-Huggins entropy equation (Flory, 1942), which is derived for athermal mixtures of molecules of different sizes.

In a mixture of multisegment molecules, consideration has to be given not only to the effect of the size of each multisegment molecule but also that contact occurs between the segments. If the molecule has r segments, and makes on average zq contacts, then the resulting nonideality may be taken into consideration by means of the surface fraction, θ_i, which is, in fact, a measure of the surface contact sites:

$$\theta_i = \frac{x_i q_i}{\sum_i x_i q_i} = \frac{x_i z q_i}{\sum_i x_i z q_i} \qquad (12)$$

The structural parameters are nonadjustable, and their values depend only on the molecular structures.

Results obtained by Lee (1987) indicated that the combinatorial part of the UNIQUAC equation was not properly evaluated. Lee's optimized results on aqueous sucrose solutions have further shown that the original r and q were overestimated and that the "proper values" might be significantly smaller.

The improper estimation of the structural parameters has a pronounced effect on the representation of the excess functions. Incorrect structural parameters would directly result in improper representation of the excess entropy, and incorrect energetic parameters would give poor representation of the excess enthalpy function. This explains why both excess entropy and enthalpy were not properly represented in previous studies (Choudhury, 1987; Lee, 1977), even though the excess Gibbs energy function was well fitted by optimization techniques. The proper representation of the excess Gibbs energy was in fact due to the fortuitous cancellation of the excess enthalpy and excess entropy (Fredenslund and Rasmussen, 1985).

Possible Deficiencies in the Assumption and Evaluation of the Structural Parameters. Previous attempts at modifying the structural parameters have been purely empirical. For instance, Lee (1987) and Saravacos et al. (1989) used the experimental data to find the optimum values of r_i and q_i for sucrose molecules; Kikic et al. (1980) introduced the exponent $\frac{2}{3}$ to r_i with $r_i^{2/3}$; Anderson and Prausnitz (1978) proposed an empirical surface parameter, q', in order to get better fit for systems containing water or alcohol molecules.

The definition of the structural parameters r_i and q_i are the following (Abrams and Prausnitz, 1975):

$$r_i = \frac{V_{wi}}{V_{ws}} \tag{13}$$

$$q_i = \frac{A_{wi}}{A_{ws}} \tag{14}$$

where V_{wi} and A_{wi} are the van der Waals volume and surface area of molecule i, calculated from the work of Bondi (1968), and V_{ws} and A_{ws} the van der Waals volume and surface area of the standard segment. Abrams and Prausnitz (1975) selected a methylene group (CH_2) of a linear polyethylene molecule of infinite length as one standard segment. The volume and surface area of the standard segment, as calculated by Abrams from Bondi's value, are 15.17 cm^3/mol and 2.5×10^9 cm^2/mol, respectively.

The choice of a methylene segment as the standard segment is perfectly legitimate if hydrocarbon mixtures are concerned. It would also be a reasonable choice in the case of a nonpolar organic mixture. But if the mixture is, for instance, a mixture of carbohydrate (a polyhydroxy compound) and water (a highly polar molecule), the choice would be questionable.

Although the normalization does not severely influence the calculation of the surface and segment fractions (θ_i and ϕ_i), the normalized surface area (q_i) enters into both the combinatorial and the residual parts of UNIQUAC as a scaling factor. As a result, the normalization has a significant influence on the calculation of the excess Gibbs energy. A proper choice of normalizing factors or standard segment is thus critical.

Second, the definition of the structural parameters assumes that r_i and q_i are independently determined from van der Waals volume and surface area. This contradicts the relation defining the bulk factor l_i (Abrams and Prausnitz, 1975),

$$l_i = \frac{z}{2}(r_i - q_i) - (r_i - 1) \tag{15}$$

or

$$zq_i = (z - 2)\, r_i + 2(1 - l_i) \tag{16}$$

This relation clearly indicates that r_i and q_i are related linearly. If r_i and q_i are determined independently, the bulk factor l_i will become an adjustable parameter without any physical significance. But, according to Guggenheim (1952) and Vera et al. (1977), l_i is not an adjustable parameter, but should represent a structural condition of the molecule.

Group Assignment for Carbohydrate Molecules. According to Bondi (1968), the van der Waals volume and surface area of a molecule are found by summation of the corresponding group properties, that is,

$$V_{wi} = \sum_k \upsilon_k V_{wk} \tag{17}$$

and

$$A_{wi} = \sum_k \upsilon_k A_{wk} \tag{18}$$

where k = 1,2,..., N (number of groups in molecule i); υ_i = number of groups of kind k in molecule i; V_{wk}, A_{wk} = van der Waals volume and surface area of group k, respectively; and V_{wi}, A_{wi} = van der Waals volume and surface area of molecule i, respectively.

Since the structural parameters r_i and q_i are defined in terms of the van der Waals volume and surface area, a similar additivity principle is applied to calculate r_i and q_i using corresponding group values (R_k and Q_k). That is,

$$r_i = \sum_k \upsilon_k R_k \tag{19}$$

$$q_i = \sum_k \upsilon_k Q_k \tag{20}$$

where R_k and Q_k are the r and q values of group k, respectively.

The fundamental assumption made in the above calculations is that a molecule is considered as an ensemble of identifiable structural units corresponding to the main chemical groups, and the geometrical parameters (r and q) of the molecule are assessed from the volumes and surfaces of the individual groups using a simple additivity rule.

A chemical basis may give an appropriate measure to define convenient chemical groups for interaction, but it does not give meaningful definition for structural units from which a geometrical property is estimated. For instance, on a chemical basis, a mannitol molecule has six OH groups. The six OH groups might behave similarly chemically, but regarding the physical extent of the group, they should be different: the OH group at the end of the molecule has greater spatial extent than the one in the middle of the molecule. Hence, to distinguish this structural difference, it may be more sensible to represent the mannitol molecule by 2 (CH_2OH) and 4 (CHOH) groups than by 6 (OH), 2 (CH_2), and 4 (CH) groups.

Modified Evaluation Methods for the Structural Parameters. In view of the analysis presented in the last chapter, the most critical problem that confronts UNIQUAC is the evaluation of the structural parameters r and q. Quasi-lattice models assume the existence of a regular lattice. Each lattice site is occupied by an equal-sized segment of molecule i and is surrounded by z (the lattice coordination number) nearest neighbor sites. Each molecule i has r_i segments and zq_i nearest neighbor sites (or contact sites). Thus, a relation between zq_i and r_i can be obtained by direct counting. For a straight or branched open-chain molecule, Guggenheim (1952) showed that r_i is related to q_i by (Guggenheim, 1952)

$$\frac{1}{2} z(r_i - q_i) = r_i - 1 \tag{21}$$

or

$$zq_i = r_i(z - 2) + 2 \tag{22}$$

Vera et al. (1977) proposed an improved method for evaluating r and q, which eliminates most of the problems pointed out in the previous chapter. In 1985, Abusleme and Vera further extended the method to calculations in the group approach.

By combining Abram's defining equations for r_i and q_i and Guggenheim's relation for r_i and q_i, Vera et al. (1977) obtained the following relation:

$$A_{wi} = \frac{(z - 2)}{z} \frac{A_{ws}}{V_{ws}} V_{wi} + \frac{2A_{ws}}{z} \tag{23}$$

and showed that the above relation between A_{wi} and V_{wi} is in fact theoretically linear, with a common slope for all families of compounds, but a unique intercept for each family. That is,

$$A_{wi} = k_1 V_{wi} + k_2 \tag{24}$$

where k_1 and k_2 are constants.

An explanation for this linearity, based on molecular considerations, has been given by Vera et al. (1977). By a least-square fit over 43 of the more common open-chain compounds, branched or unbranched (including alkanes, alcohols, aldehydes, ketones, nitriles, acids and esters), Vera et al. (1977) found that

$$zq_i = 0.4228 V_{wi} + 2(1 - l_i) \tag{25}$$

Thus, the values of q_i can be evaluated without any particular assumption about the values of V_{ws} and A_{ws}.

Vera et al. (1977) also found that the van der Waals volume V_{wi} as calculated by Bondi (1968) is closely proportional to the liquid molar volume at 20°C. A least-square fit over 33 compounds, including alkanes, alcohols, esters, ketones, alkylchlorides, and aromatics, gives

$$V_{wi} = 0.554 V_i^L (20°C) \tag{26}$$

This gives an alternative to calculate V_{wi}. Equation 25 can be rewritten as

$$zq_i = 0.2342\ V_i^L\ (20°C) + 2(1 - l_i) \quad (27)$$

As a result, sets of self-consistent values of z, A_{ws}, and V_{ws} are obtained, which are reproduced in Table 1.1. Conceptually, these new sets of normalizing factors (V_{ws} and A_{ws}) can be visualized as the van der Waals volume and surface area of an optimum arbitrary standard segment.

Table 1.2 shows the calculated r and q values of some common carbohydrate molecules using Vera's method (Abusleme and Vera, 1985).

The "Building Block" Method

Concept. Carbohydrates are conveniently classified into three major groups: mono-, oligo-, and polysaccharides. Monosaccharides are the simplest carbohydrate units and are generally considered as the building blocks for oligo- or polysaccharides. Structurally, a monosaccharide unit can be considered as a carbon chain closed by an ether group (–O–) to form a ring on which hydroxyl groups are attached. The rings can be further linked together by some other ether groups to form a series of rings.

Despite the variable nature of carbohydrate molecules, only a few basic structural units are found repeatedly joined together if the molecule is divided segment by segment. These basic units can be classified into three categories: (1) the linkage unit, (2) the hydroxl units, and (3) the alkyl units. Only one kind of linkage unit is found, the ether group (–O–). The hydroxyl units form the backbone of the molecule; the CHOH group is the most common unit found in the ring, while the CH_2OH group is mostly found attached to the ring as a side group. A COH group may also exist as a segment of the ring structure in some carbohydrates such as fructose.

Table 1.1 Self-Consistent Sets of Values of z, A_{ws}, and V_{ws}

z	A_{ws} (10^9 cm^2/mol)	V_{ws} (cm^3/mol)
6	1.88	9.46
8	2.50	14.19
10	3.13	18.92
12	3.76	23.65

Source: Adapted from Vera et al. (1977).

Table 1.2 Evaluation of r_i and q_i of Some Carbohydrate Molecules by Vera's Method

Compound	V_{wi}	l_i	r_i	q_i
Xylose	65.94	1	3.49	2.79
Fructose	77.31	1	4.09	3.27
Glucose	77.31	1	4.09	3.27
Sucrose	155.67	2	8.23	6.38
Raffinose	234.03	3	12.37	9.49

Source: Abusleme and Vera (1985).

Occasionally, some alkyl groups may be incorporated into the molecule. On this basis, Table 1.3 summarizes the basic units found in carbohydrate molecules.

Estimation methods for the R and Q values of the new structural units. Since r is a well-defined parameter, the original definition of r by Abrams and Prausnitz (1975) is used to calculate the R values of the proposed structural units, and the r value of a molecule is then calculated by the group-contribution method.

$$R_k = \frac{V_{wk}}{V^*} \tag{28}$$

and

$$r_i = \sum_k \upsilon_k R_k \tag{29}$$

where k = 1,2,...,N (number of groups in molecule i); r_i, R_k = r values

Table 1.3 Basic Structural Units for Carbohydrate Molecules

Type	Structural unit	Comment
Linkage	—O—	As link between rings or in a ring
Hydroxyl	CH_2OH	Essential backbone units
	CHOH	
	COH	
Alkyl	CH_2	Nonhydroxyl backbone units, exist occasionally
	CH	
	C	

of molecule i and group k, respectively; V_{wk} = van der Waals volume of group k; V^* = normalizing factor for R_k (or r_i); υ_k = number of groups of kind k in molecule i.

With one exception, values of van der Waals volume given by Bondi (1968)—the current best estimate for van der Waals volume, are used to calculate V_{wk}. Bondi's value for the –O– group (ether) in an aliphatic chain is 3.7 cm^3/mol. Based on two studies by Kehiaian et al. (1971, 1978), Abusleme and Vera (1985) confirmed that 10.23 cm^3/mol is the best optimal value for the van der Waals volume of the –O– group. In this study, V_w (–O–) = 10.23 cm^3/mol was used in the calculation of R(–O–).

Instead of 15.17 cm^3/mol, the normalizing factor calculated by Abrams and Prausnitz (1975) using a methylene group as a standard segment, 18.92 cm^3/mol is used for V^*. This is an optimal value calculated by Vera et al. (1977). Conceptually, V^* stands for the van der Waals volume of an optimum arbitrary standard segment.

Because of the nonpolar nature of the alkyl groups, the Q values of these groups are simply calculated by the method proposed by Abusleme and Vera (1985), which has been shown to give the best estimates for nonpolar compounds.

For evaluation of Q values of the polar groups (the linkage and the hydroxyl units), an optimization procedure is required to find the best values (Lui, 1989). This involves (1) the selection of structural homogeneous molecule, which allows the determination of the Q value of a particular structural unit unambiguously, (2) the determination of the optimized q values of the selected molecules, and (3) the calculation of the Q values of the structural units from the optimized q according to the representation. Using this procedure, values of Q for the specified units are presented in Table 1.4.

Representation and Prediction of the Thermodynamic Properties of Aqueous Carbohydrate Solutions Using the New Parameters

The five agencies carbohydrate systems selected for this validation process are the aqueous solutions of:

1. Xylose (monosaccharide; pentose)
2. Glucose (monosaccharide; hexose; structural isomer of fructose)
3. Fructose (monosaccharide; hexose; structural isomer of glucose)
4. Sucrose (disaccharide)
5. Raffinose (oligosaccharide)

Table 1.4 R and Q Values of the Building Blocks

Type	Structural unit	R	Q
Linkage	—O—	0.541	0.581
Hydroxyl	CH_2OH	0.966	1.025
	CHOH	0.783	0.889
	COH	0.601	0.737
Alkyl[a]	CH_2	0.541	0.431
	CH	0.358	0.182
	C	0.176	0.0

[a] The R and Q values of the alkyl groups are calculated as described in text.

The r_2 and q_2 values of these carbohydrate molecules are calculated using the R and Q values given in Table 1.4; the r_1 and q_1 values of the water molecule are set at 0.73 and 1.19 (Vera et al., 1977), respectively. The new structural parameters of the selected carbohydrate molecules are presented in Table 1.5. The two adjustable energy parameters (a_{12} and a_{21}) are simply determined by optimization using both the G^E and H^E data.

As shown in Fig. 1.2, the carbohydrate compounds lie perfectly on the same straight line with only a slight discrepancy found between the two isomers, glucose and fructose. This clearly indicates that the building block method does give theoretically consistent values.

By studying the (A_{12},A_{21}) values published in the DECHEMA Data Series (Gmehling and Onken, 1977), Luecke (1983) confirmed that there

Table 1.5 Parameters of Five Simple Carbohydrates by Building Block Method[a]

Compound	r_2	q_2	a_{12}	a_{21}
Xylose	4.214	4.568	206.86	−172.19
Glucose	4.998	5.344	231.29	−180.51
Fructose	4.998	5.413	182.96	−150.87
Sucrose	9.686	9.809	121.52	−111.68
Raffinose	14.374	14.433	122.88	−106.68

[a] Note: subscripts 1 and 2, respectively, refer to water and the carbohydrate molecule concerned.

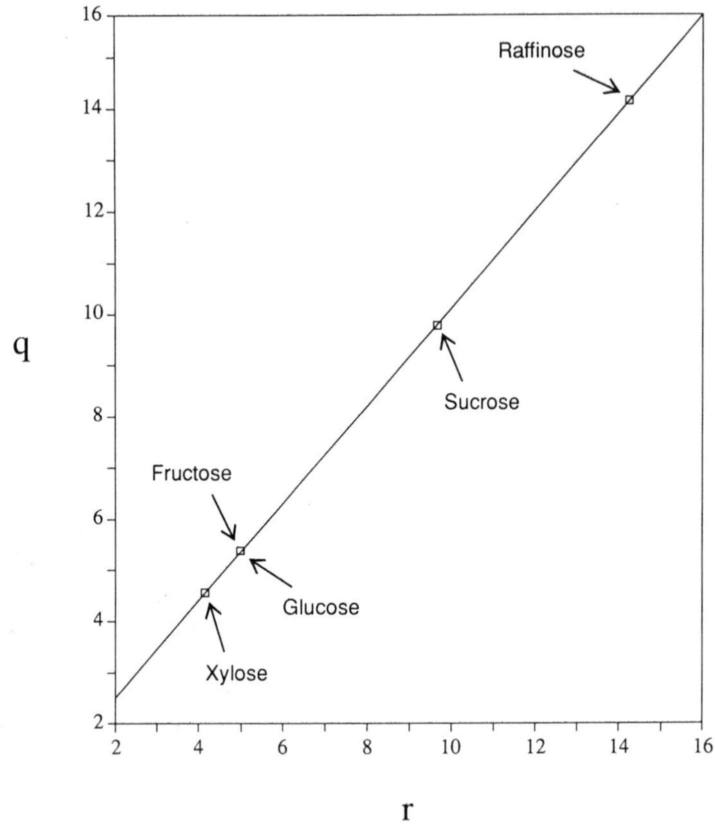

Fig. 1.2 Correlation between r and q for carbohydrates.

is a high degree of correlation between interaction (or energy) parameters in the UNIQUAC equation, and he proposed the following equation:

$$(A_{12} + c)(A_{21} + c) = 2 \times 10^6 \tag{30}$$

where c is a constant, and $A_{ji} = (u_{ji} - u_{ii}) = R\ a_{ji}$. For example, in a water-ethanol system, c is found to be 1290 when A_{12} and A_{21} are given in cal/mol.

In our case, both a_{12} and a_{21} are known, hence, c can be solved for each aqueous carbohydrate system. The c values of the five aqueous carbohydrate systems are given in Table 1.6.

Table 1.6 A_{12}, A_{21} and c Values of Aqueous Carbohydrate Systems

System	a_{12} [K]	a_{21} [K]	A_{12} [cal/mol]	A_{21} [cal/mol]	c^a [cal/mol]
Xylose	206.9	−172.2	411.1	−342.2	1429
Glucose	231.3	−180.5	459.6	−358.7	1422
Fructose	183.0	−150.9	363.6	−299.8	1421
Sucrose	121.5	−111.7	241.5	−221.9	1423
Raffinose	122.9	−106.7	244.2	−212.0	1416

[a] Mean c = 1422.

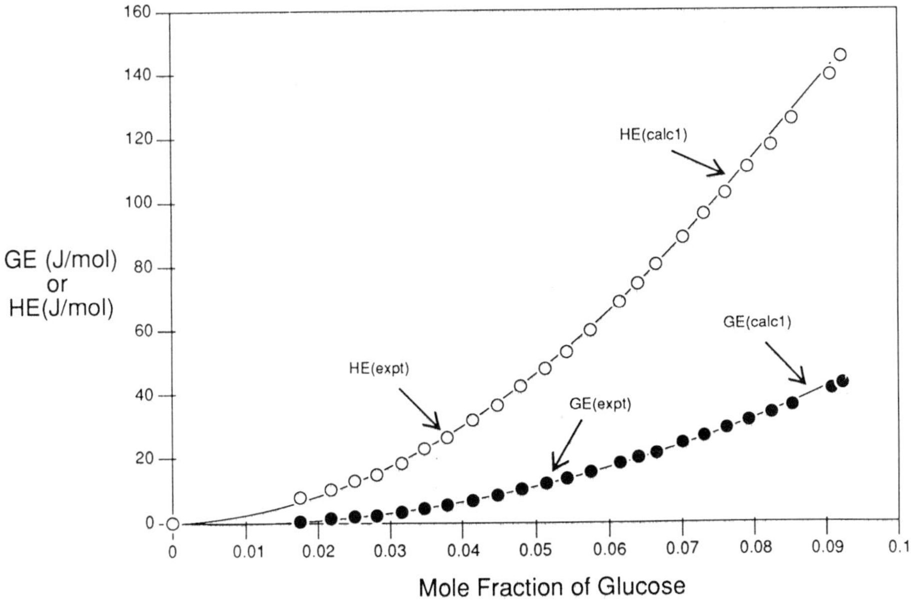

Fig. 1.3 Simultaneous representation of excess Gibbs energy, GE, and excess enthalpy, HE, for the system water-glucose at 25°C. Predicted values calculated from the UNIQUAC equation with revised parameters by the building block method. (*Experimental data adapted from Taylor and Rowlinson, 1955; Stokes and Robinson, 1966.*)

Representation of the excess properties. It is well known that the original UNIQUAC equation is unable to represent the excess Gibbs energy and excess enthalpy simultaneously with satisfactory results (Skjold-Jorgensen et al., 1980; Renon, 1985), and it generally fails to give an accurate representation of the thermodynamic properties of highly nonideal system (Anderson and Prausnitz, 1978; Nath and Bender, 1981; Brandani and Evangelista, 1984; Zou and Prausnitz, 1987). It is thus critical to test the performance of the equation using the new sets of parameters evaluated by the building block method. The simultaneous representations of the excess Gibbs energy and excess enthalpy of the five aqueous carbohydrate solutions (water-xylose, water-glucose, water-fructose, water-sucrose, and water-raffinose systems) at 25°C are shown in Figs. 1.3 to 1.5. The resulting simultaneous representation is remarkably successful (%RMSD is well within 4%) (Table 1.7).

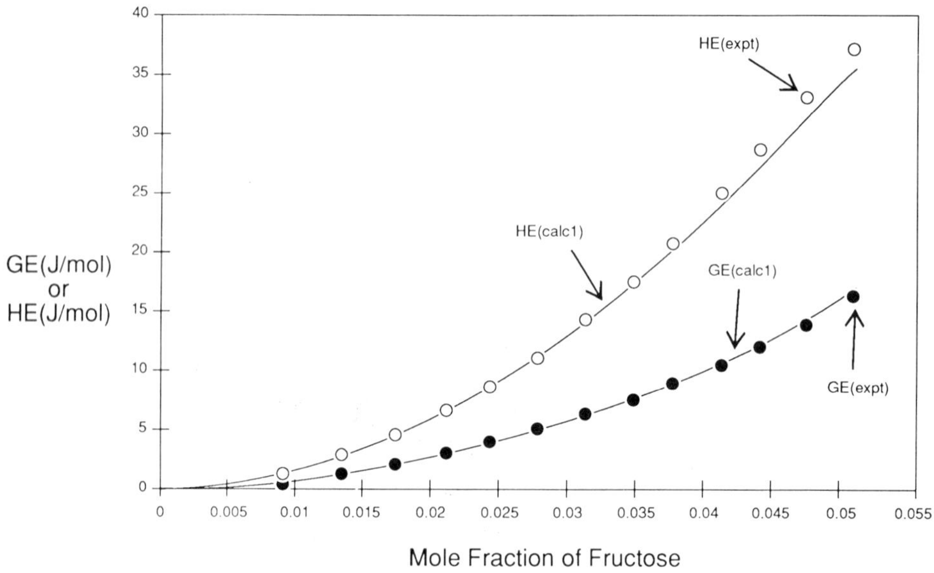

Fig. 1.4 Simultaneous representation of the excess Gibbs energy, GE, and excess enthalpy, HE, for the system water-fructose at 25°C. Predicted values calculated from the UNIQUAC equation with revised parameters by the building block method. (*Experimental data adapted from Barone et al., 1981, 1986.*)

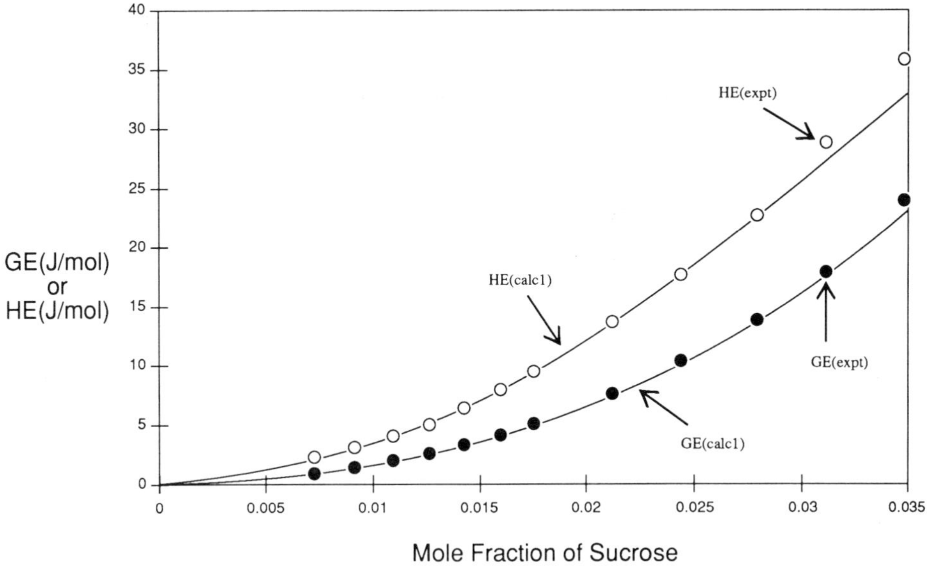

Fig. 1.5 Simultaneous representation of excess Gibbs energy, GE, and excess enthalpy, HE, for the system water-sucrose at 25°C. Predicted values calculated from the UNIQUAC equation with revised parameters by the building block method. *(Experimental data adapted from Barone et al., 1981; Savage and Wood, 1976; Robinson and Stokes, 1961.)*

Table 1.7 Mean Deviations (%RMSD) of Simultaneous Representation of G^E and H^E Data[a]

	Building block method	
System	G^E	H^E
Xylose	2.13	0.94
Glucose	3.00	3.55
Fructose	2.95	3.06
Sucrose	1.12	3.93
Raffinose	3.50	3.66

[a] Note:

$$\%RMSD = \frac{\sum_{i=1}^{n} \left(1 - \frac{Y_i^{calc}}{Y_i^{expt}}\right)^2}{n} \times 100$$

where Y_i = G^E or H^E data; n = number of data points.

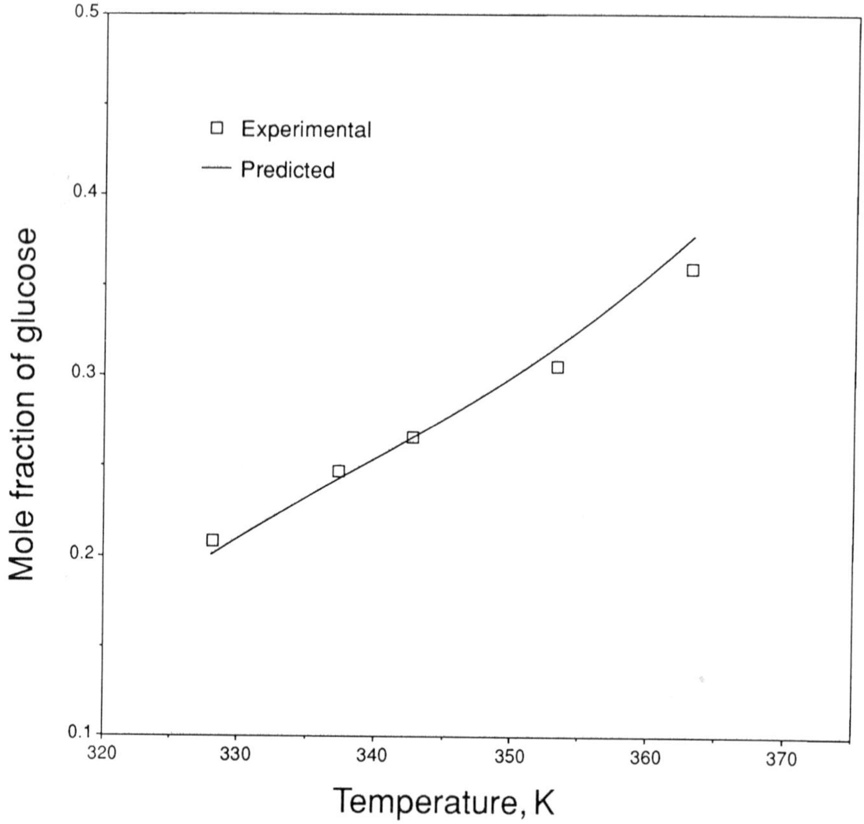

Fig. 1.6 Predicted and experimental solubility of glucose in water. (*Experimental data adapted from Honig, 1953.*)

Prediction of selected properties of glucose, fructose, and sucrose solutions. The results of these predictions are presented in Figs. 1.6 to 1.9 and show excellent results. As might be expected, since the parameters A_{12} are not temperature dependent, we observe slight deviations in the solubilities.

CONCLUSION

The UNIQUAC equations with the proposed modifications represents a definite improvement over the existing equations which can be found in the literature.

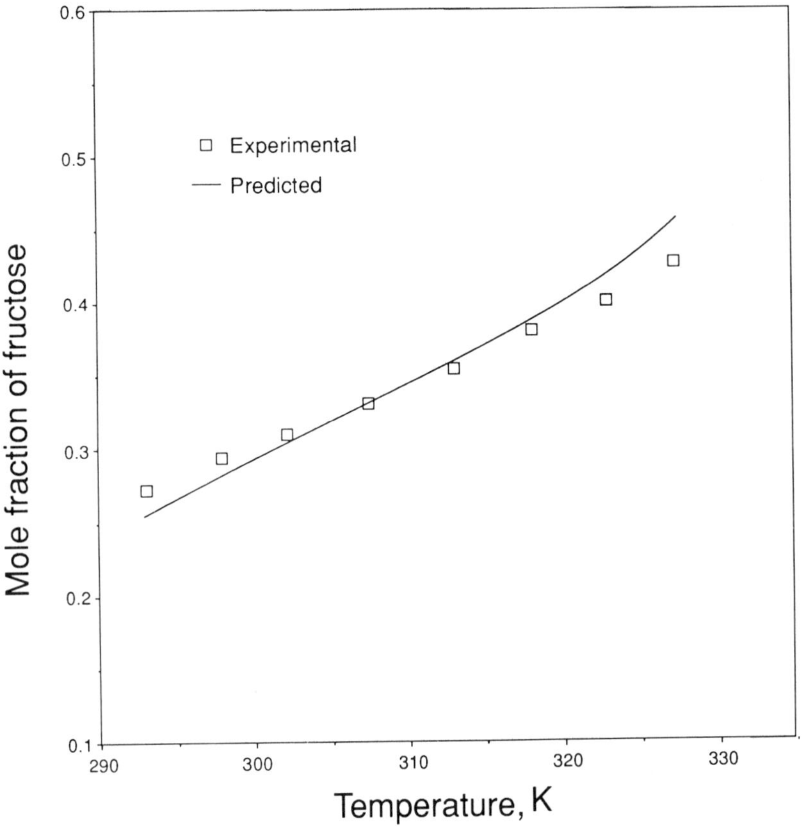

Fig. 1.7 Predicted and experimental solubility of fructose in water. (*Experimental data adapted from Honig, 1953.*)

Its extension to ionic solutions (Christensen et al., 1983) and its successful use by Lee (1987) indicates as well that the potential is there for having a powerful predictive tool for food solutions. It should be noted, however, that we are still dependent on experimental data for the determinations of the energy-related parameters. However, this has also improved because of existing correlations between the parameters.

ACKNOWLEDGMENT

The author acknowledges the support of the Natural Sciences and Engineering Research Council of Canada for the support given to this work over the years.

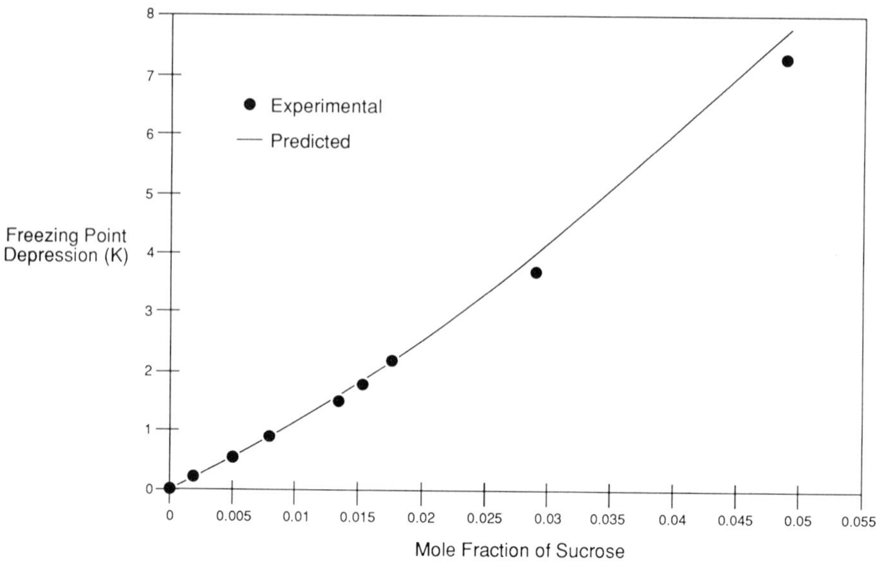

Fig. 1.8 Predicted and experimental freezing point depression of aqueous sucrose solutions. (*Experimental data from Honig, 1953.*)

SYMBOLS

a_i	activity of component i
a_w	water activity
A_w	van der Waals surface area
a_{ij}	UNIQUAC adjustable binary energy parameter
c	a constant in Eq. 30
D	objective function
G	Gibbs energy
H	enthalpy
K	equilibrium constant
L^-	relative partial molar enthalpy
I_i	bulk factor of component i
M	molecular weight
m	molality
P	pressure
q	UNIQUAC surface parameter

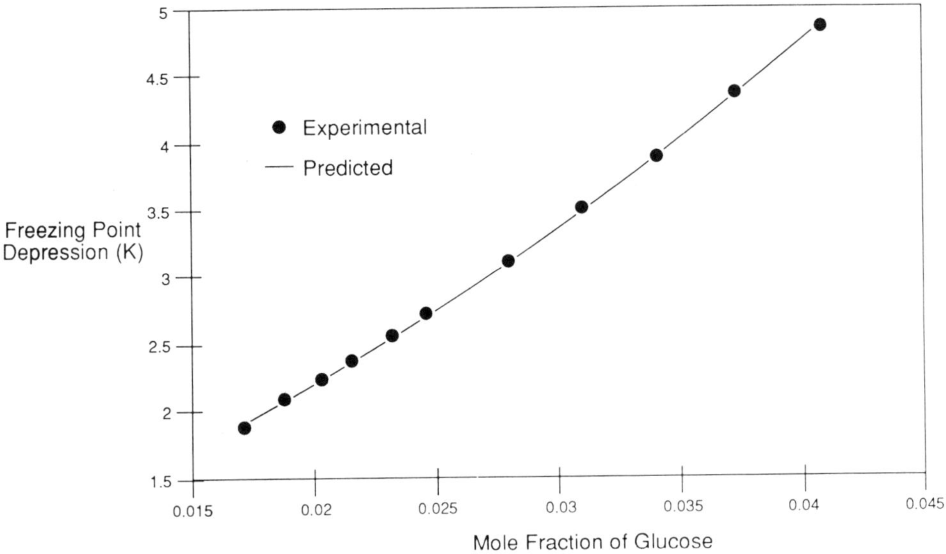

Fig. 1.9 Predicted and experimental freezing point depression of aqueous glucose solutions. Experimental data adapted from CRC (1983–1984) tables on freezing point depression.

Q_k	surface parameter of group k
r	UNIQUAC size parameter
R_k	size parameter of group k
R	universal gas constant
S	entropy
S^2	objective function
T	temperature in K
T_B	boiling temperature
T_F	freezing temperature
U	energy content
u_{ij}	UNIQUAC binary interaction parameter
V	volume
V^*	normalizing factor for R_k
V_w	van der Waals volume
x	mole fraction
Y_i	data point i for any measurement concerned
z	coordination number

Greek Letters

γ	activity coefficient
Δ	change
θ	component surface area fraction
μ	chemical potential or partial molar Gibbs energy
υ	stoichiometric coefficient
π	osmotic pressure
τ	UNIQUAC energetic parameter
ϕ	osmotic coefficient
ρ_i	component volume fraction

Superscripts

E	excess
ID	ideal state
L	liquid state
M	mixing
$\cdot$	pure state
$\circ$	standard state
∞	an infinite dilution
*	unsymmetric convention
-	partial molar quantity
(m)	molality standard state per kg solvent
(m)′	molality standard state per mole
(x)	mole fraction standard state

Subscripts

i	component i
k	group k
r	response r
s	standard segment
1	solvent
2	solute
f	fusion
m	melting

REFERENCES

Abrams, D. S. and Prausnitz, J. M. (1975). Statistical thermodynamics of liquid mixtures: a new expression of the excess Gibbs energy of partly or completely miscible systems. *AlChE J.* 21(1): 116.

Abusleme, J. A. and Vera, J. H. 1985. The quasi-chemical group solution theory for organic mixtures. *Fluid Phase Equil.* 22: 123.

Acree, W. E., Jr. 1984. *Thermodynamic Properties of Nonelectrolyte Solutions*. Academic Press, Orlando, FL.

Allaneau, B. 1979. Influence des lipides sur la retention des constituants volatiles: aspects thermodynamiques. Thesis dissertation ENSIA, Massy.

Anderson, T. F. and Prausnitz, J. M. 1978. Application of the UNIQUAC equation to calculation of multicomponent phase equilibria. 1. Vapor-liquid equilibria. 2. Liquid-liquid equilibria. *Ind. Eng. Chem. Process Des. Dev.* 17(4): 552; 17(4): 561.

Barone, G., Cacace, P., Castronuovo, G., and Elia, V. 1981a. Excess enthalpies of aqueous solutions of mono- and oligo-saccharides at 25°. *Carbohydr. Res.* 91: 101.

Barone, G., Cacace, P., Castronuovo, G., Elia, V., and Iapelli, F. 1981b. Excess thermodynamic properties of aqueous solutions of L-fucose and L-Rhamnose at 25°. *Carbohydr. Res.* 93: 11.

Barone, G., Cacace, P., Castronuovo, G., and Elia, V. 1981c. Interactions in aqueous solutions of urea-like compounds. *J. Chem. Soc. Faraday Trans. I* 77: 1569.

Barone, G., Cacace, P., Castronuovo, G., and Elia, V. 1982. Excess thermodynamic properties of aqueous solutions of hydrophillic non-electrolytes. *Gazz. Chim. Ital.* 112: 153.

Barone, G., Castronuovo, G., Doucas, D., Elia, V., and Mattia, C. A. 1983a. Excess enthalpies of aqueous solutions of monosaccharides at 298.15 K:Pentases and 2-deoxy sugars. *J. Phys. Chem.* 87: 1931.

Barone, G., Cacace, P., Castronuovo, G., Elia, V., and Lepore, U. 1983b. Excess enthalpies of aqueous solutions of methyl-D-aldopyranosides at 25°C. *Carbohydr. Res.* 115: 15.

Barone, G., Cacace, P., Castronuovo, G., and Elia, V. 1983c. Excess thermodynamic properties of myo-inositol and polyhydric alcohols in water at 25°C. *Carbohydr. Res.* 119: 1.

Barone, G., Castronuovo, G., Elia, V., and Savino, V. 1984. Excess free energies and entropies of aqueous solutions of methyl-D-pyranosides at 25°C. *J. Solution Chem.* 13: 209.

Barone, G., Cacace, P., Castronuovo, G., Elia, V., Muscetta, M., and Stassinopoulou, K. H. 1985. Thermodynamics of aqueous solutions of isometric sugars. *Fluid Phase Equil.* 20: 177.

Barone, G., Castronuovo, G., Elia, V., Celotto, D., and Santorelli, G. 1986. Excess Gibbs free energy of aqueous solutions of carbohydrates and polyhydric alcohols at 25°C. *Calorimetrie et Analyse Thermique* 17: 464.

Bondi, A. 1968. *Physical Properties of Molecular Crystals, Liquids and Glasses.* John Wiley & Sons, Inc., New York.

Brandani, V. and Evangelista, F. 1984. The UNIQUAC associated-solution theory: Vapor-liquid equilibria of binary systems containing one associating and one inert or active component. *Fluid Phase Equil.* 17: 281.

Brown, W. B. 1957. The statistical thermodynamics of mixtures of Lennard-Jones molecules. I. Random mixtures. *Phil. Trans.* 250: 175; The second-order theory of conformal solutions. *Proc. Roy. Soc. A.* 240: 651.

Chao, K. C. and Greenkorn, R. A. 1975. *Thermodynamics of Fluid—an Introduction to Equilibrium Theory.* Marcel Dekker, Inc., New York.

Chen, C. S. 1987a. Relationship between water activity and freezing point depression of food systems. *J. Food Sci.* 52: 433.

Chen, C. S. 1987b. Calculation of water activity and activity coefficient of sugar solutions and some liquid foods. *Lebensm.-Wiss. u.-Technol.* 20: 64.

Choudhury, G. S. 1987. Thermodynamic modelling of complex aqueous solutions. Ph.D. Thesis, University of Alberta, Edmonton, Canada.

Choudhury, G. S. and Le Maguer, M. 1986. Prediction of water activity in glucose solution using the UNIFAC model. In *Food Engineering and Process Applications,* p. 171, LeMaguer, M., and Jelen, P. (ed.). Elsevier, New York.

Christensen, C., Sander, B., Fredenslund, A. A., and Rasmussen, P. 1983. Towards the extension of UNIFAC to mixtures with electrolytes. *Fluid Phase Equil.* 13: 297.

Chuang, L. and Toledo, R. T. 1976. Predicting the water activity of multicomponent systems from water sorption isotherms of individual components. *J. Food Sci.* 41: 922.

Cukor, P. M. and Prausnitz, J. M. (1969). *Intl. Chem. Eng. Symp.* Ser. No. 32 (Instn. Chem. Engrs., London) 3: 88.

Derr, E. L. and Deal, C. H. 1969. Analytical solutions of groups: Correlation of activity coefficients through structural group parameters. I. *Chem. E. Symp. Ser.* (No. 32), London 3: 40.

Dunning, W. J., Evans, H. C., and Taylor, M. 1951. The vapor pressures of concentrated aqueous sucrose solutions up to the pressure of 760 mm. *J. Chem. Soc.* 2363–72.

Eyring, H. and Marchi, R. P. 1963. Significant structure theory of liquids. *J. Chem. Educ.* 40: 562.

Fast, J. D. 1970. *Entropy,* 2nd ed. McMillan and Co., London.

Ferro Fontan, C. and Chirife, J. 1981. Technical note: a refinement of Ross's equation for predicting the water activity of non-electrolyte mixtures. *J. Food Technol.* 16: 219.

Flemr, V. 1976. A note on excess Gibbs energy equations based on the local composition concept. *Collect. Czech. Chem. Commun.* 41: 3347.

Flory, P. J. 1941. Thermodynamics of highpolymer solutions. *J. Chem. Phys.* 9: 660.

Flory, P. J. 1942. Thermodynamics of high polymer solutions. *J. Chem. Phys.* 10: 51.

Flory, P. J. 1953. *Principles of Polymer Chemistry.* Cornell University Press, Ithaca, New York.

Franks, F. 1977. Solvation and conformational effects in aqueous solutions of biopolymer analogues. *Philos. Trans. Roy. Soc.* (London), Ser. B. 278: 33.

Fredenslund, A. A., Gmehling, J., and Rasmussen, P. 1977. *Vapor-Liquid Equilibria using UNIFAC,* A Group Contribution Method. Elsevier, Amsterdam.

Fredenslund, A. A., Jones, R. L., and Prausnitz, J. M. 1975. Group-contribution estimation of activity coefficients in nonideal liquid mixtures. *AlChE J.* 21: 1086.

Garrod, J. E. and Herrington, T. M. 1970. Apparent molar volumes and temperatures of maximum density of dilute aqueous sucrose solutions. *J. Phys. Chem.* 74: 363.

Gmehling, J. and Kolbe, B. 1983. Limitations of modern expressions for the excess Gibbs energy. *Fluid Phase Equil.* 13: 227.

Gmehling, J. and Onken, U. 1977. DECHEMA Chemistry Data Series, Vol. I, part 1, *Vapor-Liquid Equilibrium Data Collections.* DECHEMA, Frankfurt, Germany.

Gmehling, J., Onken, U., and Arlt, W. 1981. DECHEMA Chemistry Data Series, Vol. I, Parla, *Vapor-Liquid Equilibrium Data Collection.* DECHEMA, Frankfurt, Germany.

Gonzalez, C. and Van Ness, H. C. 1983. Excess thermodynamic functions for ternary systems, 9. total-pressure data and G^E for water/ethylene glycol/ethanol at 50°C. *J. Chem. Eng. Data* 28: 410.

Grover, D. W. and Nicol, J. M. 1940. The vapor pressure of glycerol solution at 20°C. *J. Soc. Chem. Ind.,* Lond. 59: 175.

Gucker, F. T., Jr., Pickard, H. B., and Plank, R. M. 1939. A new microcalorimeter: the heats of dilution of aqueous solutions of sucrose at 20 and 30° and their heat capacities at 25°C. *J. Am. Chem. Soc.* 61: 459.

Guggenheim, E. A. 1935. The statistical mechanics of regular solutions. *Proc. Roy. Soc.* (London) A. 148: 304.

Guggenheim, E. A. 1952. *Mixtures.* Oxford University Press, Oxford.

Hanks, R. W., Tan, R. L., and Christensen, J. J. 1978. Limits on the simultaneous correlation of G^E and H^E data by the NRTL, LEMF and Wilson's equations. *Thermochim. Acta* 23: 41.

Henley, E. J. and Seader, J. D. 1981. *Equilibrium-Stage Separation Operations in Chemical Engineering.* John Wiley & Sons, New York.

Hildebrand, J. H. 1929. Solubility. XII. Regular solutions. *J. Am. Chem. Soc.* 51: 66.

Hildebrand, J. H., Prausnitz, J. M., and Scott, R. L. 1970. *Regular and Related Solutions—The Solubility of Gases, Liquids, and Solids.* Van Nostrand Reinhold Company, New York.

Hildebrand, J. H. and Scott, R. L. 1962. *Regular Solutions.* Prentice-Hall, New York.

Honig, P. 1953. *Principles of Sugar Technology,* Vol. 1. Elsevier, New York.

Huggins, M. L. 1941. Solutions of long chain compounds. *J. Chem. Phys.* 9: 440.

Huggins, M. L. 1942. Thermodynamic properties of solutions of long-chain compounds. *Ann. N.Y. Acad. Sci.* 43: 1.

Karel, M. 1975. Stability of low and intermediate moisture foods. In *Freeze-Drying and Advanced Food Technology.* Academic Press, London.

Kehiaian, H. V., Grolier, J. P. E., and Benson, G. C. 1978. Thermodynamics of organic mixtures. A generalized quasichemical theory in terms of group surface interactions. *J. Chim. Phys. Chim. Biol.* 75: 1031.

Kehiaian, H. V., Sosnkowska-Kehiaian, K. and Hryniewicz, R. 1971. Enthalpy of mixing of ethers with hydrocarbons at 25°C and its analysis in terms of molecular surface interactions. *J. Chim. Phys. Phys. Chim. Biol.* 68: 922.

Kikic, I., Alessi, P., Rasmussen, P., and Fredenslund, A. A. 1980. On the combinatorial part of the UNIFAC and UNIQUAC models. *Can. J. Chem. Eng.* 58: 253.

Labuza, T. P. 1975. Oxidative changes in foods at low and intermediate moisture levels. In *Water Relations of Foods*. Academic Press, London.

Lebert, A. and Richon, D. 1984. Study of the influence of solute (n-alcohols and n-alkanes) chain length on their retention by purified olive oil. *J. Food Sci.* 49: 1301.

Lee, N. S. 1987. A thermodynamic model for aqueous food solutions. M. Sc. thesis, University of Alberta, Edmonton, Canada.

Le Maguer, M. 1981. A thermodynamic model for terpene-water solutions. In *Water Activity: Influences on Food Quality,* L. B. Rockland and G. F. Stewart (Ed.), p. 347. Academic Press, New York.

Le Maguer, M. 1985. Solution thermodynamics and the starch-water system. In *Properties of Water in Foods in Relation to Quality and Stability,* Simatos, D. and Mutton, J. L. (Ed.), p. 133. Martinus Nijhoff Publishers, Boston.

Le Maguer, M. 1987. Mechanics and influence of water binding on water activity. In *Water Activity: Theory and Applications to Food,* L. R. Beuchat (Ed.), p. 1. Marcel Dekker, Inc., New York.

Le Maguer, M. 1989. Thermodynamic properties for water removal processes in solid and liquid foods. In *Food Properties and Computer-aided Engineering of Food Processing Systems,* R. P. Singh and A. G. Medina (Ed.), p. 157. Kluwer Academic Publishers, Dordrecht, The Netherlands. NATO ASI Series.

Lerici, C. R., Piva, M., and Rosa, M. D. 1983. Water activity and freezing point depression of aqueous solutions and liquid foods. *J. Food Sci.* 48: 1667.

Loncin, Marcel. 1988. Activity of water and its importance in preconcentration and drying of foods. In *Preconcentrations and Drying of Food Materials,* S. Bruin (Ed.), p. 15. Elsevier Science Publishers, B. V., Amsterdam.

Luecke, R. H. 1983. Correlations in published values of UNIQUAC interaction parameters. *Fluid Phase Equil.* 14: 373.

Lui, K. K. 1989. Optimization of the UNIQUAC parameters for aqueous carbohydrate solutions. Ph.D. thesis, University of Alberta, Edmonton, Canada.

Lumry, R. and Rajander, S. 1970. *Enthalpy-Entropy Compensation Phenomena in Water Solutions and Small Molecules,* M. Goodman (Ed.). Interscience, New York.

Malanowski, S., Skjold-Jorgensen, S., Rasmussen, P. and Fredenslund, A. A. 1981. Simultaneous representation of binary VLE, LLE and HE data using the UNIQUAC model. *Chem. Eng. Sci.* 36(10): 1727.

Maurer, G. and Prausnitz, J. M. 1978. On the derivation and extension of the UNIQUAC equation. *Fluid Phase Equil.* 2: 91.

McDermott, C. and Ashton, N. 1977. Note on the definition of local composition. *Fluid Phase Equil.* 1: 33.

Money, R. W. and Born, R. 1951. Equilibrium humidity of sugar solutions. *J. Sci. Food Agric.* 2: 180.

Monfort, J. P. and Rojasr, M. D. L. 1978. A study of simplified molecular models in phase equilibrium prediction. *Fluid Phase Equil.* 2: 181.

Nath, A. and Bender, E. 1981. On the thermodynamics of associated solutions I & II. *Fluid Phase Equil.* 7: 275.

Nath, A. and Bender, E. 1983. Isothermal vapor-liquid equilibria of binary and ternary mixtures containing alcohol, alkanolamine, and water with a new static device. *J. Chem. Eng. Data* 28: 370.

Norrish, R. S. 1966. An equation for the activity coefficients and equilibrium relative humidities of water in confectionery syrups. *J. Fd. Technol.* 1: 25.

Novak, J. P., Matous, J., and Pick, J. 1987. *Liquid-Liquid Equilibria.* Elsevier, Amsterdam.

Oishi, T. and Prausnitz, J. M., 1978. Estimation of solvent activities in polymer solutions using a group-contribution method. *Ind. Eng. Chem. Process Des. Dev.* 17(3): 333.

Orye, R. V. and Prausnitz, J. M. 1965. Multicomponent equilibria with the Wilson equation. *Ind. Eng. Chem.* 57(5): 18.

Prausnitz, J. M. 1969. *Molecular Thermodynamics of Fluid-Phase Equilibria.* Prentice-Hall, Englewood Cliffs, NJ.

Prausnitz, J. M. 1977. State-of-the-art review of phase equilibria. In *Phase Equilibria and Fluid Properties in the Chemical Industry,* T. S. Storvick and S. I. Sandler, (Ed.), p. 11. ACS Symp. Ser. 60. ACS, Washington, DC.

Prausnitz, J. M., Lichtenthaler, R. N., and de Azevedo, E. G. 1986. *Molecular Thermodynamics of Fluid-Phase Equilibria,* 2nd ed. Prentice-Hall, Englewood Cliffs, NJ.

Prigogine, I., Bellemans, A., and Mathot, V. 1957. *The Molecular Theory of Solutions.* North-Holland Publishing Company, Amsterdam.

Prigogine, I. and Defay, R. 1954. *Chemical Thermodynamics.* Longmans, Green & Co., London.

Rasmussen, P. 1983. Development of G^E models including group contributions. *Fluid Phase Equil.* 13: 213.

Redlich, O. and Kister, A. T. 1948. Algebraic representation of the thermodynamic properties and the classification of solutions. *Ind. Eng. Chem.* 40: 345.

Reid, R. C., Prausnitz, J. M., and Sherwood, T. K. 1977. *The Properties of Gases and Liquids,* 3rd ed. McGraw-Hill, New York.

Renon, H. 1985. NRTL: An empirical equation or an inspiring model for fluid mixture properties. *Fluid Phase Equil.* 24: 87.

Renon, H. and Prausnitz, J. M. 1968. Local compositions in thermodynamic excess functions for liquid mixtures. *AIChE. J.* 14: 135.

Robinson, R. A. and Stokes, R. H. 1959. *Electrolyte Solutions*. 2nd ed. Butterworths, London.

Robinson, R. A. and Stokes, R. H. 1961. Activity coefficients in aqueous solutions of sucrose, mannitol and their mixtures at 25°C. *J. Phys. Chem.* 65: 1954.

Ross, K. D. 1975. Estimation of water activity in intermediate moisture foods. *Food Technol.* 29(3): 26.

Rowlinson, J. S. 1969. *Liquids and Liquid Mixtures,* 2nd ed. Butterworth, London.

Ruegg, M. and Blanc, B. 1981. The water activity of honey and related sugar solutions. *Lebensm.-Wiss. u.-Technol.* 14: 1.

Sandler, S. I. 1977. *Chemical and Engineering Thermodynamics,* Ch. 7. John Wiley & Sons, New York.

Saravacos, G. D., Karathanos, V., and Marino-Kovris, D. 1990. Volatility of fruit aroma compounds in sugar solutions. In *Flavors and Off-Flavors*. Proceedings of the 6th. International Flavor Conference, Rethymnon, Crête. 5–7 July, 1989. C. Charalambous (Ed.), p. 729. Elsevier Science Publishers, B. V. Amsterdam.

Scatchard, G., Hamer, W. J., and Wood, S. E. 1938. Isotonic solutions. I. The chemical potential of water in aqueous solutions of sodium chloride, potassium chloride, sulphuric acid, sucrose, urea and glycerol. *J. Am. Chem. Soc.* 60: 3061.

Scott, R. L. 1956. Corresponding states treatment of nonelectrolyte solutions. *J. Chem. Phys.* 25: 193.

Scott, W. J. 1957. Water relations of food spoilage microorganisms. *Adv. J. Biol. Sci.* 6: 549.

Seaton, N. A. and Glandt, E. D. 1985. Coordination number and local composition in binary mixtures of spherical adhesive particles. *Fluid Phase Equil.* 23: 165.

Skjold-Jorgensen, S., Rasmussen, P. and Fredenslund, A. A. 1980. On the temperature dependence of the UNIQUAC/UNIFAC models. *Chem. Eng. Sci.* 35: 2389.

Skjold-Jorgensen, S., Rasmussen, P., and Fredenslund, A. A. 1982. On the concentration dependence of the UNIQUAC/UNIFAC models. *Chem. Eng. Sci.* 37: 99.

Sorensen, J. M., Magnussen, T., Rasmussen, P. and Fredenslund, A. A. 1979a,b, 1980. Liquid-liquid equilibrium data: their retrieval, correlation and prediction, part I-III. *Fluid Phase Equil.* 2: 297; 3: 47; 4: 151.

Sorrentino, F., Voilley, A., and Richon, D. 1986. Activity coefficients of aroma compounds in model food systems. *AlChE J.* 32(12): 1988.

Staverman, A. 1950. The entropy of high polymer solutions. *Rec. Trav. Chem. Pays-bas* 69: 163.

Stokes, R. H. and Robinson, R. A. 1966. Interactions in aqueous nonelectrolyte solutions. I. Solute-solvent equilibria. *J. Phys. Chem.* 70(7): 2126.

Stroth, L. and Schonert, H. 1977. Excess enthalpies and excess volumes of water + sucrose + mannitol at 298.15K. *J. Chem. Thermodynamics* 9: 851.

Taylor, J. B. and Rowlinson, J. S. 1955. The thermodynamic properties of aqueous solutions of glucose. *Trans. Faraday Soc.* 51: 1183.

Troller, J. A. and Christian, J. H. B. 1978. *Water Activity and Food.* Academic Press, New York.

Van Den Berg, C. and Bruin, S. 1986. Water activity and its estimation in food systems: theoretical aspects. In *Water Activity: Influences on Food Quality,* L. B. Rockland and G. F. Stewart (Ed.), p.l. Academic Press, New York.

Van Laar, J. J. 1910. Über Dämpfspannungen von binaren Gemischen. *Z. Physik Chem.* 72: 723.

Van Laar, J. J. 1913. Zur Theorie der Dämpfspannungen von binaren Gemischen. *Z. Physik Chem.* 83: 599.

Vera, J. H., Sayegh, S. G., and Ratcliff, G. A. 1977. A quasi-lattice-local composition model for the excess Gibbs free energy of liquid mixtures. *Fluid Phase Equil.* 1: 113.

Vonka, P., Novak, J. P., Suska, J., and Pick, J. 1975. An a priori analysis of temperature dependence of Wilson and NRTL equations. *Chem. Eng. Commun.* 2: 51.

Walrafen, G. E. 1966. Raman spectral studies of the effects of urea and sucrose on water structure. *J. Chem Phys.* 44: 3726.

Wang, W. and Chao, K. C. 1983. The complete local concentration model activity coefficients. *Chem. Eng. Sci.* 38(9): 1483.

Wilson, G. M. 1964. A new expression for the excess free energy of mixing. *J. Am. Chem. Soc.* 86: 127.

Wohl, K. 1946. Thermodynamic evaluation of binary and ternary liquid systems. *Trans. AlChE* 42: 215; errata: *Chem. Eng. Progress* 49: 218 (1953).

Yalkowsky, S. H. and Valvani, S. C. 1980. Solubility and partitioning I: Solubility of nonelectrolyte in water. *J. Pharmaceutical Sci.* 69(8): 912.

Zapsalis, C. and Beck, R. A. 1985. *Food Chemistry and Nutritional Biochemistry*. John Wiley & Sons, New York.

Zou, M. and Prausnitz, J. M. 1987. Vapor-liquid and liquid-liquid equilibria in binary aqueous systems. *J. Chem. Eng. Data* 32: 34.

2

Solid-Liquid Equilibrium: Crystallization in Foods

Richard W. Hartel

University of Wisconsin
Madison, Wisconsin

INTRODUCTION

Crystallization of various components in foods plays an important role in determining overall texture and physical properties. Control of crystallization processes allows development of the appropriate or required properties in the food. Understanding the physical principles underlying crystallization processes is critical to controlling crystallization. This requires an understanding of the solid-liquid equilibrium and dynamic conditions underlying crystallization mechanisms.

There are many components of foods that may crystallize under certain conditions. These include sugars, salts, fats, ice, proteins, polysaccharides, and organic acids. Control of crystallization in foods is required for development of appropriate product texture. Understanding when and to what extent these components crystallize under certain conditions is the basis for controlling operations where crystallization is of concern. For example, in confectionery products such as fondants and creams, control of sugar crystallization is crucial to produce a large number of fine crystals that give the desired texture. Similarly, tempering of chocolates requires control of crystallization rates in addition to crystallization in the proper polymorphic form to produce a chocolate that has good snap and sheen as well as good stability during storage. Improper tempering can result in dull, soft chocolates that bloom quite readily upon storage.

While controlling crystallization in foods is certainly crucial in controlling product quality, there remains an art about the process. That is, many food processes are based on recipes and procedures passed on through the generations which require an experienced operator to ensure a high rate of success. Although the understanding of crystallization processes in foods is advancing, considerable work is still required to completely control these processes.

Control of crystallization may involve either promoting or preventing crystal formation and growth. Formation and growth is controlled to yield the desired textural properties, whereas crystallization is prevented, in situations where it might occur, to produce different textural or physical properties. For example, some caramels are made so that sugar crystallizes to yield a short texture, or to give a clean bite, while others are formulated and prepared to give a chewy texture, where sugar crystallization would be a defect. Therefore, crystallization processes need to be understood in terms of both prevention and development of the appropriate size distribution of crystals.

In foods, sugar crystallization is important in confectionery products, some dairy products, refining of sugars, and in sugar coatings on cereals (Hartel and Shastry, 1991). In confectionery products, crystallization is controlled to give the proper size distribution in grained confections such as fondant, cream, fudge and some panned goods, while crystallization is considered a defect in such products as hard candy, nougats, and toffees. Dairy products where sugar crystallization may be important are sweetened condensed milk and ice cream, where small crystals are desired in the first and prevention of lactose crystallization is desired in the latter. Refined sugars include sucrose, lactose, glucose, and fructose. Refining processes are generally much different than sugar crystallization in food products since in refining, sugars crystallize from a stirred suspension. Sugar coatings on cereals require careful operations to either give a stable crystallized coating or to prevent crystallization to give an amorphous product. Sugar coatings during hard or soft panning of confections may be similar.

Crystallization of ice is one of the most important aspects of food preservation during freezing. Removal of water from solution through formation of ice aids in food preservation by lowering the water activity of the remaining solution phase and helps to prevent microbiological growth. Controlling ice crystal formation is critical to good storage quality since the size and location of ice crystals within frozen foods determines how well the original food structure was maintained. Controlling ice crystallization is also important in frozen desserts such as ice cream or frozen yogurt. Here, many small ice crystals are desired to give the creamy, smooth texture typical of fine ice cream. Control of formulation and processing conditions are important for developing this crystal size distribution (Arbuckle, 1986). In addition, ice crystallization is important in freeze concentration where a fluid food is concentrated by partial freezing of water followed by removal of pure ice. Large ice crystals of uniform size and shape are desired for efficient separation (Heist, 1979).

Lipids can also crystallize to give texture and quality aspects to foods. Products in which lipid crystallization is important include chocolates and confectionery coatings, spreadable icings, dairy products such as butter and cream, vegetable spreads (margarine), and peanut butter. An understanding of fat crystallization processes in these products plays a critical role in determining overall product quality (Sato and Garti, 1988). In addition, lipid crystallization is important in fat and oil fractionation, where lipid products with varying melting and physical properties are made (usually) from vegetable oils by melt crystallization. For most food products, the desired crystallization yields large numbers of small crystals of the proper polymorph for good product stability (as in chocolate tempering). However, in fractionation processes, larger, uniformly shaped crystals are generally required to give efficient separation.

While the thermodynamic principles of crystallization apply to each of the above processes and products, the details of thermodynamic theories vary slightly depending on whether crystallization occurs from a solution (solute crystallizes out of solvent) or from a melt (solvent crystallizes from melt out of impurities). Control of crystallization processes depends on an understanding of these principles as well as on the effects of heat and mass transfer rates during processing.

CONTROL OF CRYSTALLIZATION

Several factors affect control of crystallization during food processing. These include the thermodynamic principles underlying crystallization processes, the interaction of engineering principles of heat and mass transfer during processing, formulation factors, and postprocessing parameters and storage conditions.

The thermodynamic principles underlying crystallization processes determine the equilibrium conditions that would apply given sufficient time and energy during processing. These principles relate to solution structure and the mechanisms of crystal nucleation and growth. However, equilibrium thermodynamics is not sufficient for quantifying or controlling crystallization processes in foods since dynamic constraints often overwhelm equilibrium conditions. These dynamic constraints are related to the rates of heat and mass transfer experienced by the food during processing and the interaction of the specific components of the food. The manner by which the thermodynamic driving force for crystallization is achieved and the rate of development of this driving force determine the rates of formation and growth of crystals. This is determined by the rates of heat and mass transfer experienced by the product during processing.

The effects of other, noncrystallizing components in the product may control the rate of crystallization in many foods. These other components may affect the thermodynamic driving force, affect the heat and mass transfer rates, or affect specific crystallization mechanisms in such a way that crystallization can be dramatically altered. For example, the presence of milkfat triglycerides in milk chocolate results in different crystallization parameters for the chocolate and results in different tempering requirements. The effects of each of the numerous components likely to be present in foods, the synergistic effects of various components, and the variability of these components in product formulations make predictions of specific effects on crystallization quite difficult and results in much of the qualitative nature of crystallization in foods. Optimal tempering profiles for a new chocolate formulation cannot be predicted from first principles of crystallization but rather are developed by experience and through trial and error. This is generally true where crystallization in foods is concerned.

Once a product has been manufactured with the appropriate physical properties and texture, proper storage conditions must be maintained to enhance storage stability. Temperature and relative humidity are generally important in storage stability and also play an important role in stability of crystallization results. Storage at high relative temperatures and high relative humidities is particularly bad. For example, storage of frozen foods is enhanced by maintaining extremely low temperatures. However, even at low temperatures, thermal cycling during storage can cause considerable changes in crystallization products. The results of storing ice cream in a frost-free home freezer are easily seen as the product coarsens to the point where it is unacceptable (Keeney, 1979). Similar changes in crystal size take place during storage of chocolates, some sugar confections, and frozen foods under cycling temperature conditions.

While understanding thermodynamic principles is important in controlling crystallization in food processing, it is evident from the above discussion that other important parameters must be considered as well. Overall control of food crystallization will require complete understanding of each of these effects. This in turn will require further work in each area since there are significant gaps in our understanding in applications of engineering principles to specific food processes, in determining the specific effects of food components on crystallization processes, and how storage conditions affect changes in crystal size distributions. Even in the field of crystallization thermodynamics, there are significant gaps in understanding that result in poor control of crystallization in food processing. For example, solutions are notorious for either crystallizing when they should not or not crystallizing when they should. Our understanding of nucleation mechanisms is limited and results in poor control of this process.

SOLUTIONS

Crystallization can be broken down into several steps. The first is generation of sufficient driving force so that crystallization can occur. This is related to solution structure and development of supersaturation. Once the appropriate driving force has been attained, nucleation occurs where crystals are generated by bringing molecules together in such a way that a crystal lattice can be formed. Once a lattice has formed, further growth occurs either by molecular addition or coagulation/aggregation with other nuclei.

Solution Structure

The relation between solution structure and crystallization has been recognized for a long time. However, our understanding of this relationship is still somewhat limited, and research continues in this important area.

A number of experimental techniques have been employed to define the solution structure prior to crystallization. These can generally be divided into two groups: those that probe solute molecules and those that probe solvent molecules. These techniques generally suffer from trying to extract information on a molecular level from macroscopic data. These shortcomings result in the lack of exact understanding of solutions prior to crystallization.

Several experiments have shown that a supersaturated solution allowed to sit in a long vertical column will develop measurable density gradients (Allen et al., 1974). This has been explained as settling of molecular clusters (either heavier or lighter than the surrounding solvent solution) thought to be the precursors to nuclei. It is logical to think that molecules begin to interact as their concentration increases until eventually they interact to the point where a nucleus can be formed and become stable. However, there is no clear evidence that these density gradients relate to clusters that are nuclei precursors, since one would expect that these clusters would be observed using other techniques. There has been little evidence for these clusters from other experiments, for example those that use vibrational spectroscopy to investigate the molecular interactions (Mathlouthi et al., 1986). One possible exception utilized Raman spectroscopy to observe changes in solution structure near a growing crystal (McMahon et al., 1984). Evidence was found for clusters of molecules forming in concentrated solutions of KNO_3 with increased clustering at increased concentrations. Molecular changes are also evident in sugar solutions, for example, as concentration increases, as found using NMR spectroscopy (Richardson et al., 1987). In general, there appears to be increased interaction between sugar molecules and decreased mobility of both water and sugar molecules. However, the exact nature of these interactions cannot be gleaned from spectroscopic data.

Changes in physical properties of solutions have also been studied to investigate the onset of nuclei formation. Primary of these properties have been diffusivities and viscosities since these properties relate directly to molecular mobility and the ability to form nuclei. Data on a variety of salts and organic material have shown that diffusivities decrease dramatically at elevated concentrations (highly supersaturated) while viscosities increase substantially as well (Myerson et al., 1990). These trends can be seen figuratively in Fig. 2.1. These data indicate the decreasing mobility of molecules under these extreme conditions but cannot specifically point out the interactions required for nucleation to occur.

A recent technique for investigating molecular properties in concentrated solutions prior to crystallization employs a molecular marker whose fluorescence spectrum changes as the molecular environment (protonation of a side group) around this molecule changes (Chakraborty and Berglund, 1990). When used in sugar solutions, these molecular changes indicated changes in the relative ratio of bulk water to solvation water as sugar concentration increased. Solvation water molecules remained approximately constant in the range 10–80% (w/w) sugar, while the bulk water decreased exponentially as concentration increased. Future use of this molecular probe technique should shed some light on this confusing

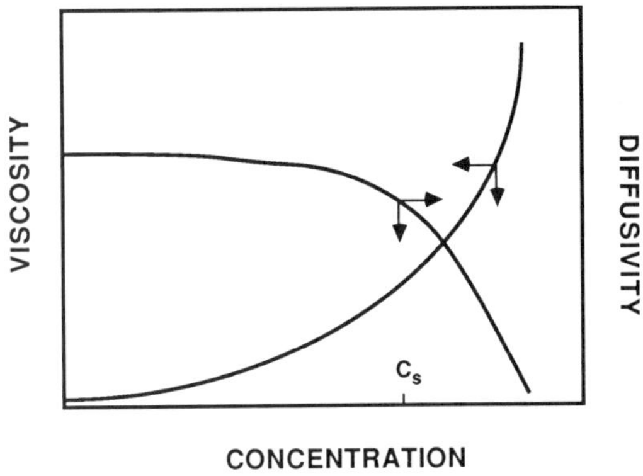

Fig. 2.1 Diffusivity and viscosity changes in supersaturated solutions.

topic of solution structure leading to crystallization and, in conjunction with spectroscopic and macroscopic techniques,should eventually lead to a full understanding of solution structure as relates to crystallization processes.

Phase Diagrams and Equilibrium Crystal Forms

Equilibrium crystal forms of substances can be detailed on a concentration-temperature plot (phase diagram) where the particular crystal form that arises at a given condition is labeled. For example, Fig. 2.2 shows the equilibrium phase diagram for D-glucose (Young, 1957). Depending on the temperature and concentration, glucose can crystallize in one of several forms including either of the two reducing sugar anomers (α or β) or as a monohydrate.

Lipid food systems crystallize in different polymorphic forms depending on the temperature and specific mixture of triglycerides (Sato, 1988). These polymorphic forms are different packing structures of triglyceride molecules resulting from different solution energetics. The exact nomenclature and numbering of these polymorphic forms is still a matter of current debate, although several standard schemes have been presented. These uncertainties are the result of the complex nature of lipid crystallization. Under the appropriate conditions, polymorphic transformations occur where the least stable (α polymorph) transform into the most stable (β polymorph). The α-to-β′ transformation is generally quite rapid while the β′-to-β transformation occurs over much longer periods of time (Sato, 1988).

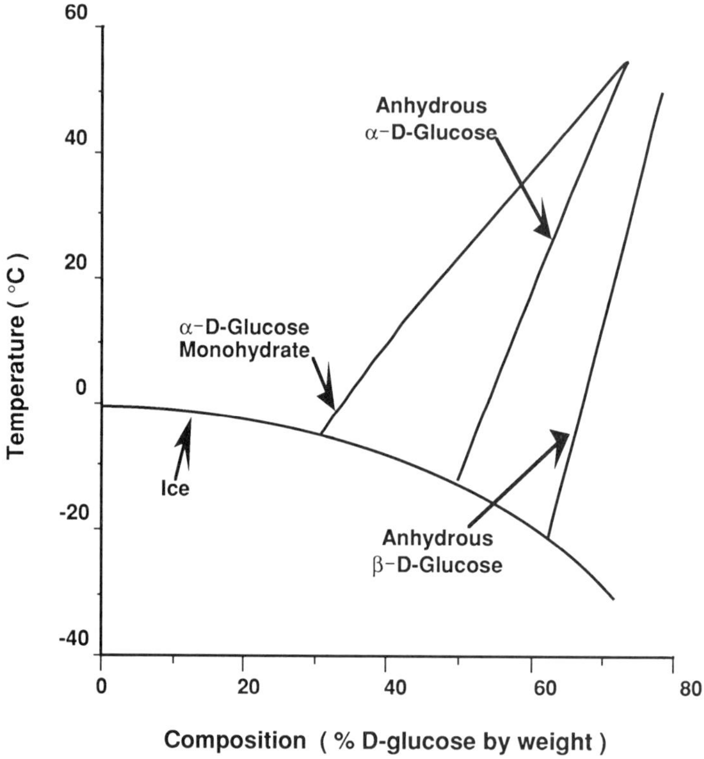

Fig. 2.2 Phase diagram for D-glucose-water. (*From Young, 1957.*)

Solubility and Supersaturation

Phase diagrams are useful when dealing with equilibrium conditions, however, the dynamic aspects of crystallization often overwhelm equilibrium concerns. In addition, phase diagrams change depending on the types and levels of impurities or additives present in the crystallization system. Therefore, these phase diagrams are of limited value when trying to predict or control crystallization processes in foods. Their main value lies in determining solubility conditions for crystallization systems, although the effects of additives or impurities are not always known.

The solubility concentration of a solute in a solvent is defined when the chemical potential of the solute in solution is identical to the chemical potential of a pure crystal of infinite size. This concentration is the maximum amount of solute allowed in solution at a given temperature. For melt systems (such as fats and water), the solid-liquid equilibrium curve is the melting point of the crystal. Small particles have a slightly elevated solubility due to surface tension effects (Mullin, 1973). This elevated solubility (known as either the Ostwald-Freundlich, Gibbs-Thompson, or Gibbs-Kelvin effect) increases at smaller sizes, however, only extremely small crystals (1–10 μm, depending on the system) exhibit this increased solubility to any measurable extent (Mullin, 1973).

Solubilities vary with type of solute, type of solvent, and temperature. For example, typical solubility curves for various sugars can be seen in Fig. 2.3 (Hartel and Shastry, 1991). Lactose exhibits the least solubility of the common sugars while fructose has the highest with near 90%. Solubilities also vary with temperature. In general, most substances exhibit increasing solubility with temperature, as in sugars, however, some salts exhibit a negative solubility curve (Mullin, 1973). Soluble salts such as calcium and magnesium carbonate decrease in solubility as temperature increases causing them to crystallize out of solution as temperature goes up.

The driving force for crystallization processes is a difference in chem-

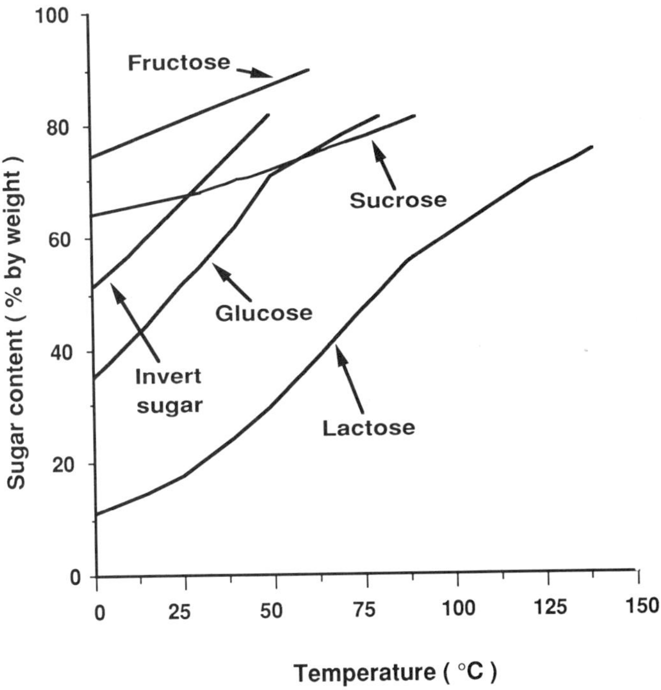

Fig. 2.3 Solubilities of various sugars. (*From Pancoast and Junk, 1980.*)

ical potential (or partial molar Gibbs free energy) between the solution phase and the crystalline solid. This chemical potential difference is satisfied by transformation of molecules from liquid to solid, or crystallization. The larger the chemical potential difference, the greater the driving force for crystallization. In solution systems (such as sugars, salts, proteins, etc.), the driving force can be related through thermodynamic theory to a concentration driving force, although this requires some approximation (Mullin and Sohnel, 1977). Thermodynamically, the driving force is the chemical potential difference.

$$\mu_{solid} - \mu_{liquid} = -\Delta\mu = \phi \tag{1}$$

Here, μ_{solid} and μ_{liquid} are the chemical potentials of solid and liquid phases, respectively, and ϕ is the chemical affinity. Chemical potential can be related to activity as

$$\mu = \mu^\circ + RT \ln (a) \tag{2}$$

where μ° is a reference state potential, R is the gas constant, T is absolute temperature, and a is activity. Depending on the nature of the solution (ideality), activities can be related to concentrations and supersaturation S then defined as concentration ratio between the actual solution condition and the solubility concentration at that temperature (C/C_s). A further approximation is generally made to relate the concentration ratio to the activity ratio. Thus,

$$\phi = RT \ln (a/a_{eq}) \approx RT \ln (C/C_s) \tag{3}$$

where a_{eq} is the activity of the equilibrium solid condition, C is concentration, and C_s is solubility concentration. At low supersaturation levels ($S < 1.1$), a Taylor series expansion of the natural logarithm of the supersaturation is often used to further simplify Eq. (3) to give

$$\phi \approx RT (S - 1) = RT (s) \tag{4}$$

where $s = S - 1 = (C - C_s)/C_s$. Thus, s or $S - 1$ may be used as an approximate supersaturation driving force at low supersaturations. Another expression for supersaturation that is often used is the concentration difference ΔC:

$$\Delta C = C - C_s \tag{5}$$

For high supersaturations, it is best to use

$$S' = \ln (C/C_s) \tag{6}$$

since the approximation for low supersaturation does not hold. If activities are known, it is best to use the ratio of activities as expressed in Eq. (3). However, activities are generally not known, even for pure substances, at high concentrations (van Hook, 1987). There is virtually no activity data available for solutions of importance in complex food formulations (Dunning, 1977).

The magnitude of supersaturation varies depending on the units used to express it. The choice of units depends on the desired results. Convenient concentration units based on weight percent are used in commercial applications, while more thermodynamically correct units (i.e., mole fractions) must be used when developing crystallization mechanisms through kinetic relationships. Some differences in standard methods of describing supersaturation are shown in Table 2.1 for sucrose at different temperatures and concentrations. Again, the most appropriate measure of supersaturation is the ratio of activities, however, these are seldom

Table 2.1 Sucrose Supersaturations Expressed in Different Units

Solution composition	C	C_s	ΔC	S	s
70% (w/w) sucrose at 20°C					
g sucrose/kg water	2333.3000	2039.0000	294.3000	1.1443	0.1443
g sucrose/kg solution	700.0000	670.9000	29.1000	1.0434	0.0434
g sucrose/L solution	943.1000	891.7000	51.4000	1.0576	0.0576
Mole sucrose/L solution	2.7550	2.6073	0.1477	1.0566	0.0566
Mole fraction sucrose	0.1093	0.0969	0.0124	1.1277	0.1277
80% (w/w) sucrose at 20°C					
g sucrose/kg water	4000.0000	2039.0000	1961.0000	1.9617	0.9617
g sucrose/kg solution	800.0000	670.9000	129.1000	1.1924	0.1924
g sucrose/L solution	1129.4000	891.7000	237.7000	1.2666	0.2666
Mole sucrose/L solution	3.2990	2.6073	0.6917	1.2653	0.2653
Mole fraction sucrose	0.1738	0.0969	0.0769	1.7932	0.7932
80% (w/w) sucrose at 40°C					
g sucrose/kg water	4000.0000	2381.0000	1619.0000	1.6800	0.6800
g sucrose/kg solution	800.0000	704.2000	95.8000	1.1360	0.1360
g sucrose/L solution	1120.2400	963.5429	156.6971	1.1626	0.1626
Mole sucrose/L solution	3.2727	2.8174	0.4553	1.1616	0.1616
Mole fraction sucrose	0.1738	0.1114	0.0624	1.5607	0.5607

known for food solutions. Thus, approximations must be made, resulting in potential errors when attempting to relate crystallization kinetics to the thermodynamic driving force. These errors are described in detail for sugar solutions by Mullin and Sohnel (1977).

Supersaturations can be created in solution systems in one of three ways, as demonstrated schematically in Fig. 2.4 (Hartel and Shastry, 1991). The example of crystallizing sugar from solution will be used to demonstrate these principles. One way to get aqueous sugar solutions to become supersaturated is to heat the water, add sugar, and then cool to a point where the concentration has exceeded the solubility concentration at that temperature. This follows line AB in Fig. 2.4. This technique is used for making rock candy at home. Another possibility is to allow some of the water to evaporate, resulting in an increased sugar concentration. This process follows line AC in Fig. 2.4 and is typical of sugar refining operations where crystallization is controlled in a vacuum evaporation pan. Another alternative is to add a second solvent in which the sugar is insoluble. The resulting mixture has a new solubility curve, and the total concentration may now be above the saturation concentration of the mix-

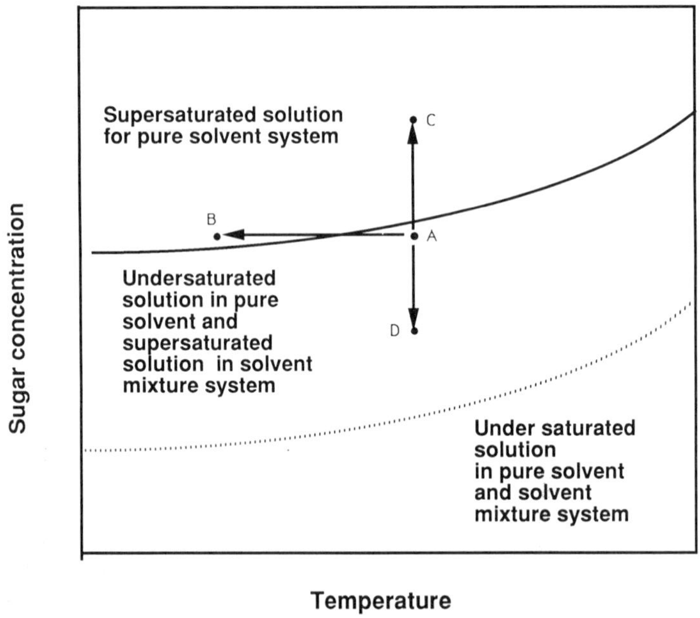

Fig. 2.4 Development of supersaturation in sucrose solutions by cooling (line AB), by evaporation (line AC), and by salting-out (line AD). (*From Hartel and Shastry, 1991.*)

ture. This is shown in Fig. 2.4 by line AD, where the concentration drops slightly due to a dilution effect, and the new solubility curve is shown by the dashed line.

For melt systems, the thermodynamic driving force is proportional to the temperature difference between the crystallizing system and the melting point. This expression is given as (Doremus, 1985)

$$\phi = \Delta H_f(T_f - T)/T_f \tag{7}$$

where ΔH_f is the latent heat of formation, T_f is the equilibrium freezing (or melting) point, and T is the actual solution temperature. This, again, is an approximation based on the thermodynamic driving force of chemical potential difference, which depends on the mechanism of crystallization and the structure of the solid-liquid interface (Kirwan and Pigford, 1969). In many situations, the crystallization rate is related to the subcooling (T_f − T) raised to some power where the value of the exponent depends on the mechanism of growth.

Metastability

Many systems, particularly substances that form glasses, form solutions at low supersaturations that are stable to crystal formation for long periods of time. This is a dynamic condition since, thermodynamically, there exists a driving force for crystallization. However, the rate of nucleation of these low supersaturations is quite low, and the time scale for the nucleation event can be upwards of years. By definition, a solution that is stable to nucleation for some reasonable time scale of operation is called metastable. The extent of the metastable zone depends on the system under study and the types and extent of other components present. For pure sucrose, the metastable region has been fairly well delineated due to its importance in sugar refining. This is represented schematically in Fig. 2.5. It has been generally accepted that pure sucrose solutions under stagnant conditions can withstand a supersaturation ratio (C/C_s) of nearly 1.3 (for concentrations in grams sucrose per 100 grams solution) before spontaneous nucleation occurs (van Hook, 1959). Between supersaturation ratios of 1.2 and 1.3, however, nucleation may be forced by either agitation or addition of sucrose crystals as seeds. Vasatko and Smelik (1968) outlined the metastability zones for sucrose, glucose, and fructose, while supersolubility limits for lactose have been presented by Hunziker and Nissen (1927). Understanding these limits is crucial to control of crystallization processes in foods, where concentrations may be extremely high.

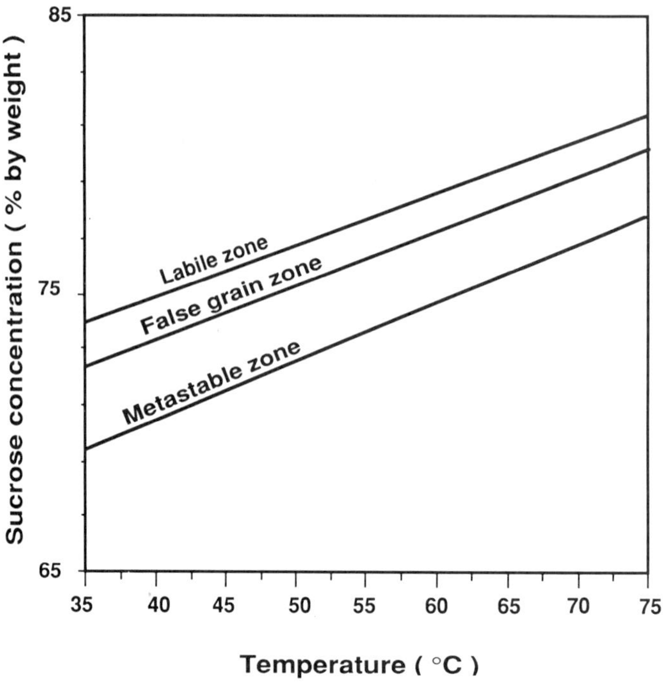

Fig. 2.5 Supersolubility curve for pure sucrose. (*From van Hook, 1959.*)

Sugar solutions held below the supersolubility limit (limit for spontaneous nucleation) will eventually nucleate as predicted by nucleation theory. However, the time scale for nucleation may be orders of magnitude longer than the available processing time. If sucrose solutions at supersaturation ratio of 1.1 are held for sufficient time, nucleation will ultimately occur. At S = 1.1, these times are on the order of 150–200 hr, while at S = 1.2, 20–30 hr are required, and at S = 1.3, only a few minutes are required for nucleation (van Hook and Bruno, 1949). This demonstrates the dynamic aspects of metastability and nucleation. Other factors, such as agitation rates, presence and types of impurities, presence and activity of nucleation centers, and temperature, also are important in determining when a solution will nucleate or when it will be metastable.

Mutarotation

An aspect of importance during crystallization of some food components is the multiple forms that may exist in solution. In particular, reducing sugars can be found in several enantiomeric forms. Lactose, for example, occurs in either the α or β anomers depending on the steric position of the reducing group (Nickerson, 1974). In solution, these anomers attain an equilibrium condition based on temperature, pH, and presence of other components. The conversion between these forms is called mutarotation and may become a limiting factor to crystallization processes under certain conditions. At temperatures below 93.5°C, where α-lactose is least soluble, the equilibrium crystalline form is α-lactose monohydrate; above this temperature, β-lactose crystallizes in the anhydrous form. As α-lactose molecules are removed from solution during crystallization, mutarotation must occur to maintain equilibrium in solution. In general, this mutarotation rate is rapid, however, there are occasions, particularly for crystal growth onto large surface areas, when mutarotation may limit the rate of growth. The effects of mutarotation on growth will be discussed in greater detail in a subsequent section.

NUCLEATION

The formation of crystals from solution, or nucleation, can occur by one of several mechanisms. These are categorized as either primary or secondary based on the mechanism of crystal formation. In primary nucleation (either homogeneous or heterogeneous), crystal nuclei are formed when molecules come together to form a stable cluster, whereas secondary nucleation occurs due to the presence of existing crystals of the desired form.

Homogeneous Nucleation

Homogeneous nucleation is a primary mechanism where nuclei are formed by a molecular accretion process. This is the classical nucleation process where molecules come together either singly or in small groups and eventually a cluster builds up to the critical size where it becomes a stable crystal nucleus. This process generally requires high supersaturations or subcoolings before nucleation can occur.

The free energy change associated with the phase transformation is related to two processes: formation of the crystal surface and formation of the bulk crystal. The free energy of formation of the crystal surface increases with surface area (size squared) while the free energy of formation of the bulk crystal decreases with volume (size cubed). The resulting dependence of free energy with cluster size gives a critical size where the free energy is at a maximum, as shown in Fig. 2.6. This critical size can be estimated based on classical nucleation theory (Mullin, 1973). For crystallization from solution, this can be written as

$$r_{crit} = 2\ \sigma v/kT\ (\ln S) \tag{8}$$

where σ is the interfacial (surface) tension of the nucleus, v is the molecular volume, and k is the Boltzman constant. S should be expressed as ratio of activities, although concentrations are often used as a simplification. For melt crystallization systems, the critical size is (Boistelle, 1988)

$$r_{crit} = 2\ \sigma v T_f/\Delta H_f\ (T_f - T) \tag{9}$$

Combining the critical free energy for nucleation with the exponential rate equation allows an expression for classical homogeneous nucleation to

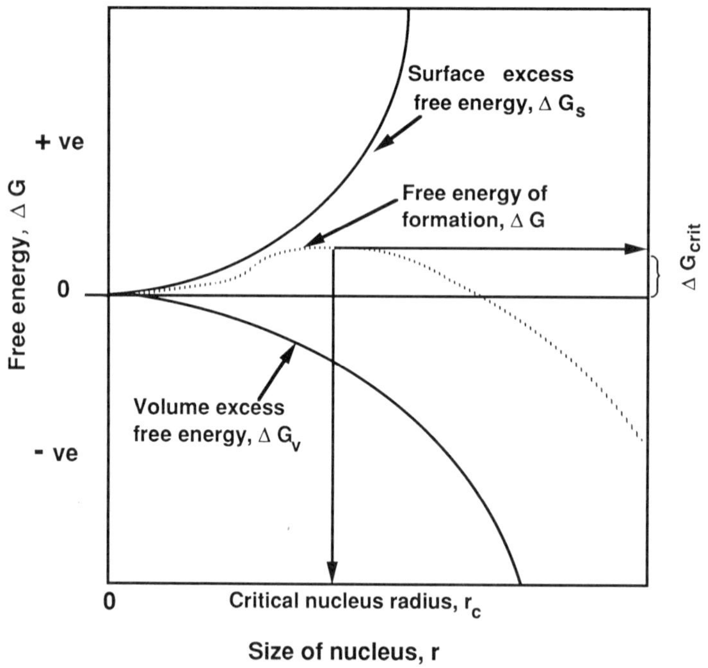

Fig. 2.6 Free energy changes during nucleation due to surface formation and bulk volume formation. (*From Hartel and Shastry, 1991.*)

be derived. For solutions, this is (Mullin, 1973)

$$J = A \exp\left[\frac{\Delta G_{crit}}{k\,T}\right] = A \exp\left[\frac{-16\pi\sigma^3 v^2}{3k^3T^3\,(\ln S)^2}\right] \tag{10}$$

where ΔG_{crit} is the critical free energy at the point of formation of a nucleus of critical size. For melts, the nucleation rate equation is

$$J = A \exp\left[-\frac{16\pi\sigma^3 v^2 T_f^2}{3kT\Delta H_f^2\,(T_f - T)^2}\right] \tag{11}$$

For viscous systems, these equations are often modified to account for volume diffusion effects through addition of another exponential term related to mobility of molecules. This has been given as (Mullin, 1973)

$$J = A \exp\left[-\frac{\Delta G_{crit}}{kT} + \frac{\Delta G_v'}{kT}\right] \tag{12}$$

Here, $\Delta G_v'$ represents the free energy associated with molecular mobility due to high viscosities. This equation predicts that nucleation rates drop off at higher supersaturations or subcoolings due to the increased influence of mass transfer inhibition. Typical rate profiles for both homogeneous and heterogeneous nucleation are sketched in Fig. 2.7. As the driving force increases, the nucleation rate initially increases due to the increased supersaturation. However, at extremely high driving forces, nucleation goes to zero due to inhibition of molecular mobility. Solutions

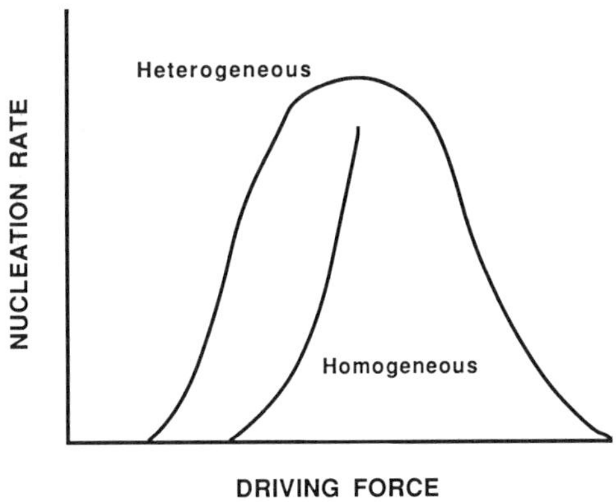

Fig. 2.7 Nucleation rates for homogeneous and heterogeneous nucleation.

that are brought to this high level of supersaturation before nucleation can occur (since nucleation is a dynamic process that requires sufficient time for molecular orientation) are stable to nucleation for long periods of time despite the thermodynamic driving force tending to push the system back towards equilibrium (crystallization). This situation results in formation of a stable amorphous phase known as the glass transition (Chapter 3).

Heterogeneous Nucleation

In heterogeneous nucleation, the process of molecular orientation is enhanced by the presence of a foreign surface (dust particles or microscopic structures in the vessel wall) which orients molecules preferentially so that a crystal lattice is more easily formed. The activity of a foreign surface to nucleate a crystal depends on the match between the structure of the nucleating site and the crystal lattice and is generally related to the relative interfacial tensions between the three phases (crystal, solution, and solid surface).

The rate of heterogeneous nucleation is often based on the homogeneous rate equation modified by the activity of the foreign surface towards nucleation. This is given as (Mullin, 1973)

$$J = A \exp\left[-\frac{\Delta G'_{crit}}{kT}\right] \tag{13}$$

where

$$\Delta G'_{crit} = \phi \, \Delta G_{crit} \tag{14}$$

$$\phi = (2 + \cos\theta)(1 - \cos^2\theta)/4 \tag{15}$$

$$\cos\theta = \frac{\sigma_{sl} - \sigma_{cs}}{\sigma_{cl}} \tag{16}$$

Here, θ is the angle formed between crystal and solid surface, σ_{sl}, σ_{cs}, and σ_{cl} are surface tensions between solid-liquid, crystal-solid and crystal-liquid, respectively. Heterogeneous nucleation is, in general, considered empirically due to the difficulty in determining the specific interactions between foreign surface and crystallizing molecules. Exact prediction and control of spontaneous nucleation in foods will not be possible until all aspects of heterogeneous nucleation are understood and can be quantified in some way. For example, lactose crystallization in ice cream remains a problem, particularly in flavored ice creams where extraneous foreign solids are added (fruits, nuts, etc.). When and where this problem (sandiness) will occur is nearly impossible to predict and difficult to prevent or control due to this lack in knowledge on the fundamental principles of heterogeneous nucleation and their application.

Secondary Nucleation

Formation of new nuclei from existing crystals can occur through several mechanisms (Larson, 1981). In suspension type crystallizers, where agitation of the crystals may be severe, crystal contacts can result in formation of new nuclei. Contacts between crystals and impeller or the wall and even other crystals can occur that result in generation of new nuclei. These contacts may result in a microattrition process where microscopic pieces of the existing crystal are distributed into the solution where they become stable nuclei and grow. Alternatively, an adsorbed layer adjacent to the growing crystal surface may be disrupted by contact and growth units distributed into the solution. The exact mechanism for this type of contact nucleation is still in question although it has received considerable study in recent years.

Fluid shear may also cause disruption of crystal surfaces, particularly of dendritic crystal growths, and cause generation of new nuclei in solution. Partial dissolution at dendrite bases due to heat and mass transfer influences may also result in generation of new nuclei. This may be enhanced by particle motion through fluid shear or even crystal contacts.

Secondary nucleation effects during crystallization have not been adequately modeled from a thermodynamic viewpoint due to the variety of mechanisms and the uncertainty associated with these mechanisms. However, empirical expressions can be used that relate secondary nucleation rates to existing hydrodynamic and thermodynamic conditions during crystallization. In suspension crystallizers, of the kind used in sugar refining, freeze concentration, or fat fractionation, expressions of the form

$$B^\circ = A_{SN} \exp\left[-\frac{E_{SN}}{RT}\right] \epsilon^l M_T^j S^q \qquad (17)$$

are used (Randolph and Larson, 1988). Here, B° is the secondary nucleation rate, A_{SN} is a frequency factor, E_{SN} is the activation energy for secondary nucleation, M_T is the total mass in suspension, ϵ is the energy input from the agitator, and l, j, and q are variable parameters. Kinetic expressions of this form have been developed for sucrose (Hartel, 1980) and lactose (Shi et al., 1990).

One case where secondary nucleation is quite important is in sugar refining. Here, sugar crystals are grown from seeds under controlled supersaturations so that a uniform distribution of crystals with the same size are formed. This allows for efficient separation of crystals in the centrifuge. Conditions in the crystallizer are maintained at supersaturations as close to the forced nucleation curve in the supersolubility diagram (see Fig. 2.5) as possible to maintain the maximum growth rate. If, however, the supersaturation wanders into the higher region of metastability, where forced nucleation can occur, a shower of nuclei will form due to secondary nucleation based on crystal contacts. This is called false grain in the refining industry and results in difficulties during centrifugation as well as a product that has a wide dispersion of crystal sizes due to the formation of nuclei during the growth process. This same process can occur during ice crystallization in freeze concentration systems and during fat crystallization during fractionation.

Freeze concentration processes based on suspension crystallization of ice may also be adversely affected by secondary nucleation, since separation and washing efficiency for ice crystals depends on their average size and size distribution. A wide size distribution and smaller average size resulting from high rates of secondary nucleation will cause inefficient separation and reduce economics of freeze concentration. Operation within a region of low subcoolings where secondary nucleation is minimized is important for efficient operation of freeze concentration processes.

Secondary nucleation is also expected to be important in fat fractionation processes (either melt or solvent) although this has not been very well documented. The dendritic structure of fat crystals should be quite prone to shearing or collision-induced breakage. Some recent data on crystallization of butterfat (Grall and Hartel, 1991) shows particle size and numbers to be a strong function of stirring speed, suggestive of secondary nucleation. However, much further work will be necessary to verify and quantify this phenomenon.

GROWTH

Crystal growth occurs by a series of steps in which the molecules of the crystallizing species transform from the liquid phase into the solid crystalline lattice. These steps may include:

- Diffusion of molecules from the bulk solution to the crystal surface
- Removal of waters of hydration or dissociation from complex molecules
- Counterdiffusion of noncrystallizing species away from the crystal surface
- Orientation of the molecules of the crystallizing species into the appropriate form for incorporation into the crystal lattice
- Diffusion of growth units or molecules along the surface to an appropriate incorporation site
- Incorporation of growth unit or molecule into the lattice structure
- Mutarotation to return the solution phase to its chemical equilibrium
- Removal of latent heat of formation released during crystallization

Any combination of these processes may affect the overall rate of growth in any particular situation. Understanding the mechanism of growth for a system requires delineation of which of these steps controls growth rates. In addition, the overall rate of growth can be calculated through knowledge of the exact mechanism for crystal growth under a variety of conditions.

In crystallization from solution, generally the major limiting steps are either bulk diffusion or surface incorporation (including diffusion along the crystal surface). The latent heat of formation for most sugar, salt, protein, etc. crystallizations is relatively small resulting in little heat transfer generation during crystal growth. On the other hand, crystallization of melt systems such as ice or fats generally have higher latent heat terms and result in more heat transfer limitation. In ice crystallization, heat removal due to phase transition is often the dominant mechanism determining overall crystal growth since approximately 144 BTU of energy must be removed as heat for every pound of ice frozen.

Growth Mechanisms

Diffusion. The concentration profiles around a growing crystal during solution growth are shown in Fig. 2.8. The driving force for diffusion of molecules from bulk solution to the crystal interface is the difference between the bulk concentration, C_b, and the interfacial concentration, C_i. The rate of diffusion then can be related to the mass transfer conditions

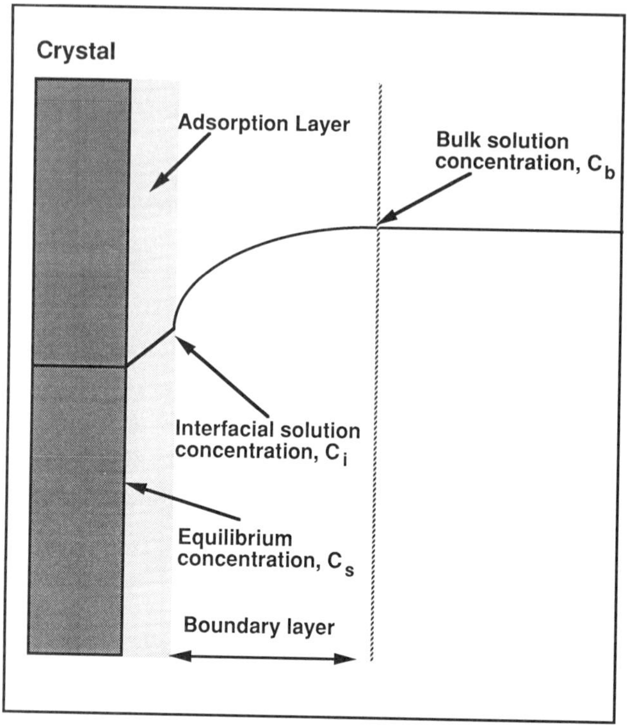

Fig. 2.8 Concentration profiles around a growing crystal. (*From Hartel and Shastry, 1991.*)

in the bulk solution and the rate of growth based on the rate of diffusion can be written as

$$G_d = k_d\,(T, D_{AB}, \mu, N, \ldots)\,[C_b - C_i] \tag{18}$$

where G_d is a linear growth rate, k_d a mass transfer coefficient, D_{AB} is molecular diffusivity of species A through B, μ is viscosity, and N is the rpm of any agitation device. The coefficient k_d in this expression is a function of temperature, bulk diffusivity, solution properties (viscosity, etc.), and the relative velocity between crystal and solution. Mass transfer coefficients can be found from correlation equations such as (Mullin, 1973; Heffels et al., 1987)

$$N_{Sh} = \frac{k_d D}{D_{AB}} = 2.0 + aN_{Re}^{b}N_{Sc}^{c} \tag{19}$$

where the Sherwood number N_{Sh} is related to the Reynolds ($N_{Re} = \rho vD/\mu$) and Schmidt ($N_{SC} = \mu/\rho D_{AB}$) numbers. Here, D is a characteristic dimension of the system, ρ is fluid density, and v is fluid velocity across the crystal.

Adsorption layer theories. Once the molecules arrive at the crystal surface, they must go through a series of steps prior to incorporation into the lattice. These steps comprise the adsorption layer theories of crystal growth. The two extreme cases represented by adsorption layer theories are two-dimensional nucleation and surface diffusion theories. In two-dimensional nucleation, a growth unit is formed on a smooth crystal surface by a similar mechanism as nucleation in the bulk solution. In this case, however, formation of the nucleus is enhanced by the presence of the existing crystal surface. In the most extreme case, a growth layer is completely filled in by two-dimensional nucleation where nuclei are formed at each incorporation site. There is no diffusion along the crystal surface in this case. Alternatively, one two-dimensional nucleus can form on a surface which then acts as an incorporation site for other growth units that arrive by diffusing along the crystal surface. Any combination of these two processes can also occur depending on the growth conditions. In general, a relatively high supersaturation is required for growth by these mechanisms, yet most crystals grow at a reasonable rate even at very low supersaturations.

This leads to the surface diffusion theories, the most common of which is based on continuous growth on a regenerating spiral dislocation (Burton et al., 1951). Here, the regenerating spiral provides a continuous incorporation site for diffusing growth units so that growth is not dependent on formation of a two-dimensional nucleus. Higher growth rates are thus predicted for low supersaturations. Growth rate kinetics based on this model can be written as (Mullin, 1973)

$$G = As^2 \tanh\left(\frac{B}{s}\right) \tag{20}$$

where A and B are parameters that depend on temperature, dislocation numbers, and dislocation activity, s is supersaturation expressed as $S - 1$ or $C/C_s - 1$. At low supersaturations, this expression simplifies to a second-order function of supersaturation, while at higher supersaturations, the dependence is linear. Crystals that grow by this mechanism typically exhibit a higher-order dependence on supersaturation at low supersaturations and a linear dependence at higher supersaturations.

Combined growth model. Diffusion theory and adsorption layer theories are typically combined in a general growth equation that relates the growth rate to a concentration driving force raised to some empirical exponent. The proportionality factor k_G is a function of temperature, agitation rate, types and levels of impurities or additives, among other factors. That is,

$$G = k_G (C_b - C_s)^m \qquad (21)$$

Determination of crystal growth kinetics requires determining m and k_G as functions of the important variable parameters.

In order to determine the mechanisms of crystal growth in a particular system, the above kinetics must be found in addition to determining the changes in crystal growth due to temperature and fluid-crystal velocity. The kinetic expression, assuming thermodynamically accurate units for expressing concentration in supersaturation have been used, can be related to the dominant mechanism of growth. However, this does not always give the specific mechanism, since an exponent of unity for m can mean either diffusion limitation or that the surface incorporation mechanism involves growth of spiral dislocations at higher supersaturations. Therefore, the change in crystal growth rate with increasing temperature and agitation rate is also necessary for determining mechanism. The parameter k_G varies with temperature with a certain activation energy (according to the Arrhenius equation) depending on the mechanism of growth. An activation energy of similar magnitude to the activation energy demonstrated by changes in diffusivity or viscosity with temperature is indicative of mass transfer limitations and suggests that diffusion is the rate-limiting step to crystal growth. Higher activation energies for k_G generally are representative of the surface incorporation step. To further verify the growth mechanism, crystal growth data at increasing fluid velocities must be taken. For a diffusion mechanism, increasing slip velocity should result in increasing growth rates since mass transfer rates are increased. If diffusion of molecules is not a concern (nor is heat transfer), then increasing slip velocity should have no effect on the crystal growth rate, and surface incorporation is the limiting mechanism. Calculations of mass transfer rates under the conditions of crystal growth can also be made to determine a growth rate based purely on diffusion. If this calculated rate is rapid compared to the measured rate, then diffusion is generally not a concern.

For sucrose crystal growth, the above experiments and calculations have been performed in sufficient detail to adequately describe the growth mechanisms over a wide range of conditions. It has been found that a combination of mass transfer and surface incorporation limit the rate of sucrose crystal growth with each playing the dominant role at different temperatures (Smythe, 1967a; van Hook, 1981). Based on the relative activation energies and the influence of agitation rate on growth, it has been determined that mass transfer rates limit the overall growth rate at higher temperatures while surface integration mechanisms limit growth at lower temperatures. At about 40°C, both mechanisms have about the same influence for sucrose crystal growth. This is an interesting result considering that, intuitively, mass transfer would be expected to be more important at lower temperatures where viscosities are higher. However, the activation energy for diffusion is low (6–8 kcal/mole) while the activation energy for surface integration is considerably higher (16–20 kcal/mole) resulting in a stronger temperature dependence for surface integration. Thus, while diffusion may be faster at higher temperatures, the relative rate of diffusion to surface integration has actually decreased, resulting in mass transfer limitation at elevated temperatures. This complete growth mechanism analysis has not been performed for very many crystallizing food components.

For melt crystallization, a general growth equation can be written as (Doremus, 1985)

$$G = B\,(T_f - T)^n \qquad (22)$$

where B is the growth coefficient and n is a variable parameter dependent on growth mechanisms. For cases where heat transfer is not a limiting factor to growth, n may be unity. However, many systems exhibit a higher than linear dependence ($n > 1$), where the exponent may depend on the crystal face being studied and the relative influence of heat transfer, mass transfer (diffusion), and surface incorporation rates (Huige, 1972). The factor B depends on relative rates of heat and mass transfer in solution. For ice crystallization, these rates have been quantified in heat and mass transfer correlation equations by Brian and Hales (1969) and Brian et al. (1969).

For food lipids, Eq. (22) is difficult to apply since there is generally a range of melting temperatures. This is due to the heterogeneous nature of the triglycerides existing in food lipids. For example, a typical, well-tempered chocolate exhibits a melting range from about 27 to 35°C, with most of the melting occurring at about 31°C (Hernqvist, 1988), whereas butterfat, an extremely heterogeneous fat, has a melting range from about −40 to 40°C. Another complicating factor is that the melting temperature of the polymorph that is crystallizing must be used rather than an average from the entire product. For these reasons, very little kinetic data based on thermodynamic driving force exists for food lipids. Generally, crystallization rates are related at specific temperatures for a given system, making comparison between systems difficult. Thus, food lipid crystallization remains mostly an empirical area of research.

Under certain circumstances, crystals grow in dendritic form, with a branched, treelike structure. This occurs typically for melt systems growing under high crystallization rates (Doremus, 1985). Although the exact reasons have yet to be formulated and quantified, it is generally considered that dendritic growth occurs due to heat transfer limitations existing during rapid growth (Gill, 1989). These heat transfer limitations are best reduced by increase of interfacial area in the form of dendrites. The stability of crystals growing under heat transfer limitations have been discussed in great detail (Mullins and Sekerka, 1963), although resolution of this question has not been fully realized.

Effects of Additives/Impurities

The presence of other components in a food formulation that affect crystal growth results in a complex situation where prediction of exact crystallization rates becomes quite difficult. Food components that may act in some way to influence crystallization rate include sugars, proteins, starches, gums, and, certainly, water. The variation in composition within these groups is quite wide and the effects of each constituent on the crystallizing species must be determined individually. However, these effects can generally be classified according to the specific interactions with the crystallizing system. Constituents may act to change the driving force for crystallization, to alter the rates of heat and mass transfer, or to act as specific growth inhibitors that act at the crystal surface or within the crystal lattice.

Driving force. Since the thermodynamic driving force for crystallization is related to the equilibrium condition between solid and liquid, any changes caused in either solubility concentration or freezing point will affect the crystallization rate. For example, the addition of other constituents to sucrose solutions may change the solubility concentration at a particular temperature. There exists a significant amount of data on solubility changes in sucrose for various component additions (van Hook, 1961; Smythe, 1967b; Mullin, 1973). However, it remains a difficult and approximate task to calculate the exact supersaturation existing in a complex food product. For example, the supersaturation driving force for sucrose crystallization in a caramel has not been calculated despite its importance in determining the stability of a caramel formulation to crystallization or graining. The effects of other sugars, proteins, and salts are difficult to quantify exactly.

In aqueous systems where water crystallizing as ice is of concern, the effects of low molecular weight constituents on freezing point may be of concern. This freezing point depression is quite important in freezing of ice cream, for example, since the draw temperature then defines the bulk subcooling and the subsequent crystallization rates. The same concern exists during freeze concentration processes (of fruit juices, for example) where increased removal of ice during concentration results in decreased driving forces as the freezing point decreases. In fact, in a batch ice crystallizer, an equilibrium will eventually be reached at high solution concentrations where no further growth of ice crystals occurs due to the thermal equilibrium. This has been documented for batch ice crystallization in skim milk (Hartel et al., 1990).

Heat and mass transfer rates. The composition of a food determines the thermal and physical properties of the liquid phase. Obviously, heat and mass transfer rates through a pure sucrose solution will be dramatically different from those through a viscous caramel mass ready to undergo crystallization. Thermal properties of importance, including thermal conductivity and specific heat, are generally reduced in food formulations compared to water since food constituents (fat, carbohydrates, proteins, etc.) have lower values than water. One place where this may not be true is in frozen foods, where thermal conductivity of ice is nearly 4 times higher than that of liquid water at the same temperature. Nevertheless, heat transfer rates can be dramatically altered in commercial food products. Mass transfer rates (diffusivities) also generally decrease due to the increased viscosity of the food product.

One case where mass transfer limitations play an important role in crystallization processes in foods is when the continuous phase is near or beyond the glass transition point (Levine and Slade, Chapter 3). In the nucleation and growth rate diagrams shown previously, this region occurs at extremely high supersaturations or subcoolings, where insufficient time for orientation of molecules into a crystalline lattice results in formation of an amorphous solid. A hard candy, for example, is made by concentrating a sugar syrup to a very high total solids level and then cooling rapidly enough that crystallization cannot occur. In this region, the nucleation is essentially zero due to the limited mobility of sugar molecules and the extremely high viscosity. Another example where crystal growth is limited by mass transfer through a glass phase occurs in ice cream and frozen desserts. Here, ice crystals are imbedded in a continuous matrix of extremely viscous syrup composed of sugar, salt, protein, and fat. At low freezer temperatures (−20°C), the continuous phase is a near amorphous solid that gives the characteristic hardness of a well-frozen ice cream. Fluctuations in temperature result in changes in molecular mobility and can result in ice crystal growth or dissolution depending on the direction of the temperature change. Stabilizers in ice cream serve, in part, to increase viscosity of this continuous phase and decrease changes in crystal size due to thermal cycling, thereby reducing heat shock and resulting in longer shelf life (Keeney, 1979).

Specific inhibition. Food composition may also affect crystallization rates through direct inhibition of constituents on the crystallizing species. This can occur in several ways including adsorption of molecules onto the crystal surface or incorporation of noncrystallizing material into the crystal lattice. Either of these mechanisms can have a large effect on crystal growth.

Adsorption of macromolecules to growing crystal surfaces has been well documented for both sugar and ice crystals. For sugar crystals, macromolecules such as pectin, dextrins, dextrans, and other oligosaccharides have been shown to inhibit crystal growth (Parrisch and Clarke, 1983). Decreases in growth rate of up to 30% were observed, as well as changes in crystal morphology since adsorption of these macromolecules generally depends on the molecular structure of the growing face. Adsorption may be greater on certain faces than others, resulting in changes in crystal shape as certain faces grow out. Michaels and van Kreveld (1966) found that gelatin adsorbed generally on all lactose crystal faces resulting in a nonselective inhibition of crystal growth. Low levels of gelatin (10 ppm) resulted in up to 66% decrease in lactose crystal growth rate with a slight effect on crystal face. Gelatin has also been found to inhibit the rate of ice crystal growth in sugar solutions (Smith and Schwartzberg, 1985) presumably by a surface adsorption mechanism.

Some components of foods may actually be incorporated into the crystal lattice structure and thereby cause growth inhibition and morphology changes. For sucrose, one such incorporated impurity is raffinose (a trisaccharide naturally occurring in sugar beet juice), which is incorporated into the lattice at the fructose-glucose part of the molecule (Vaccari et al., 1986). The galactose moiety remains in the solution layer surrounding the crystal and acts as a growth retarder since other sucrose molecules cannot come near enough to incorporate into the lattice. Visser (1980) has found a naturally occurring compound, present at low levels in milk, that inhibits lactose crystal growth. This phosphate-lactose complex is incorporated into the lattice and prevents or inhibits further growth. Another example of this phenomenon is mixed triglyceride crystals that occur naturally in food lipids. The heterogeneous nature of food lipids, such as cocoa butter or butterfat, results in a mixed triglyceride crystal structure. This is further emphasized in systems where mixed fats are crystallized together such as in milk chocolate. Here, the milkfat triglycerides are incorporated to some extent into the cocoa butter crystal lattice. The result is a softening of the chocolate, and the conditions under which the chocolate must be crystallized or tempered must be changed to give a desirable chocolate with good snap, shine, and bloom stability.

Dissolution

A topic related to crystal growth is the inverse process of dissolution, where molecules leave the crystalline lattice to become part of the fluid. This process is based again on thermodynamic equilibrium where now the driving force, or difference in chemical potential, is negative. Many researchers have considered dissolution and growth to be purely inverse processes, where the mechanisms defining crystal growth also define dissolution. However, this is generally not true as evidenced by the relative rates of dissolution and growth, under identical thermodynamic driving forces, for sucrose. Dissolution rates are generally faster than growth rates with the relative ratio of dissolution rate to growth rate ranging between 2 and 5 for sucrose crystals (Mullin, 1973). VanHook (1974) has studied the dissolution of sucrose and determined that dissolution is always a mass transfer limited process (based on diffusion of sucrose molecules away from the surface), whereas sucrose crystal growth is dependent on both mass transfer and surface incorporation of molecules. In fact, the activation energy for the dissolution process, at constant driving force but varying temperature, is approximately equal to the activation energy of the diffusion coefficient for sucrose molecules in water, indicating the mass transfer limitation. The rate of dissolution of sugar crystals, thus, is limited by the rate of removal of sucrose molecules from the crystal surface.

Growth Rate Dispersion

Since growth is dependent on the characteristics of the crystal surface, the types, numbers, and activities of dislocations on the surface significantly affect the growth rate of each individual crystal. However, each crystal has a different dislocation density depending on the conditions under which it forms and grows. Thus, each crystal grows at a slightly different rate—some will grow faster and some slower than the average. This phenomenon is known as growth rate dispersion (GRD) and has been demonstrated to occur in a number of systems (Randolph and Larson, 1988) including lactose (Shi et al., 1989) and sucrose (Liang et al., 1987). Lactose, in particular, demonstrates a wide spread of growth rates for crystals growing in a uniform environment. A large number of nuclei generated by a contact mechanism exhibit near zero growth rates despite being in a supersaturated solution (Shi et al., 1989). The phenomenon is particularly important when considering suspension crystallization where GRD causes a spread of crystal sizes. Liang and Hartel (1990) and Liang et al. (1991) have modeled GRD in continuous crystallizers for sucrose and lactose, respectively.

SUMMARY

While the thermodynamic equilibrium aspects of crystallization are important, the dynamic aspects of crystallization determine, to a large extent, the rates and extent of nucleation and growth. The equilibrium aspects discussed here apply when sufficient time is allowed for crystallization to occur and an equilibrium condition to be attained. While these thermodynamic principles are fairly well understood, there still remain questions that have not been answered. For example, the process of heterogeneous nucleation remains a mystery with only a qualitative theory available for predicting nucleation results. It is still nearly impossible to determine exactly when a solution will nucleate under given conditions without actually doing an experiment and even then, nucleation will not be exactly reproducible. This is due to conditions and parameters that are not clearly understood.

Crystallization in foods remains an operation that relies heavily on past experience since so many specific effects, from the basic thermodynamic mechanism to the specific applications during food processing, are not very well understood. In particular, the effects of food components on crystallization processes have not been studied in sufficient detail. Further research, of both theoretical and practical nature, will be required to allow crystallization processes in foods to be a well-controlled unit operation of food processing. The benefits of this research, however, will apply to a wide variety of food industries including confectionery, dairy, cereal, fats and oils, and the frozen food industry.

SYMBOLS

a	Activity of solution
a_{eq}	Equilibrium state activity
A	Parameter in growth Eq. (20)
A	Frequency factor in primary nucleation rate equation
A_{SN}	Frequency factor in secondary nucleation rate equation
B	Parameter in growth Eq. (20)
c_p	Specific heat
C	Solution concentration
C_b	Bulk solution concentration
C_i	Concentration at crystal interface
C_s	Equilibrium or solubility concentration
ΔC	Concentration difference $(C - C_s)$
D	Characteristic dimension in Reynolds number
D_{AB}	Molecular diffusivity of species A through B
E_{SN}	Activation energy for secondary nucleation
G	Linear growth rate
G_d	Diffusion based growth rate
ΔG_{crit}	Critical free energy for homogeneous nuclei formation
$\Delta G'_{crit}$	Critical free energy for heterogeneous nuclei formation
$\Delta G'_v$	Free energy for molecular mobility
ΔH_f	Latent heat of fusion
J	Heterogeneous nucleation rate
k	Thermal conductivity
k	Boltzmann constant
k_d	Mass transfer coefficient
k_G	Kinetic coefficient for combined growth model
M_T	Suspension density (mass of crystals)
N	rpm of agitation device
N_{Re}	Reynolds number $(\rho v D/\mu)$
N_{Sc}	Schmidt number $(\mu\rho/D_{AB})$
N_{Sh}	Sherwood number $(k_d\, D/D_{AB})$
r_{crit}	Critical cluster size for nuclei formation
R	Gas constant
s	Relative supersaturation $(C - C_s/C_s)$
S	Supersaturation ratio (C/C_s)
S'	Logarithm of supersaturation ratio
T	Temperature
T_f	Freezing temperature
v	Molecular volume
v	Fluid velocity at crystal surface

Greek Letters

ϵ	Energy input from agitator
θ	Angle of formation between crystal and foreign surface
μ	Viscosity
μ	Chemical potential
μ°	Reference state chemical potential
ρ	Fluid density
σ	Interfacial tension of crystal surface
σ_{cl}	Surface tension between crystal and fluid
σ_{cs}	Surface tension between crystal and foreign surface
σ_{sl}	Surface tension between foreign surface and fluid
ϕ	Chemical affinity
ϕ'	Heterogeneous nucleation factor

REFERENCES

Allen, A. T., Wood, R. M., and McDonald, M. P. 1974. Molecular associations in the sucrose-water system. *Sugar Technol. Rev.* 2: 165.

Arbuckle, W. S. 1986. *Ice Cream,* 4th ed. Van Nostrand, New York.

Boistelle, R. 1988. Fundamentals of nucleation and crystal growth. In *Crystallization and Polymorphism of Fats and Fatty Acids,* N. Garti and K. Sato (Ed), p. 189. Marcel Dekker, New York.

Brian, P. L. T. and Hales, H. B. 1969. Effects of transpiration and changing diameter on heat and mass transfer to spheres, *A. I. Ch. E. J.* 15(3): 419.

Brian, P. L. T., Hales, H. B., and Sherwood, T. K. 1969. Transport of heat and mass between liquids and spherical particles, *A. I. Ch. E. J.* 15(5): 727.

Burton, W. K., Cabrera, N., and Frank, F. C. 1951. The growth of crystals and the equilibrium structure of their surfaces, *Proc. R. Soc. London Ser. A.* 243: 299.

Chakraborty, R. and Berglund, K. A. 1990. The use of pyranine as a trace fluorescent probe to study structure in aqueous sucrose solutions. Paper #65b presented at AIChE Annual Meeting, Chicago, IL.

Doremus, R. H. 1985. *Rates of Phase Transformation.* Academic Press, Orlando, FL.

Dunning, W. J. 1977. Thermodynamics of sucrose solutions. *Z. Zuckerind.* 27(11): 727.

Gill, W. N. 1989. Heat transfer in crystal growth dynamics. *Chem. Eng. Prog.* 52: 33.

Grall, D. and Hartel, R. W. 1991. Kinetics of butterfat crystallization. Paper presented at IFT Annual Meeting, Dallas, TX.

Hartel, R. W. 1980. A kinetic study of the nucleation and growth of sucrose crystals in a continuous cooling crystallizer. Ph.D. thesis, Colorado State University, Fort Collins, CO.

Hartel, R. W. and Shastry, A. V. 1991. Sugar crystallization in food products. *Crit. Rev. Food Sci. Nutr.* 1(1): 49.

Hartel, R. W., Shi, Y., and Liang, B. 1990. Nucleation and growth of ice crystals in batch suspension crystallizer. In *Proceedings of the 11th Symposium on Industrial Crystallization,* A. Mersmann (Ed.), p. 479, GVC-VDI Gesellschaft Verfahrenstechnik und Chemieingenieurwesen, Munich.

Heffels, S. K., Weber, C. M. H., DeBruin, P. R., and DeJong, E. J. 1987. Mass transfer coefficient of diffusion at sucrose crystals in solution. *Zuckerindustrie* 112(12): 1075.

Heist, J. A. 1979. Freeze crystallization. *Chem. Eng.* 86(10): 72.

Hernqvist, L. 1988. Chocolate temper. In *Industrial Chocolate Manufacture and Use,* S. T. Beckett (Ed.), p. 159. Blackie, Glasgow.

Huige, H. J. J. 1972. Nucleation and growth of ice crystals from water and sugar solutions in continuous stirred tank crystallizers, Ph.D. thesis, Eindhoven University, Eindhoven, The Netherlands.

Hunziker, O. F. and Nissen, B. H. 1927. Lactose solubility and lactose crystal formation. II. Lactose crystal formation, *J. Dairy Sci.* 10: 139.

Keeney, P. G. 1979. Confusion over heat shock. *Food Technol.* 51(6): 116.

Kirwan, D. J. and Pigford, R. L. 1969. Crystallization kinetics of pure and binary melts. *A. I. Ch. E. J.* 15(3): 442.

Larson, M. A. 1981. Secondary nucleation: An analysis. *Chem. Eng. Comm.* 12: 161.

Liang, B. M. and Hartel, R. W. 1991. Techniques for developing nucleation and growth kinetics from MSMPR data for sucrose crystallization in the presence of growth rate dispersion, *J. Crystal Growth* 108: 129.

Liang, B. M., Hartel, R. W., and Berglund, K. A. 1987. Contact nucleation in sucrose solutions, *A. I. Ch. E. J.* 33(12): 2077.

Liang, B. M., Shi, Y., and Hartel, R. W. 1991. Growth rate dispersion effects on lactose crystal size distribution from a continuous cooling crystallizer, *J. Food Sci.*, 56(3): 848.

Mathlouthi, M., Cholli, A., and Koenig, J. L. 1986. Spectroscopic study of the structure of sucrose in the amorphous state and in aqueous solution, *Carbohydr. Res.* 147: 1.

McMahon, P. M., Berglund, K. A., and Larson, M. A. 1984. Raman spectroscopic studies of the structure of supersaturated KNO_3 solutions. In *Industrial Crystallization 84,* S. J. Jancic and E. J. deJong (Ed.), p. 229. Elsevier, London.

Michaels, A. S. and Van Kreveld, A. 1966. Influences of additives on growth rates in lactose crystals. *Neth. Milk Diary J.* 20(3): 163.

Mullin, J. W. 1973. *Crystallisation.* Butterworths, London.

Mullin, J. W. and Sohnel, O. 1977. Expressions of supersaturation in crystallization studies. *Chem. Engr. Sci.* 32: 683.

Mullins, W. W. and Sekerka, R. F. 1963. Morphological stability of a particle growing by diffusion or heat flow. *J. Appl. Phys.* 34(2): 323.

Myerson, A. S., Lo, P. Y., Kim, Y. C., and Ginde, R. 1990. Diffusion, viscosity and cluster formation in metastable solutions. In *Proceedings of the 11th Symposium on Industrial Crystallization,* A. Mersmann (Ed.), p. 847, GVC-VDI Gesellsschaft Verfahrenstechnik und Chemieingenieurwessen, Munich.

Nickerson, T. A. 1974. Lactose. In *Fundamentals of Dairy Chemistry,* B. H. Webb, A. H. Johnson and J. A. Alford (Ed.), p. 273. AVI Publ., Westport, CT.

Pancoast, H. M. and Junk, W. R. 1980. *Handbook of Sugars,* Vol. 2, p. 264. AVI Publishing, Westport, CT.

Parrisch, F. W., and Clarke, M. A. 1983. Effects of oligosaccharides and polysaccharides on sucrose crystallization, Comm. Int. Tech. Sucr. C. R. Assem. Gen., 17: 467.

Randolph, A. D. and Larson, M. A. 1988. *Theory of Particulate Processes*, 2nd Ed., Academic Press, New York.

Richardson, S. J., Baianu, I. C. and Steinberg, M. P. 1987. Mobility of water in sucrose solutions determined by deuterium and oxygen-17 nuclear magnetic resonance measurements, *J. Food Sci.* 52(3): 806.

Sato, K. 1988. Crystallization of fats and fatty acids. In *Crystallization and Polymorphism of Fats and Fatty Acids,* N. Garti and K. Sato (Ed.), p. 227. Marcel Dekker, New York.

Sato, K. and Garti, N. 1988. Crystallization and polymorphic transormation: An introduction. In *Crystallization and Polymorphism of Fats and Fatty Acids,* N. Garti and K. Sato (Ed.), p. 3. Marcel Dekker, New York.

Shi, Y., Hartel, R. W., and Liang, B. M. 1989. Formation and growth of lactose nuclei under contact nucleation conditions, *J. Dairy Sci.* 72(11): 2903.

Shi, Y., Liang, B. M., and Hartel, R. W. 1990. Crystallization kinetics of α-lactose monohydrate in a continuous cooling crystallizer, *J. Food Sci.* 55(3): 817.

Smith, C. E. and Schwartzberg, H. G. 1985. Ice crystal size changes during ripening in freeze concentration. *Biotech. Prog.* 1: 11.

Smythe, B. M. 1967a. Sucrose crystal growth. I. Rate of crystal growth in pure solutions, *Austral. J. Chem.* 20: 1087.

Smythe, B. M. 1967b. Sucrose crystal growth. II. Rate of crystal growth in the presence of impurities. *Austral. J. Chem.* 20: 1097.

Vaccari, G., Mantovani, G., Sgualdino, Aquilano, D., and Rubbo, M. 1986. The raffinose effect on sucrose morphology and kinetics. *Sugar Technol. Rev.* 13(3): 133.

van Hook, A. 1959. Nucleation in supersaturated solutions. In *Principles of Sugar Technology. Crystallization,* P. Honig (Ed.), p. 2. Elsevier, New York.

van Hook, A. 1961. *Crystallization: Theory and Practice*. Reinhold Publ., New York.

van Hook, A. 1974. The dissolution of sucrose crystals. *Int. Sugar J.* 76(905): 131.

van Hook, A. 1981. Growth of sucrose crystals: A review. *Sugar Technol. Rev.* 8(1): 41.

van Hook, A. 1987. The thermodynamic activity of concentrated sugar solutions. *Zuckerindustrie* 112(7): 596.

van Hook, A. and Bruno, A. J. 1949. Nucleation and growth in sucrose solutions. *Faraday Discuss. Soc. Chem.* 5: 112.

Vasatko, J. and Smelik, A. 1968. The crystallization of sucrose, D-glucose and D-fructose from supersaturated solutions of the metastable zone. *Zucker* 21(6): 144.

Visser, R. A. 1980. A natural crystal growth retarder in lactose. *Neth. Milk Dairy J.* 34: 255.

Young, F. E. 1957. D-glucose-water phase diagram. *J. Phys. Chem* 61: 616.

3

Glass Transitions in Foods

Harry Levine and Louise Slade

Nabisco Brands, Inc.
East Hanover, New Jersey

INTRODUCTION

What is a glass? What is a glass transition? Why is the temperature at which a glass transition occurs (Tg) so important to the processing and storage stability of so many different types of foods? Why is the effect of water as a plasticizer on Tg of such widespread relevance to food products and processes? Why are considerations of nonequilibrium glassy solid and rubbery liquid states in foods, rather than equilibrium phases, more germane to issues of food quality and safety? Why are the kinetics of heat/moisture processes for foods and of deteriorative changes in food systems during storage more often appropriately interpreted in terms of the Williams-Landel-Ferry (WLF), rather than the Arrhenius, equation? What is the "food polymer science" approach, with its central concepts of "glass dynamics" and "water dynamics," and why has this research approach proved so useful to the study of glasses and glass transitions in foods? And why has there been, since the early 1980s, such interest in these questions, and such increasing research activity, especially in the last few years, in this area of food science and technology? These are the questions we will address in this chapter. The answers will be illustrated by a review that will emphasize (1) the theoretical principles from the field of synthetic polymer science that are applicable to studies of glass transitions in foods and (2) a broad overview focusing primarily on the most recent experimental studies by a number of groups that have been especially active in this growing field of research.

PHYSICOCHEMICAL PRINCIPLES FROM SYNTHETIC POLYMER SCIENCE: FOUNDATION OF THE "FOOD POLYMER SCIENCE" APPROACH

During the 1980s, the value of a polymer science approach to the study of glasses and glass transitions, and of their importance to structure-property relationships, in food materials, products, and processes has been increasingly recognized by a growing number of food scientists. In this respect, food science has followed the compelling lead of the synthetic polymers field. Researchers who have advocated the study of glasses and glass transitions in food systems, and whose groups have been most active in this field, are listed in Table 3.1.

As reviewed in earlier publications (Slade and Levine, 1984, 1987a, b, 1988a, b, c, d, 1989, 1991a, b, 1992), the research discipline of "food polymer science" emphasizes the fundamental and generic similarities between synthetic polymers and food molecules, and provides a new interpretive (rather than theoretical) and experimental framework for the study of food systems that are kinetically constrained. Based on established structure-property relationships from the field of synthetic polymer science (Table 3.2), this innovative discipline has developed to unify structural aspects of foods, conceptualized as kinetically metastable, completely amorphous or partially crystalline, homologous polymer systems, with functional aspects, dependent on mobility and viewed in terms of "water dynamics" and "glass dynamics" (Levine and Slade, 1988a, 1989b, c; Slade et al., 1989, Slade and Levine, 1991a, b). These unified concepts have been used to explain and predict the functional properties of food materials during processing and product storage (Slade and Levine, 1984, 1987a, b, 1988a, b, c, d, 1989, 1991a, b, 1992). Key elements of this interpretive approach to investigations of structure-function relationships in food systems, with demonstrated relevance to moisture management and water relationships, include recognition of (Levine and

Table 3.1 Studies of Glasses and Glass Transitions in Foods

Researcher(s)	Refs.
Franks	Franks et al. (1977), Franks (1982, 1983a, b, 1985, 1986a, b, 1989, 1990, 1991), Franks and Grigera (1990), Finegold et al. (1989), Hatley et al. (1991)
Blanshard	Blanshard (1986, 1987, 1988), Blanshard and Franks (1987), Blanshard et al. (1991), Marsh and Blanshard (1988), Kalichevsky et al. (1991)
van den Berg	van den Berg (1981, 1986, 1991)
Karel and Roos	Karel and Flink (1983), Karel (1985, 1986, 1989), Karel and Langer (1988), Karel and Saguy (1991), Roos (1987), Paakkonen and Roos (1990), Roos and Karel (1990, 1991a, b, c, d, e), Shimada et al. (1991)
Lillford	Ablett et al. (1986, 1988, 1991a, b), Edwards et al. (1987), Lillford (1988), Attenburrow et al. (1989, 1990), Davies et al. (1991), Izzard et al. (1991)
Reid	Reid (1985, 1990), Reid and Hsu (1991), Lim and Reid (1991)
Morris, Ring, and Smith	Ring et al. (1987), Ring and Whittam (1991), Orford et al. (1989, 1990), Hutchinson et al. (1989), Morris (1990), Smith (1990), I'Anson et al. (1988, 1990), Ollett and Parker (1990), Ollett et al. (1990), Warburton et al. (1990), Noel et al. (1990), Whittam et al. (1991)
Hoseney	Hoseney et al. (1986), Hoseney (1991), Hoseney and Rogers (1990), Yost and Hoseney (1986), Doescher et al. (1987), Zeleznak and Hoseney (1987a, b)
Simatos, Le Meste, and Huang	Simatos and Karel (1988), Simatos et al. (1989), Simatos and Blond (1991), Le Meste and Duckworth (1988), Le Meste and Huang (1991), Le Meste et al. (1990, 1991a, b), Blond (1989), Matthiesen et al. (1991)
Maurice and Biliaderis	Maurice et al. (1985, 1991), Biliaderis et al. (1985, 1986a, b, c), Biliaderis and Galloway (1989), Biliaderis and Zawistowski (1990), Biliaderis and Seneviratne (1990a,b), Biliaderis (1990, 1991), Caldwell et al. (1990)
Schenz	Schenz et al. (1984, 1991)
Lund	Chungcharoen and Lund (1987), Lund (1989)
Zobel	Zobel (1988), Zobel et al. (1988)
Williams	Soesanto and Williams (1981)
Slade and Levine	Slade (1984), Slade and Levine (1984, 1987a, b, 1988a, b, c, d, 1989, 1991a, b, 1992), Slade et al. (1987, 1989), Levine and Slade (1986, 1988a, b, c, d, 1989a, b, c, d, 1990, 1991), Levine et al. (1991), Cole et al. (1983, 1984)

Table 3.2 Selected References on Structure-Property Relationships

A. Textbooks: Flory (1953), Wunderlich (1973, 1976, 1980, 1990), Cowie (1973), Ferry (1980), Rowland (1980), Turi (1981), Sears and Darby (1982), Billmeyer (1984), Sperling (1986)

B. Journal articles and book chapters: Kauzmann (1948), Williams, Landel, and Ferry (1955), Walton (1969), Brydson (1972), Sharples (1972), Hardy et al. (1973), Haward (1973), Roberts and White (1973), Flory (1974), Petrie (1975), Baro et al. (1977), Buchanan and Walters (1977), Starkweather (1980), Fuzek (1980), Johnson et al. (1980), Moy and Karasz (1980), Bair (1981), Wunderlich (1981), Shalaby (1981), Batzer and Kreibich (1981), Keinath and Boyer (1981), Gaeta et al. (1982), Fried et al. (1982), Eisenberg (1984), Graessley (1984), Ellis et al. (1984), Jin et al. (1984), Boyer et al. (1985), Matsuoka et al. (1985), Robertson (1985), Hiltner and Baer (1986), Domszy et al. (1986), Mandelkern (1986), Blum and Nagara (1986), Alfonso and Russell (1986), Cheng and Wunderlich (1987), Angell (1988), Burchard (1988), Ellis (1988), Aubin and Prud'homme (1988), Sichina (1988), Cheng (1989), Murthy et al. (1989), Ehlich and Sillescu (1990)

Slade, 1988a, 1989b, c; Slade et al., 1989; Slade and Levine, 1991a, b):

1. The behavior of foods and food materials as classical polymer systems, and that the behavior is governed by dynamics rather than energetics
2. The importance of the characteristic temperature Tg, at which the glass-rubber transition occurs, as a physicochemical parameter that can determine processibility, product properties, quality, stability, and safety of food systems
3. The central role of water as a ubiquitous plasticizer of natural and fabricated amorphous food ingredients and products
4. The effect of water as a plasticizer on Tg, and the resulting non-Arrhenius, diffusion-limited behavior of amorphous polymeric, oligomeric, and monomeric food materials in the rubbery liquid state at T > Tg
5. The significance of nonequilibrium glassy solid and rubbery liquid states (as opposed to equilibrium thermodynamic phases) in most "real world" food products and processes, and their effects on time-dependent structural and mechanical properties related to quality and storage stability

A glass is operationally defined as an amorphous (i.e., noncrystalline) solid (Haward, 1973). In fact, a glass is actually an undercooled liquid of such high viscosity ($\eta > 10^{10}$–10^{14} Pa·s (Ferry, 1980; Wunderlich, 1981; Sperling, 1986)) that it exists in a metastable, mechanical solid state, in which it is capable of supporting its own weight against flow due to the force of gravity (Haward, 1973). A glass forms when a typical liquid, with a disordered molecular structure, is cooled to a temperature generally about 100°C (for many pure liquids) below its equilibrium crystalline melting temperature (Tm) or freezing point, at a cooling rate sufficiently high to avoid crystallization of the liquid (Ferry, 1980). This solidification process, known as vitrification (Luyet, 1960), results in the "freezing in" or immobilization of the disordered structure of the liquid state (Wunderlich, 1981), such that the resulting glassy solid is spatially homogeneous, but without any long-range lattice order, and is incapable of exhibiting any long-range, cooperative relaxation behavior (e.g., translational mobility) on a practical time scale (Ferry, 1980; Wunderlich, 1990).

A glass transition in amorphous systems is a temperature-, time- (or frequency-), and composition-dependent, material-specific change in physical state, from a glassy mechanical solid to a rubbery viscous liquid (capable of flow in real time) (Ferry, 1980; Wunderlich, 1990). In terms of thermodynamics, the glass transition is operationally defined as a second-order transition (Sperling, 1986; Wunderlich, 1990) (as opposed to a first-order transition, e.g., crystalline melting) and denoted by (1) a change in slope of the volume expansion (a first derivative of the free energy), (2) a discontinuity in the thermal expansion coefficient, and (3) a discontinuity in the heat capacity (a second derivative of the free energy) (Kauzmann, 1948; Wunderlich, 1990). The glass transition is also operationally defined, based on mechanical properties, in terms of a mechanical relaxation process such as viscosity (Ferry, 1980). Figure 3.1 shows that as the temperature (T) is lowered from that of the low viscosity liquid state above Tm, where familiar Arrhenius kinetics apply, through a temperature range from Tm to Tg, a completely different, nonlinear form of kinetics, with an extraordinarily large temperature dependence (Angell, 1988; Franks and Grigera, 1990), becomes operative (Roberts and White, 1973). Then, at a temperature where cooperative mobility becomes limiting, a state transition occurs, typically manifested as a three-orders-of-magnitude change in viscosity, modulus, or mechanical relaxation rate (Eisenberg, 1984; Sperling, 1986; Noel et al., 1990). At this glass transition temperature, the viscosity of a liquid is $\sim 10^{12}$ Pa·s (10^{13} Poise), and the structural relaxation time [as determined calorimetrically, e.g., by differential scanning calorimetry (DSC)] for such a liquid is about 200 sec (Angell, 1988). A "mechanical" glass transition can be defined by combinations of temperature and deformation frequency for which, as tem-

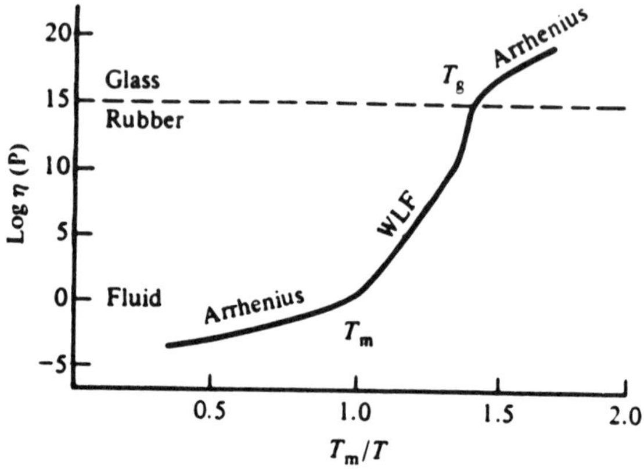

Fig. 3.1 Viscosity as a function of reduced temperature (Tm/T) for glassy and partially crystalline polymers. (*Reproduced, with permission, from Levine and Slade, 1988*a.)

perature decreases, sufficiently large numbers of mobile units (e.g., small molecules or backbone-chain segments of a macromolecule) become cooperatively immobilized (in terms of large-scale rotational and translational motions) during a time comparable to the experimental period (Buchanan and Walters, 1977; Angell, 1988; Green and Angell, 1989; Hosea et al., 1990), such that the material becomes a mechanical solid capable of supporting its own weight against flow. Perhaps the most important distinction between the dimensionally extended (alpha) relaxations, which give rise to the glass transition as translational motions become constrained at Tg, and the small-scale (beta and gamma) relaxations, for which small-scale rotational motions do *not* become constrained as T falls below Tg, is the cooperative nature of the alpha relaxations (Ferry, 1980; Wunderlich, 1981). Arrhenius kinetics become operative once again in the glassy solid at T < Tg, but the rates of all diffusion-limited processes are much lower in this high viscosity solid state than in the liquid state at T > Tg (Levine and Slade, 1988a). In fact, the difference in average relaxation times between the two Arrhenius regimes (T < Tg and T > Tm) is typically more than 14 orders of magnitude (Slade and Levine, 1988b).

The genesis of a polymer science approach to the study of glasses and glass transitions in foods dates back at least to 1966 and a seminal review by White and Cakebread (1966) of glassy states in certain sugar-containing food products. They recognized (1) the importance of the glassy state, and of the glass transition temperature and its location relative to the temperature of storage (either ambient or subzero), in a variety of aqueous food systems, including but not limited to boiled sugar candies, and (2) the critical role of water as a plasticizer of food glasses, and the quantitative Tg-depressing effect of increasing content of plasticizing moisture, whereby Tg of a particular glass-forming solute-water mixture depends on the corresponding content of plasticizing water (Wg) in that glass at its Tg (Levine and Slade, 1986, 1988a; Slade and Levine, 1988a). White and Cakebread (1966; Cakebread, 1969) were apparently the first food scientists to allude to the broader implications of nonequilibrium glassy and rubbery states to the quality, safety, and storage stability of a wide range of glass-forming aqueous food systems. Evidently, outside a small community of candy technologists, (1) the work of White and Cakebread, (2) the even earlier (but somewhat better known by food technologists) work by Makower and Dye (1956) on the kinetics of crystallization of amorphous sucrose and glucose (from the water-plasticized, rubbery state at room T > Tg), (3) more recent work by other candy technologists such as Herrington and Branfield (1984) on the physicochemical properties of sugar glasses, and (4) the broader relevance of this body of work to the field of food science and technology went largely unnoticed until the mid-1980s. Since that time, other workers (Tables 3.1 and 3.3) have helped to advance, with increasing momentum, concepts and approaches based on recognition of the importance of, and application of the principles underlying, nonequilibrium glassy and rubbery states in foods. The cur-

Table 3.3 Recent Publications Dealing with the Practical Significance of Glass Transition to Food Ingredients, Products, and Processes

A. General: Flink (1983), Wittwer and Tomka (1984), Saleeb and Pickup (1985, 1989), Chan et al. (1986), Atkins (1987), Russell (1987a), Gould and Christian (1988), Lineback and Rasper (1988), Quinquenet et al. (1988), BeMiller (Hill, 1989), Cairault et al. (1989), Fujio and Lim (1989), Russell and Oliver (1989), Shi and Seib (1989), Stepto and Dobler (1989), Hosea et al. (1990), Johnston-Banks (1990), Johnson et al. (1990), Kararli and Catalano (1990), Kararli et al. (1990), Mita (1990), Pikal and Shah (1990), Roozen and Hemminga (1990), Roper and Koch (1990), Rubin et al. (1990), Yoshida et al. (1990), Chang and Baust (1991), Cocero and Kokini (1991), Eliasson (1991), Gaines et al. (1991a, b), Hartel and Shastry (1991), Liu and Lelievre (1991), MacDonald and Lanier (1991), Shukla (1991), Watase et al. (1991)

B. Chapters in *Water Relationships in Foods* (Plenum, New York, 1991): Franks; Blanshard et al.; van den Berg; Karel and Saguy; Lim and Reid; Hoseney; Simatos and Blond; Le Meste et al.; Maurice et al.; Schenz et al.; Biliaderis; Slade and Levine; Given; Nishinari et al.; Tomka; Wasylyk and Baust

rent state of accelerating activity and interest in this area is illustrated by the fact that 16 chapters in a recent book on water relationships in foods (Table 3.3, part A) include discussions of glasses and glass transitions in food systems. Another strong indication of the emergence of this subject is the fact that a new graduate course, "Advanced Topics in Food Science: Glass Transitions, Tg, Aw, and the Physical Properties of Foods," was offered for the first time in the summer of 1990 by Profs. T. Labuza and E. A. Davis at the University of Minnesota's Food Science Department.

In previous reports and reviews (Slade and Levine, 1984, 1987a, b, 1988a, b, c, d, 1989, 1991a, b, 1992), we have described how the recognition of the key elements of the food polymer science approach and their relevance to the behavior of a broad range of different types of foods (e.g., intermediate-moisture, low-moisture, frozen, starch-based, gelatin-, gluten-, and other protein-based foods) and corresponding aqueous model systems has increased markedly during the 1980s. We illustrated the perspective afforded by using this conceptual framework and demonstrated the technological utility of this new approach to understand and explain complex behavior, design processes, and predict product quality, safety, and storage stability, based on fundamental structure-property relationships of food systems viewed as homologous families (i.e., monomers, oligomers, and high polymers) of partially crystalline glassy polymer systems plasticized by water. Referring to the food polymer science approach, John Blanshard (personal communication, 1987) has stated that "it is not often that a new concept casts fresh light across a whole area of research, but there is little doubt that the recognition of the importance of the transition from the glassy to the crystalline or rubbery state in foodstuffs, though well known in synthetic polymers, has opened up new and potentially very significant ways of thinking about food properties and stability." In a recent lecture on historical developments in industrial polysaccharides, James BeMiller (Hill, 1989) has echoed Blanshard's words by remarking that a key point regarding the future of polysaccharide research and technology is "the potential, already partly realized, in applying ideas developed for synthetic polymers to polysaccharides; for example, the importance of the glassy state in many polysaccharide applications." More praise for the food polymer science approach has come in a recent *Cereal Foods World* column on food technology entitled "A New Development in Carbohydrate Research," in which Triveni Shukla (1991) says "a recent breakthrough in fundamental research on starches [translated from basic concepts of synthetic polymers] concerns glass transition behavior and the glass transition temperature, Tg."

The technological importance of the glass transition in amorphous polymers and of the characteristic temperature at which it occurs (Tg) is well known in synthetic polymer science (Ferry, 1980; Rowland, 1980; Sears and Darby, 1982). Eisenberg (1984) has stated that "the glass transition is perhaps the most important single parameter which one needs to know before one can decide on the application of the many non-crystalline [synthetic] polymers that are now available." Especially in the last several years, a growing number of food scientists have increasingly recognized the practical significance of the glass transition as a physicochemical event that can govern food processing, product properties, quality, safety, and stability (Tables 3.1, 3.3).

This recognition has gone hand-in-hand with an increasing awareness of the inherent nonequilibrium nature of most food products and processes (Tables 3.1 and 3.3), as exemplified by intermediate-moisture foods (IMFs), in which amorphous carbohydrates (polymeric and/or monomeric) and proteins are major functional components (Slade and Levine, 1988a). Selected examples of food systems whose behavior is governed by dynamics far from equilibrium and of practical problems of food science and technology posed by their nonequilibrium nature include (Levine and Slade, 1988a, 1989b; Slade et al., 1989; Slade and Levine, 1991a, b): (1) water-vapor sorption/desorption hysteresis in concentrated food polymer systems; (2) graininess and iciness in ice cream, reduced survival of frozen enzymes and living cells, reduced activity and shelf-stability of freeze-dried proteins; (3) caking of dry powders, sugar bloom on chocolate; (4) recipe requirements for gelatin desserts; (5) cooking of cereals and grains; (6) expansion of bread or collapse of cake during baking, effects of flour and sugar on cookie baking; (7) staling of baked products; and (8) effects of sugar-water glasses and rubbers on texture and storage stability of cookies (Table 3.4).

Thermal and thermomechanical analysis methods are particularly well suited for studying such nonequilibrium systems, and for defining structure-function relationships, e.g., of synthetic amorphous polymers, from measurements of their thermal and mechanical properties (Turi, 1981; Wunderlich, 1990). DSC and dynamic mechanical analysis (DMA) have become established methods for characterizing the kinetic (i.e., time- or frequency-dependent) transition from glassy solid to rubbery liquid states that occurs at Tg in completely amorphous and partially crystalline, synthetic and natural polymer systems (Fuzek, 1980; Wunderlich, 1990), including many food materials (Ma and Harwalkar, 1990). These methods and DMTA (dynamic mechanical thermal analysis) (Davies et al., 1991; Kalichevsky et al., 1991), a variation of DMA, have been the ones of choice in most recent investigations of glasses and glass transitions in food systems. In other studies (Table 3.5), complementary methods have also been employed to measure glass transitions in food ingredients and

Table 3.4 Recent Publications Highlighting the Nonequilibrium Nature of Food Products and Processes

Nonequilibrium products and processes	Refs.
Sorption/desorption hysteresis	van den Berg (1981, 1986, 1991), Levine and Slade (1988a, 1989b), Lillford (1988), Cairault et al. (1989), Slade et al. (1989), Myers-Betts and Baianu (1990), Slade and Levine (1991a)
Graininess and iciness in ice cream, reduced survival of frozen enzymes and living cells, reduced activity and shelf-stability of freeze-dried proteins	Cole et al. (1983, 1984), Levine and Slade (1986, 1988c, d, 1989c, d, 1990), Blanshard and Franks (1987), Franks (1989, 1990)
Caking of dry powders, sugar bloom on chocolate	Flink (1983), Karel and Flink (1983), Levine and Slade (1986, 1988b, 1989a)
Recipe requirements for gelatin desserts	Slade and Levine (1987a), Levine and Slade (1988a)
Cooking of cereals and grains	Slade and Levine (1984, 1987b, 1988b, c, d, 1989), Reid and Charoenrein (1985), Burros et al. (1987), Paton (1987), Mestres et al. (1988), Levine and Slade (1989b), Knutson (1990)
Expansion of bread or collapse of cake during baking, effects of flour and sugar on cookie baking	Blanshard (1986), Slade et al. (1989), Levine and Slade (1989b), Slade and Levine (1992)
Staling of baked products	Slade (1984), Miles et al. (1985b), Hoseney (1986), Slade and Levine (1987b, 1988d, 1989), Slade et al. (1987), Clark et al. (1989), Gidley and Bulpin (1989)
Effects of sugar-water glasses and rubbers on texture and storage stability of cookies	Slade and Levine (1991a, b, 1992), Gaines et al. (1991a, b)

products and/or aspects of molecular mobility and diffusivity related to the effects of glass transitions in aqueous food glasses and rubbers. These methods include thermomechanical analysis (TMA), thermal-stimulated current/relaxation map analysis (TSC/RMA) spectroscopy, electron spin resonance (ESR) spectroscopy, nuclear magnetic reasonance (NMR) spectroscopy, Brillouin scattering spectroscopy, mechanical spectrometry, dynamic rheometry, and Instron analysis. Probably the most powerful instrumental approach to the study of glass transitions in foods would involve combined use of several of these different but complementary experimental techniques (Noel et al., 1990). For example, in a recent study by Kalichevsky et al. (1991) of the glass transition of amylopectin (the branched polymer of starch) at low moisture contents, DSC, DMTA, NMR, and Instron analyses were utilized. Similarly, Cocero and Kokini (1991) used combined mechanical spectrometry and DSC measurements in a study of the glass transition, at low moisture contents, of the glutenin component of wheat gluten protein; Rubin et al. (1990) investigated vitrification of honey, using DSC and NMR; and Johnson et al. (1990) examined the interactions of starch and sugar-water, using ESR and DSC.

Table 3.5 Studies Employing Methods Other Than DSC or DMA to Investigate Glass Transitions and Their Effects in Food Ingredients and Products

Method	Refs.
Thermomechanical analysis (TMA)	Biliaderis et al. (1986a, 1986c), Caldwell et al. (1990), Biliaderis (1990, 1991), Le Meste and Huang (1991), Le Meste et al. (1991a), Maurice et al. (1991), Schenz et al. (1991)
Thermal stimulated current/relaxation map analysis (TSC/RMA) spectroscopy	Matthiesen et al. (1991)
Electron spin resonance (ESR)	Le Meste and Duckworth (1988), Simatos and Karel (1988), Johnson et al. (1990), Roozen and Hemminga (1990), Le Meste et al. (1990, 1991b)
Nuclear magnetic resonance (NMR)	Simatos and Karel (1988), Rubin et al. (1990), Given (1991), Kalichevsky et al. (1991)
Brillouin scattering spectroscopy	Hosea et al. (1990)
Mechanical spectrometry	Masi (1989), Cocero and Kokini (1991)
Dynamic rheometry	Mita (1990)
Instron analysis	Attenburrow et al. (1990), Ollett et al. (1990), Davies et al. (1991), Kalichevsky et al. (1991)

Results of such studies have graphically demonstrated "the fact that the glass transition is not observed at a unique temperature, but is related to the frequency [or time scale (Slade and Levine, 1988c, 1991a; Orford et al., 1989; Hosea et al., 1990; Noel et al., 1990; Ring and Whittam, 1991)] of the measurement technique" (Kalichevsky et al., 1991) and that "the various techniques are sensitive to different degrees of molecular mobility" (Kalichevsky et al., 1991).

The polymer science-based interpretive approach to thermal analysis studies of structure-function relationships in food systems (Slade and Levine, 1984, 1987a, b, 1988a, b, c, d, 1989, 1991a, b, 1992) emphasizes insights gained by an appreciation of the fundamental similarities between synthetic amorphous polymers and glass-forming aqueous food materials with regard to their thermal, mechanical, and structural properties. Based on this approach, DSC results have been used to demonstrate that product quality and stability often depend on maintaining foods in kinetically metastable, dynamically constrained, time-dependent glassy and/or rubbery states, and that these non-equilibrium physical states determine the time-dependent thermomechanical, rheological, and textural properties of food materials (Slade and Levine, 1984, 1987a, b, 1988, a, b, c, d, 1989, 1991a, b, 1992).

Plasticization, and its modulating effect on the temperature location of the glass transition, is another key technological aspect of synthetic polymer science (Sears and Darby, 1982). In that field, the classical definition of a plasticizer is "a material incorporated in a polymer to increase the polymer's workability, flexibility, or extensibility" (Sears and Darby, 1982). Characteristically, the Tg of an undiluted polymer is much higher than that of a typical low molecular weight, glass-forming diluent. As the diluent concentration of a solution increases, Tg decreases monotonically, because the average MW of the homogeneous polymer-plasticizer mixture decreases, and its free volume (i.e., the volume of the mixture that is not occupied by molecules) increases (Ferry, 1980). A polymer science approach to the thermal and mechanical analysis of both model and real food systems involves recognition of the critical role of water as an effective plasticizer of amorphous polymeric, oligomeric, and monomeric food materials. Sears and Darby (1982) have stated unequivocally that "water is the most ubiquitous plasticizer in our world." Karel (1985) has noted that "water is the most important . . . plasticizer for hydrophilic food components." It has become well documented, in large part through DSC studies, that plasticization by water depresses the Tg [and the melt viscosity (Smith, 1990) and elastic modulus (Ablett et al., 1986; Hutchinson et al., 1989; Ollett et al., 1990; Davies et al., 1991)] of completely amorphous and partially crystalline food ingredients, and that this Tg depression may be advantageous or disadvantageous to product processing, functional properties, and storage stability (Levine and Slade, 1988a, 1989b; Slade et al., 1989; Slade and Levine, 1991a, b). Recently, there has been expanding interest in the importance of the effect of water as a plasticizer of many different food materials and other biopolymers, including starch and its component polymers, amylose and amylopectin; starch hydrolysis products (SHPs); low molecular weight sugars and polyhydric alcohols; nonstarch polysaccharides; gluten and one of its component polymers, glutenin; gelatin; collagen; elastin; lysozyme and other enzymes; other proteins; and the semi-crystalline cellulose and amorphous hemicellulose and lignin components of wood (Table 3.6).

Atkins (1987) has succinctly stated the important observation that "water acts as a plasticizer, dropping the Tg of most biological materials from about 200°C [for anhydrous polymers, e.g. starch, gluten, gelatin (Levine and Slade, 1988a)] to about −10°C or so [under physiological conditions of water content], without which they would be glassy" [in their native, in vivo state]. The latter Tg of about −10°C is characteristic of high molecular weight biopolymers at or above moisture contents, near 30%, corresponding to physiological conditions. This fact has been reported for many polymeric carbohydrates and proteins, including starch, gluten, and gelatin (Levine and Slade, 1988a), hemicelluloses (Kelley et al., 1987), and elastin (Atkins, 1987). Elastin epitomizes a case in which this subzero Tg is critical to healthy physiological function (Hoeve and Hoeve, 1978). Elastin exists as a completely amorphous, water-plasticized, covalently cross-linked (via disulfide bonds), network-forming polymer system whose viscoelastic properties have been likened to those of wheat gluten (Edwards et al., 1987; Ablett et al., 1988; Noel et al., 1990). In its role as a major fibrous structural protein of skin, ligaments, and arteries, elastin exists in vivo as a rubbery liquid, which demonstrates classical rubberlike elasticity (Ferry, 1980) only as long as its Tg remains below 0°C because of a water content of 0.40 g/g protein (Hoeve and Hoeve, 1978). In contrast, in the pathologic condition of arteriosclerosis (hardening of the arteries), elastin becomes a glassy solid at body temperature, because its water content decreases to 0.17 g/g and its Tg correspondingly increases to 40°C (Hoeve and Hoeve, 1978), as revealed by the so-called "glass curve" of Tg vs water content for elastin reported by Kakivaya and Hoeve (1975).

A unified conceptual approach to research on glasses and glass transitions in aqueous food polymer systems, based on principles translated from synthetic polymer science, has enhanced our qualitative understanding of structure-function relationships in a wide variety of food ingredients and products (Slade and Levine 1984, 1987a, b, 1988a, b, c, d, 1989, 1991a,

Table 3.6 Studies of Effects of Water as a Plasticizer in Food and Biological Materials

Starch, amylose, amylopectin: van den Berg (1981, 1986, 1991), Slade (1984), Slade and Levine (1984, 1987b, 1988c, d, 1989, 1991a, b, 1992), Wittwer and Tomka (1984), Biliaderis et al. (1985, 1986a, b, c), Maurice et al. (1985), Ablett et al. (1986), Blanshard (1986, 1987, 1988), Yost and Hoseney (1986), Chungcharoen and Lund (1987), Ring et al. (1987), Russell (1987a), Slade et al. (1987), Zeleznak and Hoseney (1987a, b), Levine and Slade (1988a, 1989b), Lillford (1988), Lineback and Rasper (1988), Marsh and Blanshard (1988), Zobel (1988), Zobel et al. (1988), Attenburrow et al. (1989), Biliaderis and Galloway (1989), Hutchinson et al. (1989), Lund (1989), Orford et al. (1989), Russell and Oliver (1989), Shi and Seib (1989), Stepto and Dobler (1989), Biliaderis (1990, 1991), Biliaderis and Seneviratne (1990a, b), Biliaderis and Zawistowski (1990), Ghiasi and Skarra (1990), I'Anson et al. (1990), Johnson et al. (1990), Knutson (1990), Morris (1990), Noel et al. (1990), Ollett et al. (1990), Roper and Koch (1990), Smith (1990), Warburton et al. (1990), Eliasson (1991), Given (1991), Kalichevsky et al. (1991), Liu and Lelievre (1991), Roos and Karel (1991e), Tomka (1991), Whittam et al. (1991)

Starch hydrolysis products: Cole et al. (1983, 1984), Flink (1983), Karel and Flink (1983), Saleeb and Pickup (1985, 1989), Levine and Slade (1986, 1988a, b, c, 1989a, b, c, 1990, 1991), Karel and Langer (1988), Orford et al. (1989), Levine et al. (1991), Lim and Reid (1991), Ring and Whittam (1991), Roos and Karel (1991b, e), Slade and Levine (1991a, b)

Low molecular weight sugars: White and Cakebread (1966), Cakebread (1969), MacKenzie (1977), Soesanto and Williams (1981), Franks (1982, 1985, 1986a, 1989, 1990, 1991), Herrington and Branfield (1984), Schenz et al. (1984, 1991), Karel (1985, 1986, 1989), Chan et al. (1986), Blanshard and Franks (1987), Roos (1987), Karel and Langer (1988), Levine and Slade (1988a, b, c, d, 1989a, b, c, d, 1990), Simatos and Karel (1988), Slade and Levine (1988a, b, 1991a, b, 1992), Blond (1989), Finegold et al. (1989), Green and Angell (1989), Orford et al. (1989, 1990), Simatos et al. (1989), Williams and Carnahan (1989, 1990), Caldwell et al. (1990), Franks and Grigera (1990), Noel et al. (1990), Reid (1990), Roos and Karel (1990, 1991a, c, d, e), Roozen and Hemminga (1990), Rubin et al. (1990), Ablett et al. (1991a, b), Hatley et al. (1991), Izzard et al. (1991), Karel and Saguy (1991), Le Meste and Huang (1991), Levine et al. (1991), Matthiesen et al. (1991), Maurice et al. (1991), Shimada et al. (1991), Simatos and Blond (1991)

Polyhydric alcohols: Reid (1985), Levine and Slade (1988a, b, c, d, 1989a, b, c, 1990), Quinquenet et al. (1988), Slade and Levine (1988b, 1991a), Franks and Grigera (1990), Ablett et al. (1991a)

Nonstarch polysaccharides: Kararli and Catalano (1990), Paakkonen and Roos (1990), Yoshida et al. (1990), Nishinari et al. (1991), Slade and Levine (1991a, b)

Table 3.6 Continued

Gluten: Slade (1984), Ablett et al. (1986, 1988), Hoseney et al. (1986), Doescher et al. (1987), Edwards et al. (1987), Levine and Slade (1988a, 1989b), Lillford (1988), Attenburrow et al. (1989, 1990), Fujio and Lim (1989), Slade et al. (1989), Hoseney and Rogers (1990), Davies et al. (1991), Given (1991), Hoseney (1991), Slade and Levine (1991a, 1992)
Glutenin: Cocero and Kokini (1991)
Gelatin: Jolley (1970), Yannas (1972), Borchard et al. (1980), Marshall and Petrie (1980), Slade and Levine (1984, 1987a, 1991a), Tomka (1986), Levine and Slade (1988a), Slade et al. (1989)
Collagen: Yannas (1972), Batzer and Kreibich (1981), Levine and Slade (1988a), Slade et al. (1989), Slade and Levine (1991a)
Elastin: Kakivaya and Hoeve (1975), Hoeve and Hoeve (1978), Hoeve (1980), Atkins (1987), Levine and Slade (1988a), Slade et al. (1989), Slade and Levine (1991a)
Lysozyme: Bone and Pethig (1982, 1985), Poole and Finney (1983, 1984), Morozov and Gevorkian (1985), Franks (1988), Levine and Slade (1988a), Lillford (1988), Slade et al. (1989), Myers-Betts and Baianu (1990), Slade and Levine (1991a)
Other enzymes: Poole and Finney (1984), Morozov and Gevorkian (1985), Levine and Slade (1988a), Slade et al. (1989), Slade and Levine (1991a)
Other proteins: Le Meste and Duckworth (1988), Le Meste et al. (1990, 1991b), Levine and Slade (1988a), Slade et al. (1989), Slade and Levine (1991a)
Cellulose, hemicellulose, and lignin: Salmen and Back (1977), Kelley et al. (1987), Ostberg et al. (1990)

b, 1992). Lillford and his coworkers at Unilever have advocated a related "materials science approach" to studies of (1) the influence of water on the mechanical behavior of dough and batter before, during, and after baking (Ablett et al., 1986), (2) the mechanical properties of solid food foams as affected by "the plasticizing action of water" (Attenburrow et al., 1989), (3) the influence of plasticization by water on the mobility and structural properties of nonequilibrium, kinetically metastable food polymer systems (Lillford, 1988), and (4) the plasticizing effect of water on the mobility, rheological/mechanical properties, and glassy/rubbery behavior of heat-set wheat gluten networks (Edwards et al., 1987; Attenburrow et al., 1990; Davies et al., 1991). Similarly, in a recent review of structure-property relationships in starch and another paper on starch gelatinization, Zobel (1988; Zobel et al., 1988) has cited concepts used to characterize synthetic polymers and advocated this approach to provide increased understanding of the amorphous state and its role in determining physical properties of native and gelatinized starches. Others who have recently applied a synthetic polymers or materials approach to characterize the glass transition, melting, crystallization, annealing, or gelation/network formation behavior of food polymers such as starch or gluten are listed in Table 3.7, part A. In recent years, many others have also supported the perspective based on the fundamental behavioral similarities between synthetic polymer-plasticizer and food molecule-water systems. Pertinent references are listed in Table 3.7, part B.

A central theme of our "food polymer science" approach focuses on the effect of water as a plasticizer on the glass transition and resulting

Table 3.7 Studies Employing a Synthetic Polymers or Materials Approach to the Characterization of Food Molecule–Water Systems

A. Researchers studying starch or gluten
Blanshard: (1986, 1987, 1988), Edwards et al. (1987), Marsh and Blanshard (1988), Kalichevsky et al. (1991)
Hoseney: (1991), Hoseney et al. (1986), Yost and Hoseney (1986), Doescher et al. (1987), Zeleznak and Hoseney (1987a, b), Hoseney and Rogers (1990)
Morris: (1990), Ring et al. (1987), I'Anson et al. (1990)
Ring: Ring et al. (1987), Orford et al. (1989), Noel et al. (1990), Ring and Whittam (1991), Whittam et al. (1991)
Smith: (1990), Hutchinson et al. (1989), Orford et al. (1989), Ollett et al. (1990), Warburton et al. (1990)
Biliaderis and Maurice: Biliaderis et al. (1985, 1986a, b, c), Biliaderis and Galloway (1989), Biliaderis and Zawistowski (1990), Biliaderis and Seneviratne (1990a, b), Biliaderis (1990, 1991)
Lund: (1989), Chungcharoen and Lund (1987)
Russell: (1987a), Russell and Oliver (1989)
Eliasson: (1991)
Lelievre: Liu and Lelievre (1991)
Tomka: (1991), Wittwer and Tomka (1984)
Fujio: Fujio and Lim (1989)
Mita: (1990)
Masi: (1989)
Kokini: Cocero and Kokini (1991)
B. Other studies
Karel (1985, 1986, 1989), Le Meste and Duckworth (1988), Simatos and Karel (1988), Cairault et al. (1989), Finegold et al. (1989), Franks (1989), Simatos et al. (1989), Franks and Grigera (1990), Johnson et al. (1990), Le Meste et al. (1990, 1991a, b), Ollett and Parker (1990), Orford et al. (1990), Roos and Karel (1990, 1991a, b, c, e), Roozen and Hemminga (1990), Yoshida et al. (1990), Given (1991), Izzard et al. (1991), Karel and Saguy (1991), Le Meste and Huang (1991), Lim and Reid (1991), Matthiesen et al. (1991), Reid and Hsu (1991), Shukla (1991), Simatos and Blond (1991), van den Berg (1991)

diffusion-limited behavior of water-soluble or water-miscible (collectively referred to as water-compatible) and water-sensitive amorphous materials or amorphous regions of partially crystalline materials (Levine and Slade, 1988a, 1989b; Slade et al., 1989; Slade and Levine, 1991a, b). Plasticization, on a molecular level, leads to increased intermolecular space or free volume, decreased local viscosity, and concomitant increased mobility (Ferry, 1980). Plasticization implies intimate mixing, such that a plasticizer is homogeneously blended in a polymer, or a polymer in a plasticizer. Note that a true solvent, capable of cooperative dissolution of ordered crystals and having high thermodynamic compatibility and miscibility at all proportions, is always also a plasticizer, but a plasticizer is not always a solvent (Sears and Darby, 1982). Water-compatible food polymers such as starch, gluten, and gelatin, for which water is an efficient plasticizer but not necessarily a good solvent, exhibit essentially the same physicochemical responses to plasticization by water as do many water-compatible synthetic polymers (Ellis, 1988) and many readily soluble monomeric and oligomeric carbohydrates (Levine and Slade, 1986, 1988a, b; Slade and Levine, 1988b, 1991a, b; Orford et al., 1989, 1990; Franks and Grigera, 1990; Noel et al., 1990; Davies et al., 1991; Roos and Karel, 1991d, e; Whittam et al., 1991). This fact demonstrates two underlying precepts of the food polymer science approach: (1) synthetic amorphous polymers and glass-forming aqueous food materials are fundamentally similar in behavior, and (2) food ingredients can be viewed generically as members of homologous families of completely amorphous or partially crystalline polymers, oligomers, and monomers, soluble in and/or plasticized by water (Levine and Slade, 1988a). The series from glucose through malto-oligosaccharides to the amylose and amylopectin components of starch exemplifies such a homologous polymer family.

On a theoretical basis of structure-property relationships for synthetic polymers, the functional properties of food materials during processing and product storage can be successfully explained and often predicted (Slade and Levine, 1984, 1987a, b, 1988a, b, c, d, 1989, 1991a, b, 1992). The food polymer science approach unifies structural aspects of foods, viewed as completely amorphous or partially crystalline polymer systems [the latter based on the classical "fringed micelle" morphological model (Flory, 1953; Wunderlich, 1973; Billmeyer, 1984) shown in Fig. 3.2 (Slade and Levine, 1987a)], with functional aspects, dependent on mobility and described in terms of the integrated concepts of "water dynamics" and "glass dynamics." Through this unification, the appropriate kinetic description of the nonequilibrium thermomechanical behavior of food systems has been illustrated in the context of a "dynamics map," shown in Fig. 3.3 (Slade and Levine, 1988b). This map was derived from a generic

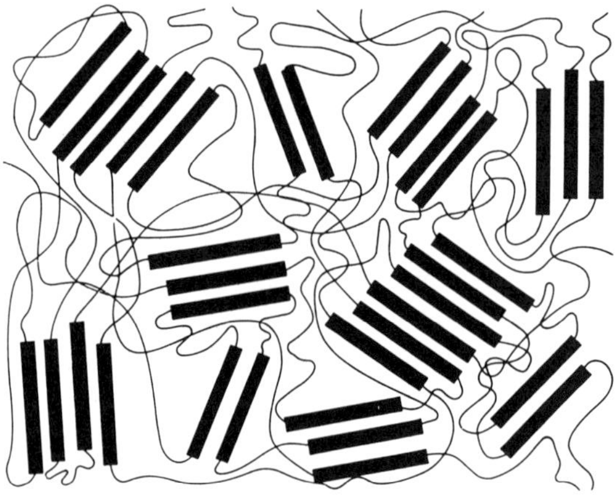

Fig. 3.2 "Fringed micelle" model of the crystalline-amorphous structure of partially crystalline polymers. (*Reproduced, with permission, from Slade and Levine, 1987a.*)

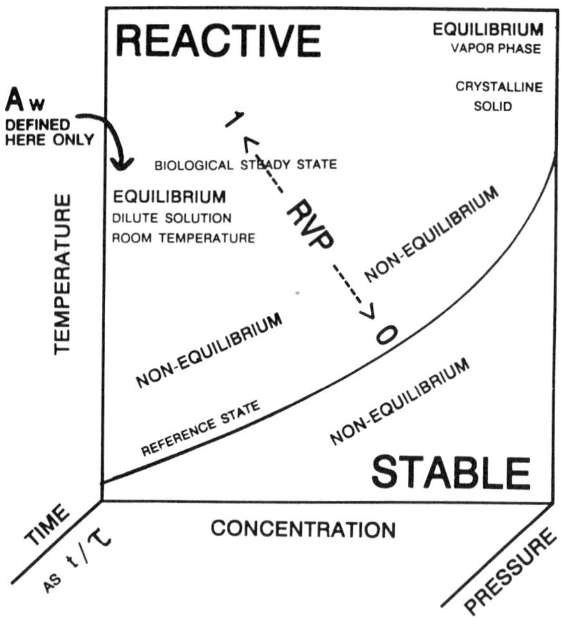

Fig. 3.3 A four-dimensional "dynamics map," with axes of temperature, concentration, time, and pressure, which can be used to describe mobility transformations in nonequilibrium glassy and rubbery systems. (*Reproduced, with permission, from Slade and Levine, 1988b.*)

solute-solvent state diagram (Franks et al., 1977; MacKenzie, 1977), in turn based originally on a more familiar equilibrium phase diagram of temperature vs. composition. The dynamics map, like the "supplemented state diagram" (MacKenzie, 1977), is complicated by the attempt to represent aspects of both equilibrium and nonequilibrium thermodynamics in a single figure. The primary distinction at atmospheric pressure is that equilibrium regions are completely described as shown in two dimensions of temperature and composition, with no time dependence, while nonequilibrium regions emphatically require the third dimension of time, expressed as t/τ, where τ is a relaxation time. In this way, the time dependence of a dynamic process can be defined in terms of the relationship between the experimental time scale and the time frame of the relaxation process undergone by the system. The established principle of time-temperature superpositioning (Sichina, 1988) has been extended to define "mobility transformations" in terms of the critical variables of time, temperature, and moisture content (with pressure as another variable of potential technological importance) (Slade and Levine, 1988b). The dynamics map has been used (Slade and Levine, 1988b) to describe mobility transformations in water-compatible food polymer systems that exist in kinetically metastable glassy and rubbery states (Levine and Slade, 1988a, 1989b; Slade et al., 1989; Slade and Levine, 1991a, b) always subject to conditionally beneficial or detrimental plasticization by water. For example, the kinetics of starch gelatinization and retrogradation have been explained in terms of mobility transformations, as illustrated by locating on the dynamics map the alternative pathways of complementary plasticization by heat and moisture (Slade and Levine, 1988b, d, 1989; Levine and Slade, 1989b). The map domains of moisture content and temperature, traditionally described with only limited success using concepts such as "water activity" (Aw) and "bound water" to interpret and explain sorption isotherms and sorption hysteresis, have been treated alternatively in terms of water dynamics (Slade and Levine, 1988a, 1991a, b; Levine and Slade, 1988a). As the name implies, water dynamics focuses on (1) the mobility and eventual "availability" of the plasticizing diluent (be it water alone or an aqueous solution) and (2) an interpretive approach to understanding how to control the mobility of the diluent in glass-forming food systems that would be inherently mobile, unstable, and reactive at $T > Tg$ and moisture content $(W) > Wg$. This concept (along with that of glass dynamics) has provided an innovative perspective on the moisture management and structural stabilization of IMF systems (Slade and Levine, 1988a, 1991a) and the "cryostabilization" of frozen, freezer-stored, and freeze-dried aqueous glass-forming food materials and products (Cole et al., 1983, 1984; Levine and Slade, 1986, 1988a, b, c, d, 1989a, b, c, d,

1990). This perspective has received considerable support from many other investigators (Table 3.8, part A).

Glass dynamics deals with the time- and temperature-dependence of relationships among composition, structure, thermomechanical properties, and functional behavior (Levine and Slade, 1988a). As its name implies, glass dynamics focuses on (1) the glass-forming solids in an aqueous food system, (2) the Tg of the resulting aqueous glass that can be produced by cooling to T < Tg, and (3) the effect of the glass transition and its Tg

Table 3.8 Publications in Which
(A) Support Is Expressed for the Perspective Provided by the Concepts of Water Dynamics and Glass Dynamics and
(B) The Applicability of WLF Kinetics to Food Systems Is Discussed

(A)	Blanshard (1986, 1987, 1988), Blanshard and Franks (1987), Chungcharoen and Lund (1987), Russell (1987a), Gould and Christian (1988), Karel and Langer (1988), Le Meste and Duckworth (1988), Lillford (1988), Lineback and Rasper (1988), Marsh and Blanshard (1988), Quinquenet et al. (1988), Simatos and Karel (1988), Zobel (1988), BeMiller (Hill, 1989), Biliaderis and Galloway (1989), Cairault et al. (1989), Franks (1989, 1990, 1991), Finegold et al. (1989), Karel (1989), Lund (1989), Orford et al. (1989), Russell and Oliver (1989), Shi and Seib (1989), Simatos et al. (1989), Biliaderis (1990, 1991), Biliaderis and Zawistowski (1990), Biliaderis and Seneviratne (1990a), Caldwell et al. (1990), Franks and Grigera (1990), Hosea et al. (1990), Hoseney and Rogers (1990), I'Anson et al. (1990), Johnson et al. (1990), Johnston-Banks (1990), Kararli and Catalano (1990), Kararli et al. (1990), Le Meste et al. (1990, 1991b), Morris (1990), Myers-Betts and Baianu (1990), Noel et al. (1990), Ollett and Parker (1990), Paakkonen and Roos (1990), Pikal and Shah (1990), Reid (1990), Roos and Karel (1990, 1991a, b, c, d, e), Roozen and Hemminga (1990), Rubin et al. (1990), Warburton et al. (1990), Ablett et al. (1991a, b), Chang and Baust (1991), Cocero and Kokini (1991), Eliasson (1991), Gaines et al. (1991a, b), Given (1991), Hatley et al. (1991), Izzard et al. (1991), Kalichevsky et al. (1991), Karel and Saguy (1991), Le Meste and Huang (1991), Lim and Reid (1991), Liu and Lelievre (1991), MacDonald and Lanier (1991), Matthiesen et al. (1991), Maurice et al. (1991), Nishinari et al. (1991), Reid and Hsu (1991), Ring and Whittam (1991), Schenz et al. (1991), Shimada et al. (1991), Shukla (1991), Simatos and Blond (1991), van den Berg (1991), Wasylyk and Baust (1991), Watase et al. (1991)
(B)	Levine and Slade (1986, 1988b, 1989d), Simatos and Karel (1988), Simatos et al. (1989), Biliaderis (1990, 1991), Franks and Grigera (1990), Morris (1990), Reid (1990), Roos and Karel (1990, 1991a, d, e), Given (1991), Karel and Saguy (1991), Lim and Reid (1991), Shimada et al. (1991), Simatos and Blond (1991)

on processing and process control, via the relationships between Tg and the temperatures of individual processing steps (which may be deliberately chosen to be first above and then below Tg) (Slade and Levine, 1991a, b). This concept emphasizes the operationally immobile, stable, and unreactive situation (actually one of kinetic metastability) that can obtain during product storage (of a practical duration) at temperatures below Tg and moisture contents below Wg. It has been used to describe a unifying concept for interpreting "collapse" phenomena (Levine and Slade, 1986, 1988b), which govern, e.g., the time-dependent caking of amorphous food powders during storage. Collapse phenomena in completely amorphous or partially crystalline food systems, investigated extensively by Karel and coworkers (To and Flink, 1978; Karel and Flink, 1983; Flink, 1983; Karel, 1985, 1986, 1989; Roos, 1987; Karel and Langer, 1988; Roos and Karel, 1990, 1991a, b, c; Paakkonen and Roos, 1990; Karel and Saguy, 1991; Shimada et al., 1991), are diffusion-limited consequences of a material-specific structural and/or mechanical relaxation process (Levine and Slade, 1986). The microscopic and macroscopic manifestations of these consequences occur in real time at a temperature ~20°C above that of an underlying molecular state transformation (Slade and Levine, 1988b; Roos and Karel, 1991a, b). This transformation from kinetically metastable amorphous solid to unstable amorphous liquid occurs at Tg (Levine and Slade, 1986). The effect of plasticization (leading to increased free volume and mobility in the dynamically constrained glass) by water on Tg is a key aspect of collapse and its mechanism (Levine and Slade, 1988b).

A general physicochemical mechanism for collapse has been described (Levine and Slade, 1986), based on occurrence of the material-specific structural transition at Tg, followed by viscous flow in the rubbery liquid state (Flink, 1973). The mechanism was derived from WLF free volume theory for (synthetic) amorphous polymers (Williams et al., 1955; Ferry, 1980). It has been concluded that Tg is identical to the phenomenological transition temperatures observed for structural collapse (Tc) and recrystallization (Tr) (Levine and Slade, 1986). The non-Arrhenius kinetics of collapse and/or recrystallization in the high viscosity rubbery state are governed by the mobility of the water-plasticized polymer matrix (Levine and Slade, 1986). These kinetics depend on the magnitude of $\Delta T = T - Tg$ (Table 3.8, part B), as defined by a temperature-dependent exponential relationship derived from WLF theory. Glass dynamics has proved a useful concept for elucidating the physicochemical mechanisms and kinetics of structural/mechanical changes involved in various melting, annealing, and (re)crystallization processes (Levine and Slade, 1988a). Such phenomena are observed in many partially crystalline food polymers and processing/storage situations, including, e.g., the gelatinization and retrogradation of starches (Slade, 1984; Slade and Levine, 1987b, 1988c). Glass dynamics has also been used to describe the viscoelastic behavior of amorphous polymeric network-forming proteins such as gluten and elastin (Slade et al., 1989).

KEY ELEMENTS OF THE FOOD POLYMER SCIENCE APPROACH

"Fringed Micelle" Structural Model

The "fringed micelle" model shown in Fig. 3.2 was originally developed to describe the morphology of partially crystalline synthetic polymers. It is a particularly useful way of viewing a three-dimensional network composed of microcrystallites (with crystalline melting temperature, Tm) that crosslink amorphous regions (with glass transition temperature, Tg) of flexible-coil chain segments (Jolley, 1970). In pure homopolymers, for which Tg is always at a lower temperature than Tm (Wunderlich, 1981), the amorphous domains can exist in a glassy solid physical state at $T < Tg$ or in a rubbery liquid state at $Tg < T < Tm$ (Levine and Slade, 1988a). The model is especially applicable to synthetic polymers that crystallize from an undercooled melt or concentrated solution to produce a metastable network of relatively low degree of crystallinity. Typically, such polymers contain small crystalline regions of only about 100 Å (Wunderlich, 1973). Thus, the model has also often been used to describe the partially crystalline structure of aqueous gels of biopolymers such as starch and gelatin (Jolley, 1970; Wunderlich, 1973; Guilbot and Godon, 1984; Slade, 1984; Slade and Levine, 1984, 1987a, b; Zobel, 1988; Lund, 1989), in which the amorphous regions contain plasticizing water and the microcrystalline regions, which serve as physical junction zones, are crystalline hydrates. The model has also been used to describe the partially crystalline morphology of frozen aqueous food polymer systems, in which case the ice crystals represent the "micelles" dispersed in a continuous amorphous matrix (the "fringe") of solute-unfrozen water (UFW) (Levine and Slade, 1988a). An important feature of the model, as applied to high molecular weight polymer systems such as starch [both native granular and gelatinized (Slade and Levine, 1988c)] and gelatin (Slade and Levine, 1987a), concerns the interconnections between crystalline and amorphous domains. A single long polymer chain can have helical (or other ordered) segments located within one or more microcrystallites that are covalently linked to flexible-coil segments in one or more amorphous regions (Jolley,

1970). Moreover, in the amorphous regions, chain segments may experience random intermolecular "entanglement couplings" (Graessley, 1984), which are topological interactions rather than covalent or noncovalent chemical bonds (Mitchell, 1980). Thus, in their thermomechanical behavior in response to plasticization by water and/or heat, the crystalline and amorphous domains are neither independent of each other nor homogeneous (Wunderlich, 1981).

The Dynamics Map

The key to our perspective on concentrated, water-plasticized food polymer systems relates to recognition of the fundamental importance of the dynamics map mentioned earlier. As shown in Fig. 3.3, the major area of the map (i.e., the area surrounding the reference state in two dimensions and projecting into the third, time, dimension) represents a nonequilibrium situation corresponding to the temperature-composition region of most far-reaching technological consequence to aqueous food systems (Slade and Levine, 1988b). The critical feature in the use of this map is identification of the glass transition as the reference state (Slade and Levine, 1988a, b; Franks and Grigera, 1990), a conclusion (Levine and Slade, 1988a; Ollett and Parker, 1990) based on WLF theory for glass-forming polymers. This line of demarcation (representing the glass curve of Tg vs composition) (Levine and Slade, 1986; Simatos and Karel, 1988; Franks, 1989) provides a basis for describing the nonequilibrium thermomechanical behavior of water-compatible polymeric materials in glassy and rubbery states, in response to changes in moisture content, temperature, and time (Levine and Slade, 1988a). Mobility is the transcendent principle underlying the definition of the glass transition as the appropriate reference state (Slade and Levine, 1988a), because mobility is the key to all transformations in time (or frequency), temperature, and composition between different relaxation states for a technologically practical system (Slade and Levine, 1988b).

The interdependent concepts of water dynamics and glass dynamics embodied in the dynamics map have provided insights into the relevance of the glassy reference state to functional aspects of a variety of food systems (Levine and Slade, 1988a; Slade and Levine, 1988b). For example, the kinetics of all constrained relaxation processes, such as translational and rotational diffusion, which are governed by the mobility of a water-plasticized polymer matrix in glass-forming systems, vary (from Arrhenius to WLF-type) between distinct temperature/structure domains, which are divided by this glass transition (Levine and Slade, 1988a, 1989b;

Slade and Levine, 1988b, 1991a, b). Thus, while familiar Arrhenius kinetics are applicable below Tg in the glassy solid state of very low mobility and very slow diffusion (the domain of glass dynamics, labeled STABLE in Fig. 3.3), WLF kinetics (Ferry, 1980) are applicable above Tg in the viscoelastic, rubbery liquid state of accelerating mobility and diffusion (the domain of water dynamics, labeled REACTIVE in Fig. 3.3) (Slade and Levine, 1988b). The WLF equation (Williams et al., 1955; Ferry, 1980) defines the kinetics of molecular-level relaxation processes, which will occur in practical time frames only in the rubbery state above Tg, in terms of an exponential, but non-Arrhenius, function of ΔT above this boundary condition (Levine and Slade, 1986, 1989b; Franks and Grigera, 1990; Lim and Reid, 1991; Roos and Karel, 1991a, d). The highest mobility and most rapid diffusion occur in the region above a second set of reference lines, the equilibrium liquidus and solidus curves, which demark the upper boundary of the WLF region where Arrhenius kinetics again apply (Levine and Slade, 1989c). Within the WLF region, kinetics accelerate according to the WLF equation from the extremely steep temperature dependence of WLF kinetics just above Tg to the moderate temperature dependence of Arrhenius kinetics on approaching Tm (Slade and Levine, 1988b). The WLF equation describes the profound range of kinetics between Tg and Tm, with correspondingly profound implications for process control, product quality, safety, and shelf life (Levine and Slade, 1989d; Slade and Levine, 1991a). Sperling (1986) has remarked that "for a generation of [synthetic] polymer scientists and rheologists, the WLF equation has provided a mainstay both in utility and theory." It should be noticed in Fig. 3.3 that Aw is correctly defined only in the region of the map corresponding to a dilute solution at equilibrium at room temperature. In contrast, the actual measured relative vapor pressure (RVP) of an IMF (nonequilibrium) system would approach zero in the limit of the glassy reference state at temperatures below Tg and moisture contents $<$ Wg, but would increase toward 1.0 with increasing temperature above Tg and increasing moisture content above Wg (Slade and Levine, 1991a, b).

One particular location among the continuum of Tg values along the reference glass curve in Fig. 3.3 results from the behavior of water as a *crystallizing* plasticizer and corresponds to an operationally invariant point [called Tg′ (Franks et al., 1977)] on a state diagram for any particular solute. Tg′ represents the solute-specific subzero Tg of the maximally freeze-concentrated, amorphous solute/UFW matrix surrounding the ice crystals in a frozen solution (Franks et al., 1977; Schenz et al., 1984; Levine and Slade, 1986; additional refs. in Table 3.9, part A). A survey of recently reported literature values of Tg′ for many different low mo-

Table 3.9 Publications Dealing with (A) Tg′, and (B) the Importance of State Diagrams and of the Tg′-Cg′ Point

(A)	Franks et al. (1977), Franks (1982, 1983b, 1985, 1986a, b, 1989, 1990, 1991), Schenz et al. (1984, 1991), Slade (1984), Slade and Levine (1984, 1987a, b, 1988a, b, c, d, 1989, 1991a, b, 1992), Reid (1985, 1990), Levine and Slade (1986, 1988a, b, c, d, 1989a, b, c, d, 1990, 1991), Blanshard and Franks (1987), Lillford (1988), Marsh and Blanshard (1988), Karel (1989), Russell and Oliver (1989), Slade et al. (1989), Biliaderis (1990), Caldwell et al. (1990), Franks and Grigera (1990), Noel et al. (1990), Paakkonen and Roos (1990), Williams and Carnahan (1990), Ablett et al. (1991a, b), Blanshard et al. (1991), Given (1991), Hatley et al. (1991), Izzard et al. (1991), Karel and Saguy (1991), Le Meste and Huang (1991), Levine et al. (1991), Lim and Reid (1991), MacDonald and Lanier (1991), Matthiesen et al. (1991), Maurice et al. (1991), Reid and Hsu (1991), Wasylyk and Baust (1991)
(B)	Schenz et al. (1984, 1991), Slade (1984), Slade and Levine (1984, 1987a, b, 1988a, b, c, d, 1989, 1991a, b, 1992), Levine and Slade (1986, 1988a, b, c, d, 1989a, b, c, d, 1990, 1991), van den Berg (1986), Blanshard and Franks (1987), Lillford (1988), Simatos and Karel (1988), Blond (1989), Franks (1989, 1990), Green and Angell (1989), Karel (1989), Simatos et al. (1989), Biliaderis (1990), Caldwell et al. (1990), Franks and Grigera (1990), Noel et al. (1990), Williams and Carnahan (1990), Ablett et al. (1991a, b), Blanshard et al. (1991), Given (1991), Hatley et al. (1991), Izzard et al. (1991), Le Meste and Huang (1991), Lim and Reid (1991), MacDonald and Lanier (1991), Matthiesen et al. (1991), Reid and Hsu (1991), Roos and Karel (1991d, e), Simatos and Blond (1991)

lecular weight carbohydrates, as well as corresponding Tg values ("dry" Tg) for the anhydrous solutes, is presented in Table 3.10. As illustrated in the schematic state diagram shown in Fig. 3.4 (Slade and Levine, 1991a), Tg′ corresponds to, and is determined by, the point of intersection of the kinetically determined glass curve for homogeneous solute-water mixtures and the nonequilibrium extension of the equilibrium liquidus curve for Tm of ice (Levine and Slade, 1986, 1988a, b, c, d, 1989a, b, c, d, 1990, 1991). This solute-specific location defines the composition of the glass that contains the maximum *practical* amount of plasticizing moisture [called Wg′, expressed as g UFW/g solute or weight % (w%) water, or alternatively designated in terms of Cg′, expressed as w% solute (Levine and Slade, 1986, 1988a)] and represents the transition from concentrated fluid to kinetically metastable, dynamically constrained solid that occurs on cooling to T < Tg′ (Slade and Levine, 1988b). In this homogeneous, freeze-concentrated, solute-water glass, the water represented by Wg′ is not "bound" energetically, but rather rendered unfreezable in

Table 3.10 Tg′ and "Dry" Tg Values for Low Molecular Weight Polyhydroxy Compounds

Polyhydroxy compound	MW	Tg′ (°C) [refs. a–d]	Other Tg′ values [refs.]	Dry Tg (°C) [refs. e–g]	Other dry Tg values [refs.]
Ethylene glycol	62.1	−85			
Propylene glycol	76.1	−67.5			
1,3-Butanediol	90.1	−63.5			
Glycerol	92.1	−65	−65 [h–k], −95 [q]	−93	−93 [j, k], −78 [q]
Erythrose	120.1	−50			
Threose	120.1	−45.5			
Erythritol	122.1	−53.5			
Thyminose (deoxyribose)	134.1	−52			
Ribulose	150.1	−50			
Xylose	150.1	−48	−48 [j–l], −47 [h, i]	9.5	9 [j–l], 13 [o, p]
Arabinose	150.1	−47.5	−48 [h, i], −48.5 [l]		4 [o]
Lyxose	150.1	−47.5		8	
Ribose	150.1	−47	−47 [h–k], −49 [y]	−10	−10 [j, k], −11 [o]
Arabitol	152.1	−47	−48 [l]		
Ribitol	152.1	−47			
Xylitol	152.1	−46.5	−46.5 [j, k]	−18.5	−19 [o], −39 [j, k]
Methyl riboside	164.2	−53			
Methyl xyloside	164.2	−49			
Quinovose (deoxyglucose)	164.2	−43.5			
Fucose (deoxygalactose)	164.2	−43			
Rhamnose (deoxymannose)	164.2	−43			27 [o]

Talose	180.2	−44		11.5	
Idose	180.2	−44			
Psicose	180.2	−44			
Altrose	180.2	−43.5		10.5	
Gulose	180.2	−42.5			
Glucose	180.2	−43	−43 [h–l, r, s], −36.5 [m]	31	21 [l], 29 [v, w], 30 [x], 31 [k, aa], 33 [z], 37 [x], 38 [o, p], 39 [j, k, u]
Fructose	180.2	−42	−42 [h–l, s, t]	11, 100	7 [o], 13 [j, k, u, y], 17 [l], 100 [k]
Fructose:Glucose (1:1 blend)		−42.5	−42.5 [j, k]	20	20 [j, k], 21 [u], 13 [o]
Galactose	180.2	−41.5	−42[h, i], −40.5 [j, k]	~30, 110	32 [o, p], 110 [k]
Allose	180.2	−41.5			
Sorbose	180.2	−41			
Mannose	180.2	−41	−41 [j, k]	30	30 [j, k], 36 [o]
Tagatose	180.2	−40.5			
Inositol	180.2	−35.5	−36 [h, i]		
Mannitol	182.2	−40			
Galactitol	182.2	−39			
Sorbitol	182.2	−43.5	−43.5 [j, k], −43 [h, i], −44 [l]	−2	−2 [bb], −3 [j–l], 0 [o]
2-o-Methyl fructoside	194.2	−51.5			
β-1-o-Methyl glucoside	194.2	−47			
3-o-Methyl glucoside	194.2	−45.5			
6-o-Methyl galactoside	194.2	−45.5			
α-1-o-Methyl glucoside	194.2	−44.5			

Table 3.10 Continued

Polyhydroxy compound	MW	Tg′ (°C) [refs. a–d]	Other Tg′ values [refs.]	Dry Tg (°C) [refs. e–g]	Other dry Tg values [refs.]
1-o-Methyl galactoside	194.2	−44.5			
1-o-Methyl mannoside	194.2	−43.5			
1-o-Ethyl glucoside	208.2	−46.5			
2-o-Ethyl fructoside	208.2	−46.5			
1-o-Ethyl galactoside	208.2	−45			
1-o-Ethyl mannoside	208.2	−43.5			
Glucoheptose	210.2	−37.5			
Mannoheptulose	210.2	−36.5			
Glucoheptulose	210.2	−36.5			
Perseitol (mannoheptitol)	212.2	−32.5			
1-o-Propyl glucoside	222.2	−43			
1-o-Propyl galactoside	222.2	−42			
1-o-Propyl mannoside	222.2	−40.5			
2,3,4,6-o-Methyl glucoside	236.2	−45.5			
Isomaltulose (palatinose)	342.3	−35.5			
Nigerose	342.3	−35.5			
Cellobiulose	342.3	−32.5			
Isomaltose	342.3	−32.5	−32 [l]		78 [o]

Sucrose	342.3	−32	−32 [h–l, n, r, t, cc, dd, ee, ff], −33 [gg, hh], −35 [ii], −40 [q], −41 [mm, nn], −41.5 [oo], −46 [kk, nn]	52	52 [k, l, cc], 57 [j, k, u, jj–ll], 66 [z], 67 [v], 70 [o]
Gentiobiose	342.3	−31.5			
Laminaribiose	342.3	−31.5			
Turanose	342.3	−31		52	
Mannobiose	342.3	−30.5		90	
Melibiose	342.3	−30.5			95 [o]
Lactulose	342.3	−30			
Maltose	342.3	−29.5	−29.5 [j, k], −30 [h, i, l], −41 [pp]	43	43 [j, k, l], 70 [z], 87 [qq], 91 [x], 92 [p], 95 [o]
Maltulose	342.3	−29.5			
Trehalose	342.3	−29.5	−29.5 [j, k], −30 [h, i, l]	79	79 [z], 77 [k]
Cellobiose	342.5	−29		77	77 [j, k]
Lactose	342.3	−28	−28 [h, i]		101 [jj, ll, rr]
Maltitol	344.3	−34.5			
Isomaltotriose	504.5	−30.5			
Panose	504.5	−28			
Raffinose	504.5	−26.5	−26 [h, i]		
Maltotriose	504.5	−23.5	−23.5 [j, k], −23 [h, i], −24 [l]	76	76 [j–l], 134 [o], 135 [x]
Nystose	666.6	−26.5		77	
Stachyose	666.6	−23.5			
Maltotetraose	666.6	−19.5		111.5	147 [x]
Maltopentaose	828.9	−16.5		125	165 [x]
α-Cyclodextrin	972.9	−9			

Table 3.10 Continued

Polyhydroxy compound	MW	Tg′ (°C) [refs. a–d]	Other Tg′ values [refs.]	Dry Tg (°C) [refs. e–g]	Other dry Tg values [refs.]
Maltohexaose	990.9	−14.5		134	175 [o, x]
Maltoheptaose	1153.0	−13.5		138.5	

References:
[a] Levine and Slade (1986).
[b] Levine and Slade (1988b).
[c] Levine and Slade (1988c).
[d] Levine and Slade (1988d).
[e] Slade and Levine (1988b).
[f] Slade and Levine (1991a).
[g] Levine and Slade (1989a).
[h] Franks (1985).
[i] Franks (1986a).
[j] Franks (1990).
[k] Franks and Grigera (1990).
[l] Franks (1989).
[m] Franks (1982).
[n] Franks (1986b).
[o] Orford et al. (1990).
[p] Noel et al. (1990).
[q] Ablett et al. (1991a).
[r] Maurice et al. (1991).
[s] Rubin et al. (1990).
[t] Schenz et al. (1991).
[u] Finegold et al. (1989).
[v] Soesanto and Williams (1981).
[w] Chan et al. (1986).
[x] Orford et al. (1989).
[y] Wasylyk and Baust (1991).
[z] Green and Angell (1989).
[aa] Ollett and Parker (1990).
[bb] Quinquenet et al. (1988).
[cc] Blanshard and Franks (1987).
[dd] Caldwell et al. (1990).
[ee] Schenz et al. (1984).
[ff] Hatley et al. (1991).
[gg] Lim and Reid (1991).
[hh] Reid and Hsu (1991).
[ii] Izzard et al. (1991).
[jj] Roos and Karel (1991a).
[kk] Roos and Karel (1991c).
[ll] Roos and Karel (1990).
[mm] Williams and Carnahan (1990).
[nn] Roos and Karel (1991d).
[oo] Ablett et al. (1991).
[pp] Roos and Karel (1991e).
[qq] Roos and Karel (1991b).
[rr] Shimada et al. (1991).

a practical time frame due to the immobility imposed by the extremely high local viscosity of ~10^{12} Pa s at Tg′ (Franks, 1982, 1983a, 1985, 1986a, b, 1989, 1990, 1991; Franks and Grigera, 1990; Levine and Slade, 1988a, 1989b; Slade et al., 1989; Slade and Levine, 1991a, b). Marsh and Blanshard (1988) have recently documented the technological importance of freeze-concentration in such situations and the practical implication of water being a readily crystallizable plasticizer, characterized by a high value of Tm/Tg ratio ≃ 2 (Soesanto and Williams, 1981; Slade and Levine, 1988b; Wunderlich, 1990). A theoretical calculation (Marsh and Blanshard, 1988) of the Tg of a typically dilute (i.e., 50%) wheat starch gel fell well below the measured value of about −5 to −7°C for Tg′ (Slade, 1984; Slade and Levine, 1984, 1987b), because the calculation, based on free volume theory, did not account for the formation of ice and freeze-concentration that occurs below about −3°C. Recognition of the practical limitation of water as a plasticizer of water-compatible solutes, due to the phase separation of ice (Slade and Levine, 1988b, 1991a), reconciled the difference between theoretical and measured values of Tg (Marsh and Blanshard, 1988). Moreover, the theoretical calculations supported the measured value of ≃27% water (Slade, 1984; Slade and Levine, 1984, 1987b) for Wg′, the maximum practical water content of an aqueous wheat starch glass. The calculated water content of the wheat starch glass with

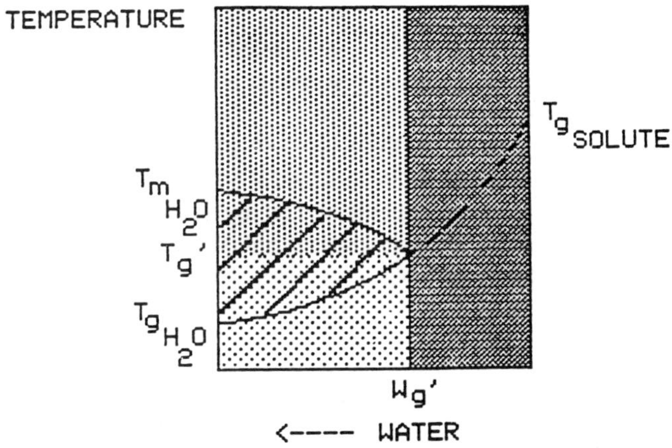

Fig. 3.4 Schematic state diagram of temperature vs. wt% water for an aqueous solution of a hypothetical, glass-forming, small carbohydrate (representing a model frozen food system), illustrating how the critical locations of Tg′ and Wg′ divide the diagram into three distinguishable structure-property domains. (*Reproduced, with permission, from Slade and Levine, 1991a.*)

Tg of about −7°C is ≈28% (Marsh and Blanshard, 1988), and 26% for a measured Tg (i.e., Tg′) of −5°C (Roos and Karel, 1991e).

In recent years, many workers have lent their strong support to the advocacy of the importance of state diagrams, and of the criticality of the solute-specific, invariant Tg′–Cg′ as the focal point of such dynamics maps, to understanding structure-property relationships of food molecule-water systems (Table 3.9, part B). It should also be noted that Tg′ corresponds to the subzero Tg mentioned by Atkins (1987) as being characteristic of many water-plasticized, rubbery biopolymers in vivo.

Effect of Molecular Weight on Tg

For pure synthetic polymers, in the absence of diluent, Tg varies with MW in a characteristic and theoretically predicted fashion, which has a significant impact on resulting mechanical and rheological properties (Ferry, 1980; Levine and Slade, 1989b). For a homologous series of amorphous linear polymers, Tg increases with increasing number-average MW ($\overline{M}n$) [due to decreasing free volume (Ferry, 1980) contributed by chain ends], up to a plateau limit for the region of entanglement coupling in rubberlike viscoelastic random networks [typically at $\overline{M}n = 1.25 \times 10^3$ to 10^5 (Graessley, 1984)], then levels off with further increases in $\overline{M}n$ (Ferry, 1980; Billmeyer, 1984). Below the entanglement $\overline{M}n$ limit, there is a theoretical linear relationship between increasing Tg and decreasing inverse $\overline{M}n$ (Sperling, 1986). [For polymers with constant values of $\overline{M}n$, Tg increases with increasing weight-average MW ($\overline{M}w$), due to increasing local viscosity (Slade and Levine, 1988b). This contribution of local viscosity is said to be especially important when comparing different MWs in the low molecular weight range (Ferry, 1980).] The difference in three-dimensional morphology and resultant mechanical and rheological properties between a collection of nonentangling, low molecular weight polymer chains and a network of entangling, high molecular weight, randomly coiled polymer chains can be imagined as analogous to the difference between masses of elbow macaroni and spaghetti (Levine and Slade, 1989b). For synthetic polymers, $\overline{M}n$ at the boundary of the entanglement plateau often corresponds to about 600 backbone-chain atoms (Sperling, 1986). Since there are typically about 20–50 backbone-chain atoms in each polymer segmental unit involved in the cooperative translational motions at Tg (Brydson, 1972), entangling high polymers are those with at least about 12–30 segmental units per chain (Levine and Slade, 1989b). Figure 3.5 (Sperling, 1986) illustrates the characteristic dependence of Tg on $\overline{M}n$ (expressed in terms of the degree of polymerization, DP) for several ho-

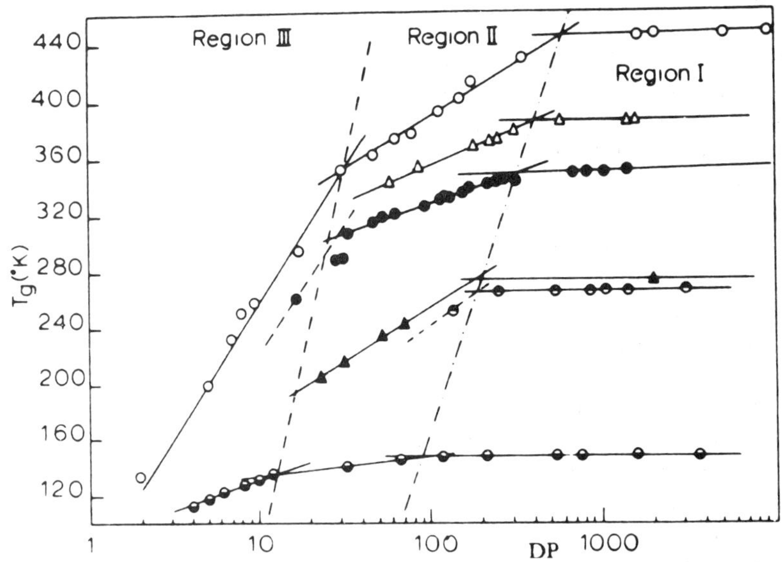

Fig. 3.5 Plot of Tg as a function of log DP (degree of polymerization, a measure of $\overline{M}n$), for poly(alpha-methyl-styrene) (open circles); poly(methylmethacrylate) (open triangles); poly(vinyl chloride) (solid circles); isotactic polypropylene (solid triangles); atactic polypropylene (circles, top half solid); and poly(dimethylsiloxane) (circles, bottom half solid). (*Reproduced, with permission, from Sperling, 1986.*)

mologous series of synthetic amorphous polymers. In this semi-log plot, the Tg values for each polymer reveal three distinguishable intersecting linear regions: (III) a steeply rising region for nonentangling small oligomers; (II) an intermediate region for nonentangling low polymers; and (I) the horizontal plateau region for entangling high polymers (Shalaby, 1981). From extensive literature data for a variety of synthetic polymers, it has been concluded that this three-region behavior is a general feature of such Tg vs log $\overline{M}n$ plots, and it has been demonstrated that data in the nonentanglement regions II and III show the theoretically predicted linear relationship between Tg and inverse $\overline{M}n$ (Shalaby, 1981).

Within a homologous family of food polymers [e.g., from the glucose monomer through maltose, maltotriose, and higher malto-oligosaccharides (e.g., maltodextrins) to the amylose and amylopectin high polymers of starch], Tg′ increases in a characteristic fashion with increasing solute MW (Levine and Slade, 1986, 1988b, d), as indicated by the data in Table 3.10. This finding has been shown to be in full accord with the established variation of Tg with MW for homologous families of pure synthetic amorphous polymers (Ferry, 1980; Billmeyer, 1984; Sperling, 1986), described above. The insights resulting from this finding have proved pivotal to the characterization of structure-function relationships in many different types of completely amorphous and partially crystalline food polymer systems (Slade and Levine, 1984, 1987a, b, 1988a, b, c, d, 1989, 1991a, b, 1992). The relationship between Tg′ and solute MW has been established from DSC measurements on over 150 different food carbohydrates (Levine and Slade, 1986, 1988b, c, d). Eighty-four (see Table 3.10) were small carbohydrates (referred to as polyhydroxy compounds, PHCs) of known, monodisperse MWs (i.e. $\overline{M}w/\overline{M}n = 1$). These PHCs represented a comprehensive but nonhomologous series of mono-, di-, and small oligosaccharides and their derivatives, including many common sugars, polyols, and glycosides, covering a MW range of 62–1153 daltons. Ninety-one others were starch hydrolysis products (SHPs), including monomeric, oligomeric, and high-polymeric carbohydrates, representing a homologous family of glucose polymers. These SHPs represented a spectrum of commercial products (including modified starches, dextrins, maltodextrins, corn syrup solids, and corn syrups), with polydisperse MWs (i.e., $\overline{M}w/\overline{M}n >> 1$), covering a very broad range of dextrose equivalent [DE, where DE $= 100/(\overline{M}n/180.2)$] values from 0.3 to 100.

Tg′ results for the commercial SHPs demonstrated precisely the same Tg vs. $\overline{M}n$ behavior as described above for synthetic amorphous polymers. Tg′ values for these SHPs (of polydisperse MWs ranging from 180 for glucose to about 60000 for a 360-DP polymer) thus demonstrated their classical behavior as a homologous family of amorphous glucose oligomers and polymers (Levine and Slade, 1986, 1988b). The plot of Tg′ vs. solute $\overline{M}n$ in Fig. 3.6 (Levine and Slade, 1986, 1988c, 1990) exhibits the same three-region behavior as shown in Fig. 3.5: (I) the plateau region indicative of the capability for entanglement coupling by high-polymeric SHPs of DE ≤ 6 and Tg′ ≥ -8°C; (II) the intermediate region of nonentangling, low-polymeric SHPs of $6 < \text{DE} < 20$; and (III) the steeply rising region of nonentangling, small SHP oligomers of DE > 20 (Levine and Slade, 1989b). Recent experimental results that confirm the Tg′ vs. solute $\overline{M}n$ behavior illustrated in Fig. 3.6, for a widely available series of maltodextrins of DE 5–36, have been reported by Lim and Reid (1991) and Roos and Karel (1991e). The plot of Tg′ vs. $1/\overline{M}n$ (inset of Fig. 3.6), with a linear correlation coefficient $r = -0.98$, demonstrates the theoretically predicted linear relationship for all the SHPs in regions II and III, with DE values > 6. The plateau region evident in Fig. 3.6 identified a lower limit of $\overline{M}n \simeq 3000$ ($\overline{DP}n \simeq 18$) for entanglement leading to viscoelastic network formation (Flory, 1953, 1974) by such polymeric SHPs in the

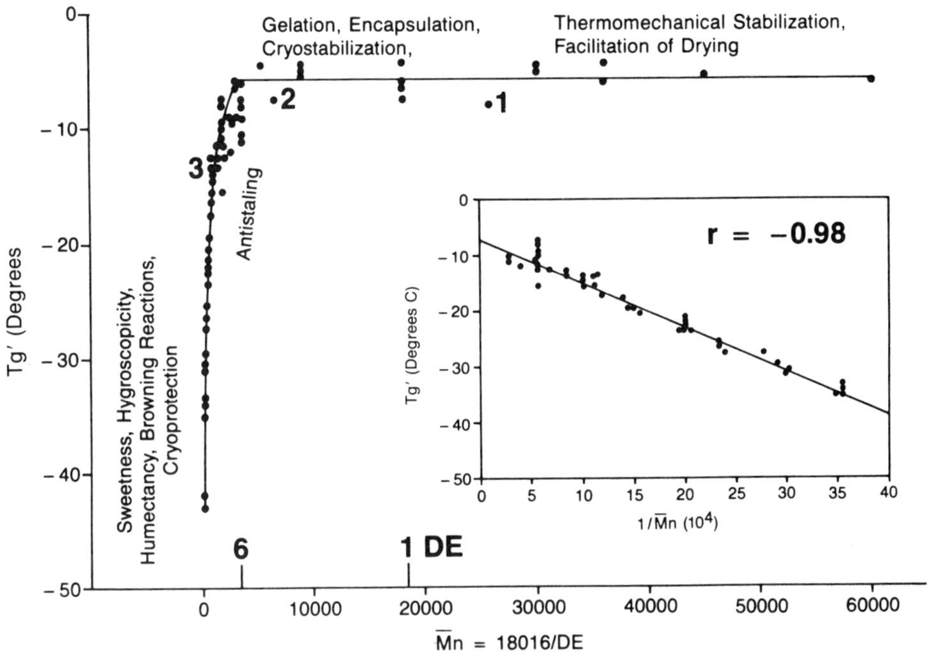

Fig. 3.6 Variation of the glass transition temperature, Tg′, for maximally frozen 20 wt% solutions against $\overline{\text{Mn}}$ (expressed as a function of DE) for commercial SHPs. DE values are indicated by numbers marked above x-axis. Areas of specific functional attributes, corresponding to three regions of the diagram, are labeled. Inset: Plot of Tg′ vs. 1/$\overline{\text{Mn}}$ (×10,000) for SHPs with $\overline{\text{Mn}}$ values below entanglement limit, illustrating theoretically predicted linear dependence. (*Reproduced, with permission, from Levine and Slade, 1986.*)

freeze-concentrated glass formed at Tg′ and Cg′ (Levine and Slade, 1986). This $\overline{\text{Mn}}$ is within the typical range of 1250–19000 for minimum entanglement MWs of many pure synthetic amorphous linear high polymers (Graessley, 1984). The corresponding $\overline{\text{DPn}}$ of about 18 is within the range of 12–30 segmental units in an entangling high-polymer chain, thus suggesting that the glucose repeat in the glucan chain (with 23 atoms/hexose ring) may represent the mobile backbone unit involved in cooperative solute motions allowed only at and above Tg′ (Levine and Slade, 1989b). The entanglement capability has been suggested to correlate well with various functional attributes (see the labels on the plateau region in Fig. 3.6) of low DE SHPs, including a predicted (Levine and Slade, 1986) and subsequently demonstrated (Levine and Slade, 1988b) ability to form thermoreversible, partially crystalline gels from aqueous solution. This gelling ability has been empirically observed and intensively investigated in the past (Richter et al., 1976a, b; Braudo et al., 1979, 1984; Reuther et al., 1984; Vorwerg et al., 1988; German et al., 1989; Radosta et al., 1989; Radosta and Schierbaum, 1990; Schierbaum et al., 1990). It has been suggested (Levine and Slade, 1988a) that SHP gelation occurs by a mechanism involving crystallization-plus-entanglement in concentrated solutions undercooled to T < Tm, as described later. Encapsulation is another predicted functional attribute of entangling, low DE SHP glass-formers that has been demonstrated in various food-product applications. Stabilization of volatile liquid flavors and aromas or other labile ingredients in low moisture food products, via encapsulation in glassy matrices for which the formulation often includes maltodextrins and/or other glass-forming SHPs or carbohydrates of high MW and Tg, has been an area of much recent research, development, and patent activity (To and Flink, 1978; Flink, 1983; Karel and Flink, 1983; Karel, 1985, 1986, 1989; Saleeb and Pickup, 1985, 1989; Karel and Langer, 1988; Levine and Slade, 1988b, 1989a; Noel et al., 1990; Levine et al., 1991; Ring and Whittam, 1991; Shimada et al., 1991).

In contrast to commercial SHPs, the series of quasi-homologous, monodisperse PHCs, including a homologous set of malto-oligosaccharides from glucose up to DP 7, manifested Tg′ values that fall below the Tg′ limit defined by SHPs for entanglement and the onset of viscoelastic rheological properties, and were incapable of gelling from solution (Levine and Slade, 1988b, d). The plot of Tg′ vs. MW for all the PHCs in Fig. 3.7 (Levine and Slade, 1988b), drawn conventionally as a smooth curve through all the points (Billmeyer, 1984), can be visualized as representing two intersecting linear regions (III for MW < 300 and II for 300 < MW < 1200) (Levine and Slade, 1989b). From the fair linearity of the Tg′ vs. 1/MW plot for all the data in the inset of Fig. 3.7, it was concluded that these diverse low-MW sugars, polyols, and glycosides show no evidence of entanglement in the freeze-concentrated glass at Tg′. The corresponding results for glucose and malto-oligosaccharides of DP 2–7 [shown in Fig. 3.8 (Levine and Slade, 1988a)] demonstrated a better linear correlation (r = −0.99) for a plot of Tg′ vs. 1/MW (inset of Fig. 3.8). This linear dependence (Wunderlich, 1981; Roos and Karel, 1991e) of the Tg′ results for the malto-oligosaccharides in aqueous solution exemplified the theoretical glass-forming behavior (i.e., diluent-free Tg vs. 1/MW) characteristic of a homologous family of nonentangling, linear, monodisperse oligomers (Ferry, 1980; Billmeyer, 1984). For all the PHCs—none larger than a heptamer of MW 1153—the main plots in Fig. 3.7 and 3.8 show that region I, representing the entanglement plateau where Tg remains

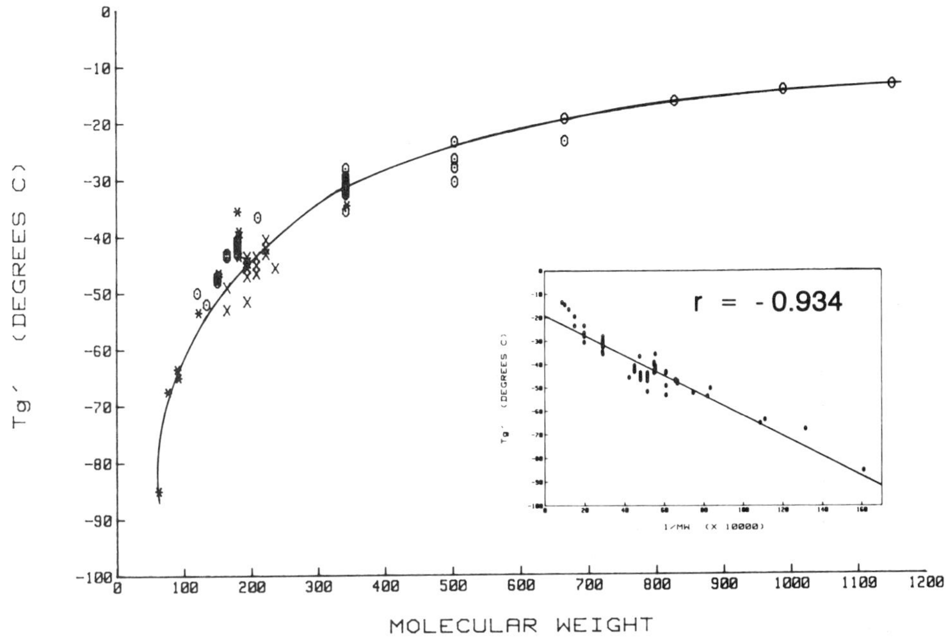

Fig. 3.7 Variation of the glass transition temperature, Tg′, for maximally frozen 20 wt% solutions against MW for sugars (o), glycosides (x), and polyols (*). Inset: Plot of Tg′ vs. 1/MW (×10,000), illustrating theoretically predicted linear dependence. *(Reproduced, with permission, from Levine and Slade, 1988b.)*

constant with increasing MW, has not been reached, in accord with the MW (and corresponding DP) range cited above as the lower limit for polymer entanglement.

As explained elsewhere (Slade and Levine, 1991a), if the relative shapes of a series of polymer-diluent glass curves are similar within a polymer series, increases in MW (of the diluent-free polymer) lead to proportional increases in both Tg and Tg′ (Slade and Levine, 1988b). This fact has been recently demonstrated by the aqueous glass curves for maltose, maltotriose, and maltohexaose published by Orford et al. (1989, 1990), coupled with the Tg′–Wg′ values for these oligosaccharides (Levine and Slade, 1986, 1988b), as illustrated in Fig. 3.9 (Slade and Levine, 1991a), and also by the aqueous glass curves and Tg′ – Cg′ values for several oligomeric maltodextrins reported by Roos and Karel (1991b). Prior to this confirmation, it had been assumed (Levine and Slade, 1989a) that a plot of Tg vs. MW for dry PHCs or SHPs would reflect the same

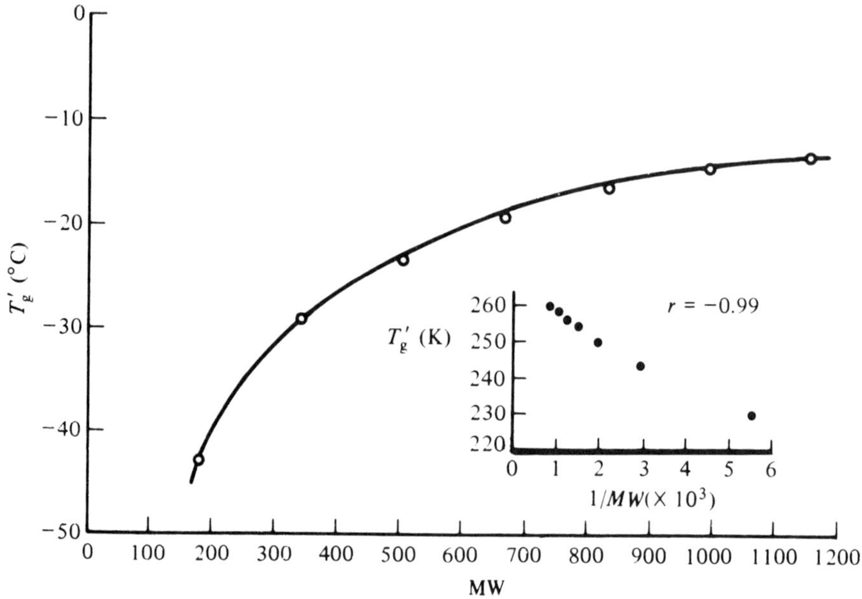

Fig. 3.8 Variation of the glass transition temperature, Tg′, for maximally frozen 20 wt% solutions against MW for a homologous series of malto-oligosaccharides from glucose through maltoheptaose. Inset: plot of Tg′ vs. 1/MW (×1000) of solute, illustrating theoretically predicted linear dependence. (*Reproduced, with permission, from Levine and Slade, 1988a.*)

fundamental behavior as that of Tg′ vs. solute MW shown in Fig. 3.6–3.8. Earlier evidence supporting this assumption had been provided by To and Flink (1978), who reported a plot of Tc vs. DP for a series of low-moisture, fractionated SHP oligomers of 2 ≤ DP ≤ 16 [i.e., nonentangling (Levine and Slade, 1986)], similar in shape to the plot of Tg′ vs. MW for the nonentangling PHCs in Fig. 3.7. It had been pointed out that Tc for low-moisture SHPs, which increases monotonically with increasing DP, represents a good quantitative approximation of "dry" Tg (Levine and Slade, 1986). The basic assumption was verified for the homologous series of glucose and its pure malto-oligomers of DP 2–7, as illustrated in Fig. 3.10 (Levine and Slade, 1989a). The plot of Tg vs. MW in Fig. 3.10a shows that dry Tg increases monotonically with increasing MW of the monodisperse sugar, from Tg = 31°C for glucose [in good agreement with several other published values (Soesanto and Williams, 1981; Chan et al., 1986; Orford et al., 1989) shown in Table 3.10] to Tg = 138.5°C for maltoheptaose (Levine and Slade, 1989a). The plot in Fig. 3.10a shows the

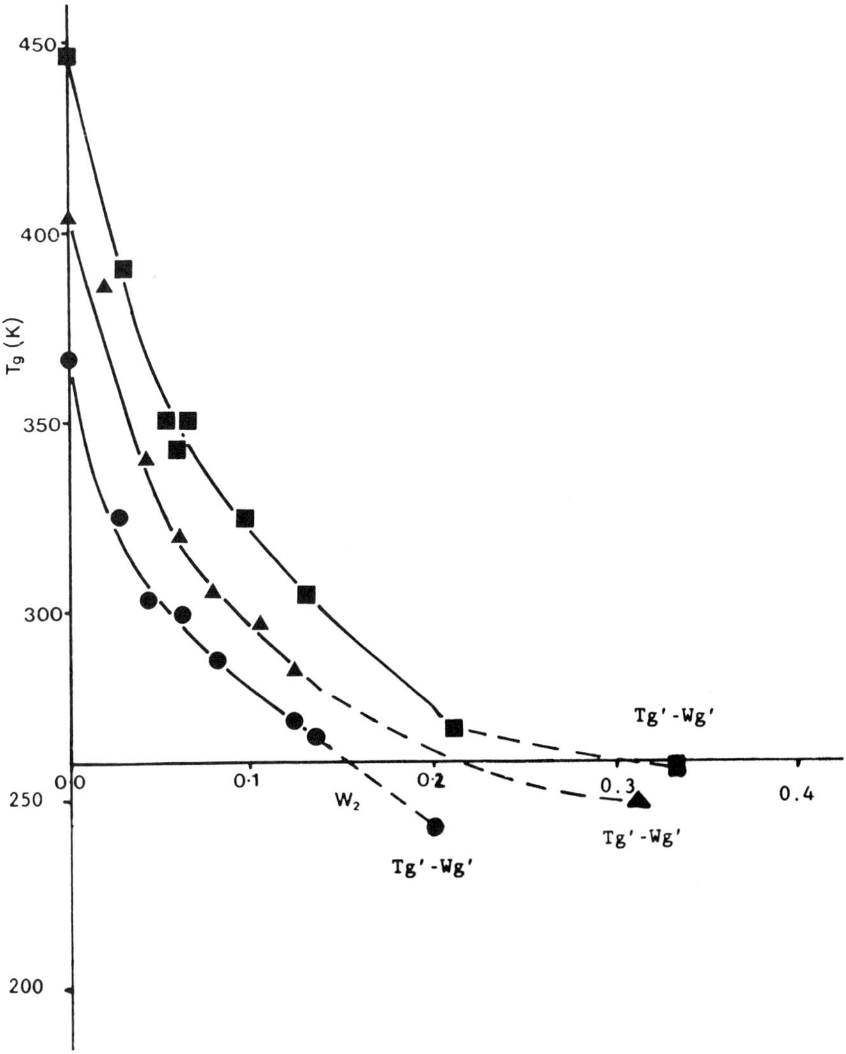

Fig. 3.9 Graph of glass transition temperature (Tg) vs. mass fraction (w_2) of water for maltose (●); maltotriose (▲); and maltohexaose (■). (Reproduced, with permission, from Orford et al., 1989.) The solid-line curves from Orford et al. (1989) are extrapolated (dashed lines) to the appropriate Tg'–Wg' points (Levine and Slade, 1986, 1988b) for these saccharides. (*Reproduced, with permission, from Slade and Levine, 1991a.*)

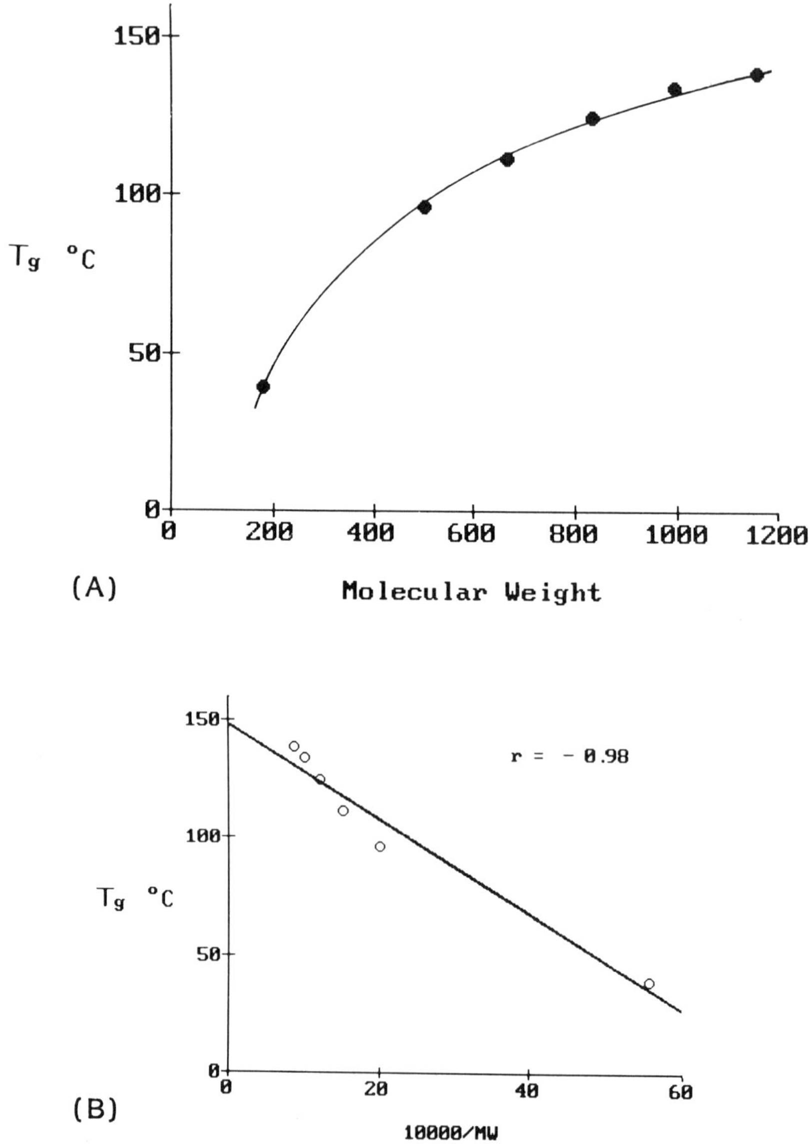

Fig. 3.10 Variation of the glass transition temperature, Tg (measured for dry powders), against (A) MW and (B) 10000/MW, for a homologous series of malto-oligosaccharides from glucose through maltoheptaose. (*Reproduced, with permission, from Levine and Slade, 1989a.*)

same qualitative curvature (and absence of an entanglement plateau) as the corresponding Tg′ plot in Fig. 3.8, and the plot of dry Tg vs. 1/MW in Fig. 3.10b shows the same linearity and r value as the corresponding Tg′ plot in the inset of Fig. 3.8. The results shown in Fig. 3.10a were subsequently corroborated by Orford et al. (1989; Ring and Whittam, 1991), who recently reported a similar curve of dry Tg vs. DP for glucose and its malto-oligomers of DP 2–6. These workers also reported (Orford et al., 1989) a plot of Tg vs. DP for amylose, amylopectin, and malto-oligomers of DP 2–13, all containing 13% water, which likewise corroborated the results shown in Figs. 3.6, 3.7, 3.8, and 3.10 and our consequent finding (Levine and Slade, 1986, 1988b) of $\overline{\mathrm{DPn}} \simeq 18$ at the boundary of the entanglement plateau for this homologous family of amorphous glucose polymers.

The general correlation between increasing saccharide MW and increasing Tg (Levine and Slade, 1988a, b, 1989a) was further confirmed in recent studies of the Tg behavior of amorphous mixtures of low molecular weight carbohydrates reported by Finegold et al. (1989) and Orford et al. (1990). Results of their studies demonstrated that measured Tg values of various dry binary mixtures of mono- and di- or oligo-saccharides can be roughly approximated from a mole fraction-weighted average of the individual dry Tg values of each saccharide in a given amorphous blend. However, beyond the general correlation between saccharide MW and dry Tg, it is worth noting, as indicated by the dry Tg data in Table 3.10, that Tg can vary substantially, even within a series of saccharides of the same MW. For example, Green and Angell (1989) have recently pointed out, for the disaccharides sucrose, maltose, and trehalose, that while their dry Tg values vary significantly, their "wet" Tg values, over a broad range of sugar/water ratios, vary even more so, with trehalose showing by far the highest "wet" Tg values, such that only trehalose solutions, at water contents less than about 4 moles water/mole disaccharide, would be in a glassy state at ambient conditions. Importantly, these workers have related this extraordinary glass-forming ability of trehalose to its unique efficacy as a dehydration protection agent in both in vivo and in vitro biopreservation systems (Green and Angell, 1989). The potential utility of this glass-forming functionality of trehalose in related food-dehydration processes and products has also been noted (Slade and Levine, 1991a).

Water Plasticization

It is well known that water, acting as a plasticizer, affects the Tg of completely amorphous polymers and both the Tg and Tm of partially crystalline polymers (Rowland, 1980; Levine and Slade, 1988a, 1989b; Slade et al., 1989; Slade and Levine, 1991a, b). Water is a "mobility enhancer," in that its low MW leads to a large increase in mobility, due to increased free volume and decreased local viscosity (Ferry, 1980), as moisture content is increased from that of a dry solute to a solution (Slade and Levine, 1988a, b). The direct plasticizing effect of increasing moisture content at constant temperature is equivalent to the effect of increasing temperature at constant moisture and leads to increased segmental mobility of chains in amorphous regions of glassy and partially crystalline polymers, allowing, in turn, a primary structural relaxation transition at decreased Tg (Rowland, 1980; Sears and Darby, 1982). State diagrams illustrating the extent of this Tg-depressing effect have been reported for a wide variety of synthetic and natural, water-compatible, glassy and partially crystalline polymers (Rowland, 1980; Levine and Slade, 1988a, 1989b; Slade et al., 1989; Slade and Levine, 1991a, b). In such diagrams, the smooth "glass curve" of Tg vs. w% composition shows the dramatic effect of water on Tg especially at low moisture contents (i.e., $\lesssim$10 wt% water) (Levine and Slade 1988a; Slade and Levine, 1991a). In this region, Tg typically decreases by about 5–10°C/wt% water (Levine and Slade, 1988a), from the neighborhood of 200°C for the anhydrous polymer (Atkins, 1987). An example of such a state diagram is shown in Fig. 3.11 (van den Berg, 1986), which depicts the amylopectin of freshly gelatinized starch as a typical water-compatible, completely amorphous polymer that exhibits a Tg curve from about 125°C for pure anhydrous starch to about −135°C, the Tg of pure amorphous solid water (Mayer, 1988), passing through Tg′ at about −5°C (and Wg′ $\simeq$ 27 wt% water) (Slade, 1984; Slade and Levine, 1984). Figure 3.11 shows the Tg of starch decreasing by about 6°C/wt% water for the first 10 wt% moisture, in good agreement with another published glass curve for starch, calculated from free volume theory (Blanshard, 1988; Marsh and Blanshard, 1988). A more recently published glass curve for completely amorphous amylopectin samples with moisture contents in the 10–25 wt% range, measured by DSC (Kalichevsky et al., 1991), shows Tg continuing to decrease by about 6.7°C/wt% water in this range. Like the curve in Fig. 3.11, this experimental curve can be smoothly extrapolated to the Tg′–Wg′ point for starch. Similarly, the glass curve for water-compatible, amorphous gluten (Fujio and Lim, 1989) in Fig. 3.12 (Hoseney et al., 1986) shows a decrease in Tg from >160°C at <1 wt% water to 15°C at 16 wt% water, a depression of about 10°C/wt% water in this moisture range. The plasticizing effect of water on gluten continues at higher moisture contents, until Tg falls to Tg′ $\simeq$ −7.5°C and Wg reaches Wg′ $\simeq$ 26 wt% water ($\simeq$0.35 g UFW/g gluten) (Slade et al., 1989). As in the glass curve for starch in Fig. 3.11,

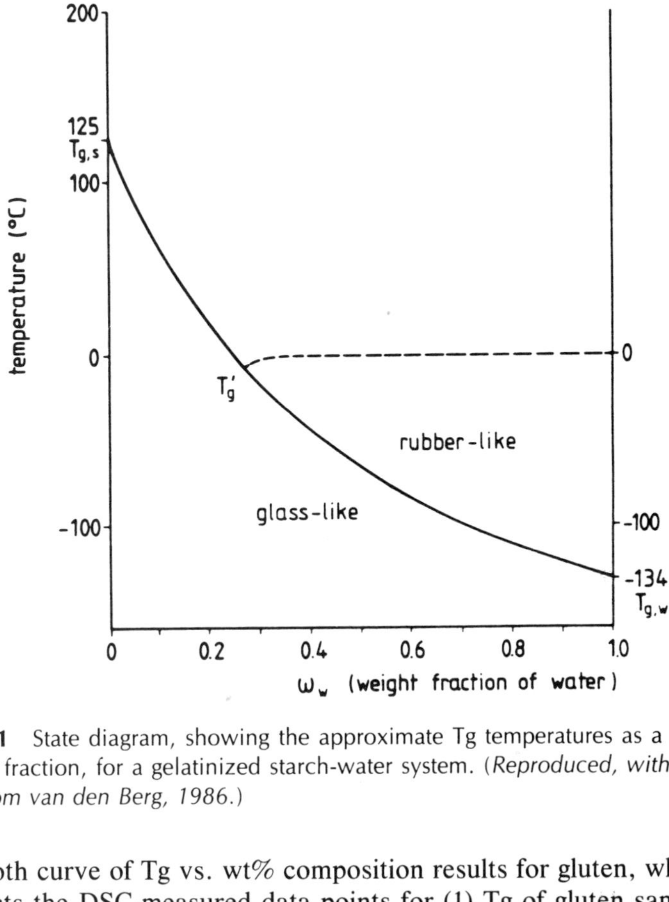

Fig. 3.11 State diagram, showing the approximate Tg temperatures as a function of mass fraction, for a gelatinized starch-water system. (*Reproduced, with permission, from van den Berg, 1986.*)

a smooth curve of Tg vs. wt% composition results for gluten, when one connects the DSC-measured data points for (1) Tg of gluten samples at <2 wt% moisture (Fujio and Lim, 1989), (2) Tg of commercial vital wheat gluten powder at 6 wt% "as is" moisture [from Slade's, (1984) original report of a glass transition in gluten], (3) Tg′–Wg′ (Slade et al., 1989), and (4) Tg of water (Mayer, 1988). Another recently published glass curve, this one for the glutenin component of gluten, likewise shows a monotonic decrease in Tg from 110–130°C at 4 wt% water to about 20°C at 14 wt% water, a depression of about 9–11°C/wt% water in this moisture range (Cocero and Kokini, 1991).

Since the first reports of a glass transition in starch (van den Berg, 1981, 1986; Slade, 1984; Slade and Levine, 1984, 1987b, 1988c; Maurice et al., 1985) and in gluten (Slade, 1984; Levine and Slade, 1988a; Slade et al., 1989), and recognition by these workers of the importance of the glass transition temperature, and its dependence on water content, to the

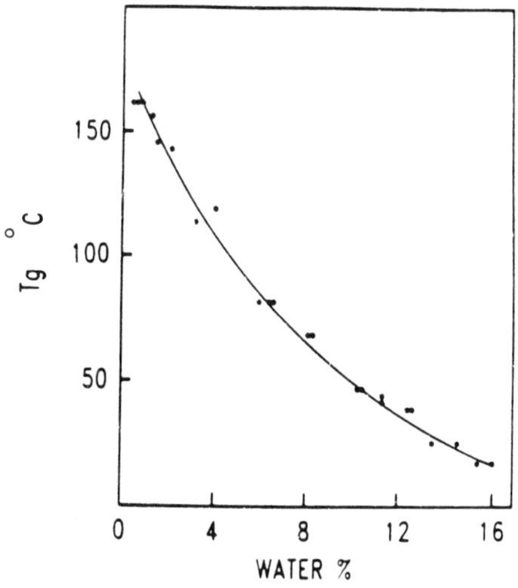

Fig. 3.12 Change in Tg as a function of moisture for a hand-washed and lyophilized wheat gluten. (*Reproduced, with permission, from Hoseney et al., 1986.*)

structure-function relationships of starch and gluten in baked products and other wheat flour-based foods, the aqueous glass curves of these two major food polymers, and the effect of plasticization by water on their Tgs, have been actively investigated by many other workers (see Table 3.11A, B). Since starch and gluten are the major storage polymers of wheat endosperm, and they are synthesized, stored in the mature seed, and hydrolyzed by enzymes of germination in the same temperature and moisture environment, it appears biochemically and physiologically logical that their aqueous glass curves should be similar, if not identical. That is, since conditions of synthesis, storage, and use are identical for the two storage polymers, equivalent performance should reflect equivalent glass curves. In fact, the results and conclusions of many of the studies cited in Table 3.11 seem to support this suggestion. Parts A and B of this table list references to studies of the Tg of starch, gluten, and/or their component polymers and to methods used to determine them. Taken together, these Tg results appear to fall in the same region of temperature and moisture content; i.e., a Tg around "room temperature" (about 15–25°C) at a moisture content around 20 wt%. The technological implications of such a Tg and moisture content to the processing and product properties of starch- and/or gluten-based food products (e.g., baked goods) are obviously profound (Table 3.11C).

In recent years, the plasticizing effect of water on the Tg of starch has

Table 3.11 Publications Dealing With Glass Transition in Starch and Gluten

A. Methods used to determine Tg of starch, amylose, and/or amylopectin

Calculated from theory: Blanshard (1986, 1988), Marsh and Blanshard (1988)

Measured by DSC: Slade (1984), Slade and Levine (1984, 1987b, 1988c), Wittwer and Tomka (1984), Maurice et al. (1985), Yost and Hoseney (1986), Zeleznak and Hoseney (1987a), Zobel et al. (1988), Orford et al. (1989), Stepto and Dobler (1989), Biliaderis (1990), Noel et al. (1990), Roper and Koch (1990), Warburton et al. (1990), Roos and Karel (1991e), Tomka (1991), Whittam et al. (1991)

Measured by DSC and TMA: Biliaderis et al. (1986a), Biliaderis (1991)

Determined from DSC and NMR data: Given (1991)

Measured by DSC, DMTA, and NMR: Kalichevsky et al. (1991)

Measured mechanically (as T at which characteristic orders-of-magnitude decrease in modulus is observed): Noel et al. (1990), Ollett et al. (1990), Smith (1990), Kalichevsky et al. (1991)

Other methods: van den Berg (1981, 1986)

B. Methods used to measure Tg of gluten or glutenin

DSC: Slade (1984), Hoseney et al. (1986), Doescher et al. (1987), Levine and Slade (1988a), Fujio and Lim (1989), Slade et al. (1989), Hoseney and Rogers (1990), Noel et al. (1990), Hoseney (1991)

DMA: Mita (1990)

Mechanically (as T at which characteristic orders-of-magnitude decrease in modulus is observed): Attenburrow et al. (1990), Davies et al. (1991)

DSC and mechanical spectrometry: Cocero and Kokini (1991)

C. Effects of Tg for starch and/or gluten on processing and product properties

Blanchard (1986), Slade and Levine (1987b, 1991a), Levine and Slade (1988a, 1989b), Shi and Seib (1989), Slade et al. (1989), Hoseney and Rogers (1990), Noel et al. (1990), Warburton et al. (1990), Cocero and Kokini (1991), Davies et al. (1991), Given (1991), Hoseney (1991), Kalichevsky et al. (1991).

also been utilized as the basis of a potentially important technological application involving the manufacture of biodegradable, thermoplastic polymers from native granular starches (Wittwer and Tomka, 1984; Stepto and Dobler, 1989; Roper and Koch, 1990; Tomka, 1991), which can be used as packaging materials in place of synthetic plastics. By means of thermomechanical processing under controlled conditions of high pressure, high temperature, and low moisture, native granular starch [a partially crystalline glassy polymer system (Slade, 1984; Slade and Levine, 1984, 1987b)] can be transformed via heat/moisture/pressure plasticization into a homogeneous polymer-plasticizer melt above its Tg and Tm. The melt can then be extruded or injection-molded (Wittwer and Tomka, 1984; Stepto and Dobler, 1989; Roper and Koch, 1990) to produce a completely amorphous, thermoplastic resinlike material with useful gas-barrier properties (Tomka, 1991).

The plasticizing effect of water on the Tg of three other glass-forming food materials is illustrated and compared in the state diagrams shown in Fig. 3.13 (Slade and Levine, 1991a) (adapted from Kelley et al., 1987, and Quinquenet et al., 1988). Hemicellulose (Kelley et al., 1987), an amorphous component of wood, is a typical water-compatible biopolymer, with a dry Tg of about 200°C that is dramatically depressed (by more than 15°C/wt% water for the first 10% water) to a Tg around −10°C [i.e., Tg′ (Slade and Levine, 1991a)] at about 30% moisture (Atkins, 1987). Hemicellulose, like starch, gluten, gelatin, and elastin, exhibits the charac-

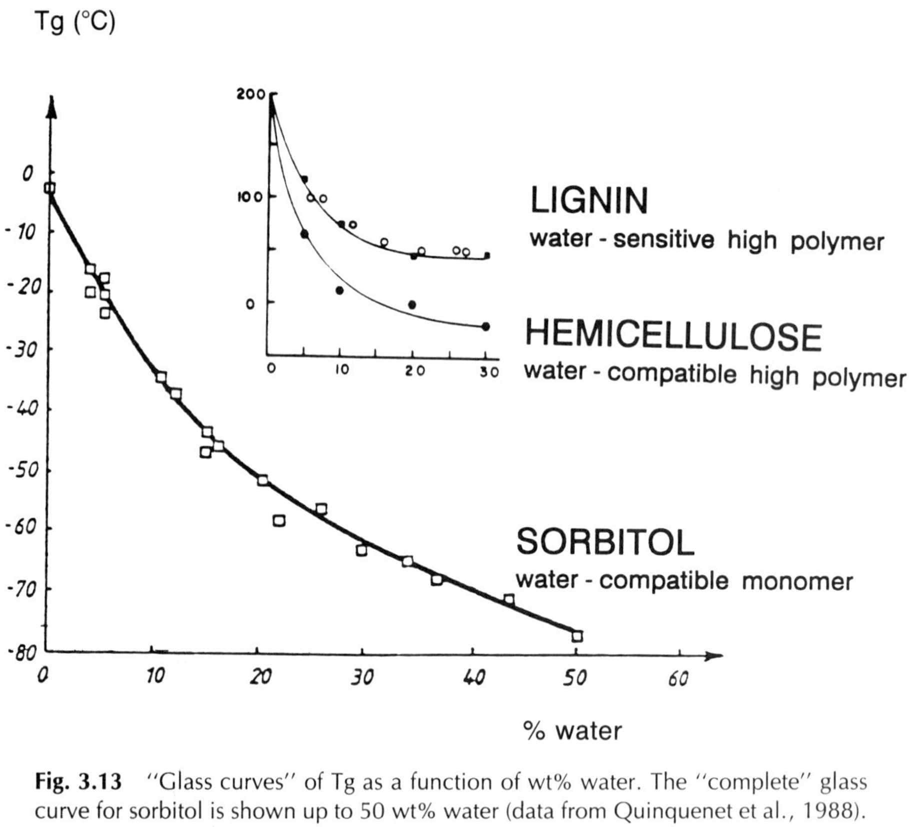

Fig. 3.13 "Glass curves" of Tg as a function of wt% water. The "complete" glass curve for sorbitol is shown up to 50 wt% water (data from Quinquenet et al., 1988). The "practical" glass curves for hemicellulose and lignin are shown up to 30 wt% water (data from Kelley et al., 1987). (*Reproduced, with permission, from Slade and Levine, 1991a.*)

teristic behavior common to all water-compatible, glass-forming solutes (Levine and Slade, 1988a): the *practical* limit to the extent of plasticization (i.e., depression of Tg by water) is determined by phase separation of crystalline ice below 0°C, so that the minimum Tg achievable during slow cooling in a practical time frame is the solute-specific Tg′ (with the corresponding maximum content of plasticizing moisture, Wg′) (Levine and Slade, 1986, 1988b; Slade and Levine, 1988b, 1991a). Accordingly, the glass curve for hemicellulose in Fig. 3.13 is typical of the "practical glass curve for a water-compatible solute," which levels off at Tg′ < 0°C, rather than continuing along the monotonic descent of the "complete glass curve" (Slade and Levine, 1991a) to Tg of water itself (e.g., in Fig. 3.11). In contrast to hemicellulose, lignin, the other major amorphous component of wood, typifies a high polymer that is only water-sensitive rather than water-compatible (Levine and Slade, 1988a; Slade and Levine, 1991a). Its glass curve (Kelley et al., 1987) also starts at about 200°C for the dry solid and decreases by more than 10°C/wt% water for the first 10% water. But the glass curve for lignin in Fig. 3.13 is typical of the "practical glass curve for a water-sensitive solute"; it levels off at a lower moisture content and at a temperature well above 0°C (Slade and Levine, 1991a). Lignin exhibits the characteristic behavior common to all water-sensitive, glass-forming solutes [e.g., synthetic high polymers that are relatively hydrophobic, such as polyethylene, polyvinyl acetate (Johnson et al., 1980), and nylons (Starkweather, 1980)] (Levine and Slade, 1988a; Slade and Levine, 1991a): the *practical* limit to its extent of plasticization by water is determined by its much more limited water-solubility and thermodynamic compatibility, leading to phase separation of liquid water (as "clusters" of water molecules) above 0°C, which would subsequently freeze on further cooling to 0°C (Levine and Slade, 1988a; Slade and Levine, 1991a). Thus, the minimum Tg achievable during cooling of a lignin-water mixture is not Tg′, but some higher Tg > Tm of ice (about 50°C for lignin, as shown in Fig. 3.13), because Tg cannot be depressed to Tg′ by clustered water in a separate (nonplasticizing) liquid phase (Levine and Slade, 1988a; Slade and Levine, 1988b, 1991a).

In contrast to the two high molecular weight biopolymers represented in Fig. 3.13, sorbitol is a water-compatible, glass-forming, monomeric polyol of low MW. As shown in Fig. 3.13, the glass curve for quench-cooled, completely amorphous sorbitol-water mixtures (Quinquenet et al., 1988) begins at a much lower temperature [Tg = −2°C for anhydrous sorbitol (Quinquenet et al., 1988; Levine and Slade, 1989a)], because of the low MW of this solute (Levine and Slade, 1988a), and shows an extent of plasticization of sorbitol at low moisture of about 3–4°C/wt% water (Slade and Levine, 1991a). The glass curve shown for sorbitol in Fig. 3.13

is the "complete glass curve" (Slade and Levine, 1991a) up to 50 wt% water, resulting from quench-cooling to avoid phase separation of ice at water contents > Wg′ and would continue smoothly down to Tg of pure amorphous water at about −135°C (Quinquenet et al., 1988), as do the "complete" glass curves of all water-compatible solutes, regardless of MW (Levine and Slade, 1986, 1988a; Slade and Levine, 1988a, b, 1991a). If these sorbitol-water mixtures had been cooled more slowly, so that ice formation and maximum freeze-concentration of the solute could have occurred during the experimental time frame, they would have been expected to manifest the "practical" glass curve for sorbitol (Slade and Levine, 1991a), with invariant values of minimum Tg = Tg′ ≃ −43.5°C and maximum Wg = Wg′ ≃ 19 wt% water ≃ 0.23 g UFW/g sorbitol (Levine and Slade, 1988b).

The glass curves (solid lines) shown in Fig. 3.9 for three malto-oligosaccharides at low moistures (i.e., W < Wg′) are noteworthy for their similarity to the glass curves for starch (Fig. 3.11), hemicellulose, sorbitol (Fig. 3.13), and gluten (Fig. 3.12) (Slade and Levine, 1991a). At moisture contents ≲ 10 wt%, maltose, maltotriose, and maltohexaose manifest extents of plasticization of about 9, 11, and 12.5°C/wt% water, respectively (Orford et al., 1989; Ring and Whittam, 1991), values within the range previously reported to be typical of many other water-compatible food oligomers and high polymers (Levine and Slade, 1988a; Slade et al., 1989). Most recently, further experimental confirmations of such plasticization results for various common food carbohydrates have come from Franks and Grigera (1990), Roos and Karel (1990, 1991a, b, d, e), and Orford et al. (1990). Franks and Grigera (1990) reported extents of plasticization at low moistures of from ~3–15°C/wt% water for a series of PHCs, including sorbitol, sucrose, maltose, maltotriose, and maltohexaose. Roos and Karel (1990, 1991a, b, d, e) obtained similar results in the same range for sucrose, lactose, maltose, and a series of commercial maltodextrins. Orford et al. (1990) extended their recent studies to ternary aqueous mixtures of several low molecular weight carbohydrates and again obtained similar results. In agreement with the DSC data of Orford et al. (1989) for the three malto-oligomers in Fig. 3.9, Franks and Grigera (1990) also showed an increasing extent of plasticization with increasing solute MW, and so did Roos and Karel (1991e) for maltose and a series of commercial maltodextrins, apparently reflecting a consequence of increasing MW difference [and thus free volume difference (Slade and Levine, 1988b, 1991a)] between plasticizer and solute (discussed further below). We added the dashed portions to the solid lines in Fig. 3.9 to show that the measured glass curves of Orford et al. (1989) can be smoothly extrapolated to the corresponding DSC-measured Tg′–Wg′ points (Levine and Slade, 1986,

1988b) on these state diagrams, at least for the cases of maltose and maltotriose (Slade and Levine, 1991a). At moisture contents > Wg′, the Tg curves (i.e., "complete" glass curves for quench-cooled solute-water blends) would be expected to continue to extrapolate smoothly, but with more gradual curvature, to the Tg at about −135°C for pure amorphous water (Slade and Levine, 1991a). Such behavior has been frequently illustrated previously for various low molecular weight carbohydrates [see discussion elsewhere (Slade and Levine, 1991a)], e.g., as shown in Fig. 3.14 (Slade and Levine, 1988b) by actual state diagrams reported for fructose, glucose, and sucrose, which are essentially identical in appearance to the schematic state diagram in Fig. 3.4 and the actual state diagram for starch in Fig. 3.11.

Figure 3.14 shows experimental data for the glass curves of glucose, fructose, sucrose (Slade and Levine, 1988b), and a 40,000 MW polyvinyl pyrrolidone (PVP-40) (Levine and Slade, 1986). These state diagrams were constructed from DSC-measured values of (1) "dry" Tg (i.e., Tg of the anhydrous solute) (Slade and Levine, 1988b; Levine and Slade, 1989a), (2) Tg′ and Cg′ (Levine and Slade, 1986, 1988b), coupled with (3) Tg of pure amorphous water, (4) Tm of pure ice, and (5) the equilibrium (MacKenzie, 1977) and nonequilibrium portions of the liquidus curve (Franks, 1982, 1985; Slade and Levine, 1988b). Figure 3.14 demonstrates that the maximum *practical* (i.e., spatially homogeneous) dilution of each amorphous solute corresponds to a particular glass in each continuum of

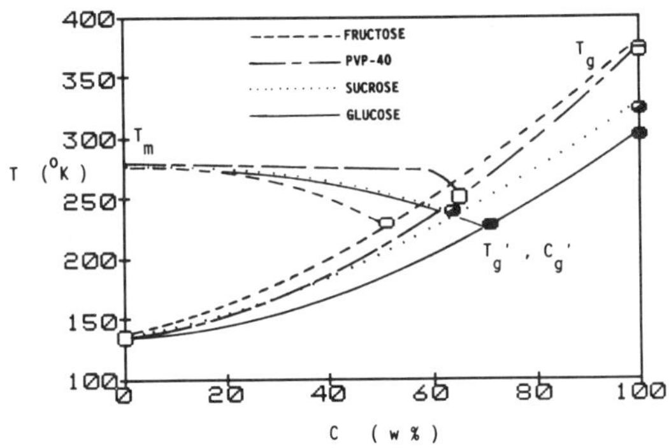

Fig. 3.14 Solute-water state diagrams of temperature vs. concentration for fructose, glucose, sucrose, and polyvinyl pyrrolidone (PVP-40), which illustrate the effect of water plasticization on the experimentally measured glass curves, and the location of the invariant point of intersection of the glass curve and the nonequilibrium portion of the liquidus curve at Tg′ and Cg′, for each solute. (*Reproduced, with permission, from Slade and Levine, 1988b.*)

glassy compositions (Slade and Levine, 1988b). As described elsewhere (Slade and Levine, 1988b, 1991a), alternative paths, such as drying by evaporation or freeze-concentration (Franks, 1982; Levine and Slade, 1988b), lead to the same operationally invariant wt% composition (Cg′), with its characteristic Tg′. If one examines the effect of decreasing plasticizer content, one sees that the elevation of Tg, due to increased solute concentration, dramatically affects the shape of the nonequilibrium, very nonideal portion of the liquidus curve. In other words, the extreme departure from the equilibrium liquidus curve for each of these solutes is related to the shape of the corresponding glass curve (Slade and Levine, 1988b). The locus of Tg′ on a state diagram depends on both free volume and local viscosity, and therefore on the inverse $\overline{\text{Mn}}$ and inverse $\overline{\text{Mw}}$, respectively (Ferry, 1980), of the dynamically constrained, kinetically metastable solution (Slade and Levine, 1988b). Thus, it has been suggested that the anomalous shape of the extrapolated liquidus curve is a consequence of the system's approach to the immobile, glassy domain, rather than the cause of the particular location of the glass at Tg′ (Levine and Slade, 1988c, 1989c, 1990; Slade and Levine, 1988b). The anomalous shape of the liquidus [described elsewhere (Franks, 1982, 1985)] reflects the nonequilibrium melting behavior of ice and the improbably low values of apparent RVP of the solution that result from the constrained approach to the glassy domain, which represents the limiting range of relaxation rates compared to the time frame of observation (Slade and Levine, 1988b, 1991a). Equally anomalous values have been observed for the RVPs of aqueous supra-glassy solutions (i.e., at T >> Tg, and >T of the rubbery region) of PHCs at ambient temperature (Levine and Slade, 1988a; Slade and Levine, 1988a). In both of these situations, apparent RVPs are often inappropriately referred to as Aw's (see Fig. 3.3), even though they are clearly nonequilibrium values, *controlled by*, rather than *controlling*, the long relaxation times of the solute-water system (Slade and Levine, 1988b, 1991a).

It is important to notice that the three-point glass curves in Fig. 3.14 are all characteristically smoothly and continuously curved over the entire range of solute-diluent wt% compositions, a point alluded to in the earlier discussion of the glass curves in Fig. 3.9, 3.11, 3.12, and 3.13. Obviously, the shapes of the glass curves in Fig. 3.14 are determined by the particular locations of the Tg′–Cg′ and dry Tg points for each solute (Slade and Levine, 1988a). The glass curve for fructose-water is smoothly curved, only because it was drawn using the higher of two dry Tg values (i.e., 100°C) measured for fructose (see Table 3.10) (Slade and Levine, 1988b). If the lower dry Tg value of 11°C had been chosen instead, the resulting glass curve for fructose-water would not be smooth (Slade and Levine,

1991a). Rather, it would have a break [or cusp (Aubin and Prud'homme, 1988)] in it at Tg′–Cg′, such that the portion from Tg of water to Tg′–Cg′ would have a different curvature than the other portion from Tg′–Cg′ to the lower dry Tg. While unusually shaped glass curves, which exhibit a cusp in Tg as a function of composition (i.e., the Tg-composition variation is not monotonic), have been reported in the synthetic amorphous polymer literature for both miscible polymer-polymer and miscible polymer-diluent blends, such a cusp is generally manifested *only* when Tg is plotted vs. volume (rather than weight) fraction, and then *only* when one of the blend components is a high polymer with MW above the entanglement limit (Aubin and Prud'homme, 1988). We know of only one report of such a cusp in a glass curve of Tg vs. wt% composition [in that case, actually attributed to partial phase separation (i.e., immiscibility) in polymer-diluent mixtures (Fried et al., 1982)], but none of a cusp in any glass curve for a miscible solute-diluent blend (such as fructose-water) in which the solute MW is well below the entanglement limit. Other cases of maxima or minima in Tg-composition plots are ordinarily attributed to specific associations or complex formation occurring at stoichiometric compositions (Hofer et al., 1989). Thus, the smooth glass curve for fructose-water in Fig. 3.14 has been suggested to represent one of several pieces of supporting evidence (Slade and Levine, 1991a) for the choice of the higher dry Tg of fructose as the one that, in conjunction with the agreed location of Tg′ (Franks, 1985, 1986a; Levine and Slade, 1988b; Rubin et al., 1990; Schenz et al., 1991; Slade and Levine, 1991a) [and its corresponding Cg′–Wg′ composition (Franks, 1985, 1986a; Levine and Slade, 1988b; Slade and Levine, 1991a)], determines the thermomechanical properties and thereby controls the mobility-related kinetic behavior of fructose-water systems in nonequilibrium glassy, rubbery, and supraglassy states (Slade and Levine, 1991a).

Notably, since the DSC-measured Tg of fructose was first reported recently (Slade and Levine, 1988b), the question of its temperature location, especially in comparison to the much-studied and better-established Tg of another monosaccharide, glucose [see Table 3.10 and discussion elsewhere (Slade and Levine, 1991a)], has been a subject of much subsequent discussion in the literature (Finegold et al., 1989; Noel et al., 1990; Ollett and Parker, 1990; Orford et al., 1990). Likewise, since the DSC-measured estimate for Cg′ of fructose [which is quite different from the somewhat more widely agreed Cg′ of glucose (Slade et al., 1988b, 1991a)] was first determined (Levine and Slade, 1988b) and then corroborated in other reports (Franks, 1985, 1986a, 1989, 1990), its numerical value has also been questioned (Schenz et al., 1991). Various other workers (see Table 3.12A) have recently questioned the simple experimental

Table 3.12 Recent Publications Discussing Values of Tg′ and Cg′ and/or Methods of Determining Those Values for Low Molecular Weight Carbohydrates

A. Methods of determining Cg′ values: Blond (1989), Noel et al. (1990), Orford et al. (1990), Ablett et al. (1991a, b), Hatley et al. (1991), Izzard et al. (1991), Reid and Hsu (1991), Roos and Karel (1991c, d, e), Schenz et al. (1991), Simatos and Blond (1991)

B. Tg′ value of −32°C for sucrose: Schenz et al. (1984), Franks (1985, 1986a, b, 1989, 1990), Blanshard and Franks (1987), Levine and Slade (1988b), Slade and Levine (1988b), Caldwell et al. (1990), Blanshard et al. (1991), Izzard et al. (1991), Hatley et al. (1991), Lim and Reid (1991), Maurice et al. (1991), Reid and Hsu (1991), Schenz et al. (1991)

C. Cg′ values for sucrose: MacKenzie (1977), Franks (1983b, 1985, 1986a, b, 1989, 1990), Blanshard and Franks (1987), Levine and Slade (1988b), Slade and Levine (1988b), Williams and Carnahan (1989, 1990), Franks and Grigera (1990), Ablett et al. (1991a, b), Blanshard et al. (1991), Hatley et al. (1991), Izzard et al. (1991), Le Meste and Huang (1991), Reid and Hsu (1991), Roos and Karel (1991c, d)

D. Tg′ as the temperature at the midpoint of the transition: Franks et al. (1977), Franks (1982, 1985, 1986a, 1989, 1990), Schenz et al. (1984, 1991), Reid (1985, 1990), Levine and Slade (1986, 1988a, b, c, d), Blanshard and Franks (1987), Caldwell et al. (1990), Franks and Grigera (1990), Williams and Carnahan (1990), Ablett et al. (1991a, b), Blanshard et al. (1991), Hatley et al. (1991), Izzard et al. (1991), Lim and Reid (1991), Maurice et al. (1991), Reid and Hsu (1991), Wasylyk and Baust (1991)

DSC method that we advocate (i.e., measurement of the ice content of a single, maximally freeze-concentrated 20 wt% solution) for the practical estimation of Cg′ for low molecular weight food carbohydrates (Levine and Slade, 1986, 1988b; Slade and Levine, 1991a). With regard to Cg′ values reported in the literature, the present state of affairs, which we have characterized as one of much recent activity and unsettled controversy (Slade and Levine, 1991a), is exemplified by the situation for sucrose. Despite fairly widespread agreement (see Table 3.12B) on the Tg′ value of −32 ± 1°C (see Table 3.10) first reported by Schenz et al. (1984), various Cg′ values reported for sucrose (most of them taken from a large number of different versions of the state diagram for sucrose-water, each of which contrasts to a greater or lesser extent with the one shown in Fig. 3.14) cover a broad range from about 64 to 83 wt% (see Table 3.12C). As reviewed elsewhere (Slade et al., 1989; Slade and Levine, 1991a), the arguments for and against particular Cg′ values and methods of measurement are complex. Special attention is drawn to some of the most recent, studied analyses of this issue by Hatley et al. (1991), Ablett et al. (1991b), and Reid and Hsu (1991). It is interesting to note, e.g., that while Ablett et al. (1991a) have advocated a Cg′ value for sucrose of ~80 wt% and criticized as incorrect our reported value of 64 wt%, they have pointed out that *no* devitrification and resulting ice formation on rewarming are routinely observed for completely vitrified DSC samples with sucrose concentrations >68 wt%. This latter finding confirmed earlier work by Williams and Carnahan (1989), whose results showed the critical effect of sample cooling/warming history on the ability to form ice in, and on the consequent Cg′ value that one might determine for, a 65 wt% sucrose solution. They also demonstrated the complete absence of freezable water (on a DSC time scale) in a 70 wt% sucrose solution, when heterogeneous nucleation of ice, either at the solution/air interface or at fracture faces of cracks formed in the brittle glass (which was, otherwise, kinetically stable in the *absence* of cracking) produced by rapid cooling to T << Tg, was deliberately prevented (Williams and Carnahan, 1990). Cracking of brittle sucrose-water (70:30 w:w) glasses after sub-Tg cooling was also noted as a common occurrence in the Brillouin scattering study by Hosea et al. (1990). These workers also reported a clearing of their cloudy, 70% sucrose sample at −15°C, which suggests to us that the phase separation they observed during warming of the sample through its glass transition region involved eutectic crystallization of both sucrose and ice [since the eutectic melting temperature, Te, for sucrose-water is −14°C (MacKenzie, 1977)]. Hatley et al. (1991) have recently concluded that "Wg′, the residual unfrozen water content of a maximally freeze-concentrated carbohydrate-water system, cannot be accurately determined from frozen solutions, i.e., by any technique that involves freezing." Despite the careful reasoning behind this paradoxical-sounding statement, we believe that the issue of appropriate Cg′–Wg′ values and method of measurement boils down to a simple question of practical relevance, i.e., what is most relevant to the actual practice of a specific food process and/or product (Slade and Levine, 1991a).

To avoid confusion, it should be noted that Roos and Karel (1991c, d, e), in several of their recent publications, have used Tg′ and Tm′ to denote the onset and endpoint temperatures of the single glass transition of a maximally freeze-concentrated solute/UFW matrix, the midpoint of which, by convention, has been designated Tg′ by many other previous workers in the field of cryotechnology (see Table 3.12D). Roos and Karel's (1991c, d, e) values of Tg′ (see Table 3.10, e.g., for sucrose, onset = −46°C and midpoint = −41°C) and Tm′ for sucrose and various other food carbohydrates correspond to values reported in the past for Tantemelting (Tam) and Tincipient melting (Tim), which were ill-defined and are now obsolete terms favored by some of the earliest workers in this field (Luyet, 1960; MacKenzie, 1977; Simatos and Karel, 1988; Simatos and Blond, 1991) to likewise describe the onset and endpoint temperatures of the single, solute-specific glass transition at Tg′ (Slade and Levine, 1991a). In the context of the widely agreed Tg′ value of −32°C for sucrose, it is interesting to note that a recent confirmation of this value has come from ESR spectroscopy measurements, of rotational correlation times for diffusion of spin-probes, after rewarming of a rapidly cooled 20 wt% sucrose solution, performed by Roozen and Hemminga (1990). They reported that at a temperature in very good agreement with the Tg′ value of −32°C measured by DSC (Schenz et al., 1984), "an extreme increase in rotational mobility" of ESR spin-probes [said to be due to a dramatic increase in free volume and decrease in viscosity (Roozen and Hemminga, 1990)], located in the freeze-concentrated amorphous matrix rather than in the ice lattice, has been observed. Le Meste and Huang (1991) have recently used a modified TMA method to study frozen sucrose solutions, and have reported results that could be construed as agreeing with both sides of the issue of the Tg′ value for sucrose. They found that (1) on heating a maximally freeze-concentrated sucrose/UFW/ice sample, the Tg′ glass transition appeared to begin around −46°C, but softening and flow were observed only above −32°C; (2) on heating nonmaximally freeze-concentrated solutions, devitrification began at Tg and continued up to −32°C, where melting of ice crystals began; and (3) with dilute solutions, ice recrystallization occurred in the temperature range from −32 to −10°C.

According to the prevailing view in the current synthetic polymer literature, the predominant contribution to the mechanism of plasticization of water-compatible glassy polymers by water at low moisture content derives from a free volume effect (Ellis et al., 1984; Jin et al., 1984; Ellis, 1988). Free volume theory (Ferry, 1980) provides the general concept that free volume is proportional to inverse $\overline{Mn}$, so that the presence of a plasticizing diluent of low MW leads to increased free volume, allowing increased backbone-chain segmental mobility. [Note that Sears and Darby (1982) have stated that "free volume is considered thermodynamically as a solvent."] The increased mobility is manifested as a decreased Tg of the binary polymer-diluent glass (Sears and Darby, 1982; Marsh and Blanshard, 1988). For synthetic amorphous high polymers, it is well known that the ability of a diluent to depress Tg decreases with increasing diluent MW (Boyer et al., 1985), as predicted by free volume theory. These facts are illustrated in Fig. 3.15 (Ferry, 1980), which shows a series of glass curves for solutions of polystyrene with various compatible organic diluents that can be undercooled without crystallizing. These smooth curves

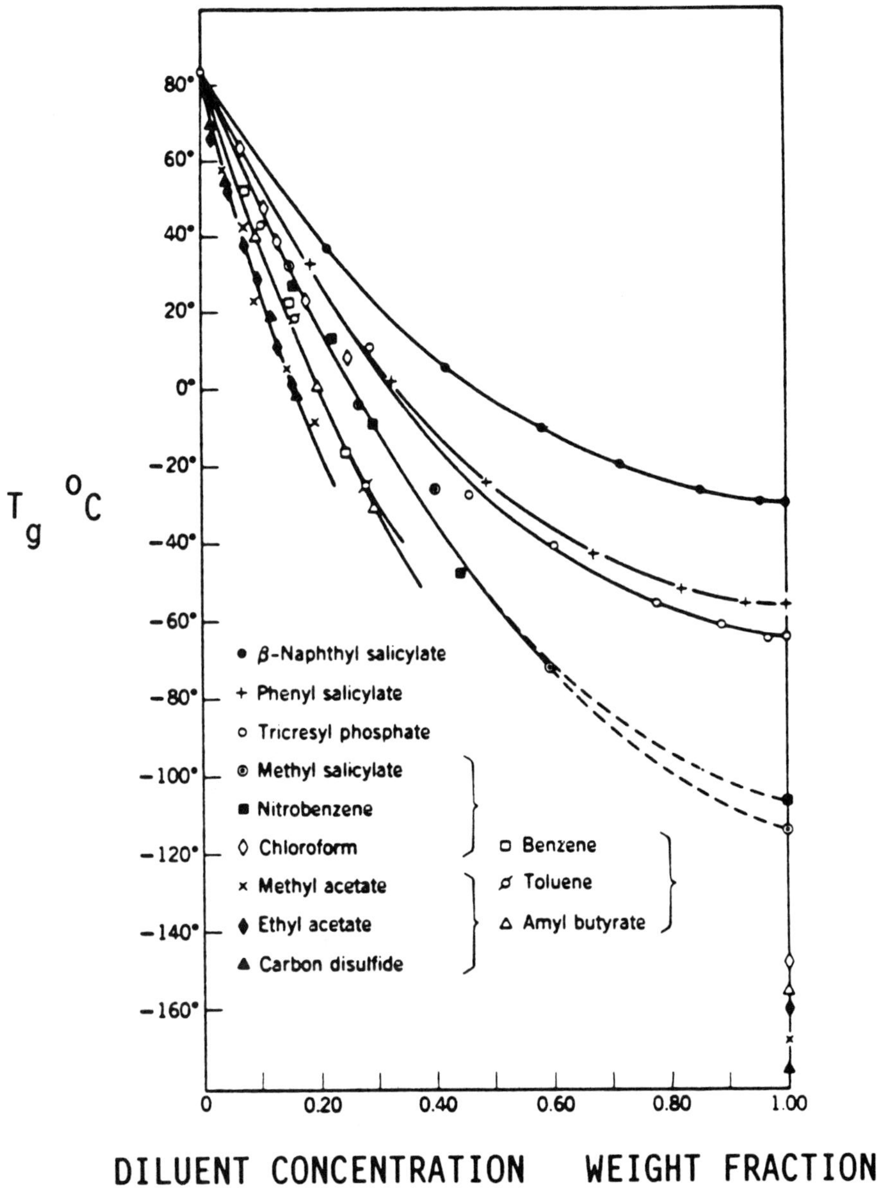

Fig. 3.15 Glass transition temperatures of polystyrene solutions with various diluents of low molecular weight, plotted against weight fraction of diluent. (*Reproduced, with permission, from Ferry, 1980.*)

illustrate the characteristic plasticizing effect of low molecular weight, glass-forming diluents of low Tg on a typical polymer of higher Tg: Tg decreases monotonically with increasing concentration (weight-fraction) of diluent, because the $\overline{M}$w of the homogeneous polymer-plasticizer mixture decreases and its free volume increases (Sperling, 1986). Figure 3.15 also shows that at a given weight-fraction of diluent, Tg of the mixture increases with increasing MW of the diluent (generally over the entire set of diluents, but rigorously within a homologous series), because the Tgs of the neat diluents likewise generally increase with increasing MW and decreasing free volume (Sperling, 1986).

In contrast, the effect of synthetic polymer plasticization by a crystallizing diluent has been illustrated by Tg results for blends of poly(vinyl chloride) (PVC) with a terpolymeric organic plasticizer that is able to crystallize on undercooling, as shown in Fig. 3.16 (Bair, 1981). In this interesting case of a polymer and plasticizer with more nearly equal MWs, while the diluent depresses the Tg of the polymer in the typical fashion, the polymer simultaneously depresses the crystallization temperature (Tcr) of the plasticizer. Thus, with increasing PVC concentration in the blend, Tcr of the plasticizer decreases as Tg of the blend increases. Upon cooling, crystallization of the plasticizer can no longer occur, within a realistic experimental time frame, in the region (on the state diagram in Fig. 3.16) of temperature and blend composition where the extrapolated crystallization curve intersects the glass curve at a particular point, which can be designated as Tg′ (Slade and Levine, 1988b). Below a critical diluent concentration (i.e., the composition of the glass at Tg′), crystallization on cooling of the plasticizer, which would be readily crystallizable if pure, essentially ceases at an incomplete extent, due to the immobility imposed by the vitrification of the glass-forming plasticizer-polymer blend. Wunderlich (1980) has described several other cases of the same type of behavior for binary mixtures of a synthetic amorphous polymer and its crystallizable monomer, as illustrated in Fig. 3.17 (Hardy et al., 1973). In each case, "the monomer liquidus curve was observed as usual. At its intersection with the Tg vs. concentration curve of the macromolecule, which shows a decreasing Tg with increasing amount of monomer (plasticization), the whole system becomes glassy without crystallization of the monomer. The polymer-rich side of the phase diagram remains thus a single-phase region throughout, while the monomer-rich samples change on cooling from a liquid solution to a two-phase system which consists of liquid solution and crystalline monomer at higher temperature, and changes at lower temperature to glassy solution and crystalline monomer" (Wunderlich, 1980). The analogy between this behavior (as exemplified in Fig. 3.16 and 3.17) of nonaqueous high-polymer systems, with their

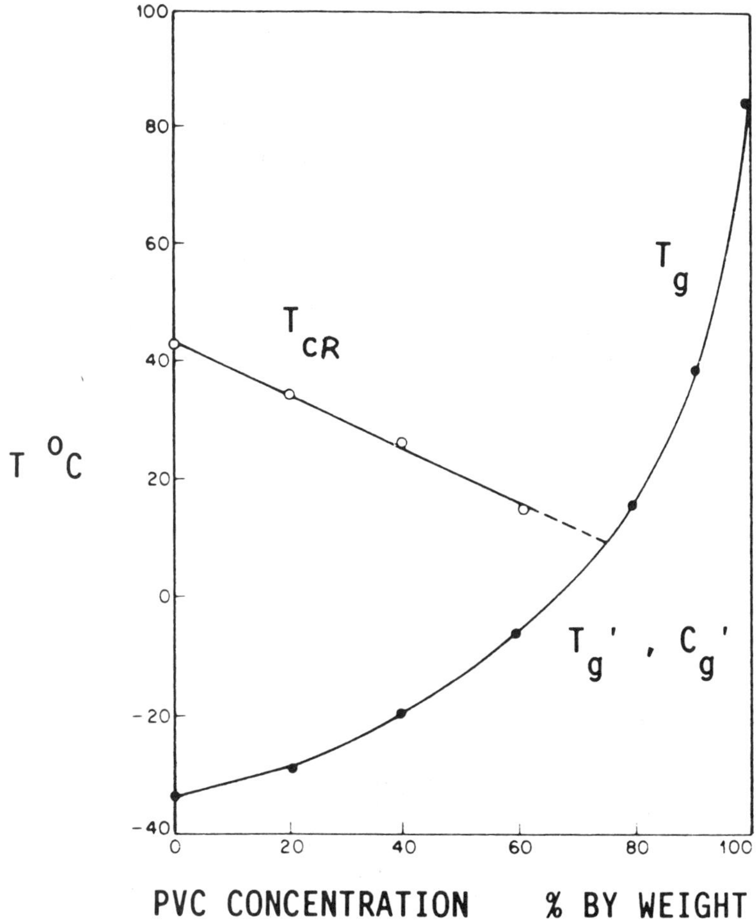

Fig. 3.16 Crystallization and glass temperatures in terpolymeric plasticizer (TP)/polyvinyl chloride (PVC) blends as a function of PVC concentration. (*Reproduced, with permission, from Bair, 1981.*)

characteristic Tg′ and corresponding composition Cg′ (MKr in Fig. 3.17), and the general behavior of aqueous glass-forming systems of water-compatible solutes (illustrated by the state diagrams in Fig. 3.4, 3.11, and 3.14) is important and fundamental to interpreting the nonequilibrium behavior of food polymer systems in the general context of the dynamics map and mobility transformations (Slade and Levine, 1988b). Even more importantly, the truly striking similarity between the state diagrams in

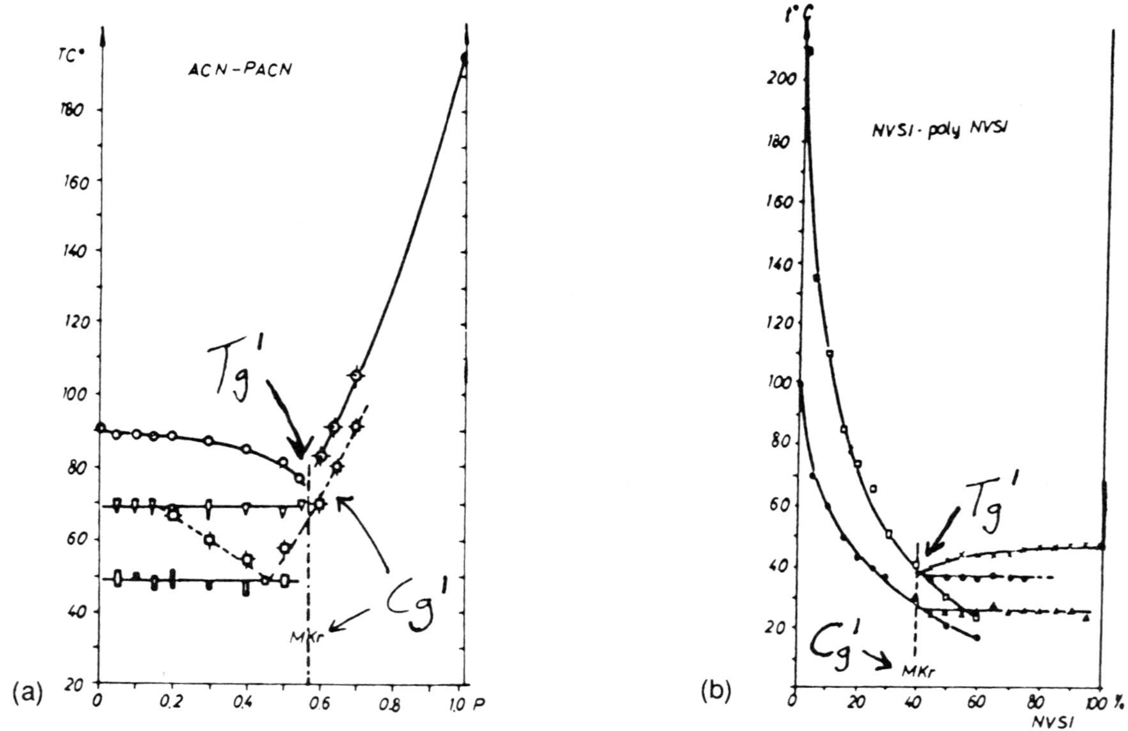

Fig. 3.17 (a) Phase diagram of acenaphthylene and polyacenaphthylene system. (b) Phase diagram of *N*-vinyl-succinimide and poly(*N*-vinyl-succinimide) system. (*Reproduced, with permission, from Hardy et al., 1973.*)

Fig. 3.17, for synthetic polymer systems in which the crystallizable diluent is a liquid organic monomer, and those in Figs. 3.4, 3.11, and 3.14, for aqueous model systems of food molecules in which the crystallizable diluent is water, illustrates the essentially identical behavior of synthetic amorphous polymers and glass-forming food molecules (Levine and Slade, 1988a).

Reports from the polymer literature (Ellis et al., 1984; Jin et al., 1984; Ellis, 1988) have demonstrated that the effectiveness of water as a plasticizer of synthetic polymers (Rowland, 1980; Levine and Slade, 1988a) (by analogy with the effectiveness of typical low molecular weight organic plasticizers, as shown in Fig. 3.15) primarily reflects water's low molar mass. Ellis and coworkers, in explaining water's plasticizing ability, discount older concepts of specific interactions, such as disruptive water-polymer hydrogen bonding in polymer hydrogen-bonded networks, or plasticizing molecules becoming "firmly bound" to polar sites along a polymer chain. Although hydrogen bonding certainly affects solubility parameters and contributes to compatibility of polymer-water blends (Sears and Darby, 1982), it has been convincingly shown that polymer flexibility does not depend on specific hydrogen bonding to backbone polar groups (Buchanan and Walters, 1977). Rather, the relative size of the mobile segment of linear backbone (Sperling, 1986), and thus the relative $\overline{M}w$ of its blend with water, governs the magnitude of plasticization and so determines Tg (Buchanan and Walters, 1977). To negate the older arguments for site-specific hydrogen bonding, NMR results have been cited which clearly indicate that water molecules in polymers with polar sites have a large degree of mobility (Ellis et al., 1984; Jin et al., 1984). In this context, mobility is defined in terms of translational and rotational degrees of freedom for molecular diffusion on a time scale of experimental measurements (Slade and Levine, 1988b). Franks (1982, 1983a, b, 1985, 1986a, b, 1989, 1990, 1991; Franks and Grigera, 1990) has advocated a similar view and presented similar evidence to try to dispel the popular (Labuza, 1985) but outdated (Simatos and Karel, 1988) myths about "bound" water and "water-binding capacity" in glass-forming food polymers or low molecular weight materials (Levine and Slade, 1988a; Slade et al., 1989). For example, proton NMR has been used to test the accessibility of water with reduced mobility in the crystalline regions of retrograded wheat starch gels (Wynne-Jones and Blanshard, 1986). Such gels are partially crystalline, with B-type hydrated crystalline regions in which water molecules constitute an integral structural part of the crystal unit cell (French, 1984; Imberty and Perez, 1988). NMR results have shown that all the water in such a starch gel can be freely exchanged with deuterium oxide (Wynne-Jones and Blanshard, 1986).

Most recently, Ellis (1988) has reported results of a comprehensive DSC study which show that several diverse synthetic "amorphous polyamides in pure and blended form exhibit a monotonic depression of Tg as a function of water content," and which "lend further credence to the simple and straightforward plasticizing action of water in polar polymers irrespective of their chemical and physical constitution." These results have helped to confirm the conclusions (Levine and Slade, 1988a; Slade et al., 1989) that (1) the behavior of hydrophilic polymers (both synthetic and natural) with aqueous diluents is precisely the same as that of nonpolar synthetic polymers (e.g., polystyrene in Fig. 3.15) with organic diluents, and (2) water-compatible food polymers such as starch, gluten, gelatin, and elastin, for which water is an efficient plasticizer but not necessarily a good solvent, exhibit the same physicochemical responses to plasticization as do many water-compatible synthetic polymers (e.g. PVP). As mentioned earlier, the excellent agreement between the measured value of Tg′ (Slade and Levine, 1984, 1987b) and the theoretical value of Tg for an aqueous wheat starch gel with $\gtrsim$27% moisture, recently calculated from free volume theory for synthetic polymers (Blanshard, 1988; Marsh and Blanshard, 1988), lends support to these conclusions. As detailed earlier, a characteristic extent of plasticization at low moisture, typically in the range of about 5–10°C/wt% water (as shown for starch in Fig. 3.11 and gluten in Fig. 3.12), but occasionally somewhat less than 5°C/wt% (e.g., sorbitol in Fig. 3.13) or as much as 20°C/wt% (e.g., hemicellulose in Fig. 3.13), has been shown to apply to a wide variety of water-compatible glassy and partially crystalline food monomers, oligomers, and high polymers (Levine and Slade, 1988a, 1989b; Orford et al., 1989, 1990; Slade et al., 1989; Franks and Grigera, 1990; Roos and Karel, 1990, 1991a, b, d, e; Cocero and Kokini, 1991; Ring and Whittam, 1991; Slade and Levine, 1991a, b). One can deduce the following general relationship (which would be more applicable within a homologous polymer family than between families) between solute MW and extent of plasticization by water. With increasing solute MW, dry Tg increases (up to the boundary of the entanglement plateau), and so then does the difference between the Tg values of the dry solute and water. Thus, the higher a solute's dry Tg, the more highly curved will be the "complete" glass curve for that solute-water system. Since Cg′ also tends generally to increase with increasing solute MW within a homologous polymer family (Levine and Slade, 1986, 1988b), the steepness of the portion of the resulting glass curve from dry Tg to Tg′–Cg′ also tends to increase with increasing solute MW. Thus, with increasing solute MW, the extent of plasticization of the solute at low moisture contents would tend to increase, from lower extents for monomers to higher extents for oligomers and high polymers (e.g., from 3–4°C/wt% for sorbitol to 12.5°C/wt% for maltohexaose), as has been seen by Orford et al. (1989), Franks and Grigera (1990), and Roos and Karel (1991e), but not previously explained on a conceptual basis of state diagrams.

In partially crystalline polymers, water plasticization occurs only in the amorphous regions (Starkweather, 1980; van den Berg, 1981; Gaeta et al., 1982; Jin et al., 1984; Murthy et al., 1989). In linear synthetic polymers with anhydrous crystalline regions and a relatively low capacity for water in the amorphous regions [e.g., nylons (Starkweather, 1980)], the % crystallinity affects Tg, such that increasing % crystallinity generally leads to increasing Tg (Jin et al., 1984; Wunderlich, 1990). This increase is due primarily to the stiffening or "antiplasticizing" effect of disperse microcrystalline crosslinks, which leads to decreased mobility of chain segments in the interconnected amorphous regions (Gaeta et al., 1982; Wunderlich, 1990). The same effect is produced by covalent crosslinks (Jin et al., 1984), which, when produced by radiation, occur only in amorphous regions (Ellis et al., 1984). In polymers with anhydrous crystalline regions, only the amorphous regions are accessible to penetration and therefore plasticization by water (Ellis et al., 1984; Jin et al., 1984; Murthy et al., 1989).

Similar phenomena are observed in partially crystalline polymers with hydrated crystalline regions, such as starch and gelatin (Maurice et al., 1985; Slade and Levine, 1987a; Levine and Slade, 1988a). In native starches, hydrolysis by aqueous acid or enzymes, at T < Tm, can occur initially only in amorphous regions (French, 1984). Similarly, acid hydrolysis of retrograded starch progresses in amorphous regions, leading to increased relative crystallinity (or even increased absolute crystallinity, due to crystal growth) of the residue (French, 1984). Dehumidification of granular starch proceeds most readily from initially mobile amorphous regions, leading to nonuniform moisture distribution (van den Berg, 1981). In partially gelatinized starches, dyeability by a pigment increases with increasing amorphous content (Kuge and Kitamura, 1985). The effective Tg that immediately precedes and thereby determines the temperature of gelatinization (Tgelat) in native starch depends on the extent and type (B vs. A vs. V polymorphs) of crystallinity in the granule (but not on amylose content) and on total moisture content and moisture distribution (Slade, 1984; Slade and Levine, 1984, 1987b, 1988c; Maurice et al., 1985). For normal and waxy (i.e., all amylopectin) starches, Tgelat increases with increasing % crystallinity (Juliano, 1982), an indirect effect due to the disproportionation of mobile short branches of amylopectin from amorphous regions to microcrystalline "micelles," thereby increasing the average MW and effective Tg of the residual amorphous constituents (Slade and Levine, 1988c), because these branches are unavailable to serve as "internal" plasticizers (Sears and Darby, 1982). The same effect of % crystallinity on Tg is seen in partially crystalline amylose or native granular starches and in partially recrystallized amylopectins at low moisture contents in the range < Wg′ (van den Berg, 1981; Zeleznak and Hoseney, 1987a; Orford et al., 1989; Kalichevsky et al., 1991): (1) Tg of the partially crystalline material is higher than that of the corresponding, completely amorphous sample with the same total moisture content; and (2) at the same moisture content, Tg increases with increasing % crystallinity of the material. Two other related phenomena are observed as a result of the nonuniform moisture distribution in situations of overall low moisture content for polymers with hydrated crystalline regions (Levine and Slade, 1988a): (1) atypically high Tg/Tm (in degrees K) ratios >> 0.80 but, of course, < 1.0 (Wunderlich, 1980; Soesanto and Williams, 1981), in contrast to the characteristic range of 0.5–0.8 for many partially crystalline synthetic polymers (Brydson, 1972; Wunderlich, 1990); and (2) a pronounced apparent depressing effect of water on Tm (Lelievre, 1976; Slade, 1984; Slade and Levine, 1984, 1987a, b; Maurice et al., 1985; Biliaderis et al., 1985, 1986a, b, c; Zobel, 1988; Zobel et al., 1988; Biliaderis and Galloway, 1989; Biliaderis, 1990, 1991; Biliaderis and Seneviratne, 1990a, b; Biliaderis and Zawistowski, 1990; Whittam et al., 1991) as well as Tg, such that both Tg and Tm decrease with increasing moisture content.

It is important to note that plasticization of the amorphous regions (e.g., the backbone segments and branch points of amylopectin molecules) of native granular starches by sorbed water is neither instantaneous nor simultaneous with the initial swelling caused by water uptake (Slade and Levine, 1991a). It has recently been demonstrated by Aguerre et al. (1989) that "the uptake of water takes place between the concentric layers" of the granule, leading to "interlamellar expansion of the starch granule structure." This sorbed water must subsequently diffuse from the interlamellar spaces to the amorphous regions of the granule before plasticization of the polymer molecules or chain segments in those amorphous regions can begin (Yost and Hoseney, 1986; Slade and Levine, 1991a). The salient point is that, while water is unquestionably an effective plasticizer of starch and all other water-compatible food polymers, the mere *presence* of water does *not* attest that plasticization has occurred (Slade and Levine, 1991a). Thus, when one reads the following from Hoseney (1986),

> When starch is placed in water, the granule is freely penetrated by water, or for that matter, by most small molecules. The starch can hold about 30% of its dry weight as moisture. The granule swells

> slightly; its increase is generally considered to be about 5%. The volume change and water absorption are reversible, and heating the system to just below its gelatinization temperature will not bring about any other changes.

one must be careful not to equate rapid penetration (e.g., through preexisting channels and voids) with rapid plasticization by water. In fact, plasticization at T < the initial Tg of a material is slow, but it occurs more rapidly at T > the initial Tg (Slade and Levine, 1991a). This was graphically demonstrated by some remarkable swelling studies of PVC described by Sears and Darby (1982). It was found that several "common PVC plasticizers . . . would not swell an unplasticized PVC sheet at room temperature in two years' time. . . . Yet, all of the plasticizers swelled the rigid PVC sheet at 76°C, the approximate Tg of the resin. When the same PVC was hot-compounded with a plasticizer, cooled, and then immersed in various other plasticizers, it would imbibe more plasticizer."

To put the modern concept of water plasticization in a more familiar context of the older, more traditional literature on "bound" and "unfreezable" water and on water sorption by food polymers at low moisture (Levine and Slade, 1988a; Slade et al., 1989; Slade and Levine, 1991a), the earliest-sorbed water fraction is most strongly plasticizing, always said to be "unfreezable" in a practical time frame, and often referred to as "bound." The later-sorbed water fraction is said to be freezable, referred to as "free," "mobile," or "loosely bound," and is either weakly or nonplasticizing, depending on the degree of water compatibility of the specific polymer. As mentioned earlier, the degree of water compatibility relates to the ability of water to depress Tg to Tg′, and to the magnitude of Wg′ (Levine and Slade, 1988a; Slade and Levine, 1988a, b). Regardless of context, a key fact about the "freezability" of water relates to the homogeneous process for the prerequisite nucleation step of ice crystallization (Franks, 1982). Even at temperatures as low as −40°C [the homogeneous nucleation temperature for ice in pure water (Franks, 1982)], a minimum on the order of 200 water molecules must associate within a domain of about 40 Å in order to form a critical nucleus that will grow spontaneously into an ice crystal (Franks, 1982). Thus, within any food material at low moisture, clusters of water molecules of lower density than about 200 molecules/40 Å would certainly require temperatures below −40°C or heterogeneous catalysts for nucleation to occur (Levine and Slade, 1989c).

The solute-specific, invariant quantity of unfrozen water captured in the glass that forms at Tg′, defined as Wg′ (Levine and Slade, 1986), is traditionally referred to by many food scientists and technologists as one measure of "bound" water (Labuza, 1985). However, "bound" water, with respect to either frozen or room-temperature food systems, is a misnomer that has persisted for at least the last 30 years, despite constant debate (Kuprianoff, 1958; Levine and Slade, 1986, 1988a, 1989b; Lillford, 1988; Simatos and Karel, 1988; Karel, 1989; Slade et al., 1989) and evermore convincing arguments that the concepts of "bound" water, "water binding," and "water-binding capacity" of a solute are incorrect, inappropriate, and misleading rather than helpful (Franks, 1982, 1983a, b, 1985, 1986a, b, 1989, 1990, 1991; Franks and Grigera, 1990). The concept of "bound" water originated in large part from a fundamental misconception that discrete "free" and "bound" physical states of water in food materials (or "free," "loosely bound," and "tightly bound" states) could provide a valid representation of water molecules in a solution at ambient temperature (Franks, 1983b; Levine and Slade, 1990). Actually, at $T > Tg'$, water molecules in a solution exist within a single physical state (i.e., liquid) characterized not by any kind of static geometry but rather by a dynamic continuum of degrees of hindered instantaneous mobility (Slade et al., 1989). In this liquid solution state, individual water molecules are only transitorally hydrogen-bonded to individual polar sites on the solute (Franks, 1983a, b; Franks and Grigera, 1990).

As explained recently (Levine and Slade, 1986, 1988c, d, 1989c, 1990; Slade and Levine, 1988b), the solute-specific value of Wg' is the maximum amount of water that can exist with that solute in a spatially homogeneous, compatible blend which, in the rubbery state, exhibits long-range cooperative relaxation behavior described by WLF kinetics, but not long-range lattice order. Further dilution beyond Wg' results in loss of cooperative mobility and onset of short-range fluid mechanics, described by Arrhenius kinetics. Thus, expression of Wg' as a water/solute number ratio [i.e., a "notional hydration number" (Franks, 1983b)] actually represents the technologically practical maximum limit for the amount of water that can act as a plasticizer of a particular solute (Franks, 1985; Levine and Slade, 1988a; Given, 1991), rather than the amount of water that is "bound" to, or whose dynamics are governed by, that solute. Part of the reason for the persistence of the concept of "bound" water in such concentrated solute systems, despite convincing evidence of its invalidity, relates to a conclusion inadvisedly extrapolated from findings for very dilute solutions. The addition of a few isolated solute molecules to pure water already causes a profound effect on the self-diffusion properties in the solution. The hindered diffusion of water molecules instantaneously in the vicinity of individual solute molecules is construed as the effect of "viscous drag"; these less-mobile water molecules are visualized to be "pulled along" with the solute during flow. But it has been demonstrated repeatedly (Franks, 1983a, b; Franks and Grigera, 1990) that the less-mobile water molecules are freely exchangeable with all of the water in the solution, leading to the inescapable, majority view that the water is not bound to the solute (Slade et al., 1989). On the other hand, in describing dilute solutions, no one has ever suggested that the solute molecules are "bound" to water molecules. When the situation is reversed, adding a few water molecules to an anhydrous solute profoundly changes the viscoelastic properties of the solute via water plasticization, which increases the free volume and decreases the local viscosity (Levine and Slade, 1990). Why then, in light of this evidence of a dramatic *increase* in the mobility of the solute, have many found it so easy to jump to the conclusion that these water molecules must be "bound" to solute molecules?

It is only recently becoming more widely acknowledged and accepted (Blanshard and Franks, 1987; Levine and Slade, 1988a, 1989c; Lillford, 1988; Simatos and Karel, 1988; Slade et al., 1989) that the so-called "bound" water corresponding to Wg′ is not energetically bound in any equilibrium thermodynamic sense. Rather, it is simply kinetically retarded, due to the extremely high local viscosity ($\sim 10^{12}$ Pa s) of the metastable glass at Tg′, and thus dynamically constrained from the translational and rotational diffusion required for ice crystal growth (Franks, 1982, 1983a, 1985, 1986a, b, 1989, 1990, 1991; Levine and Slade, 1986, 1988c, d, 1989c, 1990; Slade and Levine, 1988b; Franks and Grigera, 1990). The crucial finding that water is not "strongly bound" to polar groups on hydrophilic polymers has been demonstrated in an especially convincing fashion by the meticulous low-temperature DSC and ambient-temperature sorption studies of Pouchly and Biros (Biros et al., 1979; Pouchly et al., 1979, 1982) on the thermodynamic interaction of water with (and plasticizing effect on) hydrophilic synthetic polymers in glassy and rubbery states. This conclusion regarding the true nature of "bound" water does not mean that there are not solute-water hydrogen bonds in the glass at Tg′, only that such hydrogen bonds are the normal consequence of dissolution of a solute in water rather than the cause of the kinetic retardation that renders this water "unfreezable" in real time (Levine and Slade, 1990). The stabilizing free energy of such solute-water hydrogen bonds is no greater than for water-water hydrogen bonds in ice (Biros et al., 1979; Pouchly et al., 1979, 1982; Franks, 1983a, 1991). Analogously, for model solutions of small sugars at room temperature, results of NMR and dielectric relaxation measurements, as well as recent molecular dynamics simulations, have shown that "the residence time of a given water molecule at a solvation site [i.e. a hydroxyl group on a sugar] is extremely short, <1 ns" (Franks, 1983a, b; Franks and Grigera, 1990).

Furthermore, such results, from studies of synthetic polymers (Jin et al., 1984) and polymeric carbohydrate and protein gels (Wynne-Jones and Blanshard, 1986; Lillford, 1988) alike, have demonstrated conclusively that water molecules said to be "bound" to polar groups on such polymeric solutes are in fact highly mobile [especially compared to the mobility of water in ice (Derbyshire, 1982)] and able to exchange freely and rapidly, likewise on an NMR time scale, with other (so-called "free" or "bulk") water molecules and deuterium oxide. Other studies have concluded that "bound" water has thermally labile hydrogen bonds (Biros et al., 1979; Pouchly et al., 1979, 1982), shows cooperative molecular mobility (Hoeve, 1980), has a heat capacity approximately equal to that of liquid water rather than ice (Hoeve and Hoeve, 1978; Pouchly et al., 1979; Hoeve, 1980), and has some capability to dissolve salts (Burghoff and Pusch, 1980).

It has recently been concluded that "in the past, too much emphasis has been given to water activity and 'water binding'" (Simatos and Karel, 1988). In fact, the typical observation of two relaxation peaks (ascribed, following traditional dogma, to "free" and "bound" water) for all biological tissues and solutions that have been examined in dielectric experiments (Mashimo et al., 1987) is entirely consistent with, and exactly analogous to, the behavior of synthetic polymers with nonaqueous, non–hydrogen-bonding organic plasticizers (Sears and Darby, 1982). The traditional point of view on the "structuring" effect of solutes on water (and its association with the concept of water activity), which helped give rise to the myth of "bound" water, is rightfully being replaced (Simatos and Karel, 1988; Franks, 1991; van den Berg, 1991) by a new perspective and emphasis on the mobilizing effect of water acting as a plasticizer on solutes, which had led to a deeper qualitative understanding of structure-function relationships in aqueous food polymer systems (Slade and Levine, 1984, 1987a, b, 1988a, b, c, d, 1989, 1991a, b, 1992). For example, Labuza, who in the past has been a very well-known proponent of the concept of "bound" water (Labuza, 1985), now writes, in the context of the "water binding capacity (WBC)" of dietary fiber, that "in fact, current opinion on bound water (if there is such a thing) is that it is very different from what the expression commonly means. . . . Product development scientists should take the WBC values that currently are being bandied about with a grain of salt" (Labuza, 1989).

Williams-Landel-Ferry Theory and WLF Kinetics

At temperatures above Tg, plasticization by water affects the viscoelastic, thermomechanical, electrical, guest/host diffusion, and gas permeability properties of completely amorphous and partially crystalline polymer systems to an extent mirrored in its effect on Tg (Levine and Slade, 1988a). In the rubbery range above Tg for completely amorphous polymers or between Tg and Tm for partially crystalline polymers [in either case, typically from Tg to about Tg + 100°C for well-behaved synthetic polymers (Slade and Levine, 1988b)], the dependence of viscoelastic properties on temperature (i.e., the effect of increasing temperature on relative relaxation times) is successfully predicted (Cowie, 1973) by the WLF equation, an empirical equation whose form was originally derived from the free volume interpretation of the glass transition (Williams et al., 1955; Ferry, 1980). The WLF equation can be written as (Williams et al., 1955; Soesanto and Williams, 1981):

$$\log_{10} \frac{\dfrac{\eta}{\rho T}}{\dfrac{\eta g}{\rho g\ Tg}} = \frac{-C1(T - Tg)}{C2 + (T - Tg)}$$

where η is the viscosity or other diffusion-limited relaxation process, ρ the density, and C1 and C2 are coefficients that describe the temperature dependence of the relaxation process at temperatures above the reference temperature. For an undiluted polymer, when Tg is used as the reference temperature, C1 is proportional to the inverse of the free volume of the system at Tg, while C2 is the ratio of the free volume at Tg over the expansion coefficient of the free volume. The expansion coefficient of the free volume is the constant that describes the linear dependence, on ΔT above Tg, of the increase in free volume due to thermal expansion above Tg (i.e., the difference between the volumes of the rubbery liquid and glassy solid states). C1 and C2 are material-specific constants for an undiluted polymer (i.e., not temperature dependent), but they vary with composition in the presence of a diluent (Ferry, 1980). C1 and C2 take on the values of "universal constants" [17.44 and 51.6, respectively, as extracted from experimental data on many synthetic amorphous polymers (Williams et al., 1955)] for well-behaved polymers (Slade and Levine, 1988b). The WLF equation describes the kinetic nature of the glass transition and has been shown to be applicable to any glass-forming polymer, oligomer, or monomer (Ferry, 1980). These particular values for the "universal constants" have also been shown to apply to molten glucose (Williams et al., 1955), undercooled melts of fructose or glucose (Ollett and Parker, 1990), amorphous glucose-water mixtures (Chan et al., 1986), supersaturated sucrose solutions (Roos and Karel, 1991d), amorphous sucrose and lactose powders at low moisture (Roos and Karel, 1990, 1991a; Shimada et al., 1991), and concentrated solutions of mixed sugars (Soesanto and Williams, 1981), as examples of relevance to foods.

The equation defines mobility in terms of the non-Arrhenius temperature dependence of the rate of any diffusion-limited relaxation process occurring at a temperature T, compared to the rate of the relaxation at the reference temperature Tg, shown here in terms of how log η depends on ΔT, where $\Delta T = T - Tg$. The WLF equation is valid in the temperature range of the rubbery or undercooled liquid state, where it is typically used to describe the time-/temperature-dependent behavior of polymers (Roberts and White, 1973). The equation is based on the assumptions that polymer free volume increases linearly with increasing temperature above Tg and that segmental or mobile unit viscosity, in turn, decreases rapidly with increasing free volume (as illustrated implicitly in Fig. 3.1) (Ferry, 1980). Thus, the greater the ΔT, the faster a system is able to move (due to increased free volume and decreased mobile-unit viscosity); so the greater is the mobility, and the shorter is the relaxation time. In essence, the WLF equation and resulting master curve of log (η/ηg) vs. T − Tg (Williams et al., 1955; Soesanto and Williams, 1981) represent a mobility transformation, described in terms of a time-temperature superposition (Slade and Levine, 1988b). Such WLF plots typically show a five-orders-of-magnitude change in viscosity (or in the rates of other relaxation processes) over a 20°C interval immediately above Tg (Franks, 1985), which is characteristic of WLF behavior in the rubbery fluid range (Slade and Levine, 1988b; Noel et al., 1990; Roos and Karel, 1991d). For example, as demonstrated by Soesanto and Williams (1981), the effects of temperature and concentration on the mobility of fluids above Tg can be combined to create a single master curve, which represents the WLF equation. The viscosity data shown in Fig. 3.18 (Soesanto and Williams, 1981) were obtained for highly concentrated (>90 wt%) aqueous mixtures of fructose and sucrose. These results showed a five-orders-of-magnitude change in the viscosity of concentrated sugar solutions, over a 20°C interval near Tg, a finding in excellent accord with the behavior predicted by the quantitative form of the WLF equation, with its "universally" applicable numerical values of the coefficients C1 = 17.44 and C2 = 51.6. These results constituted the first experimental demonstration that concentrated fructose and sucrose solutions obey the WLF equation quantitatively as well as do synthetic high polymers. Similarly, it had been shown much earlier that a completely amorphous glucose melt, in the absence of diluent, has the same coefficients in the WLF equation, and thus also behaves like a typical well-behaved synthetic high polymer (Williams et al., 1955; Matsuoka et al., 1985). Furthermore, it was recently confirmed experimentally that undercooled melts of diluent-free glucose or fructose also exhibit a five-orders-of-magnitude change in viscosity over a 20°C interval above Tg (Noel et al., 1990; Ollett and Parker, 1990),

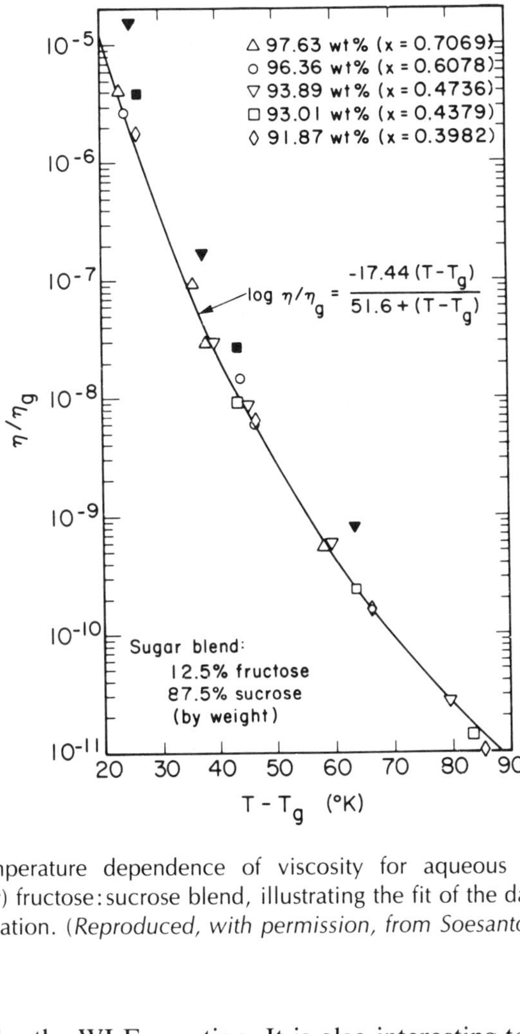

Fig. 3.18 Temperature dependence of viscosity for aqueous solutions of a 12.5:87.5 (w/w) fructose:sucrose blend, illustrating the fit of the data to the curve of the WLF equation. (*Reproduced, with permission, from Soesanto and Williams, 1981.*)

as predicted by the WLF equation. It is also interesting to note that Le Meste and Huang (1991), in a recent TMA study of frozen sucrose solutions, found that for a 20% sucrose sample, softening and concomitant flow during heating were observed to be gradual in the temperature range from −32°C (i.e., Tg′) to −10°C, but above −10°C (i.e., 22°C above Tg), sample volume decreased much more rapidly as a consequence of compression, as expected, based on WLF kinetics (Slade and Levine, 1988b).

In the context of the utility of the WLF equation, the underlying basis of the principle of time-temperature superpositioning is the equivalence between time (or frequency) and temperature as they affect the molecular relaxation processes that influence the viscoelastic behavior (i.e., the dual characteristics of viscous liquids and elastic solids) of polymeric materials and glass-forming small molecules (Ferry, 1980; Sichina, 1988; Mita, 1990). This principle is illustrated in Fig. 3.19 (Levine and Slade, 1989b), which shows a master curve of the modulus as a function of temperature or frequency for a typical partially crystalline synthetic high polymer (Graessley, 1984). Figure 3.19 has been used to describe the viscoelastic behavior of such materials, as exemplified by a kinetically metastable gelatin gel in an undercooled liquid state, in the context of WLF theory (Borchard et al., 1980). At T > Tg, gelatin gels manifest a characteristic rubberlike elasticity (Tomka et al., 1975), due to the existence of a network of entangled, randomly coiled chains (Yannas, 1972). With increasing temperature, a gelatin gel traverses the five regions of viscoelastic behavior characteristic of synthetic, partially crystalline polymers (Yannas, 1972), as illustrated in Fig. 3.19: (1) at T < Tg, vitrified glass; (2) at T =

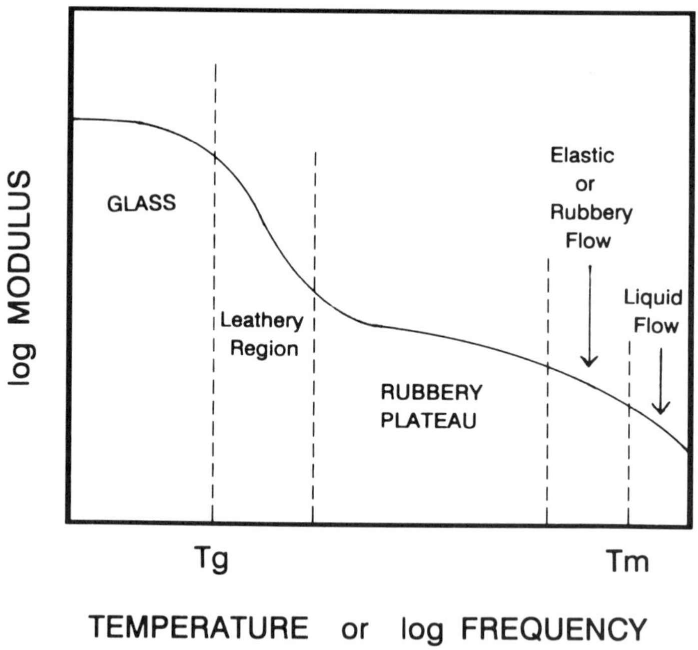

Fig. 3.19 Master curve of the modulus as a function of temperature or frequency, illustrating the five regions of viscoelastic behavior characteristic of synthetic partially crystalline polymers. (*Reproduced, with permission, from Levine and Slade, 1989b.*)

Tg, glass transition to leathery region, typically manifested as a three-orders-of-magnitude decrease in modulus; (3,4) at Tg < T < Tm, rubbery plateau to rubbery flow; and (5) at T > Tm, viscous liquid flow. It is interesting to note that at Tg < T < Tm, a gelatin gel is freely permeable to the diffusion of dispersed dyes and molecules as large as hemoglobin (Slade et al., 1989); only at T < Tg is such dye diffusion greatly inhibited (Wesson et al., 1982). The typical decrease in modulus (usually by about three or more orders of magnitude) at the glass transition, illustrated in Fig. 3.19, has also been shown in similar plots of mechanical modulus data (as functions of either temperature or moisture content) recently reported for completely amorphous, water-plasticized matrices of pure amylopectin (Kalichevsky et al., 1991), extruded starch (Ollett et al., 1990; Smith, 1990) and starch-glucose mixtures (Noel et al., 1990; Ollett et al., 1990; Smith, 1990) and elastic networks of thermoset (via covalent cross-linking) gluten (Attenburrow et al., 1990; Davies et al., 1991). Le Meste et al. (1991a) have recently used TMA to study flour-based food products, and have found that their viscoelastic properties, as functions of temperature and moisture content, correspond to those characteristic of the glassy state, the glass transition region, or the rubbery plateau.

The WLF equation is not intended for use much below Tg (i.e., in the glassy solid state) or in the very low viscosity liquid state [η <10 Pa s (Soesanto and Williams, 1981)], typically 100°C or more above Tg, where Arrhenius kinetics apply (Roberts and White, 1973; Ferry, 1980; Robertson, 1985). For partially crystalline polymers, the breadth of the temperature range of the rubbery domain of WLF behavior corresponds to the temperature interval between Tg and Tm (Wunderlich, 1976; Ferry, 1980), as illustrated in Fig. 3.1. Cheng (1989) has noted that the size of this temperature interval between Tg and Tm may be as much as several hundred degrees for synthetic high polymers. An analysis of the variation of the size of this temperature interval with the Tm/Tg ratio of representational synthetic polymers and glass-forming, low molecular weight carbohydrates has recently been reported (Slade and Levine, 1988b). This study compared the WLF behavior of kinetically metastable carbohydrate-water systems to the corresponding knowledge base for synthetic high polymers. According to the conventional description, a typical well-behaved synthetic high polymer (e.g., a representational elastomer) would manifest its Tg around 200 K in the completely amorphous state, and its Tm around 300 K in the completely crystalline state (Wunderlich, 1980), so that the ratio of Tm for the pure crystalline material to Tg for the completely amorphous material is about 1.5 (or Tg/Tm about 0.67) (Brydson, 1972; Wunderlich, 1990). Such a polymer would also have a local viscosity of about 10^{12} Pa s and a free volume fraction of about 2.5% at

Tg (Ferry, 1980). For this typical well-behaved polymer, WLF kinetics are considered to be operative in a temperature range about from Tg to 100°C above Tg (Williams et al., 1955). It can be seen that this operational definition is related to the typical Tm/Tg ratio of 1.5, since in such a case, the difference in temperature between Tg and Tm would be about 100°C. Figure 3.20A (Slade and Levine, 1988b) illustrates the conventional description of the relaxation behavior of a typical well-behaved polymer [e.g., polyvinyl acetate (Johnson et al., 1980; Matsuoka et al., 1985)], which would obey the standard form of the WLF equation with the coefficients C1 = 17.44 and C2 = 51.6. In this plot of log a_T vs ΔT [where a_T = WLF shift factor (Ferry, 1980)], the relaxation rate progresses from WLF behavior very near Tg to Arrhenius behavior at about 100°C above Tg. Within this temperature range, where technological process control would be expected, relaxation rates for WLF behavior near Tg would change by a factor of 10 for every 3°C change in temperature. In contrast, for Arrhenius behavior with familiar Q_{10} = 2 kinetics above Tm, a factor-of-10 change in relaxation rate would require a 33°C change in temperature (Slade and Levine, 1988b).

Another class of amorphous polymers has been described (Slade and Levine, 1988b) as typical but not well behaved, in the sense that they are readily crystallizable (Brydson, 1972; Haward, 1973; Ferry, 1980; Wunderlich, 1980). Highly symmetrical polymers such as poly(vinylidene chloride) and poly(vinyl cyclohexane), which manifest crystalline melting enthalpies of about 170 J/g, fit this class (Wunderlich, 1980). For such polymers, the ratio of Tm/Tg is frequently >> 1.5, so the temperature range between Tg and Tm is >> 100°C. Different WLF coefficients would be required to describe their relaxation profile, as illustrated by the plot in Fig. 3.20B drawn for C1 = 20.4 and C2 = 154.8. For a representational case of Tg $\simeq$ 200 K (with $\eta g \gtrsim 10^{12}$ Pa s, and free volume fraction $\simeq$ 2.5%) and Tm/Tg $\simeq$ 2 (Tg/Tm $\simeq$ 0.5), Tm would be $\simeq$ 400 K. Thus, there would be about a 200°C region in which relaxation rates would change from WLF behavior near Tg (in this case, by a factor of 10 for every 6°C) to Arrhenius behavior near Tm (by a factor of 10 for every 33°C) (Slade and Levine, 1988b). A notable example of a material with Tm/Tg $\simeq$ 2 is water (Soesanto and Williams, 1981; Wunderlich, 1990).

A third class of polymers, often characterized by highly unsymmetrical structures, has been described (Slade and Levine, 1988b) as atypical and poorly behaved, in that Tg is near Tm (Brydson, 1972; Wunderlich, 1980). For such polymers, with Tm/Tg << 1.5 (i.e., $\simeq$1.25, or Tg/Tm $\simeq$ 0.8), a quantitatively different form of the WLF equation would be required to describe their relaxation profile. In this case, as illustrated in Fig. 3.20C, using C1 = 12.3 and C2 = 23.3, the intercept of log a_T was plotted as

$\simeq -3$ for $\Delta T = 0$ (i.e., at Tg), in contrast to Fig 3.20A and B, where log a_T was defined as 0 at Tg. For a representational polymer in this class, Tg $\simeq$ 200 K (with $\eta g << 10^{12}$ Pa s, and free volume fraction >> 2.5%) and Tm $\simeq$ 250 K. Thus, the temperature range in which WLF kinetics would be operative is much smaller than usual. Relaxation rates would change from WLF behavior near Tg (in this case, by a factor of 10 for every 1°C) to Arrhenius behavior above Tm (by a factor of 10 for every 33°C) over a region of only about 50°C (Slade and Levine, 1988b). The synthetic polymer cited as the classic example of this behavior, which has been attributed to anomalously large free volume and high mobility at Tg, is bisphenol polycarbonate, with Tm/Tg $\simeq$ 1.18 (Brydson, 1972; Ehlich and Sillescu, 1990). This category of behavior has also been reported (Levine and Slade, 1988a, 1989a) to be exemplified by food materials such as gelatin and native starch (due to nonuniform distribution of moisture in amorphous and crystalline regions of these high polymers at low moisture) and the simple sugars fructose and galactose (due to anomalous translational free volume of these anhydrous monosaccharides) (Slade and Levine, 1988b).

The three types of behavior exemplified in Fig. 3.20A–C, in which the Tm/Tg ratio is either the typical value of 1.5, or much greater, or much less, have been compared in order to examine how the respective relaxation profiles change in the temperature interval between Tm and Tg for representational, diluent-free polymers with a common value of Tg (Slade and Levine, 1988b). As illustrated in Fig. 3.20D, this analysis revealed the critical significance of the Tm/Tg ratio for any given polymer. For a common value of Tg, different values of Tm/Tg for different polymers (e.g., food carbohydrates) can be used to compare relative mobilities at Tg and at T >> Tg (Slade and Levine, 1988b). For different values of Tg, relative mobilities can be compared based on values of the difference, Tm − Tg, rather than the ratio, Tm/Tg (Slade and Levine, 1988b). In Fig. 3.20D, the behavior of log a_T vs. ΔT is compared for different values of Tm/Tg (i.e., about 2, 1.5, and the extreme case of 1.0) to determine how mobility varies in the kinetically constrained regions of this mobility transformation map. At T >> Tg, the overall free volume for different polymers may be similar (Ferry, 1980), yet individual free volume requirements for equivalent mobility may differ significantly, as reflected in the Tm/Tg ratio. Anisotropy in either rotational mobility [which depends primarily upon free volume (Ferry, 1980)] or translational mobility [which depends primarily upon local viscosity, as well as free volume (Ferry, 1980)] may be the key determinant of a particular polymer's relaxation behavior. The glass transition is a cooperative transition (Roberts and White, 1973; Wunderlich, 1981; Angell, 1988; Cheng, 1989) resulting from

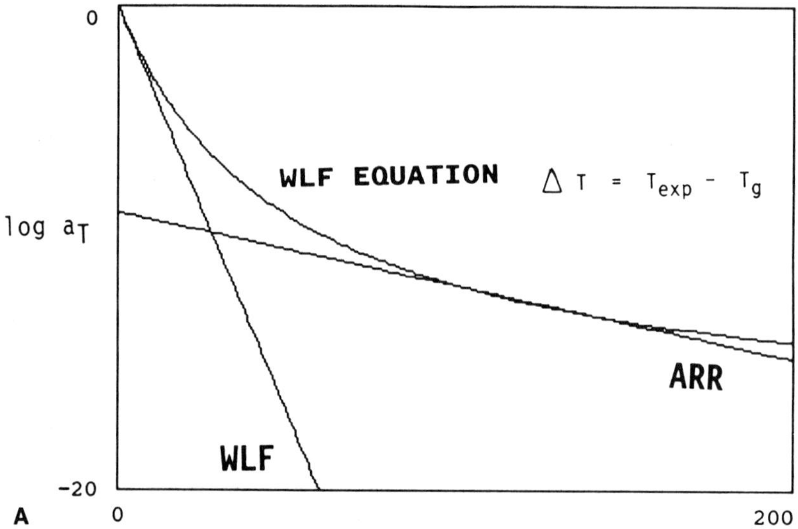

A

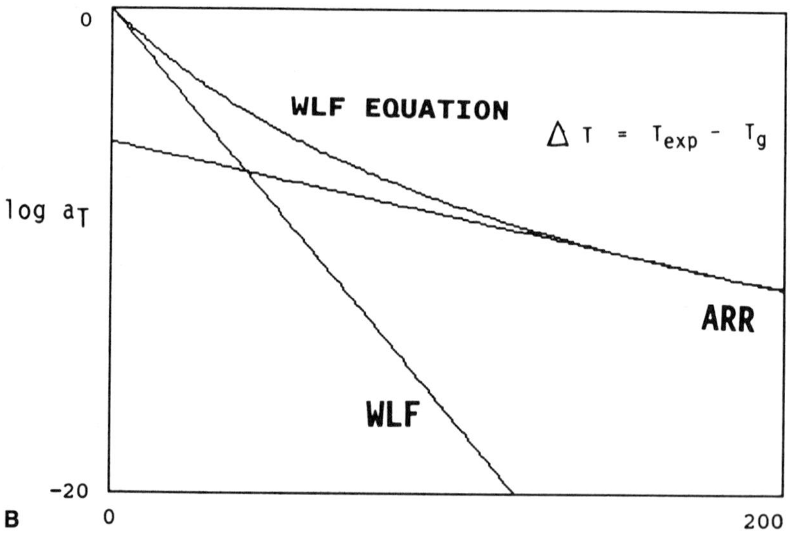

B

Fig. 3.20 WLF plots of the time-temperature scaling parameter (WLF shift factor), a_T, as a function of the temperature differential above the reference state, Tg, with the limiting regions of low and high ΔT defined by the WLF and Arrhenius kinetic equations, respectively. The curves of the WLF equation (with coefficients C1 and

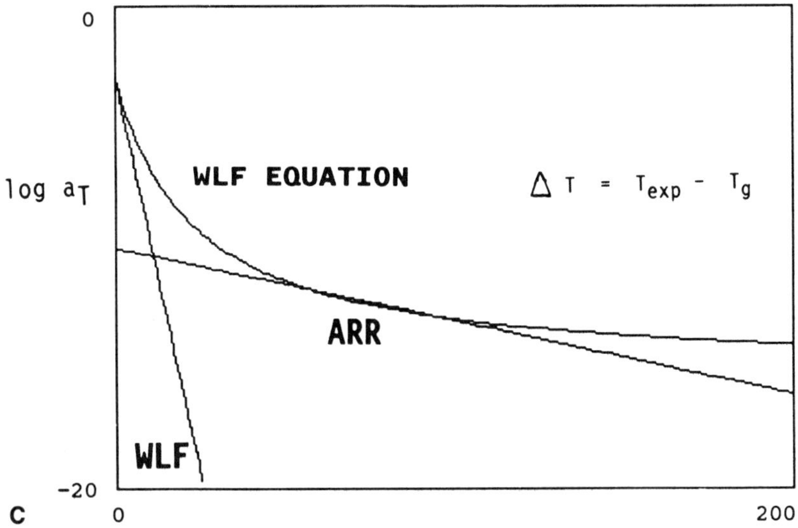

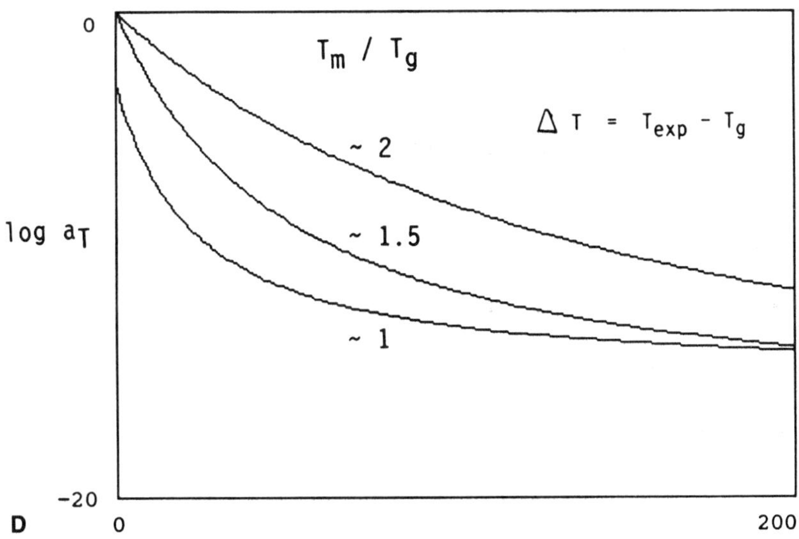

C2 as noted) illustrate the temperature dependence of the relaxation rate behavior for hypothetical polymers with Tm/Tg ratios of: (A) 1.5 (C1 = 17.44, C2 = 51.6); (B) 2.0 (C1 = 20.4, C2 = 154.8); (C) 1.0 (C1 = 12.3, C2 = 23.3); (D) 2.0, 1.5, and 1.0. (*Reproduced, with permission, from Slade and Levine, 1988b.*)

local cooperative constraints on mobility, and Tg represents a thermomechanical property controlled by the local small-molecule or segmental, rather than macroscopic, environment of a polymer. On cooling a viscous fluid of relatively symmetrical mobile units with relatively isotropic mobility, translational motions would be expected to be "locked in" at a higher temperature before rotational motions, because of the slower structural relaxations associated with the larger scale translational diffusion (Slade and Levine, 1988b). In this case, cooperative constraints of local viscosity and free volume on translational diffusion determine the temperature at which the glass transition is manifested, as a dramatic increase in relaxation times compared to the experimental time frame. However, in the case of motional anisotropy, molecular asymmetry has a much greater effect on rotational than translational diffusion, so rotational motions could be "locked in" before translational motions as the temperature is lowered (Slade and Levine, 1988b). As revealed by Fig. 3.20D, a very small ratio of Tm/Tg (i.e., close to 1.0) is accounted for by an anomalously large free volume requirement for rotational diffusion (Brydson, 1972). When the free volume requirement is so large, a glass transition (i.e., vitrification of the rubbery fluid) on cooling can actually occur even when the local viscosity of the system is relatively low (Slade and Levine, 1988b). Thus, instead of the typical "firmness' for a glass ($\simeq 10^{12}$ Pa s), such a glass (e.g., of bisphenol polycarbonate, or anhydrous fructose or galactose) may manifest a $\eta g << 10^{12}$ Pa s (Soesanto and Williams, 1981; Angell, 1988; Levine and Slade, 1988a; Slade and Levine, 1988a). [Interestingly, Ollett and Parker (1990) have recently reported a measured viscosity of $10^{4.5}$ Pa s for a fructose melt at 100°C (Noel et al., 1990), the temperature of its higher Tg (Levine and Slade, 1988a).] In such a glass, the time constant for translational diffusion may be anomalously small, indicative of high translational mobility. In contrast, in the glass of a typical well-behaved polymer, the time constant for translational diffusion would be greater than that for rotational diffusion, so an increase in local viscosity would be concomitant with a decrease in free volume (Slade and Levine, 1988b). The above analysis has pointed out the critical significance of anomalous values of Tm/Tg ratio (for the dry solute) close to 1.0 on the mobility, resultant relaxation behavior, and consequent technological process control for nonequilibrium food polymer systems (in the *presence* of water) in their supra-glassy fluid state above Tg (Slade and Levine, 1988b), in terms of the WLF kinetics of various translational diffusion-limited, mechanical/structural relaxation processes, such as gelatinization, annealing, and recrystallization of starch (Slade and Levine, 1988c).

An interesting comparison has been made between the characteristic

WLF ranges discussed above: (1a) about 100°C or more above Tg for many typical, completely amorphous, synthetic polymers, and (1b) from <50°C to 200°C or more between Tg and Tm for different categories of partially crystalline synthetic polymers; and (2) the relative breadth of the temperature range for the WLF region relevant to frozen aqueous food systems, i.e., the magnitude of the rubbery domain between Tg′ and the Tm of ice (Levine and Slade, 1989c). As illustrated schematically in Fig. 3.21 (Levine and Slade, 1989c), the rubbery domains between Tg and Tm (represented by the solid vertical bars) for pure water (about 135°C) and pure solute [e.g., about 140°C for sucrose (Levine and Slade, 1989a)] are similar in breadth to the 100–200°C span of the WLF region between Tg and Tm for many synthetic polymers (Wunderlich, 1976; Slade and Levine, 1988b). In contrast, the breadth of the temperature range between Tg′ and Te (the melting temperature for the eutectic mixture of crystalline solute and ice) in Fig. 3.21 is much smaller. For example, as reported for the specific case of a frozen sucrose solution (Lev-

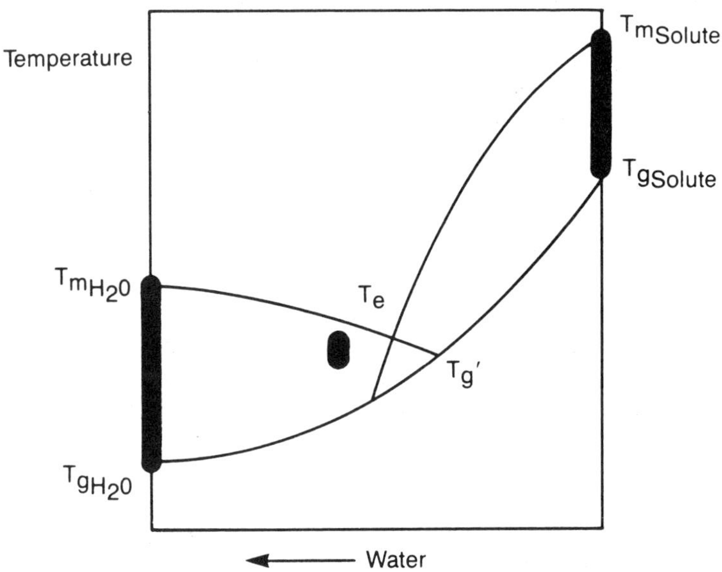

Fig. 3.21 Schematic solute-water state diagram illustrating the relative magnitudes of ΔT = Tm − Tg, the temperature range of the Williams-Landel-Ferry rubbery region, for pure solute, pure water, and the freeze-concentrated, amorphous solute-unfrozen water matrix that would exist between the Tm of ice and Tg′. (*Reproduced, with permission, from Levine and Slade, 1989c.*)

ine and Slade, 1989c), the magnitude of the WLF rubbery domain between Te (−14°C) (MacKenzie, 1977) and Tg′ (−32°C) (Schenz et al., 1984) is only 18°C. The significant implications of this fact to the texture and freezer-storage stability of frozen food products have been reviewed in detail elsewhere (Levine and Slade, 1989c, d, 1990).

Description of the time-/temperature-dependent behavior of food systems by the WLF equation requires selection of the appropriate reference Tg for any particular glass-forming material [of a given MW and extent of plasticization (Williams et al., 1955; Soesanto and Williams, 1981; Chan et al., 1986)], be it Tg for a low-moisture system (with $W < Wg'$) or Tg′ for a frozen system (with $W \geq Wg'$) (Levine and Slade, 1986, 1988b, 1989a; Roos and Karel, 1990, 1991a, d, e; Shimada et al., 1991; Slade and Levine, 1991a). For a typical, diluent-free polymer, Tg of the undercooled liquid is defined in terms of an iso-free volume state of limiting free volume (Ferry, 1980) and also approximately as an iso-viscosity state somewhere in the range from 10^{10}–10^{14} Pa s (Soesanto and Williams, 1981; Franks, 1982, 1989, 1990; Angell, 1988; Levine and Slade, 1988a; Simatos and Karel, 1988; Hofer et al., 1989; Ollett and Parker, 1990; Slade and Levine, 1991a). This iso-viscosity state refers to local, not macroscopic, viscosity. This fact constitutes a critical conceptual distinction (Slade and Levine, 1988b), as explained with regard to Fig. 3.20D.

It is interesting to consider whether the WLF equation, which predicts the dependence of the viscoelastic properties of glass-forming polymers on temperature in terms of ΔT above Tg, could be transformed to an analogous empirical equation that predicts quantitatively the dependence of polymer properties on dilution by plasticizer in terms of ΔP (for plasticizer) away from the reference Tg of the undiluted glassy polymer (Slade and Levine, 1991a). For example, one could examine the polystyrene plasticization data shown in Fig. 3.15, and calculate the extent of depression of Tg as a function of diluent concentration for different plasticizers. If the experimental temperature is maintained constant, equivalent to Tg of the undiluted polystyrene, then each extent of depression of Tg is equivalent to a ΔT. For the undiluted polymer, "ΔT above Tg" is achieved by raising the experimental temperature. For the diluted polymer, "ΔT above Tg" is achieved by keeping the experimental temperature constant and lowering Tg of the system. One could plot these "ΔT" values, achieved by plasticization, in order to transform the graph of log a_T vs. ΔT in Fig. 3.20A into a graph of log a_P vs. ΔP, by substituting the required amount of plasticizer for a given ΔT. Since the WLF equation for the effect of increased temperature on relative relaxation times is an empirical equation, it would be equally legitimate to obtain an equation for log a_P vs. ΔP.

The transformation from ΔT to ΔP would be strictly legitimate only for a system of a single monodisperse linear "polymer" (i.e., higher MW) and a single "plasticizer" (i.e., lower MW), and only if the free volume requirements for both translational and rotational mobility were the same for the "mobile segment" in the dry polymer, the diluted polymer system, and in the pure plasticizer. These conditions *might* be met in experiments in which the monomer, dimer, and low molecular weight oligomers were used as the plasticizer series for a particular linear polymer. Certainly, the conditions are not met when water is the plasticizer—not even for plasticization of small sugars, much less for starch (which is a mixture of branched and linear polymers with at least bimodal and broad MW distribution) (Slade and Levine, 1991a).

The most severe complication is that it would be essentially impossible in practice to verify experimentally the form of the WLF equation for log a_P vs. ΔP. Williams et al. (1955) were able to obtain sufficient data for so many polymers to be able to construct an empirical equation for log a_T vs. ΔT, because heat transfer is quite efficient in the solid state. It is experimentally feasible to raise the internal temperature of a sample with appropriate geometry from below Tg to well above Tg and expect that the temperature will be essentially uniform throughout the sample. Then, determination of viscosity, or self-diffusion coefficient for a reporter molecule, or other manifestation of mechanical relaxation will give usable data for construction of log a_T vs. ΔT above Tg. In contrast, mass transport is "infinitely" slow in the (glassy) solid state, and much slower in the leathery and rubbery states than can be accommodated in typical experimental time frames. Consequently, one would rarely be able to keep the experimental temperature constant and achieve ΔT values that represent uniform, steady-state moisture distribution for a sample at varying amounts of sample total moisture (Slade and Levine, 1991a). For biological high polymers, Tg of the undiluted polymer is usually so high [$\gtrsim$200°C (Atkins, 1987)] as to cause thermal degradation if this dry Tg is used as the experimental temperature (Orford et al., 1989; Noel et al., 1990; Whittam et al., 1991). If one uses experimental temperatures below dry Tg and waits for plasticized Tg to drop to and below the experimental temperature, experimental time scales will be unreasonably long (Sears and Darby, 1982; Slade and Levine, 1991a), and instantaneous moisture gradients will allow incidental chemical and physical changes to occur in the system. For small sugars with lower dry Tgs, but which are readily crystallizable [a property usually related to a large temperature interval between Tm and Tg, such that the homogeneous nucleation temperature is well above Tg (Levine and Slade, 1988a)], e.g., glucose, sucrose, and lactose, small extents of plasticization allow crystallization from the rubbery liquid state at T > Tg (Makower and Dye, 1956; White and Cakebread, 1966; Cakebread, 1969; Levine and Slade, 1988a, 1989a; Slade and Levine, 1988b, 1991a; Noel et al., 1990; Roos and Karel, 1990, 1991a, 1991e; Blanshard et al., 1991; Shimada et al., 1991), and bulk mechanical properties no longer reflect the effect of plasticizer on Tg. Possibly, the best experimental system for direct construction of log a_P vs. ΔP (to allow comparison to the indirect method of transforming log a_T vs. ΔT, based on Tg vs. ΔP data) would be a dry amorphous fructose glass plasticized by water. However, it is notoriously difficult to dry fructose syrups (Levine and Slade, 1989a), and in any event, the resulting equation could not be used to predict behavior for any other sugar.

Our feeling is that we must settle at present for a qualitative and indirect picture of log a_P vs. ΔP as a universal concept. For particular combinations of polymer and plasticizer, where the Tg vs. ΔP data are known, one can substitute, for ΔT in the WLF equation, the plasticizer concentration that produces an extent of depression of dry Tg equivalent to ΔT (Slade and Levine, 1991a).

WLF kinetics differ from Arrhenius kinetics in several important respects (Slade and Levine, 1988b; Levine and Slade, 1989c, 1990), and we have emphasized the contrast, because we believe that the qualitative differences are as influential as the quantitative differences in their impact on both experimental approach and technological significance (Levine and Slade, 1989d; Slade and Levine, 1991a). A comparison between WLF and Arrhenius kinetics begins with the recognition that the temperature dependence of microscopic relaxation parameters (including the self-diffusion coefficient, viscosity, and rotational and translational relaxation rates or times; and macroscopic processes that rely on them) changes monotonically from a steep dependence of log relaxation rate on temperature just above Tg to a shallow dependence above Tm, i.e., over a material-specific temperature range from Tg to far above Tg (Franks, 1982; Angell, 1988; Slade and Levine, 1988b; Pikal and Shah, 1990). This realization reveals two underlying diagnostic characteristics that distinguish WLF from Arrhenius kinetics (Levine and Slade, 1989d). First, the coefficient of the temperature dependence (so-called "activation energy") is defined as a constant in the expression for Arrhenius kinetics, and a plot of log relaxation rate vs. 1/T is a straight line. But the coefficient itself is temperature-dependent in the WLF expression; a plot of log relaxation rate vs. 1/T is characteristically curvilinear in the material-specific temperature range between Tg and Tm (Karel, 1989; Pikal and Shah, 1990) [as recently pointed out by Ollett and Parker (1990), who presented curvilinear plots of log η vs. 1/T for undercooled melts of fructose and glucose], approaching linearity only below Tg or above Tm (Levine and Slade, 1989c). The absolute value of the derivative of log relaxation parameter vs. 1/T increases as Tg is approached from above, and decreases abruptly to an approximately constant value as T falls below Tg, or decreases gently to an approximately constant value as T is elevated above Tm and far above Tg, where the constant value corresponds to the Arrhenius coefficient (activation energy) that characterizes the particular system and relaxation process (Levine and Slade, 1989d). The shape of the derivative profile and the temperature range over which the derivative varies are material-specific properties: typically, a range of ≳100°C for materials with a ratio of Tm/Tg ≳ 1.5, or a range of <100°C for materials with a ratio of Tm/Tg < 1.5 (Slade and Levine, 1988b). Clearly, it is the fact that the derivative varies in a material-specific and temperature-dependent fashion, rather than the particular magnitude of the derivative, that constitutes the salient feature of WLF kinetics (Levine and Slade, 1989d).

It has been noted (Slade and Levine, 1991a) that such curvilinear plots of log rate vs. 1/T (still referred to as "Arrhenius plots" by their authors) have been reported (e.g., LeBlanc et al., 1988) from studies of the storage stability (e.g., against different enzymatic activities) of various frozen food systems stored at different freezer temperatures (Duden and Scholz, 1982; LeBlanc et al., 1988; Lee et al., 1988; Baardseth and Naesset, 1989), all above the relevant Tg′ reference state of the foods in question (Levine and Slade, 1989d). Results of these studies (Duden and Scholz, 1982; LeBlanc et al., 1988; Lee et al., 1988; Baardseth and Naesset, 1989) have demonstrated that enzymatic reaction rates and corresponding rates of quality loss manifest a WLF-, rather than Arrhenius-, type temperature dependence on the ΔT of the freezer temperature above Tg′ (Levine and Slade, 1989d; Slade and Levine, 1991a). In a related vein, Lim and Reid (1991) have recently reported an excellent correlation, predicted from WLF theory and cryostabilization technology concepts (Levine and Slade, 1988c; Reid, 1990), between increasing rates of diffusion-limited deterioration reactions (e.g., model reaction systems involving enzyme hydrolysis, protein aggregation, or nonenzymatic oxidation of ascorbic acid) and increasing ΔT above Tg′. This correlation was demonstrated by results of their experimental studies of reaction kinetics in frozen aqueous model systems [formulated using a series of commercial maltodextrins with a range of previously measured Tg′ values (Levine and slade, 1986)] stored at a range of different subzero freezer temperatures. Their results also confirmed the expectation (Levine and Slade, 1988c; Reif, 1990) [based on earlier experimental results for a frozen enzyme-substrate model system (Levine and Slade, 1986)] of an essentially zero rate of reaction for each storage situation in which the Tf was below the Tg′ of the particular maltodextrin-based model system (Lim and Reid, 1991).

Second, there is no explicit reference temperature in the expression for Arrhenius kinetics, because, in fact, the implicit reference temperature is taken generically to be 0 K, regardless of the distinctive thermomechanical properties of a system, and even though Arrhenius kinetics are applicable only below Tg and above Tm (Ferry, 1980; Franks, 1982; Levine and Slade, 1988a). In contrast, the WLF equation benefits from an explicit material-specific reference temperature, which is the Tg of a component or compatible blend (Ferry, 1980). Therefore, it is critical to note that when the rate or time scale of a relaxation process can be shown to depend on a material-specific reference Tg, Arrhenius kinetics are not applicable to describe mobility transformations (time-temperature-moisture superpositions) for that process in the rubbery range from Tg to Tm, regardless of whether the average slope of the log k vs. 1/T curve can be empirically fitted by a Q_{10} = n rule and regardless of the particular magnitude of n (Levine and Slade, 1989d).

In summary, in the temperature and composition domain sufficiently above Tg, where equilibrium and steady-state thermodynamics apply, the coefficient of the temperature dependence of log relaxation rate is defined by Arrhenius kinetics to be a constant and is observed to approximate a relatively small constant value over a typical experimental range of about 20°C (Levine and Slade, 1989d). In the increasingly nonequilibrium domain of temperature and composition approaching Tg from above, the coefficient of the temperature dependence of log relaxation rate on 1/T is not a constant and increases ever more rapidly over a range of 20°C (Slade and Levine, 1988b). Typically, Arrhenius rates for aqueous systems above Tm might increase fourfold over a temperature range of 20°C (Levine and Slade, 1989d), while WLF rates near Tg would increase by 4 or 5 orders of magnitude (Franks, 1985; Slade and Levine, 1988b; Levine and Slade, 1989c). To illustrate the significance of the difference between WLF and Arrhenius kinetics, Chan et al. (1986) have noted that the dielectric relaxation behavior of amorphous glucose plasticized by water is "remarkably similar" to that of synthetic amorphous polymers in glassy and rubbery states. They showed that the rates of this mechanical relaxation process, which depends on rotational rather than translational mobility, follow the WLF equation for water-plasticized glucose mixtures in their rubbery state above Tg, but follow the Arrhenius equation for glucose-water glasses below Tg (Chan et al., 1986). Also noteworthy is Angell's (1983) pertinent observation that the temperature dependence of the transport and relaxation properties of undercooled liquid water is strikingly non-Arrhenius in the temperature range from Tm to the homogeneous nucleation temperature at −40°C (corresponding to a portion of the WLF rubbery domain shown in Fig. 3.21). This non-Arrhenius temperature dependence also typifies the case for many other viscous liquid systems that undergo restructuring processes that require the "cooperative involvement of other molecular motions" (Angell, 1983, 1988). Included in these other viscous liquid systems that exhibit non-Arrhenius behavior are concentrated aqueous solutions at subzero temperatures (Pikal and Shah, 1990), according to a suggestion by Hofer et al. (1989).

The impact of WLF behavior on the kinetics of diffusion-limited relaxation processes in water-plasticized, rubbery food polymer systems has been conceptually illustrated by the schematic curve shown in Fig. 3.22 (Levine and Slade, 1989c). Relative relaxation rates, calculated from the WLF equation with its "universal" numerical constants, demonstrate the nonlinear logarithmic relationship: for ΔT = 0, 3, 7, 11, and 21°C, corresponding relative rates would be 1, 10, 10^2, 10^3, and 10^5, respectively. These rates illustrate the five-orders-of-magnitude change, over a 20°C interval above Tg, typically shown by WLF plots, as mentioned with regard to Fig. 3.18. They are dramatically different from the rates defined by the familiar Q_{10} = 2 rule of Arrhenius kinetics for dilute solutions. As pointed out with regard to Fig. 3.20A, for Arrhenius behavior above Tm, a factor-of-10 change in relaxation rate would require a 33°C change in temperature, in comparison to a 3°C change for WLF behavior near Tg

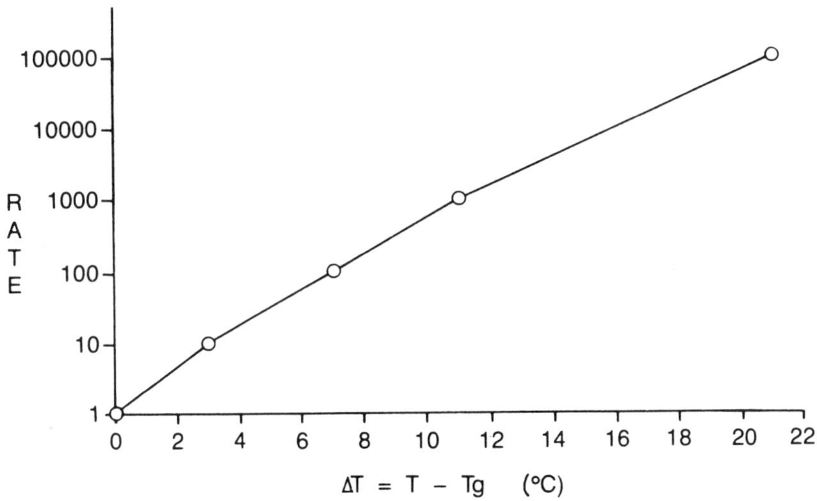

Fig. 3.22 Variation of the rate of a diffusion-limited relaxation process against $\Delta T = T - Tg$, as defined by the Williams-Landel-Ferry equation with its "universal" numerical constants of C1 = 17.44 and C2 = 51.6. (*Reproduced, with permission, from Levine and Slade, 1989c.*)

of a partially crystalline polymer of Tg/Tm = 0.67 (Slade and Levine, 1988b).

An example of WLF-governed relaxation behavior involves the kinetics of (re)crystallization (Levine and Slade, 1988a, 1989b; Slade et al., 1989). Crystallization kinetics, with regard to such common sugars as glucose, sucrose, and lactose, has been a subject with a long history of study by food scientists and technologists (Makower and Dye, 1956; White and Cakebread, 1966; Cakebread, 1969), which has seen much renewed interest in recent years (Slade and Levine, 1988b, 1991a; Levine and Slade, 1989a; Roos and Karel, 1990, 1991a; Blanshard et al., 1991; Shimada et al., 1991). Crystallization is a diffusion-limited process (Baro et al., 1977) that, on a time scale of technological significance, can only occur within the WLF rubbery domain (Wunderlich, 1976). The propagation step in the crystallization mechanism approaches a zero rate at T < Tg for an amorphous but crystallizable solute [either polymeric (Wunderlich, 1976) or monomeric], initially quenched from the melt or liquid solution state to a kinetically metastable solid state. Due to immobility in the glass, migratory diffusion of either large main-chain segments or small molecules, required for crystal growth, would be inhibited over realistic times. However, the propagation rate increases exponentially with increasing ΔT above Tg (up to Tm) (Marsh and Blanshard, 1988; Morris, 1990), due to the mobility allowed in the rubbery state. Thus, a recrystallization transition from unstable (i.e., undercooled) amorphous liquid to (partially) crystalline solid may occur at T > Tg (White and Cakebread, 1966; Karel, 1986; Phillips et al., 1986; Levine and Slade, 1988b, 1989a; Ring and Whittam, 1991), with a rate defined by the WLF equation (Levine and Slade, 1986). The facts that time-dependent recrystallization can only occur at temperatures above Tg, with a rate increasing with increasing ΔT above Tg, and manifests kinetics defined by the quantitative WLF (rather than Arrhenius) equation (Levine and Slade, 1988b) were recently confirmed in experimental studies of the recrystallization of amorphous, freeze-dried sugars (i.e., sucrose and lactose) by Roos and Karel (1990, 1991a; Shimada et al., 1991). As illustrated in Fig. 3.23 (adapted from Roos and Karel, 1990, 1991a), which combines new data for sucrose and lactose of Roos and Karel with a critical data point (i.e., the x highlighted in the oval) from the classic study of sucrose crystallization by Makower and Dye (1956), the experimental curves of the temperature dependence (in terms of ΔT above Tg) of the crystallization times for completely amorphous sucrose and lactose at low moisture contents (in the water-plasticized, rubbery liquid state at various ΔTs above Tg) fit the theoretical curve of the WLF equation (with C1 = 17.44 and C2 = 51.6), but not that of the Arrhenius equation. Other specific ex-

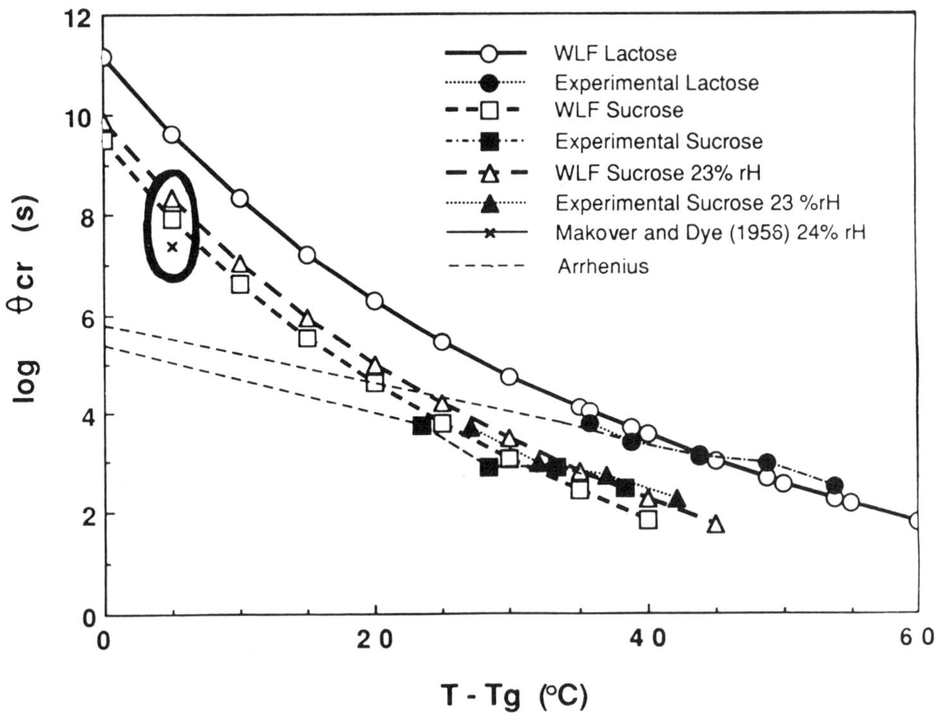

Fig. 3.23 WLF- vs. Arrhenius-type temperature dependence of the crystallization time [expressed as log θcr (sec)], as a function of $\Delta T = T - Tg$, for completely amorphous sucrose and lactose at low moisture contents. (*Adapted from Roos and Karel, 1990, 1991a.*)

amples of such a recrystallization process [i.e., a collapse phenomenon (Levine and Slade, 1988b)] include ice and solute [e.g. lactose in dairy products (Blanshard and Franks, 1987)] recrystallization in frozen aqueous systems at $T > Tr = Tg'$ (Levine and Slade, 1986).

A critical message to be distilled at this point is that the structure-property relationships of water-compatible food polymer systems are dictated by a moisture-temperature-time superposition (Starkweather, 1980; Flink, 1983; Levine and Slade, 1986). Referring to the schematic state diagram in Fig. 3.4 (which reflects the "real world" cases illustrated in Fig. 3.9 and 3.11 to 3.14) as a conceptual mobility map (which represents an extension of the dynamics map in Fig. 3.3), one sees that the Tg curve represents a boundary between nonequilibrium glassy and rubbery physical states in which various diffusion-limited processes [e.g., collapse phenomena involving mechanical and structural relaxations (Levine and Slade, 1988b)] either can (at T > Tg and W > Wg′, the high-moisture portion of the water dynamics domain corresponding to the upper-left part of Fig. 3.4, or T > Tg and W < Wg′, the low-moisture portion of the water dynamics domain corresponding to the upper-right part of Fig. 3.4) or cannot (at T < Tg, in the domain of glass dynamics corresponding to the bottom part of Fig. 3.4) occur over realistic times (Levine and Slade, 1986, 1988a; Franks, 1989, 1990). The WLF equation defines the kinetics of molecular-level relaxation processes [such as the diffusion-limited chemical and enzymatic reactions studied by Lim and Reid (1991), mentioned earlier], which will occur in practical time frames only in the rubbery state above Tg, in terms of an exponential, but non-Arrhenius, function of ΔT above this boundary condition (Levine and Slade, 1986).

As mentioned earlier and discussed in detail elsewhere (Slade and Levine, 1991a), Tg′ is the appropriate reference temperature for WLF kinetics in high-moisture food systems at temperatures above Tg′. The reference state corresponding to the glass at Tg′ can be described more completely if one recalls that Wg′ is the maximum *practical* amount of plasticizing water in such systems with water contents > Wg′ (Slade and Levine, 1991a).

Let us consider first the case of an ice-containing, solute-water system located somewhere along its liquidus curve (i.e., at some Tm > Tg′), having originated at the Tg′–Wg′ glass and undergone partial melting of ice during warming to the instant solute-unfrozen water composition (W > Wg′) of the nonice portion of the sample. For this particular case, with preexisting ice, cooling at practical rates would return the system to the Tg′–Wg′ glass (Levine and Slade, 1989c). Still, one might ask why another Tg, lower than Tg′ and corresponding to the glass with the same instant composition (W = Wg > Wg′), is not a more appropriate reference than Tg′ for describing the behavior of this system according to WLF-type kinetics. As explained elsewhere (Levine and Slade, 1989c), we consider this lower Tg to be an artificial, rather than pertinent, reference state for the ice-containing system under discussion, because the system would not have come from this lower Tg upon heating, nor would it return to this lower Tg upon cooling at practical rates.

For this reason, we consider the Tg′–Wg′ glass as *the* "practical reference state" for the above ice-containing system at Tm > Tg′ and W > Wg′ (Slade and Levine, 1991a). [An experimental validation of this point, with regard to the behavior of rubbery frozen food products during storage, has been described elsewhere (Slade and Levine, 1991a).] But what about a different case of (1) a system that is neither nucleated nor ice-containing, perhaps situated only just above its liquidus curve [i.e., at

T(c) > Tm(c)], and (2) any desired, hypothetical cooling and warming rates, perhaps supplemented by any desired manipulation of pressure in the kilobar range? Then we may consider a solute-water solution instantaneously *captured* in an unstable, ice-free condition at some point along an imaginary vertical line between its Tm and Tg. The captivation might be accomplished by (1) extremely rapid cooling to T < Tg, followed by extremely rapid heating to Tg < T < Tm, or (2) cooling to Tm > T > the homogeneous nucleation temperature (Th) > Tg, and prior to homogeneous nucleation upon further cooling to Tm > Th > T > Tg, or by cooling at 2 kilobar pressure to Tg < T < Tm. This Tg below Tg′ would then be a candidate for consideration as a WLF reference state, albeit not a practical one (Slade and Levine, 1991a).

As discussed earlier, we make a critical distinction between water-compatible and water-sensitive solutes, with regard to the temperature dependence of their practical (i.e., effective), solute-water glass curves. Because of the much more limited thermodynamic compatibility of water-sensitive solutes (e.g., nylons, PVAc, lignin) for solvating/plasticizing water, their practical glass curves are depressed much less by increasing water content and remain well above 0°C, no matter how much water is added to such a solute. The solubility parameter (Billmeyer, 1984) of water is too high at room temperature to make water an effective plasticizer of nylon (Starkweather, 1980); it decreases, and water becomes a more compatible plasticizer of nylon as T is increased above room temperature; it increases further, and water becomes incapable of plasticizing nylon as T is decreased. If water were added to nylon at a sufficiently high temperature and then quench-cooled at a sufficiently rapid rate, a lower Tg than the effective minimum value at about 6.5 wt% water (Starkweather, 1980) could be demonstrated. However, Tg of a water-sensitive solute is *never* capable of being depressed enough, so that, upon cooling at practical rates in the presence of excess water, it would reach a subzero Tg′ (and corresponding Wg′) that would be characteristic of, and specific for, each given solute-water system (Levine and Slade, 1988a). In contrast, the practical glass curve for a water-compatible solute can be depressed much more by increasing water content. Consequently, Tg of a water-compatible solute is *always* capable of being depressed enough, so that, upon cooling at practical rates in the presence of excess water, it reaches a subzero Tg′ (and corresponding Wg′) that is characteristic of, and specific for, each given solute-water system (Levine and Slade, 1988a).

An important, related symptom of this greater extent of solute-water compatibility is that Tg′ is, by definition, equal to the corresponding solute-water Tm at Wg′ (Franks, 1985; Levine and Slade, 1988a, 1989c; Roos and Karel, 1991c, d, e), so that only the amount of water in excess of this

Wg′ amount will be freezable (in a practical time frame) upon cooling at practical rates (Levine and Slade, 1986). Thus, Wg′ represents the maximum *practical* amount of plasticizing water (= ''unfreezable'' water) for a given solute, and amounts of water in excess of Wg′ would not be additionally plasticizing under practical conditions (Franks, 1985; Levine and Slade, 1986, 1988a; Slade and Levine, 1988b). [Recall that we distinguish, for this reason and based on grounds of what is practical and relevant to ice-containing, frozen aqueous systems, between the *practical* glass curve (which levels off at Tg′ for water contents > Wg′) and the *complete* glass curve (which decreases smoothly and continuously from the solute's dry Tg to Tg of pure amorphous water at −135°C) for a water-compatible solute, as illustrated earlier in Fig. 3.13.] It is significant to note that the equivalence of Wg′, as determined by DSC, with the maximum amount of highly immobile (''unfreezable'') water, as estimated by NMR, and with the maximum amount of plasticizing water that is capable of affecting the sensory textural properties of a food product has been demonstrated recently by Given (1991) from the combined NMR, DSC, and sensory analysis results of a study on a baked flour-water cracker model system. It is also noteworthy that the Wg′ of ≈27 wt% for completely amorphous, gelatinized starch (Slade, 1984; Slade and Levine, 1984, 1987b) and the water content of 27 wt% for B-type crystalline starch (Imberty and Perez, 1988) are in good agreement with the value of 26 wt% determined for the maximum water content ''in [B-type] starch granules . . . , when [native, partially crystalline, partially amorphous] starch granules are in equilibrium with excess water [at room temperature], regardless of the amount of water present'' (Kuge and Kitamura, 1985). Wg′ represents the maximum amount of water that can coexist with all of the solute, under practical conditions *at any temperature* from T < the vaporization temperature (Tvap) to T = 0 K, in a homogeneous, single phase (liquid or metastable amorphous solid) capable of cooperative, molecular mobility (at T > Tg′) (Slade and Levine, 1988b). In contrast, amounts of water in excess of Wg′ would exist in a separate, water-rich phase, either as pure ice at T < Tm, or as dilute solution at Tm < T < Tvap (Slade and Levine, 1988b). Thus, in the first case described above, the amount of water in the system that had arisen from partial melting of ice would not be solute-plasticizing water (even though it would obviously be solute-diluting and viscosity-lowering water), and so would not alter the appropriateness of Tg′ as the Tg reference state for WLF-type kinetics (Slade and Levine, 1991a).

In many critical aspects, the behavior of a water-compatible solute at a water content > Wg′ would be analogous to that of a synthetic high polymer in the presence of an excess amount (say, >30 wt%, where Dg′

= 30 wt%) of an organic plasticizing diluent (D) that can crystallize (as shown in Fig. 3.16) (Slade and Levine, 1991a). For example, if one cooled the latter system at a sufficient rate, one *could* capture a metastable, homogeneous glass with a diluent content > 30 wt% and maintain this glassy state at temperatures below the Tg of the 70− wt% polymer/30+ wt% diluent mixture. However, if one then raised the temperature to T > Tg, under conditions such that some of the diluent could and would crystallize after devitrification, then Tg′ of the remaining 70 wt% polymer/ 30 wt% diluent mixture would become the appropriate reference Tg. If the excess amount of diluent (i.e., 30+ wt% − Dg′), which had crystallized on warming, were permitted to melt on further warming to T > Tg′, this excess diluent would exist as a separate, diluent-rich phase, perturbed by the solute, but *not* plasticizing the solute (Slade and Levine, 1991a).

We have suggested that, for water-compatible solutes at low moisture contents (0 < W < Wg′), the portion of the practical glass curve between Tg of the dry solute and Tg′ serves to define a continuum of real reference states (corresponding to a continuum of practical solute-water compositions) for WLF-type kinetics (Levine and Slade, 1988a; Slade and Levine, 1988b). This suggestion has been recently validated by the results of experimental studies of the crystallization kinetics of amorphous sugars at low moisture contents by Roos and Karel (1990, 1991a; Shimada et al., 1991). In the region of the state diagram surrounding this portion of the glass curve, Tg is both defined *and* accessible on a 200-sec time scale (Angell, 1988; Johari et al., 1989), because no phase separation of water can occur, so the solute-water composition of the system remains constant and homogeneous (Slade and Levine, 1988b). In contrast, in the region of the state diagram (lower left quadrant) surrounding the portion of the glass curve between Tg′ and Tg of water, Tg is still *defined* on a 200-sec time scale, but is only accessible on a time scale many orders of magnitude smaller (Franks, 1982); otherwise, water phase-separates as ice and the remainder of the system arrives at Tg′–Wg′ (Levine and Slade, 1989c). Thus, in practice, Tg values in this region are neither real nor relevant to questions about the kinetics of diffusion-limited processes (Levine and Slade, 1989c, d). Unfortunately, it is in this same region of the state diagram, for higher-moisture situations (W > Wg′) for the diluted solute, that our interpretation of the applicable form of the WLF equation and the appropriate reference state for its use has been questioned (Simatos et al., 1989). Based on (1) the explicit description, in Chapter 17 of Ferry's 1980 book, of the variation of the WLF coefficients (C1 and C2) with increasing diluent concentration, for plasticizers that are both (a) noncrystallizing, and (b) thermodynamically compatible over the entire temperature range of dilution, and (2) our analysis (Slade and Levine, 1988b) of (a) the effect of temperature-dependent thermodynamic compatibility of the plasticizer, and (b) the effect of the variation of the magnitude of the WLF region (Tm–Tg) for crystallizing plasticizers on the WLF coefficients, we have suggested that the proper coefficients of the WLF equation must *change* with changing diluent concentration (Slade and Levine, 1988b). When the plasticizer is noncrystallizing and thermodynamically compatible over the entire temperature range, (1) the change in coefficients is continuous from 0 to 100% plasticizer, (2) the total diluent concentration defines the concentration of the plasticizer in the glassy solute-diluent blend, and (3) the Tg of the glassy solute-diluent blend is *the* reference temperature for the WLF equation with appropriate coefficients for that extent of dilution (Ferry, 1980). Most of the literature on the free-volume description of WLF kinetics deals with either undiluted synthetic polymers or dilution by such well-behaved, noncrystallizing, thermodynamically compatible plasticizers (Ferry, 1980). In contrast, when the plasticizer becomes ever less thermodynamically compatible with decreasing temperature, as for tricresyl phosphate with poly(vinyl chloride) (Sears and Darby, 1982), or in the extreme case, when the diluent, such as water, is capable of complete phase separation via crystallization, (1) the change in coefficients is only continuous from 0% diluent to the region of diluent concentration-temperature where incompatibility or phase separation is exhibited (i.e., Wg′–Tg′ for water), (2) the total diluent concentration does *not* define the concentration of plasticizer in the glassy, plasticizer-plasticized solute blend, and (3) the minimum reference Tg is that of the glass with maximum *effective* plasticizer content (Slade and Levine, 1991a). This complicated behavior, which is the rule for aqueous systems, has not been extensively described in the literature. In Chapters 11, 12, and 17, Ferry (1980) provides the bare minimum of information required to deal with systems of low molecular weight, plasticizers of marginal compatibility, and crystallizing diluents. We have expanded his analysis by demonstrating that the Tm/Tg ratio or Tm–Tg for the diluent-free solute can be used to provide normalizing guidelines for the identification of appropriate WLF coefficients (Slade and Levine, 1988b).

Another salient feature of the behavior of water-compatible solutes, at water contents both above and below Wg′, which we believe emphasizes the singularity of the Tg′–Wg′ reference state/point, involves sorption isotherm data that we have treated in terms of iso-RVP contours for glassy, rubbery, and supra-glassy substrates, as shown in Fig. 3.24 (Slade and Levine, 1991a). RVP goes from zero for a glassy substrate at temperatures and moisture contents below its effective Tg to 1 for the corresponding liquid substrate at temperatures higher than those corresponding to its rubbery state (i.e., temperatures corresponding to its supra-

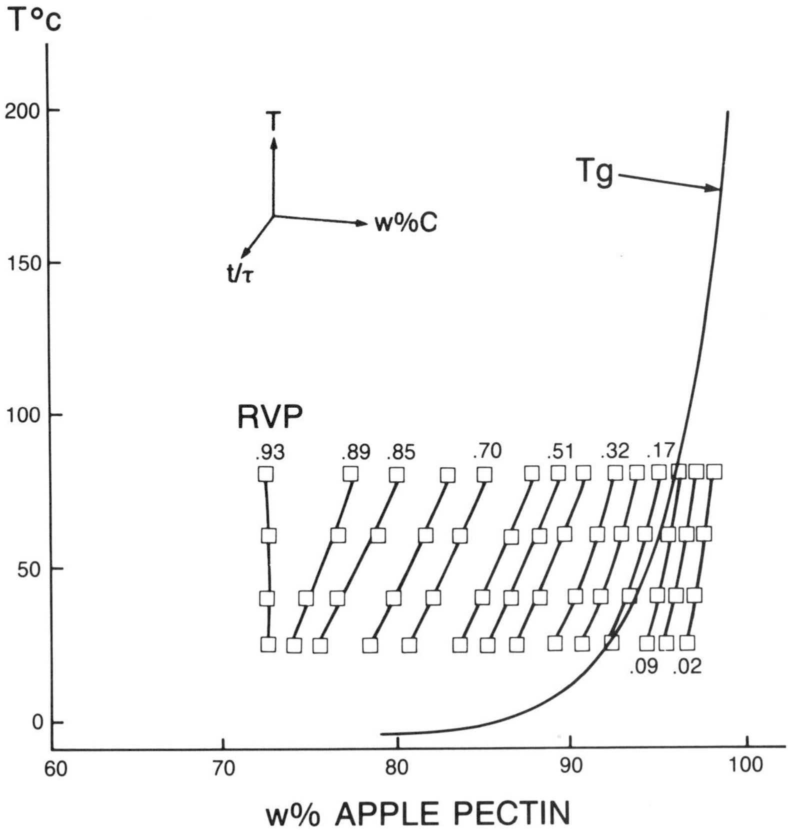

Fig. 3.24 Ad/absorption isotherm data for apple pectin (data from Weisser, 1985) transformed into a two-dimensional water/glass dynamics map of temperature vs. wt% solids, on which are compared the relative locations of a series of iso-RVP contours and a schematic, "practical" glass curve for pectin-water, based on data for hemicellulose (*from Kelley et al.*, 1987). (*Reproduced, with permission, from Slade and Levine, 1991a.*)

glassy state). However, while at temperatures (T > Tg) and moisture contents (W > Wg) corresponding to the rubbery region, RVP increases significantly with increasing water content, and the iso-RVP contours [like iso-viscosity contours (Slade and Levine, 1991a)] essentially parallel the shape of the solute-water glass curve; at temperatures > Tg′ and water contents > Wg′, RVP approaches 1, and the iso-RVP contours become vertical, i.e., lose their temperature dependence. We have interpreted this as evidence for the fact that the substrate would exist under these

T-W conditions in a supra-glassy, liquid state well above the relevant reference state, Tg′–Wg′. Importantly, if sorption were to continue and, thus, the water content of the substrate were to continue to increase to higher and higher levels above Wg′ (a situation analogous in certain vital respects to that of the partially melted, frozen aqueous system described earlier), the resultant changes in RVP, caused by this additional amount of sorbed water, would become insignificant (Slade and Levine, 1991a).

The focal point of our interpretations regarding Tg′ as the appropriate reference state for WLF-type kinetics in high-moisture systems of water-compatible solutes (Levine and Slade, 1988c) concerns the singularity of the Tg′–Wg′ point for a given solute (Levine and Slade, 1988b). For example, were it not for this singularity, how could one explain the frequently anomalous shape of the nonequilibrium extension of the equilibrium liquidus curve (Franks, 1982)? We have suggested that the Tg′–Wg′ point is not one of *accidental* intersection of the liquidus and glass curves, but rather the critical point to which the nonequilibrium portion of the liquidus curve *must* fall (Slade and Levine, 1988b). We have also noted that the shape of the liquidus curve is also determined indirectly and in part by the shape of the underlying homogeneous nucleation curve (Slade and Levine, 1988b). In the region around the Th curve, ice crystal formation (i.e., nucleation) in a dilute solution (W > Wg′) at a given T is influenced, in part, by the portion of the glass curve below the Th curve (i.e., by Tg values < Tg′). However, ice crystal growth (i.e., propagation) in that solution at a given T (and the kinetics thereof) is controlled by Tg′ (Levine and Slade, 1988c). Once the homogeneous Tg′–Wg′ glass has been formed as a consequence of maximal freezing and freeze-concentration, one would not expect that subsequent dilution of this glass (via partial ice melting at Tm > T > Tg′) would occur *uniformly* to produce a new supra-glassy, cooperative system (Slade and Levine, 1991a).

Information in Chapters 11, 12, and 17 of Ferry's 1980 book, on the effects of free volume and local viscosity in concentrated, plasticized polymers, has helped to provide a basis for understanding what determines the location of Tg′ (and corresponding Wg′) for a given solute. For synthetic high polymer-organic diluent systems, Ferry teaches that different factors/mechanisms are operative/predominant for different portions of the solute-diluent glass curve, i.e., free volume at low extents of solute dilution vs. local viscosity at lower solute concentrations (e.g., near Dg′). The simple WLF free volume theoretical approach based on the ΔT above Tg (Ferry, 1980, Chapter 11) is only applicable to the former case, i.e., low extents of solute dilution. However, our experimental results for frozen aqueous solutions of monodisperse PHCs have suggested that, in the maximally freeze-concentrated solute-UFW glass, Tg′ is *not* dependent on/determined by free volume, as evidenced by the lack of correlation between Tg′ and $1/\overline{M}n$ (of the solute-UFW) glass), but rather is dependent on/determined by local viscosity, as evidenced by the excellent linear correlation between Tg′ and $1/\overline{M}w$ (Slade and Levine, 1988b). Thus, the simple WLF free volume treatment is evidently not applicable in a straightforward way to frozen food systems, and one must instead consider the solute-specific, maximum water content of the glass (Wg′ at Tg′) and base the treatment of kinetics on this singular Tg′–Wg′ reference state (Slade and Levine, 1991a).

For the reasons described above, we have concluded that, for water-compatible solutes (monomeric, oligomeric, and polymeric, alike), at water contents ≥ Wg′, Tg′ is the appropriate Tg corresponding to the reference state for WLF-type kinetics in the rubbery liquid region at T > Tg′ (Levine and Slade, 1988c), and to the temperature boundary for Arrhenius kinetics in the glassy solid region at T < Tg′ (Levine and Slade, 1988a). In contrast to some imaginary, artificial, or purely theoretical reference state, Tg′ is the *real, practical* reference state (and Tg′–Wg′ is the reference point on the state diagram) to which such a solute-water system goes upon cooling at rates such that ice formation occurs at some T < Tm but > Tg′, and from which it subsequently comes upon heating at rates such that the melting of preexisting ice occurs in the temperature range between Tg′ and Tm (Levine and Slade, 1989c, d). This is what we mean by real and practical—cooling rates that permit ice formation, and warming rates that permit ice melting, in the rubbery region between Tg′ and Tm—and why our interpretations have been focused on this Tg′ reference state (Slade and Levine, 1991a). It should be recognized that this interpretation goes beyond the WLF theory for diluent-free, synthetic high polymers described in Chapter 11 of Ferry, 1980, and relies more on the material in Chapter 17 on concentrated, plasticized polymers.

As described earlier, there is a growing appreciation of the fact that the existence of "bound water" is a myth. However, the preceding explanation of why Tg′ is a pivotal reference temperature for descriptions of WLF behavior depends on a supplementary appreciation of four underlying precepts (Slade and Levine, 1991a): (1) rationale for selection of the particular state diagram used as a dynamics map; (2) definition of plasticizing water, in the context of the two primary requirements for an effective plasticizer; (3) identification of Wg′ as the maximum amount of plasticizing water in a practical, operational time scale; and (4) reference to melt-dilution as the effect, at temperatures above Tg′, of amounts of water in excess of Wg′. As a result of previous, incomplete discussions of these points (Levine and Slade, 1988c, 1989c), possible philosophical riddles might appear to arise, as stated by D. Reid (personal communication, 1990):

> If there is plasticizing water, and non-plasticizing, melt-diluting water, could not one equate plasticizing water with "bound water?" In what ways would plasticizing water be different from the mythical "bound water?" On the other hand, if all of the water is considered a uniform species, with no distinction between plasticizing water and melt-diluting water above Tg′, then the riddle reverts to the original question—is Tg′ the correct temperature to use as a reference for WLF kinetics above Tg′?

For considering the effects of changes in temperature or water content in the region of water content less than Wg′, there is, as alluded to earlier, a general consensus that the appropriate reference locations are the continuous set of pairs of solute-specific Tg–Wg values (such that Tg > Tg′, but Wg < Wg′) on a particular state diagram. This state diagram is a particular two-dimensional (temperature vs. water content, expressed as weight fraction or %) "snapshot" in time, which contains both equilibrium and nonequilibrium states and serves as a dynamics or mobility map (Slade and Levine, 1988b). The unique selection of this "snapshot," with its relevant set of Tg–Wg values, reflects the choice of 200 sec by polymer scientists as the time scale for the conventional definition of the glass transition (Angell, 1988; Johari et al., 1989). In terms of a DSC experiment, the glass transition temperature is defined by convention as the midpoint of the temperature range (corresponding to a time interval of 200 sec) from the onset to the completion of the relaxation that accounts for the increase in free volume that is manifested as a step-increase (discontinuity) in heat capacity (with a concomitant increase in expansion coefficient) (Wunderlich, 1981, 1990). For example, at a heating rate of 10 K/min, the glass-to-rubber relaxation is observed as a step-increase in heat capacity over a temperature range of about 33 K; at a heating rate of 1 K/min, the midpoint of the transition is expected to occur about 3 K lower, with the temperature range decreased to about 3 K (Levine and Slade, 1988a). The curved line segment, containing the reference set of Tg-Wg values on the two-dimensional map, represents a single contour line from a three-dimensional surface, at a value of 200 sec for the third (time) dimension (Slade and Levine, 1991a). This 200-sec time scale refers only to the convention used to specify a reference value of Tg; it should not be confused with the operational time scale that is relevant to the use of the dynamics map for the diagnosis of mobility. For practical application, the third dimension of the mobility map is expressed as a temporal ratio, rather than as time itself (Slade and Levine, 1988b). This operational time scale is defined as the ratio of experimental time duration (or inverse experimental frequency) to the relaxation time of the relevant underlying process, at specified conditions of temperature and moisture content. A "practical operational time scale" is any time scale for which the quotient "experimental time scale/relaxation time of the relevant process in the glassy state" approaches a value sufficiently near zero as to be "practicably" zero (Slade and Levine, 1991a).

As noted earlier, in all such discussions of the dynamics map, "mobility" refers specifically to cooperative segmental (or unit) motions, which are the underlying basis of all Tg-dependent relaxation behavior (also referred to as alpha relaxations) (Levine and Slade, 1988a), with respect to the ratios of allowed [by the local environment, where "local" refers to dimensions greater than about 10 nm (Wunderlich, 1981), but smaller than bulk sample dimensions] to required (by the size and shape of the mobile unit) free volume, local viscosity, and time (Slade and Levine, 1988b). Thus, (1) inherent requirements of free volume, local viscosity, and time for cooperative rotational or translational motions and (2) method-related requirements of sufficient mobility for detection determine the observed relationship and location of the operational values of measured transition temperatures of polymer-plasticizer blends. In contrast, small-scale motions [also referred to as beta and gamma relaxations, such as noncooperative vibrations and rotations (Levine and Slade, 1988a)] are not kinetically constrained with the same non-Arrhenius temperature dependence as Tg-dependent relaxations (Levine and Slade, 1988a) and, thus, are not diagnosed by the dynamics map. From a recent study of tracer diffusion at the glass transition, Ehlich and Sillescu (1990) have suggested that coupling of the diffusional motion of a tracer and the glass-to-rubber relaxation of a polymeric matrix "will increase with the size of the diffusant, which finally becomes a probe for monitoring the glass transition process."

Reid's philosophical question relates to the region of water content greater than or equivalent to Wg′ and the designation of Tg′–Wg′ as the appropriate reference location on the dynamics map. As already noted, we have suggested that water contents greater than Wg′ are not effective to provide further plasticization at temperatures below Tg′, due to phase separation of ice, so that Tg′–Wg′ is the glass with the maximum content of plasticizing water in a practical time scale (Levine and Slade, 1986; Slade and Levine, 1988b), and water contents greater than Wg′ lead to melt-dilution at temperatures above Tg′ (Levine and Slade, 1989c). Pure water per se exists in only two relevant circumstances: complete absence of solute, or occurrence of phase separation of ice in the presence of solute. Clearly, "plasticizing water" refers to the aqueous component of a water-solute blend in the nonice portion of a sample and should not be confused with pure water as a separate entity (Slade and Levine, 1991a).

Since there is a practical limit to the amount of plasticizing water, such that water content up to Wg′ is plasticizing, but water content greater than Wg′ is melt-diluting and nonplasticizing, the question has been raised as to how this newer description differs from the older and unacceptable description in terms of "bound" and "free" water (Slade and Levine, 1991a). The question requires an answer to prevent the equivalence of "bound water" to plasticizing water and "free water" to melt-diluting water. The primary distinction between the two descriptions is that the use of "bound" and "free" implies that there are two *types* of water, suggesting that they could be distinguished either chemically or energetically, a conclusion that is clearly contradicted by experimental observations (Franks, 1983a, b; Franks and Grigera, 1990). In contrast, plasticizing and nonplasticizing refer to the temperature- and time-dependent, operational conditions that result from different *amounts* of water, not different *types* of water. In short, "free" and "bound" unacceptably suggest two types of water that differ energetically, in contradiction to experimental observation. Plasticizing and nonplasticizing describe two regimes of operational conditions that have been observed experimentally, as discussed elsewhere (Slade and Levine, 1991a) in the context of Fig. 3.24.

Moreover, the newer description is in accord with the fact that water, outside of an ice or other crystalline lattice, is considered a uniform species, with no existence of distinguishable types (Franks, 1982). Then the question arises as to why some amounts of water could be plasticizing, and other amounts (greater than Wg′) be nonplasticizing, if water is a uniform species. The answer is based on an understanding that there are two necessary conditions for a diluent to act as an effective plasticizer: (1) thermodynamic compatibility over the entire temperature range between Tg of the pure solute and Tg of the pure diluent; and (2) the ability of the diluent to increase the free volume and decrease the local effective viscosity of the blend, compared to that of the solute alone at the same temperature (Slade and Levine, 1991a). The first condition is typically determined as a solubility parameter (a measure of thermodynamic compatibility) for polymers (Billmeyer, 1984). Thermodynamic compatibility refers to both equilibrium thermodynamic compatibility (temperature-dependent, but time-independent) and nonequilibrium thermodynamic compatibility (time- and temperature-dependent). The most effective plasticizer would not require kinetic constraint to maintain a spatially homogeneous distribution of solute and diluent. The second condition requires that, in the context of translational mobility, Tg of the diluent be less than Tg of the solute, which is usually equivalent to the linear DP of the diluent being less than the linear DP of the solute. As a consequence, the free volume (required for translational motion) of the diluent alone > that of the compatible blend > that of the solute alone at the same experimental temperature (Slade and Levine, 1991a). This understanding of the necessary conditions for effective plasticization also supports the nomination of Tg′–Wg′ as the practical reference state for systems with T > Tg′ and W > Wg′. For practical time scales, the time- and temperature-dependence of the thermodynamic compatibility of water with common food materials, such as PHCs, precludes its use as an effective plasticizer at T < Tg′. Only in shorter time scales compared to the effective relaxation time of the system (i.e., at small temperature intervals above Tg) would the family of Tg–Wg values at Wg > Wg′ and Tg < Tg′ be appropriate reference locations (Slade and Levine, 1991a). Such an appropriate use of this part of the complete glass curve as a reference contour would be the explanation of the control of ice nucleation and initial growth of critical nuclei as a function of solute concentration or pressure, before maximum freeze-concentration occurs (Slade and Levine, 1988b). In contrast, the sorption behavior mentioned earlier (Fig. 3.24) demands use of Tg′–Wg′ as the primary reference location. In a typical experimental time scale, a dramatic dependence of the apparent RVP on the underlying glass curve is already observed as a strong temperature dependence at constant moisture content for values of apparent RVP < 0.9. This condition exists at T > Tg′, when Wg′ > W > Wg. When W falls below Wg, the apparent RVP approaches zero, and the strong temperature dependence at constant W is no longer observed. When W rises above Wg′, the apparent RVP approaches 1, and again the strong temperature dependence at constant W is no longer observed (Slade and Levine, 1991a).

Crystallization/Gelation Mechanism

A classical description of crystallization as a three-step mechanism has been widely used for partially crystalline synthetic polymers crystallized, from the melt or concentrated solution, by undercooling from T > Tm to Tg < T < Tm (Jolley, 1970; Wunderlich, 1976). The mechanism is conceptually compatible with the "fringed micelle" model (Hiltner and Baer, 1986). It involves the following sequential steps, which apply universally to all crystallizable substances, regardless of MW (Wunderlich, 1976): (1) nucleation (homogeneous)—formation of critical nuclei; (2) propagation—growth of crystals from nuclei by intermolecular association; and (3) maturation—crystal perfection (by annealing of metastable microcrystallites) and/or continued slow growth (via "Ostwald ripening"). Within this universal description, flexible macromolecules are distinguished from small molecules by the possibility of nucleation by intramolecular initiation of ordered (e.g., helical) chain segments and propagation by association of chain segments for the high polymers (Wunderlich, 1976).

The thermoreversible gelation, from concentrated solution, of a number of crystallizable synthetic homopolymers and copolymers has been reported to occur by this crystallization mechanism (Boyer et al., 1985; Domszy et al., 1986; Hiltner and Baer, 1986; Mandelkern, 1986). In contrast, a different gelation mechanism, not involving crystallization and concomitant thermoreversibility, pertains to polymers in solution that remain completely amorphous in the gel state (Slade et al., 1989). Such high polymers are distinguished from oligomers by their capacity for intermolecular entanglement coupling, resulting in the formation of rubberlike viscoelastic random networks [called gels, in accord with Flory's (1974) nomenclature for disordered three-dimensional networks formed by physical aggregation] above a critical polymer concentration (Ferry, 1980). It has been demonstrated for many synthetic linear high polymers that the mechanism of gelation from concentrated solution can be distinguished by a simple dilution test (Burchard, 1988). Gelation due to entanglement in a completely amorphous polymer-diluent network can be reversed by dilution, whereas a thermoreversible, partially crystalline, polymer-diluent network gel cannot be dispersed by dilution (Levine and Slade, 1988a). Examples of food polymers that can form such amorphous entanglement gels include gluten in unoriented wheat flour dough, sodium caseinate in imitation mozzarella cheese, and casein in real cheese (Slade et al., 1989). As summarized by Mitchell (1980), "entanglement coupling is seen in most high MW polymer systems. Entanglements [in completely amorphous gels] behave as crosslinks with short lifetimes. They are believed to be topological in origin rather than involving chemical bonds." Importantly, hydrogen bonding need not be invoked to explain the viscoelastic behavior of completely amorphous gels formed from solutions of entangling polysaccharides or proteins (Slade et al., 1989).

The gelation-via-crystallization process [described as a nucleation-limited growth process (Domszy et al., 1986)] produces a metastable three-dimensional network (Mandelkern, 1986) crosslinked by "fringed micellar" (Hiltner and Baer, 1986) or chain-folded lamellar (Domszy et al., 1986) microcrystalline junction zones composed of intermolecularly associated, helical chain segments (Boyer et al., 1985). Such partially crystalline gel networks may also contain random interchain entanglements in their amorphous regions (Domszy et al., 1986). The nonequilibrium nature of the process is manifested by "well known aging phenomena" (Hiltner and Baer, 1986) [i.e, maturation (Levine and Slade, 1988a)], attributed to time-dependent crystallization processes that occur subsequent to initial gelation. The thermoreversibility of such gels is explained in terms of a crystallization (on undercooling) $\leftrightarrow$ melting (on heating to $T > Tm$) process (Domszy et al., 1986). Only recently has it been recognized that for synthetic polymer-organic diluent systems (e.g., polystyrene-toluene), such gels are not glasses (Blum and Nagara, 1986) ["gelation is not the glass transition of highly plasticized polymer" (Hiltner and Baer, 1986)] but partially crystalline rubbers (Boyer et al., 1985), in which the mobility of the diluent (in terms of rotational and translational motion) is not significantly restricted by the gel structure (Blum and Nagara, 1986). Similarly, for starch and gelatin gels, water as the diluent is highly mobile, and amounts $> Wg'$ freeze readily at subzero temperatures (Levine and Slade, 1989b). The temperature of gelation (Tgel) is above Tg (Blum and Nagara, 1986), in the rubbery fluid range up to about 100°C above Tg. Tgel is related to the flow relaxation temperature, Tfr, observed in flow relaxation of rigid amorphous entangled polymers (Boyer et al., 1985) and to Tm observed in melts of partially crystalline polymers (Hiltner and Baer, 1986). The basis for the MW-dependence of Tgel has been identified (Domszy et al., 1986) as an iso-viscous state (which may include the existence of interchain entanglements) of $\eta gel/\eta g = 10^5/10^{12} = 1/10^7$, where ηg at $Tg \simeq 10^{12}$ Pa s.

The distinction among these transition temperatures becomes especially important for elucidating how the morphology and structure of food polymer systems relate to their thermal and mechanical behavior (Slade et al., 1989). This distinction is a particularly important consideration when experimental methods involve very different time frames (e.g., mechanical measurements during compression tests or over prolonged storage; thermal analysis at scanning rates varying over four orders of magnitude; relaxation times from experiments at acoustic, microwave, or NMR frequencies) and sample preparation histories (i.e., temperature, concentration, time) (Slade et al., 1989). In the case of morphologically homogeneous, molecular amorphous solids, Tg corresponds to the limiting relaxation temperature for mobile polymer backbone-chain segments. In the case of morphologically heterogeneous, supramolecular networks, the effective "network Tg" (Slade et al., 1989; Levine and Slade, 1989b, 1991; Slade and Levine, 1991a, b) corresponds to the Tfr transition above Tg for flow relaxation (Mandelkern, 1986) of the network. For example, the ratio of Tfr/Tg varies with MW from 1.02 to 1.20 for polystyrene above its entanglement MW (Keinath and Boyer, 1981). Tfr defines an iso-viscous state of 10^5 Pa s for entanglement networks (corresponding to Tgel for partially crystalline networks) (Mandelkern, 1986). Tgel of a partially crystalline network would always be observed at or above Tfr (≡ "network Tg") of an entanglement network; both transitions occur above Tg, with an analogous influence of MW and plasticizing water (Slade et al., 1989). As an example, the effective "network Tg" responsible for mechanical firmness of freshly baked bread would be near room temperature for low extents of network formation, well above room temperature for mature networks, and equivalent to Tgel near 60°C for staled bread, even though the underlying Tg for segmental motion, responsible for the predominant second-order thermal transition, remains below 0°C at Tg′ (Slade et al., 1989; Levine and Slade, 1989b, 1991; Slade and Levine, 1991a, b).

Curiously, it has been well-established for a much longer time (Domszy et al., 1986) that the same three-step polymer crystallization mechanism describes the gelation mechanism for the classical gelling system, gelatin-water (Jolley, 1970; Wunderlich, 1980). The fact that the resulting partially crystalline gels (Marshall and Petrie, 1980) can be modeled by the "fringed micelle" structure is also widely recognized (Jolley, 1970; Wunderlich, 1976; Slade and Levine, 1984, 1987a). However, while the same facts are true with regard to the aqueous gelation of starch [i.e., retrogradation, a thermoreversible gelation-via-crystallization process that follows gelatinization and "pasting" of partially crystalline native granular starch-water mixtures (Slade and Levine, 1987b)], and despite the established importance of gelatinization to the rheological properties of bread doughs during baking (Levine and Slade, 1989b) and of retrogradation to the time-dependent texture of fresh-baked vs. aged breads (Slade and Levine, 1987b), recognition of starch (or pure amylose or amylopectin) retrogradation as a thermoreversible polymer crystallization process has been much more recent and less widespread (Table 3.13A). Blanshard (1988; Marsh and Blanshard, 1988) has applied synthetic polymer crystallization theory to investigate the kinetics of starch recrystallization and thereby gain insight into the time-dependent textural changes (i.e., staling due to firming) occurring in baked products such as bread. Similarly, Zeleznak and Hoseney (1987b) have applied principles of polymer crystallization to the interpretation of results on annealing of retrograded starch during aging of bread stored at superambient temperatures. Many of the persuasive early insights in this area have resulted from the food polymer science approach to structure-property relationships in starch of Slade and her various coworkers (Table 3.13B).

Slade et al. (see Table 3.13B) have used DSC results to demonstrate

Table 3.13 Recent Publications That Describe a Polymer Science Approach to Studies of Starch Gelatinization and/or Retrogradation

A. Retrogradation as a polymer crystallization process: Miles et al. (1985a, b), Ring (1985a, b), Ablett et al. (1986), Blanshard (1986), Ring and Orford (1986), Ring et al. (1987), Russell (1987b), I'Anson et al. (1988), Mestres et al. (1988), Zobel (1988), Clark et al. (1989), Gidley (1989), Gidley and Bulpin (1989), Lund (1989), Russell and Oliver (1989), Morris (1990), Shukla (1991)

B. Slade and coworkers: Slade (1984), Slade and Levine (1984, 1987b,1988b, c, d, 1989), Maurice et al. (1985), Biliaderis et al. (1985, 1986a), Levine and Slade (1989b)

C. Gelatinization as a nonequilibrium polymer melting process: Kuge and Kitamura (1985), Reid and Charoenrein (1985), Blanshard (1986, 1987, 1988), Yost and Hoseney (1986), Burros et al. (1987), Chungcharoen and Lund (1987), Paton (1987), Russell (1987a), Zobel (1988), Zobel et al. (1988), Lund (1989), Russell and Oliver (1989), Shi and Seib (1989), Knutson (1990), Morris (1990), Kalichevsky et al. (1991), Liu and Lelievre (1991), Shukla (1991)

D. Inappropriateness of the Flory-Huggins thermodynamic treatment of gelatinization: Burros et al. (1987), Paton (1987), Russell (1987a), Mestres et al. (1988), Lund (1989), Knutson (1990), Liu and Lelievre (1991), Ring and Whittam (1991), Whittam et al. (1991)

E. Antiplasticizing effect of sugars on gelatinization: Blanshard (1987, 1988), Chungcharoen and Lund (1987), Lund (1989), Biliaderis (1990), Biliaderis and Seneviratne (1990a, b), Hoseney and Rogers (1990), I'Anson et al. (1990), Johnson et al. (1990), Morris (1990), Eliasson (1991), Nishinari et al. (1991), Shukla (1991)

that native granular starches, both normal and waxy, exhibit nonequilibrium melting (Wunderlich, 1980), annealing, and recrystallization behavior characteristic of a kinetically metastable, water-plasticized, partially crystalline polymer system with a small extent of crystallinity. This group has stressed the significance of the conclusion, in which others have concurred (Table 3.13C), that gelatinization is a nonequilibrium polymer melting process. Gelatinization actually represents a continuum of relaxation processes (underlying a structural collapse) (Levine and Slade, 1989b) that occurs (at $T > Tg$) during heating of starch in the presence of plasticizing water and in which crystallite melting is indirectly controlled by the dynamically constrained, continuous amorphous surroundings (Slade and Levine, 1988c). That is, melting of microcrystallites, which are hydrated clusters of amylopectin branches (French, 1984; Imberty and Perez, 1988), is controlled by prerequisite plasticization ("softening" above Tg) of flexibly coiled, possibly entangled chain segments in the interconnected amorphous regions of the native granule, for which the local structure is visualized in terms of the "fringed micelle" model (Slade and Levine, 1987b). Such nonequilibrium melting in metastable, partially crystalline polymer network systems, in which the crystalline and amorphous phases are neither independent of each other nor homogeneous, is an established concept for synthetic polymers (Wunderlich, 1980; Cheng, 1989). In fact, Wunderlich (1980) and Cheng (1989) have both stressed the point that the melting of partially crystalline synthetic polymers is *never* an equilibrium process. Slade et al. have suggested (see Table 3.13B), and others have agreed (Table 3.13D), that previous attempts (e.g., Lelievre, 1976) to use the Flory-Huggins thermodynamic treatment to interpret the effect of water content on the Tm observed during gelatinization of native starch have failed to provide a mechanistic model, because Flory-Huggins theory (Flory, 1953) only applies to melting of polymers in the presence of diluent under the conditions of the equilibrium portion of the solidus curve. In this context, it is interesting to note Cheng's (1989) recent observation that multicomponent systems of solid polymers [especially those containing copolymers, such as a mixture of native normal starch (amylose + amylopectin) and water, in which amylopectin can be considered a block copolymer (Wunderlich, 1980; Slade and Levine, 1988c)] "are even more beset by non-equilibrium problems [than are single-polymer systems]. Only in the amorphous state above the glass transition can one expect (sluggish) equilibration."

An interesting and graphic illustration of the concept of nonequilibrium melting in partially crystalline synthetic polymer systems has been presented by Wunderlich (1980) and is detailed here in generic terms to help the reader better understand the applicability of this concept to the gelatinization of native granular starch. Wunderlich described the case of a synthetic block copolymer produced from comonomers A and B. Monomer A was readily crystallizable and capable of producing a high molecular weight, crystalline homopolymer of relatively low "equilibrium" Tm. In contrast, monomer B was not crystallizable and produced a completely amorphous, high molecular weight homopolymer with a Tg much higher than the "equilibrium" Tm of homopolymer A. When a minor amount of A and a major amount of B were copolymerized to produce a linear block polymer (with runs of repeat A covalently backbone-bonded to runs of repeat B, to yield a molecular structure of the type -BBBBB-AAAA-BBBBBBB-AAA-BBBBBB-), the resulting product could be made partially crystalline by crystallization from solution. Because the A and B domains were covalently linked, macroscopic phase separation upon crystallization of A was prevented, and microcrystalline "micelles" of A blocks remained dispersed in a three-dimensional amorphous network of B block "fringes." When the melting behavior of this block copolymer was analyzed by DSC, the melting transition of the crystalline A domains was observed at a temperature *above* the Tg of the amorphous B domains. The A domains were kinetically constrained against melting [by dissociation and concomitant volume expansion (Wunderlich, 1980)] at their "equilibrium" Tm by the surrounding continuous glassy matrix of B. The A domains were only free to melt (at a nonequilibrium Tm >> "equilibrium" Tm) *after* the B domains transformed from glassy solid to rubbery liquid at their Tg. Similar nonequilibrium melting behavior has been observed for solution-crystallized poly(phenylene oxide) (Cheng and Wunderlich, 1987).

The fact that water plasticization occurs only in the amorphous regions of partially crystalline, water-compatible polymers is critical to the explanation of how these metastable amorphous regions control the nonequilibrium melting behavior of the crystalline regions. The concept of nonequilibrium melting established for synthetic partially crystalline polymers has been applied to biopolymer systems such as native starch, in order to describe the mechanical relaxation process that occurs as a consequence of a dynamic heat/moisture/time treatment (see Table 3.13B). The existence of contiguous microcrystalline and amorphous regions (e.g., in native starch, the crystallizable short branches and backbone segments with their branch points, respectively, of amylopectin molecules) covalently linked through individual polymer chains creates a "fringed micelle" network. Relative dehydration of the amorphous regions to an initial low overall moisture content leads to the kinetically stable condition in which the effective Tg is higher than the "equilibrium" Tm of the hydrated crystalline regions. Consequently, the "effective" Tm (Slade and Levine, 1988c) is elevated and observed only after softening of the amorphous regions at Tg. Added water acts directly to plasticize the continuous glassy regions, leading to the kinetically metastable condition in which their "effective" Tg is depressed. Thus, the "fringe" becomes an unstable rubber at T > Tg, allowing sufficient mobility and swelling by thermal expansion and water uptake for the interconnected microcrystallites, embedded in the "fringed micelle" network, to melt [by dissociation, with concomitant volume expansion (Wunderlich, 1980)] on heating to a less kinetically constrained Tm only slightly above the depressed Tg. Recent experimental results reported by Liu and Lelievre (1991) have confirmed the earlier finding by Slade et al. (Slade, 1984; Slade and Levine, 1984, 1987b, 1988c; Maurice et al., 1985) that a Tg does, in fact, immediately precede the Tgelat of native starch. For such a melting process, use of the Flory-Huggins thermodynamic treatment to interpret the effect of water content on Tm has no theoretical basis (Lelievre, 1976; Wunderlich, 1981; Alfonso and Russell, 1986; Liu and Lelievre, 1991), because, while water as a plasticizer does affect directly the Tg and indirectly the Tm of polymers such as starch, the effect on Tm is not the direct effect experienced in equilibrium melting (i.e., dissolution) along the solidus curve. In contrast to the case of native starch, in which initial "as is" moisture is limiting, in an excess-moisture situation such as a retrograded wheat starch gel with >27 wt% water (Wg′), in which the amorphous matrix would be fully plasticized and ambient temperature would be above Tg (i.e., Tg′ ≃ −5°C), the fully hydrated and matured crystalline junctions would show the actual, lower (and closer to "equilibrium") Tm of ≃60°C for retrograded B-type starch (see Table 3.13B). [Note the analogy between the above description of nonequilibrium melting in native granular starch and the case described earlier of nonequilibrium melting in a synthetic, partially crystalline block copolymer. In this context, it is interesting that Wunderlich (1980) defines branched polymers as a special case of copolymers, using the example of a synthetic polymer with crystallizable branches, which is analogous to amylopectin.]

The description of the effect of water as a plasticizer on native starch, from the perspective of starch as a partially crystalline glassy polymer system, has been extended to the next level of complexity, i.e., three-component model systems of native starch, water, plus added sugars (Slade, 1984; Slade and Levine, 1984, 1987b). This extension is based on a recognition of gelatinization of granular starch in aqueous media as (1) a diffusion-limited, mechanical relaxation process with non-Arrhenius kinetics that depend on the mobility of the added plasticizer (Slade, 1984; Reid and Charoenrein, 1985; Slade and Levine, 1988c); and (2) a nonequilibrium melting process (as a consequence of heat/moisture treatment), which becomes cooperative and occurs at a significant rate at a characteristic gelatinization temperature corresponding to the instantaneous Tg (i.e., Tgelat = Tg) of the water-plasticized amorphous regions of amylopectin (Slade and Levine, 1988c). Gelatinization in concentrated aqueous solutions of common small sugars begins at a higher Tgelat than in water alone; a retardation effect that has been suggested to result from "antiplasticization" [as defined for synthetic polymers (Sears and Darby, 1982)] by sugar-water cosolvents, relative to the extent of plasticization by water alone (Slade, 1984; Slade and Levine, 1984, 1987b). Sugar-water, of higher average MW than water alone, causes a smaller depression of starch Tg than does pure water. In fact, isothermal treatment of starch in sugar-water, at a temperature that would result in nonequilibrium melting of amylopectin in water alone, results instead in antiplasticization by annealing and crystallite perfection (Slade and Levine, 1987b).

It has been known empirically for decades that various sugars, including sucrose, fructose, and glucose, raise the temperature of gelatinization of starch in water and delay the increase in viscosity (pasting), and that this effect increases with increasing sugar concentration (Blanshard, 1987; Chungcharoen and Lund, 1987; Lund, 1989; Eliasson, 1991). This effect of sugars on the gelatinization and pasting behavior of native and modified starches is also well known to be important to the processing and properties of food products such as baked goods, in that the effect influences the extent of wheat starch gelatinization, its retardation, or even inhibition during baking of, e.g., high-sugar cookie doughs and cake batters (Blanshard, 1986, 1988; Levine and Slade, 1989b; Slade and Levine, 1992). The elevating effect of sugars on Tgelat had been attributed in the past in part to a depression of "water activity" ("Aw") by sugars and in part to an unexplained interaction [called "sugar bridges" by Ghiasi et al. (1983) and said to involve hydrogen bonding (Spies and Hoseney, 1982)] of sugars with the amorphous areas of starch granules. Spies and Hoseney (1982) suggested that "Aw" affects Tgelat to some extent, such that Tgelat increases with decreasing "Aw" of various sugar solutions, yielding straight lines of different slope for mono-, di-, and trisaccharide sugars. However, these workers noted that the mechanism of this effect was not completely understood and that other factors were also important. There had been no successful attempt to show how the two purported aspects ("Aw" and "sugar bridges") of the effect on Tgelat might be related, or to explain the mechanism of elevation of Tgelat, prior to the description of the antiplasticizing effect of sugar-water cosolvents (Slade, 1984; Slade and Levine, 1984, 1987b).

The effect of sucrose on Tgelat of wheat starch is illustrated in Fig. 3.25 (Slade, 1984; Slade and Levine, 1987b). For convenience, Tgelat is taken as the temperature at the peak of heat uptake, as measured by DSC (Fig. 3.25 inset). Figure 3.25 shows that as the weight of sucrose is increased in a ternary mixture with constant equal weight ratio of starch and water, Tgelat increases monotonically for samples up to a 1:1:1 mixture. The DSC heat flow curve of the 1:1:1 mixture (Fig. 3.25 inset), in which 50 wt% sucrose is the added fluid outside the granules, exhibits a glass transition [evidenced by a characteristic baseline shift, associated with the free volume contribution to Tg (Wunderlich, 1981), at the leading edge of the gelatinization endotherm] at an effective Tg > 30°C higher than that seen for a 1:0:1 mixture when the added fluid is water alone. Immediately following and superimposed on the elevated glass transition is a relatively narrow crystalline melting transition, similar to that seen for native starch annealed under various time/temperature conditions (Slade, 1984; Slade and Levine, 1987b).

The effect of sucrose on Tgelat has been explained, within predictions

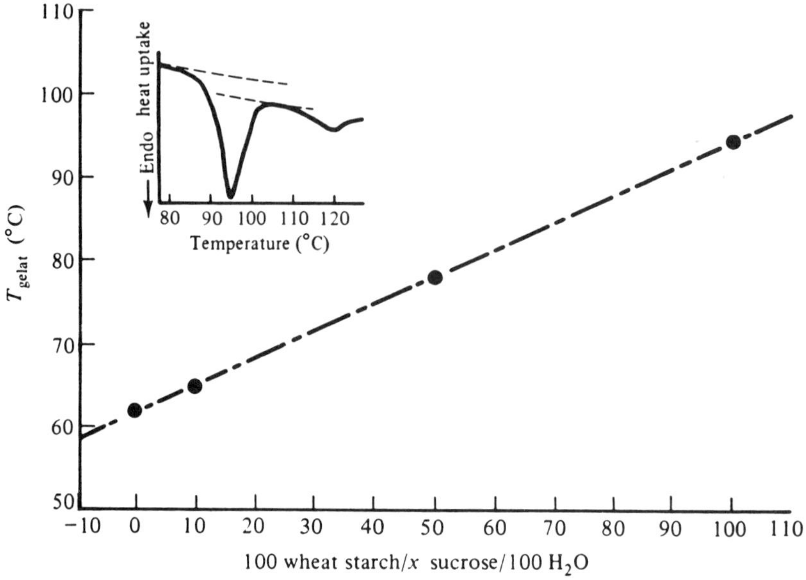

Fig. 3.25 Gelatinization temperature as a function of added sucrose content for three-component mixtures of native wheat starch:sucrose:water (100:x:100 parts by weight, where x = abscissa value). Inset: DSC heat flow curve of 100:100:100 mixture. (*Reproduced, with permission, from Slade and Levine, 1987b.*)

of the conceptual framework of starch as a partially crystalline glassy polymer, on the basis of WLF free volume theory (Slade and Levine, 1987b). If a sugar-water solution is viewed as a plasticizing cosolvent, it is evident that such a coplasticizer, of greater average MW than water alone, would be less effective in mobilizing and increasing the free volume of the amorphous fringes in the "fringed micelle" structure of a starch granule. Less effective plasticization would result directly in less depression of Tg, and thus indirectly in less depression of nonequilibrium Tm. In this sense, in comparing the efficiencies of aqueous solvents (including solutions of nonionic solutes such as sugars and polyols) as plasticizers of the glassy regions of native starch, water alone is the best plasticizer, and sugar-water cosolvents are actually *anti*plasticizers relative to water itself. By most effectively depressing the requisite Tg that initiates gelatinization, added water results in the lowest Tgelat. Increasing concentrations of a given sugar result in increasing average MWs and decreasing free volumes of the cosolvent and so increasing antiplasticization and Tgelat vs. water alone. This explanation has received considerable recent support from a number of other investigators (Table 3.13E).

Of course, increasing the concentration of a given sugar in an aqueous cosolvent also decreases "Aw" (actually RVP), but it has been established that the effective Tg of native starch in the presence of added water is independent of total sample moisture above about 30 wt% total water content (i.e., $\gtrsim$ Wg′) (Slade, 1984; Slade and Levine, 1984, 1987b; Maurice et al., 1985). Moreover, according to WLF theory, it would be expected that the extent of antiplasticization would increase with increasing $\overline{M}$w of the cosolvent, within a homologous series of cosolvent components, from monomer to dimer to oligomer to polymer. Yet, in such a case, the RVP of cosolvents at equal weight concentrations would generally increase with increasing cosolvent $\overline{M}$w. For example, for the homologous series glucose, maltose, maltotriose, 10 DE maltodextrin, RVPs of 50 wt% solutions increase from 0.85 to 0.95 (for both maltose and maltotriose) to 0.99 (Slade and Levine, 1988a).

Experimental results for the effect of this homologous series of cosolvents on Tgelat of starch have confirmed the prediction based on WLF theory, as illustrated in Fig. 3.26 (Slade, 1984; Slade and Levine, 1987b),

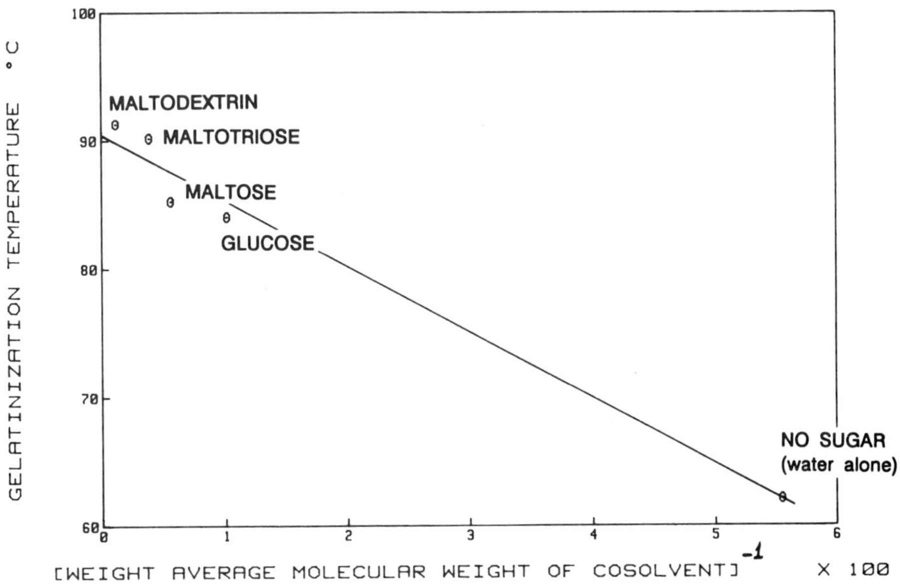

Fig. 3.26 Gelatinization temperature, Tgelat, as a function of 100/$\overline{M}$w (of cosolvent, water + glucose polymer), for three-component mixtures of native wheat starch:glucose polymer:water (1:1:1 parts by weight) and two-component mixture of native wheat starch:water (1:2 parts by weight). (*Reproduced, with permission, from Slade and Levine, 1987b.*)

and thus demonstrated that the underlying basis of this effect is not depression of "Aw" by sugars. The results in Fig. 3.26 were interpreted as follows (Slade and Levine, 1987b). As the $\overline{\mathrm{M}}$w of the antiplasticizing cosolvent increases, free volume decreases, and the resultant Tg of the cosolvent increases. Consequently, the cosolvent becomes less efficient in depressing the effective Tg that initiates the gelatinization transitions in native starch, so Tgelat increases with increasing cosolvent $\overline{\mathrm{M}}$w, despite a corresponding increase in cosolvent RVP. Further experimental confirmation of the concept of antiplasticization by sugar-water cosolvents has come from results recently reported by Biliaderis and Seneviratne (1990a, b). They showed that the nonequilibrum Tm of an amylose-glycerol monostearate crystalline complex increased with increasing sucrose concentration (Biliaderis and Seneviratne, 1990a) or with increasing cosolvent $\overline{\mathrm{M}}$w for a homologous sugar series comprising glucose and malto-oligosaccharides of DP 2-7 (Biliaderis and Seneviratne, 1990b). Eliasson (1991) recently reported similar results for the elevating effect of sucrose on the nonequilibrium Tm of an amylose-lipid crystalline complex, which she said could be satisfactorily explained based on this concept of antiplasticization. Morris (1990) and I'Anson et al. (1990) have also recently invoked this concept to explain the significant antistaling effect (i.e., inhibition of starch retrogradation via recrystallization) of ribose observed in wheat starch-ribose-water gels. The same antiplasticization concept has also recently been advanced to explain the nonequilibrium melting behavior of partially crystalline agarose-sucrose-water gels (Nishinari et al., 1991).

A comparison of the degree of elevation of Tgelat of native wheat starch, for 1:1:1 sugar-water-starch mixtures of a larger and nonhomologous series of sugars, has also been reported (Slade and Levine, 1987b). DSC results showed that Tgelat increases in the following order: water alone < galactose < xylose < fructose < mannose < glucose < maltose < lactose < maltotriose < 10 DE maltodextrin < sucrose. This same order, for fructose < glucose < maltose < sucrose, has been reported in a subsequent study of starch:sugar:water mixtures by Johnson et al. (1990), for increasing ESR correlation times of a spin-probe ESR reporter molecule [said to be indicative of decreasing rotational mobility and increasing local viscosity (Johnson et al., 1990) of the plasticizing sugar-water cosolvent in the amorphous regions of the starch granules prior to gelatinization] and for increasing DSC onset temperatures for starch gelatinization. It has been suggested that the same structure-property relationships that act as determinants of coplasticizer mobility (as opposed to "Aw") in the water dynamics domain of the dynamics map (Levine and Slade, 1988a; Slade and Levine, 1988a) appear to influence the elevation of Tgelat (Slade and Levine, 1987b). It has been concluded that no single parameter (e.g., Tg, Wg′, MW, or Tm/Tg) can explain completely the mechanism of elevation of Tgelat by sugars, but a combination of these parameters that predicts the contributions of free volume and local viscosity provides useful clues to explain why the elevating effect on Tgelat is greater for sucrose than glucose than fructose (Slade and Levine, 1987b).

As pointed out earlier with regard to the effect of water content in binary starch-water systems, previous attempts to analyze the effect of sugar content in ternary starch-sugar-water systems on the observed Tm of starch by the Flory-Huggins equilibrium thermodynamics treatment of melting-point depression by diluents (e.g., Lelievre, 1976) have been recognized to be unjustified on a rigorous theoretical basis (Slade, 1984; Blanshard, 1986, 1987; Levine and Slade, 1988a; Lund, 1989; Biliaderis and Seneviratne, 1990a). Aside from theoretical problems introduced by the facts that the interaction parameter "chi" is concentration-dependent and amylopectin microcrystallites are not a monodisperse system, Flory-Huggins theory is only appropriate to describe an equilibrium melting process. Regardless of the amount of water or sugar solution added to native starch, initial melting of a native granule remains a nonequilibrium process, in which melting of the microcrystallites is controlled by previous plasticization of the glassy regions via heat/moisture treatment to T > the effective Tg (Slade and Levine, 1987b).

Retrogradation has been described (Slade, 1984; Slade and Levine, 1984, 1987b, 1988a, c, d, 1989) as a nonequilibrium (i.e., time/temperature/moisture-dependent) polymer crystallization process in completely amorphous (in the case of waxy starches) starch-water melts. In normal stretches, retrogradation has been recently confirmed to involve both fast crystallization of amylose and slow recrystallization of amylopectin (Miles et al., 1985b; Ring and Orford, 1986; Ring et al., 1987; Russell, 1987b; I'Anson et al., 1988; Mestres et al., 1988; Russell and Oliver, 1989). Amylopectin recrystallization has been described (Slade, 1984; Slade and Levine, 1984, 1987b, 1988a, c, d, 1989) as a nucleation-limited growth process that occurs, at T > Tg, in the mobile, viscoelastic, "fringed micelle" gel network plasticized by water, and that is thermally reversible at T > Tm. This description has also been confirmed recently, for both amylopectin (Miles et al., 1985b; Ring and Orford, 1986; Ring et al., 1987) and amylose (Miles et al., 1985a, b; I'Anson et al., 1988; Clark et al., 1989; Gidley, 1989; Gidley and Bulpin, 1989), and has garnered wider support (Lund, 1989; Russell and Oliver, 1989; Morris, 1990; Shukla, 1991). The aging effects typically observed in starch gels and baked bread have been attributed (as in synthetic polymer-organic diluent gels) to time-dependent crystallization processes (i.e., maturation), primarily involving amylopectin, that occur subsequent to initial gelation (Miles et al., 1985b; Paton, 1987; Ring et al., 1987; Russell, 1987b; Marsh and Blanshard, 1988; Mestres et al., 1988; Lund, 1989; Russell and Oliver, 1989). With regard to these effects, Slade (1984) has reported that "analysis of results [of measurements of extent of recrystallization vs. time after gelatinization] by the classical Avrami equation may provide a convenient means to represent empirical data from retrogradation experiments (Jankowski and Rha, 1986; Russell, 1987b; Marsh and Blanshard, 1988; Cheng, 1989; Russell and Oliver, 1989), but some published theoretical interpretations (e.g, Kulp and Ponte, 1981) have been misleading." Complications, due to the nonequilibrium nature of starch recrystallization via the three-step mechanism, limit the theoretical utility of the Avrami parameters, which were originally derived to describe crystallization under conditions far above the glass curve (Levine and Slade, 1989b) and where details about nucleation events and constant linear growth rates were readily measurable (Wunderlich, 1976). Others have agreed with this conclusion (Miles et al., 1985a, b; Russell and Oliver, 1989; Morris, 1990) and pointed out that such an Avrami analysis allows no insight regarding crystal morphology (Russell, 1987b) and provides no clear mechanistic information (Blanshard, 1988; Cheng, 1989). Furthermore, the Avrami theory gives no indication of the temperature dependence of the crystalization rate (Marsh and Blanshard, 1988; Russell and Oliver, 1989).

It should be recalled that the same three-step crystallization mechanism also applies to low molecular weight compounds (Wunderlich, 1976; Franks, 1982), such as concentrated aqueous solutions and melts of low molecular weight carbohydrates (Slade and Levine, 1988a, b), and to recrystallization processes in frozen systems of water-compatible food materials (Levine and Slade, 1988a, b, c).

Polymer Crystallization Kinetics Theory

The classical theory of crystallization kinetics, applied to synthetic partially crystalline polymers (Wunderlich, 1976), is illustrated in Fig. 3.27 (Levine and Slade, 1989b). This theory has also been shown to describe the kinetics of starch retrogradation (Slade, 1984; Slade and Levine, 1984, 1987b; Zeleznak and Hoseney, 1987b; Blanshard, 1988; Marsh and Blanshard, 1988; Morris, 1990; Shukla, 1991) and gelatin gelation (Jolley, 1970; Slade and Levine, 1984, 1987a; Domszy et al., 1986). Figure 3.27 shows the dependence of crystallization rate on temperature within the range

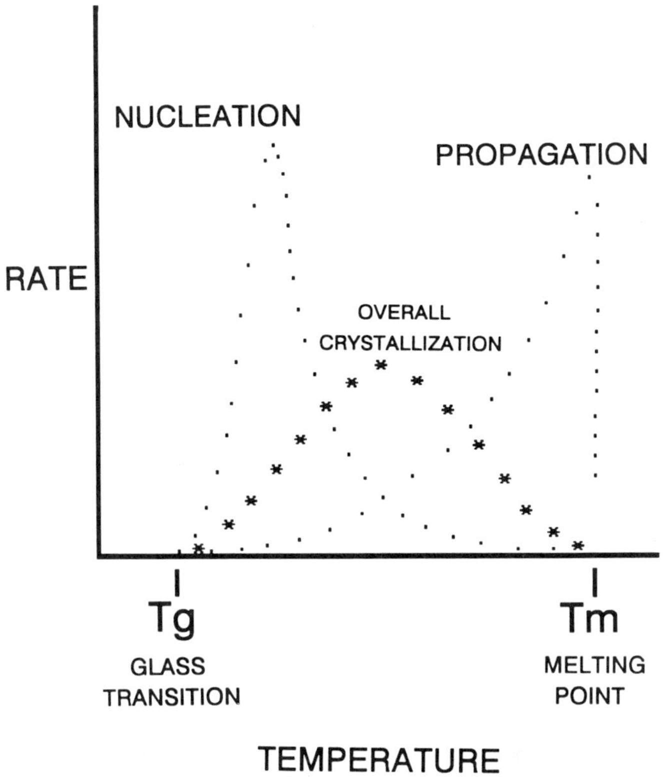

Fig. 3.27 Crystallization kinetics of partially crystalline polymers, expressed in terms of crystallization rate as a function of temperature. (*Reproduced, with permission, from Levine and Slade, 1989b.*)

Tg < T < Tm, and emphasizes the fact that gelation-via-crystallization can only occur in the rubbery (undercooled liquid) state, between the temperature limits defined by Tg and Tm (Levine and Slade, 1988a; Marsh and Blanshard, 1988). These limits, for gels recrystallized from high molecular weight gelatin solutions of concentrations up to ~65 wt% gelatin (i.e., W > Wg′ ≃ 35 wt% water), are about −12°C (= Tg′) and 37°C, respectively, while for B-type starch (or purified amylopectin) gels recrystallized from homogeneous and completely amorphous gelatinized sols or pastes containing ≳27 wt% water (= Wg′), they are about −5°C (= Tg′) and 60°C, respectively (Slade, 1984; Slade and Levine, 1984). In gelatinized potato starch:water mixtures (1:1 w/w), retrogradation has been demonstrated at single storage temperatures between 5 and 50°C

(Nakazawa et al., 1985). In retrograding potato and wheat starch gels, low-temperature storage (at 5 and 4°C, respectively) results in recrystallization to lower-Tm, less symmetrically perfect polymorphs than those produced by storage at room temperature (Nakazawa et al., 1985; Jankowski and Rha, 1986). Conversely, a higher crystallization temperature generally favors formation of the higher-Tm, more stable A-type, rather than B-type, starch polymorph (Gidley, 1987; Gidley and Bulpin, 1987). For amylopectins from waxy maize and other botanical sources, thermoreversible gelation-via-crystallization from concentrated (>10 wt% solute) aqueous solution has been observed after long-term storage at 1–5°C (Ring, 1985a; Ring and Orford, 1986; Ring et al., 1987). In baked bread, low- (4°C), intermediate- (25°C), and high-(40°C)temperature storage results in starch recrystallization manifested by corresponding lower, intermediate, and higher-Tm staling endotherms (Zeleznak and Hoseney, 1987b). In a 50% wheat starch gel, the extent of crystallization increases with decreasing storage temperature in the range 2–37°C (i.e., displays a negative temperature dependence), and the rate of recrystallization to the B-form is more rapid at 2 than at 37°C (Marsh and Blanshard, 1988). In contrast to the familiar Tm of about 60°C for thermoreversible B-type amylopectin gels with excess moisture stored at room temperature (and for stale bread) (Slade and Levine, 1987b), the corresponding Tm for thermoreversible V-type amylose gels is well above 100°C (Biliaderis et al., 1985; Ring et al., 1987), owing in part to the much higher $\overline{DP}w$ of the amylose chain segments [i.e., $\overline{DP}w \simeq 50$ vs. $\simeq 15$ for amylopectin (Ring et al., 1987)] comprising the microcrystalline junction zones. Analogously, the familiar Tm well above 100°C for various V-type lipid-amylose crystalline complexes (Biliaderis et al., 1985) is much higher than the corresponding Tm of about 70°C reported for a lipid-amylopectin crystalline complex (Slade and Levine, 1987b). These findings are fully consistent with the established relationship between increasing chain length (and MW) and increasing Tm within homologous families of partially crystalline synthetic polymers (Wunderlich, 1980; Billmeyer, 1984).

As illustrated by Fig. 3.27 and the above results on the temperature dependence of starch recrystallization, the rate of crystallization would be practically negligible at T < Tg, because nucleation is a liquid-state phenomenon [i.e., in part, a transport process through a viscous medium (Blanshard, 1988; Marsh and Blanshard, 1988)] that requires translational and orientational mobility, and such mobility is virtually disallowed (over realistic times) in a mechanical solid of $\eta > 10^{12}$ Pa s (Franks, 1982). The temperature of homogeneous nucleation (Th) can be estimated from the ratio of Th/Tm (in degrees K) (Slade and Levine, 1988b), which is typically near 0.8 for partially crystalline synthetic polymers as well as small molecules, with a reported range of 0.78–0.85 (Walton, 1969; Wunderlich, 1976). The rate of propagation goes essentially to zero below Tg, because propagation is a diffusion-limited process (Baro et al., 1977) for which practical rates also require the liquid state. At T > Tm, the rate of overall crystallization also goes to zero, because, intuitively, one realizes that crystals can neither nucleate nor propagate at any temperature at which they would be melted instantaneously.

Figure 3.27 illustrates the complex temperature dependence of the overall crystallization rate and of the rates of the separate mechanistic steps of nucleation and propagation. According to classical nucleation theory, the nucleation rate is zero at Tm and increases rapidly with decreasing temperature [and increasing extent of undercooling (Tm − T)] over a relatively narrow temperature interval, which for undiluted synthetic polymers begins at an undercooling of 30–100°C (Wunderlich, 1976). Within this temperature region, the nucleation rate shows a large negative temperature coefficient (Jolley, 1970; Marsh and Blanshard, 1988). At still lower temperatures (and greater extents of undercooling), where nucleation relies on transport and depends on local viscosity, the nucleation rate decreases with decreasing temperature and increasing local viscosity, to a near-zero rate at Tg (Wunderlich, 1976; Marsh and Blanshard, 1988). In contrast, the propagation rate increases rapidly with increasing temperature, from a near-zero rate at Tg, and shows a large positive temperature coefficient over nearly the entire rubbery range, until it drops precipitously to a zero rate at Tm (Jolley, 1970; Wunderlich, 1976; Marsh and Blanshard, 1988). The fact that the nucleation and propagation rates show temperature coefficients of opposite sign in the temperature region of intermediate undercooling has been explained (Marsh and Blanshard, 1988) by pointing out that "when the temperature has been lowered sufficiently to allow the formation of [critical] nuclei [whose size decreases with decreasing temperature (Franks, 1982; Blanshard, 1988)], the [local] viscosity is already so high that it prevents growth of crystalline material" (Baro et al., 1977). The maturation rate for nonequilibrium crystallization processes, like the propagation rate, increases with increasing temperature, up to the maximum Tm of the most mature crystals (Slade, 1984; Levine and Slade, 1988a).

As shown by the symmetrical curve in Fig. 3.27, the overall crystallization rate (i.e., the resultant rate of both the nucleation and propagation processes), at a single holding temperature, reaches a maximum at a temperature about midway between Tg and Tm, and approaches zero at Tg and Tm (Jolley, 1970; Wunderlich, 1976; Slade, 1984; Slade and Levine, 1984, 1987a, b; Blanshard, 1988; Marsh and Blanshard, 1988; Morris, 1990; Shukla, 1991). Identification of the location of the temperature of maximum crystallization rate has been described (Wunderlich, 1976) in terms of a universal empirical relationship (based on two underlying concepts) for the crystallization kinetics of synthetic high polymers. The first concept identifies a model polymer (e.g., a readily crystallizable elastomer with Tg = 200 K and Tm = 400 K) as one for which the temperature dependence of polymer melt viscosity is described by WLF kinetics (Wunderlich, 1976). [The same concept has been shown to be applicable to describe the nonequilibrium thermomechanical relaxation behavior of "typical" and "atypical" food carbohydrates in aqueous glassy and rubbery states (Slade and Levine, 1988b).] The second concept empirically defines a reduced temperature, based on Tg and Tm for typical polymers, as (T − Tg + 50 K)/(Tm − Tg + 50 K) (Wunderlich, 1976). [An analogous reduced temperature scale, based on Tg′ and Tm, has been shown to describe the rotational mobility (i.e., dielectric relaxation behavior) of concentrated aqueous sugar solutions in the supra-glassy fluid state (Slade and Levine, 1988b).] For all synthetic high polymers analyzed, the temperature position of the maximum crystallization rate, on a universal master curve like the one shown in Fig. 3.27, occurs at about 0.6 of the reduced temperature scale (Wunderlich, 1976). Low molecular weight synthetic compounds have been fitted to a similar curve, but with a different position for the maximum crystallization rate, at about 0.8 of the reduced temperature scale (Wunderlich, 1976). Based on this empirical relationship for synthetic high polymers, the calculated single holding temperature for maximum crystallization rate would be about 300 K for the model elastomer (in fact, exactly midway between Tg and Tm), −3°C for a gelatin gel with $\gtrsim$35 wt% water (a temperature made inaccessible, without detriment to product quality, due to unavoidable ice formation), 14°C for a typical B-type starch (or amylopectin) gel with $\gtrsim$27 wt% water, and 70°C for a V-type amylose gel [based on Tm = 153°C (Ring et al., 1987)] (Levine and Slade, 1989b). It has been noted (Levine and Slade, 1989b) that the calculated value of about 14°C for B-type starch is similar to (1) the empirically determined subambient temperature for the maximum rate of starch recrystallization and concomitant crumb firming during aging, reported in an excellent study of the kinetics of bread staling by Guilbot and Godon (1984), but not previously explained on the basis of the polymer crystallization kinetics theory described above; and (2) the temperature of about 5°C recently calculated from Lauritzen-Hoffman polymer crystallization kinetics theory by Marsh and Blanshard (1988) for a 50% wheat starch gel. The fact that these subambient temperatures are much closer to the operative Tg (i.e., Tg′) than to Tm (Levine and Slade, 1989b), unlike the situation depicted by the symmetrical shape of the crystallization rate curve in Fig. 3.27 that typifies the behavior of many synthetic polymers, clearly indicates that the crystallization process for B-type starch (or pure amylopectin) is strongly nucleation-limited (Slade, 1984; Miles et al., 1985b; Slade and Levine, 1987b; Marsh and Blanshard, 1988; Biliaderis and Zawistowski, 1990).

In contrast to the maximum crystallization rate achievable at a single temperature, Ferry (1948) showed for gelatin that the rate of gelation can be further increased, while the phenomenon of steadily increasing gel maturation over extended storage time can be eliminated, by a two-step temperature-cycling gelation protocol that capitalizes on the crystallization kinetics defined in Fig. 3.27. He showed that a short period for fast nucleation at 0°C (a temperature above Tg′ and near the peak of the nucleation rate curve), followed by another short period for fast crystal growth at a temperature just below Tm, produced a gelatin gel of maximum and unchanging gel strength in the shortest possible overall time. Recently, Slade has shown that a similar temperature-cycling protocol can be used to maximize the rate of starch recrystallization in freshly gelatinized starch-water mixtures with at least 27 wt% water (Slade, 1984; Slade and Levine, 1987b), resulting in a patented process for the accelerated staling of starch-based food products (Slade et al., 1987). Zeleznak and Hoseney (1987b) subsequently adopted this protocol in their study of the temperature dependence of bread staling.

CONCLUSIONS AND FUTURE PROSPECTS

In this review, we have described a food polymer science approach to the study of glasses and glass transitions in foods. This approach has been used to define structure-function relationships in food products and processes, and to assess and understand the effects of glass transitions and water plasticization on food quality, safety, and stability. We have reviewed the theoretical principles from the field of synthetic polymer science that are applicable to studies of glass transitions in foods and have presented a broad overview focusing primarily on the most recent experimental studies by a number of groups that have been especially active in this growing field of research. As a framework for this discussion, we have used the following list of questions, which we have attempted to answer or at least address.

What is a glass?

What is a glass transition?

Why is the temperature at which a glass transition occurs (Tg) so important to the processing and storage stability of so many different types of foods?

Why is the effect of water as a plasticizer on Tg of such widespread relevance to food products and processes?

Why are considerations of nonequilibrium glassy solid and rubbery liquid states in foods, rather than equilibrium phases, more germane to issues of food quality and safety?

Why are the kinetics of heat/moisture processes for foods and of deteriorative changes in food systems during storage more often appropriately interpreted in terms of the Williams-Landel-Ferry (WLF), rather than the Arrhenius, equation?

What is the food polymer science approach, with its central concepts of "glass dynamics" and "water dynamics," and why has this research approach proved so useful to the study of glasses and glass transitions in foods, and therefore gained such widespread support from so many investigators?

Why has there been, since the early 1980s, such interest in these questions, and such increasing research activity, especially in the last few years, in this area of food science and technology?

We hope we have succeeded, at least, in demonstrating to the reader that the field of food science and technology has recently enjoyed a seemingly exponential growth of interest in glasses and glass transitions in foods and in the plasticizing effect of water on Tg. This spurt of interest stems from the growing realization of the importance of glassy and rubbery states to the quality (including textural and structural characteristics), safety, and stability of foods. This state of affairs is evidenced by the fact that about 180 of the references cited in this review deal with glasses and glass transitions in food materials and systems, and more than 100 of these were published in 1989, 1990, or 1991. During the rest of the 1990s, we expect to witness even greater growth and interest in this exciting area, because it offers so many challenging questions still to be answered, while promising so many opportunities for technological advancement.

ACKNOWLEDGMENTS

We thank the following scientists, whose preprint manuscripts, cited in this chapter, were made available to us prior to their publication: S.

Ablett, C. Biliaderis, H. Bizot, J. Blanshard, A. Eliassson, F. Franks, C. Gaines, P. Given, C. Hoseney, V. Huang, M. Izzard, T. Kararli, M. Karel, J. Lelievre, M. Le Meste, P. Lillford, T. Maurice, K. Nishinari, M. Pikal, D. Reid, S. Ring, M. Roozen, T. Schenz, D. Simatos, A. Smith, I. Tomka, C. van den Berg, and R. Williams.

REFERENCES

Ablett, S., Attenburrow, G. E., and Lillford, P. J. 1986. The significance of water in the baking process. In *Chemistry and Physics of Baking*, J. M. V. Blanshard, P. J. Frazier, and T. Galliard (Ed.), pp. 30–41. Royal Society of Chemistry, London.

Ablett, S., Barnes, D. J., Davies, A. P., Ingman, S. J., and Patient, D. W. 1988. C13 and pulse NMR spectroscopy of wheat proteins. *J. Cereal Sci.* 7: 11–20.

Ablett, S., Izzard, M. J., and Lillford, P. J. 1991a. Calorimetric study of the thermal behavior of sucrose solutions. *J. Chem. Soc. Faraday Trans.* I (in press).

Ablett, S., Clark, A. H., Izzard, M. J., and Lillford, P. J. 1991b. Theoretical modelling of heat capacity-temperature data for sucrose-water systems. *J. Chem. Soc. Faraday Trans.* I (in press).

Aguerre, R. J., Suarez, C., and Viollaz, P. E. 1989. Swelling and pore structure in starchy materials. *J. Food Eng.* 9: 71–80.

Alfonso, G. C. and Russell, T. P. 1986. Kinetics of crystallization in semicrystalline/amorphous polymer mixtures. Macromolecules 19: 1143–1152.

Angell, C. A. 1983. Supercooled water. *Ann. Rev. Phys. Chem.* 34: 593–630.

Angell, C. A. 1988. Perspective on the glass transition. *J. Phys. Chem. Solids* 49: 863–871.

Atkins, A. G. 1987. Basic principles of mechanical failure in biological systems. In *Food Structure and Behaviour*, J. M. V. Blanshard and P. Lillford (Ed.), pp. 149–176. Academic Press, London.

Attenburrow, G. E., Goodband, R. M., Taylor, L. J., and Lillford, P. J. 1989. Structure, mechanics, and texture of a food sponge. *J. Cereal Sci.* 9: 61–70.

Attenburrow, G., Barnes, D. J., Davies, A. P., and Ingman, S. J. 1990. Rheological properties of wheat gluten. *J. Cereal Sci.* 12: 1–14.

Aubin, M. and Prud'homme, R. E. 1988. Analysis of the glass transition temperature of miscible polymer blends. *Macromolecules* 21: 2945–2949.

Baardseth, P. and Naesset, E. 1989. The effect of lipid-degrading enzyme activities on quality of blanched and unblanched frozen stored cauliflower estimated by sensory and instrumental analysis. *Food Chem.* 32: 39–46.

Bair, H. E. 1981. Thermal analysis of additives in polymers. In *Thermal Characterization of Polymeric Materials*, E. A. Turi (Ed.), pp. 845–909. Academic Press, Orlando, FL.

Baro, M. D., Clavaguera, N., Bordas, S., Clavaguera-Mora, M. T., and Casa-Vazquez, J. 1977. Evaluation of crystallization kinetics by DTA. *J. Thermal Anal.* 11: 271–276.

Batzer, H. and Kreibich, U. 1981. Influence of water on thermal transitions in natural polymers and synthetic polyamides. *Polym. Bull.* 5: 585–590.

Biliaderis, C. G., Page C. M., Slade, L., and Sirett, R. R. 1985. Thermal behavior of anylose-lipid complexes. *Carbohydr. Polym.* 5: 367–389.

Biliaderis, C. G., Page, C. M., Maurice, T. J., and Juliano, B. O. 1986a. Thermal characterization of rice starches: A polymeric approach to phase transitions of granular starch. *J. Agric. Food Chem.* 34: 6–14.

Biliaderis, C. G., Page, C. M., and Maurice, T. J. 1986b. Non-equilibrium melting of amylose-V complexes. *Carbohydr. Polym.* 6: 269–288.

Biliaderis, C. G., Page, C. M., and Maurice, T. J. 1986c. Multiple melting transitions of starch/monoglyceride systems. *Food Chem.* 22: 279–295.

Biliaderis, C. G. and Galloway, G. 1989. Crystallization behavior of amylose-V complexes: structure-property relationships. *Carbohydr. Res.* 189: 31–48.

Biliaderis, C. G. 1990. Thermal analysis of food carbohydrates. In *Thermal Analysis of Foods*, V. R. Harwalkar and C.-Y. Ma (Ed.), pp. 168–220. Elsevier Applied Science, London.

Biliaderis, C. G. and Zawistowski, J. 1990. Viscoelastic behavior of aging starch gels: Effects of concentration, temperature, and starch hydrolysates on network properties. *Cereal Chem.* 67: 240–246.

Biliaderis, C. G. and Seneviratne, H. D. 1990a. On the supermolecular structure and metastability of glycerol monostearate-amylose complex. *Carbohydr. Polym.* 13: 185–206.

Biliaderis, C. G. and Seneviratne, H. D. 1990b. Solute effects on the thermal stability of glycerol monostearate-amylose complex superstructures. *Carbohydr. Res.* 208: 199–213.

Biliaderis, C. G. 1991. Non-equilibrium phase transitions of aqueous starch systems. In *Water Relationships in Foods*, H. Levine and L. Slade (Ed.), pp. 251–273. Plenum Press, New York.

Billmeyer F. W. 1984. Textbook of Polymer Science, 3rd ed. Wiley-Interscience, New York.

Biros, J., Madan, R. L., and Pouchly, J. 1979. Heat capacity of water-swollen polymers above and below 0°C. *Collect. Czech. Chem. Commun.* 44: 3566–3573.

Blanshard, J. M. V. 1986. The significance of the structure and function of the starch granule in baked products. In *Chemistry and Physics of Baking*, J. M. V. Blanshard, P. J. Frazier, and T. Galliard (Ed.), pp. 1–13. Royal Society of Chemistry, London.

Blanshard, J. M. V. 1987. Starch granule structure and function: physicochemical approach. In *Starch: Properties and Potential*, T. Galliard (Ed.), pp. 16–54. John Wiley & Sons, New York.

Blanshard, J. M. V., and Franks, F. 1987. Ice crystallization and its control in frozen food systems. In *Food Structure and Behaviour*, J. M. V. Blanshard and P. Lillford (Ed.), pp. 51–65. Academic Press, London.

Blanshard, J. M. V. 1988. Elements of cereal product structure. In *Food Structure—Its Creation and Evaluation*, J. M. V. Blanshard and J. R. Mitchell (Ed.), pp. 313–330. Butterworths, London.

Blanshard, J. M. V., Muhr, A. H., and Gough, A. 1991. Crystallization from concentrated sucrose solutions. In *Water Relationships in Foods*, H. Levine and L. Slade (Ed.), pp. 639–655. Plenum Press, New York.

Blond, G. 1989. Water-galactose system: supplemented state diagram and unfrozen water. *Cryo-Lett.* 10: 299–308.

Blum, F. D. and Nagara, B. 1986. Solvent mobility in gels of atactic polystyrene. *Polym. Prepr.* 27(1): 211–212.

Bone, S. and Pethig, R. 1982. Dielectric studies of the binding of water to lysozyme. *J. Mol. Biol.* 157: 571–575.

Bone, S. and Pethig, R. 1985. Dielectric studies of protein hydration and hydration-induced flexibility. *J. Mol. Biol.* 181: 323–326.

Borchard, W., Bremer, W., and Keese, A. 1980. State diagram of the water-gelatin system. *Colloid Polym. Sci.* 258: 516–526.

Boyer, R. F., Baer, E., and Hiltner, A. 1985. Concerning gelation effects in atactic polystyrene solutions. *Macromolecules* 18: 427–434.

Braudo, E. E., Belavtseva, E. M., Titova, E. F., Plashchina, I. G., Krylov, V. L., Tolstoguzov, V. B., Schierbaum, F. R., and Richter, M.

1979. Struktur und Eigenschaften von Maltodextrin-Hydrogelen. *Starke* 31: 188–194.

Braudo, E. E., Plashchina, I. G., and Tolstoguzov, V. B. 1984. Structural characterisation of thermoreversible anionic polysaccharide gels by their elastoviscous properties. *Carbohydr. Polym.* 4: 23–48.

Brydson, J. A. 1972. The glass transition, melting point and structure. In *Polymer Science*, A. D. Jenkins (Ed.), pp. 194–249. North Holland, Amsterdam.

Buchanan, D. R. and Walters, J. P. 1977. Glass-transition temperatures of polyamide textile fibers. I: effect of water. *Textile Res. J.* 47: 398–406.

Burchard, W. 1988. Entanglement and reversible gelation for polymers of different architectures. *Prog. Colloid Polym. Sci.* 78: 63–67.

Burghoff, H. G. and Pusch, W. 1980. Thermodynamic state of water in cellulose acetate membranes. *Polym. Eng. Sci.* 20: 305–309.

Burros, B. C., Young, L. A., and Carroad, P. A. 1987. Kinetics of corn meal gelatinization at high temperature and low moisture. *J. Food Sci.* 52: 1372–1380.

Cairault, A., Bizot, H., Durning, C. J., and Roques, M. 1989. Approaching viscoelastic diffusion in polyacrylamide gels. In *Proceedings of "Drying '89"*.

Cakebread, S. H. 1969. Factors affecting the shelf life of high boilings. *Manufact. Confect.* 49: 41–44.

Caldwell, K. B., Goff, H. D., and Maurice, T. J. 1990. The use of thermal mechanical analysis to determine the influence of carbohydrates on stability of frozen dairy products. *J. Dairy Sci.* 73(supp. 1): 95.

Chan, R. K., Pathmanathan, K., and Johari, G. P. 1986. Dielectric relaxations in the liquid and glassy states of glucose and its water mixtures. *J. Phys. Chem.* 90: 6358–6362.

Chang, Z. H. and Baust, J. G. 1991. Physical aging of glassy state: DSC study of vitrified glycerol systems. *Cryobiology* 28: 87–95.

Cheng, S. Z. D. 1989. Thermal characterization of macromolecules. *J. Appl. Polym. Sci.: Appl. Polym. Symp.* 43: 315–371.

Cheng, S. Z. D. and Wunderlich, B. 1987. Glass transition and melting behavior of poly(oxy-2,6-dimethyl-1,4-phenylene). *Macromolecules* 20: 1630–1637.

Chungcharoen, A. and Lund, D. B. 1987. Influence of solutes and water on rice starch gelatinization. *Cereal Chem.* 64: 240–243.

Clark, A. H., Gidley, M. J., Richardson, R. K., and Ross-Murphy, S. B.

1989. Rheological studies of aqueous amylose gels: the effect of chain length and concentration on gel modulus. *Macromolecules* 22: 346–351.

Cocero, A. M. and Kokini, J. L. 1991. The study of the glass transition of glutenin using small amplitude oscillatory rheological measurements and DSC. *J. Rheol.* 35: 257–270.

Cole, B. A., Levine, H. I., McGuire, M. T., Nelson, K. J., and Slade, L. 1983. Soft, frozen dessert formulation. U.S. patent 4,374,154.

Cole, B. A., Levine, H. I., McGuire, M. T., Nelson, K. J., and Slade, L. 1984. Soft, frozen dessert formulation. U.S. patent 4,452,824.

Cowie, J. M. G. 1973. *Polymers: Chemistry and Physics of Modern Materials*. Intertext, New York.

Davies, A. P., Ingman, S. J., and Attenburrow, G. 1991. Plasticization and mechanical properties of heat-set wheat gluten. In *Gluten Proteins 1990*, W. Bushuk and R. Tkachuk (Ed.), pp. 21–28. Amer. Assoc. Cereal Chem., St. Paul, MN.

Derbyshire, W. 1982. Dynamics of water in heterogeneous systems with emphasis on subzero temperatures. In *Water: A Comprehensive Treatise*, F. Franks (Ed.), Vol. 7, pp. 339–430. Plenum Press, New York.

Doescher, L. C., Hoseney, R. C., and Milliken, G. A. 1987. Mechanism for cookie dough setting. *Cereal Chem.* 64: 158–163.

Domszy, R. C., Alamo, R., Edwards, C. O., and Mandelkern, L. 1986. Thermoreversible gelation and crystallization of homopolymers and copolymers. *Macromolecules* 19: 310–325.

Duden, R. and Scholz, A. 1982. Enzymatic decomposition of phospholipids in frozen parsley. *Z. Ernahrungswiss.* 21: 266–271.

Edwards, S. F., Lillford, P. J., and Blanshard, J. M. V. 1987. Gels and networks in practice and theory. In *Food Structure and Behaviour*, J. M. V. Blanshard and P. Lillford (Ed.), pp. 1–12. Academic Press, London.

Ehlich, D. and Sillescu, H. 1990. Tracer diffusion at the glass transition. *Macromolecules* 23: 1600–1610.

Eisenberg, A. 1984. The glassy state and the glass transition. In *Physical Properties of Polymers*, J. E. Mark, A. Eisenberg, W. W. Graessley, L. Mandelkern, and J. L. Koenig (Ed.), pp. 55–95. American Chemical Society, Washington, DC.

Eliasson, A.-C. 1991. A calorimetric investigation of the influence of sucrose on the gelatinization of starch. *Carbohydr. Polym.* (in press).

Ellis, T. S., Jin, X., and Karasz, F. E. 1984. The water induced plasticization behavior of semi-crystalline polyamides. *Polym. Prepr.* 25(2): 197–198.

Ellis, T. S. 1988. Moisture-induced plasticization of amorphous polyamides and their blends. *J. Appl. Polym. Sci.* 36: 451–466.

Ferry, J. D. 1948. Mechanical properties of substances of high molecular weight. *J. Am. Chem. Soc.* 70: 2244–2249.

Ferry, J. D. 1980. *Viscoelastic Properties of Polymers*, 3rd ed. John Wiley & Sons, New York.

Finegold, L., Franks, F., and Hatley, R. H. M. 1989. Glass/rubber transitions and heat capacities of binary sugar blends. *J. Chem. Soc., Faraday Trans.* I, 85: 2945–2951.

Finney, J. L. and Poole, P. L. 1984. Protein hydration and enzyme activity: role of hydration-induced conformation and dynamic changes in activity of lysozyme. *Comments Mol. Cell. Biophys.* 2: 129–151.

Flink, J. M. 1983. Structure and structure transitions in dried carbohydrate materials. In *Physical Properties of Foods*, M. Peleg and E. B. Bagley (Ed.), pp. 473–521. AVI, Westport, CT.

Flory, P. J. 1953. *Principles of Polymer Chemistry.* Cornell University Press, Ithaca, NY.

Flory, P. J. 1974. Introductory lecture—gels and gelling processes. *Faraday Disc. Chem. Soc.* 57: 7–18.

Franks, F., Asquith, M. H., Hammond, C. C., Skaer, H. B., and Echlin, P. 1977. Polymeric cryoprotectants in the preservation of biological ultrastructure. I. *J. Microsc.* 110: 223–238.

Franks, F. 1982. The properties of aqueous solutions at subzero temperatures. In *Water: A Comprehensive Treatise*, Vol. 7, F. Franks (Ed.), pp. 215–338. Plenum Press, New York.

Franks, F. 1983a. Bound water: Fact and fiction. *Cryo-Lett.* 4: 73–74.

Franks, F. 1983b. Solute-water interactions: Do polyhydroxy compounds alter the properties of water? *Cryobiology* 20: 335–345.

Franks, F. 1985. Complex aqueous systems at subzero temperatures. In *Properties of Water in Foods*, D. Simatos and J. L. Multon (Ed.), pp. 497–509. Martinus Nijhoff, Dordrecht.

Franks, F. 1986a. Metastable water at subzero temperatures. *J. Microsc.* 141: 243–249.

Franks, F. 1986b. Unfrozen water: yes; unfreezable water: hardly; bound water: certainly not. *Cryo-Lett.* 7: 207.

Franks, F. 1988. Protein hydration. In *Characterization of Proteins*, F. Franks (Ed.), pp. 127–154. Humana Press, Clifton, NJ.

Franks, F. 1989. Improved freeze-drying: An analysis of the basic scientific principles. *Process Biochem.* 24(1): R3–R7.

Franks, F. 1990. Freeze drying: From empiricism to predictability. *Cryo-Lett.* 11: 93–110.

Franks, F. and Grigera, J. R. 1990. Solution properties of low molecular weight polyhydroxy compounds. In *Water Science Reviews*, Vol. 5, *The Molecules of Life*, F. Franks (Ed.), pp. 187–289. Cambridge University Press, Cambridge, England.

Franks, F. 1991. Hydration phenomena: an update and implications for the food processing industry. In *Water Relationships in Foods*, H. Levine and L. Slade (Eds.), pp. 1–19. Plenum Press, New York.

French, D. 1984. Organization of starch granules. In *Starch: Chemistry and Technology*, 2nd ed., R. L. Whistler, J. N. Bemiller, and E. F. Paschall (Ed.), pp. 183–247. Academic Press, Orlando.

Fried, J. R., Lai, S. Y., Kleiner, L. W., and Wheeler, M. E. 1982. Experimental assessment of the thermodynamic theory of the compositional variation of Tg: PVC systems. *J. Appl. Polym. Sci.* 27: 2869–2883.

Fujio, Y. and Lim, J. K. 1989. Correlation between the glass-transition point and color change of heat-treated gluten. *Cereal Chem.* 66: 268–270.

Fuzek, J. F. 1980. Glass transition temperature of wet fibers: its measurement and significance. In *Water in Polymers*, S. P. Rowland (Ed.), pp. 515–530. ACS Symposium Series 127, American Chemical Society, Washington, DC.

Gaeta, S., Apicella, A., and Hopfenberg, H. B. 1982. Kinetics and equilibria associated with the absorption and desorption of water and lithium chloride in an ethylene-vinyl alcohol copolymer. *J Membrane Sci.* 12: 195–205.

Gaines, C. S., Kassuba, A., and Finney, P. L. 1991a. Objective measurement of cookie hardness. I. Methods assessment. *Cereal Chem.* (in press).

Gaines, C. S., Kassuba, A., Finney, P. L., and Donelson, J. R. 1991b. Attributes of laboratory-produced cookies using a typical wire-cut formulation and a sugar-snap cookie formulation. *Cereal Chem.* (in press).

German, M. L., Blumenfeld, A. L., Yuryev, V. P., and Tolstoguzov, V. B. 1989. An NMR study of structure formation in maltodextrin systems. *Carbohydr. Polym.* 11: 139–146.

Ghiasi, K., Hoseney, R. C., and Varriano-Marston, E. 1983. Effects of flour components and dough ingredients on starch gelatinization. *Cereal Chem.* 60: 58–61.

Ghiasi, K. and Skarra, L. L. 1990. Taco shell and method of manufacture. U.S. patent 4,950,490.

Gidley, M. J. 1987. Factors affecting crystalline type of native starches and model materials. *Carbohydr. Res.* 161: 301–304.

Gidley, M. J., and Bulpin, P. V. 1987. Crystallization of maltaoses as models of the crystalline forms of starch. *Carbohydr. Res.* 161: 291–300.

Gidley, M. J. 1989. Molecular mechanisms underlying amylose aggregation and gelation. *Macromolecules* 22: 351–358.

Gidley, M. J., and Bulpin, P. V. 1989. Aggregation of amylose in aqueous systems: the effect of chain length on phase behavior and aggregation kinetics. *Macromolecules* 22: 341–346.

Given, P. S. 1991. Molecular behavior of water in a flour-water baked model system. In *Water Relationships in Foods*, H. Levine and L. Slade (Ed.), pp. 465–483. Plenum Press, New York.

Gould, G. W., and Christian, J. H. B. 1988. Characterization of the state of water in foods—biological aspects. In *Food Preservation by Moisture Control*, C. C. Seow (Ed.), pp. 43–56. Elsevier Applied Science, London.

Graessley, W. W. 1984. Viscoelasticity and flow in polymer melts and concentrated solutions. In *Physical Properties of Polymers*, J. E. Mark, A. Eisenberg, W. W. Graessley, L. Mandelkern, and J. L. Koenig (Ed.), pp. 97–153. American Chemical Society, Washington, DC.

Green, J. L. and Angell, C. A. 1989. Phase relations and vitrification in saccharide-water solutions and the trehalose anomaly. *J. Phys. Chem.* 93: 2880–2882.

Guilbot, A. and Godon, B. 1984. Le pain rassis. *Cah. Nut. Diet.* 19: 171–181.

Hardy, G., Cser, F., and Takacs, E. 1973. Investigation of the binary system of a crystalline monomer and its polymer. *J. Polym. Sci.: Symp. No. 42:* 663–669.

Hartel, R. W. and Shastry, A. V. 1991. Sugar crystallization in food products. *Crit. Rev. Food Sci. Nutr.* 30: 49–112.

Hatley, R. H. M., van den Berg, C., and Franks, F. 1991. The unfrozen water content of maximally freeze-concentrated carbohydrate solutions: Validity of the methods used for its determination. *Cryo-Lett.* 12: 113–124.

Haward, R. N. 1973. The nature of polymer glasses, their packing density, and mechanical behavior. In *The Physics of Glassy Polymers*, R. N. Haward (Ed.), pp. 1–51. Applied Science, London.

Herrington, T. M. and Branfield, A. C. (1984). Physico-chemical studies on sugar glasses. I. and II. *J. Food Technol.* 19: 409–435.

Hill, M. A. 1989. *Carbohydr. Polym.* 10: 64.

Hiltner, A. and Baer, E. 1986. Reversible gelation of macromolecular systems. *Polym. Prepr.* 27(1): 207.

Hoeve, C. A. J., and Hoeve, M. B. J. A. 1978. The glass point of elastin as a function of diluent concentration. *Organ. Coat. Plast. Chem.* 39: 441–443.

Hoeve, C. A. J. 1980. The structure of water in polymers. In *Water in Polymers*, S. P. Rowland (Ed.), ACS Symp. Ser. 127, pp. 135–146. American Chemical Society, Washington, DC.

Hofer, K., Hallbrucker, A., Mayer, E., and Johari, G. P. 1989. Vitrified dilute aqueous solutions. 3. Plasticization of water's H-bonded network and the glass transition temperature's minimum. *J. Phys. Chem.* 93: 4674–4677.

Hosea, T. J. C., Ng, S. C., and Oates, C. G. 1990. A Brillouin scattering study of the glass transition in sucrose. *Food Hydrocolloids* 4: 137–147.

Hoseney, R. C. 1986. *Principles of Cereal Science and Technology.* American Association of Cereal Chemists, St. Paul, MN.

Hoseney, R. C., Zeleznak, K., and Lai, C. S. 1986. Wheat gluten: A glassy polymer. *Cereal Chem.* 63: 285–286.

Hoseney, R. C., and Rogers, D. E. 1990. The formation and properties of wheat flour doughs. *Crit. Rev. Food Sci. Nutr.* 29: 73–93.

Hoseney, R. C. 1991. Wheat gluten: rheological and gas retaining properties. In *Water Relationships in Foods*, H. Levine and L. Slade (Ed.), pp. 657–666. Plenum Press, New York.

Hutchinson, R. J., Mantle, S. A., and Smith, A. C. 1989. The effect of moisture content on the mechanical properties of extruded food foams. *J. Materials Sci.* 24: 3249–3253.

I'Anson, K. J., Miles, M. J., Morris, V. J., Ring, S. G., and Nave, C. 1988. Study of amylose gelation using synchrotron X-ray source. *Carbohydr. Polym.* 8: 45–53.

I'Anson, K. J., Miles, M. J., Morris, V. J., Besford, L. S., Jarvis, D. A., and Marsh, R. A. 1990. The effects of added sugars on the retrogradation of wheat starch gels. *J. Cereal Sci.* 11: 243–248.

Imberty, A. and Perez, S. 1988. A revisit to the three-dimensional structure of B-type starch. *Biopolymers* 27: 1205–1221.

Izzard, M. J., Ablett, S., and Lillford, P. J. 1991. A calorimetric study of the glass transition occurring in sucrose solutions. In *Food Polymers, Gels and Colloids*, E. Dickinson (Ed.), pp. 289–300. Royal Society of Chemistry, Cambridge.

Jankowski, T. and Rha, C. K. 1986. Retrogradation of starch in cooked wheat. *Starke* 38: 6–9.

Jin, X., Ellis, T. S., and Karasz, F. E. 1984. The effect of crystallinity and crosslinking on the depression of the glass transition temperature in nylon 6 by water. *J. Polym. Sci.: Polym. Phys. Edn.* 22: 1701–1717.

Johari, G. P., Hallbrucker, A., and Mayer, E. 1989. Thermal behavior of several hyperquenched organic glasses. *J. Phys. Chem.* 93: 2648–2652.

Johnson, G. E., Bair, H. E., Matsuoka, S., Anderson, E. W., and Scott, J. E. 1980. Water sorption and its effects on a polymer's dielectric behavior. In *Water in Polymers*, S. P. Rowland (Ed.), ACS Symp. Ser. 127, pp. 451–468. American Chemical Society, Washington, DC.

Johnson, J. M., Davis, E. A., and Gordon, J. 1990. Interactions of starch and sugar water measured by ESR and DSC. *Cereal Chem.* 67: 286–291.

Johnston-Banks, F. A. 1990. Gelatine. In *Food Gels*, P. Harris (Ed.), pp. 233–289. Elsevier Applied Science, London.

Jolley, J. E. 1970. The microstructure of photographic gelatin binders. *Photogr. Sci. Engn.* 14: 169–177.

Juliano, B. O. 1982. Properties of rice starch in relation to varietal differences in processing characteristics of rice grain. *J. Jpn. Soc. Starch Sci.* 29: 305–317.

Kakivaya, S. R. and Hoeve, C. A. J. 1975. The glass transition of elastin. *PNAS USA* 72: 3505–3507.

Kalichevsky, M. T., Jaroszkiewicz, E. M., Ablett, S., Blanshard, J. M. V., and Lillford, P. J. 1991. The glass transition of amylopectin measured by DSC, DMTA and NMR. *Carbohydr. Polym.* (in press).

Kararli, T. T. and Catalano, T. 1990. Stabilization of misoprostol with hydroxypropyl methylcellulose (HPMC) against degradation by water. *Pharmaceut. Res.* 7: 1186–1189.

Kararli, T. T., Hurlbut, J. B., and Needham, T. E. 1990. Glass-rubber transitions of cellulosic polymers by DMA. *J. Pharmaceut. Sci.* 79: 845–848.

Karel, M. and Flink. J. M. 1983. Some recent developments in food dehydration research. In *Advances in Drying*, Vol. 2, A. S. Mujumdar (Ed.), pp. 103–153. Hemisphere, Washington, DC.

Karel, M. 1985. Effects of water activity and water content on mobility of food components, and their effects on phase transitions in food systems. In *Properties of Water in Foods*, D. Simatos and J. L. Multon (Ed.), pp. 153–169. Martinus Nijhoff, Dordrecht.

Karel, M. 1986. Control of lipid oxidation in dried foods. In *Concentration and Drying of Foods*, D. MacCarthy (Ed.), pp. 37–51. Elsevier Applied Sci., London.

Karel, M. and Langer, R. 1988. Controlled release of food additives. In *Flavor Encapsulation*, ACS Symposium Series 370, S. J. Risch and G. A. Reineccius (Ed.), pp. 177–191. American Chemical Society, Washington, DC.

Karel, M. 1989. Role of water activity. In *Food Properties and Computer-Aided Engineering of Food Processing Systems*, R. P. Singh and A. G. Medina (Ed.), pp. 135–155. Kluwer, Dordrecht.

Karel, M. and Saguy, I. 1991. Effects of water on diffusion in food systems. In *Water Relationships in Foods*, H. Levine and L. Slade (Ed.), pp. 157–173. Plenum Press, New York.

Kauzmann, W. 1948. Nature of the glassy state and behavior of liquids at low temperatures. *Chem. Rev.* 43: 219–256.

Keinath, S. E. and Boyer, R. F. 1981. Thermomechanical analysis of Tg and T > Tg transitions in polystyrene. *J. Appl. Polym. Sci.* 26: 2077–2085.

Kelley, S. S., Rials, T. G., and Glasser, W. G. 1987. Relaxation behavior of the amorphous components of wood. *J. Materials Sci.* 22: 617–624.

Knutson, C. A. 1990. Annealing of maize starches at elevated temperatures, *Cereal Chem.* 67: 376–384.

Kuge, T. and Kitamura, S. 1985. Annealing of starch granules—warm water treatment and heat-moisture treatment. *J. Jpn. Soc. Starch Sci.* 32: 65–83.

Kulp, K. and Ponte, J. G. 1981. Staling of white pan bread: Fundamental causes. *Crit. Rev. Food Sci. Nutr.* 15: 1–48.

Kuprianoff, J. 1958. Fundamental aspects of the dehydration of foodstuffs. *Conference on Fundamental Aspects of the Dehydration of Foodstuffs*, pp. 14–23. Society of Chemical Industry, Aberdeen.

Labuza, T. P. 1985. Water binding of humectants. In *Properties of Water in Foods*, D. Simatos and J. L. Multon (Ed.), pp. 421–445. Martinus Nijhoff, Dordrecht.

Labuza, T. P. 1989. Fiber's water binding capacity. *Cereal Foods World* 34: 566–567.

LeBlanc, E. L., LeBlanc, R. J., and Blum, I. E. 1988. Prediction of quality in frozen cod fillets. *J. Food Sci.* 53: 328–340.

Lee, C. Y., Smith, N. L., and Hawbecker, D. E. 1988. Enzyme activity and quality of frozen green beans as affected by blanching and storage. *J. Food Qual.* 11: 279–287.

Lelievre, J. 1976. Theory of gelatinization in a starch-water-solute system. *Polymer* 17: 854–858.

Le Meste, M. and Duckworth, R. B. 1988. Effect of water content on the mobility of solute molecules and of protein side chains in caseinate preparations. *Intl. J. Food Sci. Technol.* 23: 457–466.

Le Meste, M., Viguier, L., Lorient, D., and Simatos, D. 1990. Rotational diffusivity of solutes in concentrated caseinates: Influence of glycosylation. *J. Food Sci.* 55: 724–727.

Le Meste, M. and Huang, V. 1991. Thermomechanical analysis of frozen sucrose solutions. Institute of Food Technologists Annual Meeting, June 2–5, Dallas, Abs. #653. *J. Food Sci.* (submitted).

Le Meste, M., Panama, J., Anderson, G., Lentz, R., and Huang, V. 1991a. Thermomechanical analysis of cereal products. American Association of Cereal Chemists Annual Meeting, Oct. 13–16, Seattle, WA.

Le Meste, M., Voilley, A., and Colas, B. 1991b. Influence of water on the mobility of small molecules dispersed in a polymeric system. In *Water Relationships in Foods*, H. Levine and L. Slade (Ed.), pp. 123–138. Plenum Press, New York.

Levine, H. and Slade, L. 1986. A polymer physico-chemical approach to the study of commercial starch hydrolysis products (SHPs). *Carbohydr. Polym.* 6: 213–244.

Levine, H. and Slade, L. 1988a. Water as a plasticizer: physico-chemical aspects of low-moisture polymeric systems. In *Water Science Reviews*, Vol. 3, F. Franks (Ed.), pp. 79–185. Cambridge University Press, Cambridge, England.

Levine, H. and Slade, L. 1988b. Collapse phenomena—a unifying concept for interpreting the behavior of low-moisture foods. In *Food Structure—Its Creation and Evaluation*, J. R. Mitchell and J. M. V. Blanshard (Ed.), pp. 149–180. Butterworths, London.

Levine, H. and Slade, L. 1988c. Principles of cryostabilization technology from structure/property relationships of water-soluble food carbohydrates—a review. *Cryo-Lett.* 9: 21–63.

Levine, H. and Slade, L. 1988d. Thermomechanical properties of small carbohydrate-water glasses and "rubbers": Kinetically-metastable systems at subzero temperatures. *J. Chem. Soc., Faraday Trans.* I, 84: 2619–2633.

Levine, H., and Slade, L. 1989a. Interpreting the behavior of low-moisture foods. In *Water and Food Quality*, T. M. Hardman (Ed.), pp. 71–134. Elsevier, London.

Levine, H. and Slade, L. 1989b. Influences of the glassy and rubbery states on the thermal, mechanical, and structural properties of doughs and baked products. In *Dough Rheology and Baked Product Texture*, H. Faridi and J. M. Faubion (Ed.), pp. 157–330. Van Nostrand Reinhold/AVI, New York.

Levine, H. and Slade, L. 1989c. A food polymer science approach to the practice of cryostabilization technology. *Comm. Agric. Food Chem.* 1: 315–396.

Levine, H. and Slade, L. 1989d. Response to the letter by Simatos, Blond, and Le Meste on the relation between glass transition and stability of a frozen product. *Cryo-Lett.* 10: 347–370.

Levine, H. and Slade, L. 1990. Cryostabilization technology: thermoanalytical evaluation of food ingredients and systems. In *Thermal Analysis of Foods*, C.-Y. Ma and V. R. Harwalkar (Ed.), pp. 221–305. Elsevier Applied Science, London.

Levine, H. and Slade, L. 1991. Polymer physicochemical characterization of oligosaccharides. In *Biotechnology of Amylodextrin Oligosaccharides*, R. B. Friedman (Ed.), pp. 219–260. ACS Symp. Ser. 458, ACS Books, Washington, DC.

Levine, H., Slade, L., Pickup, J., and van Lengerich, B. 1991. Glassy matrices containing volatile and/or labile components, and processes for preparation and use thereof. U.S. patent 5,009,900.

Lillford, P. J. 1988. The polymer/water relationship—its importance for food structure. In *Food Structure—Its Creation and Evaluation*, J. M. V. Blanshard and J. R. Mitchell (Ed.), pp. 75–92. Butterworths, London.

Lim, M. H. and Reid, D. S. 1991. Studies of reaction kinetics in relation to the Tg′ of polymers in frozen model systems. In *Water Relationships in Foods*, H. Levine and L. Slade (Ed.), pp. 103–122. Plenum Press, New York.

Lineback, D. R., and Rasper, V. F. 1988. Wheat carbohydrates. In *Wheat Chemistry and Technology*, 3rd ed., Vol. I, Y. Pomeranz (Ed.), pp. 277–372. AACC, St. Paul.

Liu, H. and Lelievre, J. 1991. A differential scanning calorimetry study of glass and melting transitions in starch suspensions and gels. *Carbohydr. Res.* (in press).

Lund, D. B. 1989. Starch gelatinization. In *Food Properties and Com-*

puter-Aided Engineering of Food Processing Systems, R. P. Singh and A. G. Medina (Ed.), pp. 299–311. Kluwer, Dordrecht.

Luyet, B. 1960. On various phase transitions occurring in aqueous solutions at low temperatures. *Ann. New York Acad. Sci.* 85: 549–569.

Ma, C. Y. and Harwalkar, V. R. 1990. *Thermal Analysis of Foods*, Elsevier Applied Science, London.

MacDonald, G. A. and Lanier, T. 1991. Carbohydrates as cryoprotectants for meats and surimi. *Food Technol.* 45(3): 150–159.

MacKenzie, A. P. 1977. Non-equilibrium freezing behavior of aqueous systems, *Phil. Trans. Royal Soc.* London B. 278: 167–189.

Makower, B. and Dye, W. B. 1956. Equilibrium moisture content and crystallization of amorphous sucrose and glucose. *J. Agric. Food Chem.* 4: 72–77.

Mandelkern, L. 1986. Thermoreversible gelation and crystallization from solution. *Polym. Prepr.* 27(1): 206.

Marsh, R. D. L. and Blanshard, J. M. V. 1988. The application of polymer crystal growth theory to the kinetics of formation of the B-amylose polymorph in a 50% wheat starch gel. *Carbohydr. Polym.* 9: 301–317.

Marshall, A. S. and Petrie, S. E. B. 1980. Thermal transitions in gelatin and aqueous gelatin solutions. *J. Photogr. Sci.* 28: 128–134.

Mashimo, S., Kuwabara, S., Yagihara, S., and Higasi, K. 1987. Dielectric relaxation time and structure of bound water in biological materials. *J. Phys. Chem.* 91: 6337–6338.

Masi, P. 1989. Study of the influence of temperature on the rheological behavior of gluten by means of dynamic mechanical analysis. In *Food Properties and Computer-aided Engineering of Food Processing Systems*, R. P. Singh and A. G. Medina (Ed.), pp. 357–362. Kluwer, Dordrecht.

Matsuoka, S., Williams, G., Johnson, G. E., Anderson, E. W., and Furukawa, T. 1985. Phenomenological relationship between dielectric relaxation and thermodynamic recovery processes near the glass transition. *Macromolecules* 18: 2652–2663.

Matthiesen, A. B., Hanson, C. E., Huang, V. T., and Le Meste, M. 1991. Thermal analysis instrument for the determination of Tg of water and sucrose solutions. Institute of Food Technologists Annual Meeting, June 2–5, Dallas, Abs. #717.

Maurice, T. J., Slade, L., Page, C., and Sirett, R. 1985. Polysaccharide-water interactions—thermal behavior of rice starch. In *Properties of Water in Foods*, D. Simatos and J. L. Multon (Ed.), pp. 211–227. Martinus Nijhoff, Dordrecht.

Maurice, T. J., Asher, Y. J., and Thomson, S. 1991. Thermomechanical analysis of frozen aqueous systems. In *Water Relationships in Foods*, H. Levine and L. Slade (Ed.), pp. 215–223. Plenum Press, New York.

Mayer, E. 1988. Hyperquenching of water and dilute aqueous solutions into their glassy states: An approach to cryofixation. *Cryo-Lett.* 9: 66–77.

Mestres, C., Colonna, P., and Buleon, A. 1988. Gelation and crystallization of maize starch after pasting, drum-drying, or extrusion cooking. *J. Cereal Sci.* 7: 123–134.

Miles, M. J., Morris, V. J., and Ring, S. G. 1985a. Gelation of amylose. *Carbohydr. Res.* 135: 257–269.

Miles, M. J., Morris, V. J., Orford, P. D., and Ring, S. G. 1985b. Roles of amylose and amylopectin in gelation and retrogradation of starch. *Carbohydr. Res.* 135: 271–281.

Mita, T. 1990. Effect of aging on the rheological properties of gluten gel. *Agric. Biol. Chem.* 54: 927–935.

Mitchell, J. R. 1980. The rheology of gels. *J. Texture Stud.* 11: 315–337.

Morozov, V. N. and Gevorkian, S. G. 1985. Low-temperature glass transition in proteins. *Biopolymers* 24: 1785–1799.

Morris, V. J. 1990. Starch gelation and retrogradation. *Trends Food Sci. Technol.* 1: 2–6.

Moy, P. and Karasz, F. E. 1980. The interactions of water with epoxy resins. In *Water in Polymers*, S. P. Rowland (Ed.), ACS Symp. Ser. 127, pp. 505–513. American Chemical Society, Washington, DC.

Murthy, N. S., Stamm, M., Sibilia, J. P., and Krimm, S. 1989. Structural changes accompanying hydration in nylon 6. *Macromolecules* 22: 1261–1267.

Myers-Betts P. A. and Baianu, I. C. 1990. Approaches to protein hydration and water activity in food proteins by NMR and vapor pressure equilibration techniques. *J. Agric. Food Chem.* 38: 1171–1177.

Nakazawa, F., Noguchi, S., Takahashi, J., and Takada, M. 1985. Retrogradation of gelatinized potato starch studied by DSC. *Agric. Biol. Chem.* 49: 953–957.

Nishinari, K., Watase, M., Williams, P. A., and Phillips, G. O. 1991. Effect of sugars and polyols on water in agarose gels. In *Water Relationships in Foods*, H. Levine and L. Slade (Ed.), pp. 235–249. Plenum Press, New York.

Noel, T. R., Ring, S. G., and Whittam, M. A. 1990. Glass transitions in low-moisture foods. *Trends Food Sci. Technol.* 1: 62–67.

Ollett, A. L. and Parker, R. 1990. The viscosity of supercooled fructose and its glass transition temperature. *J. Texture Stud.* 21: 355–362.

Ollett, A. L., Parker, R., and Smith, A. C. 1990. Deformation and fracture behavior of wheat starch plasticised with water and glucose. *J. Materials Sci.* 26: 1351–1356.

Orford, P. D., Parker, R., Ring, S. G., and Smith, A. C. 1989. The effect of water as a diluent on the glass transition behavior of malto-oligosaccharides, amylose and amylopectin. *Int. J. Biol. Macromol.* 11: 91–96.

Orford, P. D., Parker, R., and Ring, S. G. 1990. Aspects of the glass transition behavior of mixtures of carbohydrates of low molecular weight. *Carbohydr. Res.* 196: 11–18.

Ostberg, G., Salmen, L., and Terlecki, J. (1990). Softening temperature of moist wood measured by DSC. *Holzforschung* 44: 223–225.

Paakkonen, K. and Roos, Y. H. 1990. Effects of drying conditions on water sorption and phase transitions of freeze-dried horseradish roots. *J. Food Sci.* 55: 206–209.

Paton, D. 1987. DSC of oat starch pastes. *Cereal Chem.* 64: 394–399.

Petrie, S. E. B. 1975. The problem of thermodynamic equilibrium in glassy polymers. In *Polymeric Materials: Relationships Between Structure and Mechanical Behavior*, E. Baer and S. V. Radcliffe (Ed), pp. 55–118. Amer. Soc. Metals, Metals Park, OH.

Phillips, A. J., Yarwood, R. J., and Collett, J. H. 1986. Thermal analysis of freeze-dried products. *Anal. Proceed.* 23: 394–395.

Pikal, M. J. and Shah, S. 1990. The collapse temperature in freeze drying: dependence on measurement methodology and rate of water removal from the glassy phase. *Int. J. Pharm.* 62: 165–186.

Poole, P. L. and Finney, J. L. 1983. Sequential hydration of a dry globular protein. *Biopolymers* 22: 255–260.

Poole, P. L. and Finney, J. L. 1984. Sequential hydration of dry proteins. *Biopolymers* 23: 1647–1666.

Pouchly, J., Biros, J., and Benes, S. 1979. Heat capacities of water-swollen hydrophilic polymers above and below 0°C. *Makromol. Chem.* 180: 745–760.

Pouchly, J. Benes, S., Masa, Z., and Biros, J. 1982. Sorption of water in hydrophilic polymers. *Makromol. Chem.* 183: 1565–1575.

Quinquenet, S., Grabielle-Madelmont, C., Ollivon, M., and Serpelloni, M. 1988. Influence of water on pure sorbitol polymorphism. *J. Chem. Soc., Faraday Trans.* I, 84: 2609–2618.

Radosta, S., Schierbaum, F., Reuther, F., and Anger, H. 1989. Polymer-water interaction of maltodextrins—Part I: Water vapor sorption and desorption of maltodextrin powders. *Starke* 41: 395–401.

Radosta, S. and Schierbaum, F. 1990. Polymer-water interaction of maltodextrins. Part III: Non-freezable water in maltodextrin solutions and gels. *Starke* 42: 142–147.

Reid, D. S. 1985. Correlation of the phase behavior of DMSO/NaCl/water and glycerol/NaCl/water as determined by DSC with their observed behavior on a cryomicroscope. *Cryo-Letters* 6: 181–188.

Reid, D. S. and Charoenrein, S. 1985. DSC studies of starch-water interaction in gelatinization process. In *Proceedings 14th NATAS Conference*, pp. 335–340. NATAS, San Francisco.

Reid, D. S. 1990. Optimizing the quality of frozen foods. *Food Technol.* 44(7): 78–82.

Reid, D. S. and Hsu, J. 1991. The estimation of the concentration of the maximally freeze-concentrated matrix in a frozen system. *J. Food Sci.* (submitted).

Reuther, F., Damaschun, G., Gernat, C., Schierbaum, F., Kettlitz, B., Radosta, S., and Nothnagel, A. 1984. Molecular gelation mechanism of maltodextrins investigated by wide-angle X-ray scattering. *Coll. Polym. Sci.* 262: 643–647.

Richter, M., Schierbaum, F., Augustat, S., and Knoch, K. D. 1976a. Method of Producing Starch Hydrolysis Products for Use as Food Additives. U.S. patent 3,962,465.

Richter, M., Schierbaum, F., Augustat, S., and Knoch, K. D. 1976b. Method of Producing Starch Hydrolysis Products for Use as Food Additives. U.S. patent 3,986,890.

Ring, S. G. 1985a. Observations on crystallization of amylopectin from aqueous solution. *Int. J. Biol. Macromol.* 7: 253–254.

Ring, S. G. 1985b. Studies on starch gelation. *Starke* 37: 80–83.

Ring, S. G. and Orford, P. D. 1986. Recent observations on retrogradation of amylopectin. In *Gums and Stabilizers for the Food Industry* 3, G. O. Phillips, D. J. Wedlock, and P. A. Williams (Ed.), pp. 159–165. Elsevier Applied Science, London.

Ring, S. G., Colonna, P., I'Anson, K. J., Kalichevsky, M. T., Miles, M. J., Morris, V. J., and Orford, P. D. 1987. Gelation and crystallization of amylopectin. *Carbohydr. Res.* 162: 277–293.

Ring, S. G. and Whittam, M. A. 1991. Linear dextrins—solid forms and aqueous solution behavior. In *Biotechnology of Amylodextrin Oligosaccharides*, R. B. Friedman (Ed.), ACS Symp. Ser. 458, pp. 273–293. ACS Books, Washington, DC.

Roberts, G. E. and White, E. F. T. 1973. Relaxation processes in amorphous polymers. In *The Physics of Glassy Polymers*, R. N. Haward (Ed.), pp. 153–222. Applied Science, London.

Robertson, R. E. 1985. Segmental mobility in the equilibrium liquid below the glass transition. *Macromolecules* 18: 953–958.

Roos, Y. 1987. Effect of moisture on the thermal behavior of strawberries studied using DSC. *J. Food Sci.* 52: 146–149.

Roos, Y. and Karel, M. 1990. DSC study of phase transitions affecting the quality of dehydrated materials. *Biotechnol. Prog.* 6: 159–163.

Roos, Y. and Karel, M. 1991a. Plasticizing effect of water on thermal behavior and crystallization of amorphous food models. *J. Food Sci.* 56: 38–43.

Roos, Y. and Karel, M. 1991b. Phase transitions of mixtures of amorphous polysaccharides and sugars. *Biotechnol. Prog.* 7: 49–53.

Roos, Y. and Karel, M. 1991c. Phase transitions of amorphous sucrose and frozen sucrose solutions. *J. Food Sci.* 56: 266–267.

Roos, Y. and Karel, M. 1991d. Amorphous state and delayed ice formation in sucrose solutions. *Int. J. Food Sci. Technol.* (in press).

Roos, Y. and Karel, M. 1991e. Effects of water and molecular weight on glass transitions in amorphous carbohydrates and carbohydrate solutions. *J. Food Sci.* (in press).

Roozen, M. J. G. W. and Hemminga, M. A. 1990. Molecular motion in sucrose-water mixtures in the liquid and glassy state as studied by spin probe ESR. *J. Phys. Chem.* 94: 7326–7329.

Roper, H. and Koch, H. 1990. The role of starch in biodegradable thermoplastic materials. *Starke* 42: 123–130.

Rowland, S. P. 1980. *Water in Polymers*, ACS Symp. Ser. 127. American Chemical Society, Washington, DC.

Rubin, C. A., Wasylyk, J. M., and Baust, J. G. 1990. Investigation of vitrification by NMR and DSC in honey: a model carbohydrate system. *J. Agric. Food Chem.* 38: 1824–1827.

Russell, P. L. 1987a. Gelatinization of starches of different amylose/amylopectin content—DSC study. *J. Cereal Sci.* 6: 133–145.

Russell, P. L. 1987b. Aging of gels from starches of different amylose/amylopectin content studied by DSC. *J. Cereal Sci.* 6: 147–158.

Russell, P. L. and Oliver, G. 1989. The effect of pH and NaCl content on starch gel aging. A study by DSC and rheology. *J. Cereal Sci.* 10: 123–138.

Saleeb, F. Z. and Pickup, J. G. 1985. Fixing volatiles in an amorphous substrate and products therefrom. U.S. patent 4,532,145.

Saleeb, F. Z. and Pickup, J. G. 1989. Fixation of volatiles in extruded glass substrates. U.S. patent 4,820,534.

Salmen, N. L. and Back, E. L. 1977. Influence of water on Tg of cellulose. *Tappi* 60: 137–140.

Schenz, T. W., Rosolen, M. A., Levine, H., and Slade, L. 1984. DMA of frozen aqueous solutions. In *Proceedings 13th Annual Conference, North American Thermal Analysis Society*, A. R. McGhie (Ed.), Sept. 23–26, Philadelphia, pp. 57–62.

Schenz, T. W., Israel, B., and Rosolen, M. A. 1991. Thermal analysis of water-containing systems. In *Water Relationships in Foods*, H. Levine and L. Slade (Ed.), pp. 199–214. Plenum Press, New York.

Schierbaum, F., Reuther, F., Braudo, E. E., Plashchina, I. G., and Tolstoguzov, V. B. (1990). Thermodynamic parameters of the junction zones in thermoreversible maltodextrin gels. *Carbohydr. Polym.* 12: 245–253.

Sears, J. K. and Darby, J. R. 1982. *The Technology of Plasticizers*. Wiley-Interscience, New York.

Shalaby, S. W. 1981. Thermoplastic polymers. In *Thermal Characterization of Polymeric Materials*, E. A. Turi (Ed.), pp. 235–364. Academic Press, Orlando.

Sharples, A. 1972. Crystallinity. In *Polymer Science*, A. D. Jenkins (Ed.), pp. 251–321. North Holland, Amsterdam.

Shi, Y. C. and Seib, P. 1989. Properties of wheat starch compared to normal maize starch. In *Wheat Is Unique*, Y. Pomeranz (Ed.), pp. 215–234. Amer. Assoc. Cereal Chem., St. Paul, MN.

Shimada, Y., Roos, Y., and Karel, M. 1991. Oxidation of methyl linoleate encapsulated in amorphous lactose-based food model. *J. Agric. Food Chem.* 39: 637–641.

Shukla, T. P. 1991. A new development in carbohydrate research. *Cereal Foods World* 36: 251–252.

Sichina, W. J. 1988. Predicting mechanical performance and lifetime of polymeric materials. *Amer. Lab.* 20(1): 42–52.

Simatos, D. and Karel, M. 1988. Characterizing condition of water in foods: physico-chemical aspects. In *Food Preservation by Moisture Control*, C. C. Seow (Ed.), pp. 1–41. Elsevier Applied Science, London.

Simatos, D., Blond, G., and Le Meste, M. 1989. Relation between glass transition and stability of a frozen product. *Cryo-Lett.* 10: 77–84.

Simatos, D. and Blond, G. 1991. DSC studies and stability of frozen foods. In *Water Relationships in Foods*, H. Levine and L. Slade (Ed.), pp. 139–155. Plenum Press, New York.

Slade, L. 1984. Starch properties in processed foods: staling of starch-based products. American Association of Cereal Chemists Annual Meeting, 9/30–10/4, Minneapolis, MN, Abs. #112.

Slade, L. and Levine, H. 1984. Thermal analysis of starch and gelatin. In *Proceedings 13th Annual Conference, North American Thermal Analysis Society*, A. R. McGhie (Ed.), p. 64. Sept. 23–26, Philadelphia, PA.

Slade, L. and Levine, H. 1987a. Polymer-chemical properties of gelatin in foods. In *Advances in Meat Research*, Vol. 4—*Collagen as a Food*, A. M. Pearson, T. R. Dutson, and A. Bailey (Ed.), pp. 251–266. AVI, Westport, CT.

Slade, L. and Levine, H. 1987b. Recent advances in starch retrogradation. In *Industrial Polysaccharides*, S. S. Stivala, V. Crescenzi, and I. C. M. Dea (Ed.), pp. 387–430. Gordon and Breach Science Publishers, New York.

Slade, L., Altomare, R., Oltzik, R., and Medcalf, D. G. 1987. Accelerated staling of starch-based food products. U.S. patent 4,657,770.

Slade, L. and Levine, H. 1988a. Structural stability of intermediate moisture foods—a new understanding? In *Food Structure—Its Creation and Evaluation*, J. R. Mitchell and J. M. V. Blanshard (Ed.), pp. 115–147. Butterworths, London.

Slade, L. and Levine, H. 1988b. Non-equilibrium behavior of small carbohydrate-water systems. *Pure Appl. Chem.* 60: 1841–1864.

Slade, L. and Levine, H. 1988c. Non-equilibrium melting of native granular starch: I. Temperature location of the glass transition associated with gelatinization of A-type cereal starches. *Carbohydr. Polym.* 8: 183–208.

Slade, L. and Levine, H. 1988d. Thermal analysis of starch. In *1988 CRA Scientific Conference*, pp. 169–244. Corn Refiners Assoc., Washington, DC.

Slade, L. and Levine, H. 1989. A food polymer science approach to selected aspects of starch gelatinization and retrogradation. In *Frontiers in Carbohydrate Research*-1: *Food Applications*, R. P. Millane, J. N. BeMiller, and R. Chandrasekaran (Ed.), pp. 215–270. Elsevier Applied Science, London.

Slade, L., Levine, H., and Finley, J. W. 1989. Protein-water interactions: water as a plasticizer of gluten and other protein polymers. In *Protein Quality and the Effects of Processing*, R. D. Phillips and J. W. Finley (Ed.), pp. 9–124. Marcel Dekker, New York.

Slade, L. and Levine, H. 1991a. Beyond water activity: Recent advances based on an alternative approach to the assessment of food quality and safety. *CRC Crit. Revs. Food Sci. Nutr.* 30: 115–360.

Slade, L. and Levine, H. 1991b. A polymer science approach to structure/property relationships in aqueous food systems: non-equilibrium behavior of carbohydrate-water systems. In *Water Relationships in Foods*, H. Levine and L. Slade (Ed.), pp. 29–101. Plenum Press, New York.

Slade, L. and Levine, H. 1992. Structure-function relationships of cookie and cracker ingredients. In *The Science of Cookie and Cracker Production*, H. Faridi (Ed.), Ch. 2 (in press). Van Nostrand Reinhold/AVI, New York.

Smith, A. C. 1990. Studies on the physical structure of starch-based materials in the extrusion cooking process. Extrusion and Rheology of Foods Symposium, Rutgers University, New Brunswick, NJ, July 31.

Soesanto, T. and Williams, M. C. 1981. Volumetric interpretation of viscosity for concentrated and dilute sugar solutions. *J. Phys. Chem.* 85: 3338–3341.

Sperling, L. H. 1986. *Introduction to Physical Polymer Science*. Wiley-Interscience, New York.

Spies, R. D. and Hoseney, R. C. 1982. Effect of sugars on starch gelatinization. *Cereal Chem.* 59: 128–131.

Starkweather, H. W. 1980. Water in nylon. In *Water in Polymers*, S. P. Rowland (Ed.), pp. 433–440. ACS Symp. Ser. 127, American Chemical Society, Washington, DC.

Stepto, R. F. T. and Dobler, B. 1989. Method of producing destructurized starch. European patent application EP 0326517A1.

To, E. C. and Flink, J. M. 1978. "Collapse," a structural transition in freeze-dried carbohydrates. I–III. *J. Food Technol.* 13: 551–594.

Tomka, I., Bohonek, J., Spuhler, A., and Ribeaud, M. 1975. Structure and formation of the gelatin gel. *J. Photogr. Sci.* 23: 97–103.

Tomka, I. 1986. Thermodynamic theory of polar polymer solutions. *Polymer Prepr.* 27(2): 129.

Tomka, I. 1991. Thermoplastic starch. In *Water Relationships in Foods*, H. Levine and L. Slade (Ed.), pp. 627–637. Plenum Press, New York.

Turi, E. A. 1981. *Thermal Characterization of Polymeric Materials*. Academic Press, Orlando, FL.

van den Berg, C. 1981. Vapour sorption equilibria and other water-starch interactions; A physico-chemical approach. Doctoral thesis, Agricultural University, Wageningen, The Netherlands.

van den Berg, C. 1986. Water activity. In *Concentration and Drying of Foods*, D. MacCarthy (Ed.), pp. 11–36. Elsevier Applied Science, London.

van den Berg, C. 1991. Food-water relations: progress and integration, comments and thoughts. In *Water Relationships in Foods*, H. Levine and L. Slade (Ed.), pp. 21–28. Plenum Press, New York.

Vorwerg, W., Schierbaum, F. R., Reuther, F., and Kettlitz, B. 1988. Formation of thermally reversible networks from starch polysaccharides. In *Biological Synthesis of Polymer Networks*, O. Kramer (Ed.), pp. 127–139. Elsevier Applied Science, London.

Walton, A. G. 1969. Nucleation in liquids and solutions. In *Nucleation*, A. C. Zettlemoyer (Ed.), p. 225. Marcel Dekker, New York.

Warburton, S. C., Donald, A. M., and Smith, A. C. 1990. The deformation of brittle starch foams. *J. Materials Sci.* 25: (in press).

Wasylyk, J. M. and Baust, J. G. 1991. Vitreous domains in an aqueous ribose solution. In *Water Relationships in Foods*, H. Levine and L. Slade (Ed.), pp. 225–234. Plenum Press, New York.

Watase, M., Nishinari, K., and Hatakeyama, T. 1991. DSC study on properties of water in concentrated agarose gels. *Food Hydrocolloids* (in press).

Weisser, H. 1985. Influence of temperature on sorption equilibria. In *Properties of Water in Foods*, D. Simatos and J. L. Multon (Ed.), pp. 95–118. Martinus Nijhoff, Dordrecht.

Wesson, J. A., Takezoe, H., Yu, H., and Chen, S. P. 1982. Dye diffusion in swollen gels by forced Rayleigh scattering. *J. Appl. Phys.* 53: 6513–6519.

White, G. W. and Cakebread, S. H. 1966. The glassy state in certain sugar-containing food products. *J. Food Technol.* 1: 73–82.

Whittam, M. A., Noel, T. R., and Ring, S. G. 1991. Melting and glass/rubber transitions of starch polysaccharides. In *Food Polymers, Gels and Colloids*, E. Dickinson (Ed.), pp. 277–288. Royal Society Chemistry, Cambridge.

Williams, M. L., Landel, R. F., and Ferry, J. D. (1955). Temperature dependence of relaxation mechanisms in amorphous polymers and other glass-forming liquids. *J. Amer. Chem. Soc.* 77: 3701–3706.

Williams, R. J. and Carnahan, D. L. 1989. Association between ice nuclei and fracture interfaces in sucrose:water glasses. *Thermochim. Acta* 155: 103–107.

Williams, R. J. and Carnahan, D. L. 1990. Fracture faces and other interfaces as ice nucleation sites. *Cryobiology* 27: 479–482.

Wittwer, F. and Tomka, I. 1984. Polymer composition for injection moulding, European patent application 0118240A2.

Wunderlich, B. 1973. *Macromolecular Physics*, Vol. 1—*Crystal Structure, Morphology, Defects*. Academic Press, New York.

Wunderlich, B. 1976. *Macromolecular Physics*, Vol. 2—*Crystal Nucleation, Growth, Annealing*. Academic Press, New York.

Wunderlich, B. 1980. *Macromolecular Physics*, Vol. 3—*Crystal Melting*, Academic Press, New York.

Wunderlich, B. 1981. The basis of thermal analysis. In *Thermal Characterization of Polymeric Materials*, E. A. Turi (Ed.), pp. 91–234. Academic Press, Orlando.

Wunderlich, B. 1990. *Thermal Analysis*. Academic Press, Boston.

Wynne-Jones, S. and Blanshard, J. M. V. (1986). Hydration studies of wheat starch, amylopectin, amylose gels and bread by proton magnetic resonance. *Carbohydr. Polym*. 6: 289–306.

Yannas, I. V. 1972. Collagen and gelatin in the solid state. *J. Macromol. Sci.-Revs. Macromol. Chem*. C7: 49–104.

Yoshida, H., Hatakeyama, T., and Hatakeyama, H. 1990. Phase transitions of the water-xanthan system. *Polymer* 31: 693–698.

Yost, D. A. and Hoseney, R. C. 1986. Annealing and glass transition of starch. *Starke* 38: 289–292.

Zeleznak, K. J. and Hoseney, R. C. 1987a. The glass transition in starch. *Cereal Chem*. 64: 121–124.

Zeleznak, K. J. and Hoseney, R. C. 1987b. Characterization of starch from bread aged at different temperatures. *Starke* 39: 231–233.

Zobel, H. F. (1988). Starch crystal transformations and their industrial importance. *Starke* 40: 1–7.

Zobel, H. F., Young, S. N., and Rocca, L. A. 1988. Starch gelatinization: An X-ray diffraction study. *Cereal Chem*. 65: 443–446.

4

Surface Thermodynamics, Protein Adsorption, and Biofilm Development

Joseph McGuire and Viwat Krisdhasima

Oregon State University
Corvallis, Oregon

INTRODUCTION

From a thermodynamic standpoint, the extent of biological adhesion on a solid material could ideally be determined by surface energetics, i.e., the surface tensions of the contact material, liquid medium, and adhering particles involved. In natural circumstances, however, microbial adhesion is to some extent influenced by the interfacial behavior of dissolved protein. Surface chemical influences on protein adsorption thus constitute an important area for study. Many experimental results have indicated that adsorbed protein changes its conformation over time, and the rate of conformational change varies among proteins and from one surface to another. These surface-induced conformational changes are suspected to influence protein reactivity with whole cells and spores, as they have been observed to affect platelet and white cell adhesion from blood. Understanding the physical chemistry of surfaces and the nature of competitive adsorption in multiprotein systems is necessary for quantifying biofilm development. Although data for construction of a predictive model are not abundant, past and planned protein adsorption equilibrium and kinetic experiments hold promise for revealing the relative influences of solid surface, liquid medium and molecular properties on biofilm formation, as well as providing direction for its control.

SURFACE ENERGETICS

The Nature of Surface Energy

Work must be done on a pure liquid to increase its surface area. It is therefore inferred that surface molecules are in a state of higher free energy than their counterparts in the bulk. The free energy change associated with the isothermal, reversible formation of a liquid surface is termed the excess surface free energy—the excess free energy that the molecules possess by virtue of their being in the surface (Aveyard and Haydon, 1973). This higher energy state of surface molecules arises in part because surface molecules have fewer nearest neighbor interactions than bulk molecules. Since surface molecules are subject to an inward attraction normal to the surface, the surface itself is in a state of lateral tension, giving rise to the concept of liquid surface tension. For a pure liquid in equilibrium with its vapor, the surface tension and the excess surface free energy are numerically equal.

Similarly, the concepts of interfacial tension and excess interfacial free energy can be described at the interface between two pure, immiscible liquids. In principle, all of this applies to solid surfaces as well. However, absolute solid surface thermodynamic properties in general and solid-liquid interfacial energetics in particular are difficult to measure.

Evaluation of Properties Related to Solid Surface Energy

There is much interest in relating the surface properties of materials to the responses they evoke upon contact with fluid food. Such relationships would be useful as they may be used to quantify interactions that occur between a fluid food and a contact surface, such as biofilm formation on membrane, heat exchange, or food packaging surfaces.

The development of models that help to describe contact surface influences on interfacial activity depends upon unambiguous determination of properties related to solid surface energy, γ_S. To date there is no standard method for food contact surface characterization, i.e., measurement of γ_S. Often, materials are characterized under conditions that are not representative of the environment in which they must perform, i.e., an aqueous biological environment. As a consequence, there are inconsistencies among correlations reported between surface properties of materials and their tendency to interact with constituents in fluid foods.

Contact angle measurements are the basis for one of the most sensitive, yet simple techniques for describing surface energetics and thermodynamics. Contact angle data have been used extensively to evaluate solid surface properties related to surface energy. They have also been used to provide an index of material cleanliness. The derivation of these techniques and their development are discussed in most surface chemistry texts (Aveyard and Haydon, 1973; Hiemenz, 1986). Following is a brief review of basic contact angle principles, and commonly used methods for contact surface characterization.

The contact angle. The contact angle, θ, formed when a drop of liquid contacts a solid surface, is shown in Fig. 4.1. The subscripts S, L, and SL refer to the solid-vapor, liquid-vapor, and solid-liquid interfaces, respectively. The contact angle formed by a sessile drop on a given surface is routinely measured with a contact angle goniometer. This instrument consists of a light source, illuminating a stage on which the liquid drop/solid system rests, and a telescope. The drop is viewed through the telescope; cross hairs in the telescope and a 360° scale around the eyepiece enable accurate measurement of the contact angle. Young's equation (a force balance), written in its simplest form for the drop of liquid at equilibrium on a homogeneous, plane surface is

$$\gamma_S = \gamma_{SL} + \gamma_L \cos \theta \tag{1}$$

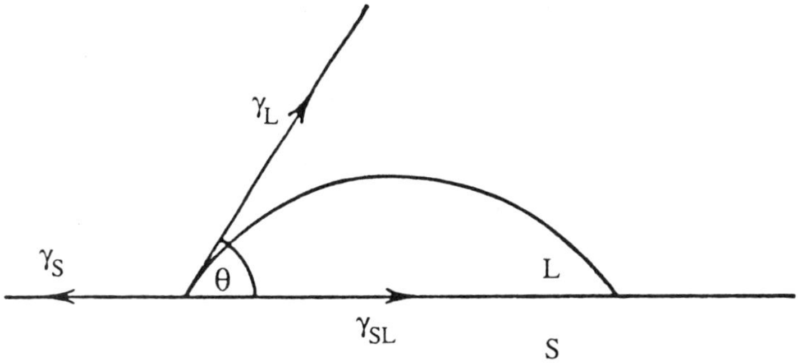

Fig. 4.1 The contact angle. (*From McGuire and Kirtley, 1988, with permission.*)

The Dupré equation (an energy balance) defines the work of adhesion, W_a, required to part the liquid from the solid:

$$W_a = \gamma_S + \gamma_L - \gamma_{SL} \quad (2)$$

The quantity γ_{SL} cannot be measured experimentally. Therefore, the Young and Dupré equations are combined to yield the following relationship for the work of adhesion between the solid and the liquid:

$$W_a = \gamma_L(1 + \cos\theta) \quad (3)$$

Rigorously, Eq. (3) applies only to a system at equilibrium. Consequently, the liquid must be saturated with the solid, and the vapor and homogeneous solid surface must be at adsorption equilibrium. This is not often the case, and values of θ for a given system may vary widely. Additionally, it should be noted that direct application of Eq. (3) to common engineering materials may yield misleading results, as these surfaces exhibit energetic heterogeneities and other surface irregularities.

Several methods are available for the determination of θ. Depending on the nature of the solid and test liquid(s) used, and the environment in which the solid surface behavior will actually be monitored, some methods result in more useful relationships than others.

The most common methods used to identify differences in surface constitution for materials used in biological fluid contact are sessile drop techniques. These are classified as advancing angle techniques, since during drop formation and contact, the liquid advances along a part of the surface with which it has not previously been equilibrated. Alternatively, the line of solid-liquid contact may occur at a position on the solid that has previously been immersed in the test liquid, i.e., the liquid recedes or is pulled back across that part of the surface with which it has reached equilibrium. Obtained this way, θ is referred to as the receding contact angle.

Use of θ in evaluation of surface energetics

Critical Surface Tension. Adopting the most common sessile drop procedure, θ is determined for each of a series of homologous liquids in contact with a solid test surface. The cosines of the angles are then plotted against the liquid/vapor surface tension of the corresponding liquids. Zisman (1964) developed this method extensively and found empirically that, for a given surface, a rectilinear fit to the data is most often obtained. The intercept of a line drawn through the data at the $\cos\theta = 1$ axis is termed the critical surface tension (γ_c) for the particular solid. Figure 4.2

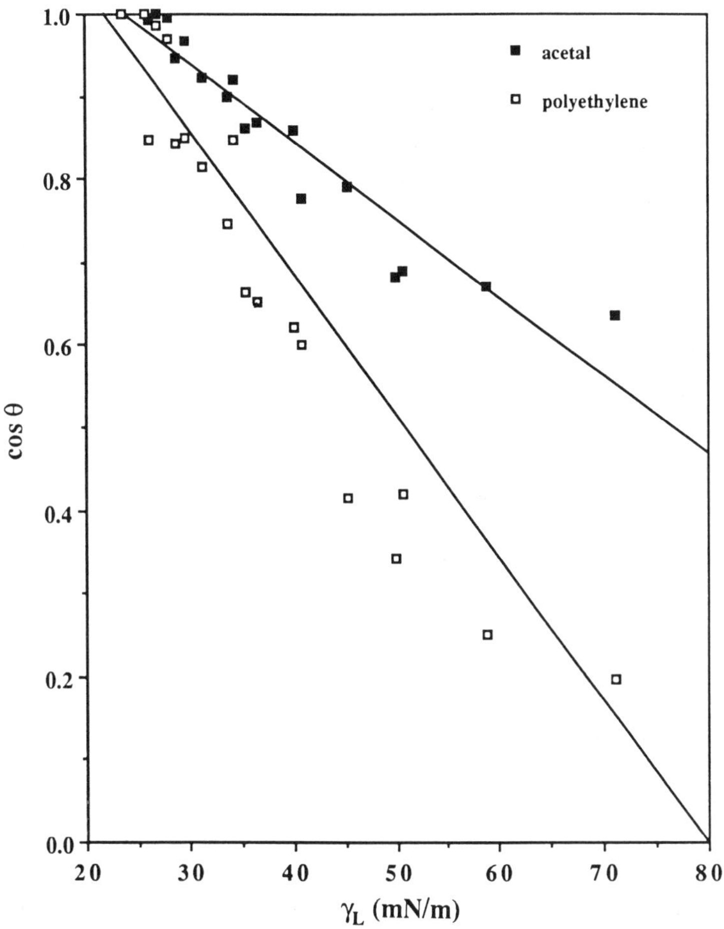

Fig. 4.2 A Zisman plot constructed with contact angle data recorded for an acetal and a polyethylene surface.

illustrates use of the Zisman plot method to evaluate γ_c for the surfaces of acetal (γ_c = 23.7 mN/m) and a commercially prepared polyethylene (γ_c = 21.8 mN/m). In these cases, each surface was contacted with drops from aqueous solutions consisting of varying concentrations of methanol or ethanol. The critical surface tension is suggested to be both a function of surface properties of the solid alone, and related to the true surface energy of the solid. However, since the data points fall within a rectangular band, different values of the critical surface tension may be determined for a given material. Additionally, as suggested by Andrade (1985a), the method can be highly misleading when examining the properties of solids that must perform in aqueous environments. This is in part due to surface restructuring that may occur under water as a more energetically favorable state is attained. Water is often used as a test liquid in the Zisman plot method, and this serves to cloud data interpretation as well. Particularly in the case of hydrophilic surfaces, water rarely falls within the band of data determined with the other test liquids commonly used. This is in part due to water penetration into the surface facilitated by its small molecular volume, and in part due to the surface restructuring that may take place upon water-solid contact.

In less commonly used receding contact angle experiments, reasonable correlations have been achieved between surface energetics of polymers and both protein adsorption and cell adhesion (Young, 1982). This is likely due to surface characterization having been performed on water-equilibrated surfaces, i.e., the surfaces were characterized in a state they would experience if contacted with an aqueous (fluid food) system. Receding angle data recorded with liquids other than water, however, are representative of the solid equilibrated with that liquid only and do not represent surface properties at the solid-water interface.

Polar and Dispersive Contributions to Surface Tension. To more quantitatively treat contact angle data, Fowkes (1964) introduced the London dispersion force contribution to the surface free energy, γ^d. Liquid surface tension may then be expressed as

$$\gamma_L = \gamma_L{}^p + \gamma_L{}^d \tag{4}$$

where superscripts p and d refer to polar and dispersive force components. The polar contribution to surface tension represents the sum of all forces that may be referred to as nondispersive (e.g., hydrogen bonding, electrostatic and dipole-dipole interactions). At the interface between any liquid and another in which the intermolecular attraction is entirely due to London dispersion forces, the only appreciable interfacial interactions to which these liquids are subject are the London dispersion forces. Moreover, Fowkes showed that the geometric mean of the dispersion force attractions should predict the magnitude of the interaction between dissimilar materials. Therefore, if only dispersive interactions are present at a liquid-liquid interface, the effect of interfacial attraction on the tension in the interface can be predicted by the geometric mean of the dispersion force components of the surface tension of two liquids 1 and 2 as $(\gamma_1{}^d\gamma_2{}^d)^{1/2}$. The tension in the interfacial region of liquid 1 is then $\gamma_1 - (\gamma_1{}^d\gamma_2{}^d)^{1/2}$, and that in the interfacial region of liquid 2 is $\gamma_2 - (\gamma_1{}^d\gamma_2{}^d)^{1/2}$. The interfacial tension is the sum of the tensions in both regions and is given by

$$\gamma_{12} = \gamma_1 + \gamma_2 - 2(\gamma_1{}^d\gamma_2{}^d)^{1/2} \tag{5}$$

By analogy with Eq. (2) (the Dupré equation), the energy required per unit area to separate the liquid-liquid interface is expressed as

$$W_a = \gamma_1 + \gamma_2 - \gamma_{12} \tag{6}$$

and, since W_a may also be divided into dispersive and polar components, i.e., $W_a = W_a{}^d + W_a{}^p$ (Fowkes, 1972), then

$$W_a{}^d = 2(\gamma_1{}^d\gamma_2{}^d)^{1/2} \tag{7}$$

Kaelble (1970) applied these concepts to liquid contact with low energy solids and developed a method for determining both the dispersive and polar components of γ_S from contact angle data (i.e., he assumed that γ_S could be expressed as $\gamma_S{}^p + \gamma_S{}^d$) as follows. Given a pure liquid for which γ_L, $\gamma_L{}^p$, and $\gamma_L{}^d$ are known, if θ is the contact angle between the liquid and some solid, the interaction can be described in terms of the reversible work of adhesion W_a, as

$$\gamma_L(1 + \cos\theta) = W_a = (W_a{}^d + W_a{}^p) = 2(\gamma_L{}^d\gamma_S{}^d)^{1/2} + 2(\gamma_L{}^p\gamma_S{}^p)^{1/2} \tag{8}$$

Contact angle data from two liquids, therefore, would provide two equations containing the unknown values $\gamma_S{}^d$ and $\gamma_S{}^p$. For each set of n pure liquids on a given surface, n equations may be written where any and all pairs of equations can be solved simultaneously for $\gamma_S{}^d$ and $\gamma_S{}^p$. Finally, the mean values ($\bar{\gamma}_S{}^d$ and $\bar{\gamma}_S{}^p$) are determined by averaging the results, i.e.,

$$\bar{\gamma}_S{}^d = \frac{1}{m}\Sigma\gamma_S{}^d \qquad \text{and} \qquad \bar{\gamma}_S{}^p = \frac{1}{m}\Sigma\gamma_S{}^p \tag{9}$$

where m is the number of pairs of simultaneous equations solved in the computation. Kaelble's method involving Eqs. (8) and (9) is mathematically convenient, and it has been adopted as more quantitative than the Zisman plot method (Nyilas et al., 1977; McGuire and Swartzel, 1987). Problems associated with respect to interpretation of the data arise from the fact that there is no theoretical basis for equating $W_a{}^p$ to the geometric mean of the polar components of solid and liquid surface tension (Fowkes, 1972).

The Hydrophilic-Hydrophobic Balance of a Surface. Although convenient, Eq. (8) is not correct and is best rewritten

$$\gamma_L(1 + \cos\theta) = W_a = 2(\gamma_L^d \gamma_S^d)^{1/2} + W_a^p \quad (10)$$

The value of the polar contribution to the work of adhesion depends upon the polar character of both the solid and the liquid that are in contact, i.e.,

$$W_a^p = f(\gamma_S^p, \gamma_L^p) \quad (11)$$

However, there is no way of calculating γ_S^p directly, and the functional representation of Eq. (11) is not known. In hope that development of an expression for W_a^p would contribute to a sound theoretical basis for calculation of γ_S, effort has been dedicated to providing direction for identification of W_a^p and its use in theoretically sound surface quantification.

One major criterion must be satisfied for experimental determination of γ_S by contact angle methods: the system defined by the drop and solid surface must be in equilibrium. Furthermore, if being assessed for contact with biological media the surface must be characterized while in equilibrium with water. These conditions are met if aqueous solutions are chosen as test liquids, and if contact angles are measured on water-equilibrated surfaces. Such is the case with respect to receding angle experiments performed with aqueous solutions. Advancing angle techniques could presumably be used if contact angles are recorded in a controlled environment of water-saturated air.

Test Liquid Characterization and Evaluation of γ_S^d. In working with Eqs. (10) and (11) to evaluate an index of solid surface hydrophobicity, evaluation of both the dispersive and polar component of test liquid surface tension is essential. For a given solid-liquid contact, if the surface tension of either the solid or the liquid has only a dispersive component (i.e., γ_S^p or $\gamma_L^p = 0$), then the interaction between the two can be entirely attributed to dispersive forces (Fowkes, 1964; 1972). For this case, Eq. (10) may be written:

$$W_a = \gamma_L(1 + \cos\theta) = 2(\gamma_L^d \gamma_S^d)^{1/2} = W_a^d \quad (12)$$

i.e., the work of adhesion is totally due to dispersive interactions. Rearranging,

$$\gamma_L^d = \frac{\gamma_L^2(1 + \cos\theta)^2}{4\gamma_S^d} \quad (13)$$

γ_L and θ are readily measurable. Consequently, given a solid surface of known $\gamma_S{}^d$ with which no polar interactions may take place, the dispersive component of liquid surface tension, $\gamma_L{}^d$, may be found with a single contact angle measurement. $\gamma_S = \gamma_S{}^d$ for paraffin wax. Obviously, evaluation of $\gamma_S{}^d$ requires knowledge of $\gamma_L{}^d$. Fortunately, a number of liquids exist (particularly hydrocarbons) whose surface tension is totally due to dispersive forces. For these liquids, $\gamma_L = \gamma_L{}^d$, where γ_L may be measured by conventional methods.

A different rearrangement of Eq. (12) yields

$$\cos\theta = \frac{2(\gamma_L{}^d\gamma_S{}^d)^{1/2}}{\gamma_L} - 1 \qquad (14)$$

Choosing a series of nonpolar liquids, for any solid a plot of cos θ vs. $(\gamma_L{}^d)^{1/2}/\gamma_L$ $(= 1/\gamma_L{}^{1/2})$ should be a straight line with slope $2(\gamma_S{}^d)^{1/2}$ intercepting the ordinate at -1. Armed with a "diagnostic solid" for which $\gamma_S = \gamma_S{}^d$ (e.g., paraffin), test liquids exhibiting both polar and nonpolar character may be characterized with use of Eq. (13).

The Relationship Between $W_a{}^p$ *and* $\gamma_L{}^p$. With test liquids of known γ_L, $\gamma_L{}^d$, and $\gamma_L{}^p$, and with methodology for evaluation of $\gamma_S{}^d$, evaluation of solid surface influences on $W_a{}^p$ should be possible according to Eq. (11). Rearranging equation (10),

$$W_a{}^p = \gamma_L(1 + \cos\theta) - 2(\gamma_L{}^d\gamma_S{}^d)^{1/2} \qquad (15)$$

Equation (15) can be used to evaluate $W_a{}^p$ for each test liquid put in contact with a given material. Having evaluated $\gamma_L{}^p$ for each test liquid (where $\gamma_L{}^p = \gamma_L{}^d$), a plot of $W_a{}^p$ vs. $\gamma_L{}^p$ can be constructed. The relationship between $W_a{}^p$ and $\gamma_L{}^p$ has been suggested to be unique for each material; differences among materials are attributed to differences in $\gamma_S{}^p$. Figure 4.3 shows a plot of $W_a{}^p$ vs. $\gamma_L{}^p$ for the surfaces of acetal and a commercially prepared polyethylene, using the same data as that of Fig. 4.2. A straight line fit to the data is usually obtained. The magnitude of the slope generally agrees with the expected polar character of each solid surface. The negative values of the intercept have been suggested to be a measure of π_S, the reduction in surface energy of the solid resulting from adsorption of vapor from the diagnostic liquid (Dann, 1970). For solids that interact by dispersion forces only, all theoretical and experimental evidence would predict that adsorption of high energy material cannot reduce the surface energy of a low energy material. For example, the surface energy of paraffin cannot be reduced by adsorption of water. The fact that a given liquid has a contact angle greater than 0° on a given

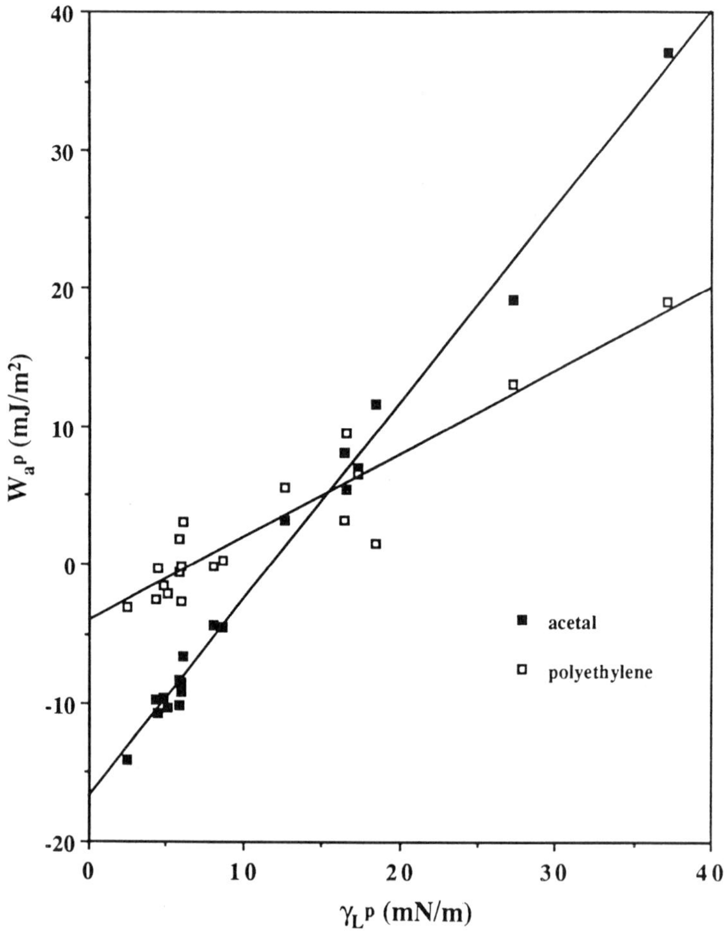

Fig. 4.3 The linear relationship observed between the polar component of the work of adhesion and the polar component of contact liquid surface tension for an acetal and a polyethylene surface.

low energy solid surface is evidence that the liquid surface energy is greater than that of the solid and π_S should be zero.

Two important points warrant consideration with respect to comparison of the methods used to construct Figs. 4.2 and 4.3, and Kaelble's method. First, $W_a{}^p$ is not proportional to $(\gamma_L{}^p)^{1/2}$; rather, it is directly proportional to $\gamma_L{}^p$. Second, the graphical method illustrated in Fig. 4.3 yields a generally better fit to the data than the Zisman plot method of Fig. 4.2 used to evaluate γ_c. In particular, in the Zisman plot of Fig. 4.2, the coefficient of determination (r^2) for acetal = 0.914, and r^2 for polyethylene = 0.899. The plot of Fig. 4.3 produced an r^2 for acetal = 0.989, and an r^2 for polyethylene = 0.917.

The value of the polar component of the work of adhesion between any given solid surface and water ($W_a{}^p{}_{water}$) can be recorded with the following equation:

$$W_a{}^p{}_{water} = k(\gamma_L{}^p{}_{water}) + b \tag{16}$$

where k and b are the slope and intercept, respectively, of a plot of $W_a{}^p$ vs. $\gamma_L{}^p$. $W_a{}^p{}_{water}$ provides an index of surface hydrophilicity (McGuire et al., 1990); since it can be calculated with the straight line equation above, actual contact angle data for water need not be used. This avoids rather serious problems associated with using pure water as a diagnostic liquid (Andrade, 1985). With reference to Fig. 4.3, $W_a{}^p{}_{water}$ for acetal was determined to be 36.2 mJ/m^2; that for polyethylene was found to be 18.3 mJ/m^2. It is of some interest to note that for these two materials, $W_a{}^p{}_{water}$ proved to provide a more discriminating indication of surface chemistry than γ_c.

In practice, contact angle data for water only occasionally fall on the straight line determined by the $W_a{}^p$ vs. $\gamma_L{}^p$ relationship, just as it only occasionally falls on the line constructed in a Zisman plot. McGuire (1990) reported values of $W_a{}^p{}_{water}$ calculated for 20 different surfaces by Eq. (16) and calculated by a single application of Eq. (15) for water only. Experimentally determined values of $W_a{}^p{}_{water}$ were found to agree with calculated values for relatively high-energy surfaces, and fell below those calculated by Eq. (16) for lower-energy surfaces.

Finally, let us comment on current thoughts about what are referred to here as polar interactions. Fowkes (1985) suggests acknowledgment of acid-base interactions rather than polar interactions, in which two polar groups may interact only when one is acidic and the other basic. Acknowledging the complexity of fluid foods, the constituents of which exhibit acidic sites, basic sites, and sites of different relative acidities and basicities, it is suggested that general polar interactions will always be present if the solid surface exhibits any degree of polar character. However, more work in this area is probably warranted.

An Equation of State Approach to Measurement of γ_S. Ward and Neumann (1974) indicated that an equation of state relationship must exist between the solid/liquid, solid/vapor, and liquid/vapor interfacial tensions of a two-component, three-phase system (such as that illustrated in Fig. 4.1), i.e.,

$$\gamma_{SL} = f(\gamma_S, \gamma_L) \tag{17}$$

Equation (17), if explicitly formulated and combined with equation (1) (Young's equation), involves four variables γ_S, γ_L, γ_{SL}, and θ. Both γ_L and θ are measured easily, enabling immediate determination of γ_S and γ_{SL}. Neumann et al. (1974) empirically obtained an explicit formulation of Eq. (17), using experimental contact angle data and liquid/vapor interfacial tensions:

$$\gamma_{SL} = \frac{(\gamma_S{}^{1/2} - \gamma_L{}^{1/2})^2}{1 - 0.015\,(\gamma_S\,\gamma_L)^{1/2}} \tag{18}$$

Combination of Eq. (18) with Eq. (1) yields

$$\cos\theta = \frac{(0.015\,\gamma_S - 2.00)(\gamma_S\gamma_L)^{1/2} + \gamma_L}{\gamma_L[0.015\,(\gamma_S\gamma_L)^{1/2} - 1]} \tag{19}$$

Equation (19) enables calculation of γ_S from the readily measurable quantities γ_L and θ.

Calculation of γ_S with the equation of state approach is associated with several important, practical limitations. First, the assumption of an ideal solid surface underlying Eq. (1) will normally not be met by real solids. However, Neumann and Good (1972) showed that advancing contact angles on smooth but energetically heterogeneous solid surfaces are uniquely determined by the low-energy portions of the solid surface; such angles are designated Young contact angles, θ_Y. Contact angles measured for use in Eq. (19) must represent Young contact angles, i.e., they must be thermodynamically characteristic for the surface tension of the substrate and not influenced by, e.g., surface roughness. Second, in formulating the equation of state, adsorption of diagnostic liquid vapor onto the solid surface was assumed to be insignificant. Neumann et al. (1974) indicated that the problem can become significant as the contact angle decreases, and for a set of values of θ_Y measured on a given solid surface, deviation among values of γ_S computed with Eq. (19) may well be attributable to adsorption and equilibrium spreading pressure. Finally, the equation of state approach was developed only for low-energy surfaces, for which γ_S is much less than the surface tension of water (Neumann et al., 1974). Indeed, inspection of Eq. (19) shows that complete spreading occurs (i.e., $\theta = 0$) when $\gamma_S = \gamma_L$. Since water has the highest surface tension of liquids commonly used in contact angle measurement, Eq. (19) applied to any solid surface for which $\gamma_S \geq \gamma_{water}$ would result in complete spreading of all diagnostic liquids and provide no reliable value for γ_S. Neumann et al. (1974) indicated, however, that Eq. (19) should be considered valid for any $\gamma_S < \gamma_{water}$.

Surface Thermodynamics and Microbial Adhesion

The fundamental mechanisms governing microbial adhesion are not well understood in a quantitative sense, although the ability of microorganisms to adhere to inert surfaces has been the focus of numerous investigations. Current theories that address bacterial adhesion to solid surfaces propose multiple mechanisms based on both observed and theoretical cell-surface interactions. The diversity of theory reflects the dynamic structural and biochemical heterogeneity of bacterial cells. Busscher and Weerkamp (1987) proposed that cell-surface interactions fall into three categories based on separation distances between cells and surfaces. At distances greater than 50 nm only attractive van der Waals forces are of consequence, whereas at 10–20 nm electrostatic charges come into play along with initial dehydration events. At distances of less than 1.5 nm specific interactions occur that lead to irreversible adhesion. It is generally believed that microbial cell surface characteristics such as charge, hydrophobicity, and surface free energy are responsible for adhesion and determine its resultant tenacity. Busscher and Weerkamp (1987) suggest that the first true cell-surface contact may be mediated by specific microbial appendages that have surface characteristics different from the rest of the cell and that are more conducive to adhesion. Indeed, it is rather well accepted that nonspecific adhesion of bacteria is largely mediated by exopolysaccharides. Once adhesion is complete, however, the bacteria continue to produce exopolysaccharides. Adherent cells become enmeshed in the matrix, and divide to form extensive microcolonies and eventually thick films. Individual cells in such a biofilm are protected to a great extent from the effects of antibacterial agents such as surfactants, biocides, and antibiotics. In general, the levels of such antibacterial agents required to kill adhered cells can be 50–100 times higher than those required to kill their counterparts in the bulk (Costerton, 1991).

From a purely thermodynamic standpoint, the extent of biological adhesion on a solid material could be determined by surface energetics, i.e., the surface energies of the contact material, liquid medium, and adhering particles involved. Absolom et al. (1983), for example, studied the adhesion of five strains of bacteria with reference to the thermodynamic model for adhesion of small particles from a suspension to a solid surface. However, there is an abundant literature of inconsistent observations on the relationship between biological adhesion and substrate surface energy, in part because substrate surface energy is estimated with markedly different procedures among laboratories, and in part because biological adhesion is probably not completely explainable in terms of solid surface energetics. For example, Baier (1980) plotted relationships between both macromolecular and whole cell adhesion and γ_c; McGuire and Swartzel (1987) plotted a relationship between dairy protein adhesion and γ_S evaluated by Kaelble's method (Kaelble, 1970); Absolom et al. (1983, 1988) plotted relationships between bacterial adhesion, and between endothelial cell adhesion and surface energy evaluated by the equation of state approach (Neumann et al., 1974); and Schakenraad et al. (1986) plotted a relationship between fibroblast cell spreading and surface energy evaluated by a method most similar to that described by Kaelble (1970). Work by Baier (1970, 1977, 1978, 1980) has consistently featured discussion of a nonlinear relationship observed when biological adhesion is plotted as a function of critical surface tension. In particular, an intermediate range of surface tensions has been observed to correlate best with minimal biological adhesion. Materials characterized by values of surface tension higher or lower than the range associated with minimal adhesion were observed to evoke a greater interaction with a biofluid. Schakenraad et al. (1986) observed what they referred to as a characteristic sigmoidal relationship between human fibroblast cell spreading and surface energy, in both the presence and absence of serum protein. The data of Absolom et al. (1983) show a linear relationship between bacterial adhesion and surface energy, but data of Absolom et al. (1988) clearly support the presence of a minimum immediately preceding sigmoidal behavior for both porcine and bovine endothelial cell adhesion plotted as a function of surface energy. However, their findings are reported simply as increased endothelial cell adhesion occurring with increased solid surface energy.

In any event, an approach to quantifying cell adhesion based on surface thermodynamics would imply that the free energy of adhesion is minimized at equilibrium. Bacterial adhesion will be favored if it causes the free energy function to decrease, and will not be favored if it causes the function to increase. If electrostatic forces and specific receptor-ligand interactions can be neglected, the change in free energy upon bacterial adhesion can be written:

$$\Delta F_{adh} = \gamma_{BS} - \gamma_{BL} - \gamma_{SL} \tag{20}$$

where F_{adh} is the free energy of adhesion per unit surface area, γ_{BS} is the bacterium-solid interfacial tension, and γ_{BL} is the bacterium-liquid interfacial tension. Equation (18) then may be considered a generic equation of state and used to calculate any interfacial tension γ_{12} from given or predetermined interfacial tensions γ_{13} and γ_{23}, where subscripts 1, 2, and 3 refer to different phases. If so, all the required interfacial tensions in

Eq. (20) can be computed, permitting evaluation of ΔF_{adh} and prediction of the relative extent of bacterial adhesion among different surfaces.

It is instructive to consider the effect of both solid surface energy and the surface tension of the suspending medium on the relative extent of adhesion to be expected among different strains of bacteria. With Eq. (20), theoretical calculations of ΔF_{adh} for the attachment of bacteria from suspension to varous substrata can be made. Figure 4.4, a plot of ΔF_{adh} as a function of γ_S, can be made for any single species of bacterium. The input data required for such a plot are the surface tensions of the bacteria γ_B, the solid surface γ_S, and the suspending liquid medium γ_L. Theoretically, this leads to a distinction between two situations. For $\gamma_L < \gamma_B$, ΔF_{adh} decreases with increasing γ_S, predicting increasing adhesion with increasing solid surface energy. When $\gamma_L > \gamma_B$, the behavior follows an

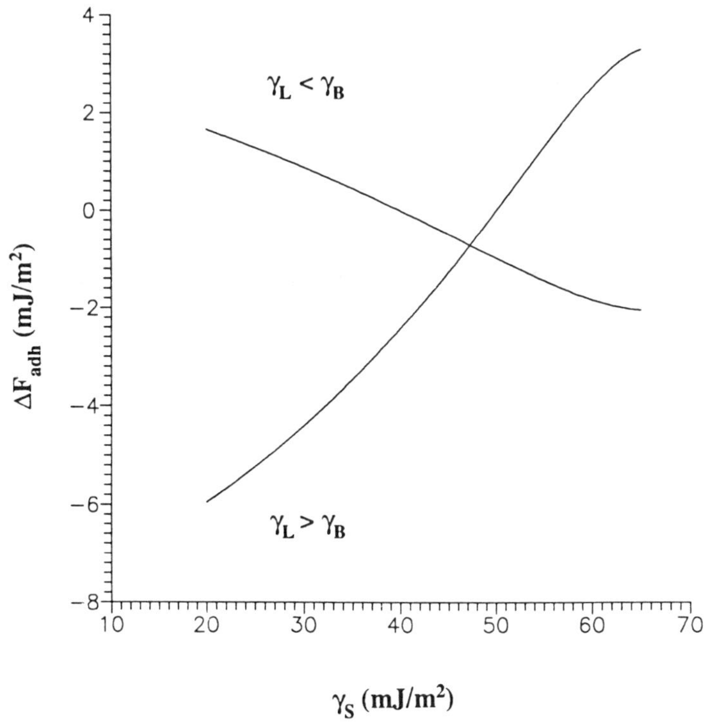

Fig. 4.4 Free energy of bacterial adhesion as a function of solid surface energy (calculated using the equation of state approach). $\gamma_B = 43$ mN/m; $\gamma_L = 40$ mN/m for $\gamma_L < \gamma_B$, and $\gamma_L = 50$ mN/m for $\gamma_L > \gamma_B$.

opposite pattern, i.e., ΔF_{adh} increases with increasing γ_S. A limiting case arises when $\gamma_L = \gamma_B$: ΔF_{adh} becomes equal to zero and independent of γ_S. This can also be seen as a direct consequence of the equation of state. If $\gamma_B = \gamma_L$, then $\gamma_{BL} = 0$. Also $\gamma_{BS} = \gamma_{SL}$, dictating that $\Delta F_{adh} = 0$. Under this limiting condition, bacterial adhesion does not depend on solid surface energy and should in general be absent if it is indeed valid to assume that other effects, such as electrostatic and specific biological interactions, can be neglected. Finally, it must be noted that according to Eq. (18), problems will arise when the quantity $0.015[\gamma_S\gamma_L]^{1/2}$ approaches unity. Figure 4.4 was constructed for a hypothetical case in which $\gamma_B = 43$ mN/m, and $\gamma_L = 40$, and then 50 mN/m. Theoretically, the shape of the plots in Fig. 4.4, both constructed for cases in which the equation of state is completely valid, should be indicative of the ΔF_{adh} vs. γ_S relationship under more realistic conditions (Absolom et al., 1983).

Measurement and careful interpretation of surface thermodynamic properties is an absolute requirement for quantifying adhesion. In natural circumstances, however, microbial adhesion must to some extent be influenced by the interfacial behavior of dissolved protein as well. Although cell-surface contact may be mediated by specific microbial appendages, some researchers argue that exudates of microbial origin might actually adhere to a preadsorbed protein film (Tosteson and Corpe, 1975; Bryers, 1987). Surface chemical influences on protein adsorption thus constitute an important area for study. Many experimental results have indicated that adsorbed protein changes its conformation over time, and the rate of conformational change varies among proteins and from one surface to another. Understanding both the physical chemistry of surfaces and the nature of competitive adsorption in multiprotein systems is thus necessary for quantifying biofilm development.

PROTEINS AT INTERFACES

The Significance of Protein Adsorption

Toward gaining a better understanding of the interfacial events that occur when a solid material is brought into contact with a biofluid, much effort has been devoted to study of protein adsorption. In addition to its apparent role in mediating microbial and spore adhesion, protein adsorption is involved in several instances of technical interest. In the food and pharmaceutical industries, proteins due to their heat sensitivity and high content in some fluid foods play a major role in the fouling of membrane surfaces used in biomolecular fractionation and fouling of heat exchange surfaces (Arnebrant and Nylander, 1986). An understanding of protein behavior at solid surfaces is of great importance in the development of new biosensors and chromatographic methods (Lesins and Ruckenstein, 1988). Protein adsorption is also of interest to clinicians since it is a primary event when foreign surfaces are exposed to blood or living tissue (Gölander and Kiss, 1988); serum protein behavior at the site of implantation of a foreign device is to a large extent related to the biocompatibility of the device.

In particular, protein adsorption kinetics and equilibrium (as it is described by an isotherm) are of interest because, interpreted with reference to a molecular model of interfacial behavior, they can provide information about the conformation of adsorbed protein molecules. Protein interactions with solid surfaces have been studied for decades, and several thorough reviews are available (Andrade, 1985b; Norde, 1986; Horbett and Brash, 1987). However, surveying the literature, one can find contradictions in reported results among laboratories. In this treatment, then, only the "generally well understood" results are summarized, to provide a basis for understanding the challenges researchers must address in this area.

Features of Protein Behavior at Interfaces

Surface-induced conformational changes. Comparisons of denaturation energetics and adsorption energetics suggest that adsorption-induced conformational changes are highly probable (Dillman and Miller, 1973; Andrade et al., 1984). Adsorbed protein experiencing such a change in conformation is depicted schematically in Fig. 4.5. Conformational changes occur to minimize interfacial free energy. Additional noncovalent bonds between the protein and surface can lead to a more "spread" configuration among protein molecules, and can assist displacement of those less tightly bound. Although there have been many hypotheses associated with adsorption-induced conformational changes in protein, little direct data is available. However, there is a large amount of experimental evidence supporting the concept that proteins exist in multiple states on a surface.

Morrissey and Stromberg (1974) used infrared difference spectroscopy to study protein conformational states adopted on silica particles. This method allows deduction of conformational state by providing information on the fraction of carbonyl groups actually bound to the surface. Their findings suggested that with respect to fibrinogen, more significant protein-surface hydrogen bonding is experienced at low solution concentra-

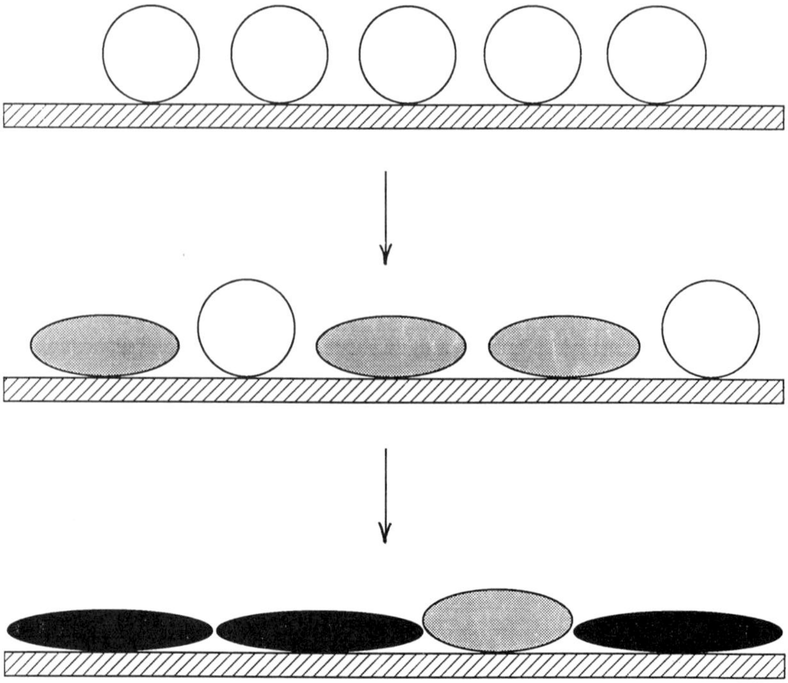

Fig. 4.5 Conformational changes experienced by adsorbed protein, resulting in multiple noncovalent bonds with the surface and a more "spread" configuration among protein molecules.

tions relative to high solution concentrations. They also found that spectra for prothrombin and bovine serum albumin were unaffected by solution concentration, indicating that the internal bonding of serum albumin and prothrombin is sufficient to prevent changes in their structure while adsorbed.

Andrade et al. (1984) used intrinsic ultraviolet total internal reflection fluorescence (UV TIRF) spectroscopy to monitor conformational changes experienced by human plasma fibronectin on hydrophobic and hydrophilic silica. Intrinsic UV TIRF spectroscopy follows adsorption by providing indirect information on the local environment of tryptophan residues. A disadvantage of this method, however, is that UV photochemical changes may be observed and interpreted as surface-induced. They reported that protein adsorbed on hydrophobic and hydrophilic surfaces showed fluorescence maxima at 326 and 321 nm, respectively, whereas the fluorescence maximum exhibited by fresh protein in solution is 321 nm. These results indicate that the adsorbed human plasma fibronectin had undergone some conformational change on the hydrophobic surface.

Another method that has lent support to the multiple states theory involves comparison of adsorption isotherms obtained from single-step and successive addition of protein. Jönsson et al. (1987) used this method to investigate the adsorption behavior of secretory fibronectin (HFN) and immunoglobulin G (IgG). They found differences between isotherms constructed on hydrophobic silica using the single-step and the successive addition methods for each protein. These results were taken to indicate virtual irreversibility of the adsorption process due to surface-dependent conformational changes occuring over time. For hydrophilic silica, the successive addition isotherm of HFN approached that obtained by single step-addition; at long equilibration times they were observed to eventually coincide. They suggested that HFN adsorbed to hydrophilic silica tends to undergo conformational changes to a lesser degree if at all, relative to HFN adsorbed to hydrophobic silica. Recently, similar experiments were done by Wahlgren and Arnebrant (1990). They studied the adsorption of β-lactoglobulin onto silica, methylated silica, and polysulfone surfaces preadsorbed with protein from a 0.1% β-lactoglobulin solution. The results indicated that values of adsorbed mass from a 12% protein solution (the highest concentration studied) onto both preadsorbed methylated silica and preadsorbed polysulfone surfaces were of the same range as those due to adsorption from a 0.1% β-lactoglobulin solution to each bare surface.

Elwing et al. (1988) used ellipsometry to study conformational changes experienced by the protein complement factor 3 on hydrophilic and hydrophobic silicon surfaces. They measured greater values of adsorbed mass on hydrophobic surfaces than on hydrophilic surfaces. Protein molecules are assumed, in general, to change conformation to a greater extent on hydrophobic surfaces relative to hydrophilic surfaces. This is due to the presence of hydrophobic interactions between the solid surface and hydrophobic "pockets" in the protein molecule. This interaction gives the molecule an extended structure, covering a relatively large area of the surface. The repulsive force normally acting between native protein molecules is probably decreased for such conformationally changed molecules (Elwing et al., 1988). This means that one should expect a greater number of adsorbed molecules per unit area on a hydrophobic surface. On a hydrophilic surface the interaction will in general be quite different. Forces acting between the surface and the molecule may be smaller in magnitude, and since the resulting conformational change will be smaller, a larger repulsive force will be present between the molecules. Therefore, the packing of adsorbed molecules will not be as close as that on a hydrophobic surface, and, in general, a smaller number of adsorbed molecules will be found on a hydrophilic surface, with each molecule occupying a smaller area on the surface.

DeFeijter et al. (1978) studied the adsorption of various proteins at air-water interfaces and proposed that conformational changes are related to the availability of adsorption sites on the contact surface. They observed that the thickness of an adsorbed κ-casein film decreased while its adsorbed mass increased for incubation times greater than 4 hours, indicating that the adsorbed κ-casein spreads out on the surface, with the time scale of the process being on the order of several hours. This effect was observed at bulk concentrations at or below 0.0001% by weight but not at higher protein concentrations. They suggested that the extent to which adsorbed κ-casein molecules change their conformation depends on the number of molecules adsorbed at the surface. For low bulk concentrations, where the surface is not completely occupied, conformational rearrangement of the adsorbed molecules is considerable; at high bulk concentrations, where the surface is completely occupied, the shape of the adsorbed molecules is hardly affected.

Walton and Maenpa (1979) applied transmission circular dichroic spectroscopy to proteins eluted after different adsorption contact times on quartz plates. They showed the extent of denaturation was directly related to contact time with the surface, for contact times less than 10 days. In other experiments, Soderquist and Walton (1980) used circular dichroism to study conformational characteristics of plasma proteins desorbed from polymer surfaces. They found that the desorbed molecules were changed in structure compared to their structures prior to adsorption. The desorption process itself depends upon the incubation time of the protein with the substrate: the longer the incubation time, the slower the desorption rate. They explained that each molecule on the surface undergoes a structural transition as a function of time that optimized protein-surface interaction.

Horbett and Brash (1987) reviewed a large body of evidence supporting the existence of multiple states of adsorbed protein. Some strong evidence includes numerous observations indicating the presence of weakly and tightly bound proteins: for example, rinsing a surface after protein contact does not remove all of the protein, yet some of the remaining protein is removable during a second, longer buffer rinse. Also, decreases in SDS elutability of proteins from an adsorbed layer are observed as the protein-surface contact time increases.

Bilayer adsorption. Many workers have reported that protein can adsorb onto a surface in more than one layer. Arnebrant et al. (1985) studied adsorption of β-lactoglobulin and ovalbumin on hydrophilic and hydrophobic chromium surfaces using ellipsometry and potential measurements. On clean hydrophilic surfaces the ellipsometric results showed that a thick, highly hydrated layer is obtained, which can be partially removed by aqueous buffer rinsing. Changes in the electrode potential, simultanously measured relative to a standard calomel electrode during both adsorption and rinsing, were also observed for this surface. The results suggest that the protein adopts a bilayer formation on the surface, with the bottom layer unfolded and attached by strong polar bonds to the surface, and not removable by rinsing. Figure 4.6 illustrates such a bilayer formed on a hydrophilic surface. Rinsing experiments showed that the upper protein layer is loosely attached, which indicates that the outer layer has a structure closer to that of its native state. This adsorption behavior was described in terms of surface-induced conformational changes and charge interactions between the protein molecule and surface by Arnebrant et al. (1985) as follows.

In a protein molecule, there are always some polar amino acid side chains that can interact strongly with a surface, even if both the protein and surface are negatively charged. Such binding is expected to result in unfolding of the protein and irreversible adsorption. The consequence of this would be the probable exposure of hydrophobic loops into aqueous solution; adsorption of a second protein layer can therefore reduce the interfacial free energy. In the case of a negatively charged metal surface, it should be expected that positive groups such as the N-terminal and lysine residues bind with direct contacts to the surface. Such charge com-

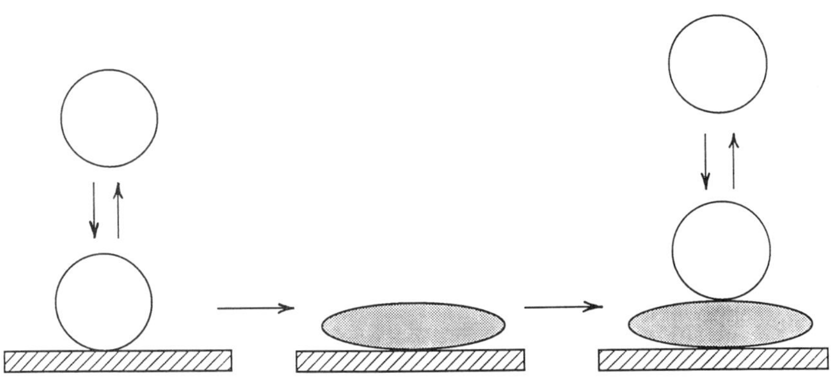

Fig. 4.6 Protein bilayer formation on a solid surface: the bottom layer is unfolded and not removable by rinsing; the conformation of protein in the upper layer approximates its native state, and protein in the upper layer is loosely attached.

pensation may explain why electrical effects of this bottom monolayer were observed to be relatively weak, compared with the upper layer. Ionic interaction, in addition to hydrophobic interaction between the top and bottom layers, can link groups in protein layers. Thus, for the case of a negatively charged surface in contact with protein molecules with a negative net charge, the upper layer will tend to orient its positive residues downwards and the negatively charged side chains outward from the surface. Due to the bulkiness of the protein molecule, a stronger dipolar effect might therefore be expected for the upper layer. In the case of protein exposed to a hydrophobic metal surface, recorded values of thickness and refractive index were found to be consistent with the formation of a monolayer.

Bilayer adsorption is not only detected by ellipsometry, but also by other methods such as fluorescence spectroscopy. Walton and Maenpa (1979) used fluorescence spectroscopy to study the behavior of bovine serum albumin adsorption on particles of copolypeptide. They found that the change in fluorescence is not only due to adsorbed protein, but also to loosely bound protein in close proximity to the surface. The concentration of loosely attached molecules is at least as high as that of strongly attached molecules.

Bilayer adsorption seems to depend upon substrate surface properties, but other evidence reveals that increased polymerization of protein causes formation of a second layer. Arnebrant and Nylander (1988) used ellipsometry to study the effect of divalent metal ions (zinc and calcium) on the adsorption of insulin to chromium and titanium surfaces. Self-association of insulin molecules in aqueous solution to form oligomeric units is influenced by specific interactions with divalent metal ions. They observed that the adsorbed mass of hexamer insulin per unit areas was comparable to the calculated value for a hexagonal close-packed monolayer of hexamers. However, adsorbed mass was greater when the protein in solution was mainly in the form of dihexamers. This higher adsorbed amount may be interpreted as the formation of a second layer, which can be removed easily be rinsing.

Adsorption equilibrium. A great deal is known about how various conditions affect the adsorbed mass of protein. Numerous protein adsorption isotherms have been constructed and compared on the basis of temperature, pH, ionic strength, conformational stability of the protein in solution, and solid surface charge and hydrophilic-hydrophobic balance. The effects of protein conformational stability and solid surface properties are perhaps best revealed in conjunction with pH and ionic strength influences.

Many workers have studied the effect of pH (e.g., Norde and Lyklema, 1978; Bagchi and Birnbaum, 1981; Penners et al., 1981; Suzawa et al., 1982; Koutsoukos et al., 1983; Van Dulm and Norde, 1983; Shirahama and Suzawa, 1985; Lee and Ruckenstein, 1988) and in general have shown that maximum adsorption occurs at or near the isoelectric point. The isoelectric point is that pH at which the net charge on the protein molecule is zero; at this point, intramolecular electrostatic respulsion is minimized, and the protein molecule will be in its most compact form (Lee and Ruckenstein, 1988).

The degree to which pH actually affects protein adsorption is determined to a large extent by the conformational stability of the particular protein molecule. Norde and Lyklema (1978) have observed that plateau amounts of adsorbed protein are virtually independent of pH for bovine pancreas ribonuclease (RNase), whereas those for human plasma albumin (HPA) vary by as much as a factor of two. Relative to RNase, HPA exhibits a high degree of conformational adaptability, which allows this protein to change its structure as conditions in the solution change. The changes in structure for HPA are reflected in the observed changes in adsorbed mass onto a given surface as a function of pH. Conversely, the lack of change in the RNase protein structure in response to changing pH leads to the adsorbed mass being independent of changing solution conditions. Goat and rabbit immunoglobulin Gs have a known structural form in the shape of a "T," with pH affecting adsorption by changing the degree of repulsion in the protein's "arms" (Bagchi and Birnbaum, 1981). Triple-helical collagen monomers are a fairly stiff rod over most of their length; as a result of this stability, pH and ionic strength have little effect on their adsorptive behavior (Penners et al., 1981).

The extent to which ionic strength affects protein adsorption is a function of the role electrostatics plays in the adsorption driving force. The interaction between electrically charge bodies is described by the DLVO theory. Lesins and Ruckenstein (1988) reviewed pertinent features of the DLVO theory, which can be briefly summarized as follows. Two components comprise the total free energy associated with an adsorbate-adsorbent interaction: (1) an electrostatic interaction and (2) a dispersion contribution to the van der Waals interaction. Ionic strength affects the electrostatic interactions to a greater extent than it does the van der Waals interactions. As ionic strength is increased, the surface charge of the protein becomes increasingly shielded, thus reducing electrostatic effects and leading to a more globular protein configuration (Lee and Ruckenstein, 1988). Previous work has generally shown an increase in the amount of protein adsorbed as ionic strength increases (Norde and Lyklema, 1978; Penners et al., 1981; Van Dulm and Norde, 1983; Jönsson et al, 1987;

Lee and Ruckenstein, 1988), but decreases in adsorbed amount have also been observed (Koutsoukos et al., 1983; Jönsson et al., 1987). Electrostatic interactions are not the only important interactions that take place during protein adsorption, and such varying observations for the adsorbed mass as a function of ionic strength are not entirely unexpected.

Hydrophobic interactions contribute to the entropic driving force observed for the adsorption of proteins from an aqueous solution (Lee and Kim, 1974; Lee and Ruckenstein, 1988). Entropy is increased due to freedom gained by water molecules resulting from partial dehydration of the solid surface and the protein "surface," and in part to structural changes that occur in the less rigid, partially dehydrated protein molecule. Since hydrophobic surfaces have a low affinity for water molecules, the protein will change conformation to remove contact of these regions with the aqueous environment by either "burying" the hydrophobic region internally or associating with similar regions on the solid surface.

In summary, one might conclude that the effect of pH and ionic strength on protein adsorption is dependent upon which type of interactions predominate (e.g., electrostatic, hydrophobic, or van der Walls interactions). Consider a negatively charged surface in contact with a protein solution. If electrostatics comprise the major interactions, adsorbed mass should be greater at pH values below the isoelectric point relative to pH values above it. Below the isoelectric point, the protein and surface are of opposite charge, while both the protein and solid surface are negatively charged at pH values greater than the isoelectric point. As ionic strength increases, the electrostatic interaction is reduced due to "shielding" by counterions; consequently, increasing the ionic strength should decrease adsorbed mass at pH values less than the isoelectric point and increase the adsorbed mass at greater values of pH. The relationship between adsorbed mass and changes in pH and ionic strength is intimately tied to protein conformational stability. In general, pH and ionic strength conditions that lead to a less stable conformation for the protein in solution will lead to an increased adsorbed mass, assuming that the protein molecule is more stable on the solid surface (Luey et al., 1991).

Another observation of importance is that protein adsorption is often an apparently irreversible process. The adsorbed mass remains constant or decreases very little when the solution in contact with the solid surface is depleted of proteins. This irreversibility is more pronounced as protein-surface contact time increases, as depicted in the hypothetical protein adsorption isotherms drawn in Fig. 4.7. Region 1 of Fig. 4.7 corresponds to short contact times, for which adsorption is essentially reversible. Slow, surface-induced conformational changes take place in region 2, and in region 3 adsorption is essentially irreversible. It is observed, however,

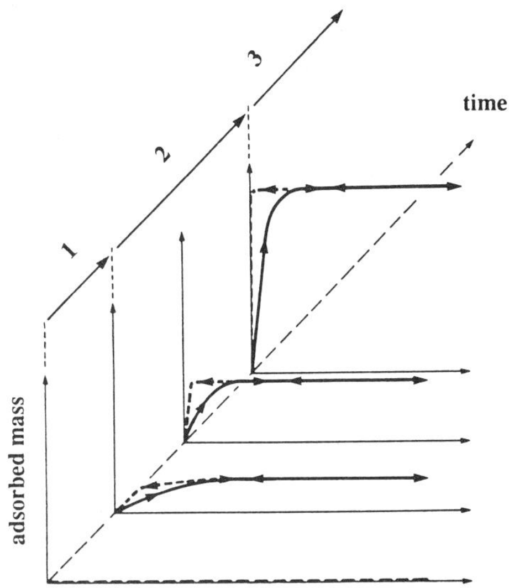

Fig. 4.7 Time-dependent irreversibility of protein adsorption illustrated with hypothetical adsorption (—) and desorption (····) isotherms.

that protein molecules "desorb" from the surface in the presence of other protein molecules. This is explained in terms of "exchange" reactions that take place on the surface (e.g., see Lundström and Elwing, 1990).

Adsorption kinetics. In considering the kinetics of any interfacial process, the question of transport vs. reaction control must be addressed. Protein adsorption at the interface is controlled not only by the intrinsic kinetic rate, which is a function of protein, solution and surface properties, but also by the rate of transport of macromolecules from the bulk solution across the concentration boundary layer near the interface. These transport phenomena are characterized by diffusion as well as hydrodynamics in a flow system.

Proteins are large molecules, and can include a number of side chains. Consequently, their diffusion coefficients are generally two orders of magnitude less than those of smaller electrolytes such as sodium chloride (Colton et al., 1981; Cussler, 1989). Diffusion coefficients may vary widely among proteins depending on their concentration and the electrostatic condition of the solution (Cussler, 1989). The initial adsorption rate of protein molecules onto a solid surface is transport-limited either at low or high concentration (Lok et al., 1982; Iordanski et al., 1983). This indicates that the rate of protein binding to the surface is faster than the rate of protein transport from solution to the surface. The diffusion limitation exists as long as there is a significant concentration gradient near the solid surface (Wojciechowski and Brash, 1990). Thus, the period of transport-limited adsorption should vary according to protein concentration and protein, solution, and surface properties. With careful design of a system to minimize the transport-limited period, an intrinsic kinetic rate can be estimated.

Much work has been done on simulation of protein adsorption by accounting for adsorption kinetic mechanisms and diffusional processes. Generally, the diffusion coefficients were assumed to be constant without regard to solution conditions. The diffusion coefficient of a protein is a strong function of solution properties such as pH, ionic strength, and concentration itself. In order to refine such simulation models, mathematical expressions to predict protein diffusion coefficients as a function of solution conditions will be needed.

In any event, changes in adsorbed mass as a function of time under various conditions have been plotted, and explanations have been proposed. Still, relatively little is known about the nature of the adsorbed layer, and predictive models to describe any aspect of adsorption as a function of protein and interfacial properties are lacking. Protein adsorption is characterized by the likely presence of a time dependence in the development of bonds with the surface, a time dependence in the lateral mobility of the protein molecules, and time-dependent conformational changes; it is thus very difficult to describe mathematically.

Many experimental observations have indicated that a major portion of the final adsorbed amount had been adsorbed within the first few minutes of contact (Soderquist and Walton 1980; Andrade et al., 1984). Soderquist and Walton (1980) used solution spectroscopy to monitor the kinetics of protein adsorption. They proposed that there are three distinct processes contributing to the kinetics of uptake of protein on polymeric surfaces. First, rapid and reversible adsorption of the proteins occurs in a short period of time. Up to 50–60% surface coverage there is a random arrangement of adsorbed molecules, but then some form of surface transition occurs that is probably in the direction of surface ordering, thereby allowing further protein uptake. Second, each molecule on the surface undergoes a structural transition as a function of time that occurs in the direction of optimizing protein-surface interaction. Third, the probability of desorption decreases with an increase in the period of incubation, and protein slowly adsorbs more or less irreversibly.

Lundström (1985) presented a dynamic model of protein adsorption on solid surfaces. The model, describing the fractional surface coverage of adsorbed molecules as a function of time, is based on the assumptions that a protein molecule may change conformation after adsorption and may desorb from the surface, as depicted in Fig. 4.8. Additionally, the rate of changing conformation was assumed to be dependent upon the availability of adsorption sites. The model was mathematically manipulated to obtain the steady-state fractional surface coverages; however, an expression describing the kinetic behavior cannot be obtained analytically. The model was compared with the case of Langmuir-like adsorption of a molecule in two different orientations, and the case of irreversible adsorption. Plateau values of isotherms constructed with the Langmuir model were reached at lower concentrations than were those constructed with the Lundström model.

Competitive Adsorption and Bulk-surface Protein Exchange Reactions. Some experimental observations indicate that adsorbed protein molecules undergo exchange with protein dissolved in solution. Elwing et al. (1987a, b) used specific antibodies to detect exchange reactions of protein antigens adsorbed to surfaces exhibiting a gradient in hydrophobicity. They observed that exchange reactions take place more readily on hydrophilic than on hydrophobic regions. They also pointed out that the results obtained were probably influenced by several factors, including the amount of specific antibodies in the antibody preparation, and the number of antigen determinants involved in the binding reaction.

Recently, Lundström and Elwing (1990) presented an expanded version of earlier models (Lundström et al., 1987) to allow for bulk-surface ex-

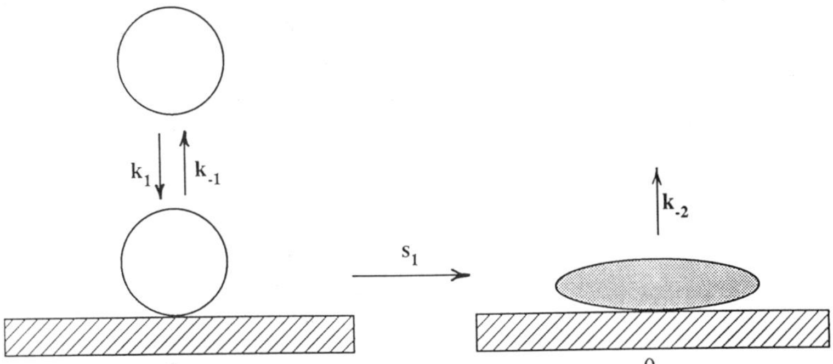

Fig. 4.8 A simple mechanism for protein adsorption in which a protein molecule may change conformation after adsorption and desorb from the surface. θ_1 and θ_2 refer to the fractional surface coverages of protein in states 1 (native) and 2 (unfolded).

change reactions among proteins in one- and two-component solutions. The interfacial events associated with adsorption and exchange from a single-component solution, from which they developed a kinetic model, are shown in Fig. 4.9. The mechanism depicted in Fig. 4.9 allows for a protein to undergo either an irreversible binding to the surface or exchange with dissolved protein, having made an initial change in conformation. Their work featured manipulation of the equations describing the fractional surface coverage of protein in specific states, and simulations of total surface coverage as a function of equilibrium concentration, and as a function of time. Due to the complexity of the model, involving several intermediate states of adsorbed protein, an expression describing the kinetic behavior of molecules in each state cannot be obtained analytically. However, the model was used to numerically simulate the change in fractional surface coverage of protein in the specific states as a function of time by assuming relative values of pertinent rate constants.

Currently, there is no adequate experimental methodology to directly monitor changes in these fractional surface coverages. Because of this limitation, a simpler model that can be statistically compared with the

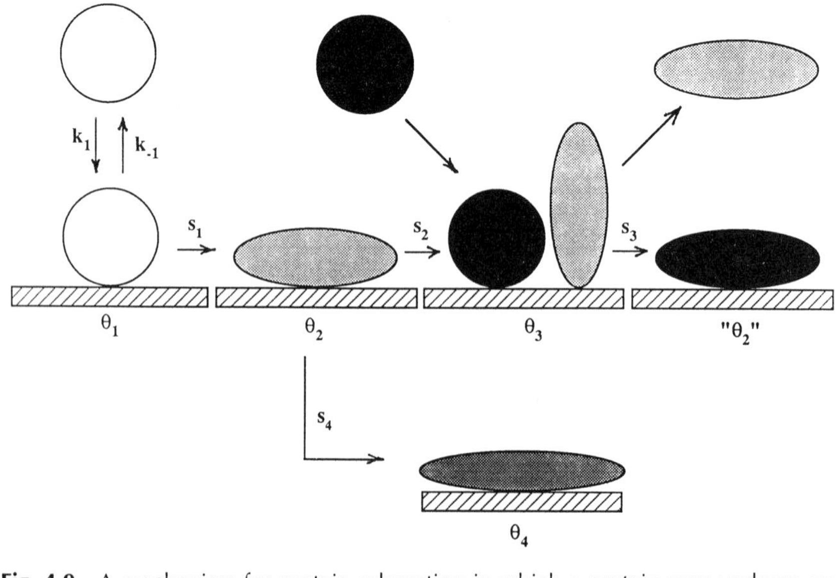

Fig. 4.9 A mechanism for protein adsorption in which a protein may undergo an irreversible binding to the surface (state 1—state 2—state 4) or exchange places with dissolved protein (state 1—state 2—state 3—"state 2"). The θ_i refer to the fractional surface coverages of protein in the various states.

available data is needed. Such a model would enable individual rate constants to be related to contact surface, solution, and protein properties.

PROTEIN SURFACE ACTIVITY AND MICROBIAL ADHESION

In clinical research, basic questions relating to mechanisms of biocompatibility, e.g., how to enhance retention of surface-bound proteins to reduce thrombosis on biomaterials, and how to optimally select cell attachment proteins to encourage cell growth in culture as well as vascular healing in implants, are in large measure addressed as protein adsorption phenomena. Platelet interaction with foreign surfaces, initiation of blood coagulation, and cell adhesion are all recognized as being related to how serum proteins behave at a particular surface.

Industrially relevant biofluids usually contain many different kinds of proteins. Often, the course of events occurring when a surface is contacted with a biofluid differs from what experiments with single component protein solutions or simplified simulants of the real fluid would predict. This is largely due to the fact that we presently know so little about the molecular basis for protein behavior at solid surfaces. Study of the actual molecular contributions to protein adsorptive behavior would address a long-standing challenge in biological adhesion research. Unlike contact surface and solution property influences on adhesion, however, the molecular origins of protein surface activity are not readily observable. In particular, we need to gain the ability to relate the relative abundance of various proteins in a mixture to what their relative abundance and average conformational state will be on a given contact surface. For example, fibrinogen and other proteins that bind to surfaces undergo conformational changes after adsorption, causing them to change their reactivity with platelets. The rate of conformational change varies among proteins and from one surface to another, perhaps explaining a major feature of why platelets and whole cells interact differently with different materials. Typically, the solution and surface concentrations of protein are different; there is also a likely time-dependent redistribution of protein. Many factors appear to be important, but the present state of knowledge in this area is of little help in the formulation of a generalized predictive model of competitive adsorption and biofilm development. Determining the factors that affect relative adsorption affinity among proteins for contact surfaces in a quantitative way seems to comprise a sound approach and is an important goal in future research.

Quantifying the Molecular Origins of Protein Surface Activity

Some comparative protein adsorption studies have been recently reported. For example, lysozyme, ribonuclease, and α-lactalbumin adsorption to hydrophilic and hydrophobic surfaces was studied by Shirahama et al. (1990), using a combination of reflectometry and streaming potential measurements. In studying sequential adsorption of two proteins, these techniques are most useful for providing an indication of whether a given protein will form an additional adsorbed layer, or will displace a given preadsorbed protein. Unfortunately, the relative importance of variations in molecular weight, charge, structural stability, shape, etc. on the observed adsorptive behavior is not clear.

Horsely et al. (1987) made a comparison of hen and human lysozyme isotherms constructed for three different surfaces, and were able to attribute differences in their adsorptive behavior to several subtle differences between the two molecules. Kato and Yutani (1988) studied the surface activity of six mutants of tryptophan synthase α-subunits (produced by amino acid substitution at a single position in the protein's interior) at the air-water interface and could attribute differences in surface tension, foaming, and emulsion properties to the free energy of denaturation of these proteins. The use of such synthetic mutants holds promise as a very attractive approach to gaining a quantitative understanding of protein behavior at solid surfaces, as it becomes possible to study the effect of a chosen, controlled molecular property on adsorption (Horbett, 1988).

Surface activity is a cumulative property of a protein, influenced by many factors: among them are size, charge, thermodynamic stability, unfolding rate, and solubility. Initially, an interface will accommodate protein molecules that have the highest rate of arrival and are most abundant in solution. In time, the adsorbed molecules are gradually displaced by others that have a higher affinity. The final composition of the adsorbed layer is thus determined by the concentrations of the various kinds of proteins in the solution and their intrinsic affinities (Shirahama et al., 1990). The number and combination of amino acids in proteins allows for multiple noncovalent bonds with a surface; certainly, unfolding of the protein can enhance the opportunity of these bonding events. The overall stability of the protein molecule is thus a key variable since it determines the likelihood of unfolding, and extent of conformational change. Past theoretical and experimental work would seem to support the hypothesis that a protein's initial attachment to a surface is most affected by solid surface and protein "surface" properties, and that unfolding, resulting in an increase in binding sites and therefore, ability to displace less tightly bound protein, is most affected by solid surface properties and protein stability. Under controlled conditions, these effects should be distinguishable and quantifiable. Using genetic variants or synthetic mutants of a single protein to measure adsorption kinetics with reference to a simple kinetic model, it should become possible to quantify the influence of protein stability on adsorption. This could be done by relating this molecular property to its effect on reaction rate constants describing initial protein attachment and unfolding at a solid surface.

This approach can be illustrated for the simple mechanism for protein adsorption sketched in Fig. 4.8, setting $k_{-2} = 0$. In describing adsorption kinetics, i.e., adsorbed mass as a function of time, it incorporates the most essential features of the model of Lundström and Elwing (1990). That is, an initial reversible step is followed by a conformational change generating an irreversibly adsorbed species. Neglecting the influence of diffusion, one can derive an expression for adsorbed mass as a function of time, protein concentration, and reaction rate constants by first writing equations describing the time-dependent fractional surface coverage of protein in each of the two states pictured—one reversibly adsorbed (θ_1) and one irreversibly adsorbed (θ_2)—as follows:

$$\frac{d\theta_1}{dt} = k_1C(1 - \theta_1 - \theta_2) - (k_{-1} + s_1)\theta_1 \qquad (21)$$

and

$$\frac{d\theta_2}{dt} = s_1\theta_1 \qquad (22)$$

Also, at any time, the total surface coverage is described by

$$\theta = \theta_1 + \theta_2 \qquad (23)$$

In solving Eqs. (21) through (23), it will be assumed that the protein molecule occupies the same surface area in states 1 and 2. The equations can then be solved analytically for θ_1 and θ_2,

$$\theta_1 = c_1 \exp(r_1t) + c_2 \exp(r_2t) \qquad (24)$$

and

$$\theta_2 = c_2 \exp(r_1t) + c_4 \exp(r_2t) + c_5 \qquad (25)$$

where the roots r_1 and r_2 (both negative numbers) for each equation above are given by

$$r_1 = \frac{-(k_1C + k_{-1} + s_1) + \sqrt{(k_1C + k_{-1} + s_1)^2 - 4\, s_1k_1C}}{2} \tag{26}$$

and

$$r_2 = \frac{-(k_1C + k_{-1} + s_1) - \sqrt{(k_1C + k_{-1} + s_1)^2 - 4\, s_1k_1C}}{2} \tag{27}$$

Addition of Eqs. (24) and (25) yields an expression for θ as a function of time:

$$\theta = \theta_1 + \theta_2 = A_1 \exp(r_1t) + A_2 \exp(r_2t) + A_3 \tag{28}$$

The surface coverages obey the initial condition that at $t = 0$, $\theta_1 = \theta_2 = \theta = 0$, as well as the final conditions: at $t = \infty$, $\theta_1 = 0$ and $d\theta_2/dt = 0$.

Nonlinear regression performed on adsorption kinetic data for a given protein-surface contact would identify the parameters A_1, A_2, A_3, r_1, and r_2. The parameter A_3 corresponds to θ_2, and A_1 and A_2 are unknown functions of protein concentration and rate constants. Incorporating the known values for r_1 and r_2 into Eqs. (26) and (27) yields two relationships among the unknown values k_1, k_{-1}, and s_1; these relationships correspond to a specific protein-surface interaction. In particular, simple algebraic manipulation of Eqs. (26) and (27) yields the following two equations with the three unknown rate constants:

$$(r_1 + r_2) = -(k_1C + k_{-1} + s_1) \tag{29}$$

and

$$r_1r_2 = s_1k_1C \tag{30}$$

Figure 4.10 shows two representative plots of kinetic data recorded in our laboratory for β-lactoglobulin adsorption to hydrophilic (Fig. 4.10a) and hydrophobic silicon (Fig. 4.10b). Silanized silicon surfaces were pre-

Fig. 4.10 Adsorption kinetics exhibited by β-lactoglobulin at (a) hydrophilic and (b) hydrophobic silicon surfaces.

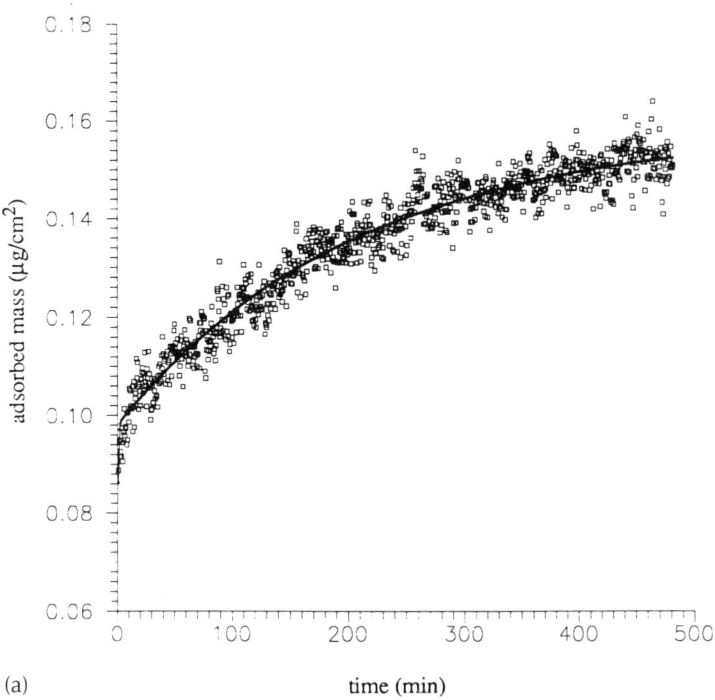
adsorbed mass (µg/cm²)
0.18
0.16
0.14
0.12
0.10
0.08
0.06
0
100
200
300
400
500
time (min)

(a)

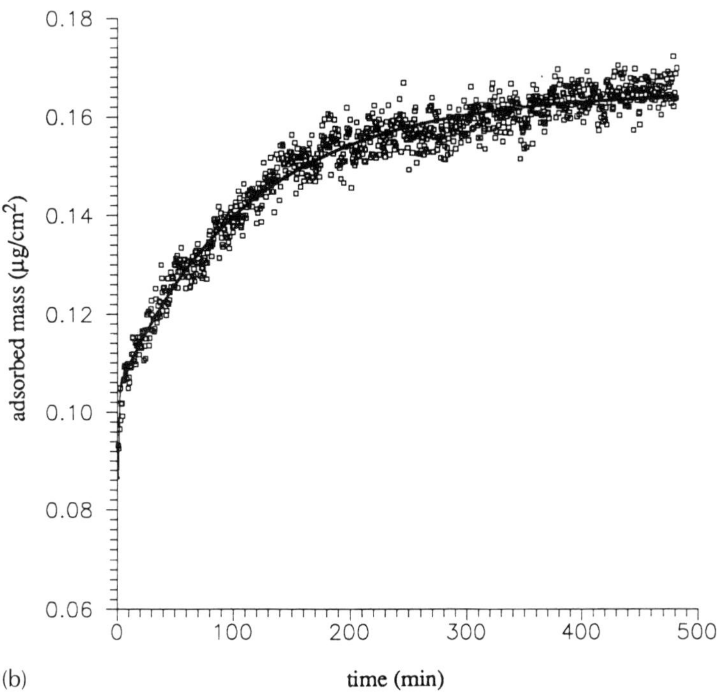
adsorbed mass (µg/cm²)
0.18
0.16
0.14
0.12
0.10
0.08
0.06
0
100
200
300
400
500
time (min)

(b)

pared according to methods similar to those described by Jönsson et al. (1982); in the cases of Fig. 4.10a and b, silanization was experimentally carried out in solutions of 0.010 and 0.100% dichlorodimethylsilane in xylene, respectively. β-Lactoglobulin, dissolved at a concentration of 1 mg/ml in phosphate buffer (pH 7.0), was contacted with each type of modified silicon surface in the absence of stirring and adsorbed mass was recorded continuously with in situ ellipsometry (McCrackin et al., 1963; Azzam and Bashara, 1977; Cuypers et al., 1983; Jonsson et al., 1985). The solid line through each set of data was generated with a nonlinear regression analysis using Eq. (28). As expected, plateau values of adsorbed mass increased with increasing surface hydrophobicity (Arnebrant and Nylander, 1986; Arnebrant et al., 1987). This is consistent with the observation that protein adsorption is an entropically driven process. It is also important to note here that the recorded kinetic data differ markedly in initial slope, indicating that the adsorption process is largely kinetically controlled in this case. Regression analysis of the data used to construct Fig. 4.10, interpreted with reference to Eqs. (29) and (30), indicates that for adsorption to hydrophilic silicon, the magnitude of $(k_1C + k_{-1} + s_1) = 2.09$ and that of $s_1k_1C = 1.00 \times 10^{-2}$. For adsorption to hydrophobic silicon, $(k_1C + k_{-1} + s_1) = 1.75$ and $s_1k_1C = 1.55 \times 10^{-2}$. Without a third relationship among the rate constants, their absolute values remain unknown. However, this particular set of data would support a hypothesis that s_1 and k_1 increase with increasing solid surface hydrophobicity, while the tendency of adsorbed protein to desorb (represented by k_{-1}) decreases.

Work featuring kinetic tests with genetic variants of a single protein could presumably be used to relate individual values of k_1, k_{-1} and s_1 to molecular properties. In particular, k_1 and k_{-1} can be hypothesized to be most affected by solid surface and protein "surface" properties, and s_1 most affected by solid surface properties and protein stability. Suppose independent kinetic tests are performed on a given surface with each of two stability variants of a single protein; the kinetic plots might resemble those shown in Fig. 4.11 (the kinetic curves of Fig. 4.10, redrawn for illustrative purposes). Equations (29) and (30) could be written for each kinetic curve yielding four equations and four unknown variables, since k_1 and k_{-1} would be assumed not to change among the mutants, with only s_1 being affected by stability. Then, for any kinetic test performed on the same contact surface with a "surface" property-changed version of one of the mutants, Eqs. (29) and (30) can be written and solved immediately for the new values of k_1 and k_{-1}, since only k_1 and k_{-1} are assumed to change upon changing "surface" properties, with s_1 remaining constant.

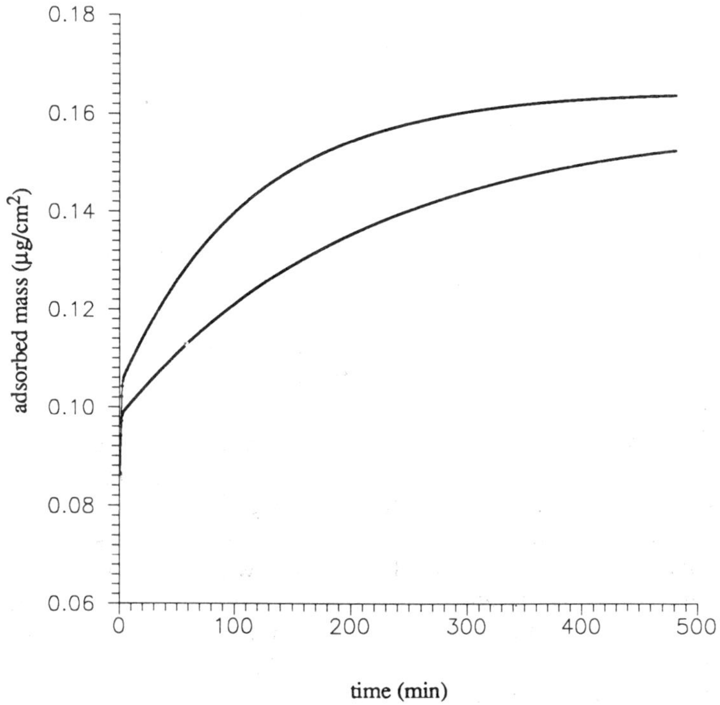

Fig. 4.11 Hypothetical adsorption kinetics exhibited by each of two stability variants of a single protein; regression analysis would yield absolute values of the rate constants k_1, k_{-1}, and s_1.

It would be important to perform this procedure with many stability variants, to best reveal the relationship between protein stability and experimentally determined s_1, and protein "surface" properties and experimentally determined k_1 and k_{-1}. With another technique (e.g., radiolabeling), protein variants could be mixed to study competitive adsorption, and to explore the possibility that s_1 could be used as an index of relative adsorption affinity in such a model system. Hopefully, competitive adsorption exhibited by mixtures of more relevant proteins could be quantitatively examined. Interpretations of previous significant work on adsorption from complex media, e.g., adsorption of fibrinogen from plasma by Slack and Horbett (1989), could be reexamined and refined as well.

In any event, with a sound idea of, e.g., stability, charge, and surface hydrophobicity influences on the eventual makeup of an adsorbed layer, deviations from model predictions could be related to properties such as molecular size and shaped in a more readily quantifiable way.

Controlling Biofilm Development

Suppose it is thought that bacterial adhesion should be enhanced in the presence of a preadsorbed, largely "unfolded" protein film. If an adsorption mechanism as simple as that shown in Fig. 4.8 adequately described protein surface behavior, then for a given surface, we would expect that if s_1 is large for the protein under study, adhesion should show an immediate increase with protein residence time on the surface (i.e., residence time preceding bacterial contact). If s_1 is small, we would expect protein residence time on the surface to have less initial effect on adhesion. Knowledge of kinetic parameters would be particularly useful in validating such a hypothesis under the more realistic circumstances of surface contact with a protein mixture. For a given surface, suppose that k_1 is small and s_1 is large for protein A, and k_1 is large and s_1 is small for protein B. Adhesion should show little change with protein residence time on the surface initially, since the surface should initially be covered by protein B, owing to its high arrival rate and tendency to maintain its conformation. But in time, protein A should displace protein B, since once attached to the surface, its tendency to unfold and make more noncovalent bonds with the surface is high; the increased number of surface contacts diminishes the likelihood of desorption or exchange with bulk protein as well. Consequently, we would expect bacterial adhesion to be enhanced with increasing protein contact time preceding the introduction of bacteria. In any event, one should expect that some knowledge of kinetic parameters governing protein surface activity should allow for definition of an optimum (i.e., minimally interacting) contact surface.

Preparation of a surface that could avoid taking part in biofilm development might come about in a variety of ways: e.g., surface coating or covalent derivitization, and immobilization of antimicrobial agents. Results presented at the 201st ACS National Meeting indicate that surfaces treated with polyethyleneoxide (PEO) actively repel proteins (Tan et al., 1991; Hubbell and Desai, 1991; Lin et al., 1991); bacterial and general cell adhesion are greatly reduced at such surfaces. The efficiency of protein repulsion was suggested to depend on the polymer surface density and the molecular weight of the polymer chains. Merrill (1991) observed that for surfaces containing no other compound than PEO, both protein adsorption and bacterial attachment were completely absent, but for "impure" PEO surfaces, protein adsorption and cell attachment occurred. In addition to the relevance of these results to biofilm control, they illustrate the interdependent nature of protein adsorption and microbial adhesion.

The subject of contact surface modification to reduce or eliminate microbial adhesion is presently being addressed with vigor. It is likely that the conceptual breakthroughs required to bring results of such research into practice will necessarily evolve in parallel with a more quantitative understanding of the nature of adsorption competition in multiprotein systems. Essential questions regarding the interplay between contact surface energetics, protein surface activity and cell attachment are certain to contribute to the subject of many future reports.

REFERENCES

Absolom, D. R., Lamberti, F. V., Policova, Z., Zingg, W., van Oss, C. J., and Neumann, A. W. 1983. *Appl. Envron. Microbiol.* 46: 90.

Absolom, D. R., Hawthorn, L. A., and Chang, G. 1988. *J. Biomed. Mater. Res.* 22: 271.

Andrade, J. D. 1985a. *Fouling and Cleaning in Food Processing*, D. Lund, E. Plett and C. Sandu, (Ed.), p. 79. University of Wisconsin, Madison.

Andrade, J. D. 1985b. *Surface and Interfacial Aspects of Biomedical Polymers. Volume 2: Protein Adsorption*, J. D. Andrade (Ed.), p. 1. Plenum Press, New York.

Andrade, J. D., Hlady, V. L., and Van Wagenen, R. A. 1984. *Pure Appl. Chem.* 56: 1345.

Arnebrant, T. and Nylander, T. 1986. *J. Colloid Interface Sci.* 111: 529.

Arnebrant, T. and Nylander, T. 1988. *J. Colloid Interface Sci.* 122: 557.

Arnebrant, T., Ivarsson, B., Larsson, K., Lundström, I., and Nylander, T. 1985. *Progr. Colloid Polymer Sci.* 70: 62.

Arnebrant, T., Barton K., and Nylander, T. 1987. *J. Colloid Interface Sci.* 119: 383.

Aveyard, R. and Haydon, D. A. 1973. *An Introduction to the Principles of Surface Chemistry*. Cambridge University Press, Cambridge, England.

Azzam, R. and Bashara, N. 1977. *Ellipsometry and Polarized Light*. North Holland Publ., Amsterdam.

Bagchi, P. and Birnbaum, S. 1981. *J. Colloid Interface Sci.* 83: 460.

Baier, R. E. 1970. *Adhesion in Biological Systems*, R. S. Manly (Ed.), p. 15. Academic Press, New York.

Baier, R. E. 1977. *Annals New York Acad. Sci.* 283: 17.

Baier, R. E. 1978. *Artifical Organs* 2: 422.

Baier, R. E. 1980. *Adsorption of Microorganisms to Surfaces*, G. Britton and K. C. Marshall (Ed.), p. 59. Wiley, New York.

Bryes, J. D. 1987. *Biotechnol. Prog.* 3: 57.

Buscher, H. J. and Weerkamp, A. H. 1987. *FEMS Microbiol. Rev.* 46: 165.

Colton, C. K., Smith, K. A., and deLeo, A. 1981. Paper presented at AIChE Annual Meeting, New Orleans.

Costerton, J. W. 1991. Paper no. 185, ACS National Meeting, Atlanta.

Cussler, E. L. 1989. *Diffusion: Mass Transfer in Fluid Systems*. Cambridge University Press, Cambridge, England.

Cuypers, P., Corsel, J., Janssen, M., Kop, J., Hermens, W. and Hemker, H. 1983. *J. Biol. Chem.* 258: 2426.

Dann, J. R. 1970. *J. Colloid Interface Sci.* 32: 321.

deFeijter, J. A., Benjamins, J., and Veer, F. A. 1978. *Biopolymers* 17: 1759.

Dillman, W. J., and Miller, I. F. 1973. *J. Colloid Interface Sci.* 44: 221.

Elwing, H., Welin, S., Askendal, A., Nilsson, U., and Lundström, I. 1987a. *J. Colloid Interface Sci.* 119: 203.

Elwing, H., Askendal, A., and Lundström, I. 1987b. *Progr. Colloid Polymer Sci.* 74: 103.

Elwing, H., Welin, S., Askendal, A., and Lundström, I. 1988. *J. Colloid Interface Sci.* 123: 306.

Fowkes, F. M. 1964. *Ind. Eng. Chem.* 56(12): 40.

Fowkes, F. M. 1972. *J. Adhesion* 4: 155–159.

Fowkes, F. M. 1985. *Surface and Interfacial Aspects of Biomedical Polymers*. Vol. 2: *Protein Adsorption*, J. D. Andrade (Ed.), p. 337. Plenum Press, New York.

Gölander, C. and Kiss, E. 1988. *J. Colloid Interface Sci.* 121: 240.

Hiemenz, P. C. 1986. *Principles of Colloid and Surface Chemistry*. Marcel Dekker, New York.

Horbett, T. A. 1988. *Protein Engineering* 2: 172.

Horbett, T. A. and Brash, J. L. 1987. *Proteins at Interfaces: Physicochemical and Biochemical Studies*, J. L. Brash and T. A. Horbett (Ed.), p. 1. ACS Symp. Ser. 343, Washington, DC.

Horsely, D., Herron, J., Hlady, V., and Andrade, J. D. 1987. *Proteins at Interfaces: Physicochemical and Biochemical Studies*, J. L. Brash and T. A. Horbett (Ed.), p. 290. ACS Symp. Ser. 343, Washington, DC.

Hubbell, J. A. and Desai, N. P. 1991. Paper no. 255, ACS National Meeting, Atlanta.

Iordanski, A. L., Polischuk, A. J. A., and Zaikov, G. E. 1983. *JMS—Rev. Macromol. Chem. Phys.* C23: 33.

Jönsson, U., Malmqvist, M. Rönnberg, I., and Berghem, L. (1982). *J. Colloid Interface Sci.* 90: 148.

Jönsson, U., Malmqvist, M., and Rönnberg, I. 1985. *J. Colloid Interface Sci.* 103: 360.

Jönsson, U., Lundström, I., and Rönnberg, I. 1987. *J. Colloic Interface Sci.* 117: 127.

Kaelble, D. H. 1970. *J. Adhesion*, 2: 66.

Kato, A. and Yutani, K. *Protein Engineering* 2: 153.

Koutsoukos, P., Norde, W., and Lyklema, J. 1983. *J. Colloid Interface Sci.* 95: 385.

Lee, R. and Kim, S. 1974. *J. Biomed. Mat. Res.* 8: 251.

Lee, S. and Ruckenstein, E. 1988. *J. Colloid Interface Sci* 125: 365.

Lesins, V. and Ruckenstein, E. 1988. *Biotechnol. Prog.* 4: 12.

Lin, Y. S., Hlady, V., Gölander, C. G., Wang, Y. J., and Andrade, J. D. 1991. Paper no. 262, ACS National Meeting, Atlanta.

Lok, B. K., Cheng, Y. L., and Robertson, C. R. 1983. *J. Colloid Interface Sci.* 91: 104.

Luey, J., McGuire, J., and Sproull, R. D. 1991. *J. Colloid Interface Sci.* 143: 489.

Lundström, I. 1985. *Progr. Colloid Polymer Sci.* 70: 76.

Lundström, I. and Elwing, H. 1990. *J. Colloid Interface Sci.* 136: 68.

Lundström, I., Ivarsson, B., Jönsson, U., and Elwing, H. 1987. *Polymer Surfaces and Interfaces*, W. J. Feast and H. S. Munro (Ed.), p. 201. Wiley, New York.

McCrackin, F., Passaglia, E., Stromberg, R., and Steinberg, H. 1963. *J. Res. NBS—A. Physics and Chemistry* 67A: 363.

McGuire, J. 1990. *J. Food Engrg.* 12: 239.

McGuire, J. and Kirtley, S. A. 1988. *J. Food Engrg.* 8: 273.

McGuire, J. and Swartzel, K. R. 1987. *Surf. Interface Anal.* 10: 430.

McGuire, J., Lee, E., and Sproull, R. D. 1990. *Surf. Interface Anal.* 15: 603.

Merrill, E. W. 1991. Paper no. 264, ACS National Meeting, Atlanta.

Morrissey, B. W. and Stromberg, R. R. 1974. *J. Colloid Interface Sci.* 46: 152.

Neumann, A. W. and Good, R. J. 1972. *J. Colloid Interface Sci.* 38: 341.

Neumann, A. W., Good, R. J., Hope, C. J., and Sejpal, M. 1974. *J. Colloid Interface Sci.* 49: 291.

Norde, W. 1986. *Adv. Colloid Interface Sci.* 25: 267.

Norde, W. and Lyklema, J. 1978. *J. Colloid Interface Sci.* 66: 257.

Nyilas, E., Morton, W. A., Cumming, R. D., Lederman, D. M., Chiu, T. H., and Baier, R. E. 1977. *J. Biomed. Mater. Res. Symp.* 8: 51.

Penners, G., Priel, Z., and Silberberg, A. 1981. *J. Colloid Interface Sci.* 80: 437.

Schakenraad, J. M., Busscher, H. J., Wildevuur, C. R. H., and Arends, J. 1986. *J. Biomed. Mater. Res.* 20: 773.

Shirahama, H. and Suzawa, T. 1985. *J. Colloid Interface Sci.* 104: 416.

Shirahama, H., Lyklema, J., and Norde, W. 1990. *J. Colloid Interface Sci.* 139: 177.

Slack, S. M. and Horbett, T. A. 1989. *J. Colloid Interface Sci.* 133: 148.

Soderquist, M. E. and Walton, A. G. 1980. *J. Colloid Interface Sci.* 75: 386.

Suzawa, T., Shirahama, H., and Fujimoto, T. 1982. *J. Colloid Interface Sci.* 86: 144.

Tan, J. S., Butterfield, D. E., Martic, P. A., Caldwell, K. D., and Li, J. 1991. Paper no. 184, ACS National Meeting, Atlanta.

Tosteson, T. R. and Corpe, W. A. 1975. *Can. J. Microbiol.* 21: 1025.

Van Dulm, P. and Norde, W. 1983. *J. Colloid Interface Sci.* 91: 248.

Wahlgren, M. and Arnebrant, T. 1990. *J. Colloid Interface Sci.* 136: 259.

Walton, A. G. and Maenpa, F. C. 1979. *J. Colloid Interface Sci.* 72: 265.

Ward, C. A. and Neumann, A. W. 1974. *J. Colloid Interface Sci.* 49: 286.

Wojciechowski, P. W. and Brash, J. L. 1990. *J. Colloid Interface Sci.* 104: 239.

Young, B. R. 1982. *Biomaterials: Interfacial Phenomena and Applications*, S. Cooper and N. A. Peppas (Eds.), p. 317. Adv. Chem. Ser. 199.

Zisman, W. A. 1964. *Contact Angle, Wettability and Adhesion*, F. M. Fowkes (Ed.), p. 1. Adv. Chem. Ser. 43.

5

Gels and Gelling

Allan H. Clark

Unilever Research Laboratory
Sharnbrook, Bedford, England

INTRODUCTION

The importance of gels and gelling processes to food technologists is undeniable. In food materials gelling components fulfill many functions, particularly in relation to texture and stability, and they of course significantly influence processing procedures. Their importance is especially great at the present time when the need to replace fat in foods has given rise to many new applications of water-based gelling systems.

In the present review, the nature of gels as materials will be examined closely. Although no exact definition of a gel will emerge, it is hoped that what is said will make the essential nature of gels and gel character much clearer. Attention will then turn to processes available for generating gels from aqueous solutions of polymers, colloidal particles, small-molecule solutes, etc., and to the underlying network-building mechanisms involved. Here, emphasis will be placed on the critical nature of the gelation event (at the gel point), and various theories of gelation (classical and modern) will be reviewed. At this point physical methods, available for monitoring the gelation process, will also be considered. From among these, mechanical spectroscopy will be singled out as a particularly important means of identifying the gel point and characterizing gel systems. This emphasis on mechanical behavior of gels will then be extended to include description of experiments designed to monitor gel behavior under conditions where this becomes increasingly nonlinear and includes failure.

Up to this point, the discussion will have focused on gels based on one main network-forming component. In a last short section, however, this restriction is lifted, and the treatment extended to include cases of mixed network gels where at least two such components are present. In conclusion, some attempt will be made to assess what has been achieved so far by physical gel studies and what still remains to be done.

THE NATURE OF THE GEL "STATE"

In his introductory lecture to a 1974 Symposium on Gels and Gelling Processes, P. J. Flory (1974) discussed at length what it was about materials commonly called gels that set them apart and justified a special classification. Various generally recognized attributes of gels were examined to see if they could provide a clue to the origins of unique character. For example, gels are sometimes said to contain two dispersed components (or "phases") one of which is solid (or at least, at a molecular level, through its long-range continuity of structure, imparts solid character to the gel) and the other liquid. Thus, although gels are usually considered as showing solid character overall, one of the important things about them is that some features of a liquid may also be present. Since overall solid behavior is expected, however, there seems to be no suggestion that a gel represents a new state of matter somehow intermediate between solid and liquid (although some might dispute this).

Examining things from a more molecular point of view Flory pointed out that the difficulties underlying definition of gels actually increased as one moved from a phenomenological description to a molecular one, since materials commonly recognized as examples of gels turned out to have very widely differing underlying molecular structures. In the light of these structures, gels, it seemed, could be classified in microstructural terms under four separate headings:

1. Lamellar liquid crystalline mesophases (e.g., soap gels and phospholipids)

2. Disordered covalently cross-linked polymeric networks (e.g., rubber and swollen rubbers)
3. Polymeric networks cross-linked by physical aggregation (e.g., gelatin gels)
4. Particulate networks (e.g., some globular protein gels, gels based on colloidal aggregates, etc.)

If there is any common factor, Flory concluded, linking these classes, it is structural coherence over long-distance scales leading to the mechanical attributes of a solid (i.e., very long stress relaxation times) when the material is deformed. The structures involved could vary greatly, however, in their detailed response to deformation, with the consequence that such gels could generally be expected to show the mechanical properties of complex solids (or indeed complex fluids) such as viscoelasticity and plasticity, rather than just simple Hookean behavior.

For the purposes of the present article, which aims to describe the most common gelling systems of interest to food technologists, attention will be focused principally on gel types 2, 3 and 4. Although systems of type 1 will sometimes be important in food materials and make structural (and other) contributions, these appear to be sufficiently different in principle (i.e., may not always involve a well-defined network) that they may merit separate experimental and theoretical examination. Our picture of a food gel, therefore, will be of a material containing a continuous and well-defined network assembled from particles or polymers (the "solid phase" referred to earlier), this network being embedded in a water-based solvent (the "liquid phase"). In practice, covalent networks of type 2 will not feature strongly in the discussion, as they are much less common in food materials than physical networks involving less permanent cross-links. The covalent networks (of which cross-linked gluten would be an example) can be thought of simply as limiting cases of the more temporary network-based gels discussed in the sections below.

In summary then, the difficulty about precisely defining gels relates in part to their structural diversity and complexity, if a microstructural criterion is to be developed, and partly to the continuous nature of physical variables if a phenomenological definition is sought (through mechanical behavior, for example). Thus if gels are viscoelastic solids, just how long does the longest stress relaxation time have to be for the gel description to be valid; for there is no doubt that most physically cross-linked gels have finite longest relaxation times, not the effectively infinite values required in theory for a solid? Also, if gels are recognized as "soft" solids, just how large a value for the equilibrium shear modulus (assuming relaxation times are long enough to allow us to measure such a quantity) is acceptable?

Another point of concern in gel definition might be the size of applied stress or strain within which linear viscoelastic behavior is retained. If a gel breaks down and flows at very small applied stresses, it will be perceived as a very different sort of material from a much more obviously solid and elastic material such as a concentrated gelatin "jelly." Considerations of this kind have in fact led to terms such as "strong" and "weak" to be used as further subclassifications of gels (Clark and Ross-Murphy, 1987).

Overall, it may be seen from the above discussion that even in the light of modern scientific analysis, the old adage that "gels are more readily recognized than defined" remains more than slightly true. Where we draw our boundaries in definition is to quite an extent arbitrary and subjective, so perhaps it is well to avoid the temptation to define gels at all; rather, we should consider instead the general physical attributes (as experimentally established) of materials we broadly recognize as gels and be sure where they fit into the overall spectrum of things.

PRODUCTION OF GELS—GEL NETWORK TYPES

A review like the present one should perhaps have been entitled "Gelling and Gels" since only very rarely in a technical process is a preformed gel used as a starting material. More commonly, the build-up of network structure occurs as the product-making process proceeds: the gel forming spontaneously in situ from a simple polymer solution, or particulate suspension, as externally controllable conditions of temperature or solution composition (pH, ionic strength, low molecular weight solutes, ion types, etc.) are changed. Thus, the sol-gel conversion process often (perhaps usually) involves aggregation of particles, or macromolecules, with eventual generation of a network spanning the entire container volume.

Where gelation is concerned, a very important point must be made about the underlying aggregation processes involved, which has long been asserted by most writers on the subject, and appears to contain the very essence of the phenomenon. This concerns criticality, and the idea that a gelling sol passes through a critical point, or state of critical branching, on its way to becoming a gel (Flory, 1941; Stockmayer, 1943, 1944; Gordon and Ross-Murphy, 1975). Here the idea is that aggregation may proceed for some time leading to a progressively more viscous solution, but at some critical degree of aggregation the gel point is reached, at which stage the viscosity diverges rapidly to infinity, along with the weight average molecular weight of the aggregates. At this point also, the system's (ideally equilibrium) elastic modulus changes from an effectively zero value to a growing finite result consistent with solid character. At this, the critically branched gel point, one aggregate molecule outstrips all the others in size, and subsequently becomes the gel network, or gel "phase," while the rest of the aggregated material remains distributed in size, with a much lower average molecular weight. As gelation proceeds, the lower molecular weight sol fraction reduces in amount as material bonds into the gel, and at high degrees of curing, the sol fraction disappears altogether.

The critical nature of gelation will be discussed in more detail in the next section, where theoretical descriptions of gelation are discussed, but it should be emphasized that the critical nature of branching processes leading to gelation and the existence of a gel point seem to be essential features of materials that can be justifiably called gels. Even though the mechanical attributes of the fully formed gels may range in type in the manner discussed earlier and so raise questions about gel classification, it appears that to be worthy of the title gel, these systems should have at some time in their history become critically branched and passed through a gel point.

Other types of aggregation process leading to growth of molecular weight (e.g., the Von Smoluchowski, 1917, model of diffusion-limited aggregation frequently adopted in colloid science) are well known, and have been described theoretically, but these do not show criticality. No doubt at high polymer or particle concentrations a molecule of arbitrary size can eventually be generated by such alternative mechanisms, but it remains questionable whether materials arising in this way should be regarded as gels (i.e., will satisfy the mechanical requirements for "weak" or "strong" gel classification to be described later).

In the section on theory to follow it will become clear that the critical point gel-formation mechanism essentially involves a reaction-controlled kinetic scheme, in which large particles react much more rapidly than small ones, leading eventually to explosive growth in aggregate size, while in the simplest diffusion-limited Von Smoluchowski coagulation mechanism, all particles react with roughly the same speed (or slow down with increasing size), and critical behavior is not predicted to occur.

Before closing this section on methods of generating gels and the gel point concept, it will be useful to focus in more detail on the types of molecular process that can give rise to gel networks and on the structures of the networks produced. Where physical networks are concerned, many common examples are to be found (Clark and Ross-Murphy, 1987). In gelatin gel formation, for example, a process common in food technology, random coil gelatin molecules in a cooled solution undergo a conformational change as they try to reassert the triple helical collagen structure in which they originally occurred. This reorganization occurs in a highly imperfect fashion, however, and leads to a complex network of aggregated gelatin chains, the sites of cross-linking being peptide segments locally ordered in triple helical form, and separated by lengths of disordered chain, giving the gel some of the attributes of a covalently bonded elastomer. Unlike permanently cross-linked networks, however, the cross-links in gelatin gels are thermally labile leading to gel thermoreversibility, i.e., gel "melting," and of course gelatin gel cross-links are not adequately described as point cross-links. They have finite size and extension within the network.

The second fundamental class of biopolymer, the polysaccharides (Rees, 1972; Rees and Welsh, 1977; Clark and Ross-Murphy, 1987), are particularly prolific in their capacity to form physical association networks via disorder-to-order conformational transformations. This is probably to be expected as, for several classes of polysaccharide, this property is closely related to their functions in nature where they often act as structural components (plants, seaweeds, etc.). In some cases double helices may be involved in forming junction zones, while in others, microcrystalline stacks of either double- or single-ordered strands constitute cross-links, and in at least one case (curdlan) triple helices may occur. Again, as for gelatin, the cross-links are finite in size, and potentially reversible (though high temperatures may be required). Unlike gelatin, however, in polysaccharide gels, the amount of disordered chain conformation present is usually quite low and is conformationally stiffer, making polysaccharide gels (e.g., agar and carrageenan gels) much more brittle. In addition, the polydisperse stacking of numbers of ordered chains into what amounts to very small crystalline cross-links, which is a feature of many polysaccharide gels, is not thought to be a structural feature of gels based on gelatin.

Gel networks of the association type are indicated schematically in Fig. 5.1b where they are compared with the point covalent cross-link situation shown in Fig. 5.1a. Indications of possible molecular origins of junction-zone formation in association networks are shown in Fig. 5.2, where (a) indicates the triple helical situation of gelatin, (b) the corresponding double helical mechanism often proposed for marine polysaccharides, and (c) shows single ribbons bonded into multiple chain assemblies by specific binding of cations. This "egg-box" arrangement, as it has been called, is generally accepted for alginate and pectin gels (algal and plant cell wall polysaccharides, respectively).

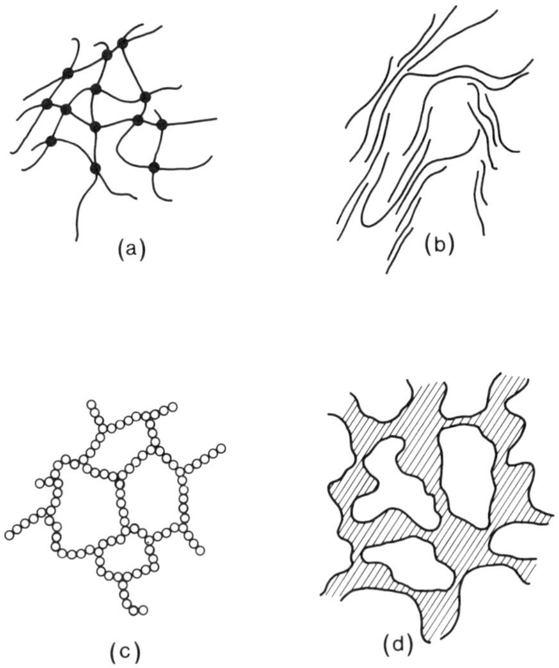

Fig. 5.1 Representations of common gel network–building mechanisms: (a) point (covalent) cross-linking; (b) extended junction zone formation (physical association networks); (c) particle association; (d) spinodal decomposition.

Point cross-linking and chain-associated junction formation are by no means the only mechanisms of aggregation and network formation by polymer molecules which have been described. Networks often described as particulate (gel classification 4 above) arise when macromolecules in compact semi-rigid conformational form, or other types of particle, aggregate into chains or random clumps. Again changes in solution conditions (temperature, pH, ionic strength, presence of small molecules, specific ions, etc.) introduce a driving force for aggregation, the bonds holding the particles together, varying considerably in origin (hydrophobic, hydrogen bonding, electrostatic, etc.) and sometimes representing a combination of types simultaneously active. Whereas colloidal systems in general can form networks of this type fairly readily, it is perhaps a less common situation for biopolymers, the gelation of globular proteins by heating (Clark and Lee-Tuffnell, 1986), and the aggregation of caseins under various conditions (Clark and Ross-Murphy, 1987), being perhaps the prime examples. Gelation of globular proteins is usually achieved by

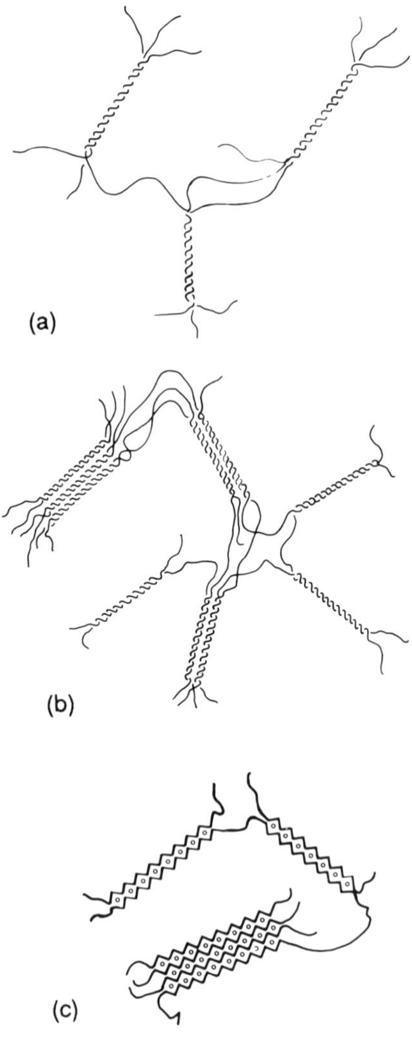

Fig. 5.2 Representations of common modes of extended junction zone formation: (a) triple helices formed in gelatin gelation; (b) double helix formation and association as in agarose gels; (c) "egg-box" formation by ion complexation as in alginate gels.

heating protein solutions (e.g., egg protein ovalbumin) above the denaturation temperature, at which point unfolded molecules aggregate in a manner determined by their state of charge, and exposure of hydrophobic and hydrogen bonding residues. Network formation of the particulate type is indicated schematically in Fig. 5.1c.

Finally, yet another mechanism of network formation which has been proposed is that of spinodal decomposition (Cahn, 1965). Incompatibility between solvent and polymer in a polymer solution can reach levels at certain temperatures and solvent compositions at which the solution becomes unstable, and spontaneously demixes to give a highly concentrated phase and a dilute one. This process proceeds via concentration fluctuations (Cahn, 1965) and, in its initial stages, leads to a periodic structure in which the concentrated phase interpenetrates the dilute phase, and seems to be made up of regions of approximately equal size. Provided that further "ripening" of this structure is inhibited by, for example, rapid gelation of the concentrated phase by one of the aggregation mechanisms already discussed (or, alternatively, glass formation, if solvent levels are very low), the resulting less fluid dispersed phase then itself constitutes a gel super-network. This mechanism has been proposed particularly for rodlike macromolecules in organic solvents (Miller et al., 1978; Tohyama and Miller, 1981), and it may also have some bearing on the formation of heterogeneous network structures in some of the more conventional water-based gels of food interest. For these last it has yet to be shown that glass-formation is important, the super-network being formed by superimposing a solvent-induced demixing process on an aggregation mechanism. The nature of the nonuniform super-network structure which can arise by spinodal decomposition is illustrated schematically in Fig. 5.1d.

Lastly, in this discussion of gel production it is worth adding some comments on the influence of shear, since in most processing situations the network-building processes that have been described take place in far from quiescent conditions. At the simplest, a uniform shear field will be involved, but more commonly the field will vary spatially. Among the issues which can be raised are: (1) whether the position of the gel point, i.e., critical degree of cross-linking, is likely to be changed by shearing; (2) whether the kinetics of aggregation which allow approach to this gel point will be altered; (3) how the size range of gel particles generated at the gel point is affected (since clearly in a shear field, the gel can no longer be treated as a single macroscopic "molecule"); and (4) whether, on removal of the shear field, cross-linking processes continue and the gel "heals." These are all important issues, and to the author's knowledge, in the biopolymer field, none of them seems to have been studied in great depth up to the present time, although related problems have been considered by colloid scientists studying shear field effects on flocculation (Swift and Friedlander, 1964).

In the gel area, based on current understanding of gelation mechanisms (see below), one might speculate that the position of the gel point (i.e., critical degree of cross-linking necessary for gelation) is unlikely to be much affected by shearing, though a change in level of intramolecular wastage of cross-links may occur and have some limited effect. It is also unlikely that for strongly cross-linked systems, the usual shear fields employed will greatly influence cross-linking kinetics, but this may be less true for more transient network situations where progress to the gel point may be slowed down or even inhibited (change in bond equilibrium position) by shearing. The size range of aggregates will clearly depend on the shear field applied, the average size reducing as the shear field increases, but no quantitative prediction seems available at present. As far as "healing" is concerned; where strong gels are involved this may sometimes occur, but in general the process will be slow and limited on account of the lack of cross-link reversibility and general dynamic character of the underlying networks. For more weakly cross-linked systems, on the other hand, where much faster processes take place, one might expect substantial association of microgel particles. In the following section attention will focus solely on theories available to describe gelation in quiescent conditions, the development of these to include shearing effects being apparently a topic for future research.

THEORIES OF GELATION

The earliest type of gelation phenomenon to be treated theoretically was that of covalent condensation of polyfunctional monomers (Fig. 5.3) either in an organic solvent or, more commonly, with the monomer itself as solvent. This early pioneering description of gelation by Flory (1941) and Stockmayer (1943, 1944) concluded that gelation corresponded to a situation where a macroscopically large branched polymer molecule became present spanning the entire volume of the sample. It was also concluded that, at the early stages of production of this gel "molecule," by no means all of the monomers originally involved formed part of this network. A "sol" fraction consisting of free monomers, and small aggregates based on these, was shown to co-exist with the network, being intimately dispersed through it. As cross-linking proceeded, this sol fraction diminished in amount, and the solid character (in terms, for example, of a measured shear modulus) became greater. Eventually, as this "curing" process moved to completion, most of the free material became connected into the gel network.

The Flory-Stockmayer treatment involved statistical arguments about the molecular weight distributions to be expected, if monomers reacted randomly, with a reaction probability determined by the overall fraction of bonding sites involved in cross-link formation at a particular stage of

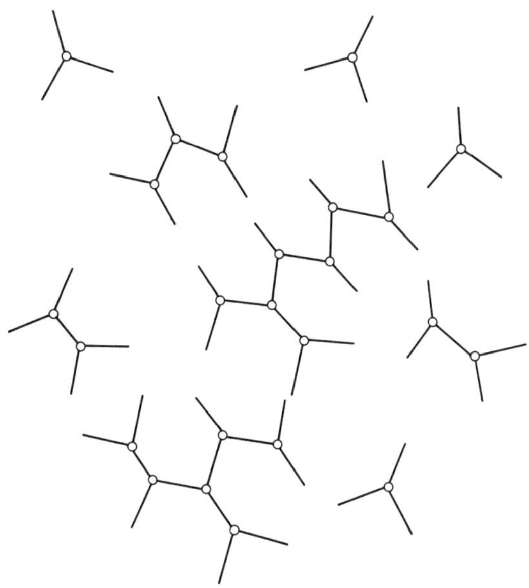

Fig. 5.3 Random f-functional (here f = 3) polycondensation model underlying Flory-Stockmayer description of gelation.

curing. The most striking conclusion of this treatment was that the emergence of the gel fraction was a sudden event which occurred only when the degree of cross-linking reached a critical value. Prior to this point, there was a lag period during which the system became increasingly viscous as simple molecular condensation took place. At the gel point, however, where the critical cross-linking threshold was reached, the viscosity diverged to infinity, there being a corresponding divergence of the weight-average molecular weight. Gelation was thus understood as a critical phenomenon, the system passing continuously, and uniformly, from liquid to solid, through a critical state where certain of its physical properties diverged. It is worth noting that in this description there is no suggestion that gelation represents a first- or second-order phase transition (in fact it has sometimes been described as a fifth-order Ehrenfest transition, see, e.g., Gordon and Ross-Murphy, 1975).

The Flory-Stockmayer treatment allowed mathematical expressions for the sol molecular weight distribution, and the amounts of sol and gel present at any stage of curing, to be derived, as well as treatments of more complex condensing systems based on more than one kind of monomer. They also developed expressions for the critical level of cross-linking necessary to reach the gel point for these situations. For example, if the system consists of identical monomers, each with f sites for condensation (f-functional monomer), the critical fraction of sites which must have reacted is predicted to be $\alpha = 1/(f - 1)$.

The Flory-Stockmayer treatment is time independent in the sense that a particular value of α is assumed to apply to a system (without detailed discussion of how long this level of cross-linking has taken to happen), and the implications of this situation are determined by statistical arguments linked to certain assumptions about the monomers, and the cross-linking process. Stockmayer did, however, give some indications of what were the essential elements of the type of cross-linking kinetics which would lead to the gel point situation. Thus, if aggregation is assumed irreversible, and is a process by which i mers and j mers react to a form p mers, the underlying implication of the critical nature of gelation is that the rate constant for such a process is proportional to the product ij. Thus, larger species react much more rapidly than small ones, and this leads to the explosive nature of gel formation, a point made in the previous section.

For physical cross-linking, this kinetic scheme must be modified to include dissociation of aggregates, but the basic conclusions of the Flory-Stockmayer approach remain valid. Today, in fact, there are many published articles available (e.g., McLeod, 1962; Dusek, 1979; Ziff, 1980; Ziff and Stell, 1980; Leyvraz and Tschudi, 1980, 1981, 1982; Cohen and Benedek, 1982; Ernst et al., 1982; Ziff et al., 1982, 1983; Van Dongen and Ernst, 1983; Clark, 1991) which discuss gelation from the kinetic viewpoint, treating both irreversible and reversible aggregation events. Where these employ the rate constant assumption mentioned above for the growth of aggregates, the Flory-Stockmayer conclusions emerge. The kinetic approach does allow alternative models to be explored, however, and the nature of kinetic concepts, such as gel times (see next section) can be examined, as well as the influence of aggregate concentration on cross-linking if a substantial amount of a supporting solvent is employed. The question of solvent is exceptionally important where physical food gels are concerned. Polysaccharide gels, in particular, may be strong solid structures but still contain over 99% water by weight.

The first discussion of how to modify the Flory-Stockmayer theory to include potential reversibility of cross-links, and the presence of large quantities of solvent, was by Hermans (1965), who considered the aggregation of carboxymethyl cellulose at various concentrations in water. This approach used a mass-action argument (Ostwald dilution law) to formulate the extent of cross-linking in relation to polymer concentration. This provided values for α to be used in the Flory-Stockmayer formulas. The idea of a critical gelling concentration emerged, since the critical level of α necessary for gel formation would be reached via mass action only when the polymer concentration, C, reached a certain value (the critical concentration C_o). By combining these ideas with those of ideal rubber theory, expressions of a general nature were derived for the gel elastic shear modulus in terms of the dimensionless quantity C/C_o. This approach has proved very useful in treating physical gels, even though sometimes one must be wary of applying the underlying equilibrium assumption where systems aggregating through strong bonds are involved.

Subsequently, the model has been refined (Clark and Ross-Murphy, 1985; Clark, 1987) to make use of more sophisticated mathematical approaches to network formation such as the cascade approach of Gordon and coworkers (Gordon, 1962; Gordon and Ross-Murphy, 1975). The use of ideal rubber theory (Flory, 1953; Treloar, 1970) has also been modified to allow for the much more rigid character of the elastically active network chains constituting such networks. These new approaches have allowed formulation of the critical concentration in terms of parameters such as molecular weight, the equilibrium constant for bond formation and the number of cross-linking sites per molecule. It has also allowed the shear-modulus concentration relationship to be formulated somewhat differently from that proposed by Hermans (but once again in a universal dimensionless form). A kinetic model (Clark, 1991) based on these principles has allowed the approach to equilibrium to be studied, and this has led to a universal formulation of the gel time (time for system to evolve to critically branched state) as a function of polymer concentration. The kinetic model also allows an examination to be made of the implications of applying the equilibrium model in a nonequilibrium situation.

The body of theory just described is sometimes referred to as "mean field" meaning that it deals with the idea of connectivity only and does not give aggregates a space-filling geometrical significance and allow them to be embedded in real space. The Flory-Stockmayer approach, and later models developed from it, thus ignore excluded volume and all the spatial implications that derive from this. An alternative theoretical approach to gelation that has been explored for quite a number of years now and that seeks to overcome the difficulties just outlined is percolation theory (Stauffer, 1976, 1979, 1981; Stauffer et al., 1982; Adam and Delsanti, 1989). There are several types of model (e.g., bond percolation, site percolation, site-bond correlated percolation, etc.), but the essence of the approach where gelation is concerned is to place monomers on a lattice (d-dimensional) and then randomly introduce a certain proportion of bonds connecting these. As this process proceeds, clusters of monomers (aggregates) develop naturally, and at a critical threshold of bonding, the gel point divergence property is observed in the form of a "molecule" spanning the entire percolation lattice. One advantage of the percolation approach on a geometrical lattice is that the formation of intramolecular bonds (cyclization) can occur quite naturally, something which is assumed to be absent in the original Flory-Stockmayer approach.

The percolation method clearly says something about cluster size distributions, and critical levels of cross-linking, as does Flory-Stockmayer theory, which has been described as a particular form of percolation, i.e., percolation on a Bethe lattice (Fisher and Essam, 1961), but the particular success claimed for it is its more correct predication of what are called critical exponents. As was said earlier, some physical properties of a system diverge at the critical point, and gelling systems are no exception, weight average molecular weight, viscosity and elastic modulus showing irregular mathematical behavior when the gel point is reached. It is usually claimed that mean field theories are unable to describe such properties accurately in the critical region. In particular, if a quantity is assumed to show a power law dependence $(p - p_c)^{\beta}$ on the difference between the value of a parameter p describing the system's state (e.g., temperature, concentration, pressure, etc.), close to the critical point and p_c, its value at the critical point, mean field theories are said to give less accurate predictions of β than, say, percolation on a lattice. For a given type of critical phenomenon, theorists expect to define a set of critical exponents for a corresponding set of physical quantities and to find that these same values recur when other (and apparently different) critical phenomena are considered. The possession of common critical exponents is a means of classifying various critical phenomena into groups, and is an example of the principle of "universality" being applied. In fact, in relation to polymer gelation, it is not yet clear whether experiment does show unequivocally that percolation exponents are more realistic than corresponding classical ones, and, indeed, one suspects that from the food technologist's point of view, this is not a vitally important issue. Gels are usually handled which have substantially passed through the gel point, and the success of theories of gelation will be judged more in terms of capacity to explain the properties of such well-cured gels than those of critically branched systems.

Before leaving percolation theory, one must say that one major advantage of this approach has been its suitability to treat particles or polymers gelling in the presence of solvent and suffering simultaneous phase separation (demixing) on account of this solvent (Coniglio et al., 1979). Here a lattice approach (site-bond correlated percolation) involving both condensing monomers and solvent molecules (Fig. 5.4) is combined with the methods of statistical mechanics to predict temperature-composition phase diagrams for reversibly gelling systems. Here solvent-polymer and polymer-polymer interaction terms provide a statistical weighting scheme

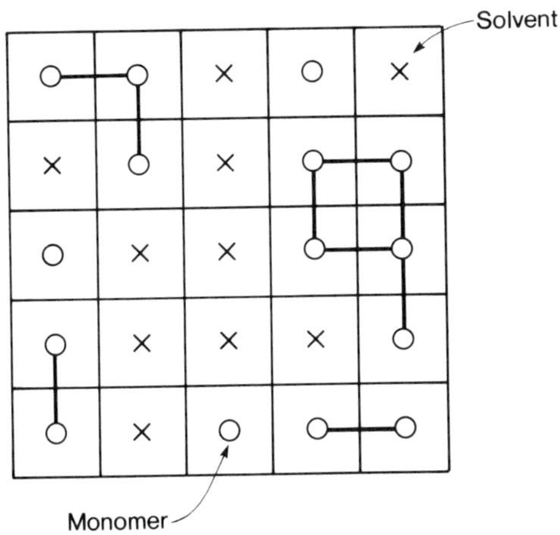

Fig. 5.4 Lattice treatment of gelation underlying the site-bond correlated percolation approach to the sol-gel transition in the presence of solvent.

for configurations and are combined with the connectivity arguments of gel theory, to produce phase diagrams of the type shown in Fig. 5.5. In these, a sol-gel line (line of critally branched gel states) combines with the usual miscibility gap of solution theory to produce a combined description of aggregation and demixing, and alternative situations occur, depending on the relationship between the point of intersection of these curves and the upper critical solution point. In the area within the miscibility gap, particularly in that part defined as the unstable spinodal region, demixing and gelling both occur, and either a heterogeneous gel structure or a suspension of microgel particles develops. If spinodal decomposition applies, the gel regions will have similar sizes. This will lead to a roughly periodic fluctuating microstructure (on a micron scale, for example).

In relation to the site-bond correlation percolation approach to gelling in a solvent it must be emphasized, that the calculations described so far in the literature refer to systems at equilibrium and to reversible processes. Real heterogeneous networks may well be strongly cross-linked and, because of this, not at all susceptible to reversible shifting across the phase diagram as external constraints vary. Nonetheless, it is interesting that experimentally established "state" diagrams for heat-set globular protein gels made under varying temperature and, more particularly, pH and ionic strength conditions are remarkably similar to those in Fig. 5.5 (Clark and Lee-Tuffnell, 1986).

In addition to percolation theory, related modern theoretical ap-

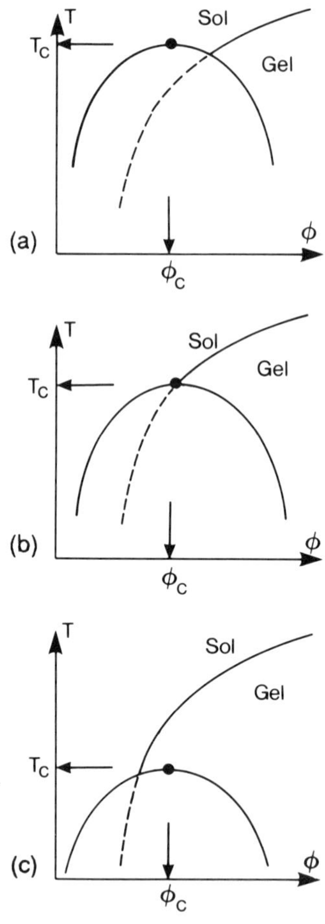

Fig. 5.5 Phase diagrams (temperature T vs. phase volume ϕ) for reversible gelling system predicted by lattice statistical mechanical treatment of Coniglio et al. (1979): (a), (b), and (c) show cases where critical solution point (T_c, ϕ_c) lies outside, on the boundary of, and within the gel region.

proaches to gelation include simulation of aggregation processes by computer. In so-called diffusion-limited aggregation (DLA) models (Witten and Sander, 1981; Witten, 1985; Meakin, 1987), monomers are projected towards a growing aggregate and stick with a certain probability. As the number of events increases, complex structures develop (Fig. 5.6), which are most easily described by the ideas of fractal geometry (Mandelbrot, 1982). Different model rules lead to different so-called fractal dimensions (the exponent relating the increase in number of particles in structure to the increase in radius). For example, the fractal dimension is 3 for a homogeneous sphere, but is usually much less than this, and fractional, for a fractal aggregate. This effect can be systematically explored. From the point of view of modeling gelation, the DLA model in its simplest form is unlikely to be appropriate on account of the limited range of aggregation events it considers, but developments of such simulations to include more general aggregate-aggregate encounters (Meakin, 1983; Jullien et al., 1987) have been described. These have been remarkably successful in reproducing the appearance of tenuous particulate networks as determined by electron microscopy.

Apart from microscopy, experimental verification of the results of such simulations is usually by scattering experiments involving light or shorter wavelength radiation, since by this approach fractal dimensions can be measured. It is also claimed that the mechanical properties of gels (e.g., modulus-concentration and ultimate properties) are also related to the fractal nature of the aggregates (Bremer et al., 1989; Walstra et al., 1991) and in particular to their fractal dimensions. At this stage it remains unclear whether such models may be used to predict absolute values of mechanical properties (the shear modulus, for example) and indeed whether they imply critical branching (i.e., a critical gel point).

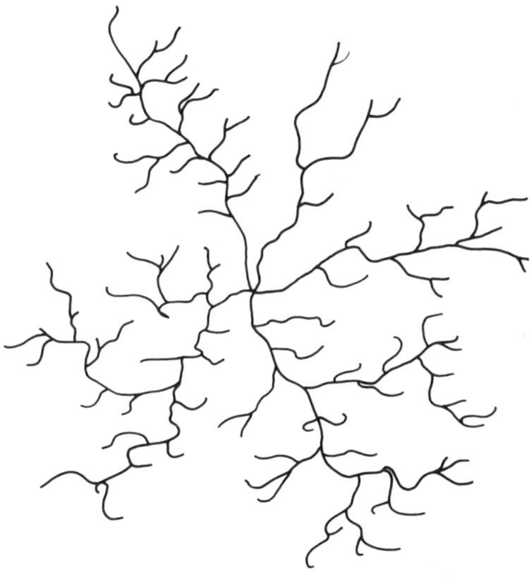

Fig. 5.6 Fractal cluster of type generated by diffusion-limited aggregation (DLA) approach.

EXPERIMENTAL MONITORING OF GELATION

In a previous section it was indicated that changes in temperature or system composition (pH, ionic strength, specific chemical additives such as certain ions, small molecules, enzymes, etc.) could set an aggregation process in motion ultimately leading to a gel, with passage through the gel point occurring on the way. Often, gelation in biopolymer solutions is sufficiently slow that the progress of such aggregation can be followed by conventional means (Clark and Ross-Murphy, 1987) such as light scattering, X-ray scattering, various forms of spectroscopy, etc. In some situations however, rapid processes are involved, and in these cases more sophisticated kinetic approaches (Norton et al., 1983) must be used (stopped-flow polarimetry, for example).

In general, however, conventional methods have yielded much useful information about molecular events underlying network-building processes, and perhaps more interestingly in relation to gel theory, they have provided details of the variation in average molecular weight and aggregate size (including fractal characteristics) which are taking place. Indeed, spectroscopic and scattering studies are largely responsible for the molecular pictures (Figs. 5.1 and 5.2, for example) that are currently available for biopolymer network formation, such as helix formation of various multiplicities, "egg-box" formation, bead-models (for protein gels), etc.

The essence of gelation is, however, the gel point transformation from liquid to solid, and this suggests that the most important attributes of the system to monitor will be mechanical ones, both before and after critical branching. There are, of course, many types of mechanical measurement and measuring devices, but when monitoring the progress of gel cure, certain principles must be adhered to. Thus, the measurement must be completed rapidly, so that a time-resolved study is possible; and it must not disturb the system outside its linear response range. It must also provide a physical quantity that is objectively significant, and a good indicator of network character, and one which can be addressed (at least to some extent) by theory.

These restrictions tend to eliminate transient methods such as creep and stress relaxation experiments, and all forms of test involving continuous scanning of applied stress or strain, and leave dynamic testing, i.e., mechanical spectroscopy (Ferry, 1980) involving oscillatory applied stress or strain, as the most promising (if not the only) option. In the strain-controlled version of this experiment, for example, tested material is subjected to a sinusoidally oscillating strain (deformation). This generates a sinusoidally oscillating stress response, which is, to some extent, out of phase with the strain. The amplitude of this stress, and the difference in phase between it and the strain, are measured; and the stress is resolved into one component (elastic response) completely in phase with the strain, and another (viscous response) 90° out of phase. The result of the measurement (when reduced to unit applied strain) is the complex shear modulus G^*, which can be expressed either in terms of the original amplitude and phase difference, or, as will be adopted here, in terms of real and imaginary parts, G' and G'', proportional to the amplitudes of the in-phase, and 90° out-of-phase, components, respectively.

The storage and loss components of the complex modulus G' and G'', so defined, depend of course on measurement frequency and, as functions of frequency, are objectively defined physical quantities amenable to theoretical interpretation. Ideally, a very low frequency of oscillation should be applied during the cure experiment to achieve measurement of, as nearly as possible, the equilibrium shear modulus, since it is this quantity which (ideally) becomes nonzero at the gel point, and enables the gel point to be identified. In practice, however, the need for time-resolution when making measurements during curing means that this last situation cannot be realized to arbitrarily low frequency. Thus, in practice, measurements are made at some chosen low frequency (1 Hz, for example) defining a "pseudo-equilibrium modulus." In cure monitoring, it is also important to keep the amplitude of the applied disturbance (stress or strain) sufficiently small that the system always responds linearly. For extremely strain-sensitive systems (e.g., physical gelling systems very close to the gel point) this can be a taxing experimental demand.

MECHANICAL SPECTROSCOPY AND GELATION

Results of monitoring G' and G'' as a function of curing time for a typical gelling system appear in Fig. 5.7. As expected, G'' is initially greater than G', indicating the starting solution to be a viscoelastic fluid of low G^*, and one which is relatively much more viscous than it is elastic. At a certain time the G' and G'' curves cross-over, and then G' rises rapidly as the system develops a much higher modulus, and shows much greater relative elasticity. After the cross-over, G'' rises too, and in many cases passes through a maximum before falling again as the network becomes more completely established, and still more elastic. The time to reach the critically branched condition (gel time t_g) is usually considered to correspond to the cross-over point (Te Nijenhuis and Winter, 1989), but in experiments that are not sensitive enough to accurately measure low values for G' and G'', the time at which the real part of the modulus starts to climb rapidly is all that is available. Finally, as Fig. 5.7 shows, at long

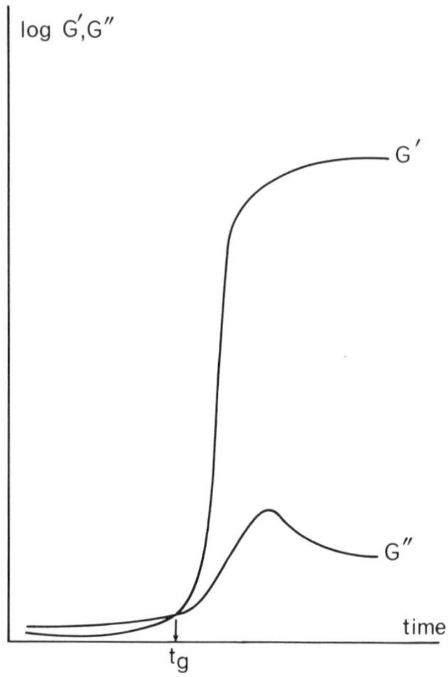

Fig. 5.7 Typical G′, G″ vs. time-cure curves obtained by application of dynamical mechanical spectroscopy to a gelling system. The cross-over point is often taken as an indicator of the gel time t_g.

times G′ and G″ will tend to plateau limits, with G″ having a limiting value much smaller than G′ (100× smaller, for example). Since the terminal variation of the components of the modulus with time may be slow, the use of logarithmic time in plots may be necessary to establish such limits with certainty.

If the experiment of Fig. 5.7 is carried out at different concentrations of gelling material under essentially the same gelling conditions, e.g., temperature, ionic strength, stoichiometric ratio of counterions, etc., the plateau height for G′ will vary considerably, usually increasing substantially with concentration. The gel time will also vary with concentration, shortening as the concentration is increased (Clark, 1991; Ross-Murphy, 1991). These trends can be analyzed graphically by making plots of log t_g and log G′ (plateau value) against log C (Figs. 5.8 and 5.9), the advantage of the log plots being that power law relationships, e.g., $t_g \propto C^{\beta}$, are immediately obvious. This type of power law should be distinguished from the critical exponent laws discussed earlier. The latter are related to variation in the vicinity of the critical point, and for G′ or G″, for example,

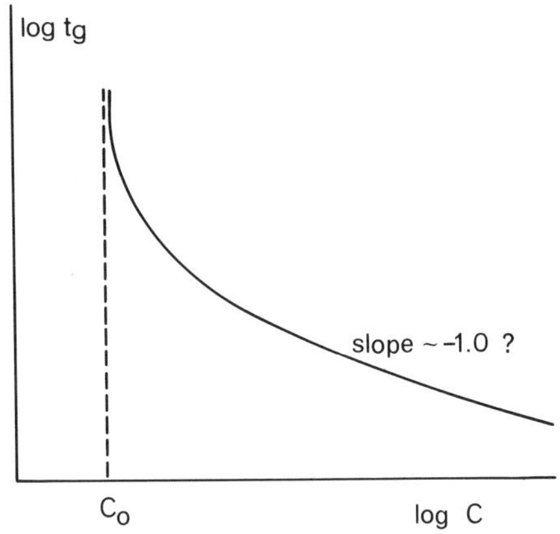

Fig. 5.8 Predicted relationship between gel time (t_g) and concentration (C) obtained by applying modified Flory-Stockmayer model to reversible gelation in a solvent medium. The critical concentration C_o is the concentration at which the gel time diverges logarithmically to infinity.

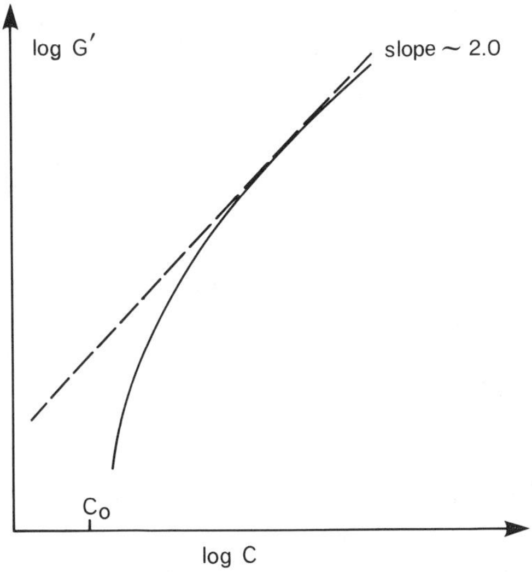

Fig. 5.9 Relationship between limiting (or plateau) shear modulus component (G′) and concentration (C) predicted by applying same model as described for Fig. 5.8. The power law relationship between these quantities is close to C^2 at high concentrations (though exponent is predicted to be smaller than 2 for certain model assumptions).

would require plotting against $C-C_o$. The power laws now discussed relate to variation of the modulus with the absolute size of the concentration, and may well be valid for concentrations much greater than critical. An example is the often-stated principle that the modulus of a gel (e.g., gelatin "jelly") varies as C^2.

If the simple kinetic model based on Flory-Stockmayer kinetics and reversible aggregation discussed earlier is adopted as a means of describing gelation, computer calculations using this approach (see, e.g., Clark, 1991) suggest mathematical forms for the gel time and gel modulus vs. concentration plots of the form shown in Figs. 5.8 and 5.9. Figure 5.8 suggests that the gel time should diverge logarithmically at the critical concentration C_o discussed earlier, and if data can be measured accurately near C_o, this plot provides (at least in principle) a means of critical concentration evaluation. Beyond this, at concentrations considerably higher than C_o, a constant power law region is expected, and the simple kinetic approach suggests that the appropriate power is -1. In practice, however, much higher negative powers have been found for, for example, gelatin and certain carrageenan systems (Oakenfull, 1984; Oakenfull and Scott, 1986; Oakenfull and Morris, 1987). Ross-Murphy (1991) has discussed this discrepancy in some detail, and explains it in terms of neglect of intramolecular cyclization (wastage of cross-links) in the kinetic model, an event that is likely to be prevalent near the gel point and lead to a loss of potential cross-linking sites. An alternative explanation put forward by Oakenfull and coworkers is that the larger negative slope contradicts the second-order kinetics underlying the modified Flory-Stockmayer approach and is consistent with junction zone formation occurring cooperatively, through the simultaneous coming together of several chains. Such high-order kinetic mechanisms seem unrealistic, however.

Figure 5.9 shows the general form expected for log G' vs. log C results. At high concentrations, a constant (or very nearly so) power law region is predicted, and although the slope can vary depending on exact details of the kinetic model assumed (Clark, 1991), the result is usually close to two. At concentrations nearer critical, the power law index is expected to vary continuously, becoming very much greater than two as C_o is approached. Thus the common idea that gel moduli vary as C^2 is not generally valid, being true only for fairly strong, concentrated gels. In fact, since measurements are often made over very limited concentration ranges, and intermediate ranges between the C_o and C^2 regions in Fig. 5.9, it is not uncommon to see a much higher power law of, for example, five quoted for gelling systems (Walstra et al., 1991). In the author's opinion it would be necessary to establish this law over a wider range of concentration than is usually done (or perhaps is possible for certain systems), before its existence could be claimed as fundamentally significant, and a clear indicator of failure of the classical mean field approach to gelation.

In terms of experimental support for the limiting modulus-concentration relationship of Fig. 5.9, one can say that unlike the case of experimental data for gel times, experimental gel modulus data supports the predictions of the kinetic model rather well (see also Clark and Ross-Murphy, 1985; Clark, 1986). Figure 5.10 shows some typical results (Clark and Ross-Murphy, 1985; Clark et al., 1990; Brown, 1991) for several well-known biopolymer systems (κ-carrageenan, agar, bovine serum albumin, and gelatin gels), and though individual curves vary considerably in position on the plot they all tend to be slightly curved with a power law of

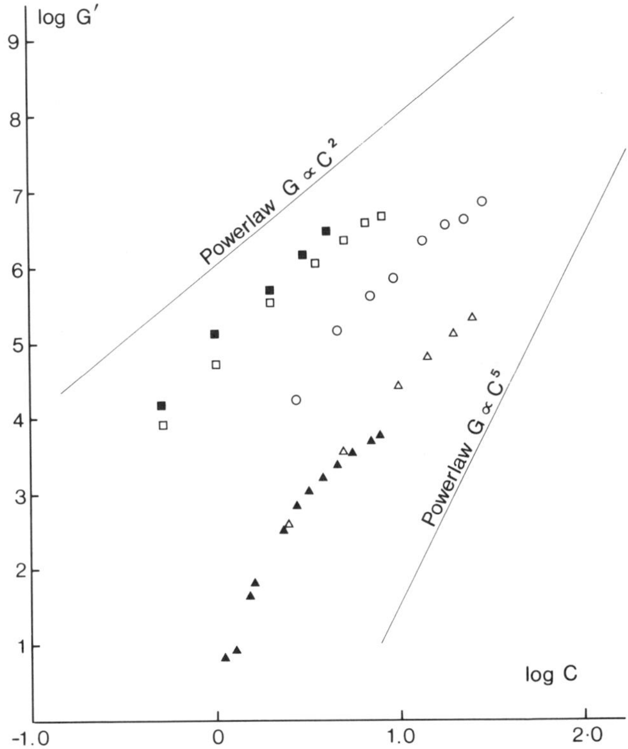

Fig. 5.10 Experimental plateau log G′ vs. log C data for common gelling systems: gelatin type 1 (▲), gelatin type 2 (△), heat-set BSA pH 7.0 (○), and agarose (□), and κ-carrageenan (■). Slopes corresponding to C^2 and C^5 powerlaw behavior are included for comparison.

roughly two at the highest concentrations studied, and a somewhat higher power at lower concentrations. Constant power law behavior is not observed. Thus, although the data is never extensive enough in concentration range (in particular, is never measured close enough to the critical concentration) to enable the very high power law behavior of Fig. 5.9 to be demonstrated, increasing power law behavior as concentration falls seems clearly verified.

A very valuable alternative to Figs. 5.8 and 5.9 (and indeed, Fig. 5.10) is to plot log t_g and log G' as functions of C/C_o, assuming that an estimate of C_o is available (from gel time results or direct fitting of modulus-concentration data as discussed below). This is a first step towards establishing universal relations for these quantities, for if results for different systems are plotted together in this way, particularly for G' vs. C data, it becomes clear that simple translation along the ordinate axis, brings all such results onto a common line or mastercurve (Fig. 5.11). Several authors have discussed this mastercurve approach, particularly its success in relation to gel-modulus concentration data (Hermans, 1965; Bikbov et al., 1979; Clark, 1991; Ross-Murphy, 1991), and have attempted to provide a mathematical form for the universal relationship involved. For example, in their cascade theory treatment, Clark and Ross-Murphy (1985) provide explicit analytical expressions, which include slightly different limiting power laws (and hence mastercurve forms) at high C, depending on the number of cross-linking sites per gelling unit.

The success of the mastercurve approach as a means of simplifying experimental data by exposing an underlying universality is shown in Fig. 5.11. Here the results of Fig. 5.10 have been reduced to mastercurve form by least-squares fitting to one of the analytical forms referred to. For the data shown, a model involving a very large number of cross-linking sites per molecule was found to give the best overall reduction of the five sets of experimental data, and for the heat-set protein gel systems in particular this is perhaps a surprising result. A model involving a small number of cross-linking sites would seem more appropriate, as the aggregating protein molecules are not likely to be grossly unfolded (Clark and Lee-Tuffnell, 1986), and would be expected to bind to neighbors through only a limited number of potential contact zones. Caution is needed, however, when attempting such a detailed interpretation of the mastercurve analysis, as in the theoretical treatment of network elasticity which underlies it (Clark and Ross-Murphy, 1985; Clark, 1987), network defect effects likely to be prevalent at high polymer concentrations (such as trapped entanglements) have been neglected. Unfortunately, the establishment of numbers of cross-linking sites appropriate to a particular molecule by other experimental means (i.e., independently of the kinetic model) is far from easy. While the number of cross-linking sites (or functionality as it

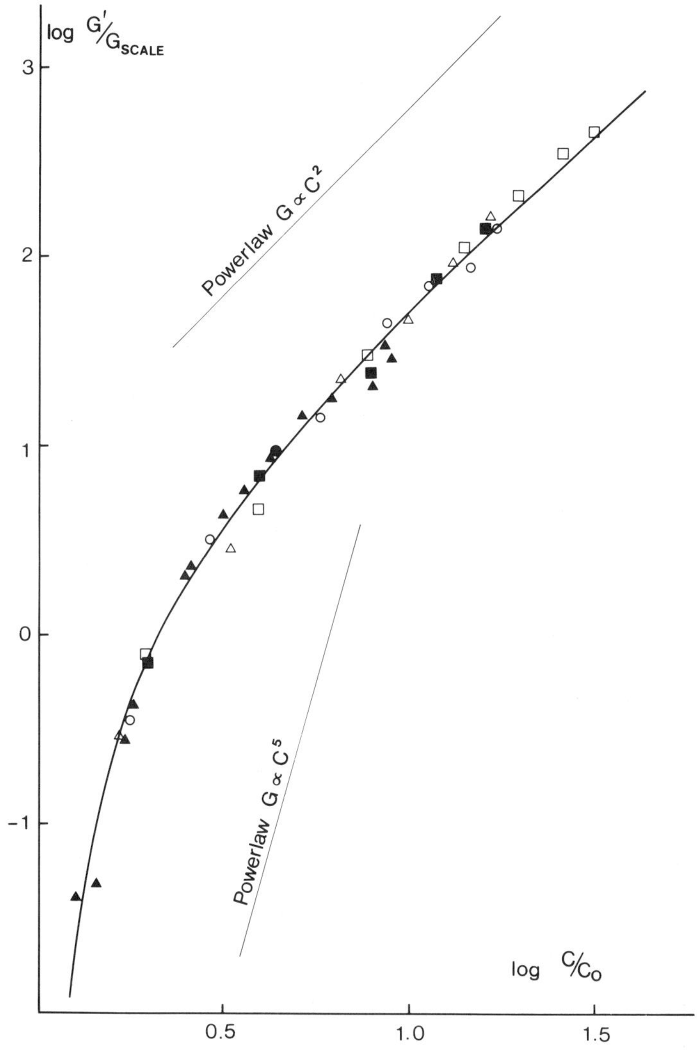

Fig. 5.11 Data of Fig. 5.10 rescaled to dimensionless form by fitting to high-f mastercurve (heavy line) of Clark and Ross-Murphy (1985).

is called in polymer literature) is a quantity fairly readily determined for molecules forming covalent bonds, unfortunately, it is far less easy to establish its value for particles aggregating physically. This remains a central element of difficulty when attempting to adjudicate experimentally between network theories of physical gelation using this concept.

At a more practical level, the G′ versus C/C_0 mastercurves formulated by Clark and Ross-Murphy are found to be extremely effective in fitting modulus-concentration data over a wide range of concentration. This provides both a means of interpolating, and extrapolating, such data, and of estimating C_o. Such estimates of critical concentration from modulus-concentration data are much easier to arrive at experimentally than from gel time measurements and of course provide a means of cross-checking these results.

Finally, in relation to the mastercurve approach, it is emphasized that the theoretical descriptions just discussed are based on kinetic models which assume complete reversibility in the making and breaking of bonds, and an eventual establishment of equilibrium at long curing times. Clark (1991) has discussed the validity of using such functions to describe experimental modulus-concentration data measured for a system under kinetic control (such as, for example, is likely to be the case for most strongly gelling systems). Good fits of the analytical mastercurves to such data can still be expected, provided the data satisfies certain measurement conditions, but the model parameters must now be treated with caution, particularly the concept of the critical concentration. In the irreversible situation the critical concentration becomes a kinetically determined quantity having little fundamental significance. In the irreversible case, in fact, all systems would ideally have zero critical concentrations in the limit of infinite time if other processes such as intramolecular cyclization (already referred to in connection with gel times) did not take place. Indeed, the failure of the reversible model to accurately predict gel time vs. concentration relationships for many strongly gelling systems is almost certainly a consequence of the failure of the equilibrium assumption. Such a breakdown is much less obvious, it seems, where modulus-concentration data are concerned.

Lastly, one must add that all of these difficulties become compounded where gelation is accompanied by phase separation, for here the critical concentration concept, and the application of the mastercurve approach require especial care. In this situation, an apparent critical concentration may be found which relates simply to a heterogeneous system's ability to span the volume available to it. At concentrations lower than this, the system may still have passed through the gel point and formed macroscopic aggregates, but these may then take on the form of microgel particles only, suspended in the solvent medium.

MECHANICAL PROPERTIES OF FULLY CURED GELS

In this section attention will focus on mechanical properties of either fully cured gels (i.e., those that have achieved the limiting plateau modulus condition) or gels which, though still unstable, are changing only very slowly with time. These properties will divide conveniently into: (1) results of measurements made at small deformation which maintain systems within their linear response region, and (2) the results of mechanical tests in which linearity is not maintained and failure can occur.

Small-Deformation Studies

Mechanical spectroscopy is also applied to limitingly cured gels to: (1) establish the frequency dependences of G′ and G″; (2) study the strain dependence of the modulus, and so establish the limits of linear behavior; and (3) study the temperature dependences of the G′(ω) and G″(ω) functions, and so characterize gel "melting" behavior (or, if the gels are not thermoreversible, to at least establish the direction of modulus change with changing temperature). These further studies of "finished" gels allow experimentalists to distinguish between so-called "strong" and "weak" gels, and to eliminate from "gel" classification certain materials that superficially, at least, seem to be entitled to such a description (Clark and Ross-Murphy, 1987).

The linear viscoelastic modulus (G′,G″) vs. frequency spectrum of a gel (strong or weak) is expected to have the general form shown in Fig. 5.12a. This indicates that G′ and G″ are expected to be roughly independent of frequency over a wide range of frequency, and G″ is expected to be very much smaller than G′. Ideally, the frequency range should extend as widely as possible, but in practice in mechanical spectrometers, the accessible range is roughly 10^{-2}–10^{2} Hz. Other approaches would be necessary (such as stress relaxation or creep) to extend this range, and such transient methods have their own particular limitations. For the ideal gel, G′ holds its value down to zero frequency, whilst G″ becomes zero, i.e., an equilibrium modulus is present, but in practice many gels may fall considerably short of this requirement, showing frequency dependence of the modulus below the lower frequency limit in Fig. 5.12a. At higher frequencies, deviations are also expected in real systems, as real materials pass through the rubber-glass transition region of the mechanical spectrum and reach the glassy plateau. For most heavily hydrated food gels this plateau is not accessed within the frequency window indicated in Fig. 5.12, hence the sometimes quoted remark (Mitchell, 1984) that the frequency spectrum of gels is "rather uninteresting."

Where a different frequency response is found over the frequency range in Fig. 5.12, one could meaningfully say (albeit a little arbitrarily) that the material is not a gel. For example, Fig. 5.12b shows the typical modulus-frequency response of an entangled polymer solution. Clearly, this ma-

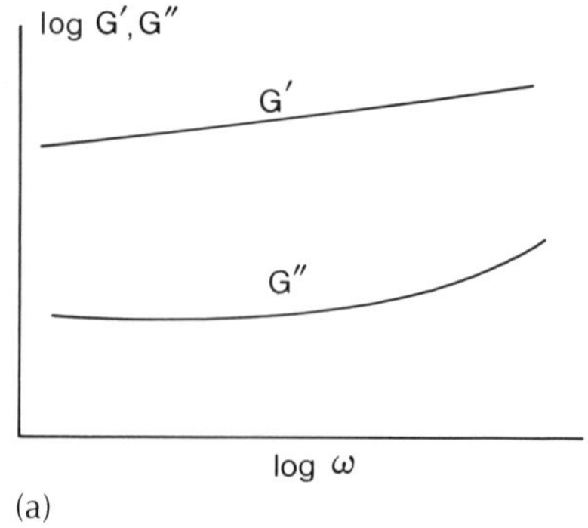

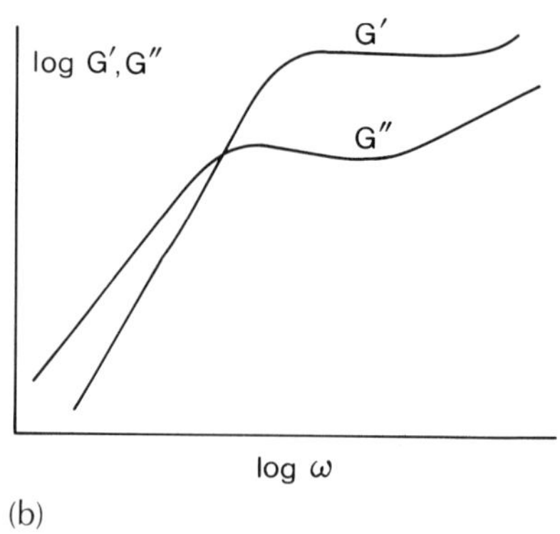

Fig. 5.12 G′, G″ vs. log frequency (log ω) responses for: (a) a gel, and (b) an entangled polymer solution. Measurements made within linear viscoelastic response region and within a frequency range ~10^{-2}–10^{2} Hz.

terial shows viscoelastic liquid behavior below a certain frequency level, and this cross-over point for the functions $G'(\omega)$ and $G''(\omega)$, plus the considerable frequency dependence of these quantities it implies, enables us to eliminate such entanglement systems as gel candidates.

A second important approach to studying gel character is to look at the strain dependence of the mechanical spectrum. Figure 5.13 contrasts typical results for weak (curve a) and strong (curve b) gels. For the former, the amplitude of the complex modulus $|G^*|$, measured at some frequency, falls off in magnitude very rapidly with strain, as the strain is increased in a dynamic mechanical experiment. The limit of linear response could be 0.5% strain, or much less, while strong gels (and indeed entanglement networks, Fig. 5.13, curve c) can be strained to much larger extents before nonlinearity develops. In a weak gel, the network is held together by linkages which are stronger and more permanent than entanglements, but which are nonetheless much much more susceptible to disruption than those in strong gel systems (or indeed in the entanglement case). Strong gels may remain very nearly linear to extensions of 20% or more, before the entire gel ruptures. Weak gels yield and flow nonlinearly rather than fail.

Turning to temperature effects, monitoring the modulus of a fully cured gel against temperature, provides a useful method of demonstrating "gel melting," and establishing the temperature range over which this occurs. In thermoreversible gels (such as gelatin) this effect is easily demonstrated by mechanical spectroscopy, as is the reformation of the network on recooling. Application of the simple reversible kinetic model (modified Flory-Stockmayer gel theory) discussed earlier enables the melting point to be related to polymer concentration and molecular weight (Clark and Ross-Murphy, 1987). Interestingly, these same relationships were discussed by Eldridge and Ferry (1954) many years ago in relation to studies of gelatin. Where gelation is thermally irreversible (heat-set protein gels, for example), measurement of the temperature dependence of the modulus is still of interest as usually it falls with increasing temperature (Richardson and Ross-Murphy, 1981a, b), in contrast to expectation for an ideal rubber network. This suggests that either network strands contribute less to the modulus at higher temperatures, or there is some reduction in cross-link density, even though this is insufficient to cause melting.

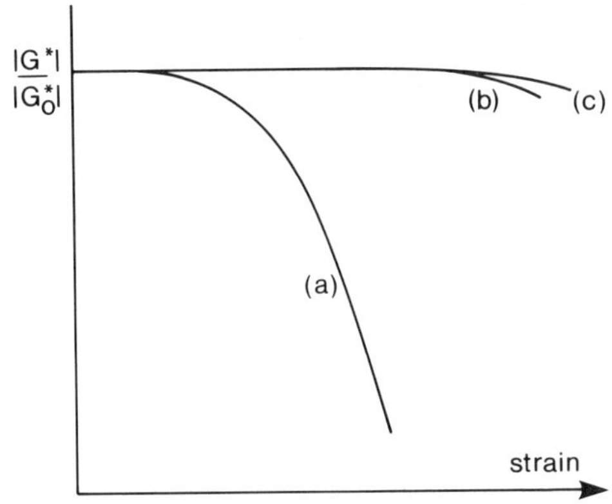

Fig. 5.13 Typical strain dependences of absolute size of dynamic modulus $|G^*|$, for situations of: (a) a "weak" gel, (b) a "strong" gel, and (c) an entanglement network.

Finally, the modulus data for fully cured gels containing varying amounts of polymer can be plotted against concentration as discussed earlier and used to test the various theoretical models currently available for this relationship.

Large-Deformation Studies

Small-deformation mechanical measurements of the linear viscoelastic behavior of gels are useful for inferring structural properties of the "unperturbed" materials (or at least for making connections between their mechanical behavior, and structural models determined by other physical techniques). They also provide the raw data for testing theoretical approaches aimed at explaining the size of the equilibrium modulus, and (in some cases) the form of the linear relaxational behavior. As we have already seen, however, it is also necessary to examine nonlinear behavior, if the important classification of gels into weak and strong categories is to be achieved. It is also important to use nonlinear testing (and failure studies) to distinguish between different types of strong gel, for within this class, large differences are found in the forms of stress-strain responses, including differences in such features as extensibility before rupture, and yielding behavior. Agar gels, for example, are quite brittle, whereas gelatin gels are more "rubbery" (McEvoy et al., 1984, 1985a). Some modified starch gels, on the other hand, yield and "flow" rather than fail, and hence show a form of plasticity.

Although less work (certainly of a rigorous scientific nature) has been published describing nonlinear gel behavior, some progress has been made both in the study of strong and weak gels. The mechanical measurement methods used usually involve transient approaches, such as application of stress and strain steps (and ramps), but dynamic methods can be employed if one is prepared to examine stress or strain responses which are no longer simple sinusoidal in form. In all of these experiments, the tensorial nature of the stress or strain functions can be restricted to simple shear, but for solids it is perhaps more common for tests to be made in uniaxial compression or extension. Of course, in addition to all of these methods, a wide range of semi-empirical approaches exist such as the application of so-called texturometers which, while they serve an important function for the food industry in product control and evaluation situations, do not provide measurements readily interpretable by theory and hence are not likely to substantially contribute to fundamental understanding.

Satisfactory large-deformation studies must therefore employ well-defined stress and strain functions, which can be measured accurately and which (generally) are uniform throughout sample bulk. It is also important to be able to control and, where necessary, account for the effects of friction between samples and measuring devices and indeed to be able to make corrections for any other instrumental effects.

Studies of strong gels satisfying these criteria are available. A particularly interesting discussion is to be found in articles by McEvoy et al. (1984, 1985a). Here agar and gelatin gels were studied in extension, gel rings (or ovals) being used, and strained on an Instron tester by means of two dowel pins, one mounted on the transducer stage and the other on the driven stage. In this way stress-strain curves could be measured at different strain rates, and the failure properties characterized by stress-to-break and strain-to-break measurements (Fig. 5.14). The gels were studied over a range of concentrations, and in some cases swollen gels were included (i.e., gels made by allowing a system to set at one concentration, and then immersing in solvent, and swelling to a new concentration). The failure parameters were plotted in the form of so-called failure envelopes (stress-to-break vs. strain-to-break plots). It was shown that both agar and gelatin gels exhibited stress-strain behavior that varied significantly from linear Hookean form [the BST equation, described by Blatz et al., (1976), fitted well] and that the results depended to some

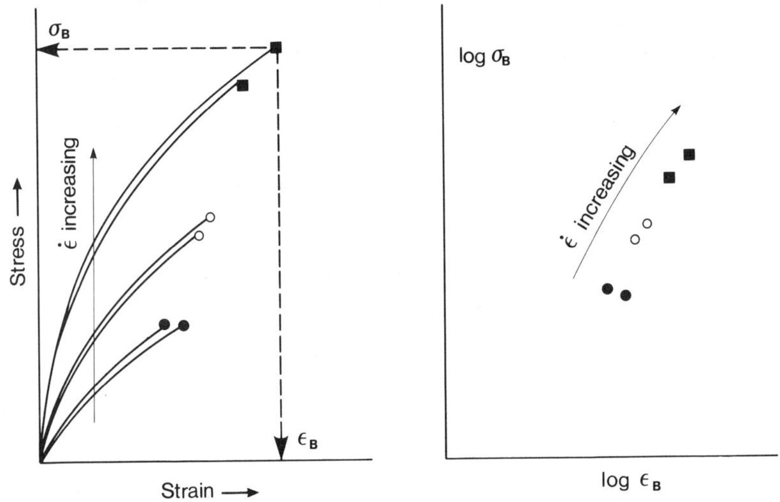

Fig. 5.14 Failure envelope measurement for strong gels as a function of increasing strain rate ($\dot{\epsilon}$). σ_B and ϵ_B are, respectively, the stress-to-break and strain-to-break. Results of two replicate experiments are shown for each strain rate.

extent on applied strain rate (i.e., viscoelasticity was present). This last effect was not large, however, and linear viscoelastic moduli extracted from the data (BST fits) agreed well with those measured by other methods (mechanical spectroscopy).

In relation to failure behavior, the failure envelope approach for the gelatin and agar gels showed that failure characteristics were strongly strain-rate dependent, and this finding was discussed within the framework of the Bueche-Halpin (1964) theory of elastomer rupture. Results for gels at different concentrations could be plotted on universal failure envelopes by suitable scaling. Differences between agar and gelatin gels were discussed in the context of the rather different network structures underlying these materials. It is interesting that neither system showed the resistance to failure characteristic of synthetic polymer rubbers, failure occurring early in the stress-strain profile, i.e., in the region where the effective modulus was falling off in value, not in the region where it was increasing again. Solvated biopolymer gels are much more brittle than conventional rubbers, which is consistent with their weaker physical cross-links and less extensible network chain properties. In this connection, the stress-strain behavior of physical polysaccharide gels (prior to failure) has recently been described theoretically by Higgs and Ball (1990) in terms of a "reel-chain" model, in which the chains in junction-zones can be pulled out of their compact unperturbed arrangement by the deformation process.

Large-deformation and failure studies of other biopolymer gel systems, such as examples based on starch and casein, have also been reported. Of particular interest is recent work by Van Vliet et al. (1991) who have obtained a number of interesting stress-strain responses (uniaxial compression) for such gels, and have described theoretical analyses of the failure properties in terms of fracture theories. Large-deformation work on starch systems (and gelatin) has also been reported by Christianson et al. (1984), who carried out experiments in both simple shear and torsion. Results were analyzed using the Mooney-Rivlin equation (see, e.g., Treloar, 1970) and attempts were made to introduce normal force measurements as a means of separately determining the constants C_1 and C_2 which appear in this equation. The value of including normal force measurements in large-deformation studies of food materials (including gels) has also been suggested in a recent review of the subject (Bagley and Christianson, 1988). In this article the authors particularly emphasize the importance of using more than one type of deformation mode when characterizing the nonlinear and failure properties of complex materials.

Turning to studies of weak gels we note that, while strong gels rupture fairly early as strain is increased, weak gels yield and flow, the steady-state stress being much less dependent on shear rate than is consistent with Newtonian behavior. In other words such materials behave as infinite viscosity solids in the unperturbed state, but become strongly shear-thinning fluids on application of an imposed strain. As this strain is increased from zero, the measured stress rises through a distinct maximum (stress overshoot effect) before falling to the steady state result (Fig. 5.15). As Richardson and Ross-Murphy (1987) have shown for solutions of xanthan (a bacterial polysaccharide) within a certain concentration range, the amount of overshoot increases rapidly with strain rate. They have also demonstrated that the sheared system requires a very considerable time to recover its original properties when shearing is stopped. Richardson and Ross-Murphy contrast these properties with those of guar solutions (entanglement networks) where the stress overshoot effect is less pronounced and where recovery time is at least two orders of magnitude

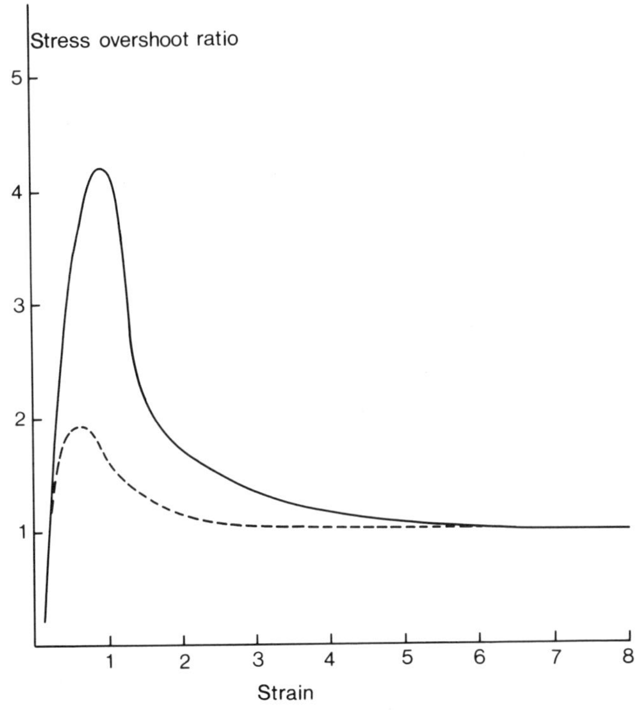

Fig. 5.15 Typical stress overshoot behavior of weak gels submitted to constant strain rate. Low strain rate (broken line); high strain rate (solid line).

shorter. In addition they have demonstrated that whereas guar solutions show Cox-Merz superpossibility (Cox and Merz, 1958) of continuous shear and dynamic viscosity ($|G^*|/i\omega$), this property is significantly violated by xanthan systems in the weak gel condition. The distinction between weak gels and entanglement systems is once again clearly demonstrated by studies of nonlinear viscoelasticity.

It is worth adding that a theoretical approach to weak gel formation and to the mechanical properties of such gels, based on the reversible cross-link approach described earlier, has recently been reported (Chu and Prud'homme, 1987). As well as describing the kinetics of network curing, and potentially the gel time and modulus-concentration behavior discussed earlier, this theory also formulates a nonlinear constitutive re-

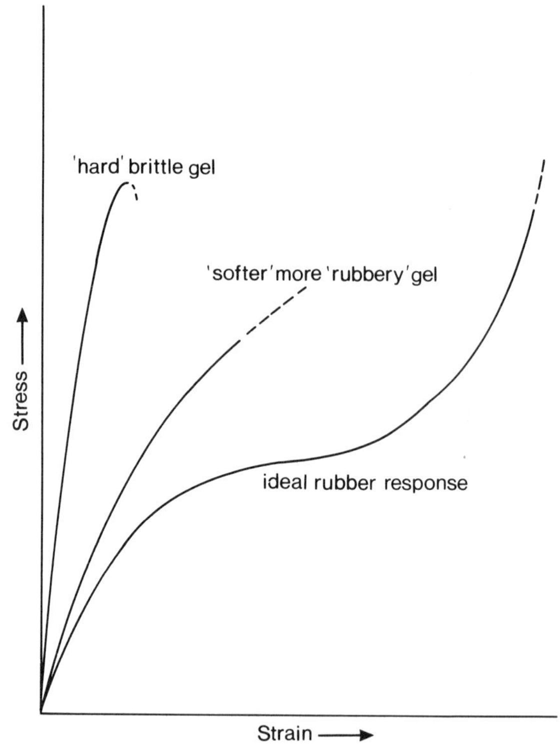

Fig. 5.16A Possible stress-strain responses of gels prior to failure. In both cases the modulus falls initially as strain increases. The response of a covalently bonded rubber is shown for comparison (though not accurately to scale). Responses are illustrative only, no reference to a particular form of mechanical test being intended.

lation which describes the viscoelasticity of weak gels including nonlinear behavior. This is a highly phenomenological approach and unfortunately is not readily amenable to molecular interpretation.

Lastly, in this section on large-deformation studies, it should be said that the distinction between strong and weak gels is not a discontinuous one. A whole range of complex intermediate behavior is likely to be met with between these extremes. Gel networks will differ greatly in the strength of connecting cross-links, and on the flexibility of the units connected together, as well as on the level of network heterogeneity. These properties will ensure a complex diversity of response (Figs. 5.15 and 5.16) in large deformation experiments, corresponding to an equally diverse range of material behavior when these systems form the substance of food materials.

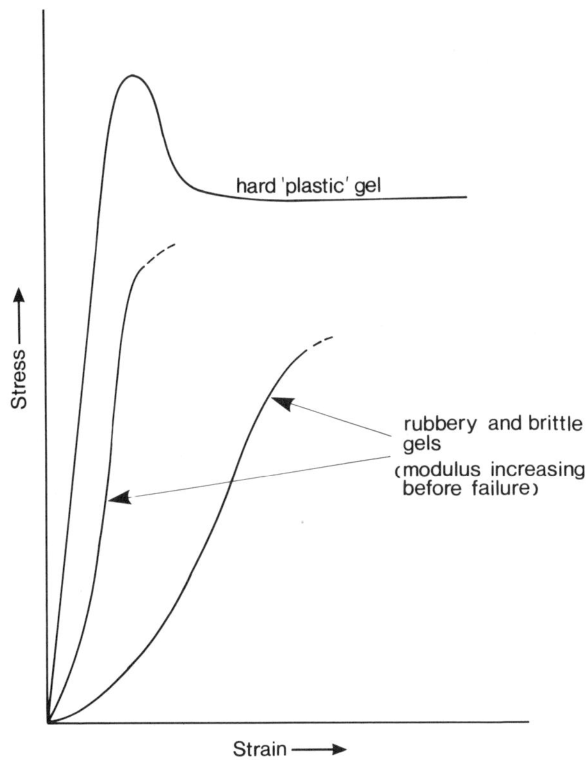

Fig. 5.16B Further stress-strain response types including situations where modulus increases before failure, and response of "plastic" gel.

OTHER PROPERTIES OF GELS

In this review much of the emphasis has been on the mechanical behavior of gels. This is natural since, as was explained in the introduction, it is in terms of specific mechanical attributes that gel behavior makes its presence most clearly felt. However, gels have other important and distinctive properties and though limits of space preclude detailed discussion of these, some summary of such additional features may be worthwhile.

One important property of a gel is its ability to swell or deswell in an appropriate solvent. Flory 1953 has provided a theoretical framework for this phenomenon in terms of solvent-polymer and polymer-polymer interactions and network elasticity. In a slightly refined form (Erman and Flory, 1986) this theory has proved valuable when interpreting some very interesting solvent effects on covalent polyacrylamide networks recently demonstrated by Tanaka and coworkers (Tanaka, 1978, 1981; Ohmine and Tanaka, 1982). Phase changes and critical behavior involving the network can be induced by changes in solvent composition and temperature, and of course there are accompanying changes in mechanical behavior. Where physical gels are concerned, the deswelling behavior of agar and gelatin gels, when subjected to pressure, and the mechanical implications of these processes have also recently been described (Lips et al., 1988).

In food applications the appearance of a gel network may also be important, since in some cases transparency may be an advantage, while in others (e.g., the simulation of fats) opacity may be required. Appearance is clearly determined by microstructural inhomogeneities of a size sufficient to scatter light. Such heterogeneities induced by solvent effects have already been discussed in relation to mechanical properties.

In other applications of gels such as their use in electrophoresis or gel filtration (i.e., in separation techniques), pore size is important and indeed the porosity of gels is a descriptive quantity measurable by pressure-induced flow through them. Closely related to this aspect is the question of the diffusion of free molecules within gel networks and their specific interactions with the network itself. This might include interaction with flavors.

MIXED GEL SYSTEMS

So far the discussion has implicitly focused on systems containing essentially one kind of network-forming component, e.g., one type of biopolymer. In food applications, gels of interest are often more complex than this, involving not only the inclusion of solid particles of various sorts as inert fillers, but often being built from more than one network-forming substance. In this connection the study of mixed gels becomes analogous to the study of composite materials in conventional synthetic polymer science. The main difference is that food gels usually involve a large amount of solvent (aqueous solution) as an extra, and sometimes quite complex, constituent. The study of the mechanical (and indeed structural) properties of these more complex gels involves very similar measurement approaches to those already discussed, but now the extent of experimentation necessary, and the range of information generated, may be somewhat greater. For example, mechanical monitoring of the gelation of mixed systems by dynamic methods (Clark et al., 1983, Doublier and Llamas, 1991) requires consideration of systems with essentially two concentration degrees of freedom, and more complex thermal regimes are often required during curing.

A principal issue in the study of mixed biopolymer systems is the question of phase compatibility of the two network-building components prior to, and during, gelation. Thermodynamic incompatibility of polymers of different types, when in solution, is a well-understood and well-documented subject (Flory, 1953) and has been addressed theoretically many years ago by extension of Flory-Huggins theory. Homogeneous solutions of unlike polymers can be stable over certain concentration ranges, but more often than not they are unstable and demix to become biphasic, the individual phases containing predominantly one of the two network building components (Fig. 5.17). If gelling also takes place after (or during) the demixing process, complex multiphasic gel microstructures (Fig. 5.17) develop, which vary from situations where one phase can be regarded as dispersed through the other, to essentially bicontinuous materials. Electron (and light) microscopical imaging of such microstructures have been reported (Clark et al., 1982; 1983), and where biopolymers of food interest are concerned the question of thermodynamic compatibility has been extensively researched (Tolstoguzov, 1986, 1988).

The complex microstructure of mixed gels, and the varied rheological character of the component phases, naturally gives rise to complex mechanical behavior, both at small and at large deformation. Clark and coworkers (Clark et al., 1983; Clark, 1987) have considered the implications of demixing and gelling in relation to the size of the linear elastic modulus of the composite produced. A model allowing estimation of this quantity from information about the gelling behavior of individual components and their tendency to demix has been described. In terms of large deformation and failure behavior, McEvoy et al. (1985b) extended the failure envelope

(a)

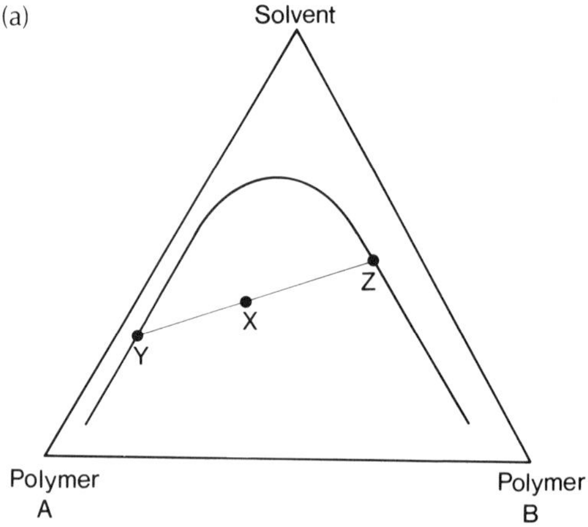

(b)

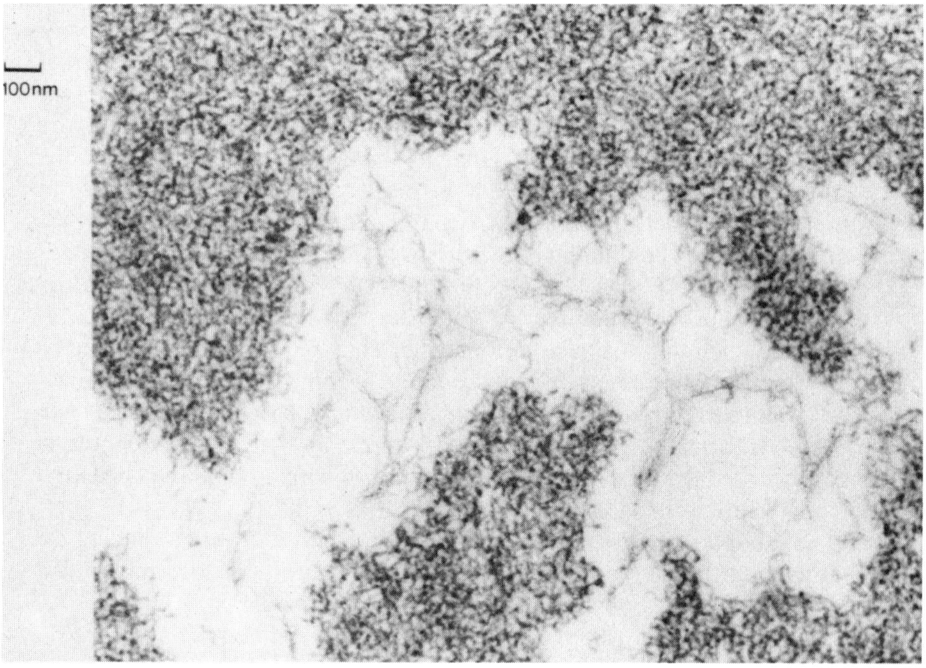

Fig. 5.17 (a) Triangular phase diagram indicating tendency of solutions of unlike polymers A and B (e.g., protein and polysaccharide) to demix spontaneously. (b) Section of electron micrograph showing regions of protein (dark stain) and polysaccharide (lighter stain) in a typical composite biopolymer gel.

treatment of individual agar and gelatin systems to treat the composite gel case. In terms of absolute prediction of behavior much remains to be done, however, and as yet the complex kinetic processes of combined phase-separation and gelling, and their implications for mechanical (and other) behavior remain obscure. Mechanisms such as spinodal decomposition for the demixing process, already discussed in connection with the generation of phase-separated gels from systems containing just one gelling component, must again be centrally important.

CONCLUSION

In this review attention has focused on the most common classes of gel structure described by Flory in 1974. For these systems in which a network component or "phase" is readily identifiable, the critical nature of the gel-forming process has been emphasized. The development of the gel from a pregel state, through the critical point, to a fully cured condition has been discussed as a kinetic process from both classical and more modern points of view. For the physical gels which are principally of relevance in foods, modification of these theories is essential to take into account the presence of solvent and the reversible nature of cross-links. In terms of the extent of this reversibility and the susceptibility of networks to disruption by imposed strain, gels have been roughly classed as "strong" or "weak," and experimental methods for making this distinction have been discussed. These relate essentially to the large deformation properties of such systems, which range from substantial extensibility followed by abrupt rupture, at the strong gel end of the spectrum, to yielding at very low applied strains and flow, at the other. Thus, in terms of their potential performance in real food processing and handling situations, the substances we call gels form a broad spectrum of material types, which is made even more complex by the multiphasic nature of most of the gels of real practical importance.

It is hoped that this review has succeeded in broadly summarizing the progress that has been made so far in the scientific study of gels (both experimentally and theoretically) and that it has indicated areas where understanding remains weak and where future research will be necessary. Thus, whereas undoubted progress has been made by experiment and theory directed at understanding the formation and structure of gels in quiescent conditions (absence of shear, for example) and of characterizing the linear viscoelastic properties of such substances, the influence of shear on their kinetic development, and the nature of their nonlinear viscoelastic and failure properties (even in the absence of shear) have only just begun to be studied rigorously. These limitations in the progress of understanding are even more obvious for mixed gels, for here all the problems associated with single component systems become compounded by the superimposed issue of kinetic demixing during gelation, and the resulting heterogeneity of microstructure.

ACKNOWLEDGMENTS

It is a pleasure to acknowledge the considerable assistance I have received from many coworkers, both current and former, within Unilever Research. These are too numerous to name individually, but I am especially indebted to Prof. S. B. Ross-Murphy, Dr. R. K. Richardson, and Mr. H. McEvoy for their help, inspiration, and encouragement over many years. I am also indebted to Mrs. L. Linger for help in the preparation of the manuscript.

REFERENCES

Adam, M. and Delsanti, M. 1989. *Contemp. Phys.* 30: 203.

Bagley, E. B. and Christianson, D. D. 1988. *Food Structure—Its Creation and Evaluation*, J. M. V. Blanshard and J. R. Mitchell (Ed.), p. 401. Butterworths, London.

Bikbov, T. M., Grinberg, V. Ya., Antonov, Yu. A., Tolstoguzov, V. B., and Schmandke, H. 1979. *Polym. Bull.* 1: 865.

Blatz, P. I., Sharda, S. C., and Tschoegl, N. W. 1976. *Trans. Soc. Rheol.* 18: 145.

Bremer, L. G. B., Van Vliet, T., and Walstra, P. 1989. *J. Chem. Soc., Farad. Trans. 1* 85: 3359.

Brown, C. R. T. 1991. Unilever Research Laboratory, Bedford U.K., private communication.

Bueche, F. and Halpin, J. C. 1964. *J. Appl. Phys.* 35: 36.

Cahn, J. W. 1965. *J. Chem. Phys.* 42: 93.

Christianson, D. D., Navickis, L. L., Bagley, E. B., and Wolf W. J. 1984. *Gums and Stabilizers for the Food Industry 2*, G. O. Philips, D. J. Wedlock and P. A. Williams, (Ed.), p. 123. Pergamon Press, New York.

Chu, A. S. and Prud'homme, R. K. 1987. *J. Non-Newtonian Fluid Mech.* 22: 207.

Clark, A. H., Richardson, R. K., Robinson, G., Ross-Murphy, S. B., and Weaver, A. C. 1982. *Proc. Food Nutrit. Sci.* 6: 149.

Clark, A. H., Richardson, R. K., Ross-Murphy, S. B., and Stubbs, J. M., 1983. *Macromolecules* 16: 1367.

Clark, A. H. and Ross-Murphy, S. B. 1985. *Br. Polym. J.* 17: 164.

Clark, A. H. and Lee-Tuffnell, C. D. 1986. *Functional Properties of Food Macromolecules*, J. R. Mitchell and D. A. Ledward (Ed.), p. 203. Elsevier Applied Science, London.

Clark, A. H. and Ross-Murphy, S. B. 1987. *Adv. Polymer Sci.* 83: 57.

Clark, A. H. 1987. *Food Structure and Behavior*. J. M. V. Blanshard and P. J. Lillford (Ed.), p. 13. Academic Press, London.

Clark, A. H., Ross-Murphy, S. B., Nishinari, K., and Watase, M. 1990. *Physical Networks Polymers and Gels*, W. Burchard and S. B. Ross-Murphy, (Ed.), p. 209. Elsevier Science Publishers, London.

Clark, A. H. (1991). *Food Polymers, Gels and Colloids*, E. Dickinson (Ed.), p. 322. Royal Society and Chemistry, London.

Cohen, R. J. and Benedek, G. B. 1982. *J. Phys. Chem.* 86: 3696.

Coniglio, A., Stanley, H. E., and Klein, W. 1979. *Phys. Rev. Letts.* 42: 518.

Cox, W. P. and Mertz, E. H. 1958. *J. Polymer Sci.* 28: 619.

Doublier, J. L. and Llamas, G. 1991. *Food Polymers Gels and Colloids*, E. Dickinson (Ed.), p. 349. Royal Society of Chemistry, London.

Dusek, K. 1979. *Polymer Bull.* 1: 523.

Eldridge, J. E. and Ferry, J. D. 1954. *J. Phys. Chem.* 58: 992.

Erman, B. and Flory, P. J. 1986. *Macromolecules* 19: 2342.

Ernst, M. H., Hendriks, E. M., and Ziff, R. M. 1982. *J. Phys. A: Math. Gen.* 15: L743.

Ferry, J. D. 1980. *Viscoelastic Properties of Polymers, 3rd. ed.* John Wiley, New York.

Fisher, M. E. and Essam, J. W. 1961. *J. Math. Phys.* 2: 609.

Flory, P. J. 1941. *J. Amer. Chem. Soc.* 63: 3083: 3091: 3096.

Flory, P. J. 1953. *Principles of Polymer Chemistry*. Cornell University Press, Ithaca, NY.

Flory, P. J. 1974. *Farad. Discuss. Chem. Soc.* 57: 7.

Gordon, M. 1962. *Proc, Roy. Soc (London) Ser. A* 268: 240.

Gordon, M. and Ross-Murphy, S. B. 1975. *Pure Appl. Chem.* 43: 1.

Hermans, J. R. 1965. *J. Polym. Sci. Part A* 3: 1859.

Higgs, P. G. and Ball, R. C. 1990. *Physical Networks Polymers and Gels*,

W. Burchard and S. B. Ross-Murphy (Ed.), p. 185. Elsevier Science Publishers, London.

Jullien, R. Botet, and Mors, P. M. 1987. *Farad. Discuss. Chem. Soc.* 83: 125.

Leyvraz, F. and Tschudi, H. R. 1980. *J. Phys. A: Math. Gen.* 13: 1867.

Leyvraz, F. and Tschudi, H. R. 1981. *J. Phys. A: Math. Gen.* 14: 3389.

Leyvraz, F. and Tschudi, H. R. 1982. *J. Phys. A: Math. Gen.* 15: 1951.

Lips, A., Hart, P. M., and Clark, A. H. 1988. *Food Hydrocolloids* 2: 141.

Mandelbrot, B. 1982. *The Fractal Geometry of Nature*. Freeman, New York.

McEvoy, H., Ross-Murphy, S. B., and Clark, A. H. 1984. *Gums and Stabilisers for the Food Industry 2*, G. O. Philips (Ed.), p. 111. Pergamon Press, Oxford, England.

McEvoy, H., Ross-Murphy, S. B., and Clark, A. H. 1985a. *Polymer* 26: 1483.

McEvoy, H., Ross-Murphy, S. B., and Clark, A. H. 1985b. *Polymer* 26: 1493.

McLeod, J. B. 1962. *Quart. J. Math. Oxford* 13: 119.

Meakin, P. 1983. *Phys. Rev. Lett.* 51: 1119.

Meakin, P. 1987. *Farad. Discuss. Chem. Soc.* 83: 113.

Miller, W. G., Kou, L., Tohyama, K., and Voltaggio, V. 1978. *J. Polym. Sci: Polymer Symposium* 65: 91.

Mitchell, J. R. 1984. *Proc. Int. Workshop on Plant Polysaccharides: Structure and Function*, C. Mercier and M. Rinaudo (Ed.), p. 93. INRA/CNRS.

Norton, I. T., Goodall, D. M., Morris, E. R., and Rees, D. A. 1983. *J. Chem. Soc. Farad. Trans. 1* 79: 2501.

Oakenfull, D. 1984. *J. Food Sci.* 49: 1103.

Oakenfull, D. and Scott, A. 1986. *Gums and Stabilisers for the Food Industry 3*, G. O. Philips, D. J. Wedlock, and P. A. Williams (Ed.), p. 465. Elsevier Science Publishers, London.

Oakenfull, D. and Morris, V. J. 1987. *Chem. Ind.* 16: 201.

Ohmine, I. and Tanaka, T. 1982. *J. Chem. Phys.* 77: 5725.

Rees, D. A. 1972. *Chem. Ind.* 19: 630.

Rees, D. A. and Welsh, E. J. 1977. *Angew. Chem. Int. Ed. Engl.* 16: 214.

Richardson, R. K. and Ross-Murphy, S. B. 1981a. *Int. J. Biol. Macromol.* 3: 315.

Richardson, R. K. and Ross-Murphy, S. B. 1981b. *Brit. Polymer J.* 13: 11.

Richardson, R. K. and Ross-Murphy, S. B. 1987. *Int. J. Biol. Macromol.* 9: 257.

Ross-Murphy, S. B. 1991. *Food Polymers, Gels and Colloids* E. Dickinson (Ed.), p. 357. Royal Society of Chemistry, London.

Stanley, H. E., Family, F., and Gould, H. 1985. *J. Polym. Sci: Polymer Symposium* 73: 19.

Stauffer, D. 1976. *J. Chem. Soc., Farad. Trans. II* 72: 1354.

Stauffer, D. 1979. *Phys. Rep.* 54: 1.

Stauffer, D. 1981. *Pure Appl. Chem.* 53: 1479.

Stauffer, D., Coniglio, A., and Adam, M. 1982. *Adv. Polym. Sci.* 44: 103.

Stockmayer, W. H. 1943. *J. Chem. Phys.* 11: 45.

Stockmayer, W. H. 1944. *J. Chem. Phys.* 12: 125.

Swift, D. L. and Friedlander, S. K. 1964. *J. Coll. Interf. Sci.* 19: 621.

Tanaka, T. 1978. *Phys. Rev. Lett.* 40: 820.

Tanaka, T. 1981. *Sci. Am.* 244: 110.

Te Nijenhuis, K. and Winter, H. H. 1989. *Macromolecules* 22: 411.

Tohyama, K. and Miller, W. G. 1981. *Nature* 289: 813.

Tostoguzov, V. B. 1986. *Functional Properties of Food Macromolecules*, J. R. Mitchell and D. A. Ledward (Ed.), p. 385. Elsevier Science Publishers, London.

Tolstoguzov, V. B. 1988. *Food Hydrocolloids* 2: 339.

Treloar, L. R. G. 1970. *The Physics of Rubber Elasticity*, 2nd Ed. Clarendon Press, Oxford, England.

Van Dongen, P. and Ernst, M. H. 1983. *J. Phys. A: Math Gen.* 16: L327.

Van Vliet, T., Luyton, H., and Walstra, P. 1991. *Food Polymers, Gels and Colloids*, E. Dickinson (Ed.), p. 392. Royal Society of Chemistry, London.

Von Smoluchowski, M. 1917. *Z. f. Physik. Chemie* 92: 129.

Walstra, P., Van Vliet, T., and Bremer, L. G. B. 1991. *Food Polymers Gels and Colloids*, E. Dickinson (Ed.), p. 369. Royal Society of Chemistry, London.

Witten, T. A. Jr., and Sander, L. M. 1981. *Phys. Rev. Lett.* 47: 1400.

Witten, T. A. Jr. 1985. *J. Polym. Sci.: Polymer Symposium* 73: 7.

Ziff, R. M. and Stell, G. 1980. *J. Chem. Phys.* 73: 3492.

Ziff, R. M. 1980. *J. Stat. Phys.* 23: 241.

Ziff, R. M., Hendriks, E. M., and Ernst, M. H. 1982. *Phys. Rev. Letts.* 49: 593.

Ziff, R. M., Ernst, M. H., and Hendriks, E. M. 1983. *J. Phys. A: Math. Gen.* 16: 2293.

6

Emulsions

Ganesan Narsimhan

Purdue University
West Lafayette, Indiana

INTRODUCTION

Emulsions are dispersions of one liquid phase in the form of fine droplets in another immiscible liquid phase. The two immiscible phases are usually oil and water, so emulsions can be broadly classified as oil-in-water or water-in-oil emulsions depending on the dispersed phase. In food systems, a much broader definition can be employed to include systems in which the continuous phase is semi-solid or a gel and the system also contains solid particles (crystals) and air. Some typical food emulsions are shown in Table 6.1. Milk, cream, ice cream, butter, margarine, salad dressing and meat emulsions are examples.

Since formation of emulsion involves dispersion of one phase in another in the form of fine droplets, it requires creation of large interfacial area. The change in the free energy ΔG during emulsion formation is

$$\Delta G = \gamma \Delta A - T \Delta S \tag{1}$$

Table 6.1 Typical Food Colloids

Food	Type of emulsion	Method of preparation	Mechanism of stabilization
(1) Milk	O/W	Natural product	Protein membrane
(2) Cream	A + O/W	Centrifugation	As (1) + particle stabilization of air
(3) Ice cream	A + O/W	Homogenization	As (2) + ice network
(4) Butter and margarine	W/O	Churning and in votator	In crystal network
(5) Sauces	O/W	High-speed mixing and homogenization	By protein and polysaccharide
(6) Fabricated meat products	O/W	Low-speed mixing and chopping	Gelled protein matrix
(7) Bakery products	A + O/W	Mixing	Starch and protein network

O = oil; A = air; W = aqueous phase.
Source: Darling and Birkett, 1986.

where γ is the interfacial tension, ΔA is the increase in the interfacial area, T is the temperature and ΔS is the change in the entropy due to dispersion of one phase into a large number of droplets. Excepting for ultra low values of γ, not usually encountered in food emulsions, ΔG, is usually positive, therefore emulsions are thermodynamically unstable. In other words, emulsions eventually will break due to coalescence of droplets. As the droplets approach each other due to relative motion brought about by Brownian and gravitational forces, they are attracted to each other by van der Waals forces. Without a stabilizer, the continuous phase between the two colliding droplets drains, leading to their coalescence. Emulsifiers and stabilizers employed in food emulsions, however, slow down coalescence considerably, thereby providing long shelf-life (kinetic stability). Proteins are used as stabilizers in most food emulsions. They tend to adsorb at the oil-water interface, thereby modifying interparticle forces, and so provide kinetic stability to the emulsion. A wide variety of small molecule surfactants are also used for the same purpose. Adsorption of these introduce electrostatic and steric repulsive forces between the droplets, thereby providing an energy barrier to collisions between droplets. Adsorption, especially of proteins, also provides surface mechanical properties that prevent rupture of the drop interface. Overall stability can also be achieved by addition of polysaccharides to the continuous phase. They increase the viscosity of the continuous medium and, therefore reduce rates of droplet collisions and rates of drainage of the continuous phase between the colliding droplets. It is necessary to understand mechanisms of stabilization to develop formulations with desirable texture and shelf-life. Since modification of interparticle forces is one of the most important mechanisms of stabilization, the nature of these forces should be understood. Flocculation of droplets resulting from their encounter precedes coalescence. Coalescence is brought about by drainage and rupture of the thin continuous phase film between two droplets. Surface mechanical properties of the interface, so important to thin film drainage and stability, are greatly influenced by the competitive adsorption in mixed emulsifier systems consisting of protein, polysaccharide and other small molecule surfactants. The present review attempts to highlight important aspects of different mechanisms of stabilization in food emulsion, but is not comprehensive; therefore the reader is referred to other reviews on this subject for details. A brief description of the factors involved in the control of drop size during emulsion formation is also given. The main focus of this review is on basic processes and their physics insofar as they relate to control of shelf life of emulsions. No attempt has been made to cover the technology of different products, different experimental techniques for characterizing emulsions or different types of emulsification equipment.

COLLOIDAL FORCES BETWEEN DROPLETS

As pointed out earlier, an important factor that affects the stability of emulsions is the nature of interparticle forces between emulsion drops. As the individual droplets of the dispersed phase in an emulsion approach each other, they are attracted to each other by van der Waals forces. In the absence of suitable stabilizer, the liquid film between them rapidly drains, resulting in coalescence of the drops. The overall rate of coalescence can be slowed down by modifying the forces between the droplets. This is accomplished by adding a stabilizer which adsorbs at the droplet surface, changing them in such a way to introduce a repulsive force which provides an energy barrier to coalescence. Proteins such as milk, egg and soya proteins, triglycerides, and polar lipids such as lecithin are commonly used stabilizers in food emulsions. Repulsive forces between the droplets with adsorbed proteins may be electrostatic, steric, and hydration forces. These interaction forces are influenced by the extent of adsorption of proteins and emulsifier, pH, ionic strength, etc. It is important to quantify these interactions to better understand the role of stabilizers on emulsion stability.

Van der Waals Interaction

Van der Waals interaction is the result of permanent dipole–permanent dipole, permanent dipole–induced dipole, and induced dipole–induced dipole interactions between molecules. Of these, induced dipole–induced dipole interactions, known as dispersion forces, are experienced by all molecules and have their origin from the rapidly fluctuating dipole moment of a neutral molecule which polarizes other neighboring molecules, thus giving rise to an attractive interaction. Assuming pairwise additivity, the net interaction potential between two emulsion droplets of radii R_1, and R_2 is given by

$$V_a = -\frac{A}{6}\left[\frac{2R_1R_2}{s^2 + 2R_1s + 2R_2s} + \frac{2R_1R_2}{s^2 + 2R_1s + 2R_2s + 4R_1R_2} + \ln\left(\frac{s^2 + 2R_1s + 2R_2s}{s^2 + 2R_1s + 2R_2s + 4R_1R_2}\right)\right] \quad (2)$$

where A is the effective Hamaker constant, which depends on the polarizabilities of droplets and the intervening continuous medium, and s is the shortest surface-to-surface separation between the two droplets. When the distance of separation is much smaller than the radii of the droplets, the above equation simplifies to:

$$V_A \simeq -\frac{AR_1R_2}{6s(R_1 + R_2)} \qquad R_1, R_2, \gg s \quad (3)$$

Van der Waals attraction is a fairly long-range interaction, with the attractive force increasing significantly at shorter distances of separation. A macroscopic continuum theory for the evaluation of attractive force has been developed by Lifshitz (1956) and Dzyaloshinnsky et al. (1961). When the size of the emulsion droplet is comparable to the wavelength λ of propogation of electromagnetic radiation in the intervening medium, which is typically around 100 nm, dipoles shift during propogation, causing retardation effects, which diminish attraction and cannot be neglected. The Hamaker constant is then a function of the distance of separation between the droplets and can, in principle, be determined using Lifshitz (1956) theory, leading to a rather unwieldy expression for the retarded van der Waals interaction. However, the retarded interaction between two equal-sized emulsion droplets of radius R can be approximated by (Schenkel and Kitchener, 1960)

$$V_A = -\frac{AR}{12s}(1 + 1.77b)^{-1} \qquad b \leq 2$$

$$V_A = -\frac{AR}{12s}(0.490b^{-1} - 0.145b^{-2} + 0.017b^{-3}) \qquad b \leq 0.5 \tag{4}$$

where

$$b = 2\pi v_0 s/c \tag{5}$$

c being the velocity of light. Retardation always reduces the attractive force, increasingly so at larger distances of separation, and as pointed out earlier, becomes important for distances of separation greater than around 100 nm.

For emulsion droplets (1) dispersed in a continuous medium (2), the effective Hamaker constant A_{212} is given by,

$$A_{212} = A_{11} + A_{22} - 2A_{12} \tag{6}$$

where A_{ij} (i,j = 1,2) refers to the interaction between materials i and j. It is reasonable to approximate A_{12} as the geometric mean of A_{11} and A_{22}. Consequently,

$$A_{212} = (A_{11}^{1/2} - A_{22}^{1/2})^2 \tag{7}$$

The effective Hamaker constant is very sensitive to the polarizability of the continuous medium, as can be seen from Fig. 6.1. The net force is always attractive, becoming zero only when the dispersed and the continuous phases have the same polarizabilities. For a food emulsion of O/W type, the effective Hamaker constant for interaction between a pair of oil droplets is in the range of 2–5 × 10^{-21} J. For food emulsions, the van der Waals attraction has to be corrected to account for the effect of adsorbed macromolecular layers. Vold (1961) and Vincent (1973) have presented equations for evaluating these correction factors. Generally, adsorbed layers tend to reduce the net attractive force, the extent of reduction being dependent on the nature and the thickness of the layer.

The above equations can be employed to evaluate the van der Waals interaction when the emulsion droplets are undeformed. When the drop-

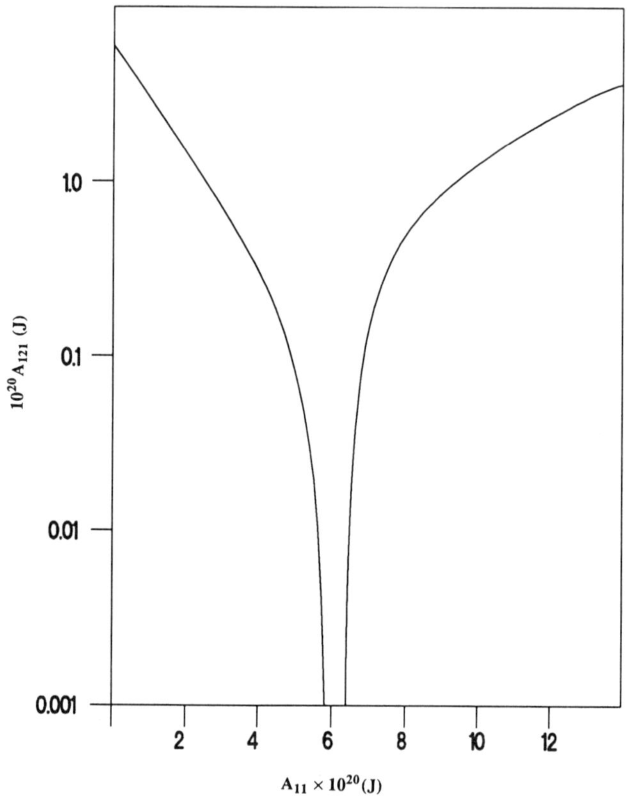

Fig. 6.1 Effect of an intervening medium on interaction of two semi-infinite bodies.

lets are close to each other, however, they deform and are separated by a thin film of the continuous phase. In such cases, the equation for van der Waals interaction of flat plates can be employed, i.e.

$$V_A = -\frac{A}{12\pi s^2} \tag{8}$$

where s is the thickness of thin film, and the van der Waals interaction V_A is now expressed per unit area of the film interface.

Electrostatic Interactions

In food emulsions, electrostatic interactions are provided by ionic emulsifiers and proteins which, upon adsorption at the oil-water interface, impart charge to the emulsion droplets. Electroneutrality requires that the net charge on the emulsion droplet be counterbalanced. Consequently, oppositely charged ions (counterions) in the continuous phase are preferentially attracted to the emulsion droplet, whereas ions of the same charge (coions) are repelled. Because of thermal motion, the neutralization occurs over a region in the vicinity of the droplet known as the "electrical double layer." This double layer consists of an inner compact "Stern layer" and an outer "diffuse layer." Every emulsion droplet is surrounded by an electrical double layer, as shown in Fig. 6.2. When two droplets approach each other, they experience electrostatic repulsion because of the overlap of double layers (Fig. 6.2). This repulsion can be viewed as manifestation of an increase in free energy or, equivalently, as osmotic pressure created by the difference in the ion concentration in the overlap region compared to bulk. When two emulsion droplets approach, several situations may arise. If the approach is slow, the ions on the surface of the droplet will be in equilibrium with the ions in the bulk, so the surface *potential* of the approaching droplets can be assumed constant. On the other hand, if the approach is very rapid, the ions do not have sufficient time during an encounter to reach equilibrium so that the surface *charge* of the approaching droplets can be assumed constant. Though real situations will lie between the above two extremes, for Brownian and

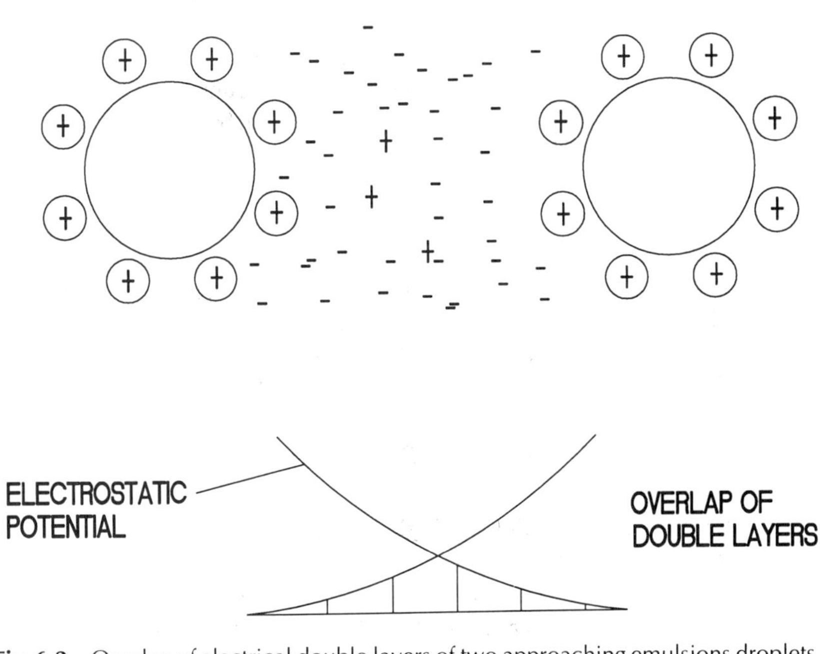

Fig. 6.2 Overlap of electrical double layers of two approaching emulsions droplets.

gravitational motions of droplets, assumption of constant surface potential during encounter may be reasonable.

The thickness of the double layer depends on the ionic strength of the continuous medium, the valence number of the counter ion, and the dielectric constant of the medium, ϵ, and is given by

$$\frac{1}{\kappa} = \left[\frac{\epsilon_o \epsilon k T}{e^2 \sum z_i^2 n_{io}}\right]^{1/2} \tag{9}$$

where κ, the Debye-Hückel parameter, is inversely proportional to double-layer thickness, ϵ_o is the permittivity of vacuum, k is the Boltzmann's constant, T is the temperature, e is the elementary charge, and z_i and n_{io} are the valence number and bulk number concentration of the ith ion, respectively. As can be seen from Fig. 6.3, double-layer thickness decreases with increase in the ionic strength as well as with the valence number of ions for symmetrical electrolytes. For asymmetrical electrolytes, however, the valence number of counterion dictates the thickness of the double layer. Because of oil's low dielectric constant, the double-layer thickness in oil is much smaller than that in water. As a result, electrostatic repulsion in water-in-oil emulsion has a much smaller range, and therefore is not an effective stabilizing factor. For the same reason, the capacity of internal double layer within oil droplets in case of oil-in-water emulsions is very small and can be neglected in the calculations.

When the double layer around the droplets is sufficiently thin (i.e., κR is sufficiently large) Derjaguin's procedure (Derjaguin and Landau, 1941) can be employed to evaluate electrostatic repulsion between two spherical equal-sized emulsion droplets for low surface potentials, i.e.,

$$V_R = 2\pi\epsilon_o\epsilon R\psi_o^2 \ln\{1 + \exp(-\kappa s)\} \tag{10}$$

where ψ_o is the surface potential of the droplet. When the double layer around the droplet is very extensive ($\kappa R < 5$), Derjaguin's procedure begins to break down. Verwey and Overbeek (1948) developed an approximate procedure for low surface potentials to give

$$V_R = 2\pi\epsilon_o\epsilon R\psi_o^2 \exp(-\kappa s) \tag{11}$$

More exact solutions, valid for higher surface potentials, can be found elsewhere (Verwey and Overbeek 1948). Similar equations for the case of constant surface charge are given by Wiese and Healy (1970). The surface potential ψ_o is usually taken to be the zeta potential, the potential

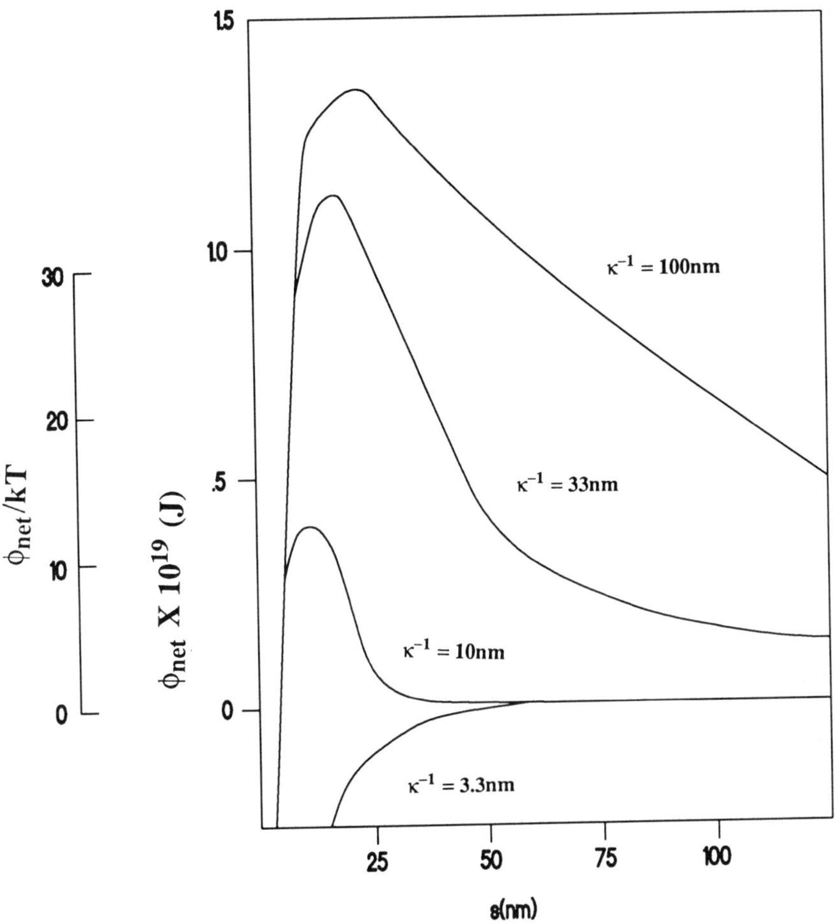

Fig. 6.3 Plot of ϕ_{net} vs. s, the separation of surfaces, for two spheres of equal radius (100 nm). Curves are drawn for different values of κ with constant values of A (10^{-19} J) and ψ_0 (25.7 mV).

at the plane of shear. Such an approximation may be reasonable if the interfacial adsorbed layer consists only of small molecular weight surfactants. If proteins are adsorbed at the interface, however, the thickness of adsorbed layer can be several nanometers (Fig. 6.4), since the adsorbed protein molecule is present in the form of loops and trains. Consequently, zeta potential will not be a true measure of the surface potential in such systems since the plane of shear is located far from the surface. Moreover,

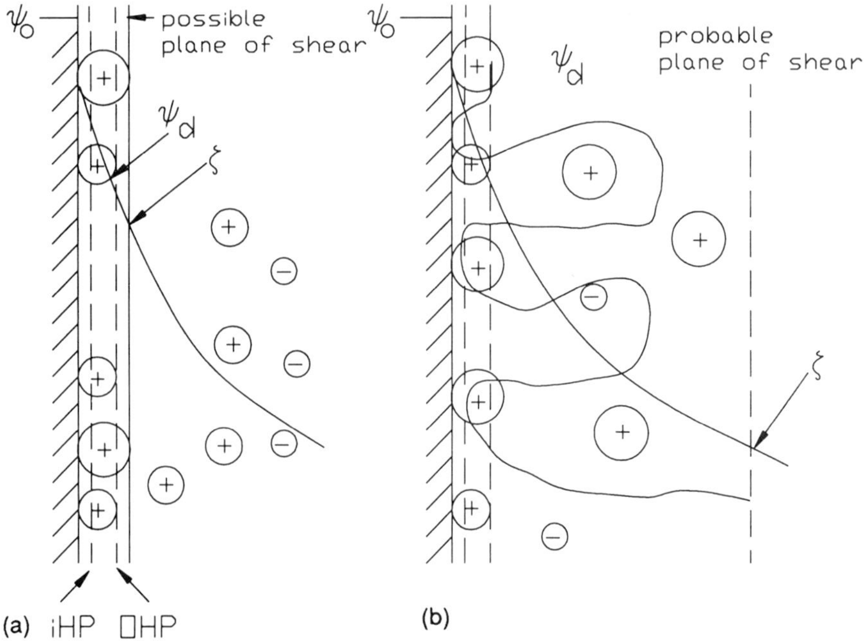

Fig. 6.4 Schematic representation of the structure of the electrical double layer in the (a) absence and (b) presence of adsorbed homopolymer. (*From Tadros and Vincent, 1983.*)

adsorption of macromolecule may change the surface charge density through modification of adsorption of potential determining ions. Also, the dielectric constant of the inner protein layer will be different. No comprehensive theory capable of accounting for the above effects is yet available. However, the zeta potential, ζ, and ψ_d, the potential at the Stern plane, can be related through the Guoy-Chapman equation (Tadros and Vincent, 1983):

$$\tanh \frac{ze\zeta}{4kT} = \tanh \frac{ze\psi_d}{4kT} \exp[-k(\delta - d)] \qquad (12)$$

where δ is the thickness of the interfacial layer and d is the thickness of the inner layer.

Forces Associated with Adsorbed Macromolecules

In food emulsions, proteins are usually employed to obtain stability. Because of the presence of hydrophobic and hydrophilic functional groups, proteins tend to adsorb at oil-water interfaces. The adsorbed protein molecule orients itself at the interface so hydrophobic functional groups point into the oil phase, whereas hydrophilic functional groups are exposed to the aqueous medium. Consequently, in oil-in-water emulsion, for example, the hydrophobic functional groups adsorb at the oil droplet surface in the form of trains with the hydrophilic functional groups protruding into the aqueous medium in the form of loops and trains as shown schematically in Fig. 6.4. When two emulsion droplets with adsorbed macromolecules approach, they experience a steric interaction (which, though usually a repulsive force, can, in principle, be attractive) as a result of the overlap of the adsorbed layers (Fig. 6.4).

It is convenient to visualize the overlap of adsorbed layers as occurring in two different domains. These domains delimit different regions in which different phenomena can clearly be identified as responsible for steric interaction. These zones depend on the relative magnitudes of the thickness of the adsorbed macromolecular layer and the distance between the two emulsion droplets. A schematic representation of these domains is shown in Fig. 6.5 (Napper, 1983). If the distance of separation between the droplets is more than twice the thickness L_s of the adsorbed layer, the droplets do not experience any steric interaction since the adsorbed layers do not overlap. If the distance between the droplets is between one and two spans of adsorbed macromolecule, i.e., $L_s \leq d \leq 2L_s$, L_s and d respectively, being the thickness of the adsorbed layer and the distance between the two droplets (Fig. 6.5b), the adsorbed segments overlap. This regime is referred to as the interpenetrational regime (Napper, 1983). In the interpenetrational regime, steric interaction is created mainly by the increase in the concentration of adsorbed segments in the overlap region. This increase in the concentration of macromolecular segments is accomplished at the expense of solvent molecules. If the interaction between the adsorbed segments and the solvent is favorable (i.e., if the solvent is "good"), then interpenetration of adsorbed segments should be thermodynamically unfavorable and therefore should result in an increase in the free energy. In other words, for good solvency conditions, interpenetration of adsorbed segments should result in a repulsive force. On the other hand, if the interaction between the adsorbed segments and the solvent is unfavorable, interpenetration is thermodynamically favorable, and therefore should result in a decrease in the free energy and an attractive force of interaction. The overall process that determines steric interaction in this domain is the free energy of mixing since the adsorbed segments are demixed with the solvent in the interpenetrational zone.

When the distance of separation between the two droplets is less than the thickness of the adsorbed layer (i.e., $d \leq L_s$), loops and tails of ad-

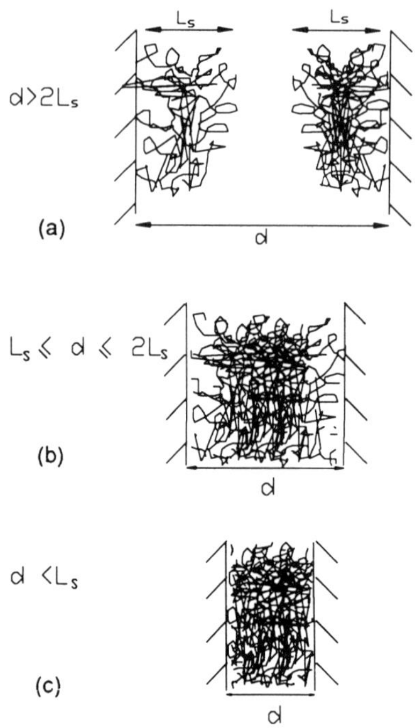

Fig. 6.5 Schematic representation of the three domains of close approach for sterically stabilized flat plates; (a) the noninterpenetrational domain ($d > 2L_S$); (b) the interpenetrational domain ($L_S \leq d \leq 2L_S$); (c) interpenetration plus-compression domain ($d < L_s$). *(From Napper, 1983.)*

sorbed macromolecules protruding from one emulsion droplet are compressed by the second droplet. Of course, interpenetration of adsorbed segments still occurs. This regime is referred to by Napper (1983) as the interpenetration-plus-compression domain (Fig. 6.5c). Compression of loops and tails by a droplet surface results in a loss of configurational entropy of adsorbed segments. This loss of configurational entropy is manifested as a repulsive force, irrespective of the nature of interaction between the adsorbed segments and the solvent. The net free energy of interaction is therefore given by

$$\Delta G = \Delta G_{mix} + \Delta G_{el} \tag{13}$$

where the two right-hand terms refer to the contributions from mixing and elastic free energies, respectively.

Mackor (1951) was the first to propose a theory for steric stabilization. His model pictured adsorbed segments as rigid rods of length L, flexibly attached to the particle surface with ball joints (Fig. 6.6). He assumed that steric repulsion arose solely due to loss of configurational entropy when two particles with adsorbed segments approached each other. The free energy of interaction can therefore be written as

$$\Delta G_R = G_R(h) - G_R(\infty) = -TS_R(h) + TS_R(\infty) \tag{14}$$

where $G_R(h)$ is the Gibbs free energy, when the separation distance is h and $G_R(\infty)$ is its value at infinite separation, $S_R(h)$ and $S_R(\infty)$ are corresponding values of entropy, and T is the absolute temperature. The loss of configurational entropy can be evaluated from Boltzmann's relationship $S = k \ln W$, where W is the number of possible configurations that a rigid rod can assume. For low surface coverage of adsorbed macromolecule, the free energy of interaction has been shown by Mackor (1951)

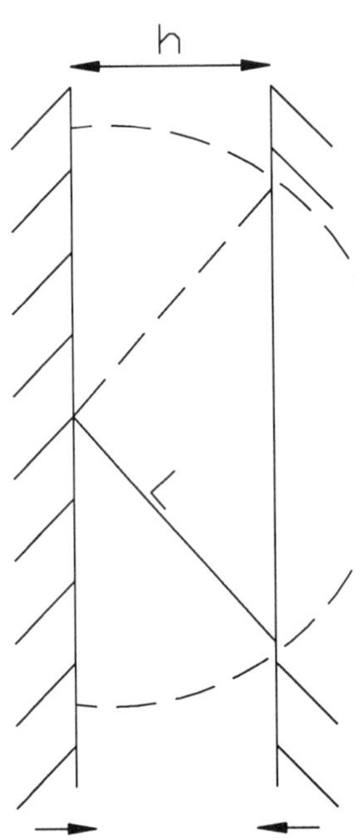

Fig. 6.6 Diagrammatic representation of the loss of configurational entropy of overlapping adsorbed layers according to the theory of Mackor (1951).

to be

$$\Delta G_R = N_S\theta kT\left(1 - \frac{h}{L}\right) \tag{15}$$

where $N_S\theta$ is the number of segments per unit area of the surface, θ being the fractional surface coverage, and h is the distance of separation between the two neighboring surfaces (Fig. 6.6).

The main drawback of Mackor's model is that it completely disregards the contribution from mixing free energy and therefore predicts only steric repulsion irrespective of the nature of interaction between the adsorbed segments and the solvent. Consequently, Mackor's (1951) model cannot predict flocculation of sterically stabilized dispersions upon heating. Other more refined models, e.g., those of Clayfield and Lumb (1966a, b, 1968, 1974a, b, c) and Jackel (1964), suffer from the same drawback.

The importance of the free energy of mixing on the steric interaction was first recognized by Fischer (1958). The mixing free energy in the lens-shaped overlap layer can be written as

$$\Delta G_M = 2\int_{\Delta V} \pi_E \, dV \tag{16}$$

where ΔV is the total volume of the overlap region and π_E, the excess osmotic pressure of macromolecular segments, is given by

$$\pi_E = RTB_2c_2^2T \ldots \tag{17}$$

where R is the gas constant, B_2 is the second virial coefficient used to express the deviation from ideality in the osmotic pressure of the solute, and c_2 is the segment concentration. Virial coefficients higher than the second are ignored in Eq. (17). From Eqs. (16) and (17) one obtains

$$\Delta G_M = 2RT\int_{\Delta V} B_2c_2^2 dV \tag{18}$$

Fischer (1958) assumed that the segment density distribution in the overlap region is uniform so that

$$\Delta G_M = 2RTB_2c_2^2\Delta V \tag{19}$$

The validity of such a simplifying assumption will be discussed later. From geometric considerations, ΔV, the volume of the overlap region, is given by

$$\Delta V = (2/3)\,\pi\left(8 - \frac{H_0}{2}\right)(3a + 2\delta + H_0/2) \qquad (20)$$

where δ is the thickness of the adsorbed layer and H_0 is the minimum distance of separation between two spherical particles of radius a. The second virial coefficient B_2 of the adsorbed segments is given by

$$B_2 = \frac{0.5 - \chi}{\overline{V}_1 \rho_2^{\,2}} \qquad (21)$$

where χ is the Flory-Huggins parameter, $\overline{V}_1$ is the molar volume of the solvent, and ρ_2 is the density of adsorbed segments. Equation (19) can now be written as

$$\Delta G_M = (4\pi c_2^{\,2}/3\overline{V}_1\rho_2^{\,2})kT(0.5 - \chi)(\delta - H_0/2)^2(3a + 2\delta + h_0/2) \qquad (22)$$

The above expression can be recast into a more convenient form as

$$\Delta G_M = 4\pi a\omega^2 N_A(\overline{V}_2^{\,2}/\overline{V}_1)(0.5 - \chi)kT(1 - H_0/2L_S)^2 \qquad (23)$$

where L_S, the span of the adsorbed macromolecular segments, can be set equal to δ. In the above equation, ω is the weight of adsorbed segments per unit area and $\overline{V}_2$ is the partial specific volume of adsorbed segments. Note that the free energy of steric interaction depends on the Flory-Huggins χ parameter. If the solvent is athermal (i.e., $\chi = 0.5$), the free energy of steric interaction is zero, since the second virial coefficient of the adsorbed segments is zero (i.e., the solution is ideal). If $\chi < 0.5$, the segment-solvent interaction is favorable (i.e., "good solvent"), the free energy of steric interaction is positive. For "poor" solvent ($\chi > 0.5$), however, the free energy of interaction is negative.

Fischer's treatment does not account for the elastic component of free energy in the interpenetrational and compression zone. Consequently, for constant segment density distributions, Eqs. (22) and (23) are valid only when $L_S \le H_0 \le 2L_S$. For $H_0 \le L_S$, Smitham et al. (1975) showed that

$$\Delta G_M = 2\pi akT\omega^2(\overline{V}_2^{\,2}/\overline{V}_1)N_A(0.5 - \chi)S \qquad (24)$$

where $S = 3\ln(L_S/H_0) + 2(H_0/L_S) - 1.5$

Similar expressions for the free energy of steric interaction have been derived for different segment density distributions (Napper, 1983). The general expression is given by (24) where the function S takes on different functional forms for different segment density distributions.

The predicted steric interaction energy between two polystyrene latex particles stabilized by polyvinyl alcohol in water for different segment density distributions is shown in Fig. 6.7 (Smithiam and Napper, 1979). The variation of steric interaction with distance is fairly insensitive to the type of segment density distribution for Gaussian, radial Gaussian, and constant segment density distributions. For an exponential distribution, however, the predicted steric interaction falls off much more rapidly than for the others, possibly because of rapid depletion in the segments away from the surface.

Expressions for steric interactions mainly have been derived based on an assumed segment density distribution of the adsorbed macromolecules in the vicinity of emulsion droplets. This segment density distribution, however, depends on various factors such as the molecular weight, segment-segment, and, most importantly, segment-solvent interactions of the adsorbed macromolecule. Because of the complexity of the structure of protein molecules commonly employed in the stabilization of food emul-

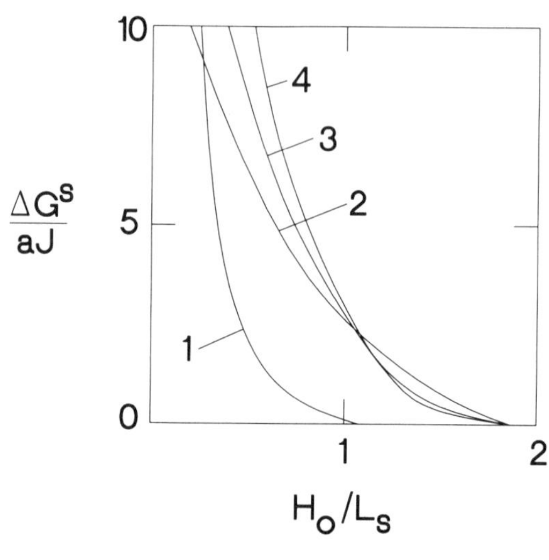

Fig. 6.7 The distance dependence of the steric interaction energy for two spheres of equal radius stabilized by polymer layers with different segment density distribution functions: 1, exponential; 2, constant; 3, Gaussian; 4, radial Gaussian. (*After Smitham and Napper, 1979.*)

sions, it is extremely difficult to predict its segment density distribution. Only for synthetic polymers with uniform structure have attempts been made to predict the segment density distributions (Meir, 1967; Hesselink et al., 1971); these have been based on the solution of diffusion equation for random flight chains terminally attached to a surface. It is therefore necessary to obtain the segment density distribution of proteins near the emulsion droplets independently through experimental measurements.

Forces Associated with Free Macromolecules

Water-soluble polymers such as gums, pectins, and gelatin can cause agglomeration when present at very low concentrations. Polysaccharides also can induce agglomeration in food emulsions (Dickinson and Euston, 1990). Free macromolecules (i.e., macromolecules that do not adsorb onto surfaces of emulsion droplets) give rise to attractive or repulsive forces when their concentration between two emulsion droplets changes because of geometric factors. Consider, for example, free macromolecules in solution between two emulsion droplets that are far apart. The conformation of a macromolecule between the two droplets will be the same as that in the bulk (Fig. 6.8a). When the distance between the two droplets becomes comparable to the radius of gyration of the macromolecule, however, the macromolecule cannot take its natural confirmation because of geometric constraints caused by the emulsion droplets. For favorable interaction between the solvent and the macromolecule, such changes in the conformation of the macromolecule (Fig. 6.8b) result in an increase in the free energy of the system, thereby giving rise to a repulsive force. If the two emulsion droplets are able to overcome this repulsive force and come even closer (Fig. 6.8c), the macromolecule is completely squeezed out of the region between the two droplets. Demixing of macromolecule and the solvent should lead to further increase in the free energy. At this point, the region between the two emulsion droplets consists only of a microreservoir of solvent. Further decrease in the distance between the emulsion droplets, however, squeezes the solvent into the bulk. This should result in a decrease in the free energy since mixing of the solvent with the macromolecules in the bulk is energetically favorable. Consequently, the potential energy diagram for interaction of two emulsion droplets exhibits an energy barrier very much like electrostatic interaction as shown in Fig. 6.9. As pointed out above, at very small separations, the droplets experience attraction. This attraction between two droplets separated only by solvent can also be viewed as due to excess osmotic force pushing the droplets towards each other. The height of the energy barrier as well

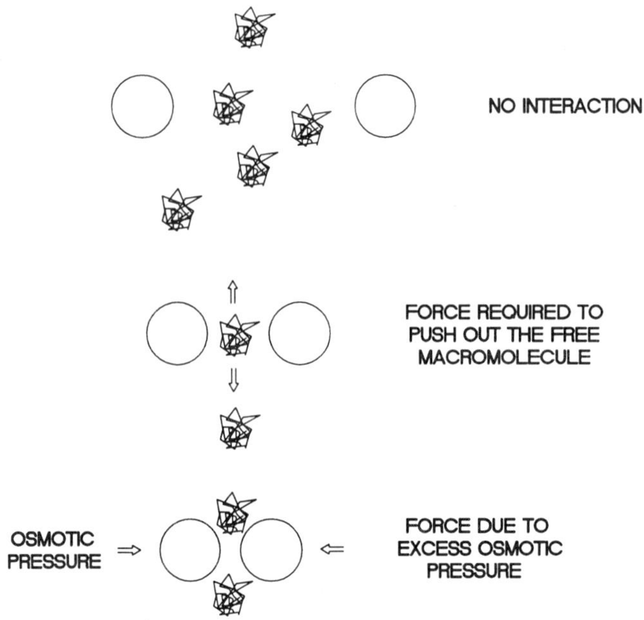

Fig. 6.8 Forces associated with free macromolecules. (*From Napper, 1983.*)

as the depth of the potential well depend on the concentration of free macromolecule in solution. At low concentration, the energy barrier will be small ($<5\ kT$) and allow flocculation of emulsion droplets. This phenomenon is known as depletion flocculation, since it is due to depletion of macromolecules in the region between the droplets. At higher macromolecular concentration, however, kinetic stability is imparted to the emulsion because the energy barrier is sufficiently large. This phenomena is usually referred to as depletion stabilization.

The attractive potential due to depletion flocculation has been evaluated by Asakura and Oosawa (1954) by considering the macromolecules as rigid spheres, of diameter Δ, with no interactions between them and particles (droplets) of radius a to give

$$\Delta G = 0 \qquad d \geq 2a + \Delta$$

$$\Delta G = -\pi_o(\pi/12)\{2(2a + \Delta)^3 - 3(2a + \Delta)^2 d + d^3\} \qquad 2a \leq d \leq 2a + \Delta \quad (25)$$

where d is the distance of separation between the two particles and π_o, the limiting Van't Hoff osmotic pressure of the dissolved polymer, is given by

$$\pi_o = kT\, N_2/V \tag{26}$$

N_2 and V being the number of polymer molecules and volume, respectively. Further refinements have been made by Asakura and Oosawa (1958) and Sieglaff (1959) to account for the flexibility of the macromolecule and its second virial coefficient. It has been shown that flexible macromolecules provide stronger interparticle force than rigid spherical macromolecules such as globular proteins. Charge on the molecule has also been shown to increase the interparticle force.

Combination of Forces

The combination of different interparticle forces discussed above determines the stability of emulsion. For electrocratically stabilized emul-

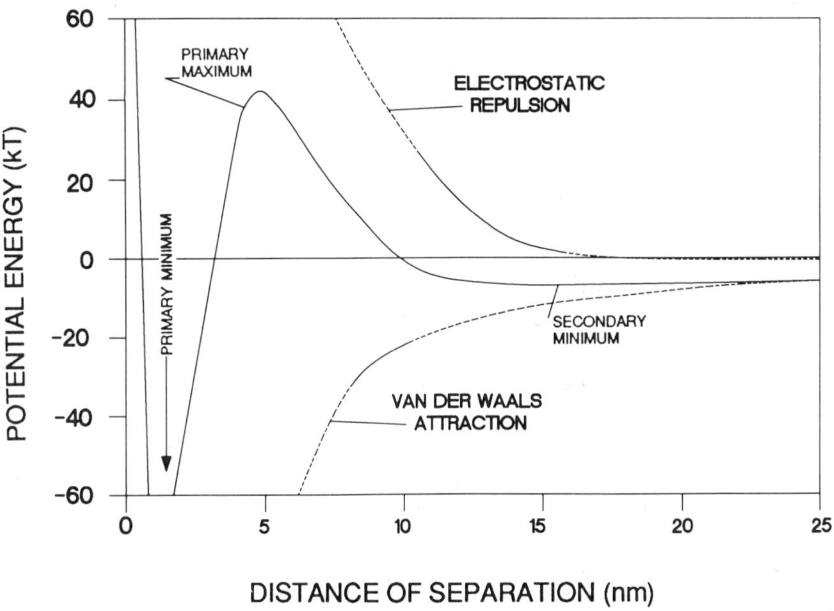

Fig. 6.9 Potential energy vs. distance curve for a pair of electrostatically stabilized droplets; also shows the separate contributions of the electrostatic and van der Waals components. (*From Parker, 1982.*)

sion, the important interparticle forces are van der Waals attraction and double-layer repulsion. At very small and large distances of separation, van der Waals attraction predominates over double-layer repulsion, so droplets experience a net attractive potential. At intermediate distances of separation, however, electrostatic repulsion predominates. Therefore, the overall potential exhibits a primary minimum at short separation, an energy barrier at intermediate separations, and possibly a shallow secondary minimum at large separation, as can be seen in Fig. 6.9. The stability of emulsion depends on the height of the energy barrier. If the energy barrier is greater than 25 kT, the relative motion between the two approaching emulsion droplets will not possess sufficient energy to overcome the energy barrier. Consequently, the two droplets will not experience a close encounter. On the other hand, for energy barriers of the order of kT, the two droplets can come together in the primary minimum, leading subsequently to their coalescence. The height of the energy barrier depends on surface potential, ionic strength, and particle size. The sensitivity of the energy barrier to the surface potential is shown in Fig. 6.10. Up to a surface potential of about 18 mV there is no energy barrier. For surface potentials higher than around 18 mV, however, the energy barrier increases dramatically with surface potential. The rate of exponential decay of double layer potential increases as κ increases [see Eq. (11)], and κ is higher at higher ionic strength. Therefore, the double layer is thinner at higher salt concentrations, resulting in smaller energy barriers, as can be seen in Fig. 6.3. Consequently, increases in salt concentration tend to destabilize the emulsion. At a certain salt concentration, known as critical flocculation concentration, the energy barrier disappears so that the emulsion tends to flocculate spontaneously. The critical flocculation concentration is smaller for electrolytes with a higher valence number of counterions and is proportional to the inverse of sixth power of counterion valence number. This relationship is known as the Schulze-Hardy rule. For emulsions stabilized by macromolecules such as proteins, steric interactions also come into the picture. As pointed out earlier, the steric interaction is greatly influenced by the quality of the solvent, i.e., the nature of segment-solvent interaction. For favorable interactions ($\chi < 0.5$), steric repulsion predominates over van der Waals interactions at short distances of separation so that there is *no* primary minimum, as can be seen from Fig. 6.11. As a result, the emulsion is thermodynamically stable since the droplets cannot experience a close encounter necessary for coalescence. When the solvent is worse ($\chi > 0.5$), however, the overall interaction potential exhibits a potential well at short distance of separation, because of which the droplets can flocculate (see Fig. 6.11). These remarks regarding the overall potential of sterically stabilized emulsion

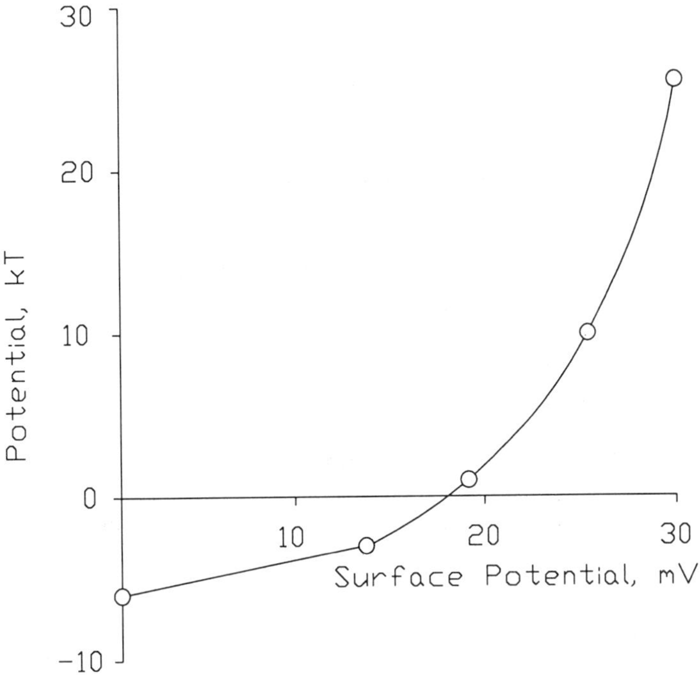

Fig. 6.10 The maximum of the resulting (the total) potential is critically dependent on the surface potential. (*From Friberg, 1976.*)

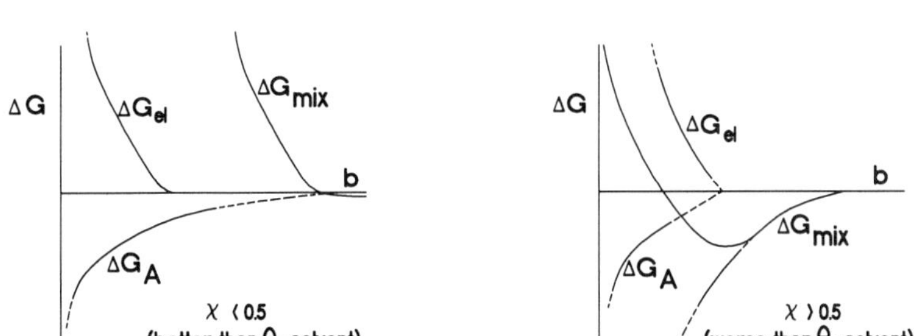

Fig. 6.11 The interaction of free energy–separation curves for two droplets stabilized by polymer. (*From Tadros and Vincent, 1983.*)

are valid only for undeformed droplets. When the droplets are very close, they deform, thus modifying the overall interaction. Moreover, rearrangement of loops and trains of adsorbed segments may also occur, leading further to changes in the overall potential. The sensitivity of flocculation of sterically stabilized emulsion to the quality of solvent was demonstrated by March and Napper (1977) in their investigation of flocculation behavior of O/W emulsions stabilized by block and graft copolymers. They observed that the amount of cream layer increased because of flocculation above a critical flocculation temperature. Typical experimental data for the thickness of cream layer at different temperatures for toluene droplets stabilized by polyethylene oxide are shown in Fig. 6.12. The thickness of the cream layer decreased upon cooling, indicating that the phenomenon is reversible. The increase in the cream layer was shown to be due to flocculation. The flocculated drops were found to be spherical with minimal deformation. The critical flocculation temperature corresponded to the θ temperature of the stabilizing polymer. At the θ temperature, the second virial coefficient of the macromolecule vanishes. Consequently, the solution behaves like an "ideal" solution. Therefore, θ temperature is a transition temperature at which the solvent changes from "good" to "poor." Therefore, flocculation above the critical flocculation temperature is probably due to a shift in the medium's solvency from "good" to "worse." The results of the investigation for various polymers are shown in Table 6.2.

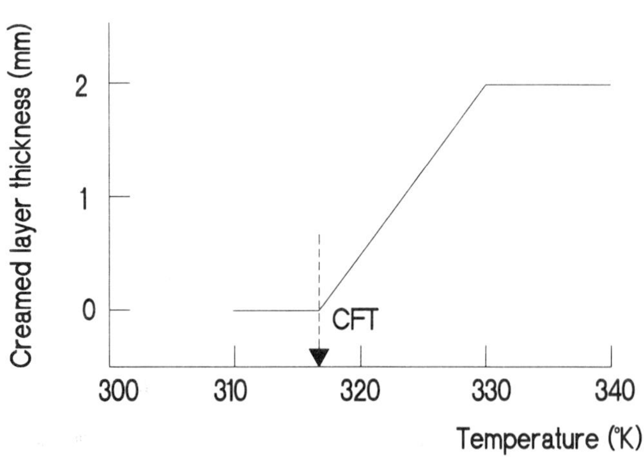

Fig. 6.12 Location of the CFT for a sterically stabilized emulsion. (*From Tadros and Vincent, 1983.*)

Table 6.2 CFT Values for Emulsions

Type	Stabilizing moiety	Anchor polymer	Internal phase	External phase	CFT (K)	θ (K)
O/W	PEO	PVAc	Toluene	0.39 mol dm^{-3} $MgSO_4$	318 ± 2	318 ± 2
O/W	PVA	PVAc	Toluene	2.0 mol dm^{-3} NaCl	297 ± 2	299 ± 2
O/W	PAM	PVAc	Toluene	2.1 mol dm^{-3} $(NH_4)_2SO_4$	297 ± 2	298 ± 2
O/W	PAA	PVAc	Toluene	2.0 mol dm^{-3} HCl	298 ± 2	879 ± 52
O/W	PLM	PAA	Water	*n*-Pentanol	301 ± 3	303 ± 5
W/O	PS	PEO	Water	Cyclohexane	308 ± 3	307 ± 1

Source: Tadros and Vincent, 1983.

FLOCCULATION

Because of van der Waals attraction, two colliding emulsion droplets can "flocculate" to form a doublet. Even though the emulsion droplets belonging to a doublet maintain their identity, they move as one kinetic unit. Since the individual droplets in a "floc" are separated by a thin film of intervening continuous phase, rupture of these films can lead to coalescence. Consequently, flocculation can be viewed as a precursor to coalescence and eventual breakdown of the emulsion. It is, therefore, important to predict rates of flocculation to understand emulsion stability. The rate of flocculation, in turn, depends both on interparticle forces as well as relative motion between droplets. The three main mechanisms of particle loss are creaming, Brownian flocculation, and sedimentation flocculation as schematically shown in Fig. 6.13. In creaming, droplets either rise or sediment out of the system due to the density difference. For O/W emulsion, oil being lighter than water, oil droplets rise to the top to form a cream layer. Encounters between emulsion droplets occur because of relative Brownian motion between the droplets. Such encounters may result in flocculation if the interactions between the droplets are favorable. Gravity-induced flocculation occurs because of differential rates of cream-

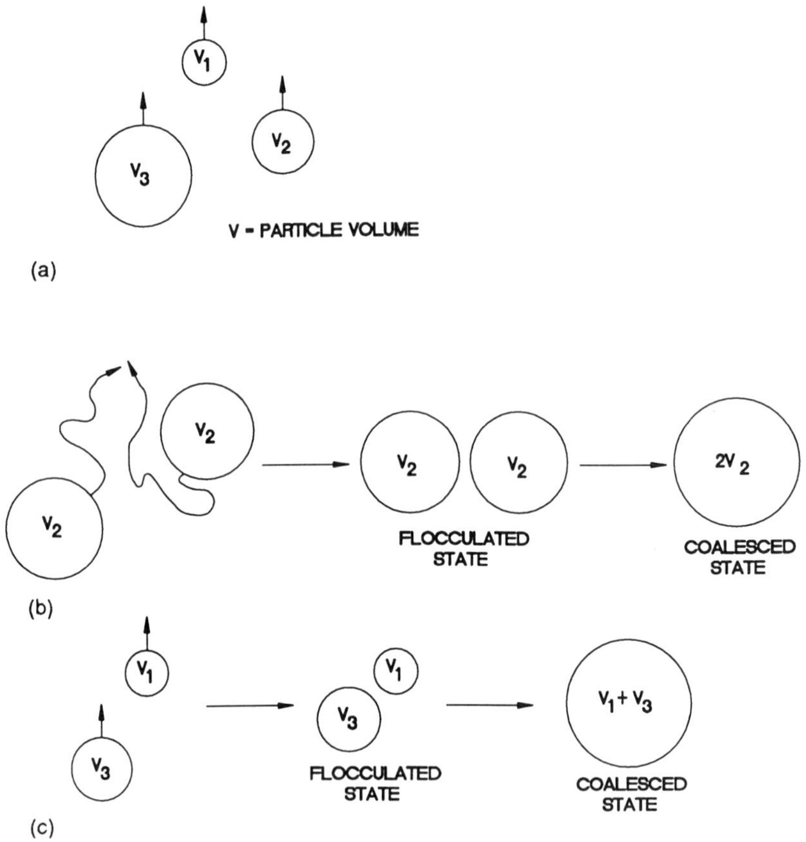

Fig. 6.13 Colloidal breaking mechanisms in quiescent media: (a) creaming, (b) Brownian flocculation, (c) sedimentation flocculation. (*From Melik and Fogler, 1988.*)

ing of droplets of different sizes. Larger droplets sweep up smaller droplets in their path, often resulting in flocculation.

Brownian Flocculation

Smoluchowski (1917) considered Brownian flocculation of particles in the absence of interparticle forces. Such analysis describes flocculation of emulsion droplets when no stabilizers such as surfactants or proteins are used to modify the inter-particle forces, and gives rise to the following expression for J_f, the rate of flocculation,

$$J_f = 8\pi a D n_o \tag{27}$$

where a is the radius of the particle, n_o is the initial particle number concentration, and D, the diffusion coefficient of the particle, is given by

$$D = \frac{kT}{6\pi\mu a} \tag{28}$$

where k is the Boltzmann's constant, T is the absolute temperature, and μ is the viscosity of the continuous medium. From Eqs. (27) and (28), one can see that such (fast) coagulation is independent of particle size. The half-life $\tau_{1/2}$, i.e., the time for the original number concentration to decrease by a factor 2, is given by

$$\tau_{1/2} = \frac{3\pi}{4n_o kT} \tag{29}$$

Typical values of $\tau_{1/2}$ for O/W emulsion, for different emulsion drop sizes, and volume fractions ϕ are shown in Table 6.3 (Tadros and Vincent, 1983). The number concentration $n_o = 3\phi/9\pi a^3$. Therefore $\tau_{1/2}$ is smaller for larger volume fractions and small drop sizes. For typical drop sizes encountered in food emulsions, $\tau_{1/2}$ are extremely small (of the order of msec). It is, therefore, apparent that the emulsion would break immediately if no stabilizers were employed to modify the interparticle forces. (The values of half-life reported in Table 6.3 are for neutrally buoyant emulsions.) Even though the half-life of Brownian flocculation for large

Table 6.3 Half-lives of Emulsions of Different Phase Volumes and Droplet Sizes

	ϕ (%)[a]			
a (μm)	10^{-3}	1	10	50
0.1	76 sec	76 msec	7.6 msec	1.5 msec
1	21 hr	76 sec	7.6 sec	1.5 sec
10	4 months	21 hr	2 hr	25 min

[a] Volume percent dispersed phase in mixture.
Source: Darling and Birkett, 1986.

droplets is very high, the predominant mechanism of particle loss for such large particles is creaming and sedimentation flocculation.

The effect of energy barriers on the kinetics of flocculation can be accounted for by considering the interparticle forces in evaluating rates of collisions between droplets. Such analysis by Fuchs (1934) gave the following expression for the rate of slow flocculation J_s:

$$J_s = 8\pi a D n_o / W \tag{30}$$

where the stability ratio W is given by

$$W = 2a \int_{2a}^{\infty} \exp\left(\frac{V_T}{kT}\right) \frac{dr}{r^2} \tag{31}$$

r being the center-to-center distance between two droplets. In Eq. (31) V_T, the net interaction potential between two emulsion droplets, can be evaluated using the appropriate equations for different contributions, as discussed in the previous section. By employing the method of steepest descent, Reerink and Overbeek (1954) have evaluated the following approximate expression for the stability ratio:

$$\log W = A - B \frac{a\gamma^2}{z^2} \log c \tag{32}$$

where A and B are constants, c is the electrolyte concentration, z is the valence number of electrolyte, and γ is given by,

$$\gamma = \frac{\exp(ze\psi_o/2kT) - 1}{\exp(ze\psi_o/2kT) + 1} \tag{33}$$

Equation (32) predicts that a log-log plot of W vs. c should yield a straight line in the slow flocculation region. Experimental measurements seem to validate this prediction. However, the experimental results of the effect of particle size are not consistent with Eq. (32) and indicate that the slope of the straight line, rather than being proportional to the particle size, is fairly insensitive to particle size. No satisfactory explanation for such a behavior has yet been given. Since the energy barrier decreases as electrolyte concentration increases (see Fig. 6.3), W should be independent of c at high electrolyte concentrations. Therefore, a plot of log W vs. log c should consist of two straight lines intersecting at critical flocculation

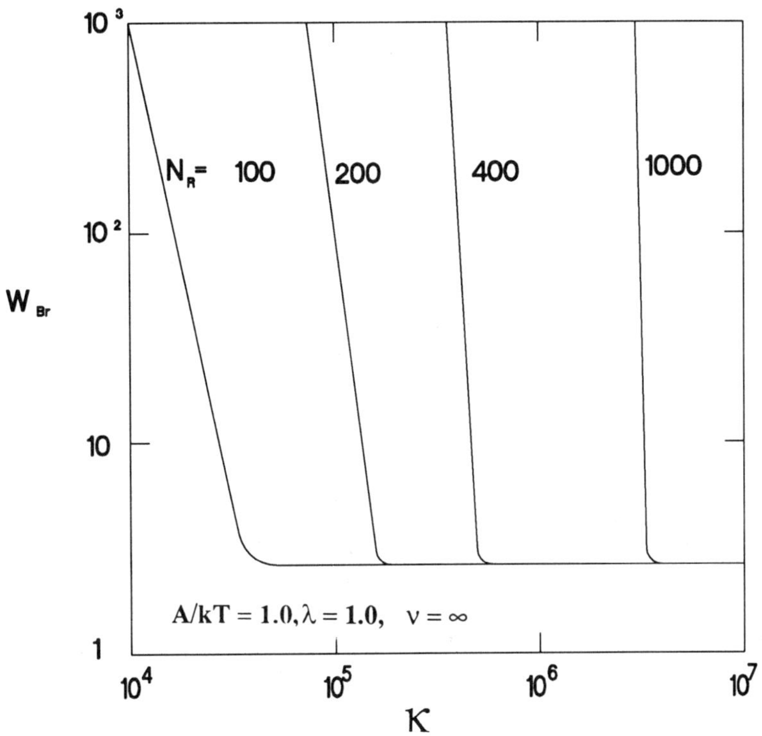

Fig. 6.14 Effect of electrolyte concentration on the Brownian flocculation stability factor for various surface potentials. (*From Melik and Fogler, 1988.*)

concentration. Plots for various surface potentials is shown in Fig. 6.14. The critical flocculation concentration ($c \alpha \kappa^2$) increases as surface potential of the droplets increases ($\psi_o \alpha N_R{}^{1/2}$), as can be seen from Fig. 6.14. It should be noted, however, that the stability ratio W at high electrolyte concentrations (large values of κ) is slightly larger than unity. This is a consequence of resistance due to hydrodynamic interaction not accounted for in Smoluchowski's analysis. Expressions for hydrodynamic interaction between two droplets of different sizes are given by Spielman (1970). Theoretical plots of stability ratio vs. electrolyte concentration for symmetrical electrolytes of different valence numbers indicate, for example, that the critical flocculation concentration for 2:2 electrolyte is much smaller than that for 1:1 electrolyte, which is consistent with the Schulze-Hardy rule.

Creaming

The creaming or settling velocity of an emulsion droplet of radius a is given by the Stokes equation:

$$U = \frac{2a^2 \mid \Delta\rho \mid g}{9\mu} \tag{34}$$

where $\Delta\rho$ is the density difference between the two phases and g is the acceleration due to gravity. As can be seen from the above equation, the creaming velocity is a strong function of droplet size. Creaming is reduced for smaller emulsion droplets, higher viscosity of the medium, and smaller density difference.

Equation (34) is derived under the assumption that the emulsion droplets essentially behave as solid spheres. Such an assumption is usually valid when droplets have a high viscosity. Also, when sufficiently high concentrations of surfactants and especially globular proteins are adsorbed at the droplet interface, they provide surface viscosity and surface elasticity (Dickinson and Stainsky, 1982) high enough to make droplets behave like solid particles. For pure emulsion droplets without any surfactant, the interface is completely mobile and the creaming velocity is given by the following expression, due to Hadamard and Rybczynski (Levich 1962):

$$U = \frac{2a^2 \mid \Delta\rho \mid g(\mu_d + \mu_c)}{3\mu_c(3\mu_d + 2\mu_c)} \tag{35}$$

where μ_c and μ_d refer to the viscosities of the continuous medium and the droplet, respectively. The creaming velocity predicted by Eq. (35) is greater than that predicted by Stokes equation (34). The actual creaming velocity will lie between the two limits, though Stoke's equation is a fairly good approximation. Since the dispersion medium in many food emulsions is non-Newtonian, the apparent viscosity of the medium will depend on the shear rate. In emulsions containing polysaccharides such as pectin and carrageenan, the dispersion medium is pseudoplastic. This tends to reduce the creaming velocity considerably since the apparent viscosity of the dispersion medium is extremely high at small shear rates.

At high volume fractions, the creaming velocity is reduced by hydrodynamic interaction between the droplets. For dilute suspension of spherical particles, the hydrodynamic interaction can be accounted for to give (Batchelor 1972)

$$\overline{U} = (1 - 6.55\phi)U \tag{36}$$

where $\phi(\leq 2\%)$ is the volume fraction of the dispersed phase and U is given by Eq. (34). Hydrodynamic interactions are very important at higher volume fractions, though no reliable equation is available for the prediction of creaming velocity.

Gravity-Induced Flocculation

The rate of collision G_{ij} between emulsion droplets of radii a_i and a_j due to gravity is given by Reddy and Fogler (1981):

$$G_{ij} = \frac{8\pi \Delta\rho}{g\mu W_{ij}} (a_i^2 - a_j^2)(a_i + a_j)^2 \tag{37}$$

where W_{ij} refers to the stability ratio. The above expression is derived assuming that the particle paths are straight lines, the collision efficiency is unity, and the stability ratio is identical to that of Brownian coagulation. Since the Brownian coagulation frequency B_{ij} is given by

$$B_{ij} = 2kT \frac{(a_i + a_j)^2}{3\mu W_{ij} a_i a_j} \tag{38}$$

the ratio of gravitational collision frequency G_{ij} and Brownian collision frequency B_{ij} is (Reddy and Fogler, 1981)

$$\frac{G_{ij}}{B_{ij}} = \frac{4\pi \Delta\rho g}{3kT} a_i a_j (a_i^2 - a_j^2) \tag{39}$$

This ratio goes through a maximum at a certain particle size ratio $R_{ij} = a_i/a_j$ and is given by

$$\left(\frac{G_{ij}}{B_{ij}}\right)_{\max} = \frac{8\pi \Delta\rho g}{9(3^{1/2})kT} a_i^4 \tag{40}$$

The stability of an emulsion to gravity-induced flocculation depends on the particle size, size ratio, density difference, viscosity, surface potential, and electrolyte concentration. Melik and Fogler (1988) presented a stability plane for flocculation, as shown in Fig. 6.15. The effect of electrolyte concentration on the stability of the emulsion can be examined from the figure. For example, for a 2-μm emulsion drop, a size ratio of 0.5, $\Delta\rho = 0.2$ g/cc and a Hamaker constant of 10^{-14} erg, $N_G \approx 50$. For high electrolyte concentrations, i.e., $\kappa > 10^3$, the droplets are unstable

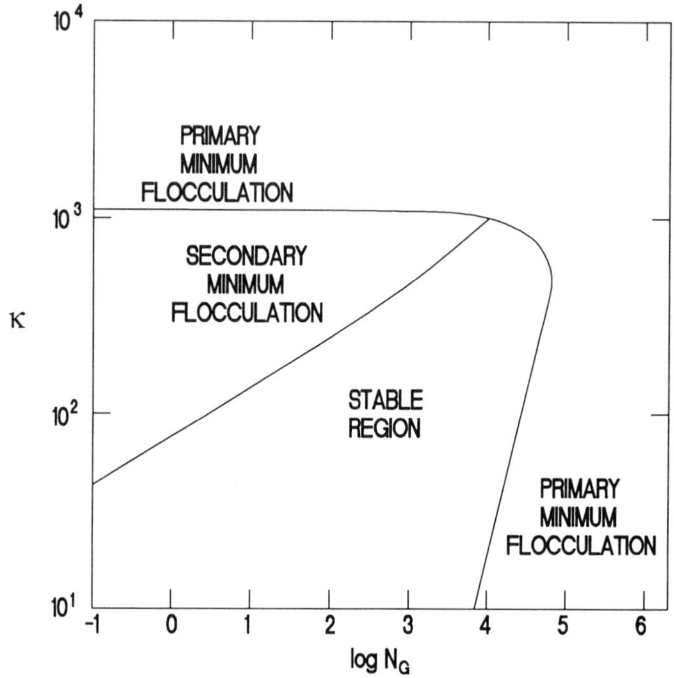

Fig. 6.15 Gravity-induced flocculation stability plane. (*From Melik and Fogler, 1988.*)

to gravitational flocculation. For $\kappa < 10^3$, however (i.e., at moderate electrolyte concentrations), the energy barrier is large enough to prevent flocculation in the primary minimum, but the secondary minimum is deep enough to permit flocculation. For even lower electrolyte concentrations i.e., $\kappa < 10^2$, flocculation is prevented completely, and the droplets are stable.

Melik and Fogler (1988) identified different regimes of dominant particle loss mechanisms, which are shown in Fig. 6.16 for two different values of $g\Delta\rho$. In region I, i.e., for emulsion droplets smaller than 0.15 μm, creaming and gravity-induced flocculation will be negligible and droplets disappear primarily due to Brownian flocculation. The rate of Brownian flocculation, of course, will depend on the strength of electrostatic repulsion and the total number concentration of droplets. As expected, the upper limit of emulsion drop size does not depend on the size ratio $\lambda(a_1/a_2)$. In region II, gravity-induced flocculation will be negligible compared to Brownian flocculation. Creaming will be important in this region.

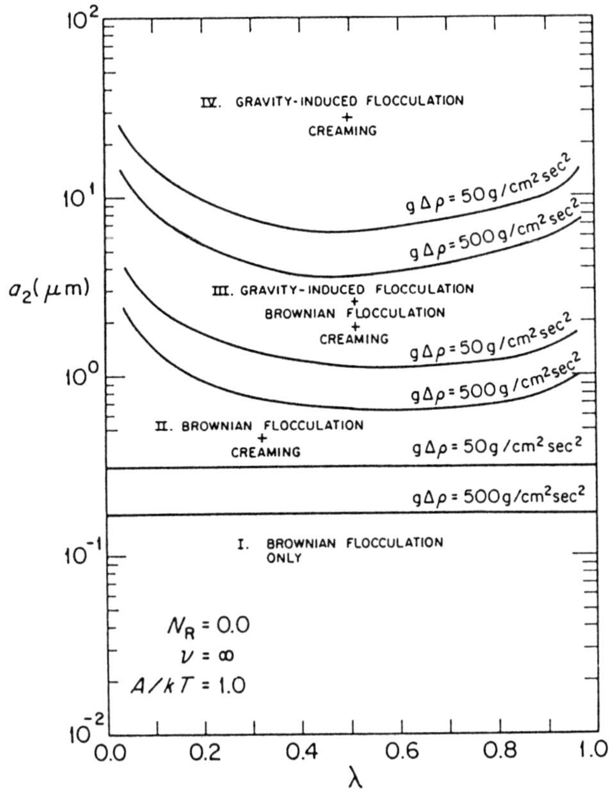

Fig. 6.16 Regimes of dominant particle loss mechanisms in quiescent media. (*From Melik and Fogler, 1988.*)

For intermediate drop sizes that fall into region III, all three mechanisms will be important. For very large emulsion droplets, i.e., droplets larger than 20 μm, Brownian flocculation will be negligible and droplets will be lost mainly by gravity-induced flocculation and creaming.

Simultaneous Flocculation and Creaming

Melik and Fogler (1984, 1988) formulated a general convection diffusion equation to account for different particle loss mechanisms. They solved this equation along with a population balance equation to describe the spatial and temperal variations of particle size distributions. Melik and Fogler provide details of the procedure. Typical variation of particle size

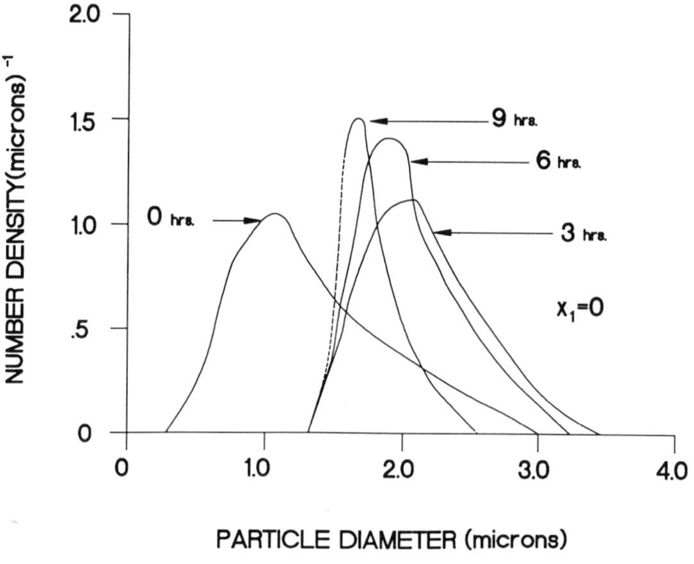

Fig. 6.17 Changes in the particle size distribution with time at $x_1 = 0$ for the case of simultaneous flocculation + creaming. (*Melik and Fogler, 1988.*)

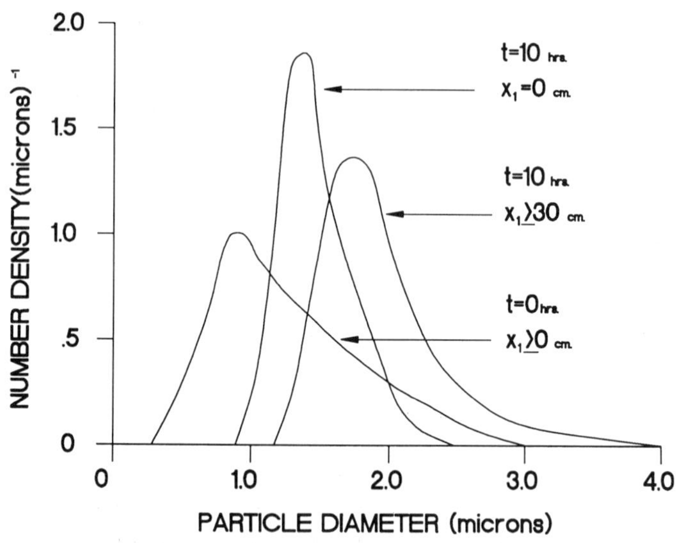

Fig. 6.18 Particle size distribution at various positions for the case of simultaneous flocculation + creaming. (*From Melik and Fogler, 1988.*)

distribution at the bottom of the container ($x_1 = 0$) at different times is shown in Fig. 6.17. For the first 3 hours, the particle size distribution shifts to the right because of flocculation. As the particle size increases, however, the rate of flocculation decreases exponentially. Consequently, creaming becomes comparable to flocculation for larger particles. As a result, for times >3 hours, larger particles preferentially cream out (rise), resulting in the shift of the particle size distribution to smaller sizes. Also, the distribution becomes increasingly narrower because of low rates of creaming and flocculation as a result of decreasing number concentration. Variations of particle size distributions at two different locations are shown in Fig. 6.18. Initially the particle size distribution is independent of position. Because of creaming, the particle size distribution at a height

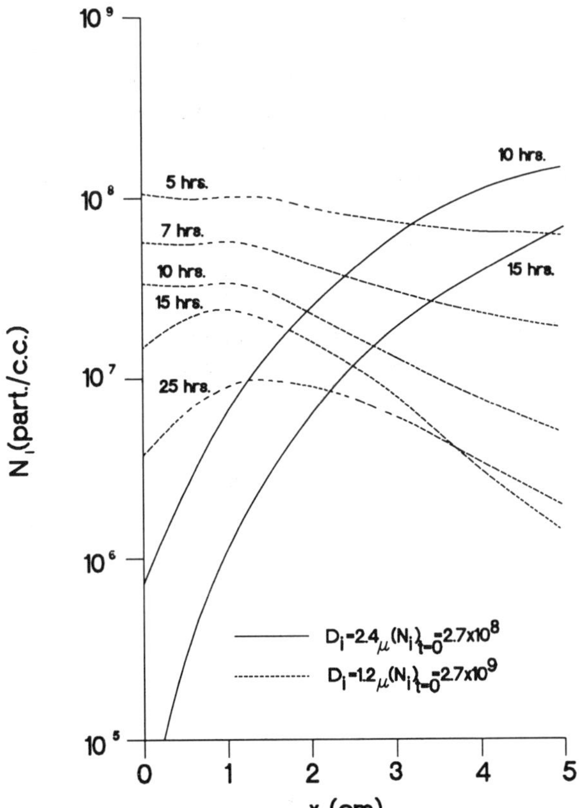

Fig. 6.19 Concentration of particles of specified size as a function of position for the case of simultaneous flocculation + creaming. (*From Melik and Fogler, 1988.*)

of 30 cm becomes coarser than the distribution at the bottom. The concentration profiles of particles of two different sizes is shown in Fig. 6.19. As expected, the concentration of larger particle increases in the direction of creaming. For smaller particles, however, the concentration at short times decreases slowly in the direction of creaming. At longer times, there is a shallow maximum in the concentration profile because of competing effects of flocculation and creaming.

COALESCENCE

When two emulsion droplets approach each other, they tend to flocculate (either at the primary or secondary minimum) because of van der Waals attraction. Unlike solid particles, droplets tend to deform upon close approach and form plane parallel faces separated by thin liquid films (Fig. 6.20b). In large droplets, the deformed faces form a dimple (Fig. 6.20c), because of which the intervening film tends to be thinnest at the periphery. The continuous-phase liquid separating the two droplets drains because of capillary forces resulting from the radius of curvature of the droplets. The drainage will continue until the thin film either ruptures, resulting in coalescence, or reaches an equilibrium thickness as a result of balance of forces acting across the film. If the two droplets are of different liquids, rupture of thin film leads to engulfment. The equilibrium film is subjected to thermal and mechanical perturbations; its stability when perturbed depends on the nature of interaction forces. Because of the increase in the

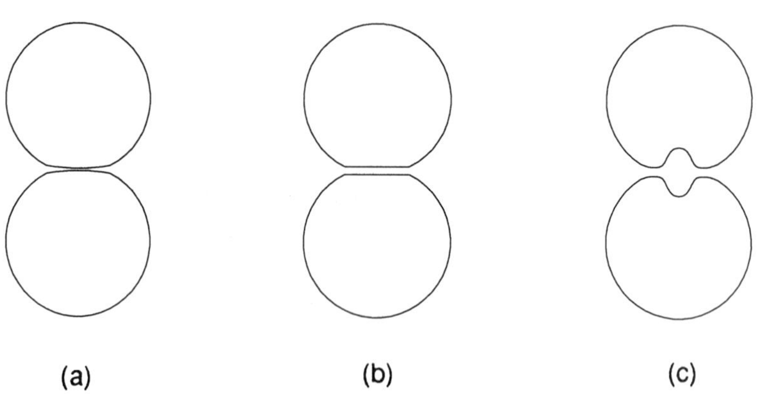

Fig. 6.20 Close approach of two liquid droplets: (a) distortion without dimpling, (b) formation of flat lamela, (c) distortion with dimpling.

interfacial area due to deformation, the surface concentration of surfactants tends to decrease in the planar interface. Moreover, movement of the interface as a result of film drainage leads to a gradient of surface concentration. The resulting Gibbs-Marangoni effect tends to stabilize the thin film by retarding film drainage. Also, surfactant depletion due to surface expansion is partly compensated by its adsorption from bulk. The stability of thin film to rupture, i.e., resistance to coalescence, depends strongly on the mechanical properties of the interface. For examples, adsorption of proteins tends to impart stability to thin films by providing high surface elasticity and surface viscosity.

Thin Film Drainage

The process of coalescence can be split into (1) mutual approach of two droplets to form a plane parallel film, (2) drainage of the film, and (3) rupture of the film. Stages (1) and (3) are instantaneous so that the lifetime of an emulsion droplet subsequent to collision is governed by stage (2). The geometry of a symmetrical plane parallel film is shown in Fig. 6.21. The pressure p_o inside the thin film is almost equal to the pres-

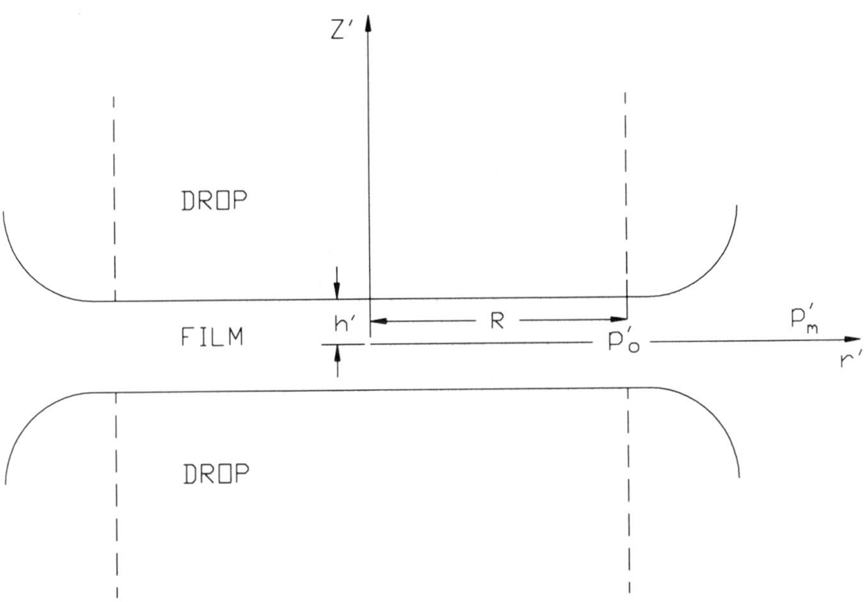

Fig. 6.21 Schematic of thin films between two droplets. (*From Wasan and Malhotra, 1986.*)

sure inside the emulsion droplet. The pressure p_m inside the meniscus is smaller than the pressure inside the droplet by the capillary pressure σ/R, σ and R being the interfacial tension and the drop radius, respectively. Consequently, there is a pressure gradient Δp which tends to drain the continuous phase liquid from the thin film to the meniscus. When because of drainage, the film thickness becomes of the order of a few hundred nanometers, the faces of the plane parallel film will experience an additional pressure due to interactions between the two faces of the film. This disjoining pressure Π, defined as the pressure that tends to "disjoin" or pull the two faces of the film apart, has contributions due to van der Waals, electrostatic, and steric interactions. The pressure drop Δp responsible for film drainage, is now given by

$$\Delta p = \frac{\sigma}{R} - \Pi \tag{41}$$

where

$$\Pi = \Pi_{VW} + \Pi_{el} + \Pi_{st} \tag{42}$$

and

$$\Pi_{VW} = \frac{-A}{6\pi h^3} \tag{43}$$

where A is the effective Hamaker constant and h is the film thickness.

$$\Pi_{el} = 2n_o kT \left\{ \cosh\left(\frac{e\psi_o}{kT}\right) - 1 \right\} \tag{44}$$

where n_o is the number concentration of electrolyte, e is the elementary charge, ψ_o is the potential at the center of the film, k is the Boltzmann's constant, and T is the temperature. ψ_o can further be related to the surface potential ψ_s through

$$\psi_o = \frac{\psi_s}{\cosh(\kappa h/2)} \tag{45}$$

κ being the Debye-Hückel parameter given by Eq. (9). If macromolecules, such as proteins, are employed to stabilize the emulsion, assuming constant segment density distribution, Π_{st} is given by (Narsimhan, 1991) (for $L_s \le h \le 2L_s$)

$$\Pi_{st} = 2kTB_2 \frac{\Gamma^2}{L_s^2} \tag{46}$$

where B_2 is the second virial coefficient of the adsorbed segments of the macromolecule, L_s is the thickness of adsorbed layer and Γ is its surface concentration. Relating the second virial coefficient B_2 to Flory-Huggins χ parameter through Eq. (21), one obtains

$$\Pi_{st} = \frac{2kT(0.5 - \chi)\Gamma^2}{\overline{V}_1 \rho_2^2 L_s^2} \tag{47}$$

for $h \le L_s$,

$$\Pi_{st} = \frac{2kT(0.5 - \chi)\Gamma^2}{\overline{V}_1 \rho_2^2 L_s^2} + \frac{2kT\nu}{h} \tag{48}$$

where ν is the number of adsorbed macromolecules per unit area.

The disjoining pressure is a function of film thickness. A schematic of the variation of Π with film thickness is shown in Fig. 6.22. At large and small film thicknesses, the effect of van der Waals attraction is predominant, so two secondary minima may exist. At intermediate film thicknesses, electrostatic repulsion due to overlap of double layers may be dominant, so there is a maximum. At very small film thickness, there is another maximum because of steric repulsion. The draining film reaches an equilibrium film thickness h_{eq} if the net driving force for drainage is zero, i.e.,

$$\Delta p = \frac{\sigma}{R} - \Pi(h_{eq}) = 0 \tag{49}$$

i.e., $\Pi(h_{eq}) = \sigma/R$. Therefore, the equilibrium film thickness will depend on the interfacial tension, drop size and, more importantly, the nature of the interactions. If $\sigma/R > \Pi^{el}_{\max}$, $\Pi^{el}_{\max}$ being the maximum disjoining pressure due to electrostatic forces, the film will thin until it reaches the thickness x_{F_2} (see Fig. 6.22), at which point it will exhibit a jump transition to x_{F_3} but the equilibrium film thickness will be less than x_{F_3}. Such jump

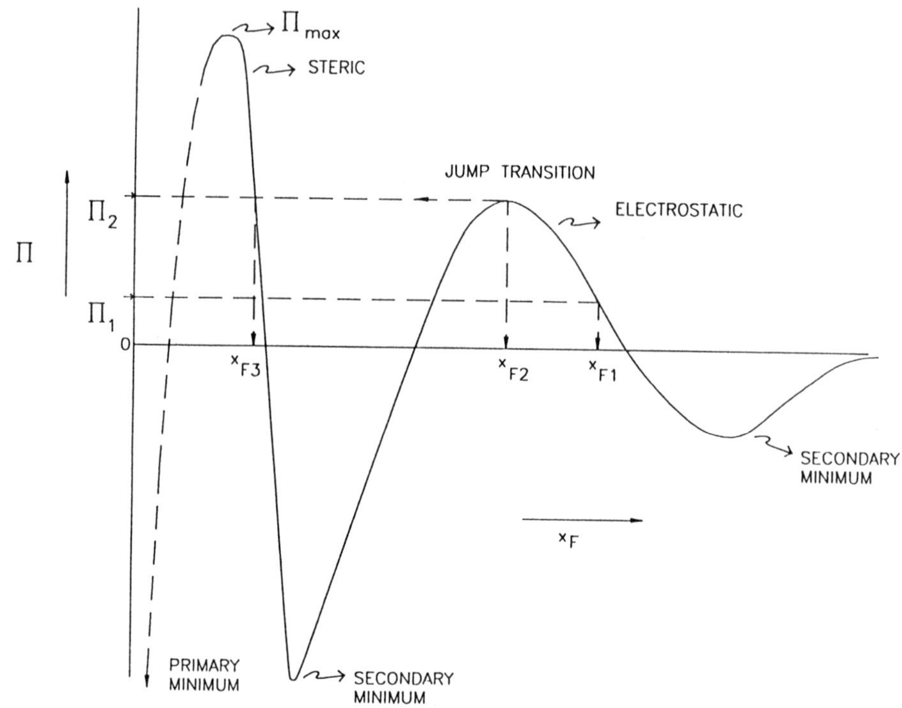

Fig. 6.22 Schematic of the variation of disjoining pressure Π with the film thickness X_F. *(From Narsimhan, 1991.)*

transitions in film thickness have been observed experimentally (Smith and Mitchell, 1976).

As pointed out earlier, high surfactant concentration, especially of proteins, provide high surface viscosity and surface elasticity, so the drop interface is rendered immobile. Under such conditions, the velocity of drainage V_{Re} of thin film is given by Reynolds' equation,

$$V_{Re} = \frac{2\Delta p h^3}{3\mu R^2} \tag{50}$$

μ being the viscosity of the continuous phase.

When the interface is mobile, the velocity of drainage of the film is higher than predicted by Reynolds' equation. For pure emulsions in the absence of any stabilizer, the velocity of film drainage V is given by

(Ivanov and Traykov, 1976, 1971)

$$V = \left(\frac{32 \Delta p^2}{a_1{}^2 \rho_d \mu_d R^4} \right)^{1/3} h^{5/3} \tag{51}$$

where a_1, a weak function of $h \approx 1$, and ρ_d and μ_d are the density and viscosity of the droplet, respectively. It is interesting to note that the velocity of drainage does not depend on the viscosity of the film μ. This occurs because most of the energy dissipation occurs within the droplet as a result of internal circulation created by the mobility of the interface. Also, the functional dependence of drainage velocity with respect to h and Δp is changed by the interfacial mobility.

For emulsions stabilized by a surfactant, two special cases can be considered: a surfactant soluble either in the continuous phase or in the drops. For surfactants soluble in the continuous phase, the velocity of drainage V' is given by (Ivanov, 1980)

$$\frac{V'}{V_{\text{Re}}} = 1 + 1/(\epsilon^f + \epsilon^e) \tag{52}$$

where $\epsilon^f = \dfrac{h}{h_s}$, $h_s = \dfrac{-6\mu D_s[\partial\Gamma/\partial c]}{\Gamma(\partial\rho/\partial c)}$, $\epsilon^e = \left(\dfrac{h}{h_e}\right)^{4/3}$, and $h_e = \left(\dfrac{108R^2\mu^3}{a_1\rho_d\mu_d\Delta p}\right)^{1/4}$. Even for small surfactant concentrations, $\epsilon^e \ll \epsilon^f$ so that

$$\frac{V'}{V_{\text{Re}}} \approx 1 + \frac{1}{\epsilon^f} \tag{53}$$

which is the equation for velocity of drainage of a foam film. Therefore, even a small amount of surfactant suppresses internal circulation and energy dissipation in the droplet. Consequently, V' depends on μ, *not* on μ_d, when sufficient concentration of surfactant is present. In Eq. (52), D_s is the surface diffusion coefficient of the surfactant, Γ is the surface concentration, and σ is the interfacial tension.

When the surfactant is only soluble in the droplet, it does *not* affect the velocity of film drainage (Ivanov, 1980), which is the same as that for a system without any surfactant, i.e., the velocity of film drainage V'' for this case is given by Eq. (51). The velocities of thinning of the film for two systems differing only in the solubility of the surfactant may differ by orders of magnitude, since

$$\frac{V'}{V''} = \epsilon^{e} \ll 1 \tag{54}$$

Since a surfactant that is soluble only in the drop phase does not stabilize an emulsion, i.e., change the rate of thin film drainage, the emulsion will coalesce quickly and invert to form an emulsion in which the surfactant is soluble in the continuous phase, consistent with Bancroft's rule.

Drainage time of thin films for different concentrations of the surfactant sodium lauryl sulphate (SLS) measured by Rao et al. (1982) is shown in Fig. 6.23. As expected, the surface tension decreased with increase in SLS concentration, eventually becoming constant above the critical micelle concentration (CMC). The drainage time initially increased with SLS concentration below the CMC because of the decrease in capillary pressure, the driving force for film drainage. The critical thickness decreased

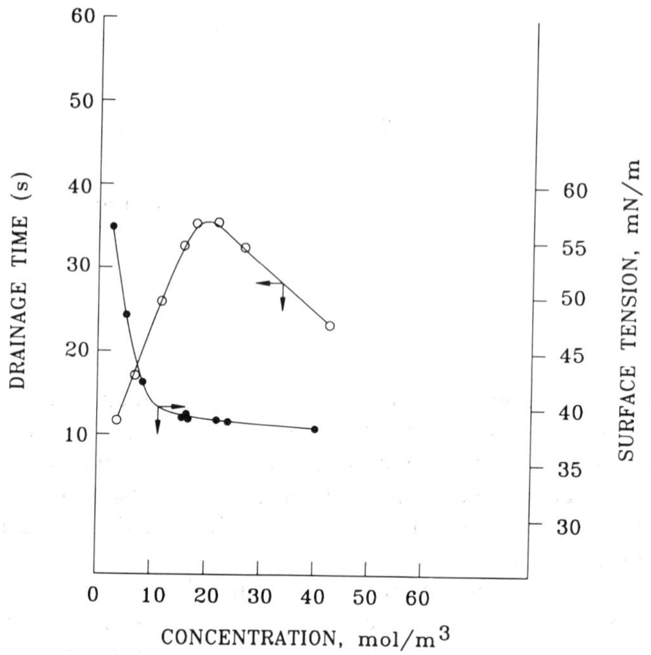

Fig. 6.23 Drainage time and surface tension as a function of SLS bulk concentration. 'K' = ∞, C_{el} = 0.25 kmol/m^3 NaCl, R = 9 × 10^{-1} mm. (*From Rao et al., 1982.*)

as well. At the CMC, the film surface is saturated with the surfactant. Any expansion of the surface leads to an increase in the surface tension because of the decrease in the surface concentration. Above the CMC, this surface depletion of surfactant due to surface expansion is quickly rectified by diffusion of surfactant from the bulk. Consequently, the drainage time was found to decrease with SLS concentration above CMC. Experimental data of drainage time of emulsion films as a function of film radius obtained by Manev et al. (1984), shown in Fig. 6.24, indicate that drainage time is proportional to $R^{0.8}$. The power is much smaller than 2, predicted by Reynolds equation, indicating that the interface is mobile. The effect of surface elasticity, E_S, on drainage time is shown in Fig. 6.25 (Wasan and Malhotra, 1986). The elasticity number E_S is a measure of relative importance of surface concentration gradient caused by surface expansion and bulk diffusion, which tends to nullify such gradient. At high elasticity number ($E_S > 10^6$), the velocity of thinning is given by the Reynolds equation and therefore is independent of interfacial viscosity.

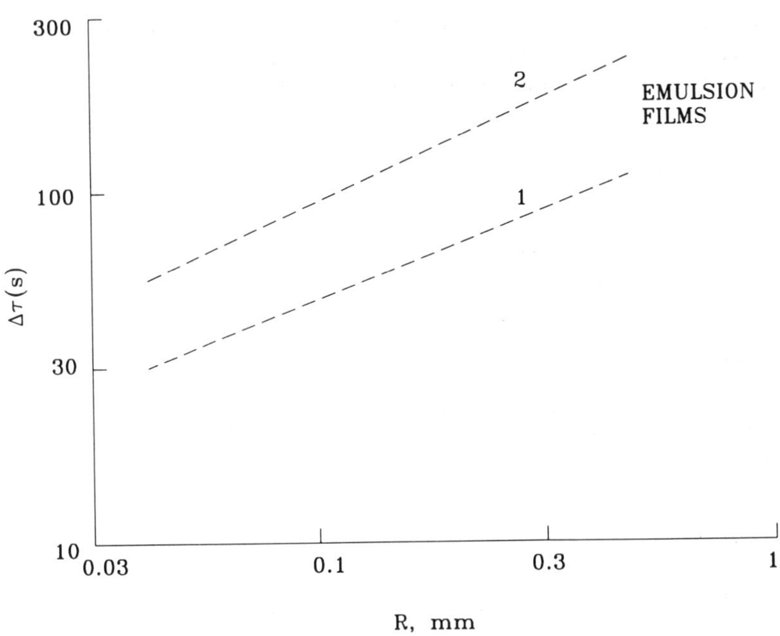

Fig. 6.24 Drainage time Δt from a thickness $h_0 = 200$ nm to a constant thickness $h = 50$ nm for emulsion films as a function of film radius R. (1) 5.6×10^{-4} kmol/m^3 SLS + 0.3 kmol/m^3 NaCl, (2) 4.3×10^{-4} kmol/m^3 SLS + 0.1 kmol/m^3 NaCl. (*From Manev et al., 1984.*)

For smaller E_S, however, the time of drainage increases with interfacial viscosity, eventually becoming equal to that predicted by Reynolds equation at sufficiently large interfacial viscosity. Such behavior is due to the velocity of drainage being much higher at lower interfacial viscosities because of increased mobility of the interface. Note that the dependence of drainage time on interfacial viscosity is more pronounced at smaller elasticities. Fischer and Mitchell (1989) measured the thickness of draining thin films between triglyceride droplets stabilized by the proteins lysozyme and β-casein. Film thickness as a function of time is shown in Fig. 6.26 for lysozyme-stabilized film. The film drains to a constant metastable thickness and remains at this thickness for times that can vary from minutes to hours before eventual abrupt collapse, leading to droplet coalescence. The metastable film thickness decreased with increase in salt concentration and agreed with the values calculated from a simple balance of capillary, van der Waals and electrostatic forces. Fischer and Mitchell (1989) found that the lifetime of the metastable film depended on its age and thickness; thicker films had greater lifetimes. For adsorbed films less than 5 minutes old, coalescence occurred within a few seconds. On the other hand, the lifetime of the draining film was many hours for adsorbed films greater than 2 hours old. This was attributed to the mechanical strength of the interface, being higher for aged films. In contrast,

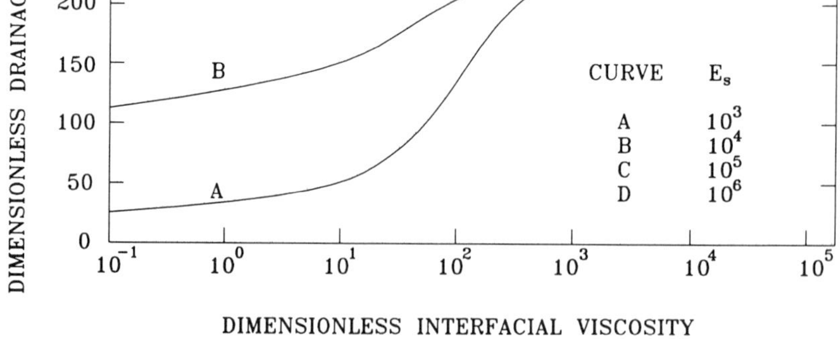

Fig. 6.25 Effect of interfacial viscosity number, V_s, on drainage time, t for various values of elasticity number, E_s. (*From Wasan and Malhotra, 1986.*)

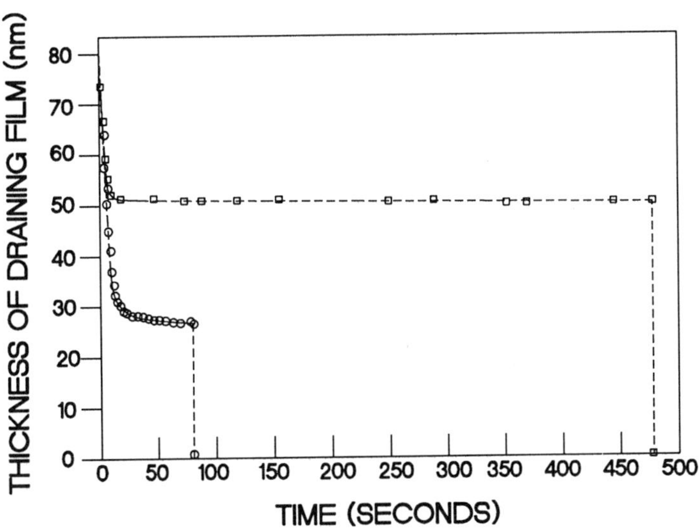

Fig. 6.26 Thickness of draining film as a function of time for 10^{-3}% w/w lysozyme in aqueous solution between triglyceride droplets. Squares, aqueous solution contains 2×10^{-3} M NaCl; circles, aqueous solution contains 10^{-2} M NaCl. (*From Fisher and Mitchell, 1989*).

the thickness of films stabilized by β-casein decreased monotonically for several minutes, after which they collapsed, as shown in Fig. 6.27. Collapse of fresh films generally resulted in immediate coalescence. For aged films, however, collapse occurred at a film thickness of 39 ± 4 nm, which decreased the film thickness abruptly to 4 ± 2 nm, a thickness corresponding to a monolayer of β-casein. The monolayers of the two interfaces remained in adhesive contact for times ranging from 0.25 to 10 min.

Thin Film Stability

Drainage of thin liquid films by itself cannot lead to coalescence since the ratio of the viscous to inertial forces becomes unbounded as films become infinitesimally thin. Consequently, coalescence has to be preceded by film rupture. Stability of thin films to external perturbations has been investigated extensively (see Scheludko, 1962; Vrij, 1966; Ivanov et al., 1970; Felderhoff, 1968; Ruckenstein and Jain, 1974).

Fluctuations of film thickness can be represented as a superposition of an infinite number of waves of different frequencies and amplitudes. External perturbation gives rise to (1) a restoring force caused by the local

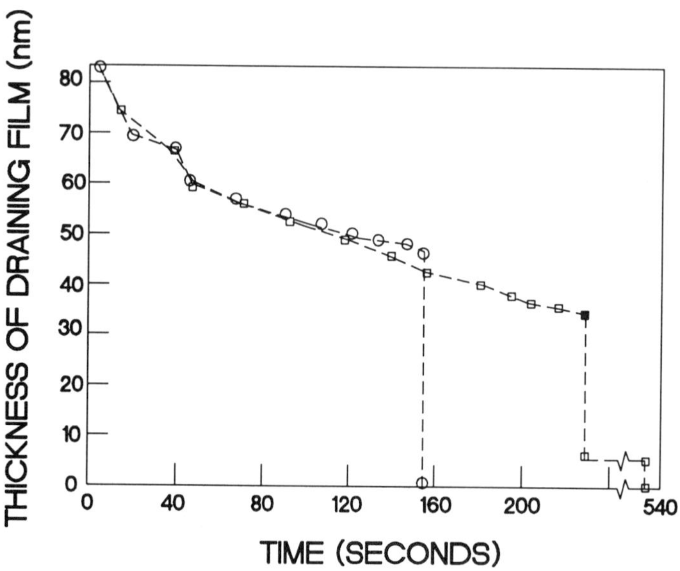

Fig. 6.27 Thickness of the draining film as a function of time for an aqueous solution of β-casein (10^{-3}% w/w). (○-○-○), Collapse of draining film leads to coalescence; (□-□-□), collapse of draining film leads to adhesion, with coalescence occurring ca. 320 sec later. (*From Fischer and Mitchell, 1989.*)

capillary pressure, which tends to smooth the film surface; (2) a negative disjoining pressure, which tends to deform the film further; and (3) a positive disjoining pressure due to electrostatic and steric forces, which tends to stabilize the film. There exists a film thickness below which perturbations grow leading to eventual rupture of a film. A necessary condition for spontaneous growth of perturbations is

$$\frac{d\pi_A}{dh} > \frac{d\pi_{el}}{dh} + \frac{d\pi_{st}}{dh} + \frac{d\pi_\gamma}{dh} \tag{55}$$

where π_A, π_{el}, and π_{st} refer to van der Waals, electrostatic, and steric contribution of disjoining pressure and h is the film thickness. π_γ refers to local variations in the disjoining pressure due to local rises in interfacial tension resulting from local fluctuation in thickness, and is given by

$$\pi_\gamma = \frac{64\gamma\Delta h}{\lambda^2} \tag{56}$$

where Δh is the fluctuation amplitude of wavelength λ. Scheludko (1962) derived the following expression for h_{crit}, the critical thickness of a film where only van der Waals forces act,

$$h_{crit} = \left(\frac{A_H \lambda^2}{128\pi\gamma}\right)^{1/4} \tag{57}$$

A_H being the Hamaker constant.

For large film thicknesses, $\pi_A \ll \pi_\gamma$, Vrij (1966) derived the following expression for critical thickness:

$$h_{crit} = 0.268 \left(\frac{A_H{}^2 R^2}{\gamma\pi_\gamma f}\right)^{1/2} \tag{58}$$

where R is the radius of the film and f is a factor that depends on h. For small thicknesses, where $\pi_\gamma \ll \pi_A$ (Vrij, 1968);

$$h_{crit} = 0.22 \left(\frac{A_H R^2}{\gamma f}\right)^{1/4} \tag{59}$$

Both Scheludko's and Vrij's expressions for critical thickness have the same form for small film thickness. Ivanov et al. (1974) also derived expressions for the critical thickness of rupture of a draining film accounting for the growth of perturbations during film drainage.

MacRitchie (1976) postulated that an energy barrier associated with the formation of a "hole" of critical radius in both interfaces must be overcome for the rupture of thin films. Formation of a "hole" requires compressional energy to create a clean interface of certain dimensions. The compressional energy barrier ΔG^* is given by

$$\Delta G^* = \frac{(\delta A_S)^2 \epsilon}{2A_s} \tag{60}$$

where A_S is the droplet surface area, δA_S is twice the critical area, and the modulus ϵ is defined by

$$\epsilon = \frac{-d\pi_S}{d\ ln\ A_x} \tag{61}$$

π_S being the surface pressure. MacRitchie (1976) pointed out that the formation of clean interface due to compression of adsorbed monolayer of macromolecules, such as protein, depends on the local value of ϵ during an encounter between two colliding droplets, which is influenced by the rearrangement and desorption of adsorbed loops and trains. MacRitchie's approach is likely to be valid for small droplets, where Brownian aggregation is predominant and coalescence occurs during collisions of short duration (Dickinson and Stainsby, 1982). Coalescence due to drainage and rupture of thin films, on the other hand, will generally occur for larger droplets.

The dependence of critical thickness of rupture of thin film on the concentration of sodium lauryl sulfate, as measured by Rao et al. (1982), is shown in Fig. 6.28. The critical thickness of rupture decreased with increasing SLS concentration, reaching a constant value above the CMC.

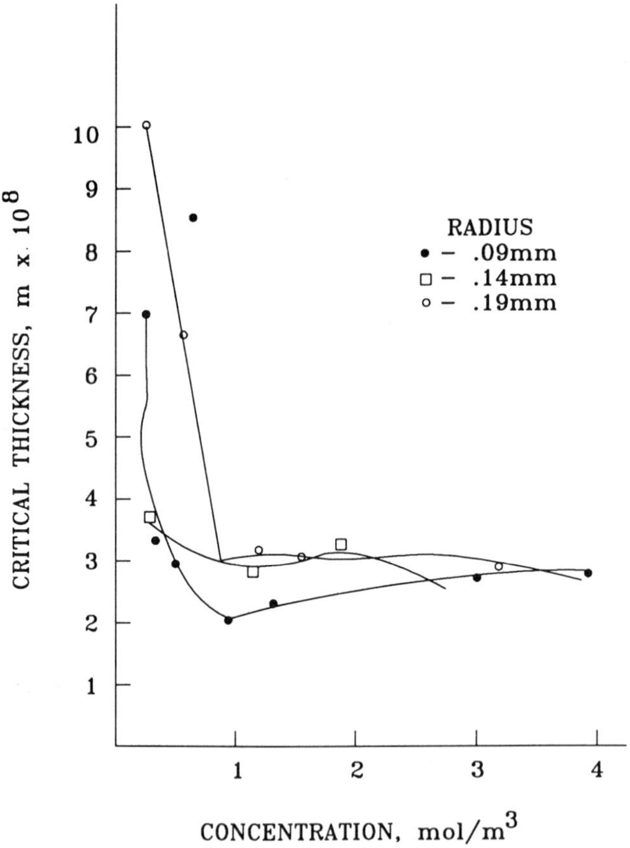

Fig. 6.28 Critical thickness as a function of SLS bulk concentration C_{el} = 0.1 kmol/m^3 NaCl. (*From Rao et al., 1982.*)

Since the amplitude of surface waves responsible for film rupture are large at low SLS concentrations, the film ruptures at a large critical thickness. At higher SLS concentrations, the waves damp so the film can drain to a much smaller thickness before rupture. Critical thickness of rupture also decreased as film radius decreased. As pointed out earlier, lysozyme stabilized films between two triglyceride droplets drained to an equilibrium thickness, before eventual collapse leading to coalescence. Films stabilized by β-casein, however, drain until they reach a critical thickness and then collapse. Globular proteins such as lysozyme, therefore, provide better stability for thin films, mainly due to better mechanical properties of adsorbed protein layer.

Coalescence of Drops

There have been many experimental investigations of coalescence of droplets at planar interfaces. Such studies provide information not only about the stability of droplets to coalescence, but also about mechanisms of coalescence. In most such experiments, a liquid droplet is released into an immiscible liquid and allowed to come to rest at a plane interface of the same liquid as the droplet. The coalescence time is recorded. The median coalescent time ($\tau_{1/2}$) of sufficiently large sample of droplets is a measure of drop stability. A detailed review of such investigations can be found in Jeffreys and Davies (1971). Figure 6.29 is a typical plot of fraction of droplets with lifetime exceeding time t for orange oil droplets at an oil-water interface in the presence of carboxymethyl cellulose as reported by Siber and Mizrai (1975). Coalescence is not observed for times shorter than t_0, after which coalescence occurs as a statistical process. Gillispie and Rideal (1956) derived a semi-empirical relationship to describe such behavior,

$$\ln \frac{N}{N_0} = -K_c(t - t_0)^{3/2} \tag{62}$$

N_0 is the total number of drops, N is the number with lifetime greater than t, and K_c is a coalescence constant. That coalescence is not observed prior to t_0 indicates that a certain time is needed for the film to drain to a critical thickness and that coalescence is extremely unlikely for film thicknesses greater than the critical thickness. Charles and Mason (1960) demonstrated through high-speed photography that the rupture of thin film does not always occur either at the same position or at the same film thickness. Stability of droplets to coalescence was found to decrease at

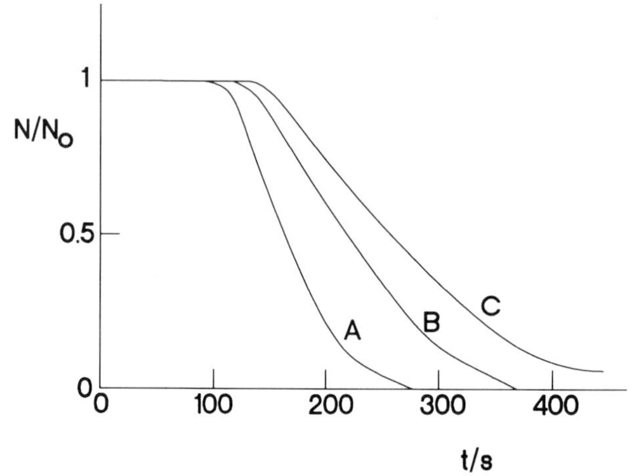

Fig. 6.29 Lifetimes of orange-oil droplets at an interface of oil and an aqueous solution of carboxymethylcellulose (CMC) at 25°C. The fraction N/N_0 of drops with lifetime exceeding t is plotted against t for various concentrations of CMC: A, 10^{-5} wt%; B, 10^{-3} wt%; C, 0.1 wt%. (*From Silber and Mizrai, 1975.*)

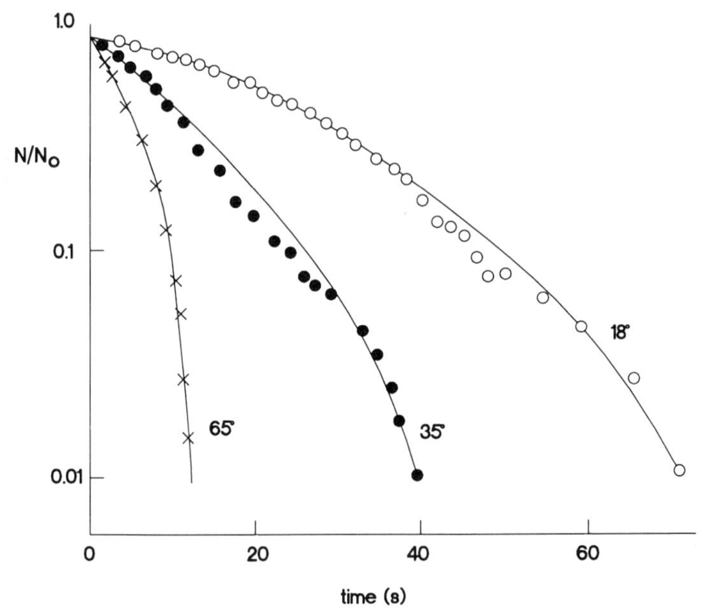

Fig. 6.30 Influence of temperature on coalescence rate. (*From Gillespie and Rideal, 1956.*)

higher temperature, as shown in Fig. 6.30 (Gillispie and Rideal, 1956). Even though the average lifetime of a droplet indicates its stability, quantitative analysis of drainage and stability requires knowledge of exact shape of the droplet as it approaches the interface. A detailed account of equations for drainage and stability of drops is given by Tadros and Vincent (1983). In a recent study, Dickinson et al. (1988) investigated coalescence stability of emulsion droplets in protein solutions at planar interface for β-casein, κ-casein, and lysozyme. A typical plot of average coalescence time as a function of drop diameter for lysozyme is shown in Fig. 6.31. The average coalescence time decreased as drop size increased. Also, no difference in the coalescence time was observed for 20-min emulsion droplets stabilized by β-casein, κ-casein, and lysozyme. When droplets were aged for 72 hours, however, differences in the stabilizing ability of these proteins were lysozyme > κ-casein > β-casein. As shown in Fig. 6.32, the surface shear viscosity of the lysozyme ad-

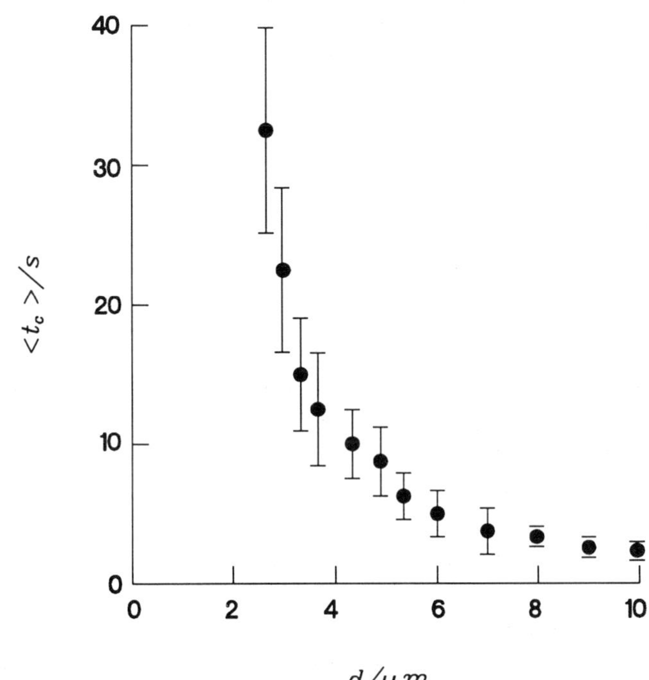

Fig. 6.31 Average coalescence time $\langle t_c \rangle$ as a function of droplet diameter *d* for a 20 minute old lysozyme system. Bars on point refer to standard deviations about the mean. (*From Dickinson et al., 1988.*)

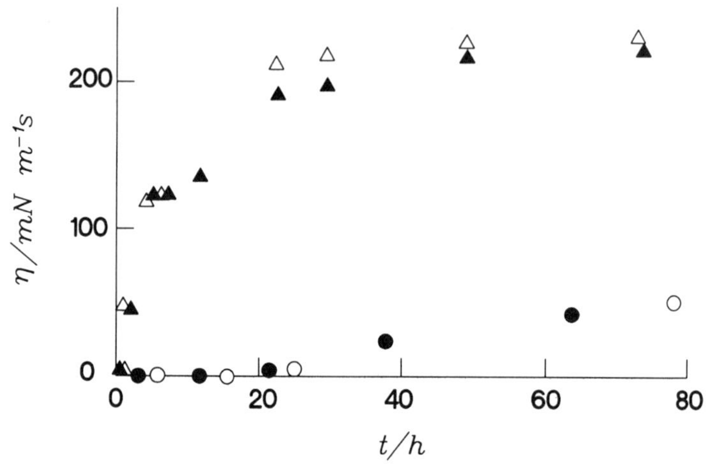

Fig. 6.32 Time-dependent surface shear viscosity of protein films adsorbed at the *n*-hexadecane-water interface (10^{-4} wt%, pH 7, ionic strength 0.1 mol dm^{-3}, 25°C). The viscosity η is plotted against adsorption time *t*: ○, ●, κ-casein; △, ▲, lysozyme. Values for β-casein are not shown, being indistinguishable (on this scale) from η = 0. (*From Dickinson et al., 1988.*)

sorbed layer was much higher than that of κ-casein, the surface shear viscosity of β-casein adsorbed layer being almost zero (Dickinson et al., 1988). At small times, however, the surface shear viscosity of lysozyme adsorbed layer was almost zero, possibly due to a lag in its adsorption. It is therefore evident that the observed differences in the stabilizing action of the three proteins are directly related to the mechanical properties of the interfacial layer. Stability was also found to be highly sensitive to ionic strength and bulk protein concentration. Stability was enhanced at lower ionic strengths because of increased electrostatic repulsion. Probability of coalescence was lower for higher protein concentration and after aging of the interfacial layer.

Stabilization of Emulsions by Solid Particles

Finely divided solids have long been known to stabilize emulsions. Such emulsions are sometimes referred to as pickering emulsions. When an emulsion is stabilized by solid particles, the phase that wets the solid particle is usually the continuous phase. Stabilization by solid particles is important in food emulsions since monoglycerides, the commonly employed emulsifier, crystallizes, thus forming solid particles at the interface

(Friberg, 1976). The mechanism of stabilization by solid particles is not well understood. However, in order for stabilization to occur, the surface of the droplet should be covered by a monolayer of fine solid particles; i.e., the solid particles should be much finer than the droplets for stabilization to occur. It is believed that these solid particles prevent drainage of thin film of continuous phase liquid between the droplets. It has been shown by Schulman (1954) that emulsions are most stable when the contact angle of the solid particle is close to 90°.

Stabilization of emulsion by solid particles can best be explained in terms of free energy considerations. The free energy change in moving a spherical solid particle of radius a and contact angle θ from the interface into the phase that is preferentially wetted by the particle is given by (Tadros and Vincent, 1982)

$$\Delta G = \pi a^2 \gamma_{o/W} (1 - \cos \theta)^2 \tag{63}$$

$\gamma_{o/W}$ being the interfacial tension of oil and water. This change in the free energy is always positive, being maximum when $\theta = 90°$ and zero when $\theta = 0$. The variation of ΔG with contact angle is shown in Fig. 6.33. As

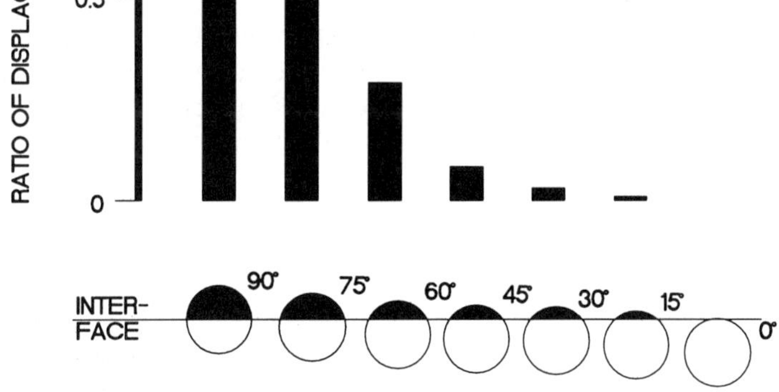

Fig. 6.33 The energy to move an adsorbed particle into the most wetting phase is strongly dependent on the contact angle between the two liquids and the solid surface. (*From Friberg, 1976.*)

pointed out earlier, displacement of solid particles from the interface is necessary for coalescence. Consequently, coalescence is prevented by the increase in the free energy for particle displacement.

Fat Crystallization

In many food oil-in-water emulsions, fat crystallization occurs. Formation of fat crystals in the oil phase destabilizes emulsions. In dressings, for example, crystallization of oil phase leads to visible oil separation, whereas some crystallization of fat is beneficial for whippability in creams and ice creams (Darling and Birkett, 1987). Van Boekel (1980) investigated the mechanism of emulsion destabilization by fat crystals and concluded that destabilization occurs due to penetration of interfacial adsorbed layer of surfactant by fat crystals leading to bridging of droplets. Because of the small radius of curvature at the point of contact, bridges grow rapidly due to interfacial forces, thus forming a strong sintered bond between the droplets (Darling and Brikett, 1987). The mechanism of penetration of interfacial layer by crystals is shown schematically in Fig. 6.34. Because of the radius of curvature of small emulsion droplets, fat remains supercooled below its melting point. This supercooling eventually leads to rapid nucleation of fine crystals. Penetration of the interfacial layer by these crystals depends on the wettability of the continuous phase by the crystals. If the crystals preferentially wet the oil phase, no penetration occurs. Crystal size, shape, and orientation are also important factors. Larger crystals are more likely to penetrate the interfacial layer. Mechanical deformation of crystals, which favor stabilization, depends on the composition and the type (α or β) of crystals. Thickness of the interfacial adsorbed layer also plays an important role in the destabilizing action of crystals. As will be pointed out later, low molecular weight surfactants tend to displace proteins from interfaces thus reducing the thickness of the interfacial layer, thereby leading to reduction in emulsion stability.

Liquid Crystalline Structures at Interfaces

At the surface of emulsion droplets, the three-component system—oil, surfactant, and water—can form ordered liquid crystalline structures due to weak interactions between liquid molecules and emulsifiers. Such liquid crystalline structures are encountered in food emulsions. The importance of liquid crystalline structures in the stability of food emulsions has been demonstrated by Friberg and coworkers (1969, 1971). Different types of such ordered structures are shown in Fig. 6.35. The binary water + emulsifier system forms a hexagonal I mesophase (M) when water and emulsifier concentrations are about equal. At low emulsifier concentrations, a micellar solution (L_1) is formed. Adding oil results first in the formation of lamellar mesophase (N) and then a hexagonal II mesophase (RM). At high concentrations, the ternary mixture forms a reversed micellar solution. The type of liquid crystalline structure depends on the composition, type of emulsion (O/W or W/O) and temperature. For example, lamellar structures are favored in O/W emulsions, whereas hexagonal II mesophases are favored in W/O emulsions (Krog and Lauridsen, 1976).

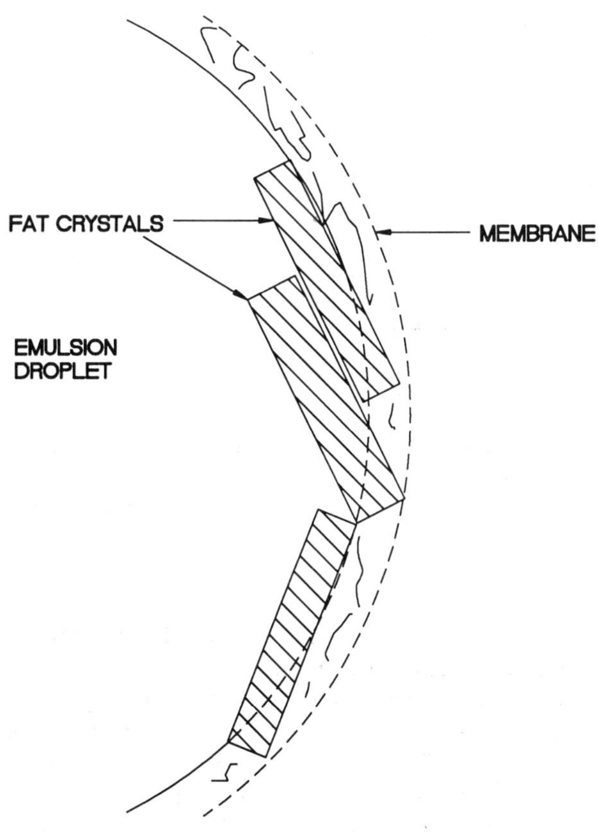

Fig. 6.34 Schematic representation of fat crystals penetrating the interfacial membrane at the surface of an emulsion droplet. (*From Darling and Birkett, 1986.*)

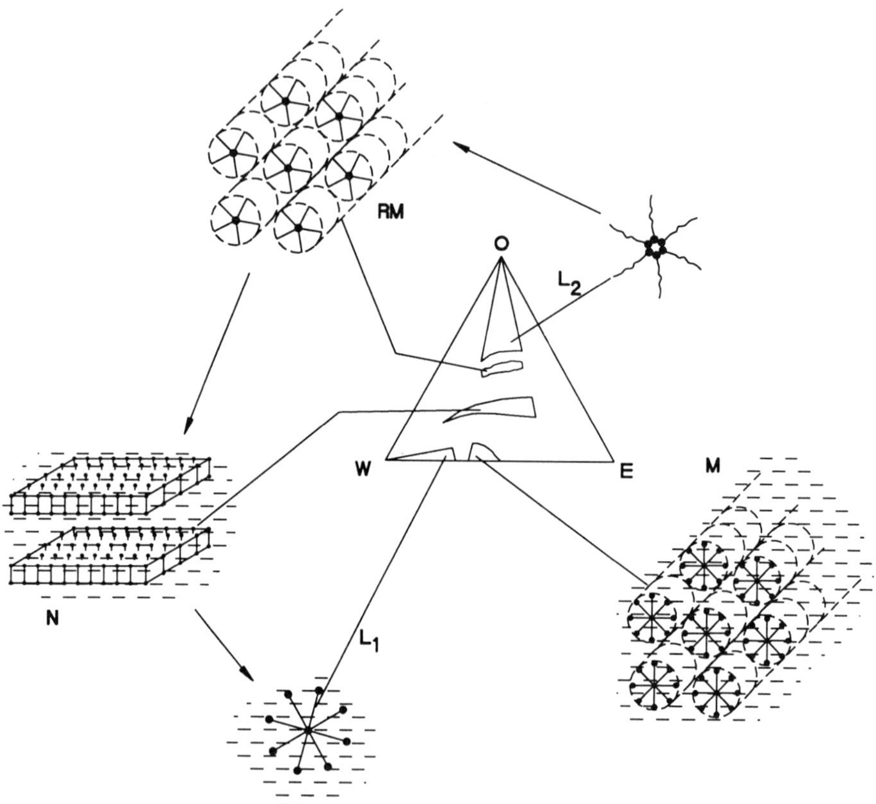

Fig. 6.35 Normal and reversed micelles and different liquid crystalline structures in a system of water-surface active substance–amphiphilic substance. L_1 and L_2 are micellar solutions, M, N, and RM are liquid crystals, with hexagonal I, lamellar, and hexagonal II structures, respectively. (*From Friberg, 1976.*)

It is important to characterize the ternary phase diagram for such systems. The importance of oil-emulsifier interaction in the formation of liquid crystalline structures has been demonstrated by Friberg et al. (1976), who showed that hexadecane + water + polyoxyethylene nonyl phenol ether resulted in coexistence of two phases, one aqueous and the other hydrocarbon. When the aliphatic hexadecane was replaced by aromatic *p*-xylene, however, a single liquid crystalline phase was formed.

Liquid crystalline structures possess bulk viscosities that are sometimes an order of magnitude higher than that of oil and water. Such high viscosities tend to substantially reduce rates of creaming and flocculation.

Presence of liquid crystalline structures at the interface provides surface mechanical properties that retard thin film drainage and rupture. High surface elasticity of such structures has been demonstrated by Boyd et al. (1976).

Stabilization of emulsion is also aided by reductions in van der Waals attractive forces due to liquid crystalline structures. Consider, for example, the steps involved in coalescence of emulsion drops having liquid crystalline structures at the interface as shown schematically in Fig. 6.36. Stage I refers to two water droplets in W/O emulsion with liquid crystalline structures separated by a distance d_o at their interfaces. Stage II refers to flocculated state where the distance between the droplets is reduced from d_o to m_o ($m_o \ll d_o$), m_o being the thickness of hydrophobic layers in the liquid crystalline phase. Stage III refers to coalescence where the water droplets are separated by an oil layer of thickness m_o. This can be attained through the removal of intervening liquid crystalline layers one

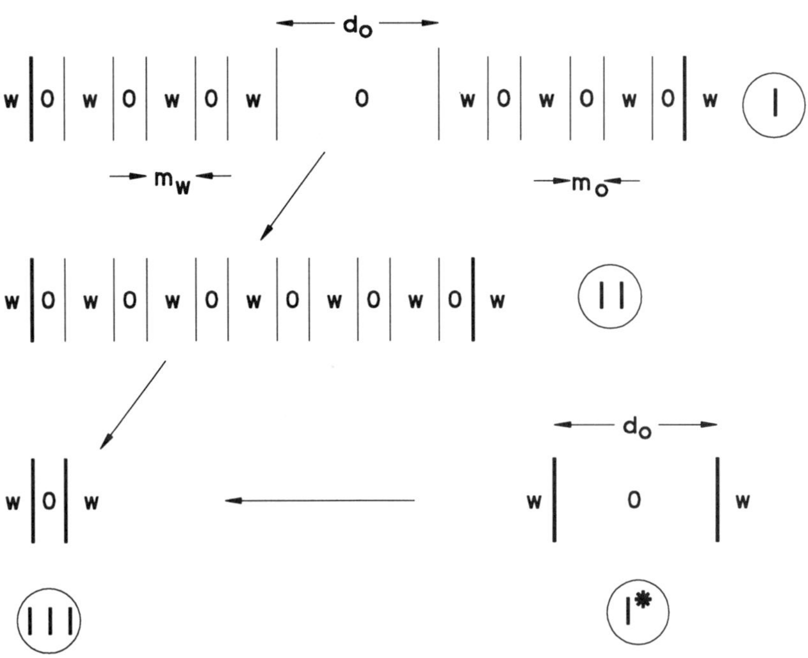

Fig. 6.36 Stages in coalescence of two water droplets covered with lamellar liquid crystal (three bilayers on each droplet surface are illustrated). The distances m_o and m_w represent the thicknesses of oil and water layers; d_O is the initial surface-to-surface separation of the covered droplets. (*From Friberg et al., 1976.*)

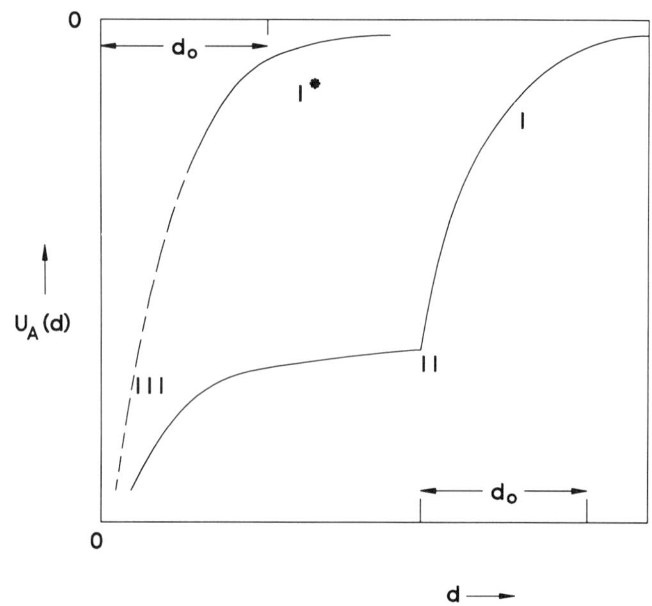

Fig. 6.37 Effect of liquid crystalline layer on dispersion interaction between two water droplets. The potential energy $U_A(d)$ is shown as a function of the particle separation d. (From Friberg et al., 1976.)

at a time. Of course, coalescence is assumed to occur due to rupture of this oil layer. Stage I* refers to two water droplets without liquid crystalline layers separated by a distance d_o. Consequently, coalescence can occur either from I → II → III or from I* → III. The variation of $U_A(d)$, the potential energy of van der Waals interaction, with d the distance of separation of water droplets, with and without liquid crystalline structures is shown in Fig. 6.37 (Friberg et al., 1976). In the flocculation stage, the variation of $U_A(d)$ for both the cases are similar. In the coalescence stage, however, $U_A(d)$ changes very slowly with distance for droplets covered with liquid crystalline layers. Consequently, the droplets experience much smaller van der Waals attractive force, thus stabilizing the emulsion.

FOOD EMULSIFIERS

Different types of surfactants are used as emulsifiers in food emulsions. Table 6.4 lists different types of emulsifiers, emulsion type, and appli-

Table 6.4 Important Food Surfactants and Their Typical Applications

General classification	Surfactant	Emulsion type	Application
Protein	Caseinate	O/W	Ice cream
	Whey protein	O/W	Cake batter
	Egg protein	O/W	Mayonnaise
Polysaccharide	Methyl cellulose	O/W	Artificial cream
	Propylene glycol alginate	O/W	Salad dressing
Small-molecule surfactants	Monoglycerides	W/O	Margarine
	Acid esters of monoglycerides	W/O & O/W	Bakery products
	Sorbitan fatty acid esters	W/O	Confectionery
	Polyoxyethylene derivatives of sorbitan fatty acid esters	O/W	Toppings
	Lecithin	O/W & W/O	Milk powder

Source: Darling and Birkett, 1986.

cations (Darling and Birkett, 1987). The most important criterion for food emulsifier is that it be nontoxic; therefore it should be approved by regulatory agencies. The emulsifiers can be broadly classified as small-molecule surfactants, polysaccharides, and proteins. Monoglycerides and their esters, fatty acids and their esters, and phospholipids are different types of small-molecule surfactants. Monoglycerides tend to adsorb at triglyceride-water interfaces. This strong adsorption along with their ability to form lamellar liquid crystalline structures makes them good stabilizing agents in food emulsions. As pointed out earlier, liquid crystalline structures stabilize emulsions through different mechanisms—increase in bulk viscosity, reduction in van der Waals attraction, and enhancement of surface mechanical properties. The surface mechanical properties of the interfacial adsorbed layer, in general, depend on the efficiency of packing of the surfactant molecules at the interface. Fatty acids and their esters can form either O/W or W/O emulsions depending on the length of the hydrocarbon chain, i.e., the HLB ratio. Long hydrocarbon chains, for example, fatty acids and their esters, tend to stabilize W/O emulsions. Phospholipids also form liquid crystalline structures, and their mechanism of stabilization is similar to that of monoglycerides. The most commonly employed phospholipid in food emulsions is lecithin, which is derived from eggs and soybean. Detailed discussions of different applications of emulsifiers in food are given by Krog et al. (1986), Jaynes (1986), and Dickinson and Stainsby (1982).

Methyl cellulose and propylene glycol alginate, and other gums, such as carboxy methyl cellulose, guar gum, and xanthan gum, are used to stabilize the emulsions. Polysaccharides stabilize the emulsions mainly because they increase the viscosity of the continuous phase, thereby retarding creaming, flocculation, thin film drainage, and coalescence. At very low concentration, they can destabilize the emulsion through depletion flocculation.

Caesinate, whey protein, egg protein, and soy protein are some of the commonly employed proteins in food emulsions. A detailed review of protein stabilized foams and emulsions is given by Halling (1981). Proteins are invariably involved in stabilizing food emulsion. Because they contain both hydrophobic and hydrophilic functional groups, proteins are inherently surface active, i.e., they tend to adsorb at air/water and oil/water interfaces. An interfacial adsorbed protein layer stabilizes the emulsion via two mechanisms. First, the adsorbed protein layer provides steric repulsion between droplets, resulting in an energy barrier to flocculation and subsequent coalescence. Second, adsorption of protein provides interfacial viscosity and interfacial elasticity high enough to retard film drainage and rupture. The surface activity and stabilizing ability of different proteins depend on their tertiary structure. Proteins, being macromolecules, tend to adsorb much more slowly than small molecule surfactants. Consequently, the kinetics of adsorption of proteins are important. In addition, competitive adsorption of mixture of proteins and surfactants is extremely important in complex interaction in such systems. These topics will be discussed later.

Hydrophile-Lipophile Balance

Hydrophile-lipophile balance (HLB) is an empirical scale devised by Griffin (1949) to enable one to choose an appropriate surfactant for formulation of emulsions. This scale is based on the relative percentage of hydrophilic and hydrophobic functional groups in the surfactant molecule. Surfactants with HLB numbers in the range 4–6 are suitable for stabilizing W/O emulsions, whereas those with HLB numbers in the range of 8–18 are suitable for O/W emulsions. A summary of HLB ranges and their applications is given in Table 6.5 (Tadros and Vincent, 1983). Table 6.6

Table 6.5 A Summary of HLB Ranges and Their Application

HLB range	Application
3–6	W/O emulsifier
7–9	Wetting agent
8–18	O/W emulsifier
13–15	Detergent
15–18	Solubilizer

Source: Tadros and Vincent, 1983.

lists HLB values for some food emulsifiers (Dickinson and Stainsby, 1982). The original experimental method for determining HLB numbers is quite laborious (Griffin, 1949). Alternative experimental methods are discussed by Tadros and Vincent (1983). Davies (1957) devised a group contribution technique for evaluating HLB of surfactant molecules by assigning group numbers to different functional groups and by using the following equation:

$$HLB = 7 + \sum_i n_H(i) - \sum_j n_L(j), \tag{64}$$

where $n_H(i)$ and $n_L(j)$ are the group numbers of hydrophilic group i and

Table 6.6 HLB Values n_{HL} of Some Food Emulsifiers

Emulsifier	n_{HL}
Sorbitan tristearate (Span 65)	2.1
Glycerol Monostearate	3.8
Sorbitan monooleate (Span 80)	4.3
Propylene glycol monolaurate	4.5
Succinic acid ester of monoglycerides	5.3
Sorbitan monopalmitate (Span 40)	6.7
Sorbitan monolaurate (Span 20)	8.6
Diacetyl tartaric acid ester of monoglycerides	9.2
Polyoxyethylene sorbitan monostearate (Tween 60)	14.9
Polyoxyethylene sorbitan monopalmitate (Tween 40)	15.6
Polyoxyethylene sorbitan monolaurate (Tween 20)	16.7
Sodium oleate	18.0
Sodium stearoyl-2-lactylate	21.0

Source: Dickinson and Stainsby, 1982.

hydrophobic group j, respectively. The group numbers of different functional groups are given in Table 6.7 (Davies, 1953). It should be noted that the group numbers of ionic functional groups listed in Table 6.7 depend on the ionic strength of the aqueous phase.

Davies (1957, 1961) related the HLB value of a surfactant molecule to the relative coalescence rates of O/W and W/O emulsions. Accounting for electrostatic and hydration energy barriers, the ratio of the rates of coalescence in W/O and O/W emulsions $K_{W/O}$ and $K_{O/W}$, respectively, is given by

$$\frac{K_{W/O}}{K_{O/W}} = \frac{A_2}{A_1} \exp\left(\frac{B\psi_o^2 + \theta \sum E_h - 2m\omega\theta}{RT}\right) \tag{65}$$

where A_2 and A_2 are the hydrodynamic collision factors for oil and water droplets, respectively. ψ_o is the surface potential, θ is the fractional surface coverage of surfactant, $\sum E_h$ is the energy barrier to hydration, m is the number of $—CH_2—$ groups in the surfactant, and ω is the energy barrier involved in the transfer of a $—CH_2—$ group into water. Substituting appropriate values for the constants, Eq. (65) reduces to (Tadros and Vincent, 1983)

$$\ln \frac{A_1 K_{W/O}}{A_2 K_{O/W}} = 2.2\theta(HLB - 7) \tag{66}$$

which relates the relative rate constants for coalescence to HLB. The collision factors in the above equation are directly proportioned to the volume fraction and inversely to the viscosity of the continuous phase. Therefore, for water and oil phases of the same viscosity and an emulsion

Table 6.7 Hydrophilic Group Numbers

Group	n_H	Group	n_H
$—SO_4Na$	38.7	—COOH	2.1
—COOK	21.1	—OH (free)	1.9
—COONa	19.1	—O—	1.3
Tertiary amine	9.4	—OH (sorbitan)	0.5
Ester (sorbitan)	6.8	$—(CH_2—CH_2—O)—$	0.33
Ester (free)	2.4	$—(CH_2—CH_2—CH_2—O)—$	−0.15[a]

[a] The negative value denotes the group is lipophilic.
Source: Davies, 1957.

of 50% volume fraction, Eq. (66) reduces to

$$\ln \frac{K_{W/O}}{K_{O/W}} = 2.2\theta(HLB - 7) \tag{67}$$

For an HLB of 7, therefore, neither a W/O nor a O/W is preferentially formed. For HLB < 7, however, $K_{W/O} < K_{O/W}$. As a result, HLB < 7 leads to the formation of a W/O emulsion.

Even though HLB values provide a rough guide for the appropriateness of a certain emulsifier, they should be applied with caution. For example, an emulsion prepared with a blend of emulsifiers is more stable than one prepared with a single emulsifier of the same HLB value. More importantly, in food emulsions, HLB does not reflect the interaction of emulsifiers with proteins. In protein-stabilized emulsions, small molecule surfactants tend to destabilize the emulsion by displacing protein from the interface.

Adsorption of Proteins

Protein molecules can lower their free energy by adsorbing at the O/W interface and exposing their hydrophobic functional groups to the oil phase and their hydrophilic groups to the aqueous phase. This decrease in the free energy is the driving force for their adsorption. Protein molecules diffuse from bulk solution to the interface. Consequently, their rate of adsorption is equal to their rate of diffusion. Experimental measurements of kinetics of adsorption of proteins by Graham and Phillips (1979) and MacRitchie et al. (1963) seem to validate diffusion controlled adsorption. Also, at low surface concentrations, globular proteins completely unfold upon adsorption, thus leading to the surface-induced denaturation (Graham and Phillips, 1979). At sufficiently high surface concentrations, however, an adsorbing protein molecule has to do work against surface pressure to clear area at the interface for its adsorption (MacRitchie, 1963). As a result, the protein molecule has to overcome an energy barrier for its adsorption. In addition, if the protein molecule is charged, its adsorption charges the O/W interface, setting up an electrical double layer in its vicinity. Consequently, an adsorbing protein molecule has to do additional work to overcome the electrostatic repulsion. The existence of energy barriers to adsorption due to surface pressure and charge have been amply demonstrated by MacRitchie and coworkers (1963). Experimental measurements of adsorption of β-casein, lysozyme, and bovine serum albumin at decane/water interface by Graham and Phillips (1979) indicate that the surface concentration for monolayer coverage for the three proteins is 2–3 mg/m^2. β-Casein, being a flexible random coil protein, was the most surface active. Schematic representations of protein structure at a fluid interface for flexible random coil and globular proteins are shown in Fig. 6.38. Flexible random coil proteins adsorb at the interface in the form of trains and loops. Adsorption of globular protein, however, involves partial penetration of the molecule, and the molecule more or less retains its original structure. The latter was especially true for lysozyme because of the presence of large number of disulfide bonds (Graham and Phillips, 1979). Monolayer coverage occurred at extremely low protein concentration. Consequently, at protein concentration employed in food emulsions, multilayer adsorption of proteins at the O/W interface occurs. Though multilayer adsorption does not decrease interfacial tension below that for monolayer coverage, it increases adsorbed protein layer thickness, and therefore extends the steric interaction over larger distances. Experimental observation of interfacial viscosity of adsorbed proteins (Graham and Phillips, 1980) showed that globular

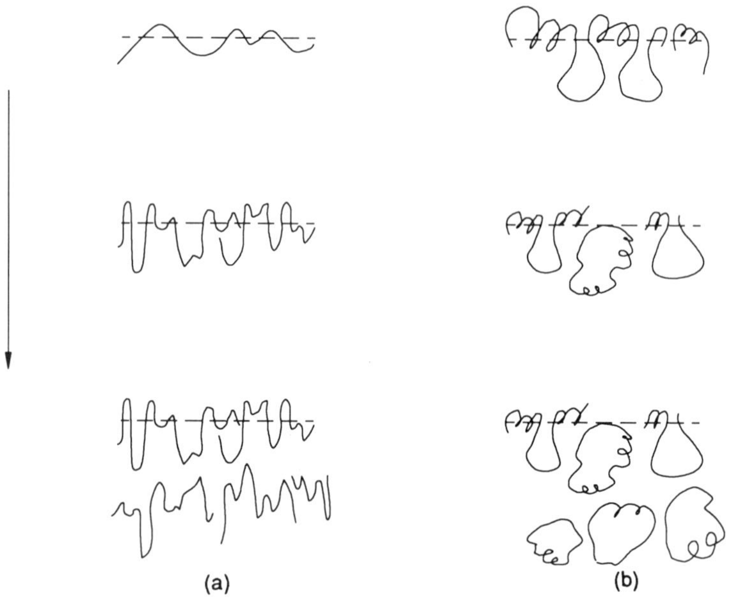

Fig. 6.38 Schematic representation of protein structure at a fluid interface of (a) flexible, random-coil proteins; and (b) globular, highly structured proteins. The arrow denotes increasing protein concentration. (*From Darling and Birkett, 1986.*)

proteins, such as bovine serum albumin and lysozyme, provided much better mechanical properties than flexible random coil proteins, such as β-casein. Therefore, flexible random coil proteins are more surface active, but globular proteins tend to provide better stability to emulsions.

Competitive Adsorption in Mixed Emulsifier Systems

In addition to proteins, mixtures of surfactants are normally used in food emulsions to give better stabilization. The greater stability with mixed surfactant systems is mainly due to increased strength of the interfacial adsorbed layer because of better packing of surfactant molecules and improved viscoelastic properties. It has been well recognized that use of cosurfactant along with the main surfactant helps prepare stable concentrated emulsions. Detailed discussion of this "self-bodying" action as well as other interactions between emulsifiers is given by Tadros and Vincent (1983) and Parker (1982).

Proteins are the most important emulsifiers in food emulsions. Small molecule emulsifiers are used along with proteins to provide controlled destabilization of emulsions and provide desirable texture by partially displacing proteins from the interface in such food systems as whipped creams, toppings, ice cream, and powdered toppings (Parker, 1982). It is therefore important to understand competitive adsorption between protein and other surfactants. Even though small molecule surfactants are, in general, more surface active than proteins and displace proteins completely from the interface when used in sufficiently high concentrations, details of competitive adsorption in specific systems are sensitive to the nature of protein-surfactant interactions. Surfactants sometimes bind to proteins and affects their adsorption, net charge, and surface hydrophobicity (Dickinson and Waskett, 1989). Protein-surfactant binding has been investigated extensively. Figure 6.39 shows typical binding isotherms for mixtures of lysozyme and sodium dodecyl sulfate (SDS) (Furushima et al., 1982). At low SDS concentration, where the binding number is less than 10, the interactions are primarily electrostatic. At high SDS concentrations, the interaction is mainly hydrophobic. There is a dramatic increase in binding number near the critical micelle concentration of SDS. Binding of cationic surfactants with proteins also exhibited a similar trend. The literature on protein-surfactant binding is reviewed by Dickinson and Waskett (1989). Figure 6.40 shows interfacial tension data (Dickinson and Waskett, 1989) for the system of SDS gelatin at a *n*-tetradecane–water interface plotted vs. SDS concentration for different gelatin concentrations. At high SDS concentrations (above the CMC), interfacial tensions

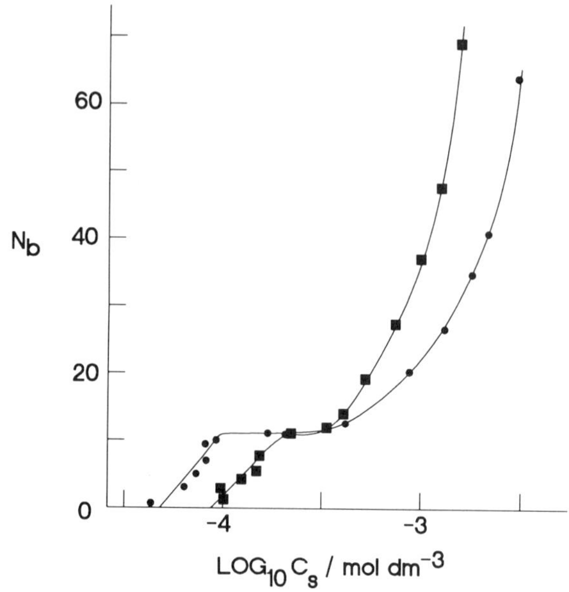

Fig. 6.39 Binding isotherms for SDS + lysozyme (25°C, pH 5.8) at two electrolyte concentrations. The binding number N_b is plotted against the logarithm of the surfactant concentration C_s; ●, 001M NaCl; ■, 0.03M NaCl. (*From Fukushima et al., 1982.*)

for SDS + gelatin at different gelatin concentrations are the same as that for pure SDS, implying thereby that the gelatin is completely displaced from the interface. At low SDS concentrations, the interfacial tension is lower for SDS + protein mixtures with higher gelatin concentrations. At intermediate SDS concentrations, interfacial tension reaches a plateau, decreasing further at high SDS concentrations. The critical micelle concentration of SDS, the concentration at which γ reaches the lower limit, was found to be higher for higher protein concentrations. According to Dickinson and Waskett (1989), the general plot of γ vs. surfactant concentration for protein surfactant mixture can be divided schematically into five regions as shown in Fig. 6.41. At very low surfactant concentrations, the interfacial tension is the same as that of protein. In region II, interfacial tension falls off rapidly, possibly due both to occupation of empty surface sites between protein train segments as well as to formation of protein-surfactant complex. In the intermediate plateau region III, the interfacial tension is constant over a wide range of surfactant concentrations including concentrations higher than CMC of pure surfactant pre-

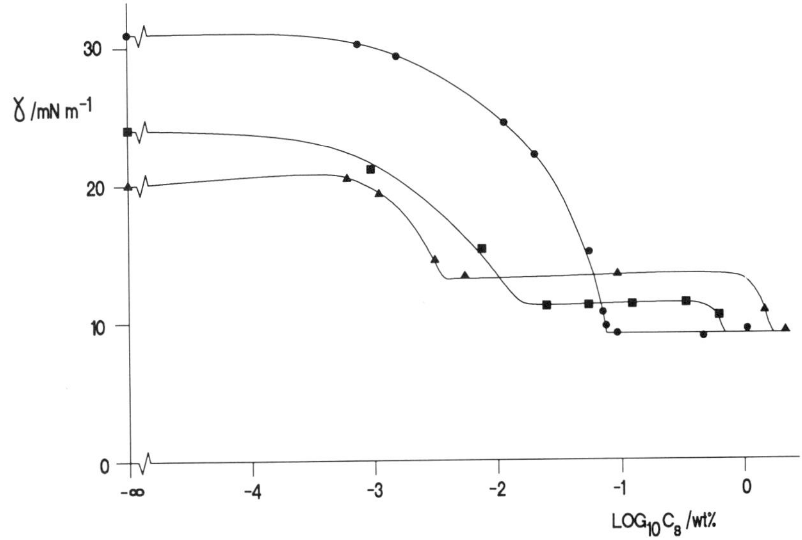

Fig. 6.40 Interfacial tension for SDS + gelatin as an *n*-tetradecane–water interface (25°C, pH 7.2, 0.005 M phosphate buffer). The tension γ is plotted against the logarithm of the surfactant concentration C_s: ●, pure SDS; ■, SDS + 0.1 wt% gelatin; ▲, SDS + 0.4 wt% gelatin. (*From Dickinson and Woskett, 1989.*)

sumably because it is energetically favorable for surfactant to bind to protein. In region IV, the interfacial tension decreases, eventually becoming equal to that of pure surfactant (region V) as a result of complete displacement of protein by surfactant. A pictorial representation of different regions is shown in Fig. 6.42 (Dickinson and Waskett, 1989). In-

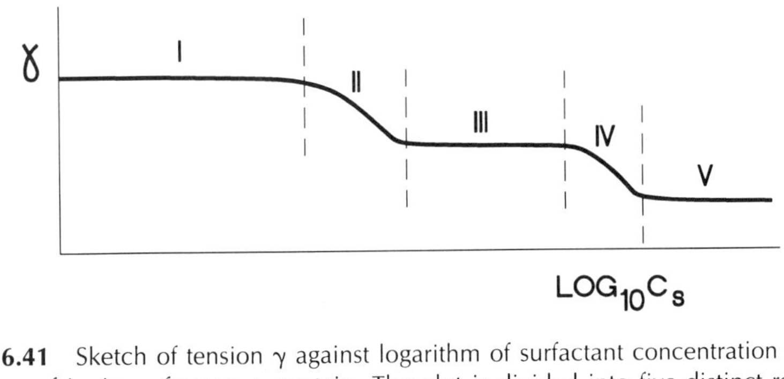

Fig. 6.41 Sketch of tension γ against logarithm of surfactant concentration C_s for system of ionic surfactant + protein. The plot is divided into five distinct regions (I–V). (*From Dickinson and Woskett, 1989.*)

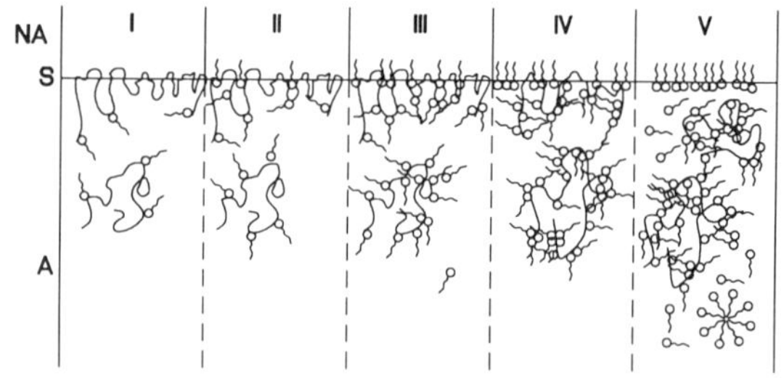

Fig. 6.42 Pictorial representation of protein-surfactant binding and competitive adsorption as a function of increasing surfactant concentration. Illustrations refer to a mixture of idealized random-chain protein molecules and simple ionic amphiphiles at the surface (S) between aqueous (A) and nonaqueous (NA) phases. (*From Dickinson and Waskett, 1989.*)

teractions between proteins and surfactants at fluid interfaces can be inferred from surface rheological measurements. Displacement of protein by surfactant at high surfactant concentrations results in a sudden drop in interfacial viscosity. Experimental results for displacement of proteins such as β-casein or β-lactoglobulin by SDS from a sunflower oil-in-water emulsion interface, obtained by de Feijter et al. (1987), are shown in Fig.

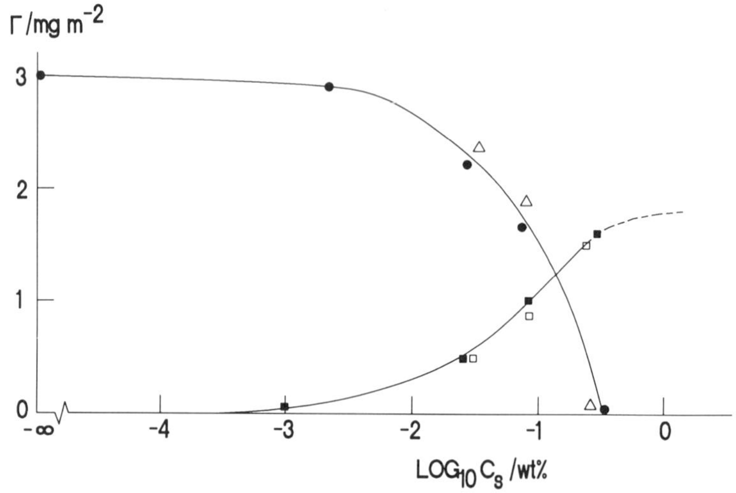

Fig. 6.43 Displacement of protein from the droplet surface by surfactant in sunflower seed oil-in-water emulsions (pH 6.5, 50 wt% oil). The surface concentration Γ is plotted against the logarithm of the (final) surfactant concentration C_S: □, △, SDS + β-casein; ■, ●, SDS + β-lactoglobulin. (*From Dickinson and Waskett, 1989.*)

6.43. The results were found to be virtually independent of the type of protein and the nature of the oil phase. Displacement curves for two dodecyl anionic surfactants, SDS and sodium dodecyl sulfonate, were very similar. The shapes of displacement curves with Tweens 20 and 80 were found to be slightly different from those for anionic surfactants.

Whereas proteins stabilize emulsions by adsorbing at the oil-water interface, polysaccharides achieve a similar effect mainly by modifying the rheological properties of the continuous medium. Polysaccharides form complexes with protein. Binding of polysaccharide to protein can be due to strong irreversible covalent bonding or to weak reversible association. In any event, such association increases the stability of an emulsion since it satisfies conflicting requirements of hydrophobicity (from protein) for strong adsorption and substantial hydrophilicity (from polysaccharide) and provide much thicker adsorbed layer for steric repulsion. Dickinson and Euston (1991) compared the stabilizing ability of a whey protein + propylene glycol alginate (PGA) complex with a mixture of whey protein + PGA of the same composition, but without any association, for *n*-tetradecane-in-water emulsion as shown in Fig. 6.44. Emulsion droplets

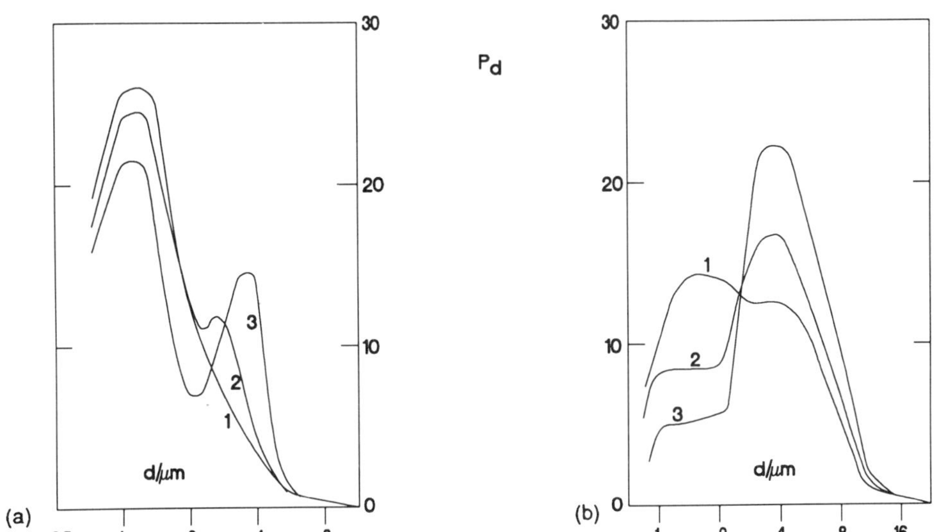

Fig. 6.44 Time-dependent droplet-size distribution of *N*-tetradecane-in-water emulsions (10 wt% oil, 0.72 wt% emulsifier, pH 3) made with (a) whey protein-PGA covalent hybrid (1 : 10) and (b) whey protein + PGA mixture (1 : 10). The volume-weighted percentage droplet probability P_d as given by the coulter counter is plotted against the droplet diameter d: (1) fresh emulsion; (2) after 48 hr; (3) after 128 hr. (*From Dickinson and Euston, 1991.*)

produced with whey protein + PGA complex were much smaller (Fig. 6.44a) and more stable with respect to coalescence than those produced with the whey protein + PGA mixture (Fig. 6.44b). Formation of a protein-polysaccharide complex is also reflected by the changes in the interfacial viscosity. Figure 6.45 shows the effect of PGA addition on the surface viscosity of sodium caseinate adsorbed layer (Dickinson and Euston, 1991). Following the addition of PGA after 2 hours, the rate of increase of surface viscosity was much higher than that for only sodium caseinate. This is believed to be due to formation of a more structured interfacial film as a result of complexation. In contrast, the system caseinate + xanthan showed no evidence of complex formation. Experimental measurement of creaming profile of *n*-tetradecane-in-water emulsion stabilized by caseinate indicates that low concentrations of xanthan tend to increase the creaming rate and hence destabilize the emulsion as shown in Fig. 6.46 (Dickinson and Euston, 1991). Such behavior is believed to be due to an increase in the emulsion drop size as a result of depletion flocculation of droplets by nonadsorbing polymer xanthan. Similar results were also observed for other polysaccharides such as dextran (Bullin et al., 1988) and succinoglycan (Cao et al., 1990).

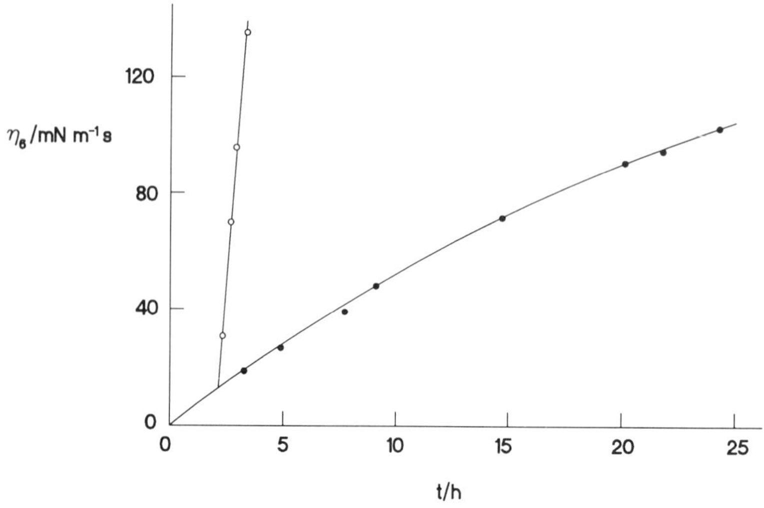

Fig. 6.45 Surface viscosity of caseinate + PGA at the *n*-hexadecane–water interface (pH 3, 25°C). The apparent surface shear viscosity η_s is plotted against the adsorption time *t*: ●, 10^{-3} wt% caseinate from $t = 0$ hr; ○, 10^{-3} wt% caseinate from $t = 0$ hr + 8×10^{-3} wt% PGA introduced at $t = 2$ hr. (*From Dickinson and Euston, 1991.*)

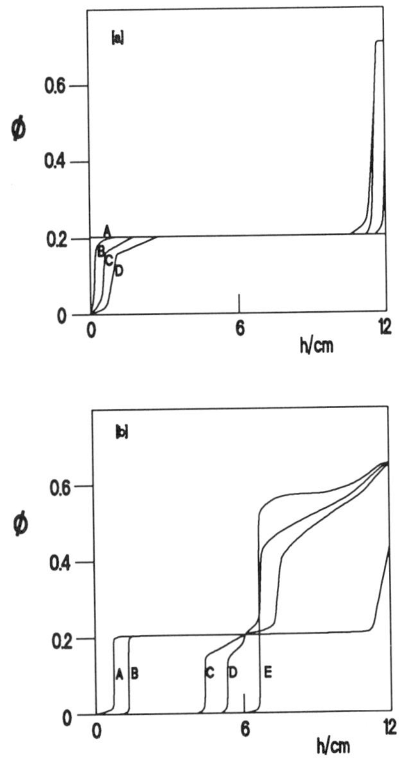

Fig. 6.46 Effect of xanthan on the rate of creaming of a *n*-tetradecane-in-water emulsion (15 wt% oil, 0.75 wt% caseinate, pH 7) as determined by ultrasonic velocity measurement at 20°C. The oil volume fraction ϕ is plotted as a function of height h: (a) emulsion with no added polymer stored for (A) 0 hr, (b) 72 hr, (c) 168 hr, and (D) 216 hr; (b) emulsion containing 0.05 wt% polymer stored for (A) 19 hr, (B) 43 hr, (C) 127 hr, (D) 154 hr, and (E) 223 hr. (*From Dickinson and Euston, 1991.*)

EMULSION FORMATION

Emulsions are formed by applying mechanical energy to a mixture of two immiscible liquid phases, so that one liquid disperses in the form of fine droplets. The mechanical energy initially disrupts the interface to form big droplets, which are subsequently broken into smaller droplets. Droplets must be deformed for their disruption. Since any deformation of the droplet leads to an increase in the Laplace pressure because of the radius of curvature, external pressure exceeding this Laplace pressure must be applied to break it. The external pressure gradient should exceed $2\gamma/r^2$ to break a spherical droplet of radius r (Walstra, 1983). Application of shear to a droplet will also deform the droplet. Breakage of droplet will occur if the applied shear stress exceeds the Laplace pressure. The smaller the droplets, the larger must be pressure gradient. Use of an emulsifier during emulsion formation decreases the energy required for emulsification since adsorption of the emulsifier at the oil-water interface decreases the interfacial tension γ. Of course, adsorption equilibrium will not be achieved during deformation and breakup of droplets, since time scales for breakup are very small (of the order of milliseconds). However, use of emulsifiers will result in interfacial tension gradients during drop deformation because of expansion of the interface. The interfacial tension gradient via the Gibbs-Marangani effect tends to prevent coalescence of droplets during their formation. As Walstra (1983) points out, most of the mechanical energy input is required to create high pressure gradients necessary for drop deformation. This energy is eventually dissipated, and only a very small fraction of the energy input is used in creating new interfacial area. In the following sections, a brief review of different mechanisms of emulsion formation is given. Details can be found elsewhere (see Walstra, 1983; Parker, 1982; Dickinson and Stainsby, 1982).

Formation of Large Droplets from the Disruption of Planar Interface

To disperse one liquid phase into another, a planar interface between two bulk phases has to be disrupted to produce drops. Such disruption can occur due to Rayleigh-Taylor instability, which is initiated by acceleration of interface perpendicular to its plane and directed from the lighter to denser phase. Maximum instability for inviscid fluids of constant interfacial tension γ occurs at a certain wavelength λ_{opt} (Walstra 1983).

$$k_{opt}^2 = \left(\frac{2\pi}{\lambda_{opt}}\right)^2 = \frac{\Delta\rho a}{3\gamma} \tag{68}$$

where k_{opt} is the corresponding wave number, $\Delta\rho$ is the density difference, and a is the acceleration of the interface. Walstra (1983) showed that only coarse droplets, of sizes of the order of millimeters, can be formed by such a process.

When the two liquid phases move with different velocities parallel to the interface, instability occurs, leading to disruption of one phase. This instability is known as Kelvin-Helmholtz instability. Small perturbations will grow (Walstra, 1983) if the wave number k satisfies Eq. (69).

$$k > \frac{a(\rho_1{}^2 - \rho_2{}^2)}{\rho_1\rho_2(u_1 - u_2)^2} \tag{69}$$

and

$$(u_1 - u_2)^2 > 2(\rho_1 + \rho_2)\,\frac{[a\gamma(\rho_1 - \rho_2)]^{1/2}}{\rho_1\rho_2} \tag{70}$$

In the above equations, ρ_1 and ρ_2 are the densities of the two phases and u_1 and u_2 are their velocities. It turns out that disturbances of any wavelength will grow if the velocity difference between the two phases exceeds about 0.1 msec^{-1}.

One phase can also be elongated in the form of a thread in the other phase, either due to mixing or due to fingering as a result of the above instabilities. Such threads are unstable to imposed disturbances. When the amplitude of the growing disturbance becomes equal to the radius of the cylindrical thread, breakage of the thread into droplets ensues. The radius, r, of the droplet, obtained from stability analysis, is

$$r = \theta\left(\frac{3\pi}{2b_{opt}}\right)^{1/3} \tag{71}$$

where θ is the diameter of the cylindrical thread and

$$b_{opt} = K_{opt}\,\theta = 2\pi\theta/\lambda_{opt} \tag{72}$$

where k_{opt} and λ_{opt} are the wavenumber and the corresponding wavelength at which the growth rate of the imposed disturbance is maximal.

Breakup of Large Droplets

The mechanism of drop breakup depends on the type of flow field the droplets are exposed to in emulsification equipment. In colloid mills, for example, droplets are exposed to high shear. In high pressure homogenizers, the droplets are exposed to a turbulent flow field. In high speed stirrers, the droplets are exposed to turbulence in the vicinity of the stirrer blades as well as shear. In ultrasonic jet generators ("liquid whistles"), the droplets break due to cavitation. Detailed description of different types of emulsification equipment can be found elsewhere (see Walstra, 1983; Dickinson and Stainsby, 1982).

In simple shear, a droplet is distorted into a prolate ellipsoid under the action of viscous forces and is oriented 45° to the direction of flow (Taylor, 1934). For large deformations, the droplet breaks if its Weber number *We* exceeds a critical value We_{crit}. *We* is defined as the ratio of viscous forces to capillary forces and is equal to $\mu_c Gr/\gamma$, G being the shear rate. Variation of We_{crit} with the viscosity ratio $q^{crit} = \mu_D/\mu_c$ is shown in Fig. 6.47. Drops do not break for $q > 4$. For $q < 4$, We_{crit} increases as q decreases. For plane hyperbolic flow, We_{crit} was smaller (Fig. 6.47), and drops can break for any viscosity ratio. For small viscosity ratios, the drop attains a slender pointed shape and small droplets detach from the pointed ends. For moderate to high q, the droplet deforms into a thread, which eventually breaks into many smaller droplets due to Raleigh instability.

Turbulent flow is characterized by velocity fluctuations, the root mean square value of which is higher for higher intensity of turbulent flow. The turbulent energy can be pictured as being distributed over eddies of different scales or sizes. The scale of an eddy refers to a distance over which fluctuations occur. Turbulent energy is essentially convected from larger to smaller eddies. Since viscous effects are important for very small eddies, energy transported to these small eddies is dissipated as a result of viscous dissipation. For high intensity of turbulence, i.e., for very high Reynolds numbers, the turbulence is locally isotropic for a range of small eddy sizes (Kolmogorov, 1941). The smallest size of eddies, or Kolmogorov scale, is given by

$$l_0 = \mu^{3/4} \rho^{-3/4} \epsilon^{-1/4} \tag{73}$$

where ϵ is the energy dissipation rate per unit volume. For eddies of size greater than l_0, the relative velocity over distance l is governed only by inertial forces and is given by

$$u(l) \sim c\epsilon^{1/3} l^{1/3} \rho^{-1/3}, \tag{74}$$

c being a constant. The turbulent velocity fluctuation $u(d)$ between two diametrically opposite sides of a droplet of diameter d leads to a pressure difference $\rho_c u^2(d)$. If this pressure difference exceeds capillary pressure, drop breakup occurs. The maximum stable drop diameter $d_{\max}$ is then given by

$$d_{\max} = c\epsilon^{-2/5} \gamma^{3/5} \rho_c^{-1/5} \tag{75}$$

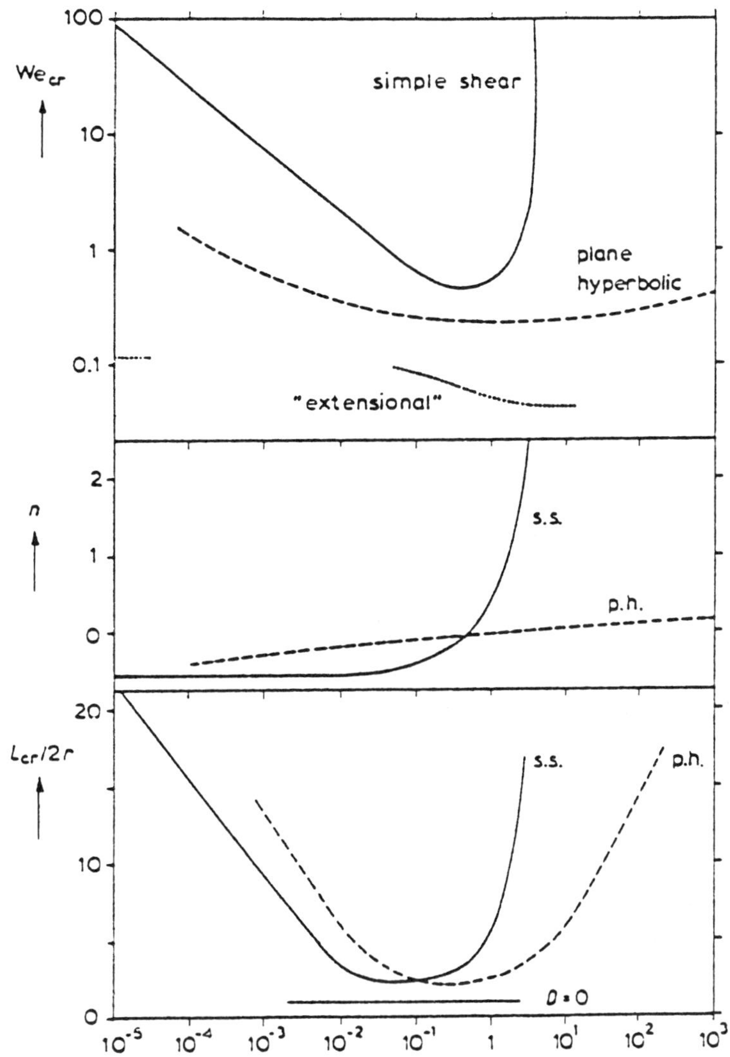

Fig. 6.47 Critical Weber number for breakup We_{cr} (i.e., the drop will break in the region above the curve); viscosity exponent $n = d \log We_{cr}/d \log q$; and the largest drop dimension at burst L_{cr} relative to original drop diameter; for various types of steady flow as a function of the viscosity ratio q. (*From Walstra, 1983.*)

Equation (75) gives the order of magnitude for drop sizes during emulsion formation in a turbulent flow field. Of course, in order for Eq. (75) to be valid, d_{max} must be $>l_0$ as given by Eq. (73), which gives the minimum drop size d_{min} for which Eq. (75) applies.

$$d_{min} = \frac{3\mu_c^2}{\gamma\rho_c} \tag{76}$$

On the other hand, if the drop sizes are smaller than l_0, viscous dissipation becomes important and the relative velocity is given by

$$u(l) = c\epsilon^{1/2} d \mu_c^{-1/2} \tag{77}$$

and the maximum stable diameter by

$$d_{max} = c\gamma\epsilon^{-1/2}\mu_c^{-1/2} \tag{78}$$

Cavitation is formation and collapse of small vapor bubbles in a liquid as a result of local pressure becoming less than the liquid's vapor pressure of liquid. Such low pressures can occur due to extremely high velocity of the liquid. Collapse of a cavity produces a shock wave, which travels through the liquid. Implosion of a cavity in the vicinity of a drop leads to suction on part of the drop and therefore to its fragmentation.

Effect of Emulsifier on Formation of Emulsions

As already pointed out, emulsifier decreases the amount of energy required for emulsification by decreasing interfacial tension. The interfacial tension will not be an equilibrium value and will depend on the rate of adsorption of emulsifier. Order of magnitude estimates of different time scales in turbulent flow (Walstra, 1983) indicate that the time scale of adsorption of emulsifier is much greater than the time scale of deformation so that under dynamic conditions the interfacial tension is close to that of the pure liquids. Because of the expansion of the interface due to drop deformation, the interfacial tension will also depend on surface dilational modulus. The surface free energy needed for enlarging a drop surface will be $\gamma\Delta A + A\Delta\gamma$, instead of just $\gamma\Delta A$, because of interfacial tension gradients along the surface. The increase in energy requirement due to this effect is small. More important, the interfacial tension gradient, as a result of the Gibbs-Marangoni effect, tends to prevent coalescence of droplets by retarding the rate of thin film drainage between coalescing drops. In addition to interfacial tension, drop size in a turbulent flow is influenced by energy dissipation rate per unit volume ϵ. In fact, there can be a tremendous variation in ϵ depending on the type of emulsification equipment and emulsification condition (from 10^4 W m^{-3} for a paddle stirrer to 10^{12} W m^{-3} for a high pressure homogenizer). On the other hand, the range of interfacial tension values is much smaller (4–40 mN m^{-3}). In fact, experimental observation of mean drop size for different types of emulsifiers and conditions by Darling and Birkett (1988) as shown in Table 6.8 indicates that surfactant chemistry does not greatly influence drop size during formation. There is very little variation in droplet size for a variety of surfactants and viscosity of the aqueous continuous phase, excepting at very high viscosity.

Table 6.8 Effect of Emulsifier on Mean Droplet Size as Determined by Nanosizer at 20°C[a]

	Aqueous phase composition	Mean droplet size (μm)	Equilibrium interfacial tension (mNm^{-1})	Aqueous phase viscosity (mPas)
1 wt%	Tween 60[b]	0.46	7.4	1.3
2 wt%	Sodium caseinate[c]	0.63	11.0	2.0
1 wt%	Sodium stearoyl lactylate[d]	0.43	11.9	5.8
2 wt%	Methyl cellulose[e] M20	1.33	9.3	25.7
2 wt%	M20 + 1 wt% Tween 60	0.57	8.3	26.5
1 wt%	Tween 60 + 47 wt% sucrose	0.36	7.4	27.0
2 wt%	Hydroxypropylmethyl cellulose HPM 450	2.23	14.7	240

[a] Emulsions were made at 20°C with 10 vol% soya bean oil by premixing in a Silverson mixer and a single-pass homogenization at 14 MPa on a Rannie 3-piston homogenizer. Interfacial tensions were measured against decane at 20°C and 0.1 sec^{-1} using a Deer Rheometer.
[b] Polyoxyethylene (20) sorbitanmonostearate (Atlas Chemical Industries).
[c] DMV spray-dried powder.
[d] SSL 2004 (2004016) (PPF International).
[e] Cellulose derivatives (CourtauldsAcetate).
Source: Darling and Birkett, 1986.

CONCLUSIONS

As is evident by now, food emulsions are complex systems whose stability is influenced by many parameters. Even though a lot of progress has been made recently in understanding phenomena involved in stabilizating emulsions, predicting long-term behavior of such systems still remains a craft because of its complexity. Well-characterized, simpler model systems should be investigated to delineate mechanisms of stabilization. While such studies illuminate different mechanisms, experimental investigation of complex ill-defined real systems are still needed to elucidate effects of interaction among various mechanisms of stabilization. Even though techniques for direct measurement of interaction forces have been employed to quantify the steric forces for synthetic polymers, very little information is available with regard to steric interaction between protein layers. It is also known that proteins modify the structure of electrical double layer because of the distribution of charged groups in adsorbed protein trains and loops and the low dielectric constant of the adsorbed protein layer. It is therefore necessary to characterize electrostatic and steric interaction between adsorbed protein layers through direct measurement. In addition, experimental measurements of the thickness of adsorbed layers at O/W interfaces will also be useful in quantifying steric interactions. Experimental data on adsorption isotherms and surface equations of state (interfacial tension vs. concentration) for proteins are scarce. Also, little information is available on adsorption isotherms of mixed emulsifier systems at O/W interfaces. It is now recognized that kinetics of adsorption of proteins and competitive adsorption of proteins and small molecule surfactants, especially displacement of proteins by small molecule surfactants, are important in stabilization of emulsions. More systematic study of these effects should be undertaken for different systems. General guidelines for selection of emulsifiers should be developed accounting for the effects of competitive adsorption. Of course, it is extremely important to characterize interfacial rheology of mixed adsorbed layers of proteins and emulsifiers, since stability of emulsion is greatly influenced by surface mechanical properties. Coalescence of emulsion droplets is not fully understood. Though studies on coalescence of large single droplets at planar interfaces are useful for a better understanding of the coalescence process, such studies should be supplemented by experimental investigation of drainage and stability of isolated, protein stabilized thin films of much smaller sizes. It is also important to characterize rates of coalescence through the measurement of evolution of drop size distribution. Finally, systematic studies are needed to elucidate the mechanism of stabilization by adsorbed particles, fat crystals, and liquid crystals.

REFERENCES

Asakura, S. and Oosawa, F. 1954. *J. Chem. Phys.* 22: 1255.

Asakura, S. and Oosawa, F. 1958. *J. Polymer Sci.* 33: 183.

Batchelor, G. K. 1972. *J. Fluid Mech.* 52: 245.

Boyd, J. V., Krog, N., and Sherman, P. 1976. *Theory and Practice of Emulsion Technology*, A. L. Smith (Ed.), p. 123. Academic Press, New York.

Bullin, S., Dickinson, E., Impeg, S. J., Narhan, S. K., and Stainsby, G. 1988. In *Gums and Stabilizers for the Food Industry*, G. O. Phillips, D. J. Wedlock, and P. A. William (Ed.), Vol. 4, p. 337. IRL Press, Oxford, England.

Cao, Y., Dickinson, E., and Wedlock, D. J. 1990. *Food Hydrocolloids* 4: 185.

Clayfield, E. J. and Lumb, E. C. 1966a. *J. Colloid Sci.* 22: 269.

Clayfield, E. J. and Lumb, E. C. 1966b. *J. Colloid Sci.* 22: 285.

Clayfield, E. J. and Lumb, E. C. 1968. *Macromolecules* 1: 133.

Clayfield, E. J. and Lumb, E. C. 1974a. *J. Colloid Interface Sci.* 47: 6.

Clayfield, E. J. and Lumb, E. C. 1974b. *J. Colloid Interface Sci.* 47: 16.

Clayfield, E. J. and Lumb, E. C. 1974c. *J. Colloid Interface Sci.* 49: 489.

Darling, D. F. and Birkett, R. J. 1986. Food colloids in practice. In *Food Emulsions and Foams*, E. Dickinson (Ed.), Special Publication No. 58, p. 1. Royal Society of Chemistry, Burlington House, London.

Davies, J. T. 1957. *Proc. Int. Congr. Surface Activity* (2d), Vol. 1, p. 426. Butterworth, London.

Davies, J. T. and Rideal, E. K. 1961. *Interfacial Phenomena.* Academic Press, New York.

De Feijter, J. A., Benjamin, J., and Tamboer, M. 1987. *Colloids Surf.* 27: 243.

Derjaguin, B. V. and Landau, L. 1941. *Acta Physiochem.* USSR 14: 633.

Dickinson, E. and Euston, S. R. 1991. Stability of food emulsions containing both protein and polysaccharide. In *Food Polymers, Gels and Colloids*, E. Dickinson (Ed.), p. 132. Special Publication No. 82, Royal Society of Chemistry, Thomas Graham House, Cambridge.

Dickinson, E., Murray, B. S., and Stainsby, G. 1988. *Chem. Soc. Faraday Trans.* 1, 84: 871.

Dickinson, E. and Stainsby, G. 1982. *Colloids in Food.* Applied Science Publishers.

Dickinson, E. and Woskett, C. M. 1989. Competitive Adsorption between proteins and small molecule surfactants in food emulsions. In *Food Colloids*, R. D. Bee, P. Richmond, and J. Mingins (Ed.), p. 74. Special Publications No. 75. Royal Society of Chemistry, Thomas Graham House, Cambridge.

Dzyaloshinsski, E. I., Lifshitz, E. M., and Pitaevskii, L. P. 1961. *Adv. Phys.* 10: 165.

Felderhoff, B. V. 1968. *J. Chem. Phys.* 48: 44.

Fischer, E. W. 1958. *Kolloid-Z.* 160: 120.

Fischer, L. R. and Mitchell, E. E. 1989. Effect of adsorbed proteins on interactions between emulsion droplets. In *Food Colloids*, R. D. Bee, P. Richmand, and J. Mingins (Ed.), p. 123. Special Publications No. 75, Royal Society of Chemistry, Thomas Graham House, Cambridge.

Friberg, S. 1971. *J. Colloid Interface Sci.* 37: 291.

Friberg, S. 1976. Emulsion stability. In *Food Emulsions*, S. Friberg (Ed.). Marcel Dekker, New York.

Friberg, S., Mandell, L., and Larsson, M. 1969. *J. Colloid Interface Sci.* 29: 155.

Fuchs, N. 1934. *Z. Phys.* 89: 736.

Fukushima, K., Murata, Y., and Tanaka, M. 1982. *Bull. Chem. Soc. Jpn.* 55: 1376.

Gillispie, T. and Rideal, E. R. 1956. *Trans. Faraday Soc.* 52: 173.

Graham, D. E. and Phillips, M. C. 1979a. *J. Colloid Interface Sci.* 70: 403.

Graham, D. E. and Phillips, M. C. 1979b. *J. Colloid Interface Sci.* 70: 427.

Graham, D. E. and Phillips, M. C. 1980. *J. Colloid Interface Sci.* 76: 240.

Griffin, W. C. 1943. *J. Soc. Cosmetic Chemists* 1: 311.

Halling, P. J. 1981. *Crit. Rev. Food Sci. & Nutrition* 155.

Hesselink, F. Th. 1971. *J. Phys. Chem.* 75: 65.

Ivanov, I. B. 1980. *Pure Appl. Chem.* 52: 1241.

Ivanov, I. B. and Dimitrov, D. St. 1974. *Colloid & Polymer Sci.* 252: 982.

Ivanov, I. B., Radoev, E., Manev, E., and Scheludko, A. 1970. *Trans. Faraday Soc.* 66: 1262.

Ivanov, I. B., and Traykov, T. T. 1976. *Intern J. of Multiphase Flow.* 397.

Jackel, K. 1964. *Kolloid-Z.Z. Polymers* 197: 143.

Jaynes, E. N. 1986. Applications in the food industry II. In *Encyclopedia of Emulsion Technology*, Vol. 2, P. Becher (Ed.). Marcel Dekker, New York.

Jeffreys, G. V. and Davies, G. A. 1971. *Recent Advances in Liquid-Liquid Extraction*. Oxford, England.

Kolmogorov, A. N. 1941a. *Dokl. Akad. Nauk. SSSR* 30:301.

Kolmogorov, A. N. 1941b. *Dokl. Akad. Nauk. SSSR* 31:99.

Kolmogorov, A. N. 1941c. *Dokl. Akad. Nauk. SSSR* 32: 16.

Krog, N. L., Rusom, T. H., and Larsson, K. 1986. Applications in the food industry 1. In *Encyclopedia of Emulsion Technology*, Vol. 2, P. Becher, (Ed.). Marcel Dekker, New York.

Levich, V. G. 1962. *Physicochemical Hydrodynamics*. Prentice-Hall, Englewood Cliffs, NJ.

Lifshitz, E. M. 1956. *Sov. Phys. JETP* 2: 73.

Mackor, E. L. 1951. *J. Colloid Sci.* 6: 492.

MacRitchie, F. 1976. *J. Colloid Interface Sci.* 56: 53.

MacRitchie, F. and Alexander, A. E. 1963a. *J. Colloid Interface Sci.* 18: 453.

MacRitchie, F. and Alexander, A. E. 1963b. *J. Colloid Interface Sci.* 18: 464.

Manev, E. D., Sazdanova, S. V., and Wasan, D. T. 1984. *J. Colloid Interface Sci.* 97: 591.

March, G. C. and Napper, D. J. 1977. *J. Colloid Interface Sci.* 61: 383.

Meir, D. J. 1967. *J. Phys. Chem.* 71: 1861.

Melik, D. H. and Fogler, H. S. 1988. Colloid stability in quiescent media. In *Encyclopedia of Emulsion Technology*, Vol. 3, P. Becher (Ed.). Marcel Dekker, New York.

Melik, D. H. and Fogler, H. S. 1984. *J. Colloid Interface Sci.* 101: 84.

Napper, D. H. 1983. *Polymeric Stabilization of Colloidal Dispersions*. Academic Press, New York.

Narsimhan, G. 1991. *Colloids and Surf.* (In press).

Parker, N. S. 1982. *CRC Crit. Reviews in Food Sci. & Nutrition* 25: 285.

Rao, A. A., Wasan, D. T., and Manev, E. D. 1982. *Chem. Eng. Commn.* 15: 63.

Reddy, S. R. and Fogler, H. S. 1981. *J. Colloid Interface Sci.* 79: 105.

Reeriink, H. and Overbeek, J. Th. G. 1954. *Discuss. Faraday Soc.* 18: 74.

Ruckenstein, E. and Jain, R. K. 1974. *Faraday Trans. II* 70: 132.

Scheludko, A. 1962. *Proc. Kon. Ned. Akad. Ventenach.* B56: 87.

Schenkel, J. H. and Kitchener, J. A. 1960. *Trans. Faraday Soc.* 56: 161.

Schulman, J. H. and Leja, I. 1954. *Trans. Faraday Soc.* 50: 598.

Sieglaff, C. L. 1959. *J. Polymer Sci.* 41: 319.

Silber, D. and Mizrai, S. 1975. *J. Food Sci.* 40: 1174.

Smith, A. L. and Mitchell, D. P. 1976. *Theory and Practice of Emulsion Technology*, A. L. Smith (Ed.), p. 61. Academic Press, New York.

Smithiam, J. B., Evans, R., and Napper, D. H. 1975. *J. Chem. Soc. Faraday Trans. I.* 71: 285.

Smithiam, J. B. and Napper, D. H. 1979. *Colloid Polymer Sci.* 257: 748.

Smoluchowski, M. 1917. *Z. Phys. Chem.* 92: 129.

Spielman, L. A. 1970. *J. Colloid Interface Sci.* 33: 562.

Tadros, T. F. and Vincent, B. 1983. Emulsion stability. In *Encyclopedia of Emulsion Technology*, Vol. 1, P. Becher (Ed.). Marcel Dekker, New York.

Taylor, G. I. 1934. *Proc. Roy. Soc. London* A146: 501.

Van Boekel, M. A. J. S. 1980. Influence of fat crystals in the oil phase on the stability of oil-in-water emulsions. Ph.D. thesis, University of Wageningen.

Verwey, E. J. W. and Overbeek, J. Th. G. 1948. *Theory of Stability of Lyophobic Colloids*. Elsevier, Amsterdam.

Vincent, B. 1973. *J. Colloid Interface Sci.* 42: 270.

Vold, M. J. 1961. *J. Colloid Sci.* 16: 1.

Vrij, A. 1966. *Disc. Faraday Soc.* 42: 23.

Walstra, P. 1983. Formation of emulsions. In *Encyclopedia of Emulsion Technology*, Vol. 1, P. Becher (Ed.). Marcel Dekker, New York.

Wasan, D. T. and Malhotra, A. K. 1986. *AIChE Symposium Series #252* 86: 5.

Wiese, G. and Healy, T. W. 1970. *Trans. Faraday Soc.* 66: 490.

7

Generation of Engineered Structures in Gels

José Miguel Aguilera

Catholic University
Santiago, Chile

INTRODUCTION

If foods were protein pills, people's lives would be much duller and the field of food technology would be insignificant. Fortunately, foods are not protein pills but highly structured, multicomponent materials whose properties are recognized and appreciated by consumers. Therefore the generation of new structures is a continuing technological undertaking. Proteins, carbohydrates, and sometimes lipids interacting in an aqueous environment determine the native or processed structure of foods. Breakdown of the food during mastication unveils those mechanical properties buried in the structure that are perceived as texture and initiates a host of physicochemical phenomena in the cavity of the mouth. The integrated relationship between chemical components–structure–textural properties is a determinant in the acceptability and desirability of many foods (Aguilera and Stanley, 1990). Consequently, a major task of modern food technology is to generate functional or engineered structures with desirable mechanical and transport properties from a limited array of ingredients.

Biopolymers (proteins and polysaccharides) and, to a limited extent, lipids are the main building blocks available to the food technologist for structure development. Since ancient times many structures have become associated with raw and processed foods. Structures in the latter are a consequence of intense and subtle interactions during processing between polymeric molecules, lipids, water, and solutes dispersed in a complex physicochemical milieu.

What is food structure? In food technology, structure relates to the organization of a number of similar or dissimilar elements, their binding into a unit, and the interrelationships between the individual elements or their groupings (Raeuber and Nikolaus, 1980). This concept differs from that proposed in materials science mainly in the scale of significance of the individual elements (Hornbogen, 1986). While in a metal alloy or in a superconductor the atomic arrangement may have a major influence on their performance, in foods the lowest boundary relevant to textural properties is usually that of associated polymeric molecules or phase discontinuities (e.g., 0.1–10 μm). The term "microstructure" will be used throughout this paper to refer to this scale of significance.

To provide a basis for discussing the design and generation of engineered structures in gels, let us at this point attempt to dissect foods into their major architectural elements. Table 7.1 presents a list of elements that characterize some common semi-solid foods. As graphically depicted in this scheme, most semi-solid foods are combinations of gel matrices (hydrated or collapsed) with microstructural elements such as fibers, fat globules, solutions, or encapsulating walls.

A gel can be defined as a three-dimensional polymeric network holding large quantities of solvent (aqueous solution) and showing mechanical rigidity (absence of fluidity during observation times). In this sense, all cellular interiors are considered to be gels (Silberberg, 1989) and most foods having a cellular structure are modified gels. Moreover, since the overwhelming majority of food products are solids containing between 50 and 90% water, they can be regarded in many ways as multicomponent gels (Tolstoguzov, 1983).

Food structuring or the generation of engineered structures may involve recombining ingredients derived from agricultural and marine raw materials into acceptable human foods (Aguilera and Stanley, 1990). The ultimate motive may vary from the utilization of cheaper stocks to increase food supplies or the extension of shelf life in less developed countries (LDCs), to the creation of new textures, cost reduction, and the design of nutritionally adequate and safer products in industrialized so-

Table 7.1 Major Microstructural Elements in Some Semi-Solid Foods

Foods	Elements	Microstructure	Relative Scale
Meats			
Raw	Fibers oriented in a protein gel matrix		100
Processed	Fat globules dispersed in protein gel matrix	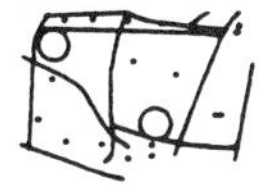	100
Dairy			
Cheeses, yogurt	Protein gel–fat globule composite	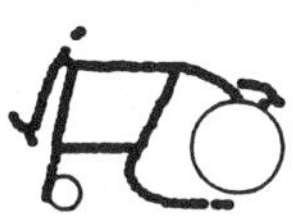	1
Cereals, pulses			
Raw, dry	Starch granules dispersed in a collapsed gel matrix	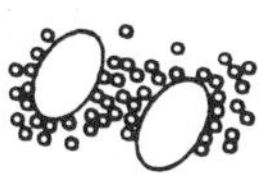	10
Cooked	Starch particulates dispersed in a protein–starch gel matrix		10
Snacks	Air dispersed in starch-protein matrix (solid foam)	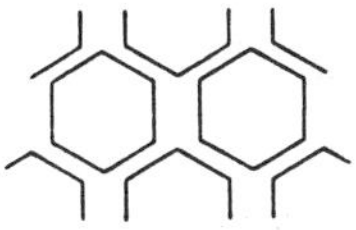	1000
Fruits/Vegetables			
Raw, fresh	Gel surrounded by skin (cellular)		100
Jams, jellies	True polysaccharide gel	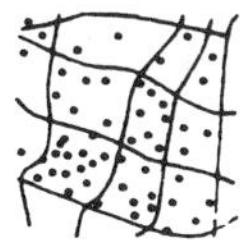	0.1

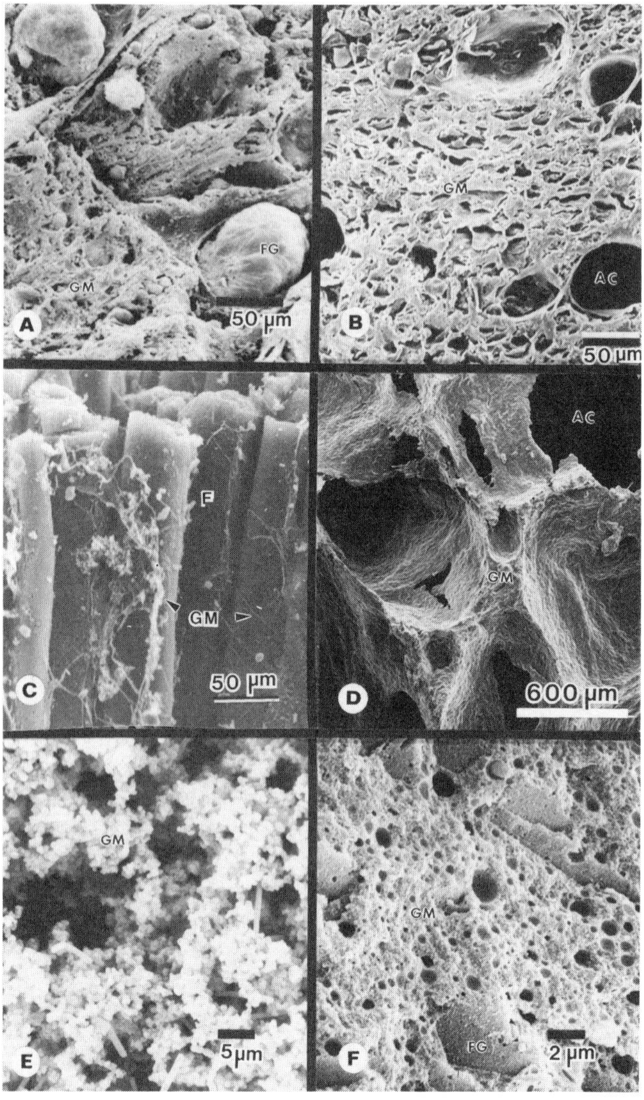

Fig. 7.1 Examples of common foods viewed as composite gels and dimensions of microstructural elements involved. (A) Commercial frankfurter; (B) commercial surimi; (C) fish; (D) bread; (E) yogurt; (F) processed Colby cheese. GM = gel matrix; FG = fat globules; F = Fibers; AC = air cells. (*Adapted from Aguilera and Stanley, 1990.*)

cieties. Whatever the purpose, gels derived from food biopolymers are extremely attractive high-moisture, semi-solid matrices.

The objectives of this chapter are to discuss the physicochemical bases by which gels can be transformed into simulated foods, to study their microstructural features, and to analyze the processing implications and textural properties of products. This review will not refer to uses of gels as binders of comminuted pieces (a subject reviewed by Pearson and Dutson, 1987).

FOODS AS COMPOSITE MATERIALS

Composites of two or more materials may be produced to attain a unique property that neither of the original components possess. Daily life applications of polymer composites with improved mechanical properties abound: rubber tires, fiberglass, plywood, and flexible packaging, among others. Our own body also utilizes composite structures when unique mechanical functions must be obtained and a few components are available. It is amazing, for example, that soft connective tissues perform in tension, as in tendons, or in compression, when part of cartilages; yet both mechanical roles result from a single basic structure—collagen fibrils embedded in a gel-like matrix (Hiltner et al., 1985).

Foods are highly heterogeneous materials. High-moisture, semi-solid foods derived from biological tissues or fluids can be modeled to a first approximation as composites, where the mechanical (or textural) properties result from the interplay of at least two microstructural elements, one of which is a gel (see Table 7.1). Meat, bread, hard cheeses, fresh fruits and vegetables, covering a wide spectrum of foods groups, can be classified, respectively, as a fibrous composite, a solid foam, a particle-reinforced composite, and a cellular material. Figure 7.1 shows the microstructural appearance of several foods leading to their proposed classification as composites. For the purpose of this presentation, simple gels will be considered as basic structures to be modified to achieve some desirable foodlike properties. The approach will consist of viewing the target foodlike structures as gel composites.

CHARACTERISTICS OF COMPOSITE MATERIALS

Composites with improved mechanical performance may be classified into three basic categories: materials containing dispersed particles, materials containing fibers, and laminates. Concrete, fiberglass, and plywood, respectively, are typical examples of these materials. Foamed and fibrous materials also have peculiar mechanical properties, with paper and polystyrene foam being representative (Fig. 7.2).

Addition of particles or fibers to polymeric matrices may produce several desirable mechanical effects: increases in stiffness, creep resistance, and fracture toughness. A fundamental difference exists between particle- and fiber-reinforced materials. In particulate reinforcement the matrix and dispersed particles share the load. In fiber-reinforced composite materials the matrix transmits the load to the fibers, which are the primary load-bearing constituents (Broutman and Krock, 1967). In both cases the improved mechanical behavior results from a complex interplay between the matrix, the filler, and the interfacial region (i.e., the state of adhesion between filler and matrix).

The most important properties of particulate materials when used as reinforcing agents are the size, shape, surface properties, and the state of distribution in the matrix. If there is no bonding between filler and matrix or if the rigidity of the filler is less than that of the matrix, the modulus of the composite will decrease with increasing amounts of filler. Models describing the effect of particles in the modulus and strength of polymeric materials have been recently reviewed by Ahmed and Jones (1990). The model most often used to predict the dependence of the modulus of a food composite (E_c) with composition is the one developed by

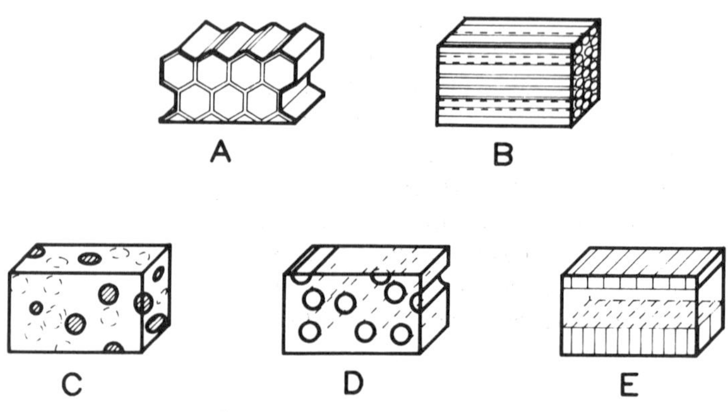

Fig. 7.2 Heterogeneous basic structures in solids: (A) solid foams; (B) fibrous solids; and the three main types of composite materials; (C) with particulates; (D) with fibers; and (E) laminates.

Hirsh for rigid inclusions in a rigid matrix (Morris, 1986; Clark, 1988):

$$E_c = k(E_p f_p + E_m f_m) + (1 - k) \frac{E_p E_m}{E_p f_m + E_m f_p}$$

where E and f are the moduli and the volume fraction of particulates, p, and the matrix, m, respectively. A lower bound, which is identified with a poorly bonded filler, is predicted for $k = 0$; for a perfectly bonded filler, $k = 1$; and the upper bound solution reduces to the additivity rule ($E_c = E_p f_p + E_m f_m$). Parameters k and $(1 - k)$ are the relative proportions of material conforming to the upper and lower bound solutions, respectively. The empirical parameter k can be determined by curve fitting (Ahmed and Jones, 1990). At a certain intermediate volume fraction, phase inversion occurs (e.g., a composite having a matrix of material *A* and particulates of *B* changes to a matrix of material *B* containing inclusions of *A*), and mechanical properties may change accordingly.

Our challenge is to creatively adapt concepts of composite materials technology and apply them soundly to food development. This has been done in the past with extrusion and fiber spinning. A concomitant task is to relate data for mechanical properties of food composites with target textural properties.

SOLID FOAMS AND SPONGES

A variety of solid foods, most notably bakery products, have a porous and open macrostructure resulting from a dispersion of gas bubbles in a matrix (Fig. 7.1D). They are referred to, somewhat loosely, as foams or sponges and are low-density, porous, thin-walled structures. While sponges are capable of imbibing large quantities of solvent without dissolving, foams do not have that property and would normally dissolve in the presence of water.

Starches and sometimes proteins can be induced to foam by two main mechanisms: (1) sudden vaporization of a solvent, usually water, as in extrusion-cooking or after dipping in hot cooking oil, or (2) in situ generation of a gaseous phase (usually carbon dioxide and/or water vapor) as in proofing and dough baking. This latter mechanism is extensively used to produce insulating foams in polymer processing and is also used in food processing. Vapor-induced puffing occurs in extrusion processes when starch-based extrudates pass through dies at temperatures greater than 100°C. A related technology, explosion-puffing, is also used to expand starchy materials and produce ready-to-eat breakfast cereals. It has also been utilized to expand vegetables prior to and during drying to facilitate rehydration. Vegetables are also puffed by high temperature heating near the end of the drying process.

During extrusion cooking starch granules become progressively hydrated, disrupted, heated, and compacted into a dough, as they move along the extruder barrel. The native semi-crystalline structure of raw starch granules is gradually lost as hydration proceeds and temperature increases until gelatinization or melting set in (Zobel, 1988). As the hot melt approaches the die, further degradation into dextrinized forms occurs. If a ternary diagram representing blends of raw, gelatinized, and dextrinized starch is constructed, starchy extrudates have physicochemical properties similar to blends positioned in the hatched area of Fig. 7.3. Porosity increases as more dextrinized forms are attained. Solid starch foams (cross-hatched area in Fig. 7.3) are expanded products, soluble in hot water, and yield pastes with low viscosity and high susceptibility to α-amylase attack (Gomez and Aguilera, 1983, 1984).

Expansion of the melt after exit at the die occurs by viscous flow induced through the combined action of two mechanisms: (1) the release of normal stresses in the extrudate (the so-called die swell phenomenon) and (2) growth of minute bubbles of superheated steam after pressure release. Expansion due to die swell is usually relatively small compared to that due to vapor expansion. Water vapor evaporation induces rapid cooling of the melt increasing the viscosity until flow at the walls of air

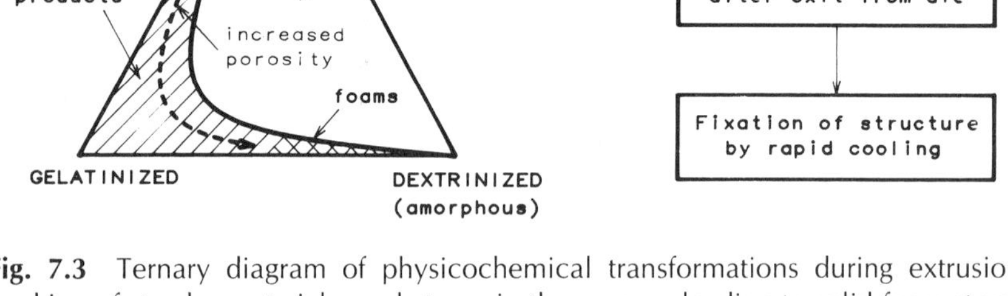

Fig. 7.3 Ternary diagram of physicochemical transformations during extrusion cooking of starchy materials, and stages in the process leading to solid foam structures.

cells ceases, resulting in a porous solid or foam. Starch in cell walls of low-moisture foams is present in a glassy state and extensively degraded to low molecular weight dextrins (Gomez and Aguilera, 1984; Colonna et al. 1989). The main sequence of transformations involved and stages of the extrusion process of starch foams are represented in Fig. 7.3.

The texture of starch-based, solid foams depends on the amylose:amylopectin ratio. High-amylose products are hard and dense, whereas high-amylopectin starches give fragile products, hence some amylose is required to resist breakage and give adequate texture (Matz, 1984). An amylose:amylopectin ratio to close to 50:50 has been recommended for optimum expansion (Chinnaswamy and Hanna, 1988). The microstructures of solid starchy foams made from corn and wheat are presented in Fig. 7.4. The expansion ratio (diameter of extrudate:diameter of the die) achieved after extrusion cooking of starch in a twin extruder

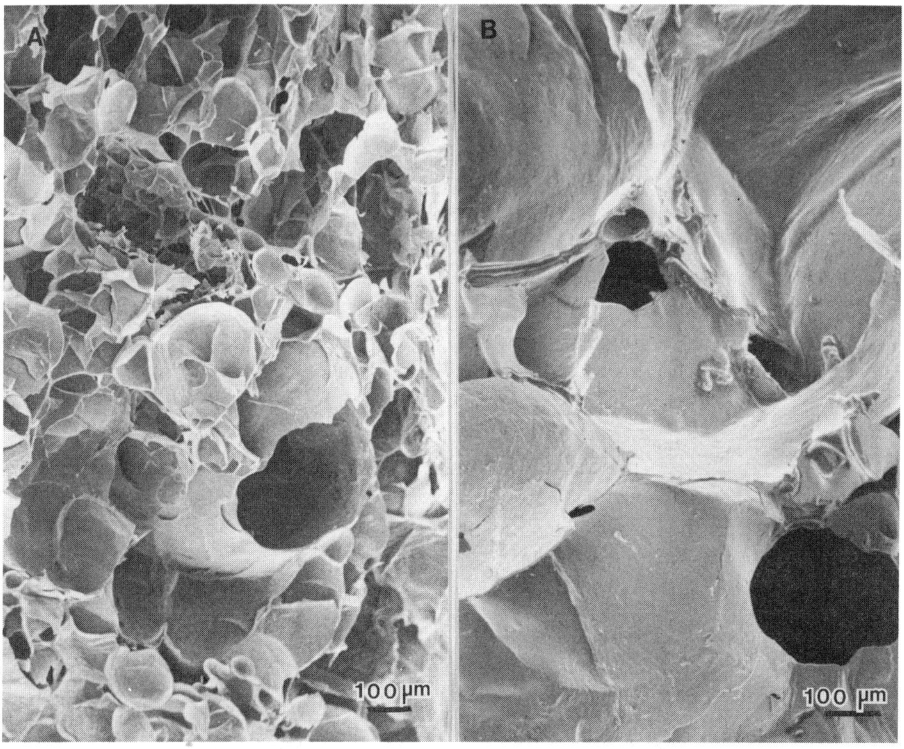

Fig. 7.4 Scanning electron micrograph starchy solid foams made by extrusion cooking: (A) corn meal; (B) wheat flour.

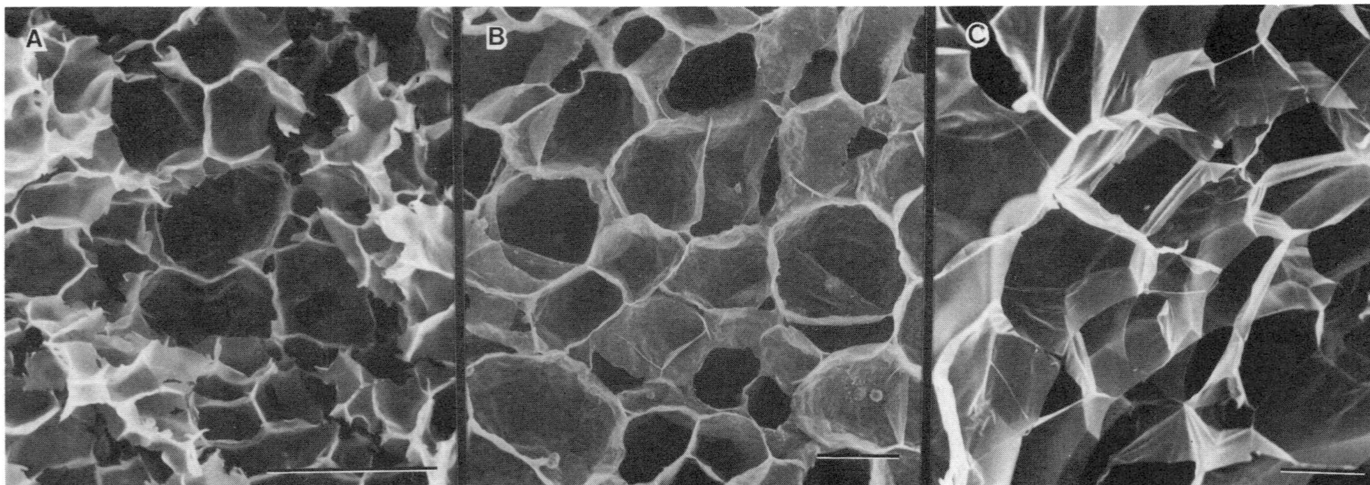

Fig. 7.5 Scanning electron micrographs of porous and cellular structures: (A) starch sponge; (B) cells in sugarbeet; (C) man-made polystyrene foam.

can be as high as 6.1, and densities of the solid foams may range between from 0.04 to 1.79 g/cm^3 (Hayter et al., 1986). Increased expansion can be attained by decreasing the die size (increase the pressure), extruding at higher temperatures (e.g., ca. 200°C), and reducing the moisture content of the dough (e.g., to 16%). Higher pressures increase the solubility of water in the polymer melt facilitating a more uniform nucleation of bubbles and resulting in a narrower pore distribution. Mechanical properties of solid foams from the perspective of foods have been amply discussed by Jeronimidis (1988), who concluded that the abundant theoretical basis available for predicting mechanical properties and failure mechanisms should be assessed for food foams and related to textural properties.

An interesting use of starchy food foams is as matrices for vacuum infusion of high melting point fats blended with other dehydrated ingredients, even small particulates (Barret and Ross, 1990). In summary, foam fabrication by extrusion cooking involves complex physicochemical principles (crystalline to amorphous transitions, nucleation and growth of bubbles, solubility of gases) and intricate rheological problems.

Freeze-texturization (to be discussed later) has been used to produce porous starch-based sponges (Torres, 1977). The process involves preparing starch or amylose dispersions, autoclaving at high temperature with agitation, cooling to an intermediate temperature and molding while hot. After molding the product is further cooled to room temperature and the gel formed is subsequently frozen. Control of the microstructure is achieved during the heating and freezing steps and to a certain extent by the formulation: sponges prepared from high-amylose corn starch have a "fibrillar crab meat-like" character while those prepared from pure amylose are "flaky meat-like" (Torres, 1987). Stronger sponges are obtained by high-temperature heating (e.g., autoclaving at 163°C) and slow freezing at temperatures slightly below the freezing point (Bice and MacMasters, 1944). Contrary to foams, sponges consist largely of gelatinized starch and thus possess high water-holding capacity and low solubility (Gomez and Aguilera, 1983). In the diagram of Fig. 7.3 sponges would be located close to the "gelatinized" vertex. Microstructural aspects of one type of starch sponge are presented in Fig. 7.5A and compared to the architectural features of the cell wall matrix of sugarbeet (Fig. 7.5B) and of an expanded polystyrene foam (Fig. 7.5C).

FIBROUS STRUCTURES

High meat prices and changing food habits led to the development of so-called textured and spun proteins derived from soybeans. The processes used were an adaptation of the manufacture of thermoplastic products and man-made textile fibers, respectively. In both processes, globular proteins in an aqueous medium must unfold into chains, stretch, and interact with each other until a heat-stable matrix is formed. Soy products textured by extrusion cooking consist of a continuous protein matrix with fibrous appearance and a discontinuous amorphous carbohydrate phase. Spun soy fibers are protein filaments which can be bound together to simulate meat. Fabrication methods for these "conventional" protein products will not be reviewed here; the reader is referred to Aguilera and Stanley (1986, 1990) for details on the biochemical and microstructural principles involved.

A novel technique to achieve a fibrous structure in protein/protein or protein/polysaccharide systems is the "spinneretless spinning" process. As described by Tolstoguzov (1988a), the process is based on three fundamental phenomena: (1) thermodynamic incompatibility in aqueous solutions of proteins or of protein/polysaccharide mixtures, leading to a two-phase system, (2) flow-induced deformation and orientation of dispersed elements in the two-phase system, and (3) gelation of one or both phases. According to Tolstoguzov (1988a) the process results in three types of products: fibers, capillary gels, or fiber-filled gels depending on whether the dispersed or the continuous phase or both are gelled (Fig. 7.6).

Thermodynamic incompatibility between polymers in solution is a re-

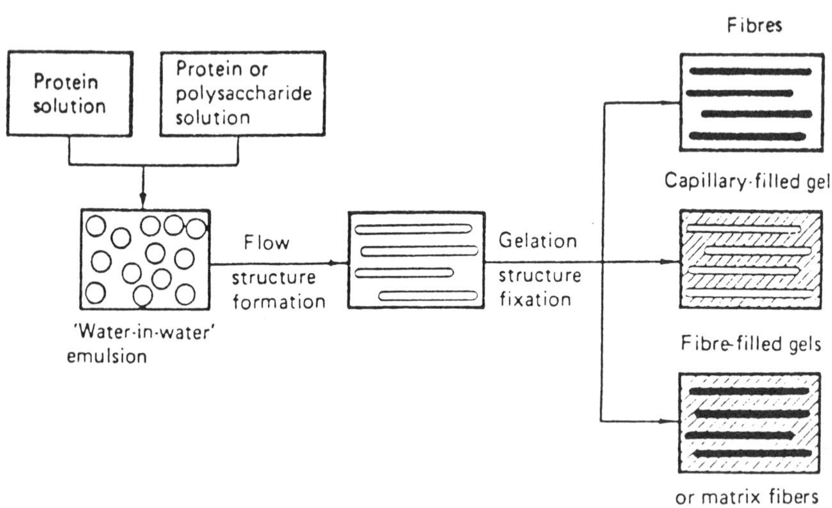

Fig. 7.6 Type of products generated by the spinneretless process. (*From Tolstoguzov, 1988a.*)

current concept in the generation of heterogeneous structures in gels. (For a thorough coverage of the thermodynamic principles, see Flory, 1953.) The Gibbs free energy of mixing, which has to be negative for spontaneous miscibility, consists of an entropy term and an enthalpy term. The entropy term, representing the variety of ways of arranging the solvent and polymer molecules, is remarkably simple: it depends on the number of moles and on the volume fractions of polymer and solvent. Hence, on a *molar* basis this term should be small for a polymer solution. The term for the enthalpy of mixing, is more complex. It involves the number of chain segments in a polymer and the interaction factors (c_i) between pair of chain segments and pair of solvent molecules and between one polymer segment and one solvent molecule. For polymers having a large number of segments, the interaction energy per mole will tend to dominate the entropy of mixing per mole.

Suppose two polymer molecules repel each other. Mixing solutions of these two polymers will result in an energetically stable state only when molecules of the two polymers are surrounded by their own kind, e.g., two phases are formed, each one containing one class of polymer. Polymers are said to be incompatible in solution. Interestingly enough, this is a common result when mixing two polymer solutions.

In particular, separation in two phases is typical of many protein solutions, including (1) solutions of proteins belonging to different classes within the Osborne classification (i.e., albumins, globulins, glutelins and prolamines), (2) systems of globular and random coil proteins, and (3) mixtures of native and denatured forms of the same proteins (Tolstoguzov, 1988b). A typical phase diagram for an aqueous two-phase system of proteins is presented in Fig. 7.7. The binodial curve separates the region of one- and two-phase systems, and the tie lines (dashed lines) connect the composition of the coexisting phases.

The actual fabrication process starts with the preparation of a two-phase aqueous polymeric system and is followed by deformation and shaping by flow through a nozzle. Gelation of either phase can be achieved by mediation of heat, metal ions, or concentration of the solution (Fig. 7.7). A wide range of mechanical properties can be expected depending on the type of composite formed, properties of individual gelling agents, and relative proportion between the phases. For example, in many cases the moduli of elasticity and the tensile strength are linear functions of the volume fraction of one of the phases and obey the additivity rule of composites. Micrographs of casein-pectin fibers show a fibrous structure similar to spun soy fibers, indicating that fibers may be used to imitate meat products (Tolstoguzov, 1988b).

Textural modification of proteins by freezing is an old process used to

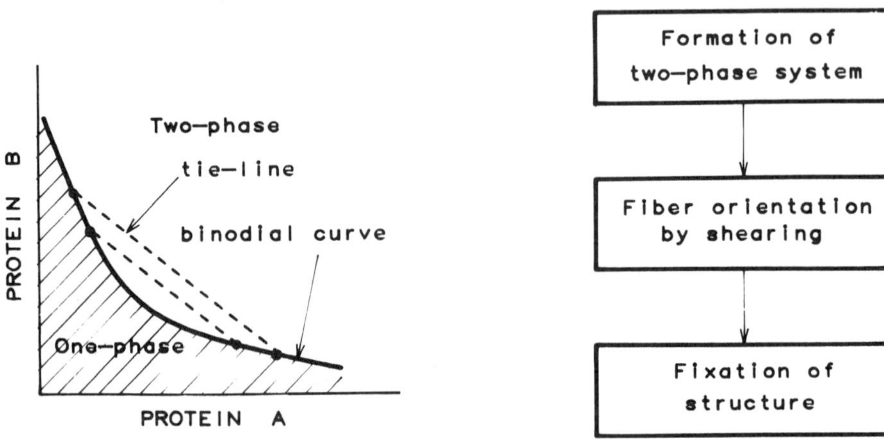

Fig. 7.7 Thermodynamic phase diagram of an incompatible binary protein solution and process stages in the spinneretless spinning process.

produce kori-tofu, a porous, spongelike product from soybean curd consumed in Japan. The physicochemical principle of freeze texturization of freeze alignment is similar to that of spinneretless spinning in that two phases are initially formed (ice and concentrated solution/suspension), in this case by cooling below the freezing point of the solution (Fig. 7.8). Fibrillar structure is induced by controlled growth and orientation of the

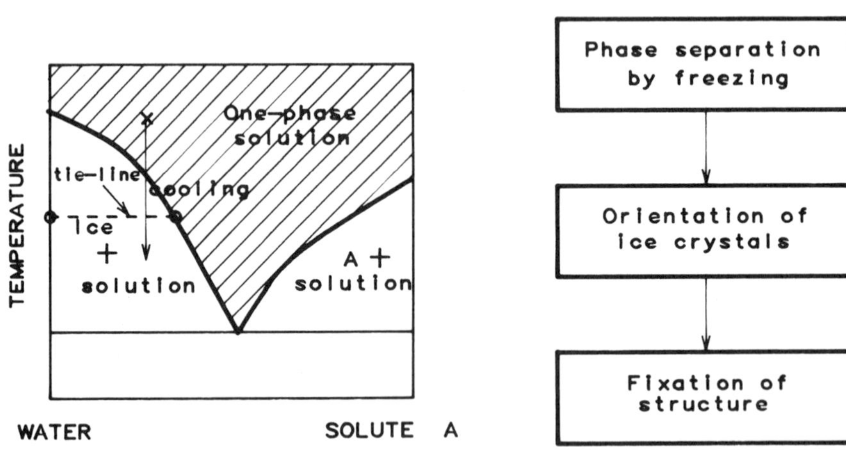

Fig. 7.8 Thermodynamic phase diagram of water-solute mixtures and process stages in freeze texturization.

ice phase favored by unidirectional freezing and is later fixed through protein-protein interactions (Lugay and Kim, 1981; Lawrence et al., 1986). Figure 7.8 shows the main stages in the freeze texturization process.

A thorough discussion of physicochemical principles involved in the freeze texturization process and a detailed comment of various patents is presented in Lillford (1985). He noted that the freezing rate is critical as to whether a spongy or fibrous structure is formed. Contrary to freeze preservation of foods, freeze texturization relies on slow freezing so that the rate of crystal growth exceeds the rate of nucleation. To provide homogeneous texture, uniform slow freezing is required. This is difficult to achieve, because heat transfer rates and solute concentrations change as freezing fronts advance. To alleviate this problem and force fiber formation, use of mechanical means to break ice crystals and induce alignment has been proposed in some patents. Fibrousness increases with decreasing solids concentration and closeness to a pH of maximum protein solubility (Lugay and Kim, 1981). Fixation of the frozen fibrillar structure is achieved by thermal denaturation, use of chemical fixatives for cross-linking or denaturation of the protein during freeze-drying. Figure 7.9 presents the fibrous microstructure of freeze-dried soy milk frozen unidirectionally at −76°C. Fiber formation by freezing is relevant not only to food technology but also in materials science, from which useful know-how can be adopted. In the production of long zirconia fibers from freezing solutions, the ratio between the temperature gradient and the

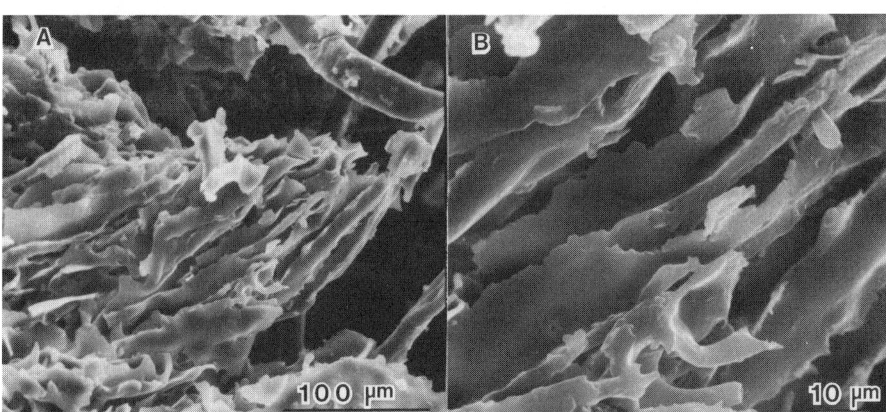

Fig. 7.9 Scanning electron micrographs of (A) soy milk after unidirectional freezing at −76°C and freeze-drying; (B) larger view of central area in A.

freezing rate was the preferred control variable used to optimize fiber length (Kokubo et al., 1988).

A "pre-gel" or micellar state favored by hydrophobic interactions between protein molecules in solution has been studied by Murray et al. (1981). The process exploits salting-in phenomena, where proteins (mainly of plant origin) are exposed to a salt solution (i.e., 0.3 M NaCl) and self-associate into micelles when the extracted mass is diluted in water. Protein micelles (1–10 μm in diameter) form fibers when injected in hot water. A possible advantage of this method over traditional fiber spinning would be an increase in the heterogeneity of the fibers, which then would mimic the complex microstructure of meat fibers.

REINFORCEMENT BY SOLID PARTICLES AND FIBERS

Many foods are composites in which a particulate material (filler) is embedded in a gel matrix, thereby affecting rheological properties. Particulate fillers are responsible for essential textural characteristics of some traditional foods. For example, the cells of cooked potatoes and legumes are gels filled with gelatinized starch granules. Filling food matrices with particulate or fibrous materials may accomplish one of three objectives: (1) mimic the structure of natural food composites, (2) accommodate an ingredient which comes in such forms, or (3) replace an expensive major component of the matrix without weakening the structure.

Particle size and interfacial interaction with the matrix have a large influence on the mechanical properties of particulate food composites (Ring and Stainsby, 1982; Brownsey et al., 1986). Whey protein gels filled with solid glass spheres having modified surfaces or with oil droplets coated by a protein layer were tested in compression and their microstructure examined (Langley and Green, 1989). Particles that were integral part of the composite (active fillers) induced several desirable characteristics: (1) high reinforcement independent of the nature of the filler, (2) good bonding revealed by fracture within the protein matrix, and (3) high extents of reinforcement (e.g., a high ratio of modulus of the composite to that of the unfilled matrix), which increased almost exponentially with the volume fraction of the filler. On the contrary, gels containing particles that did not interact with the matrix were strengthened to a lesser extent, and only when the fractional volume of particles exceeded 0.3. Fracture in compression happened adjacent to the particle surface indicating poor interfacial bonding. Gels containing oil droplets were only produced when the particle diameter was less than about 1 μm and for oil volume fractions of less than 0.42.

Surimi is a good example of an underutilized food resource transformed into structured products that are in great demand. Surimi is an intermediate gel produced by washing refined fish mince, mainly Alaska pollack, and mixing it with cryoprotectants for acceptable frozen storage. Its most important functional attribute is the gel-forming ability of the undenatured myofibrillar proteins when heated above 70°C. Further development of the expanding fish gel market is limited (1) by the decrease in stocks of white, low-fat fish, and (2) by the relatively poor efficiency of the surimi process (yields are approximately 20% based on whole fish). This has prompted studies of use of small fatty fish (pelagic fish) abundant on the Pacific coast of South America (catches of ca. 10 million tons/year) and of combinations of surimi (or kamaboko) with other gel-forming ingredients or fillers.

Potato starch has been a preferred adjunct for diluting fish gels; in fact, most of the kamaboko sold in Japan contains between 5 and 20% added starch. Addition of starch at 10% displaces the firmness vs. gelation temperature curve for surimi upwards, as shown in Fig. 7.10 (Susuki, 1981). Starch-containing surimi gels became firmer when potato starch was added as a filler at levels up to 15% (Fig. 7.10); the reinforcing effect peaked at 6% addition (Lee and Kim, 1986). These authors surmised that this effect was probably due to a combined effect of a denser protein matrix (by water partition between starch and protein) and the presence of gelatinized starch granules acting as plastic inclusions. Dilution of surimi with egg albumin weakened the gel to an extent that could not be surmounted even by addition of starch granules. Better knowledge of the fundamental interactions between gel-forming myofibrillar proteins in fish and other gelling agents or fillers would allow for a more rational search for new adjuncts for development of less costly fish gels.

Surimi has also a unique ability to "set" at low temperatures and can be used to bind food pieces while avoiding precooking or freezing. Torley and Lanier (1989) demonstrated that surimi and meat proteins are compatible and gel when added in all proportions. However, excessive extraction of myofibrillar protein in meat should be avoided since it dilutes the setting effect of surimi. These findings may allow utilization of mixtures of cheaper meat cuts and fish mince in fresh meatlike products formed by a thermo-set process.

Fiber-filled composites have not found a place yet in food development, apparently due to a lack of suitable sources of edible fibers. Intact fish fibers can be obtained by controlled disintegration of fatty fish (Rutman et al., 1988) and utilized as a low-cost extender for surimi gel. Work being performed at this laboratory on the effect of added fish fibers on the texture of surimi (Tabilo and Aguilera, unpublished data) has shown that no significant changes in penetration or compression forces occur when up to 0.5 weight fraction of fiber is added (Fig. 7.11). At higher levels of addition the strength of composites increases, but textural attributes depart from those of surimi (i.e., surimi acts as a binder). Since no reinforcement occurs, the effect of adding fibers to surimi apparently conforms to a lower-bound situation for composites, denoting poor bonding between fibers and matrix. The microstructure of a 80:20 gel-fiber composite is more porous than that of surimi and poor bonding between fibers and the matrix is corroborated (Fig. 7.12); however, textural characteristics of the composite, as measured by folding tests, were satisfactory. Factors known to affect adhesion between meat fibers and gels include the wettability of fibers (improved by absence of fat or connective tissue on the surface), type and concentration of adhesive (i.e., extracted pro-

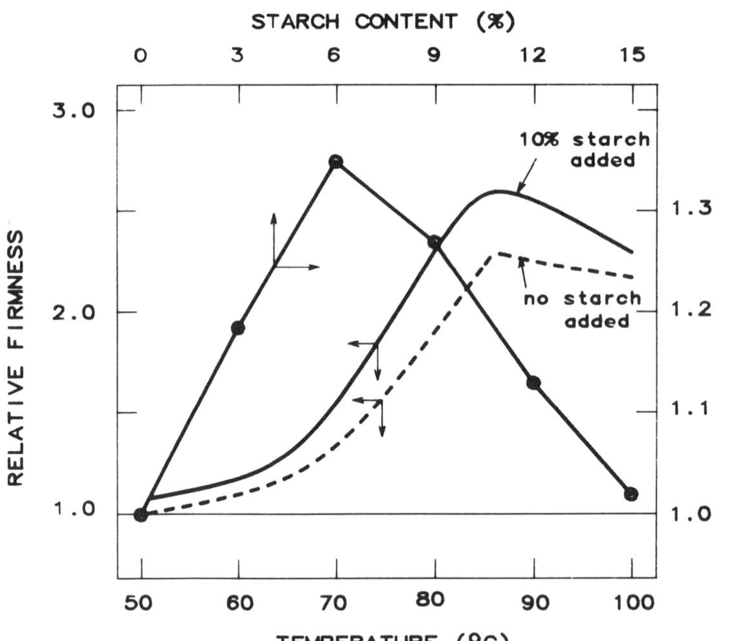

Fig. 7.10 Effect of the presence of starch granules on gelation temperature and firmness of a surimi gel. Relative firmness = firmness of composite/firmness of surimi. (*Adapted from Susuki, 1981, and Lee and Kim, 1986.*)

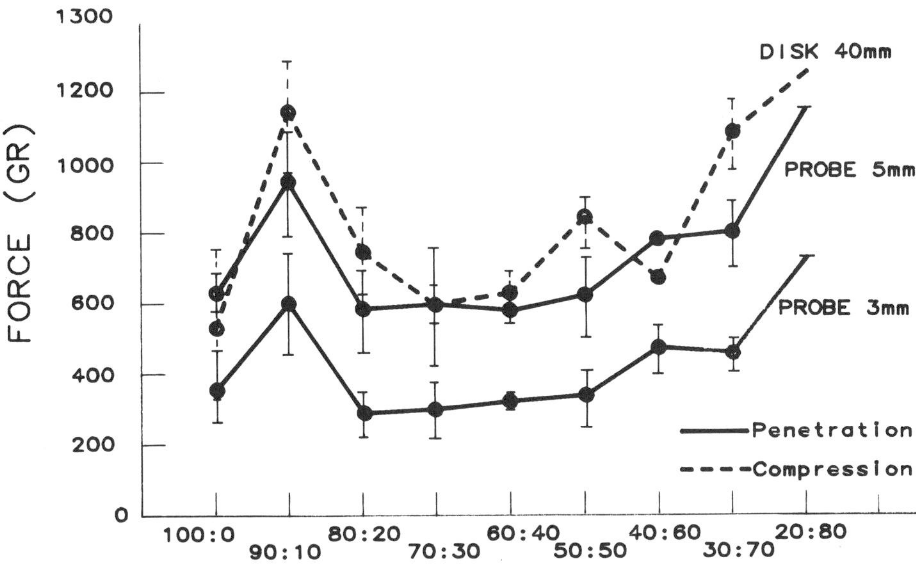

Fig. 7.11 Compression (at 20% deformation) and penetration forces (with two probe sizes) in surimi/fish fiber composites. Bars indicate ± one standard deviation. (*From Tabilo and Aguilera, unpublished data.*)

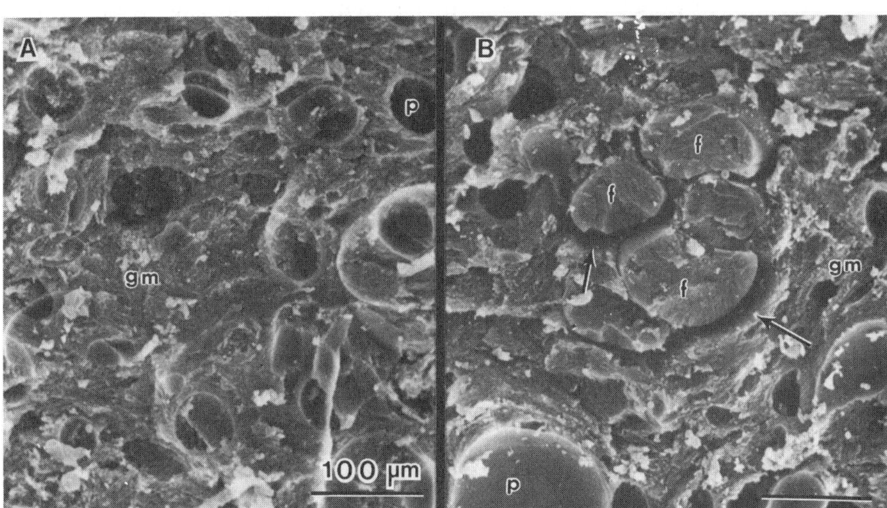

Fig. 7.12 Scanning electron micrographs of (A) surimi gel and (B) surimi/fish fiber composite, 80:20 ratio. Arrows show poor bonding between fibers and the matrix of the composite. f = fibers, gm = gel matrix; p = pore.

tein), quality of the dispersion medium (i.e., level of salts, pH), and temperature of heating (Jolley and Purslow, 1988).

A word should be said at this point regarding mechanical testing of food composites. Assays should include two limiting cases important to foods: (1) small deformations at slow speeds, conditions relevant to structural integrity and present during processing, packaging and transport, and (2) fracture under high rates of deformation, a situation relevant to texture or mouthfeel (Green et al., 1990).

MIXED DAIRY GELS

It is economically advantageous to substitute an expensive gelling agent by cheaper ones if product functionality is not impaired. This has led to an obtuse search for gel substitutes performed mostly by trial and error. Gel properties can be dramatically changed, rarely positively and usually negatively, by addition of a second gelling agent. In commercial applications one seeks for synergistic interactions or added functionality of mixed gels.

The difficulty in predicting the functional outcome of a mixed gel system stems from the many parameters involved and a lack of understanding of relevant physicochemical phenomena. When two polymers A and B are mixed together and gelled, several types of gel structures may arise, depending basically upon the thermodynamic compatibility between components and their respective mechanisms of gelation. Possible microstructures proposed for mixed protein gels (Ziegler and Foegeding, 1990) and mixed polysaccharide gels (Cairns et al., 1987) are represented in Fig. 7.13. The simplest microstructure is formed when only one polymer gels and the other remains in solution (A, single polymer network). More complex forms may occur if both polymers gel and form separate networks but interlace one another (B, interpenetrating network) or are definitely segregated (C, phase-separated or composite network). Strong interactions and reinforced microstructures are likely to be obtained if binding occurs between both networks (D, coupled network) or if monomers copolymerize in the same network (E, copolymer network).

An effort to predict the type of microstructure of a mixed gel from phase equilibrium data has been undertaken by Russian scientists (Gotlieb et al., 1988). As previously reviewed, when two thermodynamically incompatible polymers are mixed, a single or a two-phase solution may be formed depending on the relative concentrations and their relation to the binodial curve (Fig. 7.7). Several of the mixed gel microstructures shown

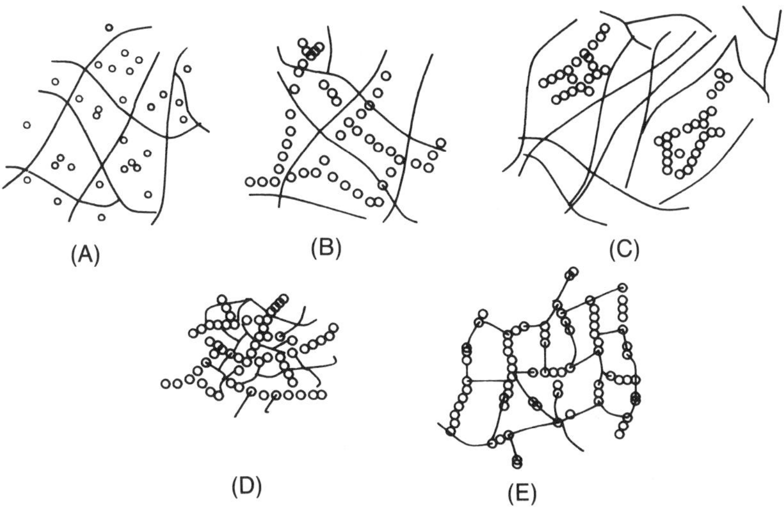

Fig. 7.13 Proposed structures for mixed gels. (A) Single polymer network; (B) interpenetrating network; (C) phase-separated or composite network; (D) coupled network; (E) copolymer network.

in Fig. 7.13 could be predicted from solubility diagrams if quasi-equilibrium conditions prevailed during gelation. For example, single polymer networks may occur in a mixture that is above the critical gel concentration of one polymer but is very diluted in the other (zones I and III, Fig. 7.14C). Phase-segregated gels would be produced if the original solution lays in the two-phase region (zone III, Fig. 7.14C). Since gelation is a kinetic process, nonequilibrium conditions are likely to predominate, changing the relative concentrations of the reacting species with time, thus making this fundamental approach of little practical value. In fact, some processes to generate engineered structures rely on differences in rates of gelation to trap one gel in another. Since the phenomenon involved occur at microscopic level (Clark et al., 1982), transmission electron microscopy has been a valuable tool in studying phase separation and its effect on the homogeneity of mixed gels. The discussion that follows will concentrate on mixed dairy gels; binary polysaccharide gels are discussed in Cairns et al. (1987) and Morris (1986); mixed protein gels are reviewed in Ziegler and Foegeding (1990).

An interesting gelling mixture is the water–casein–whey proteins system of milk. Whey proteins form gels upon heating by a denaturation-aggregation process typical of globular proteins (Mulvihill and Kinsella, 1987); casein gels are formed by aggregation of casein micelles induced

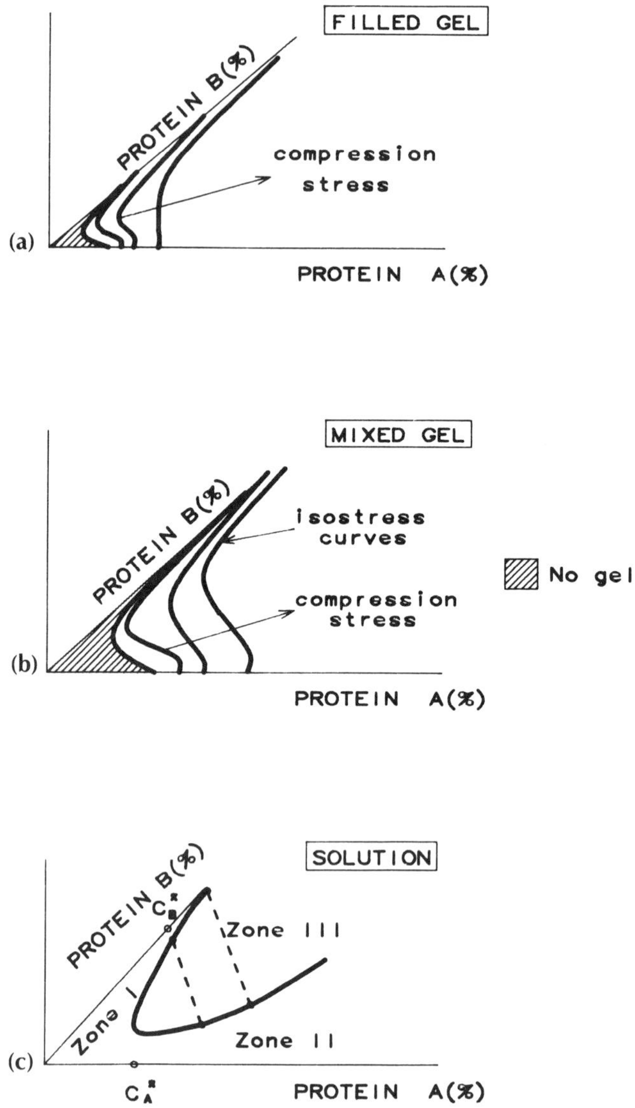

Fig. 7.14 Phase diagram of solutions of protein A and B (highly speculative) and compressibility contour plots for mixed and filled dairy gels made from whey protein (protein A) and skim milk (protein B) at pH 4.6–4.8.

either by charge neutralization during acidification (Heertje et al., 1985) or by enzymatic action. Most commercial, gelled dairy products (e.g., yogurt, cheese curd) have a structure consisting of a casein network with whey proteins remaining largely in solution. However, both protein species do interact during processing, as demonstrated by denatured β-lactoglobulin adhering to casein micelles (Davies et al., 1976), bridging of casein chains by whey-protein strands in yogurt-type gels (Modler and Kalab, 1983), and gelation of κ-casein and β-lactoglobulin complexes when heated above 70°C (Doi et al., 1983). In this latter case, an -SH/-S-S- interchange as well as a hydrophobic bond have been implied in formation of the complex.

The present large availability of skim milk powder (SMP) and refined whey concentrates (WPC) and isolates (WPI) in the world market has prompted study of mixed dairy gels. In the near future, micellar casein produced by microfiltration could also become a widely available ingredient. Formation of mixed casein–whey protein gels require controlled acidification and heating for both gelling mechanisms to operate. In model systems, this is achieved by the addition of an acid-generating source, e.g., glucono-δ-lactone (GDL), to the mixed protein solution and by heating over 70°C. Expected advantages of mixed dairy gels over single gels are wider functionality over an extended pH range.

Synergistic effects have been demonstrated for mixed gels made from SMP and WPC heated to 90°C in the presence of GDL. Yogurt-type gels having 10% total solids (TS) were firmer in a 50:50 blend than gels made from either SMP or WPI alone (Aguilera and Kessler, 1989). Results from an extension of this work to a broader protein concentration and pH ranges (SMP, 0–20%, WPI, 0–10% TS and 3.6–3.9, 4.6–4.8, 5.1–5.3, respectively) are represented as contour plots of isostress curves (curves of equal compression force at 20% deformation) in Fig. 7.14B (Aguilera and Kinsella, 1991).

Physicochemical interpretation of these results at present level of knowledge is highly speculative but is presented as an effort to stimulate future working hypotheses. A typical phase equilibrium diagram for a system protein A–protein B and water in which both proteins are incompatible in solution is shown in Fig. 7.14c; Fig. 7.14a and b represent a textural characteristic (compressibility) of mixed and filled gels made from these solutions. Four zones are clearly distinguishable in the phase diagram that may affect differently the gelation process (Fig. 7.14a, b). These zones are (1) an area close to the origin, where the protein concentration of solutions is below the critical concentrations for gel formation (C_A^* and C_B^*) and is related to the shaded areas (no-gel) in the middle and upper diagrams; (2) zone I, in which addition of small amounts of protein A to solutions predominantly of protein B strengthen the structure of gels and allow gelation to occur at lower concentrations of B (middle diagram). This case could correspond to formation of a copolymer gel (Fig. 7.13). In the case of dairy proteins, this effect is in accord with the reported compatibility between casein (protein B) and whey proteins (protein A) during gelling; (3) zone II, where mixed gels made from solutions of A having small amounts of protein B added show no reinforcement, in spite of the increased protein content. Formation of a gel matrix of A occluding a solution of B can be hypothesized in this case; and (4) an intermediate region where gels having a constant solids content present similar compression stresses. The different biochemical nature of casein and whey proteins suggests that dairy gels whose compositon fall in zone III may be phase-separated networks.

FILLED DAIRY GELS

Many dairy products like cheeses and yogurt contain fat. Fabricated filled dairy gels must accommodate the lipid phase without weakening the microstructure. Fat in engineered dairy foods affects texture and acts as a carrier for flavors, colors, and nutrients (i.e., fatty acids and vitamins).

Milk fat globules whose membranes have been modified by homogenization are regarded today as "active" building blocks for new dairy products (Buchheim, 1986). The presence of an artificial protein membrane in recombined fat globules or a modified membrane having adsorbed casein (and whey proteins) in natural fat globules changes their physicochemical and mechanical properties considerably (Aguilera and Kessler, 1988). Indeed, dairy products have been shown to acquire different rheological properties when fat globules from homogenized milk participate in acid or enzymic gelation processes (Buchheim et al., 1986). Fat globules with new protein membranes behave as large protein units. Aqueous emulsions of fat globules with adsorbed casein micelles form a weak curd when the pH drops to 4.6, as casein micelles would. When used as fillers in composites, recombined fat globules provide a shear modulus 7.5 times higher than that of natural milk fat globules (Luyten and van Vliet, 1990). Moreover, recombined fat globules made with natural or hydrogenated vegetable oils and coated with a casein or whey protein layer can be used in dairy gel analogs with minimal losses in functionality with respect to milk fat globules. The old unit operation of homogenization has now been revived as an effective means of producing "active" fat globules with new protein coatings and small particle dimensions (i.e., $<1\ \mu m$).

The reinforcing effects of protein-coated fat globules on firmness (compressive strength) of filled dairy gels and the opposed action of membraneless fat globules (emulsified with the surfactant Tween 20) or nonhomogenized natural milk fat globules are shown in Fig. 7.15. The protein adsorbed at the surface of these "active" fat globules makes them perform as polymerization nuclei for protein molecules in solution and allows formation of a copolymer network, equivalent to that shown in Fig. 7.13E. The microstructures of mixed and filled dairy gels are shown in Fig. 7.16. Mixed gels (Fig. 7.16A) have thinner chains than filled gels (Fig. 7.16B) due to the absence of fat globules in their protein networks. Actual involvement of fat globules in the network and formation of a copolymer

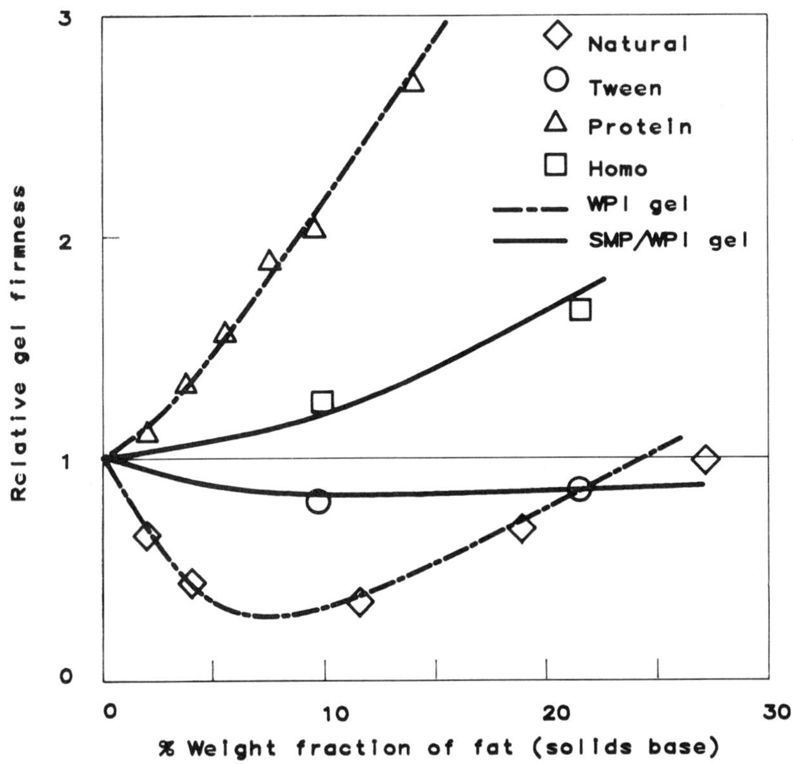

Fig. 7.15 Effect of active and non-active fillers on the compression strength of: (— - — -) whey protein gel; (———) mixed skim milk/whey protein gel. Nomenclature: ◇, washed natural fat globules; □, homogenized fat globules; ○, fat globules stabilized with Tween 20; △, reformed fat globules stabilized with whey protein.

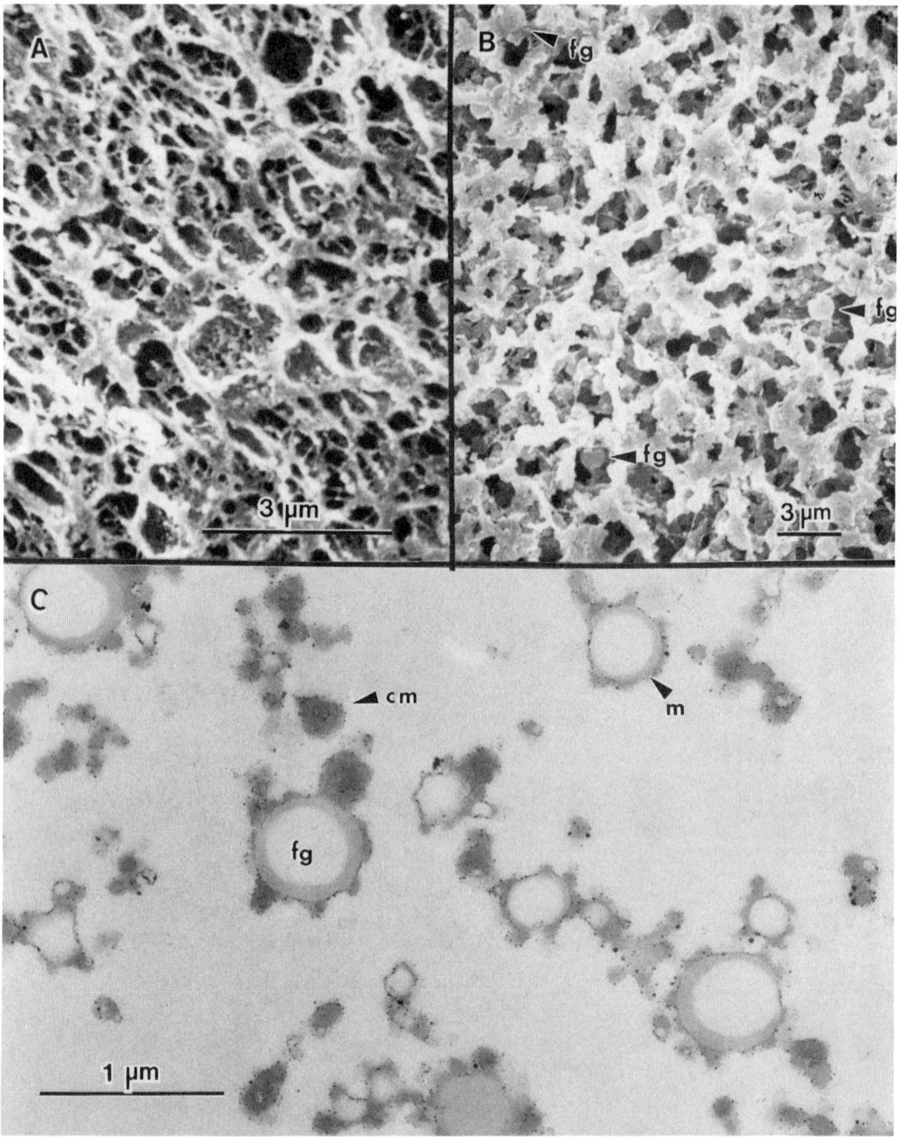

Fig. 7.16 Scanning electron micrographs of mixed and filled dairy gels (cryopreparation). (A) Mixed gel WPI(2%)/SMP(5%); (B) filled WPI(3%)/SMP(8%), arrows point at fat globules (fg); (C) transmission electron micrograph of the filled gel, fg = fat globules, cm = casein micelles, m = membrane.

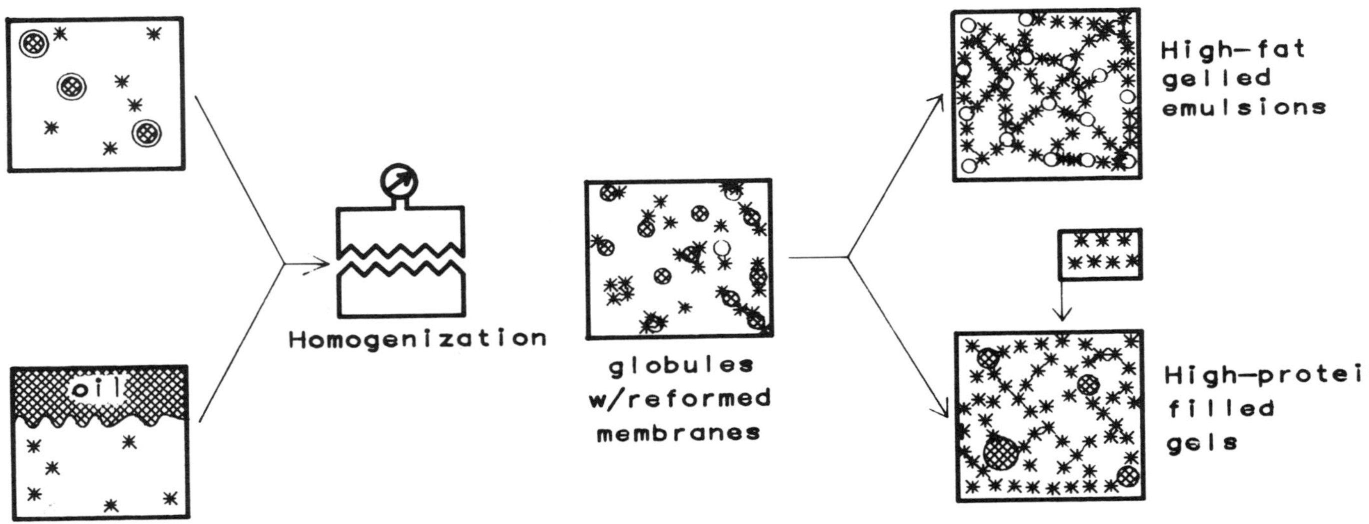

Fig. 7.17 Schematic of processes to produce high-protein filled dairy gels and high-fat gelled dairy emulsions.

is clearly demonstrated in Fig. 7.16C. Microstructural aspects of mixed and filled dairy gels are discussed by Aguilera et al. (1991).

When mixed dairy gels are filled with fat globules having reformed membranes (particle size $< 1.26\ \mu m$), the gelation boundary and the iso-stress curves are displaced toward the origin, reflecting a strengthening effect of fat globules (Fig. 7.14A). Corroboration of these results in actual products has been reported by Green et al. (1990) for cheese analogs.

Solid textures holding a vegetable oil/water emulsion even at oil volume fractions of 0.5 can be obtained by gelation of adsorbed whey proteins. Key factors are thorough homogenization step leading to a low mean fat globule size (e.g., less than 1 μm) and to protein adsorption at the oil/water interface, heating the undisturbed emulsion above 70°C, and control of the pH and ionic strength (Jost et al., 1990). It is remarkable that small vegetable oil droplets coated with whey protein can also act as nuclei on which further whey protein molecules preferentially bind during gelation (Jost et al., 1989). A schematic depicting processes to fabricate high-protein filled dairy gels and gelled dairy emulsions is presented in Fig. 7.17.

CELLULAR STRUCTURES OF FRUITS AND VEGETABLES

Technically speaking, cellular structures of fruits and vegetables are foams with closed-cell geometry, filled with a gel (see Fig. 7.5B). The presence of water stabilizes the cell walls against buckling and gives rise to a turgor pressure of the order of 1–3 MPa. Cell walls, in turn, are composites consisting of an amorphous matrix strengthened with cellulose fibers. In addition, textural requirements in fruits and vegetables call sometimes for failure across the cells (as in raw products) or for individual cell separation (as in cooked legumes). This level of complexity makes the generation of artificial cellular structures a difficult endeavor.

High-moisture, semi-solid structures of gels can provide suitable starting materials for the fabrication of simulated fruits and vegetables; in fact, this application of gelation appears frequently in the patent literature. Some advantages of fruit substitutes are uniformity in size and shape and improved color stability during storage. In this respect alginates and pectates, being able to form gels with divalent cations at ambient temperature, have become preferred materials. In several cases fruit purée or fruit pulp is also incorporated to blend in flavors and textural traits of the natural product.

Depending on the type of fruit texture to be mimicked, at least four

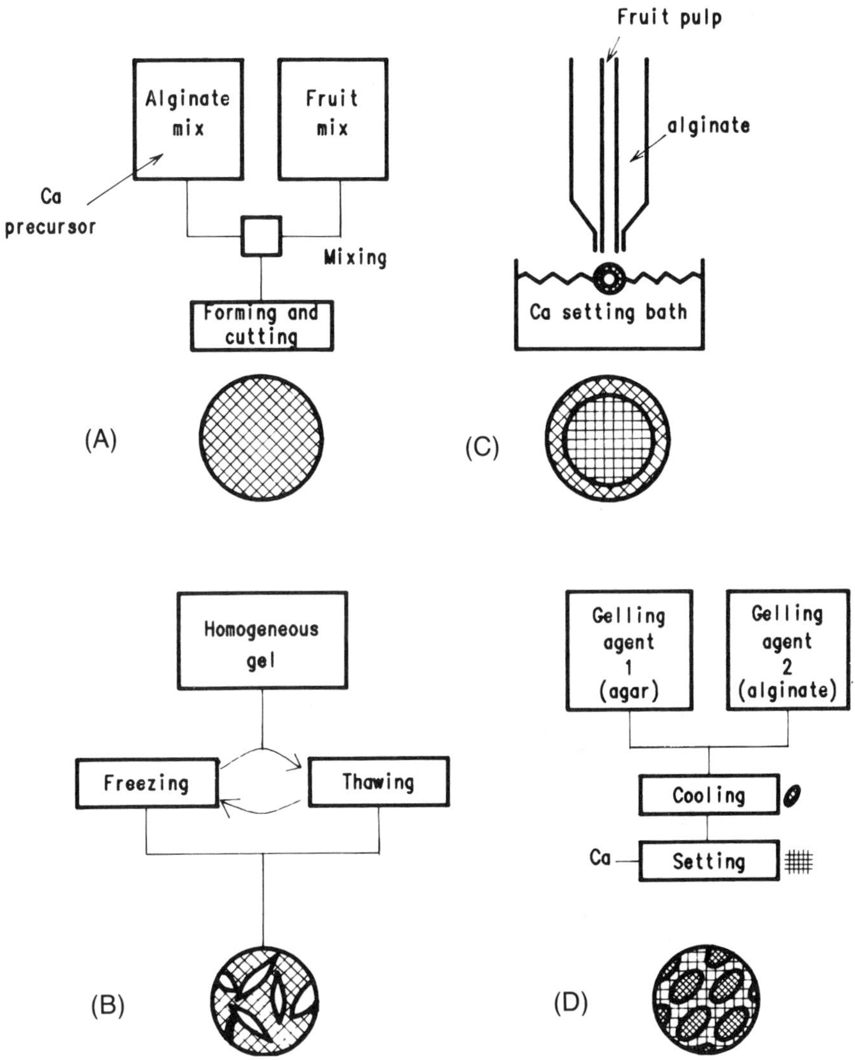

Fig. 7.18 Alternative processes for the generation of artificial fruits.

basic methods (Fig. 7.18) for generating the structure exist: (1) Analogs of fruits having close to an homogeneous texture (e.g., apples, peaches, etc.) are generally produced by gelation of alginate in the presence of calcium cations. Sime (1990) describes one such process as follows: A mix having the alginate and the source of calcium (usually anhydrous dicalcium phosphate, DCP) and another mix containing a fruit purée, sequestrant, and acid are rapidly blended in a high-shear mixer. As the two mixes are homogenized, the pH is lowered (by contact with the fruit purée) and DCP decomposes releasing calcium. The alginate gel is allowed to form under shearless conditions and the product is cut into fruitlike shapes (Fig. 7.18A). Figure 7.19 shows the microstructure of a commercial artificial fruit product of this sort. (2) A certain degree of heterogeneity in texture is achieved by subjecting an alginate gel to freeze-thaw cycles (Szczesniak, 1968). Slow freezing allows for the formation of a "skin" surrounding ice crystals which simulates cell walls. Further structural changes are induced by additives, such as other gums, malic acid, starch, fats, etc., that improve the textural characteristics of the product (Fig. 7.18B). (3) Analogs of small fruits (berries) having a distinctive outer skin can be made by dropping a solution containing a gel setting-agent (e.g., calcium chloride) into another solution containing the alginate. Thus, gelling proceeds from inside out and the interior remains liquid (Johnson, 1972). Another patent uses a co-extrusion process whereby drops of fruit purée are externally coated as they come out of a nozzle by a uniform film of alginate solution (Fig. 7.18C). Formation of the outer skin proceeds at two fronts: externally by contact of the outer surface of the co-extruded

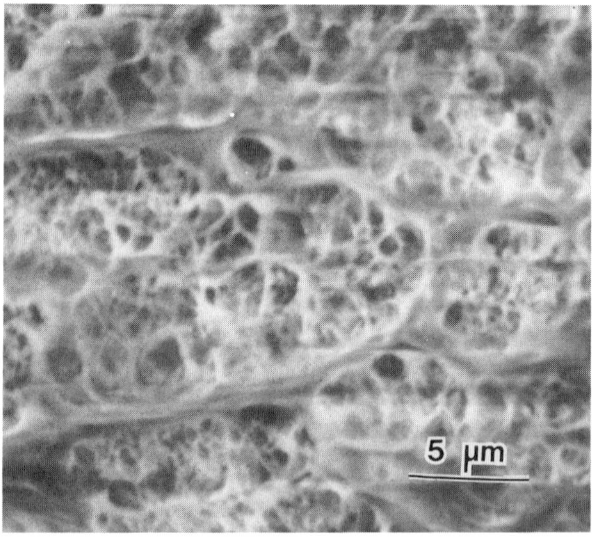

Fig. 7.19 Scanning electron micrograph (using a cryostage) of a commercial artificial fruit product based on an alginate gel. (*Courtesy of Prof. D. W. Stanley.*)

piece with a calcium setting bath and internally by gelation of the internal surface with calcium ions in the fruit purée (Sime, 1990). A method for producing artificial grain caviar achieves a similar effect described as "enveloped granules of gelatin filled with casein" (Tolstoguzov, 1983). (4) More heterogeneous textures can be achieved by allowing a dispersed gelling agent to gel first in the form of small beads and then enclosing them in the matrix of another polymer (Wood, 1975). Ideally the two gelling mechanisms should be different, for example, by using an alginate as the first gelling agent and agar as the second one (Fig. 7.18D).

Some organoleptic problems of engineered fruit structures become apparent towards the end of mastication. Poor water-holding capacity and rapid flavor release usually yields a dry and tasteless fibrous residue that is unpleasant to swallow. Addition of texture modifiers (sucrose) and some fillers such as polyvinyl alcohol and carboxymethylcellulose has been suggested to moderate the chewiness (Luh et al., 1977). Similarly, a method whereby the calcium alginate gel provides not only the basis for texture development but also a controlled-release mechanism for flavors and colors has been proposed by King (1988).

CONCLUSIONS

Gels, having a semi-solid structure composed of a polymeric network and an aqueous occluded phase, are unique food matrices for generating a wide range of microstructures in foods. Future structuring technologies must rely on a sound basic knowledge of the physicochemical phenomena underlying formation, controlled breakdown and reformation of the matrix and of the interrelation between the network and elements in the dispersed phase. Several physicochemical phenomena are involved in the generation and control of engineered structures in gels, such as interfacial aspects, phase transitions, phase equilibria, and reaction kinetics. The ultimate challenge for food technologists is to be able to relate mechanical and transport properties of man-made structures to desirable organoleptic properties of the target food.

ACKNOWLEDGMENTS

Food Biomaterials work at the Department of Chemical Engineering is funded by the United States Agency for International Development (USAID) and FONDECYT.

REFERENCES

Aguilera, J. M. and Kessler, H. G. 1988. *Milchwissenschaft* 43: 411.

Aguilera, J. M. and Kessler, H. G. 1989. *J. Food Sci.* 54: 1213.

Aguilera, J. M. and Kinsella, J. E. 1991. *J. Food Sci.* 56: 1224.

Aguilera, J. M., Kinsella, J. E., and Liboff, M. 1991. *Food Microstruct.* (submitted)

Aguilera, J. M. and Stanley, D. W. 1986. In *Food Engineering and Process Applications*, Vol 2, M. Le Maguer and P. Jelen (Ed.), p. 131. Elsevier Applied Science Publishers, London.

Aguilera, J. M. and Stanley, D. W. 1990. *Microstructural Principles of Food Processing and Engineering*. Elsevier Applied Science, London.

Ahmed, S. and Jones, F. R. 1990. *J. Mater. Sci.* 25: 4933.

Barret, A. H. and Ross, E. W. 1990. *J. Food Sci.* 55: 1378.

Bice, C. W. and MacMasters, M. M. 1947. *U. S. Patent* 2,423,475.

Broutman, L. J. and Krock, R. H. 1967. *Modern Composite Materials*. Addison-Wesley Publishing Co., Reading, MA.

Brownsey, G. J., Ellis, H. S., Ridout, M. J., and Ring, S. G. 1986. In *Gums and Stabilizers for the Food Industry 3*, G. O. Phillips, D. J. Wedlock, and P. A. Williams (Ed.), p. 525. Elsevier Applied Science Publishers, London.

Buchheim, W. 1986. *Kieler Milch. Forsch.* 38: 227.

Buchheim W., Falk G., and Hinz, A. 1986. *Food Microstruct.* 5: 181.

Cairns, P., Miles, M. J., Morris, V. J., and Brownsey, G. J. 1987. *Carbohydrate Res.* 160: 411.

Chinnaswamy, R. and Hanna, M. A. 1988. *Cereal Chem.* 65: 138.

Clark, A. H. 1988. In *Food Structure and Behaviour*, J. M. V. Blanshard and P. J. Lillford (Ed.), p. 13. Academic Press, New York.

Clark, A. H., Richardson, R., and Weaver, A. 1982. *Progr. Food Nutr. Sci.* 6: 149.

Colonna, P., Tayeb, J., and Mercier, C. 1989. In *Extrusion Cooking*, C. Mercier, P. Linko, and J. M. Harper (Ed.), p. 247. American Assoc. Cereal Chem., St. Paul, MN.

Davies, F. L., Shankar, P. A., Brooder, B. E., and Hobbs, D. G. 1976. *J. Dairy. Res.* 45: 53.

Doi, H., Ideno, S., Kuo, F. H., Ibuki, F., and Kanamori, M. 1983. *J. Nutr. Sci. Vitaminol.* 29: 679.

Flory, P. J. 1953. Ch. XIII. In *Principles of Polymer Chemistry*. Cornell University Press, Ithaca, NY.

Gomez, M. H. and Aguilera, J. M. 1983. *J. Food Sci.* 48: 378.

Gomez, M. H. and Aguilera, J. M. 1984. *J. Food Sci.* 49: 40.

Gotlieb, A. M., Plashchina, I. G., Braudo, E. E., Titova, E. F., Belavtseva, E. M., and Tolstoguzov, V. B. 1988. *Die Nahrung* 32: 927.

Green, M. L., Marshall, R. J., and Langley, K. R. 1990. in *Rheology of Food. Pharmaceutical and Biological Materials with General Rheology*, R. E. Carter (Ed.), p. 1. Elsevier Applied Science Publishers, London.

Hayter, A. L. and Smith, A. C. 1988. *J. Mater. Sci.* 23: 736.

Hayter, A. L., Smith, A. C., and Richmond, P. 1986. *J. Mater. Sci.* 21: 3729.

Heertje, I., Visser, J., and Smits, P. 1985. *Food Microstruct.* 4: 267.

Hiltner, A., Cassidy, J. J., and Baer, E. 1985. *Ann. Rev. Mater. Sci.* 15: 455.

Hornbogen, E. 1986. *J. Mater. Sci.* 21: 3737.

Jeronimidis, G. 1988. In *Food Structure—Its Creation and Evaluation*, J. M. V. Blanshard and J. R. Mitchell (Ed.), p. 59. Butterworths, London.

Johnson, E. W. 1972. U. S. Patent 3,682,654.

Jolley, P. D. and Purslow, P. P. 1988. In *Food Structure—Its Creation and Evaluation*, J. M. V. Blanshard and J. R. Mitchell (Ed.), p. 231. Butterworths, London.

Jost, R., Dannenberg, F., and Rosset, J. 1989. *Food Microstruct.* 8: 23.

Jost, R., Dannenberg, F., and Gumy, D. 1990. *Proceedings of the 23rd International Dairy Congress*, Montreal, Oct. 7–12, pp. 1481–1491.

King, A. H. 1988. In *Flavor Encapsulation*, S. J. Risch and G. A. Reineccius (Ed.), p. 122. ACS Symposium Series 370, American Chemical Society, Washington DC.

Kokubo, T., Teranishi, Y., Maki, T., and Sakka, S. 1988. *J. Mater. Sci.* 23: 1126.

Langley, K. R. and Green, M. L. 1989. *J. Texture Stud.* 20: 191.

Lawrence, R., Consolación, F., and Jelen, P. 1986. *Food Technol.* 40(3): 77.

Lee, C. M. and Kim, J. M. 1986. In *Food Engineering and Process Applications*, Vol. 1, M. Le Maguer and P. Jelen (Ed.), p. 63. Elsevier Applied Science Publishers, London.

Lillford, P. J. 1985. In *Properties of Water in Foods*, D. Simatos and J. L. Multon (Ed.), p. 543. M. Nijhoff Publisher, Dordrecht.

Lugay, J. C. and Kim, M. K. 1981. In *Utilization of Protein Resources*,

D. W. Stanley, E. D. Murray, and D. H. Lees (Ed.), p. 177. Food and Nutrition Press, Westport, CT.

Luh, N., Flink, J. M., and Karel, M. 1977. *J. Food Sci.* 42: 976.

Luyten, H. and van Vliet, T. 1990. In *Rheology of Food, Pharmaceutical and Biological Materials with General Rheology*, R. E. Carter (Ed.), p. 43. Elsevier Applied Science Publishers, London.

Matz, S. A. 1984. *Snack Food Technology*. AVI Publishing Co., Westport, CT.

Modler, H. W. and Kalab, M. 1983. *J. Dairy Sci.* 66: 430.

Morris, V. J. 1986. In *Gums and Stabilizers in the Food Industry 3*, G. O. Phillips, D. J. Wedlock, and P. A. Williams (Ed.), p. 87. Elsevier Applied Science Publishers, London.

Mulvihill, D. M. and Kinsella, J. E. 1987. *Food Technol.* 41(9): 102.

Murray, E. D., Myers, C. D., Barker, L. D., and Maurice, T. J. 1981. In *Utilization of Protein Resources*, D. W. Stanley, E. D. Murray, and D. H. Lees (Ed.), p. 158. Food and Nutrition Press, Westport, CT.

Pearson, A. M. and Dutson, T. R. 1987. *Advances in Meat Research*. Vol. 3, *Restructured Meat and Poultry Products*, Van Nostrand Reinhold Co., New York.

Raeuber, H. J. and Nikolaus, H. 1980. *J. Texture Stud.* 11: 187.

Ring, S. and Stainsby, G. 1982. *Prog. Fd. Nutr. Sci.* 6: 323.

Rutman, M., Aguilera, J. M., and Bisso, J. 1988. *J. Food Sci.* 53: 284.

Silberberg, A. 1989. In *Polymers in Aqueous Media Performance through Association*, J. E. Glass (Ed.), p. 3. Advances in Chemistry Series No. 223, American Chemical Society, Washington, DC.

Sime, W. J. 1990. In *Food Gels*, P. Harris (Ed.), p. 53. Elsevier Press, London.

Susuki, T. 1981. *Fish and Krill Protein: Processing Technology*. Applied Sci. Publishers Ltd., London.

Szczesniak, A. S. 1968. U. S. Patent 3,362,831.

Tolstoguzov, V. B. 1983. *J. Texture Stud.* 14: 183.

Tolstoguzov, V. B. 1988a. In *Food Structure—Its Creation and Evaluation*, J. M. V. Blanshard and J. R. Mitchell (Eds.), p. 181. Butterworths, London.

Tolstoguzov, V. B. 1988b. *Food Hydrocolloids* 2: 339.

Torley, P. J. and Lanier, T. C. 1989. *Pacifichem Conference*, Honolulu, Hawaii, Dec. 1989.

Torres, A. 1977. *Textural Properties of Amylose and Starch Sponges.*

M.Sc. thesis, Department of Food and Agricultural Engineering, University of Massachusetts, Amherst, Mass.

Visser, J. 1988. In *Food Structure its Creation and Evaluation*, J. M. V. Blanshard and J. R. Mitchell (Ed.), p. 197. Butterworths, London.

Wood, F. W. 1975. U. S. Patent 3,892,870.

Ziegler, G. R. and Foegeding, E. A. 1990. *Adv. Food Nutrition Res.* 34: 203.

Zobel, H. F. 1988. *Starch/Stärke* 40: 1.

8

Physicochemical Aspects of Muscle Tissue Behavior

E. Allen Foegeding and Donald D. Hamann

North Carolina State University
Raleigh, North Carolina

INTRODUCTION

The biological function of skeletal muscle is locomotion. This is accomplished by contraction and relaxation of myofibrils contained within muscle cells (fibers). To fulfill this task, skeletal muscle is designed with an off/on switching system (myoneural junction and calcium-regulated contraction), a regulated motor (actin, myosin, troponin, and tropomyosin), and a rigid support system to pull against (connective tissue and bones). This dynamic system is powered by adenosine triphosphate (ATP) and maintained by a constant turnover of the component parts (via endogenous proteinases). The precise function of specific muscles varies from

---

Paper No FS91-20 of the Journal Series of the North Carolina Agricultural Research Service, Raleigh, NC 27695-7643. The use of trade names does not imply endorsement by the North Carolina Agricultural Research Service of the products named, nor criticisms of similar ones not mentioned.

extremes of rapid, forceful contractions required to move large weights to slow, tonic contractions required in posture. Biology fulfills these specific needs by maintaining general form but substituting key elements associated with physiological need. For example, there are at least three different thin filament regulatory programs for fast twitch rabbit muscle which function in altering the sensitivity to calcium regulation of contraction (Schachat et al., 1990), and six different myosin heavy chain isoforms are found in chicken breast muscle during development (Hofmann et al., 1988).

Skeletal muscles from pork, beef, chicken, turkey, and fish species represent the most common meats used in the manufacturing of muscle foods. In meat products the molecules and structures that previously served the biological function of motion are now the basis for texture, appearance, and flavor. The quality of any particular muscle food will therefore depend on compositional factors associated with individual muscles (this being a direct reflection of the biological function), the changes that occur during the conversion from muscle to meat (rigor-associated changes), and mechanical processes and additives used in manufacturing a particular muscle food. We will address some physicochemical principles that are important in meat consumed in a form that has undergone minimal processing (anisotropic muscle foods) and those that have been chopped or ground to a small particle size such that they are considered rheologically isotropic.

ANISOTROPIC MUSCLE FOODS

Structure of Muscle

Muscle fibers (cells) are multinucleated, elongated, with conical or tapering ends. Their diameters range from 10 to 100 μm with lengths of 1–40 mm (Cassens, 1987). Contained within the fibers are the protein structures that contract, the myofibrils. Myofibrils are composed of contractile (myosin and actin), regulatory (troponin and tropomyosin), structural (alpha-actinin, M-protein, C-protein, etc.), and cytoskeletal (titin, desmin) proteins. These myofibrillar proteins are assembled into filaments and structures that form the sarcomere, the primary contractile unit of muscle.

Surrounding each fiber is the plasma membrane referred to as the sarcolemma. This structure is involved in ion transport and contains hormonal receptors and proteins important to other biological functions (Mickelson and Louis, 1985). Surrounding the sarcolemma is the basement lamina, which contains type IV collagen, laminin (a 900 kDa protein with domains that interact with other basement membrane components and the cell surfaces), and proteoglycans, with heparan sulfate as the predominant proteoglycan (Mayne and Sanderson, 1985).

Around each fiber, above the basement lamina, is the endomysium. This connective tissue is the outermost component surrounding the muscle cell. Groups of cells are collected into bundles by perimysium, and muscles are enclosed by endomysium. The predominant collagen in the epimysium and perimysium is type I, with type III as a minor component (Light and Champion, 1984). Types I and III collagen are the predominant forms found in the endomysium; minor amounts of types IV and V are also present.

Tenderness (Texture)

The human perception of texture or tenderness of meats is associated with mechanical failure properties, related to muscle structure, and the ease of lubrication or juiciness which determines the amount of mastication required before swallowing (Dransfield et al., 1984). The mechanical failure component of meat texture is determined by the force required to shear through the layers of connective tissue and muscle fiber structures. Shearing perpendicular to the long axis of the fiber requires rupturing two or three layers of connective tissue, the basement lamina, and the myofibrils. Fracture that occurs parallel to the long axis of the fiber requires, at a minimum, separation at the perimysium. Meat texture is determined primarily by the structures and associations within the connective tissue and myofibrils.

The biological function of muscle determines distribution and types of connective tissue and myofibillar structures. In general, muscles of locomotion, such as the *semitendinosus,* have more connective tissue and are tougher than those of support (e.g., *longissimus dorsi*). As animals mature the collagen fibers are stabilized by intermolecular cross-links; this is the primary reaction associated with connective tissue-toughening (Bailey, 1985). Muscle-to-muscle variation in myofibillar protein is related to fiber twitch response. In comparing fast and slow twitch fibers, there are fast and slow protein isoform differences (Gauthier, 1990), along with variation in structures such as Z-disk thickness (Dahlin et al., 1976). Therefore, for any particular cut of meat, the "inherent" meat texture will be determined by the biological function of the muscles comprising that cut and age of the animal (which will affect the amount of crosslinking within the collagen).

When an animal is stunned and exsanguated, it begins the process of converting muscle to meat. The initial phase of this process is a loss of biological homeostasis, which causes an irreversible contraction (rigor mortis) and alteration of regulators controlling endogenous proteinases. The contribution of myofibrils to meat texture is determined by the combination of sarcomere length and disruption of myofibillar structure (by proteolytic and/or osmotic changes) (Ouali, 1990). Superimposed on the above-mentioned factors is the effect of heating and processing additives on connective tissue and myofibillar proteins. Hydrolysis of proteins within these structures by endogenous or exogenous proteinases will cause weakening that can alter textural properties. The texture that is perceived when one consumes a cooked piece of meat is therefore determined by the sum of biological, post-mortem, and cooking factors. Furthermore, the many components that contribute to textural properties make it very difficult to adjust one variable while holding all others constant. This leads to approaches which determine the effects of some treatment on texture without a rigorous understanding of the mechanism(s) causing the change.

Rheological Measurement of Texture

One of the most widely used instruments for measuring mechanical failure properties of meat is the Warner-Bratzler shear (Warner, 1952). The failure points in the force-deformation curves generated from shearing through muscle(s) with the Warner-Bratzler shear can be interpreted as failure in myofibillar and connective tissue components (Moller, 1981). The data shown in Table 8.1 are representative of many studies where sensory and mechanical failure tests are used to evaluate texture. The sensory data indicates that the meat treated with NaCl and the deionized water control are less tender than the bromelain-based tenderizer treatment, however, the shear-force values were unable to show a similar significant difference between treatments (Foegeding and Larick, 1986). This could be due to the relative sensitivity of sensory and mechanical testing or differences in the way humans and the Warner-Bratzler shear cause mechanical failure. The complex nature of the Warner-Bratzler shear force-deformation pattern does not allow for rigorous evaluation of the stresses and strains associated with shearing through meat, therefore, comparison among studies is limited to those which use exactly the same testing condition.

Another approach to investigations is to use principles associated with materials and polymer science as general models for mechanical properties of meat. Rheological properties determined in tension, compres-

Table 8.1 Tenderness of *Semitendinosus* Steaks from Mature Cattle

	Sensory		Warner-Bratzler shear		
Treatment	Juiciness	Tenderness	M-Force (kg force)	C-Force (kg force)	Work
Deionized water	5.3^{a}	3.2^{b}	3.1^{a}	5.2^{a}	153^{a}
2% NaCl	5.7^{a}	3.7^{b}	3.0^{a}	5.0^{ab}	142^{ab}
Tenderizer (bromelain-based)	5.8^{a}	6.2^{a}	2.6^{a}	4.1^{b}	116^{b}

Steaks were soaked for 2 hr and heated to 65°C. Samples were 1 cm^2 × 2.5 cm; sensory scores were based on an 8-point structured scale of low (1) vs. high (8) descripters.
[ab] Mean values (5 replications) within columns with similar superscripts are not significantly different ($p > 0.05$).
Source: Foegeding and Larick, 1986.

sion, or torsion permits one to calculate fundamental properties of failure stresses (force/unit area) and strains. If failure stress, or a nonfailure stress/strain modulus, is reported in units of Pascals (Newtons/m^2) and the deformation rate is given (important due to viscoelastic effects), then relative comparisons among investigations are possible. Purslow (1985) used tensile testing to determine fracture stress (ultimate tensile strength) of beef *semitendinosus*. Meat which had been cooked at 80 ± 1°C for one hour had failure stress values of ~300 kPa parallel and ~25 kPa perpendicular to the fibers (strain rates of 1.46–7.3 min^{-1}). Moreover, analysis of the failure pattern showed that fibers fracture as individual units and the strain energy is not transferred to adjoining fiber bundles. Purslow (1985) suggested a two-phase composite model for fracture behavior of muscle. Treating meat with collagenase is a selective way to weaken the connective tissue structures without affecting the myofibrils. Addition of collagenase to beef flank steak decreases maximum stress parallel (from 92.0 to 58.3 kPa) and perpendicular (from 23.7 to 10.6 kPa) to the fibers. The collagenase treatment also causes an increase in the degree of meat anisotropy (parallel stress/perpendicular stress) (Sacks et al., 1988). The hydrolysis of collagen causes a decrease in the strain at maximum stress parallel to the fibers, however, strain perpendicular to the fibers is increased. These findings suggest that the partial hydrolysis in the perimysium makes the meat more deformable perpendicular to the fibers. Tensile testing has also been used to evaluate physical properties of frozen and thawed beef (Munro, 1983). Beef has three characteristic fracture curves that are strain rate ($0.555–111.1 \times 10^3$ sec^{-1}) and meat temperature dependent. Once meat is frozen, failure stress shows a sharp

Table 8.2 Tensile Strength of Beef Pulled Parallel to the Fiber Axis

	Temperature (°C)			
	0	−10	−20	−30
Tensile strength (kPa)	410	2820	5580	4390
Strain at max. load (%)	184	0.91	1.63	1.01

Source: Munro, 1983.

increase and strain decreases, characteristic of a brittle material (Table 8.2). The failure stresses reported in the above studies varied from 10.6 to 410 kPa for the unfrozen meat. If the variation among studies due to strain rate (viscoelastic properties) was normalized, the remaining differences can be accounted for by changes in the physical properties of the meat.

Sinusoidal compression is another mechanical testing procedure which has been used to evaluate meat texture. Kamoun and Culioli (1988) devised a compression cell that supports samples on the bottom and parallel sides so that meat deformation under compression is restricted to the one axis perpendicular to the compressive force. This cell allows for the stress-strain properties of meat to be determined longitudinally and transverse to the long fiber axis. Uncooked muscles which are high (*semimembra-*

Table 8.3 Maximum Stress at a Compression Ratio of 0.3 ($\Delta h/h_0$) of Muscle Under Sinusoidal Compression

	Maximum Stress[a]		
	Longitudinal	Transverse	Longitudinal/Transverse
Longissimus dorsi			
Uncooked	160 kPa	200 kPa	0.80
55°C[b]	380	280	1.36
90°C	780	410	1.90
Semimembranosus			
Uncooked	200	260	0.77
55°C	450	420	1.07
90°C	940	400	2.35

[a] Stress values were determined by compressing muscle at 10 Hz in a direction longitudinal or transverse to the fiber axis.
[b] Heated at 55°C or 90°C for 30 min.
Source: Kamoun and Culioli, 1988.

nosus) and low (*longissimus dorsi*) in connective tissue have a similar low degree of anisotropy (longitudinal stress/transverse stress) (Table 8.3). However, cooking makes both muscles more anisotropic. The main difference between muscles is in the increased stress required for longitudinal deformation because the stress required for deformation transverse to the fiber axis is similar between muscles (Table 8.3).

ISOTROPIC MUSCLE FOODS

Changes During Particle Size Reduction

A large variety of muscle foods are made from meat which has been ground, chopped, or flaked. These processes disrupt the connective tissue matrix and therefore are one of the simplest ways to alter the texture of muscle foods. If salt is added to meat before or during the particle reduction process, the myofibrillar structure is disrupted and free proteins become part of a dispersion which also contains fibers, fibrils, and filaments. The relative amount of each component in suspension will depend on the mechanical process used for particle size reduction and the condition of meat (i.e., temperature). The most abundant myofibrillar proteins (Table 8.4) are actin and myosin, which have monomeric long axes of 5.5 nm and ~170 nm, respectively (Elliott and Offer, 1978; Holmes et al., 1990). These proteins can be considered the smallest "functional macromolecular particles" found in processed meats. Meat batters comminuted in a bowl chopper have been shown to contain particles ranging from 1 to 30 mm (Girard et al., 1985). Particles in that size range could be pieces of muscle fibers and/or bundles of myofibrils. Meat that is flaked

Table 8.4 Protein Content (% total protein) of Chicken Leg and Breast Muscles

	Leg (Red)		Breast (White)	
Myofibrillar	56.7%		54.5%	
Actin		12.0%		11.5%
Myosin		24.9		23.9
Other soluble proteins	37.6		43.6	
Collagen	5.6		1.3	

Data for 6-week-old chickens.
Source: Khalili and Zarkadas, 1988.

can have particles ranging in thickness from 430 to 1700 μm and particle densities of 7 to 368 particles/g (Sheard et al., 1990). From a practical prospective, this means that there is a continuum of particle size, shape, and protein composition in processed meats. Therefore, certain processed meats (frankfurters, bologna, surimi) can be regarded as isotropic to rheological testing while being heterogeneous in particle composition. The isotropic rheological properties reflect the particle composition of a specific product.

Rheological Properties of Viscoelastic Fluid Meet Batters

Meat chopped to the degree that is used in manufacturing frankfurters, bologna, and surimi has particle sizes small enough that the material behaves as a viscoelastic fluid. This allows the meat to be pumped to various points in the processing line and emphasizes the importance of the flow character of each product batter. We will not go into detail on all factors affecting meat batter rheology but will discuss examples related to fat replacement in frankfurters and the rheological character of surimi. More detailed insight on factors influencing meat batter rheology can be gained from Burge and Action (1984), Hamm (1975), and Payne and Rizvi (1988).

Frankfurter batters are pseudoplastic and often have a yield stress below which they act as a solid. Apparent viscosity and the yield stress tend to increase with percent fat and decrease with added water (Payne and Risvi, 1988). Therefore, fat replacers in frankfurters must first provide a final product texture similar to the higher-fat products and maintain a batter consistency such that the processing line will perform well. The batters shown in Table 8.5 are a fat-reduced batter (containing a 90% water-carbohydrate gel that was chopped into the batter) and a control batter at ~30% fat. The goal was to replace the fat with water that will not cook out and provide the mouthfeel of fat. Tube rheometry was used to obtain flow data on both batters, and the results are shown in Fig. 8.1.

Table 8.5 Two Beef/Pork Frankfurter Batters

Component	Low-fat batter	Control batter
Protein (%)	10.7	10.1
Fat (%)	10.0	29.9
H_2O (%)	70.7	54.1
Salt (%)	2.0	2.0
Other (%)	6.6	3.9

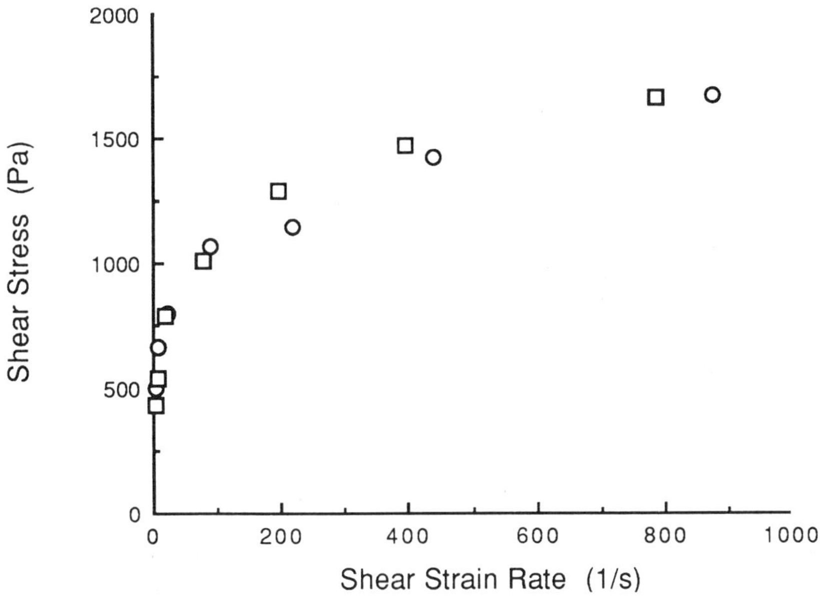

Fig. 8.1 Flow curves for 10% fat (squares) and 30% fat (circles) frankfurter batters at 10°C. Batter formulations are found in Table 8.5.

Although the water contents and fat contents were very different, the pseudoplastic flow curves for the two batters were nearly identical, showing that the fat replacement gel produced flow character at 10°C similar to the fat. The rheological model of power law without a yield stress was used because addition of a yield stress did not statistically improve the model. The respective equations for low-fat and control batters were:

$$\text{Shear stress, kPa} = 419\ (\text{shear strain rate, s}^{-1})^{0.235}\ (R^2 = 0.990) \qquad (1)$$

$$\text{Shear stress, kPa} = 494\ (\text{shear strain rate, s}^{-1})^{0.197}\ (R^2 = 0.993) \qquad (2)$$

The curves of Fig. 8.1 are in general agreement with models in the literature (Payne and Risvi, 1988).

Surimi is frozen, stabilized, water-washed minced fish that is used in making products such as imitation crab legs. Alaska pollack surimi batters for making shellfish analogs have flow characteristics somewhat different from frankfurters. Surimi (80% water, 7.5% cryoprotectant, 3% sodium chloride, 9% muscle protein, and 0.5% other) flow curves are best modeled with a yield stress (Kim et al., 1986).

$$\text{Shear stress, kPa} = 160 + 29.1\ (\text{shear strain rate, s}^{-1})^{0.73}\ (R^2 = 0.994) \tag{3}$$

The protein level in this surimi was about 1% lower that of the red meat batters of Fig. 8.1. Surimi batters made from the same lot were subjected to 15 freeze-thaw cycles to see how freeze-thaw history affects the flow properties. The surimi subjected to freeze-thaw fit the following equation:

$$\text{Shear stress, kPa} = 80 + 163\ (\text{shear strain rate, s}^{-1})^{0.473}\ (R^2 = 0.998) \tag{4}$$

Application of this equation shows that surimi subjected to freeze-thaw produced higher stress values suggesting that denaturation (i.e., protein unfolding) caused a higher suspended vòlume fraction, thus increasing the viscosity of the paste. Supporting evidence for freeze-thaw denaturation was found in the observation that final product shear stress and shear strain at fracture values were lower for the freeze-thaw surimi (Kim et al., 1986).

Thermally Induced Transformation into a Viscoelastic Solid (Gel)

Changes in rheological properties. Once meat batters are formed via chopping, they are extruded into a container (casing) and cooked to form the finished product. Cooking of meat batters forms thermally induced gels (gels formed upon heating rather than cooling) of myofibrillar protein. This viscoelastic liquid–to–viscoelastic solid conversion can be observed by changes in viscous energy loss when the sol/gel is deformed and released in a cyclic manner (Foegeding and Ramsey, 1987). If the cyclic motion is sinusoidal, the relative viscous energy lost as heat can be determined by the phase angle, by which strain lags stress (Fig. 8.2). Gelation is seen as a decrease in energy loss or phase angle and always precedes a major increase in shear modulus (G) or storage modulus (G′) (Foegeding and Ramsey, 1987; Saliba et al., 1987) (Fig. 8.2). Once the gel is formed, it is irreversible regarding its viscoelastic solid state. Rheological properties determined at both nonfailure strains (Fig. 8.3) and failure strains (Howe et al., 1991) are temperature dependent. This emphasizes the importance of temperature control when evaluating rheological (textural) properties by mechanical and sensory testing.

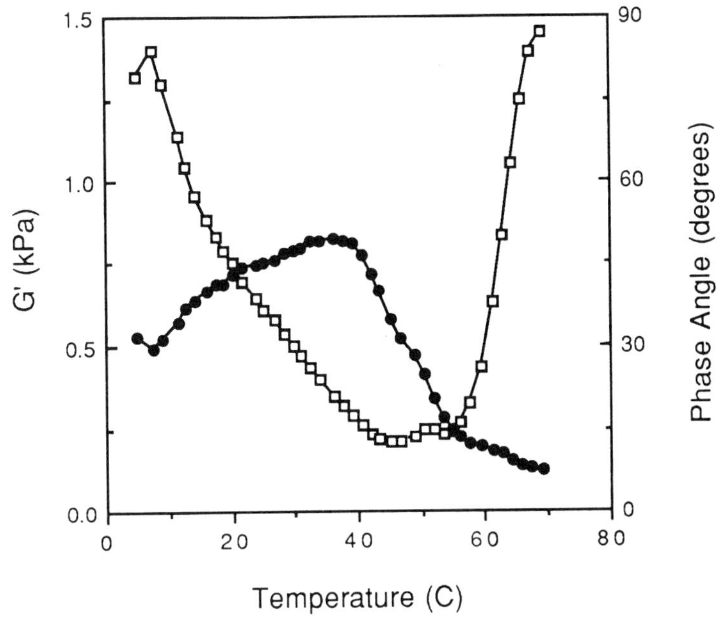

Fig. 8.2 Rheological changes in 13% protein comminuted turkey breast batters during heating at 1°C/min. Storage modulus (G′, squares) and phase angle (circles) were determined at 0.05 Hz at a maximum strain of 0.037.

Myosin gelation. The gelation of myosin has been investigated as a model system to understand one of the primary reactions responsible for textural properties of muscle foods. This topic was recently reviewed by (Asghar et al., 1985). A generalized model for thermally induced gelation is seen in Fig. 8.4. The initial change in protein structure in the sequence of events that leads to gelation is one which causes intermolecular protein interactions. This can be caused by any "structure-perturbing" agent, but we will restrict our discussion to thermally induced gelation. The myosin molecule consists of a globular "head" structure connected to a "tail" of coiled coil α-helix (Elliott and Offer, 1978). Structural changes occur in the myosin head at temperatures as low as 37°C (Burke et al., 1987). Heating myosin at 40°C causes association in the head region such that the aggregates have a "daisy wheel" appearance (Yamamoto, 1990). The daisy wheel aggregate could be one of the initial structures formed by heating, especially in myosin dispersed from the thick filaments. The myosin tail has from two (Stafford, 1985) to six (Potekhin et al., 1979) unfolding domains. Myosin rod (tail) heated at 40°C develops a rigid gel

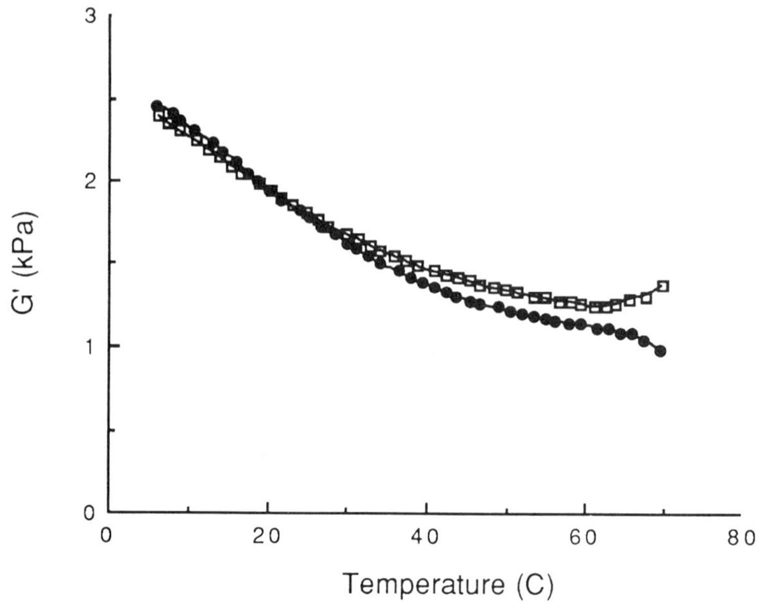

Fig. 8.3 Rheological changes in 13% protein comminuted turkey breast batters during cooling (squares) and reheating (circles). The procedure for forming gels and rheological testing is as described in Fig. 8.2.

structure coinciding, with a decrease in α-helix (Samejima et al., 1981). This indicates that rod-rod aggregates are also possible during the first phase of gelation.

When porcine myosin is gelled by isothermal heating, the gel strength (work required to fracture the gel) increases up to 55°C, then shows lower values at 60° and 70°C (Foegeding et al., 1986). This is not true for suspensions heated at 12°C/hr or 50°C/hr, where gel strength continues to

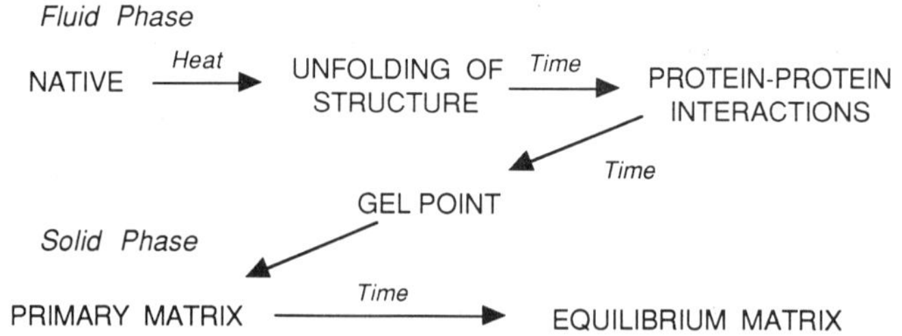

Fig. 8.4 Generalized scheme for thermally induced gelation of proteins.

increase at 60° and 70°C. It was speculated that the slower heating processes (12°C/hr or 50°C/hr) allowed for gelation of a less unfolded molecule which produced a stronger gel. This topic was recently investigated using a kinetic model for gelation (Wu et al., 1991). Changes in shear modulus of myosin suspensions heated isothermally from 43° to 80°C all fit kinetic models with orders between 1.6 and 1.8, therefore, second-order kinetics were used to analyze the data. The equation derived by Wu et al. (1991) to describe myosin gelation was

$$G = G_e\left[1 - \frac{k\,t_c + 1/A_o}{k\,t + 1/A_o}\right] \tag{5}$$

Where the equilibrium shear modulus (G_e) and rate constant (k) were determined by an iteration method, the critical time for gelation (t_c) was read directly for the shear modulus-time curve and the initial concentration of myosin (A_o) varied from 5 to 25 mg/ml.

The results seen in Table 8.6 for chicken myosin are consistent with those of Foegeding et al. (1986) for porcine myosin, in that the gel structure is weaker when formed under isothermal heating at T > 55°C. The start of gelation (T > 43°C) and the decrease in G_e at ~55°C coincide with rabbit myosin results for unfolding transitions observed by Stafford (1985)

Table 8.6 Myosin Gelation[a]

Temperature (°C)		t_c (min)	Rate const. ml/(mg min)	G_e (Pa)
43 *Did not gel (after 1 hr)*				
44		16.0	0.0024	14.8
45	T_{m1}			
46		2.0	0.026	19.0
48		0.8	0.040	21.9
50		0.6	0.065	22.3
52		0.4	0.88	13.2
53	T_{m2}			
54		0.5	0.28	15.1
56		0.7	0.23	11.7
58		0.7	0.21	9.9
60		8.0	0.39	0.7

[a] Rheological data of critical time (t_c), second-order rate constant and equilibrium shear modulus (G_e) are for chicken myosin from Wu et al. (1991). The thermal denaturation transitions (T_m) are for rabbit myosin from Stafford (1985).

($T_{m,1}$ = 45°C and $T_{m,2}$ = 53°C) and shear modulus transition of 43°C and 55°C reported by Ishioroshi et al. (1979). The variation in exact transition temperature among all studies is probably related to species-associated isoform differences (Rodgers et al., 1987). One way to interpret these results is with the scheme shown in Fig. 8.5. This model depicts reversible unfolding into two denatured states that can aggregate, irreversibly, into two different gel (precipitate) structures. This general model has been shown to apply to denaturation and irreversible aggregation of T4 lysozyme (Wetzel et al., 1990). For the myosin data seen in Table 8.6, one could assume that the transition to U_1 is the 43–45°C myosin transition and the gel matrix formed from the U_1 state has a G_e of ~22 Pa. The second unfolded state, U_2, is much less favorable to gelation and may have a critical protein concentration (minimal concentration required for gelation) less that 10 mg/ml. This is supported by the observation that thermally denatured myosin changes shape such that the radius of gyration (R_g) decreases. Myosin rod (rabbit) in the native state has a R_g of ~42 nm and after heating to 60°C the R_g has decreased to ~13 nm (Rodgers et al., 1987). This could decrease the relative contribution of individual myosins to the total length of the gel structure. The net result is that gels formed isothermally at any given temperature would have a microstructure composed of different aggregates, with the relative abundance depending on the kinetics of unfolding and aggregation. Experiments designed to test the validity of this model would benefit our understanding of the mechanism(s) by which heating treatments alter the rheological properties of gels and, hopefully, explain aspects of textural properties in muscle foods.

Rheological properties determined at failure (fracture) When isotropic meat products are consumed, as with anisotropic meats, the stress and deformation to failure have major contributions to the overall evaluation

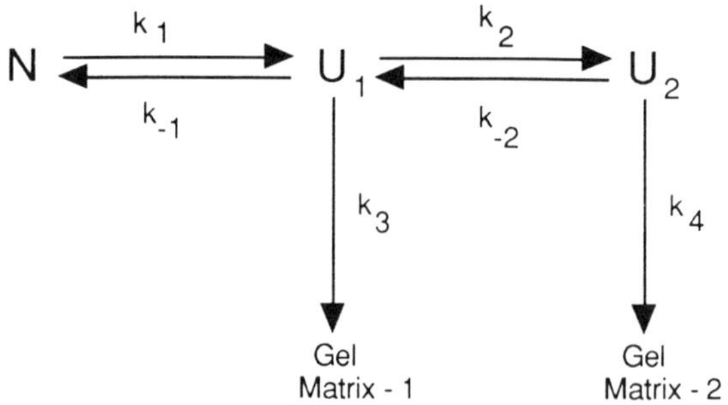

Fig. 8.5 Scheme for denaturation and gelation of a protein with two unfolded states. *(Adapted from Wetzel et al., 1990.)*

Table 8.7 Rheological Properties of Comminuted Meats

Meat	% Protein	pH	Shear stress (kPa)	Shear strain
Montejano et al. (1984)				
Surimi	15	—	76.6	2.28
Turkey thigh	16	—	41.3	1.32
Beef[a]	16	—	26.8	1.18
Pork[a]	16	—	27.4	1.18
Daum (1990)				
Turkey thigh	13	6.4	46.2	2.01
	13	5.9	31.5	1.63
Turkey breast	13	5.9	20.1	1.37
	13	6.4	43.9	2.15

[a] Pooled *longissimus dorsi, psoas major,* and *gluteus medius.*

of texture. Montejano et al. (1985) determined that for isotropic, viscoelastic food gels (comminuted fish, beef, pork, and turkey and egg white), texture profile analysis sensory notes are significantly correlated ($p < 0.01$) with the fundamental rheological properties of true shear stress (τ_{max}) and true shear strain (γ_{max}) determined at failure (fracture). The sensory notes with correlation coefficients of ≥ 0.80 were springiness, firmness, cohesiveness, denseness, coarseness, and graininess. In general, stress reflects the firmness of the gel, whereas strain is an indication of the cohesive properties. Some typical values of shear stress and shear strain at failure are seen in Table 8.7. The data from Montejano et al. (1984) show that a variety of comminuted meats can be compared on the basis of shear stress and shear strain at failure.

These physical properties will be influenced by chemical factors such as protein concentration, pH, ionic strength, and thermal processing conditions. The study of Daum (1990) (Table 8.7) showed that altering the pH causes changes in shear stress and strain, whereas Foegeding (1990) reported that altering the protein concentration causes major changes in shear stress with minimal affect on shear strain. The physicochemical mechanisms responsible for shear stress and shear strain at failure are poorly understood, and insight into these mechanisms offers the possibility of improving processed meat texture.

Meat Products as Multicomponent Gels

In order to gain an understanding into physical chemical mechanisms, it is best to start with first principals and a model system. Myosin gelation offers some insight into processed meat rheology but is somewhat limited in extrapolating to the complex mixture found in many muscle foods. The theory and approaches used in understanding multicomponent or composite gels offers a fundamental framework for which to study comminuted meat products (Tolstoguzov and Braudo, 1983; Brownsey et al., 1987). With this approach, the gelling proteins within meat would be considered the primary matrix, and any other component, including nongelling meat proteins, can be evaluated for their ability to change the rheological properties of the primary matrix. Model systems of particle filled gelatin gels have shown that the filled particle volume fraction, particle rheology, and ability to interact with the primary matrix will each influence the overall rheology of the composite system (Brownsey et al., 1987).

This does not mean that a more classical approach of studying the effects of different ingredients on rheological properties does not produce useful information. For example, the study of Foegeding and Ramsey (1987) showed that iota-carrageenan, kappa-carrageenan, and xanthan gum affect the shear stress and shear strain at failure of a low-fat frankfurter and gave some indication of which hydrocolloids would function best in low-fat products. However, in retrospect, the mechanism underlying these changes could have also been determined with additional experiments. A combination of experimental methods which determine fundamental rheological properties (stress and strain) and designs which are based in theory from polymer and materials science (i.e., multicomponent gels) would provide an excellent approach for one to further the understanding of the physicochemical mechanisms involved in meat texture.

REFERENCES

Asghar, A., Samejima, K., and Yasui, T. 1985. Functionality of muscle proteins in gelation mechanisms of structured meat products. *CRC Crit. Rev. Food Sci. Nutr.* 22: 27–107.

Bailey, A. J. 1985. The role of collagen in the development of muscle and its relationship to eating quality. *J. Animal Sci.* 60: 1580–1587.

Brownsey, G. J., Ellis, H. S., Ridout, M. J., and Ring, S. G. 1987. Elasticity and failure in composite gels. *J. Rheology* 31: 635–649.

Burge, D. L., Jr. and Action, J. C. 1984. Rheological properties of comminuted meat batters and the relationship to constituent interactions. *J. Food Technol.* 19: 719–724.

Burke, M., Zaager, S., and Bliss, J. 1987. Substructure of skeletal myosin subfragment 1 revealed by thermal denaturation. *Biochemistry* 26: 1492–1496.

Cassens, R. G. 1987. Structure of muscle. In *The Science of Meat and Meat Products*. Food & Nutrition Press, Inc., Westport, CT.

Dahlin, K. J., Allen, C. E., Benson, E. S., and Hegarty, P. V. J. 1976. Ultrastructural differences induced by heat in red and white avian muscles in rigor mortis. *J. Ultrastruct. Res.* 56: 96–106.

Daum, D. L. 1990. The effect of pH on the gelation of comminuted turkey breast and thigh meat. M.S. thesis, North Carolina Sate University, Raleigh, NC.

Dransfield, E., Francombe, M. A., and Whelehan, O. P. 1984. Relationship between sensory attributes in cooked meat. *J. Texture Studies* 15: 33–48.

Elliott, A. and Offer, G. 1978. Shape and flexibility of the myosin molecule. *J. Mol. Biol.* 123: 505–519.

Foegeding, E. A. 1990. Development of a test to predict gelation properties of raw turkey muscle proteins. *J. Food Sci.* 55: 932–936.

Foegeding, E. A., Allen, C. E., and Dayton, W. R. 1986. Effect of heating rate on thermally formed myosin, fibrinogen and albumin gels. *J. Food Sci.* 51: 104–108.

Foegeding, E. A. and Larick D. K. 1986. Tenderization of beef with bacterial collagenase. *Meat Sci.* 18: 201–214.

Foegeding, E. A. and Ramsey, S. R. 1987. Rheological and water-holding properties of gelled meat batters containing iota carrageenan, kappa carrageenan or xanthan gum. *J. Food Sci.* 52: 549–553.

Gauthier, G. F. 1990. Differential distribution of myosin isoforms among the myofibrils of individual developing muscle fibers. *J. Cell Biol.* 110: 693–701.

Girard, J. P., Dantchev, S., and Calderon, F. (1985). Relationship between temperature and degree of grinding in the cutter and the structure of frankfurter mixtures. *Fleischwirtsch.* 65: 811–814.

Hamm, R. 1975. On the rheology of minced meat. *J. Texture Studies* 6: 281–296.

Hofmann, S., Dusterhoft, S., and Pette, D. 1988. Six myosin heavy chain isoforms are expressed during chick breast muscle development. *FEBS Lett.* 238: 245–248.

Holmes, K. C., Popp, D., Gebhard, W., and Kabsch W. 1990. Atomic model of the actin filament. *Nature* 347: 44–49.

Howe, J. R., Hamann, D. D., and Park. J. W. 1991. Influence of test temperature and strain rate on fracture properties (texture) of surimi gels. Annual Meeting of The Institute of Food Technologists, Dallas, TX.

Ishioroshi, M., Samejima, K., and Yasui T. 1979. Heat-induced gelation of myosin: Factors of pH and salt concentrations. *J. Food Sci.* 44: 1280–1284.

Kamoun, M. and Culioli, J. 1988. Mechanical behavior of cooked meat under sinusoidal compression. *J. Texture Studies* 19: 117–136.

Khalili, A. D. and Zarkadas, C. G. 1988. Determination of myofibrillar and connective tissue protein contents of young and adult avian (*Gallus domesticus*) skeletal muscles and the N-methylhistidine content of avian actins. *Poultry Sci.* 67: 1593–1614.

Kim, B. Y., Hamann, D. D., and Wu, M. C. 1986. Effects of freeze-thaw abuse on the viscosity and gel forming properties of surimi from two species. *J. Food Sci.* 51: 951–956.

Light, N. and Champion, A. E. 1984. Characterization of muscle epimysium, perimysium and endomysium collagens. *Biochem. J.* 219: 1017–1026.

Mayne, R. and Sanderson, R. D. 1985. The extracellular matrix of skeletal muscle. *Collagen Rel. Res.* 5: 449–468.

Mickelson, J. R. and Louis, C. F. 1985. Components of purified sarcolemma from porcine skeletal muscle. *Arch. Biochem. Biophys.* 242: 112–126.

Moller, A. J. 1981. Analysis of Warner-Bratzler shear pattern with regard to myofibrillar and connective tissue components of tenderness. *Meat Sci.* 5: 247–260.

Montejano, J. G., Hamann, D. D., and Lanier, T. C. 1984. Thermally induced gelation of selected comminuted muscle systems—rheological changes during processing, final strengths and microstructure. *J. Food Sci.* 49: 1496–1505.

Montejano, J. G., Hamann, D. D., and Lanier, T. C. 1985. Comparison of two instrumental methods with sensory texture of protein gels. *J. Texture Studies* 16: 403–424.

Munro, P. A. 1983. The tensile properties of frozen and thawed lean beef. *Meat Sci.* 8: 43–61.

Ouali, A. 1990. Meat tenderization: Possible causes and mechanisms. A review. *J. Muscle Foods* 1: 129–165.

Payne, N. N. and Rizvi, S. S. H. 1988. Rheological behavior of comminuted meat batters. *J. Food Sci.* 53: 70–73.

Potekhin, S. A., Trapkov, V. A., and Privalov, P. L. 1979. Stepwise pattern of the thermal denaturing of helical fragments of myosin. *Biofizika* 24: 45–50.

Purslow, P. P. 1985. The physical basis of meat texture: Observations on the fracture behavior of cooked bovine *M. semitendinosus*. *Meat Sci.* 12: 39–60.

Rodgers, M. E., Karr, T., Biedermann, K., Ueno, H., and Harrington, W. F. 1987. Thermal stability of myosin rod from various species. *Biochemistry* 26: 8703–8708.

Sacks, M. S., Kronick, P. L., and Buechler, P. R. 1988. Contribution of intramuscular connective tissue to the viscoelastic properties of post-rigor bovine muscle. *J. Food Sci.* 53: 19–24.

Saliba, D. A., Foegeding, E. A., and Hamann, D. D. 1987. Structural failure and nondestructive rheological analysis of frankfurter batters: Effects of heating rates and sugars. *J. Texture Studies* 18: 241–259.

Samejima, K., Ishioroshi, M., and Yasui, T. 1981. Relative roles of the head and tail portions of the molecule in heat-induced gelation of myosin. *J. Food Sci.* 46: 1412–1418.

Schachat, F., Briggs, M., Williamson, E. K., McGinnis, H., Diamond, M. S., and Brandt, P. W. 1990. Expression of fast thin filament proteins. Defining fiber archetypes in a molecular continuum. In *The Dynamic State of Muscle Fibers*. Walter de Gruyter & Co., Berlin.

Sheard, P. R., Cousins, A., Jolley, P. D., and Voyle, C. A. 1990. Particle characteristics of flake-cut meat. *Food Structure* 9: 45–56.

Stafford, W. F. 1985. Effect of various anions on the stability of the coiled coil of skeletal muscle myosin. *Biochemistry* 24: 3314–3321.

Tolstoguzov, V. B. and Braudo, E. E. 1983. Fabricated foodstuffs as multicomponent gels. *J. Texture Studies* 14: 183–212.

Warner, K. F. 1952. Adventures in testing meat for tenderness. *Proceedings Fifth Annual Reciprocal Meat Conference,* 5: pp. 156–160. National Livestock and Meat Board, Chicago.

Wetzel, R., Perry, L. J., Mulkerrin, M. G., and Randall, L. M. 1990. Unfolding and inactivation—genetic and chemical approaches to the stabilization of T4-lysozyme and human interferon gamma against irreversible thermal denaturation. In *Protein Design and the Development of New Therapeutics and Vaccines,* J. B. Hook and G. Poste (Ed.), pp. 79–115. Plenum Publishing Co., New York.

Wu, J. Q., Hamann, D. D., and Foegeding, E. A. 1991. Myosin gelation kinetic study based on rheological measurements. *J. Agric. Food Chem.* 39: 229–236.

Yamamoto, K. 1990. Electron microscopy of thermal aggregation of myosin. *J. Biochem.* 108: 896–898.

9

Physical Chemistry of Bread Dough

R. Carl Hoseney

Kansas State University
Manhattan, Kansas

INTRODUCTION

Wheat flour is unique in its ability to form a viscoelastic dough that will retain gas and produce a light loaf of bread. The composition of wheat flour is not very different from that of flours from other cereals. However, as illustrated in Fig. 9.1, the differences in the sizes of loaves produced from 100 g of various flours are quite dramatic (He and Hoseney, 1991c).

Most bread flours contain between 11.0 and 12.0% protein. It is generally believed that the uniqueness of wheat flour is caused by the properties of its storage proteins, gluten. The gluten-protein complex is a mixture of a large number of proteins (Wrigley and Beitz, 1988) containing about an equal mixture of two basic groups: the prolamins, called gliadin, and the glutelins, called glutenin. The mixture of the gliadins and glutenins give gluten its viscoelastic properties.

The gliadins comprise many proteins (over 50 have been identified) that are single chained and of moderate molecular size. The molecular

Fig. 9.1 Loaves made from wheat, rye, rice, and corn flours, with volumes of 838 ± 11.5, 244 ± 4.8, 200 ± 9.1, and 179 ± 9.5, respectively.

weights range from about 30,000 daltons up to almost 100,000 daltons, with the majority of the proteins being in the 40,000–50,000 dalton range. When isolated, the gliadins are physically sticky and very extensible.

The glutenins are much larger. Their average molecular weight has been reported to be about 3 million daltons. Because of their large size and their tendency to undergo interchange, isolating individual glutenin proteins has proved to be essentially impossible. Therefore, we have no estimate of the number of glutenin proteins. The glutenins are multichained, and the individual peptides are linked by disulfide bonds. The peptides, or reduced glutenins, are many in number and range in molecular weight from 20,000 to 200,000 daltons. If the average molecular weight of 3 million is correct, and the average peptide is 100,000 daltons, then the average glutenin is composed of 30 peptides. In its isolated form, glutenin is resilient but not very extensible. The complex polymer appears to be linked together by the disulfide bonds mainly at the ends of the polypeptide chains. Little, if any, crosslinking appears to be present.

The amino acid composition of gluten is somewhat unique in several aspects. Differences between the amino acid composition of gliadins and glutenins are relatively minor. In both groups of proteins, the most predominant amino acid is glutamine, accounting for about 35% of the total gluten protein. Because it is in its amide form, glutamine does not contribute charge to the protein but, instead, is responsible for a high tendency for hydrogen bonding. The second most prevalent amino acid is proline, making up about 14% of the gluten. Because it has a ring structure, proline interrupts the order in protein chains. Studies have reported relatively low amounts of order in the gluten proteins. The other interesting point about the amino acid composition of the gluten proteins is the rather low amount of positively charged amino acids. Thus, the picture that becomes clear is that the gluten proteins have essentially no negative charges (the acids exist as their amides) and a low number of positive charges. Thus, the protein has a very low charge density. This results in the protein chains being able to approach each other and in a large amount of interaction between chains by secondary forces (hydrogen bonding and hydrophobic bonding). Such bindings are important in gluten's viscoelastic properties.

Although most cereal chemists agree that the uniqueness of wheat flour comes from the gluten, wheat starch also has also been shown to be somewhat unique. Wheat starch can be replaced only by rye or barley starch in bread (Hoseney et al., 1971) or in cake (Sollars and Rubenthaler, 1971). The wheat flour lipids are also unique, and their function in baking is not replaceable with other cereal lipids (Badi and Hoseney, 1976). Other partially unique features of wheat flour are the water-soluble pentosans. Although found in flour in rather small amounts (about 0.5%), they have been shown to contribute to loaf volume (Hoseney et al., 1969). Once again, only barley and rye appear to contain similar materials.

FORMATION OF DOUGH

Wheat flour, although appearing very fine, is actually made up of discrete particles that are quite large when compared to starch granules or protein molecules (Fig. 9.2). When water is added to flour, a number of different things happen that depend upon the ratio of flour to water. At high levels of water, a slurry is obtained; at low water content, the flour remains as a moist powder; but at intermediate levels of water, a thick sticky paste is formed. As this paste is mixed, several amazing changes occur. It becomes more resistent to stirring and turns into a dough that takes on a drier appearance. The changes that occur can be illustrated with a mixograph curve (Fig. 9.3). At the start of mixing, the dough offers little resistance to extension, primarily because the flour is only hydrated on the surface of the particles and, therefore, a large excess of water is present to give the dough fluidity. As the mixing proceeds, more and more of the water is used in hydrating the flour. As the protein becomes hydrated, it forms fibrils (Bernardin and Kasarda, 1973). As the level of

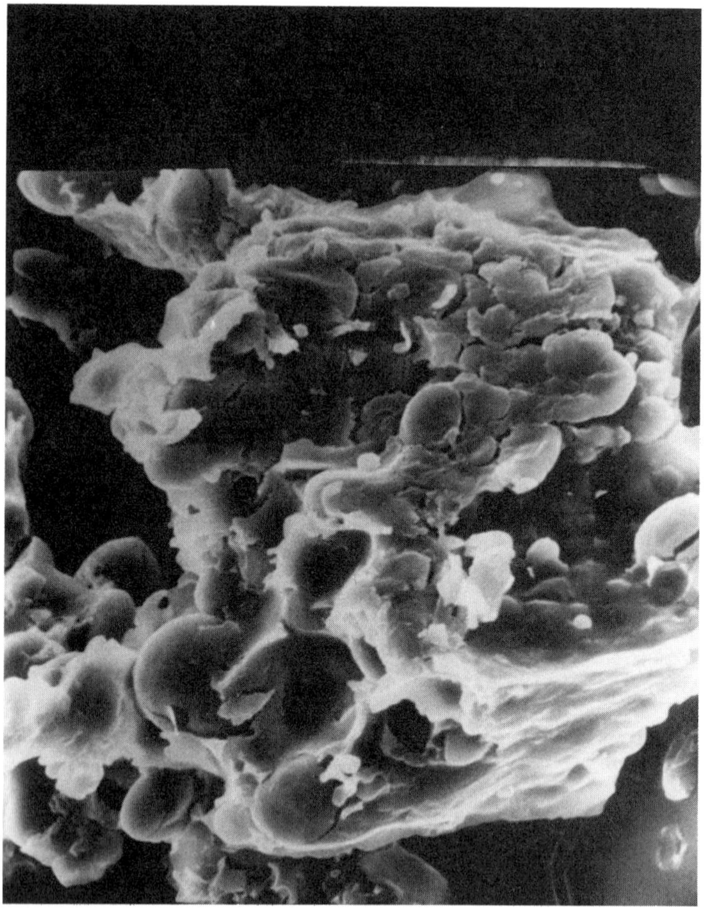

Fig. 9.2 Scanning electron photomicrograph of a hard wheat flour particle. (*From Hoseney and Seib, 1973.*)

water that can act as a plasticizer decreases and the amount of fibrils increases, the resistance to extension gradually increases. This continues until all the flour protein is hydrated. Stated in slightly different terms, the flour particles hydrate on their outer surface, then during mixing, the hydrated surface is wiped away, and the newly exposed surface is hydrated. This process is repeated until all the flour particles are worn away.

The above view of dough mixing is simplistic in that it considers mixing only as a hydration process. Although we think hydration is a very important part of mixing, it is not the only factor. If hydration were the only factor, then the mixing curve should rise to a maximum and then stay at that level during additional mixing. However, it is clear from the mixing curves (Fig. 9.3) that after reaching a peak, the curves decline, showing that the dough has less resistance to mixing. This is the classic breakdown with overmixing. Of course, some flours break down much more than others.

A complete understanding of the factors involved in overmixing is beyond the scope of this presentation. (The reader is referred to Schroeder and Hoseney, 1978; Hoseney, 1985.) In general, as a dough is mixed, it becomes more resistant to extension. With continued rapid mixing of the dough, certain disulfide bonds in the larger proteins are stressed and break (MacRitchie, 1975). The breaking of these bonds results in a free radical being formed on the sulfur (thiyl radical). Flour contains compounds that will react with those free radicals and, thus, become grafted to the protein. As a net result, the protein is reduced in size. This reduces the resistance to extension of the dough and, thereby, produces the peak on the mixing curve.

The formation of free radicals also causes a rapid and rather extensive

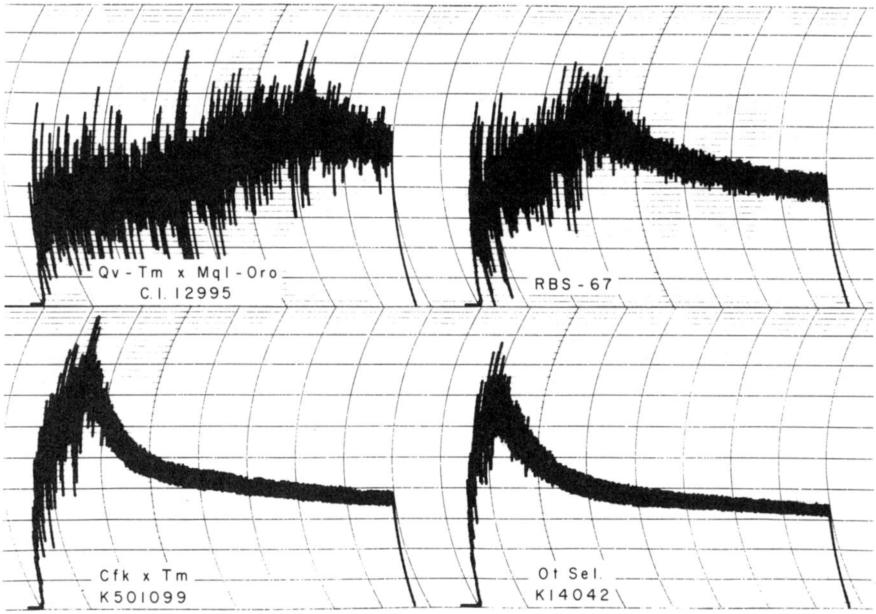

Fig. 9.3 Mixograms curves of hard wheat flours, showing long (C.I. 12995), medium (RBS-67), short (K501099), and very short (K14042) mixing times.

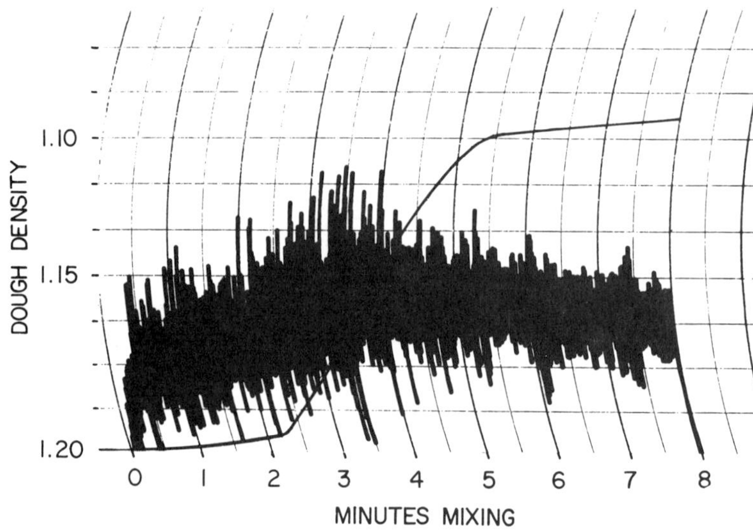

Fig. 9.4 Mixogram with a line (———) showing the change in dough density as a function of mixing.

interchange of the peptides in the glutenin proteins during mixing (Stewart and Maurtzen, 1966; Maurtzen, 1967; Jones and Carnegie, 1971). This may be a very important part of the dough development that occurs during mixing.

Another important event that occurs during dough mixing is the incorporation of air into the dough (Junge et al., 1981). This can be shown by measuring the dough density as a function of mixing time (Fig. 9.4). As dough is developed, it becomes much better at incorporating air. In fact, as shown by the mixing curve, more than half of the air incorporation occurs after the mixing peak has been reached. Of course, doughs are not mixed beyond their peak because they lose their ability to retain gas. It has also been shown that poor quality flours will incorporate more air during mixing to optimum than will good quality flours (Baker and Mize, 1946; He and Hoseney, 1991a). The air incorporated during mixing is very important because it forms the nuclei that determine the number of cells in the finished loaf of bread.

RHEOLOGICAL PROPERTIES OF DOUGH

Wheat flour dough is a viscoelastic material, which has both elastic and viscous properties. Wheat flour dough is not truly elastic but rather partially elastic. A true elastic material is a rubber band. As we stretch the band, it offers more resistance to extension. When we release the rubber band, it recoils to its original size and shape, providing we had not exceeded its elastic limit during the original deformation. Dough is elastic only if we deform it and release the deformation force rapidly. In other words, if we deform dough and hold it in that form for some time, the elastic property is lost.

For material to be elastic, two criteria must be met: the molecules must be large and they must be crosslinked. The molecular weight of raw rubber is large but it is not crosslinked. Thus, raw rubber has viscous flow properties. When sulfur is added and rubber heated, in a process called vulcanization, the rubber becomes truly elastic. Wheat flour dough appears to be more like raw rubber. The glutenin proteins are large in molecular size, but they are not crosslinked. The partial elastic properties result from either entanglement of the large molecules and/or secondary bonds (hydrogen or hydrophobic). Both the secondary bonds and entanglements

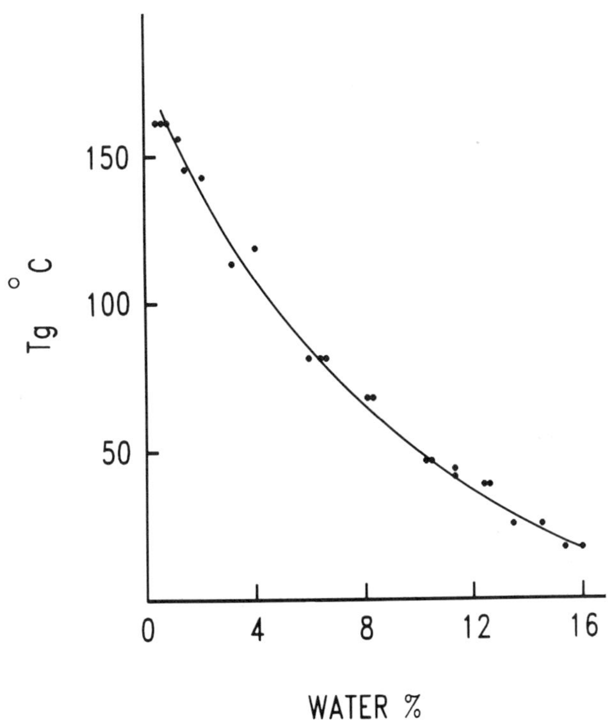

Fig. 9.5 Change in T_g as a function of moisture content of both hand-washed and commercial gluten.

lose their effectiveness with time, and, thus, the elastic properties of dough disappear with time. With the breaking of the secondary bonds, the dough is free to flow under force, even the force of gravity.

In dough, the continuous phase is the gluten protein. Therefore, many of the viscoelastic properties of the dough are the result of the properties of gluten. There have been several speculations about the factors causing the viscoelastic properties of gluten. These included the fact that under certain conditions, gluten is in a beta-turn configuration (Tatham et al., 1985). Perhaps the simplest and, therefore, most likely explanation is that gluten is a random glassy polymer (Hoseney et al., 1986). At room temperature dry gluten is glassy, but it goes through a glass transition as the moisture content reaches about 16% (Fig. 9.5). Thus, at the moisture content of dough the polymer will be rubbery (Hoseney, 1991). The uniqueness of gluten may be the fact that it has a glass transition at the correct temperature.

The starch in dough then acts as a filler for the gluten polymer. As is the case with many fillers, the starch does not appear to be an inert filler. Rheological studies show that the elastic modulus of gluten increases as we add starch, indicating that there is an interaction between the gluten and the starch (He and Hoseney, 1991a). These studies indicate that glutens from different flours interact with the same starch differently. Thus, much of the difference in quality of gluten for breadmaking may be explained by how it interacts with the starch.

GAS RETENTION OF DOUGH

Another important property of wheat flour doughs is their ability to retain gas. This property is not truly unique. It has been shown that bread with good volume can be produced from starch containing surfactants in the correct form (Rotsch 1954; Rajapaksa et al., 1983) and also with starch and various gums, such as xanthan. Of course, the bread produced is not close to that produced from wheat flour in a number of important characteristics, e.g., texture and flavor.

When bread dough is mixed, air is incorporated as mentioned above. The other formula ingredients including sugar and yeast also are mixed into a uniform mixture. Just after the dough is mixed, it is aerobic because of the oxygen in the air. However, yeast rapidly uses this oxygen and produces an anaerobic dough. Yeast is a facultative organism and can readily ferment under anaerobic conditions. If there is no sugar in the formula, the flour contains about 2% of its weight in fermentable material. As the result of fermentation, yeast produces equal molar amounts of carbon dioxide and ethyl alcohol. The reactions are exothermic, and considerable heat is produced during fermentation.

As fermentation starts, the oxygen in the aqueous phase is consumed and the yeast produces carbon dioxide. The carbon dioxide first has to saturate the aqueous phase of the dough, which gives rise to the lag seen in plots of gas production. These lags have been explained as an adaption phase in which the yeast comes out of dormancy. Saturation of the aqueous phase is probably a more accurate explanation. Once the aqueous phase is saturated, the carbon dioxide must find some place to which it can diffuse. One possibility is diffusing to the preformed gas cells that were formed when air was incorporated during mixing. These cells have lost their oxygen but are still intact because the nitrogen in air is not soluble. The other possibility is diffusing to the outer surface of the dough and escaping to the atmosphere. Of course, it most cases, this is a long distance from where the gas was produced.

The yeast and the carbon dioxide it generates cannot produce any new gas cells (Baker and Mize, 1941). From bubble mechanics, we learn that the pressure in a bubble is defined by the following formula:

$$p = \frac{2\gamma}{r}$$

where p is the pressure in the bubble, γ is the interfacial tension, and r is the radius of the bubble. This formula clearly shows that the pressure required to form a new bubble, zero radius, is infinitely large and, therefore, not possible.

Thus, once the aqueous phase is saturated, the carbon dioxide must diffuse into a preformed gas cell or to the surface of the dough. Only that gas produced near the surface of the dough can diffuse to the surface. As the carbon dioxide enters a gas cell, the pressure in the cell increases. Because the dough is viscoelastic, the dough will flow to bring the pressure to equilibrium. As a result, the dough increases in volume or, in other words, the dough is leavened.

The concept often heard in cereal chemistry that the gas is retained in a gas bubble by a gluten film or membrane is without basis. This is disproved by asking the simple question, if there is a membrane that holds the gas in, then how does the gas get into the cell? There is no magic in gas retention (Hoseney, 1984). Gas diffuses from the saturated aqueous phase into preformed gas cells. It cannot diffuse back out because the aqueous phase is saturated, and the yeast is producing more carbon dioxide to keep it saturated. The difference between doughs that retain gas, such as wheat flour doughs, and those that do not, such as corn flour

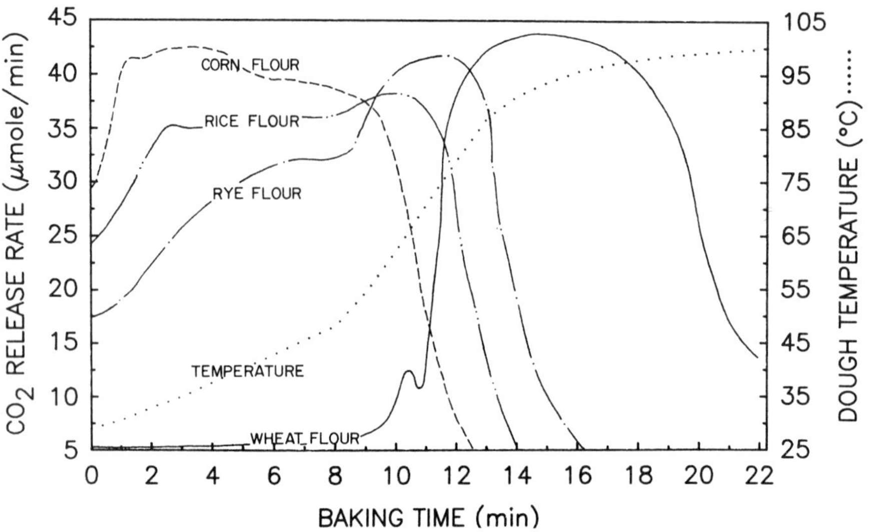

Fig. 9.6 Rate of loss of carbon dioxide during baking from doughs made from wheat, rye, rice, and corn flours.

doughs, is simply the rate of diffusion. In corn flour doughs, the continuous phase is the water and, therefore, diffusion is fast. In wheat flour, the highly hydrated gluten is the continuous phase, and gas diffusion is slow. Therefore, wheat flour dough retains gas and the corn flour dough does not (Fig. 9.6).

During fermentation, the dough is often punched or remixed. As a result of this working, some gas is lost from the dough; however, this is not the goal of the punching. As a result of leavening, the gas cells have grown large, and punching subdivides these cells into many more, smaller cells. As a result of this, the crumb grain of the bread is made much finer. Also during fermentation, the sugar close to the yeast is depleted. The rate of gas production slows because the sugar must diffuse to the yeast cell. As a result of punching, the dough is "mixed," thus getting the yeast and sugar together. This restores a rapid rate of fermentation.

RHEOLOGICAL CHANGES DURING FERMENTATION

When wheat flour dough comes from the mixer, it is tight, but within a few minutes, it relaxes into a very extensible dough. In general, the dough is slightly sticky and almost dead feeling (like clay). During fermentation, the character of the dough changes dramatically. The dough becomes much more elastic and less extensible. It becomes drier to the touch and now appears to be lively. A number of factors may be involved in those changes, but the most important is clearly the effect of yeast. To study the changes occurring during fermentation, a simple spread test was developed (Hoseney et al., 1979). With the test, it was easy to show that the products of yeast fermentation have no effect on the rheological properties of dough. The test also showed that the yeast itself, when in contact with the dough, caused the changes. The biochemical nature of those changes is not yet clear. Lubricated compression studies have reached the same conclusion (Cullen-Rafei et al., 1988). The bottom line is that the properties of the dough are changed dramatically during fermentation, and most of this change is the result of the yeast. The oxidants that are commonly used in breadmaking also exert their effects during fermentation.

DOUGH IN THE OVEN

At the end of the proofing stage, most doughs have a similar volume, and only very poor quality loaves lose appreciable gas during proofing (He and Hoseney, 1991b). The differences in loaf volume then occur during the first few minutes in the oven. The volume of a loaf of bread expands

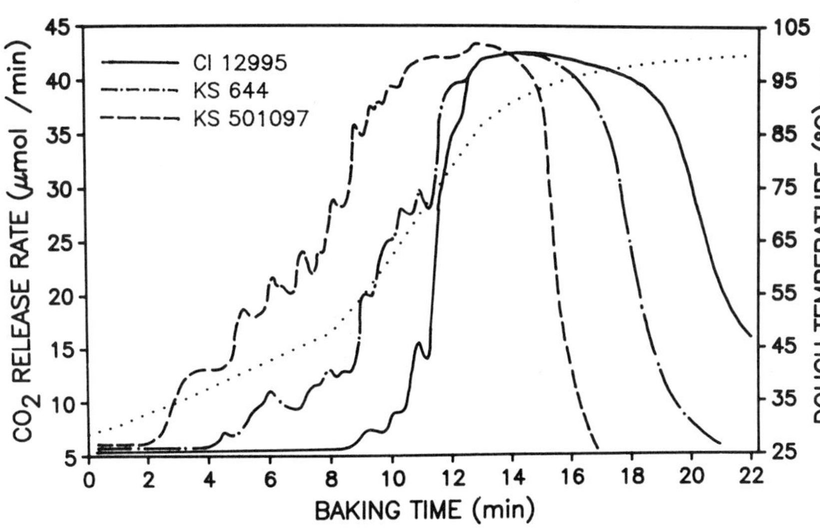

Fig. 9.7 Comparison of carbon dioxide release from CI 12995 (good quality), KS 644 (intermediate quality), and KS 501097 (poor quality) flour doughs.

rapidly after it is placed in the oven and will reach its maximum size within 6 to 8 minutes. This period of rapid increase in volume is called the oven spring stage of baking.

What causes the difference in oven spring is not clear. It could be a rheological resistance to the rapid increase in volume or the effect of temperature on the dough. Using an electrical resistance oven to heat the dough and a IR carbon dioxide detector to monitor the loss of gas, He and Hoseney (1991b) were able to show that poor quality flour lost appreciable amounts of gas during the early stages of baking, while the dough was still at a relatively low temperature. On the other hand, good quality flour produced a dough that lost relatively little gas until the temperature approached 72°C (Fig. 9.7). At that temperature, the dough appeared to suddenly lose its ability to retain gas. This is not consistent with an increase in the rate of gas diffusion but, instead, appears to indicate a rapid transition of some type.

TRANSFORMATION OF A CLOSED-CELL FOAM (DOUGH) TO AN OPEN-CELL SPONGE STRUCTURE (BREAD)

In a well-mixed bread dough, the continuous phase is the hydrated gluten protein. The discontinuous phases include the starch granules and the air cells in the dough. As discussed earlier, the dough retains gas because its diffusion to the surface is slow. As a dough is heated, a number of changes take place. Probably the most important is the change in the dough's rheology. Heating gluten to about 90°C gave only a small decrease in its elastic modulus (Dreese et al., 1988). Above 90°C, particularly with longer heating, the gluten started to polymerize (crosslink) and the elastic modulus increased. Addition of starch to the gluten resulted in a large increase in elastic modulus at about the temperature of starch gelatinization. At first glance, this might be explained as a transfer of water from the gluten to the starch, and, of course, this undoubtedly occurs. However, removing water from the system changed the elastic modulus but did not affect the loss tangent. Heating the gluten starch system changed both the elastic modulus and the tangent (Dreese et al., 1988). This was explained as an interaction between the protein and the gelatinized starch. When such a system was heated further (80°C), the elastic modulus decreased substantially, probably because of decreased hydrogen bonding between the protein and the starch.

In a complete bread formula, the starch gelatinizes at about 65°C or only slightly below the temperature at which major loss of gas occurs.

He and Hoseney (1991b) have suggested that the interaction between the protein and the starch changes the rheology of the dough such that it can no longer expand under the force of the gas. As a result, the cell walls fail, and the dough is transformed into a gas continuous sponge structure. Once the system is gas continuous, it no longer retains gas, as is shown by the rapid loss of gas. This finding is in agreement with the much earlier work of Baker (1939), who also concluded that bread was a sponge structure. The fact that bread is air continuous can easily be shown by gently blowing through a slice of bread; very little, if any, resistance is found. Being air continuous also explains why a loaf of bread does not collapse when it is removed from the oven. If it were still dough, an air discontinuous foam, it would collapse.

FIRMING OF BREAD (STALING)

Upon storage, bread becomes firm. Actually, many changes occur during bread storage, but firming appears to be the most important. Firming has been explained as a recrystallization of the amylopectin fraction of the starch in the bread. Recently, this concept has been questioned. For example, work has shown that partly dried bread firms at a very rapid rate but that recrystallization does not occur under these conditions (Rogers et al., 1988).

Bread firming and starch recrystallization are not synonymous, although both occur during bread storage (Martin et al., 1991). Factors such as presence of shortening and baking time influenced water hydration capacity of bread crumb. These factors also affected bread firmness during 5 days of storage. A proposed model of the mechanism for bread firming maintains that crosslinks (hydrogen bonds) between the continuous protein matrix and the discontinuous remnants of starch granules result in bread firming.

High levels (100 SKB units/loaf) of bacterial alpha-amylase stopped bread from firming during 5 days of storage. Supplementation with malted barley flour did not decrease bread firming rate, and the bread did not contain a large quantity of dextrins between 3 and 9 glucose units in length. However, bread supplemented with either fungal or bacterial alpha-amylase contained larger quantities of residual dextrins DP 3 to 9. Beta-amylase was effective in reducing bread firmness; however, very high levels of supplementation did not stop the bread-firming process. Bread supplemented with pullulanase firmed at a faster rate. Dextrins of a particular size are presumably responsible for the antifirming effect. A mechanism of bread firming proposed by Martin and Hoseney (1991) maintains that crosslinks between protein fibrils and starch granule remnants result in bread firming. Dextrins of a particular size interfere with crosslinks (H-bonds) between starch and protein.

REFERENCES

Badi, S. M. and Hoseney, R. C. 1976. *Cereal Chem.* 53: 733.

Baker, J. C. 1939. *Cereal Chem.* 16: 730.

Baker, J. C. and Mize, M. D. 1941. *Cereal Chem.* 19: 19.

Baker, J. C. and Mize, M. D. 1946. *Cereal Chem.* 23: 39.

Bernardin, J. E. and Kasarda, D. D. 1973. *Cereal Chem.* 50: 529.

Cullen-Refai, A., Faubion, J. M., and Hoseney, R. C. 1988. *Cereal Chem.* 65: 401.

Dreese, P. C., Faubion, J. M., and Hoseney, R. C. 1988. *Cereal Chem.* 65: 348.

He, H. and Hoseney, R. C. 1991a. *Cereal Chem.* 68: 526.

He, H. and Hoseney, R. C. 1991b. *Cereal Chem.* 68: 521.

He, H. and Hoseney, R. C. 1991c. *Cereal Chem.* 68: 334.

Hoseney, R. C. 1984. *Cereal Foods World* 29: 305.

Hoseney, R. C. 1985. *Cereal Foods World* 30: 453.

Hoseney, R. C. 1991. *Water Relationships in Foods: Advances in the 1980s and trends for the 1990s,* L. Slade and H. Levine (Ed.). Plenum Press, New York.

Hoseney, R. C. and Rogers, D. E. 1990. *Crit. Rev. in Food Sci. and Nutr.* 29: 73.

Hoseney, R. C. and Seib, P. A. 1973. *Baker's Dig.* 47(6): 26.

Hoseney, R. C., Finney, K. F., Shogren, M. D., and Pomeranz, Y. 1969. *Cereal Chem.* 46: 117.

Hoseney, R. C., Finney, K. F., Pomeranz, Y., and Shogren, M. D. 1971. *Cereal Chem.* 48: 191.

Hoseney, R. C., Hsu, K. H., and Junge, R. C. 1979. *Cereal Chem.* 56: 141.

Hoseney, R. C., Zeleznak, K., and Lai, C. S. 1986. *Cereal Chem.* 63: 285.

Jones, I. K. and Carnegie, P. R. 1971. *J. Sci. Fd Agric.* 22: 358.

Junge, R. C., Hoseney, R. C., and Varriano-Marston, E. 1981. *Cereal Chem.* 58: 338.

MacRitchie, F. 1975. *J. Polymer Sci.* 49: 85.

Martin, M. L. and Hoseney, R. C. 1991. *Cereal Chem.* 68: 503.

Martin, M. L., Zeleznak, K. J., and Hoseney, R. C. 1991. *Cereal Chem.* 68: 498.

Maurtzen, C. A. M. 1967. *Cereal Chem.* 44: 170.

Rajapaksa, D., Eliasson, A.-C., and Larsson, K. 1983. *J. Cereal Sci.* 1: 53.

Rogers, D. E., Zeleznak, K. J., Lai, C. S., and Hoseney, R. C. 1988. *Cereal Chem.* 65: 398.

Rotsch, A. 1954. *Brot und Gebäck* 8: 129.

Schroeder, L. F. and Hoseney, R. C. 1978. *Cereal Chem.* 55: 348.

Sollars, W. F. and Rubenthaler, G. L. 1971. *Cereal Chem.* 48: 397.

Stewart, P. R. and Maurtzen, C. M. 1966. *Austr. J. Biol. Sci.* 19: 1125.

Tatham, A. S., Miflin, B. J., and Shewry, P. R. 1985. *Cereal Chem.* 62: 405.

Wrigley, C. W. and Beitz, J. A. 1988. *Wheat Chemistry and Technology* Vol. 1, 3rd ed., Y. Pomeranz (Ed.). AACC, St. Paul, MN.

10

Physical and Chemical Properties of Nutrients Affecting Their Absorption and Utilization

Dennis T. Gordon and Lori Pellett

University of Missouri
Columbia, Missouri

INTRODUCTION

The science of food as it relates to the primary function of providing nutrition has progressed far beyond the quantification of individual nutrient content. The foundation of these advances, although at times obscured from immediate recognition in the scientific literature, is our understanding of the chemical and physical properties of food ingredients.

The food chemist/technologist works to understand how these properties of food ingredients affect their functionality (Szczesniak, 1990; Arciszewski, 1991; Shen and Jackson, 1991). The nutritionist needs to understand them to improve the nutritional value of food (Sipple and McNutt, 1974; Perkins and Visek, 1983). The ultimate goal of both groups is to improve food quality (Fig. 10.1). Good health and freedom from disease, as long-term goals in providing good nutrition, are thus directly related to the chemical and physical characteristics of food ingredients.

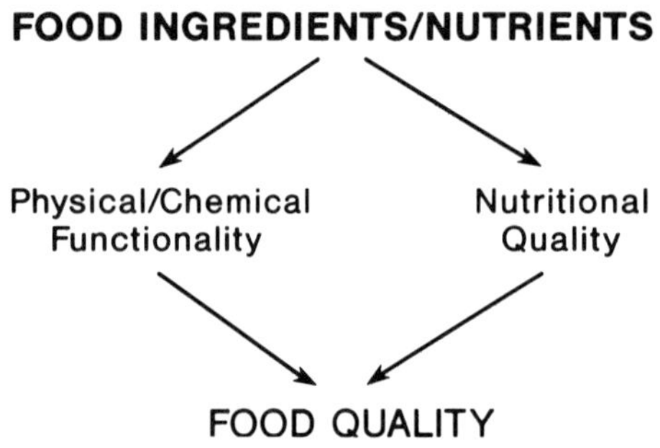

Fig. 10.1 Food ingredients are nutrients, which have physical, chemical, and nutritional properties that contribute to overall food quality.

This chapter deals specifically with the relationship between the chemical and physical properties of food ingredients and the body's ability to perform biochemical and physiological processes with them in meeting its nutritional needs. The terms "nutrient" and "food ingredient" are considered synonymous and will be used interchangeably.

A partial listing of the common physical and chemical properties that could be used to describe an element, a compound, a nutrient, or a food ingredient is presented in Table 10.1. There are also many good references which describe the physical and chemical properties of each major class of food ingredients (Lineback and Inglett, 1982; Weiss, 1983; Pomeranz, 1985; Dreher, 1987; Bauernfeind and Lachance, 1991; Goldberg and Williams, 1991).

Table 10.1 Common Physical and Chemical Properties of Nutrients/Food Ingredients[a]

Color
Molecular weight
Melting/Boiling point
Density
Optical rotation
Dissociation constant
Redox potential
Oxidation number (valence)
Coordination number
Solubility

[a] Similar chemical and physical properties can be used to describe nonfood ingredients. The list is not complete.

Three additional areas integrally related to the nutritional quality of foods and interrelated with each other are: (1) the chemical nature (referred to as speciation) and physical forms of food ingredients; (2) the actual amount of the ingested nutrient absorbed and utilized by the body (i.e., bioavailability); and (3) interactions among nutrients that both positively and negatively affect their bioavailability, which is of core importance. These three areas are briefly defined and will be discussed in greater detail with examples and a special focus on the elements known as essential minerals.

The authors wish to note that it is difficult to adequately separate nutrient chemical and physical properties in explaining their combined effect on a nutrient's ability to meet the body's nutritional needs. However, the chemical properties of a nutrient appear to be predominately related to its capacity to meet the nutritional needs of the body from a biochemical perspective. The physical form and properties of a nutrient exert a greater influence on physiological actions within the body.

NUTRIENT SPECIATION AND PHYSICAL FORM

Individual nutrients can exist in different chemical and physical forms, exhibit different properties differentiated by, and differentially affect, the body. Thus, nutrient speciation plays a vital role in achieving nutritional adequacy. The systematic determination of the chemical forms in which a nutrient can exist can be described as specie analysis or speciation.

Examples of nutrients having different chemical and physical forms are listed in Table 10.2. That these characteristics of a nutrient can determine if it will be absorbed or utilized by the body is illustrated in the case of iron.

Iron in red meats is present mainly in hemoglobin and myoglobin molecules (Francis, 1985). The iron in the porphyrin rings of these molecules may be in the iron (II) (i.e., ferrous) form (i.e., myoglobin and hemoglobin) or iron (III) (i.e., ferric) form (i.e., metmyoglobin and methemoglobin) oxidation states. Irrespective of the oxidation state of this iron, the porphyrin-bound iron in these complexes is preferentially absorbed into the intestinal mucosal cell compared to either free or nonheme iron (II) or iron (III) ions. This occurs because of the different absorptive pathways for organically bound iron in the porphyrin ring compared to nonheme iron ions (Wheby et al., 1970; Bjorn-Rasmussen et al., 1974). For this reason, the consumption of red meats is strongly advised for growing children and women (*Surgeon General's Report on Nutrition and Health*, 1988).

Table 10.2 Different Chemical/Physical Forms of Nutrients/Food Ingredients[a]

- Sugars and amino acids
  - Stereoisomers, D form; R clockwise/right
  - Stereoisomers, L form; S counter-clockwise/left[b]
- Fatty acids
  - Saturated/Unsaturated
  - *cis/trans*
- Dietary fiber
  - Insoluble
  - Soluble
- Vitamins
  - Vitamin A
    - Retinol
    - Retinal, retinaldehyde
    - Retinoic acid
    - β-Carotene (vitamin A precursor)
  - Vitamin E
    - α-Tocopherol
    - β-Tocopherol
    - γ-Tocopherol
    - δ-Tocopherol
  - Vitamin K
    - Menadione
    - Phylloquinone
    - Menaquinone-n
    - Menaquinone-7
  - Niacin
    - Nicotinic acid
    - Nicotinamide
  - Vitamin B_6
    - Pyridoxine
    - Pyridoxamine
    - Pyridoxal
    - Pyridoxic acid
  - Folic acid
    - 7,8-Dihydofolate
    - 5,6,7,8-Tetrahydrofolate
    - 5-Methyltetrahydrofolate
    - 10-Formyltetrahydrofolate
- Vitamins
  - Vitamin B_{12}
    - Cyanocobalamin
    - Hydroxocobalamin
    - Chlorocobalamin
    - Nitritocobalamin
    - Thiocyanatocobalamin
- Minerals
  - Iron
    - Organically bound heme iron
      - Hemoglobin
      - Myoglobin
    - Organically bound nonheme iron
      - Ferritin
      - Hemosiderin
    - Inorganic
      - Iron II (ferrous)
      - Iron III (ferric)
  - Zinc
    - Organic complexed with metallothionein
    - Inorganic Zinc II (Zn^{++})
  - Copper
    - Organic, complexed with
      - Metallothionein
      - Ceruloplasmin
      - Hemocyanin
    - Inorganic
      - Copper I (cuprous)
      - Copper II (cupric)
  - Selenium
    - Organic, complexed with
      - Selenomethionine, plant
      - Selenocysteine, animal
    - Inorganic
      - Selenium VI (selenate)
      - Selenium IV (selenite)
      - Selenium II (selenide)

[a] Listing contains only a representative sampling of the different chemical forms that exist for various nutrients. A biological system can differentiate among these chemical species to achieve its nutritional needs. The biopotencies of different species of nutrients are usually different.
[b] Sugars and amino acids can also be either dextrorotatory (+) or levorotatory (−).

The nonheme portion of the iron in our diet exists in complexes or as iron salts (Monsen et al., 1978). The iron in these complexes and salts must be released and solubilized before absorption can occur. It is commonly believed that the iron (II) oxidation state of iron is more soluble in the intestine and therefore more easily absorbed than the iron (III) oxidation state (Turnbull et al., 1962; Pla et al., 1973). If a nutrient (i.e., iron) cannot be solubilized in the intestine because of its chemical form (i.e., ferric hydroxide, ferric orthophosphate), it will not be absorbed into the body (Rees and Monsen, 1973; Pla et al., 1976). Solubility in the stomach and intestine remains one of the most important physical properties influencing the absorption of any nutrient. Although the concept of nutrient speciation normally refers to the micronutrients (i.e., vitamins and minerals), the basic units of the macronutrients (i.e., sugars, fatty acids, and amino acids) should also be considered. Physiological functions affected by the different physical forms of primarily macronutrients and end-processing-induced changes are listed in Table 10.3.

Monosaccharides and amino acids absorbed by the body exist in the D and L forms, respectively. Efforts are made to ensure that D forms of amino acids are not formed during food processing because of their antinutritional and potentially toxic effects (Jansen, 1981; Cheftel et al., 1985). The L forms of sugars have been proposed as noncaloric sugar substitutes (O'Brien-Nabors and Gelardi, 1986; Anderson, 1989). However, the use of L-sugars for purposes of reducing caloric density of foods or drinks may not be advisable (Herman, 1974).

Saturated fatty acids are considered one of the major dietary factors contributing to coronary heart disease (Ahrens et al., 1957; Expert Panel on Population Strategies, 1990). The *cis/trans* isomers of the same fatty acid have different physical properties, such as different melting points (Swern, 1979). *Trans* fatty acids are thought to act like saturated fatty acids in elevating blood cholesterol concentrations (Beare-Rogers, 1983; Mensink and Katon, 1990). Adding to the controversy concerning saturated fatty acids and coronary heart disease has been the report that the body differentiates between stearic acid (C18:0) and palmitic acid (C16:0). Stearic acid is claimed to be nonatherogenic compared to palmitic acid, because stearic acid has been observed not to elevate blood cholesterol concentrations (Bonanome and Grundy, 1988). And finally, the position of the fatty acids on the glycerol molecule in triglycerides has been reported to affect blood lipid concentrations (Kritchevsky, 1988). Dietary lipids, and their relationship to coronary heart disease, as briefly cited, provide a classical example of the relationship between the physical and possibly chemical properties of a food ingredient and a major health problem. However, we are not convinced that these views are wholly valid, nor does space allow us to refute possible extension of the adverse

Table 10.3 Potential Physiological Changes Resulting from Physical/Chemical Alterations to Major Classes of Food Ingredients by Common Food-Processing Procedures

Physical/Chemical alteration or processing	Physiological functions
Carbohydrates	
Milling/Particle size	Rate of ingestion
Cooking	Gastric emptying
Amylose/Amylopectin ratio	Rate of digestion
Molecular size (simple/complex)	Glycemic index
	Colonic fuels
	Satiety/Satiation
Fats/Oils	
Polyunsaturation/Saturation ratio	Blood cholesterol
Interesterification	Lipid metabolism
Hydrogenation to include *cis/trans* isomerization	
Proteins	
Denaturation	Digestibility
Cooking	Protein quality
Maillard reaction	
Dietary fiber	
Milling/Particle size	Intestinal bulk
Soluble/Insoluble	Water-holding capacity
Molecular size	Time of waste passage
	Nutrient absorption
	Colonic fuels
Vitamins and minerals	
Cooking[a]	Essential components for growth, maintenance and reproduction
Physical/Chemical forms[a]	

[a] Affects bioavailability.

role of saturated fatty acids in human health to include *trans* fatty acids. There is no scientific data to explain why saturated and unsaturated fats/oil act differently in affecting blood lipid concentrations and atherosclerosis.

BIOAVAILABILITY

After considering the nutrient content of our food supply, nutrient bioavailability is the most important aspect in achieving good nutrition, the foundation in the relationship between food science and nutrition, and the central focus of this chapter. Bioavailability is defined as the amount of an ingested nutrient absorbed and utilized by the body compared to the total amount ingested (O'Dell, 1984), as illustrated in Fig. 10.2. A nutrient may be absorbed into the body, but if it does not fulfill a biochemical or physiological function, it does not achieve biological potency. The speciation of a nutrient and its interaction with other nutrients are

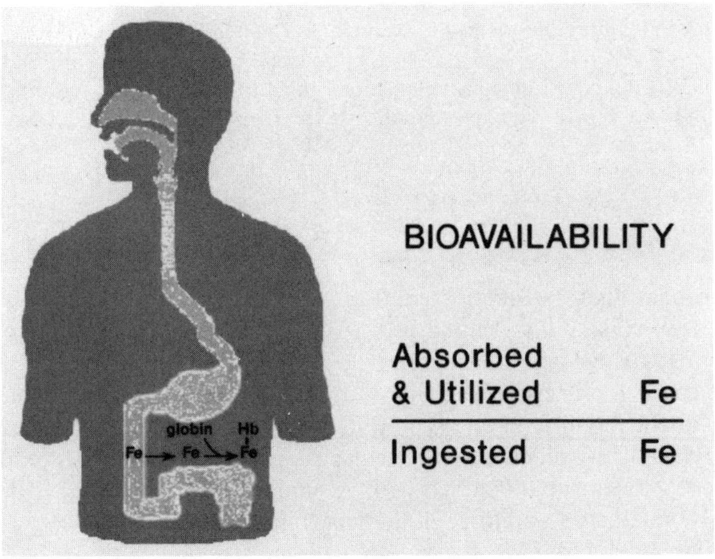

Fig. 10.2 The concept of iron bioavailability is illustrated. Iron can be ingested in a number of different physical and chemical forms in a varied diet (see Table 10.2). It is generally assumed that nonheme iron must be initially solubilized from any dietary complex in the small intestine. It is further hypothesized that this nonheme iron exists in the iron II oxidation state, which may complex with other molecules before being absorbed across the intestinal lumen. From intestinal mucosal cells, iron is transferred to the systemic blood and transported via transferrin in the iron III oxidation state to various sites where needed. For example, absorbed dietary iron can be incorporated into protoporphyrin, and the resulting complex, which now contains iron in the iron II oxidation state, is called heme. The conjugation of heme with globin protein produces one subunit, and the final product, hemoglobin (Hb-Fe), is a globular molecule made up of four subunits. The measurement of hemoglobin provides a useful index of iron status and the bioavailability of dietary iron. The ratio of absorbed and utilized iron compared to the amount ingested is defined as bioavailability.

HIGH NUTRIENT BIOAVAILABILITY

WHEN TOTAL FOOD INTAKE SHOULD BE REDUCED FOR PURPOSES OF DIETING AND WEIGHT MANAGEMENT.

WHEN TOTAL FOOD INTAKE SHOULD BE INCREASED DURING PERIODS OF ILLNESS OR LACK OF SUFFICIENT QUANTITIES.

Fig. 10.3 High micronutrient (i.e., vitamin and mineral) bioavailability should be a goal when individuals need to reduce or increase calorie intake. For example, in individuals working to decrease their food intake to manage their weight, the bioavailability of remaining micronutrients should be high. In populations where energy and micronutrient contents are marginal, the bioavailability of these limited micronutrients should also be high. Among individuals suffering from illness and disease and having limited food intake, nutrient bioavailability should be high.

the primary factors affecting nutrient bioavailability. Conditions when nutrient bioavailability should be maximized are shown in Fig. 10.3.

Almost all ingested food ingredients are utilized by the body as essential nutrients or for energy. Their basic units are absorbed almost quantitatively in the small intestine (Guyton, 1991) (Table 10.4). Aspartame, L-aspartyl-L-phenylalanine methyl ester, is a dipeptide that is completely digested and its subunits absorbed. Although the amount ingested is extremely small, the amino acids in aspartame could provide energy, and phenylalanine is an essential amino acid.

Table 10.4 Ingestion of Macromolecules and Absorption of Individual Subunits

Ingest	Absorb
Complex carbohydrates	Sugars
Proteins	Amino acids
Fats/Oils	Fatty acids
Minerals/Mineral salts	Ions
Vitamins/Conjugates	Free vitamin
Dietary fiber	Short-chain fatty acids[a]

[a] Absorption occurs after fermentation in ileum, cecum, and/or colon.
Source: Guyton, 1991.

Table 10.5 Biological Potency of Tocopherols[a]

Trival names	For normal fetal development in rats (%)	Prevention of erthrocyte hemolysis in rats (%)	Symptoms of muscular dystrophy[b] Prevention in chickens (%)	Symptoms of muscular dystrophy[b] Curative in rats[c] (%)
α-Tocopherol	100	100	100	100
β-Tocopherol	25–40	15–27	12	—
γ-Tocopherol	1–11	3–20	5	11
δ-Tocopherol	1	0.3–2.0	—	—

[a] Comparing tocopherols at equal dietary concentrations.
[b] Defined as a wasting away and atrophy of muscle mass, which can be caused or prevented with inadequate or adequate dietary vitamin E concentrations, respectively.
[c] In rats fed vitamin E–deficient diets and then supplemented with different tocopherols at equal dietary concentrations.
Source: Machlin, 1991

Exceptions are ingested food ingredients which are not absorbed but are excreted intact (e.g., sucrose polyesters) or absorbed with (e.g., sodium cyclamate) or without (e.g., saccharin) possible subsequent metabolism. Although absorbed, these latter compounds provide neither essential nutrients nor a significant source of energy.

As with speciation and physical form, and illustrating the integral relationship with them, bioavailability has historically been described with reference to the micronutrients. Thus, the four common chemical forms of tocopherol in foods differ in biopotency, as reported in Table 10.5 (Machlin, 1991). Solubility, cited previously with reference to iron, is one of the most important physical properties of a nutrient which affects its bioavailability, and in vitro determinations of nutrient solubility have been used to assess potential in vivo bioavailability (Schricker et al., 1982; Clydesdale, 1988).

INTERACTIONS

An interaction is defined as a mutual or reciprocal action or influence. In nutrition, several specific types of interactions are recognized, giving the term added meaning (Table 10.6). Testing for true interaction effects among nutrients requires properly designed studies and statistical tests.

Table 10.6 Nutrient Interactions Affecting Nutrient Bioavailability

Noncompetetive (one-sided)[a]
Antagonistic (negative)
Synergistic (positive)
Examples:
Macro/micronutrient interactions:
animal protein/iron (positive)
fructose/copper (negative)
phytic acid/zinc and copper (negative and positive)
dietary fiber/minerals (commonly claimed to be negative, but no effect).
Micro/micronutrient interactions:
copper needed for iron metabolism (synergistic-positive)
excess iron will impair copper metabolism (antagonistic-negative)
Competitive (two-sided)[b,c]
Negative
Reversible
Examples:
Iron and zinc
Zinc and copper
Multinutrient Combination of Noncompetitive and Competitive Interactions of Three or More Nutrients[d]
Iron, zinc, and copper
Copper, molybdenum, and sulfur

[a] Not true interactions, but the effect of one nutrient on another.
[b] Results from similar chemical and physical forms and properties.
[c] Classical interactions between two nutrients described as illustrated with properly designed experiments to test for interaction effects.
[d] Most important and possibly the most realistic set of interactions constantly occurring in any living system exposed to different foods.

The phrase "nutrient interaction" is sometimes used without adequate definition or description.

Noncompetitive, nonreciprocal nutrient interactions are those in which the relationship is one-sided, and they can be positive (i.e., synergistic) or negative (i.e., antagonistic). Typical is the effect of an excess of one dietary component, often a macronutrient, on the absorption and/or utilization of a micronutrient(s).

Noncompetitive micro/micronutrient interactions also fall into this category. For the proper functioning of iron in the body, copper adequacy is required, and excess has not been determined to be detrimental, an example of a positive interaction.

Conversely, excess iron appears to have a pronounced negative effect on copper metabolism, a one-sided antagonistic interaction. However, the true effects of copper on iron metabolism may not have been investigated enough to totally exclude a reverse reaction. Such one-sided noncompetitive interactions between two nutrients can usually be explained based on the chemical and/or physical properties of the nutrients.

The most common type of nutritional interaction is the competitive, i.e., reciprocal and truly two-sided, interaction, which may be either negative or reversible. Two pairs of transition elements that display competitive interactions are zinc and copper, and iron and zinc. These reciprocal interactions usually result from the similarity in the chemical properties of two elements or nutrients.

As would be expected, interactions among more than two nutrients, for example, iron, zinc, and copper, can also occur, probably because of simultaneous effects of different groups of properties of these elements. It is the future elucidation of these multinutrient interactions that holds the greatest promise in terms of acquisition of knowledge that will enable food scientists and nutritionists to improve the nutritional quality of foods.

PHYSICAL AND CHEMICAL PROPERTIES OF NUTRIENTS VERSUS PHYSIOLOGICAL EFFECTS

As mentioned, although complex molecules are ingested, the body only absorbs the simplest subunits making up food ingredients (Table 10.4). Both the subunits themselves and the macronutrients built from the subunits have a wide variety of chemical and physical properties which affect the biochemical and physiological properties of nutrient digestion, absorption, and utilization.

Complex carbohydrates, including dietary fiber, can and have been chemically and physically modified to meet food-processing needs and consumer product acceptance (Tables 10.3 and 10.7) (Rutenberg and Solarek, 1984). Physiological functions reported to be changed by ingesting different forms of complex carbohydrates and dietary fibers include rate of ingestion (Haber, et al., 1977; Duncan et al., 1983), rate of gastric emptying (Wilmshurst and Crawley, 1980; Schwartz et al., 1982), nutrient absorption (Jenkins et al., 1978, 1981), and amount of colonic fuel (Cummings et al., 1986).

Possibly one of the most investigated and interesting physiological aspects of dietary carbohydrates is their effect on blood glucose and insulin concentrations (Crapo et al., 1976; Coulston et al., 1980; Jenkins et al., 1981). Although these effects may be studied in both normal and diabetic individuals, the results of these investigations have special relevance for

Table 10.7 Starch and Starch Derivatives Available for Use in Foods Offering Different Functional Properties Based on Chemical/Physical Properties

Starches
Amylose
Amylopectin
Cycloamyloses
Starch derivatives produced by various chemical treatments
Hypochlorite oxidized starches (used in batter and breading)
Crosslinked starches (stable, high viscosity starch paste)
Starch esters (thickening, viscosity stability and clarity)
Hydroxyalkyl starches (viscosity stability, water holding at low temperature)
Starch phosphate monoesters (emulsifiers for vegetable oil in water)
Cationic starches[a]
Starch xanthates[a]
Dialdehyde starch[a]

[a] Not approved for use in foods.
Source: Rutenberg and Solarek, 1984.

diabetics whose blood glucose concentrations exceed 100 mg/dl because of insufficient insulin (Type II diabetic; non–insulin-dependent diabetes mellitus, NIDDM) or complete lack of insulin (Type I diabetic; insulin-dependent diabetes mellitus, IDDM). Because high blood glucose concentrations can lead to coma and death, practical dietary approaches for regulating blood glucose concentrations in these individuals are highly advantageous and recommended (Nutritional Recommendations and Principles, 1986, 1987; Consensus Development Conference on Diet and Exercise, 1987). Great variations in blood insulin concentrations are also to be avoided because excess insulin can drastically depress blood glucose concentrations.

Determination of the glycemic index is a method for testing the effects of different dietary carbohydrates and dietary fibers on blood glucose and insulin concentrations. This involves having individuals consume 50 g of carbohydrate in various physical forms after an overnight fast. Blood glucose concentrations are then determined every 10–20 minutes over 3–5 hours, and these data are plotted. A 50-g dose of glucose is used as the standard reference material and is usually administered first. On a second occasion (i.e., a day later), the individual is given 50 g of carbohydrate in another form (i.e., fructose, pasta, potato starch, etc). The glycemic index is calculated as the area under the curve of the test substance divided by the area under the curve produced by the standard glucose solution. The ratio multiplied by 100 is the glycemic index (Fig. 10.4). Glycemic index values determined with nondiabetic subjects for a number of carbohydrate-rich foods are reported in Table 10.8 (Jenkins et al., 1987).

Recently, a variation of this test was used to determine if differences in blood glucose concentrations could be observed among Type II diabetic individuals fed 50 g of glucose, 50 g of carbohydrate as corn flakes, or muffins containing either 50 g of high amylose starch or 50 g of low amylose (i.e., high in amylopectin) starch (Krezawski et al., 1987). The important modification made in this study compared to previous experiments of this nature was that blood glucose and insulin concentrations were assessed over a similar time period without any food, permitting each subject to serve as his own control. Use of the resulting baseline correction values thus obtained provided a more physiologically accurate determination of the effects of different dietary carbohydrates upon the gly-

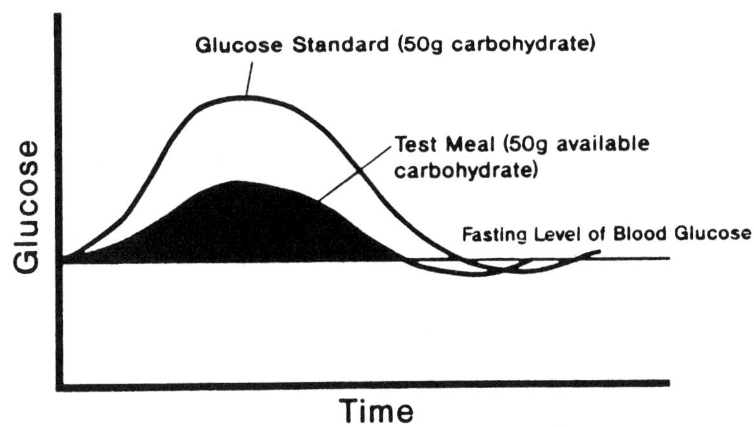

Determination of Glycemic Index GI

$$= \frac{\text{Area in Light Gray}}{\text{Area in Dark Gray and Light Gray}} \times 100$$

Fig. 10.4 Change in blood glucose concentrations over time after ingestion of 50 g of carbohydrate or a test meal containing dietary carbohydrate. Time of postprandial glucose measurements may vary from 3 to 5 hr. Blood glucose concentrations are reported in mg/dl or as the change from baseline value (i.e., Δmg/dl), which is represented by the horizontal line. The area under the horizontal line (i.e., baseline) is referred to as rebound hyperglycemia and is not used in calculating the glycemic index value.

Table 10.8 Glycemic Indices of Carbohydrate-Rich Foods in Normal, Non-diabetic Subjects[a]

	Glycemic index		Glycemic index
Breads		Legumes	
White bread	100	Baked beans	60
Whole wheat bread	99	Lima beans	115
Cereal products		Fruit	
Spaghetti	66	Apple	53
Rice, white	83	Banana	79
		Orange	66
Breakfast cereals		Sugars	
Cornflakes	119	Sucrose	86
Oatmeal (cooked)	85		
Vegetables		Dairy products	
Green peas, frozen	74	Ice cream	52
Potatoes, new boiled	81	Yogurt	52

[a] Similar values observed in diabetic individuals.
Source: Jenkins and Jenkins, 1987.

cemic index. The results from these experiments are illustrated in Figs. 10.5, 10.6, and 10.7.

From these data, it appears that the ingestion and subsequent digestion of high amylose starch results in a decrease in blood sugar over a normal postprandial period. The exact cause of this change is not known, but it is speculated that the straight-chained glucose polymers of amylose are less easily disrupted by cooking processes and are more slowly digested by carohydrases in the intestine than amylopectin (Banks et al., 1973). This would result in slower absorption of glucose into the bloodstream. The fact that high amylose starch is less easily digested has been further demonstrated in humans and in mice (Wolf et al., 1977). The determination of glycemic index values of foods may provide important information to help people with various metabolic disorders, and the test is subject to ongoing improvements (British Nutrition Foundation's Task Force, 1990).

The sucrose polyesters (Mattson and Volpenhein, 1972; Jandacek et al., 1980), used as fat substitutes, provide a now classical example of the physical nature of a food ingredient affecting its digestion. Because sucrose substituted for glycerol in a triglyceride molecule is much larger, pancreatic lipase cannot hydrolyze ester linkages connecting the fatty acids to the sucrose molecule.

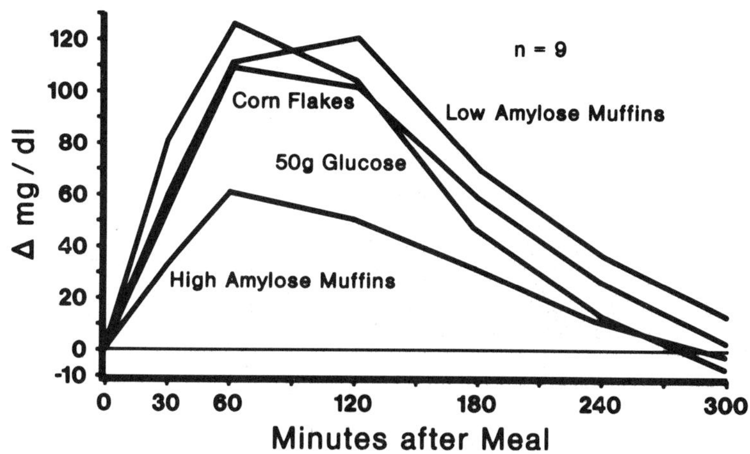

Fig. 10.5 Change in plasma glucose response (Δmg/dl) in Type II diabetics after ingestion of 50 g carbohydrate in four different forms. Each subject served as his own control, and mean fasting glucose concentrations among nine subjects were 180 ± 10 mg/dl. Only the response observed when individuals consumed 50 g high amylose starch in muffins was significantly lower ($p \leq 0.01$) compared to the three other sources of carbohydrates evaluated which included glucose, corn flakes or low amylose muffins. (*From Krezawski et al., 1987.*)

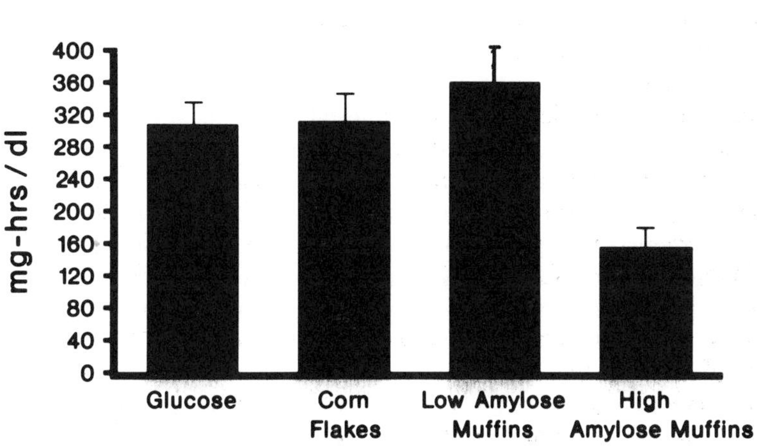

Fig. 10.6 Mean glucose areas over baseline over 5 hr illustrated in Fig. 10.5. Areas for corn flakes (25–30% amylose) and muffins with low amylose starch (0–7% amylose) were not significantly different compared to glucose control. Area for muffins with high amylose starch (65–75% amylose) was significantly lower ($p \leq 0.01$) by approximately 50% compared to the glucose control. (*From Krezawski et al., 1987.*)

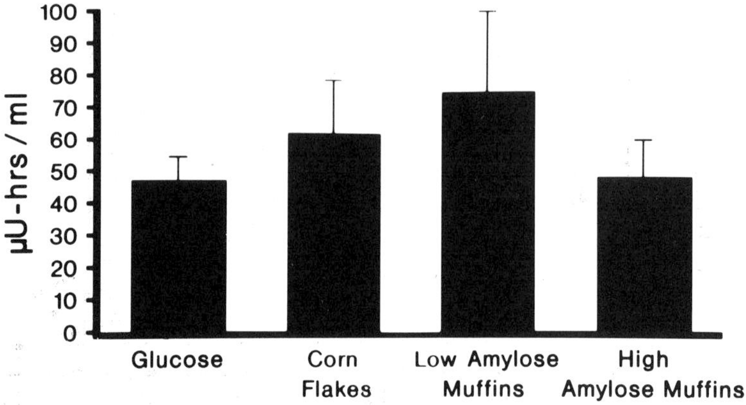

Fig. 10.7 Mean areas under insulin curves in Type II diabetics over 5 hr after consuming 50 g of carbohydrates provided in 4 different forms (see Figs. 10.5 and 10.6). Values were not significantly different among the four treatments. Serum insulin was determined by radioimmunoassay. (*From Krezawski et al., 1987.*)

Dietary fiber as a special class of complex carbohydrates provides another set of relevant examples. Dietary fiber is defined as the remnants of plant cell walls resistant to digestion in the small intestine of humans and animals (Trowell et al., 1976). Recently it has been divided into two categories based on the solubility characteristics of its components (Prosky et al., 1988). Insoluble dietary fiber consists of cellulose, hemicelluloses, and lignin, which have variable chemical compositions and physical characteristics depending upon the source (Southgate, 1969; Selvendran, 1984). A few examples of soluble dietary fibers are the polymers pectin, β-glucan, and guar gum (Furda, 1981; Selvendran, 1985). The soluble fibers also vary greatly among themselves as well as in comparison to insoluble fibers, facilitating a wide variety of applications in food systems (Glicksman, 1969, 1982). From a physiological perspective, the insoluble dietary fibers enhance the rate at which waste products pass through the intestine and promote intestinal water retention (McConnell et al., 1974; Stephen and Cummings, 1979). The soluble fibers, because of their ability to form sols and gels, are believed to regulate the rate at which nutrients are absorbed across the small intestine (Blackburn et al., 1984; Todd et al., 1990). The physiological effects attributed to insoluble and soluble dietary fibers are indicated in Table 10.9, and examples of these are illustrated in Figs. 10.8 and 10.9 for the small and large intestines, respectively.

Table 10.9 Relationships Between Total Dietary Fiber Ingestion and Intestinal and Systemic Physiological Functions

Mouth
Chewing and salivation (S & I)[a]
Stomach
Satiation and satiety (S & I)
Gastric emptying (S & I)
Small intestine
Digestion of macronutrients (S & I)
Absorption of nutrients (S & I)
Bile acid reabsorption (S & I)
Intestinal morphology (S & I)
Intestinal cell proliferation (S & I)
Colon/Large intestine
Stool weight/water content/dilution (I)
Transit time (I)
Stool frequency/laxation (I)
Microflora (S & I)
Fermentation; short chain fatty acid (SCFA) production (S & I)
Colonic cell proliferation/morphology (S & I)
Blood/Hepatic metabolism
Postprandial serum-glucose and hormone concentrations (S)
Serum lipid (e.g., cholesterol) concentrations (S)

[a] S refers to soluble dietary fibers, and I refers to insoluble dietary fibers suggested to have a major influence on systemic/physiological functions.

Dietary fibers are mainly degraded in the large intestine by fermentation (Wolin, 1981; Fleming and Arce, 1986; Cummings et al., 1989; Topping, 1991). Generally, more of the soluble dietary fibers are degraded in the large intestine than are insoluble dietary fibers. The primary products of this colonic fermentation, short-chain fatty acids (SCFA), are either utilized by the mucosal cells of the large intestine or transported to the liver. SCFA have been reported to regulate mucosal cell functioning, (Sakata et al., 1980; Kruh, 1982) and possibly inhibit cholesterol synthesis in the liver after being absorbed from the colon (Boila et al., 1981; Illman et al., 1988). That insoluble dietary fibers ferment less than soluble ones provides an excellent example of a physical property affecting biochemical, and ultimately physiological, processes.

Factors already considered in discussing effects of saturated and unsaturated fatty acids on blood cholesterol concentrations also apply for dietary fats and oils. We ingest triglycerides, of which fatty acids are an

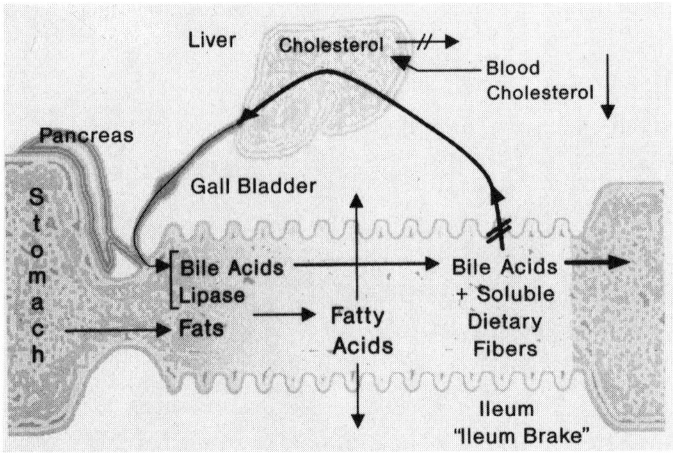

Fig. 10.8 Possible mechanism by which soluble dietary fibers lower blood cholesterol concentrations. In the ileum, soluble dietary fibers will accumulate, increasing the viscosity of the luminal contents in this area of the intestine. Bile acids, produced in the liver and transported via the bile, assist in the hydrolysis of dietary fats. The blockage of bile acid reabsorption from the ileum ("ileum brake") results in blood cholesterol being diverted to bile acid synthesis to compensate for the amount of bile acid not reabsorbed. Blood cholesterol concentrations are theorized to be reduced through this mechanism.

integral part. Numerous studies suggest that the more unsaturated the dietary lipid, the more pronounced its effect on the lowering of blood cholesterol, and possibly triglyceride concentrations (Keys et al., 1965; Mattson and Grundy, 1985). The effects on blood cholesterol and triglyceride concentrations in hyperlipoproteinemic human subjects fed diets containing vegetable or fish oils are reported in Tables 10.10 and 10.11 (Phillipson et al., 1985). In this study, subjects were fed basically identical diets, except for the dietary lipid composition. All diets provided 30% calories as fat. Cocoa butter and peanut oil were the primary fats in the control diet. Salmon and salmon oil supplements were the sources of fish oil. The vegetable diet contained safflower and corn oils. Subjects with high blood cholesterol and triglyceride concentrations were evaluated to more easily demonstrate the effects of dietary lipids on blood lipid concentrations. In both Type IIb (Table 10.10) and Type V (Table 10.11) hyperlipoproteinemic individuals, the more unsaturated the dietary lipids, the more pronounced the lowering of blood cholesterol and triglyceride concentrations. Possibly, the most unique aspect of feeding dietary

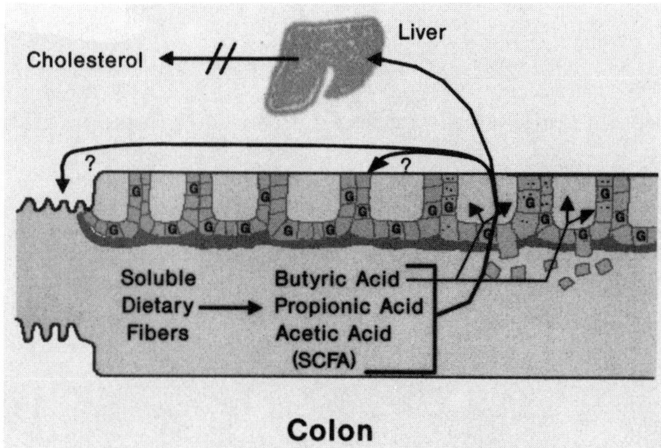

Fig. 10.9 Possible physiological effects of soluble dietary fibers (i.e., fermentable dietary fibers) mediated through the large intestine. Fermentation of soluble dietary fiber in the large intestine results in the production of short-chain fatty acids (SCFA). SCFA, principally butyric acid (C4:0), are thought to act as regulators of cell cycling and metabolism. Cells may alter their S-phase cycle (i.e., DNA replication) and/or migrate faster up the crypt and be discarded. The absorption and transfer of propionic acid (C3:0) has been suggested as a regulator of cholesterol biosynthesis in the liver.

Table 10.10 Plasma Cholesterol and Triglyceride Concentrations in Type IIb Hyperlipoproteinemic Individuals Fed Diets Containing Different Oils

	Control[a] (mg/dl)	Fish oil[a] (mg/dl)	Vegetable oil[a] (mg/dl)
Total			
Cholesterol	285 ± 8[d]	199 ± 12[b]	235 ± 6[c]
Triglyceride	374 ± 105[d]	136 ± 26[b]	258 ± 72[c]
VLDL			
Cholesterol	80 ± 22[d]	18 ± 6[b]	46 ± 18[c]
Triglyceride	304 ± 94[d]	69 ± 21[b]	177 ± 62[c]
LDL			
Cholesterol	173 ± 25[b]	156 ± 19[b,c]	149 ± 16[c]
Triglyceride	59 ± 16	64 ± 11	56 ± 9
HDL			
Cholesterol	36 ± 5[b]	33 ± 7[b]	44 ± 9[c]

[a] After 4-week dietary periods starting with control values.
[b,c,d] Mean values in a row not followed by the same letter are significantly different ($p \leq 0.05$).
VLDL = very low-density lipoproteins; LDL = low-density lipoproteins; HDL = high-density lipoproteins.
Source: Phillipson et al., 1985.

Table 10.11 Plasma Cholesterol and Triglyceride Concentrations in Type V Hyperlipoproteinemic Individuals Fed Diets Containing Different Oils

	Control[a] (mg/dl)	Fish oil[a] (mg/dl)	Vegetable oil[a] (mg/dl)
Total			
Cholesterol	377 ± 155[c]	295 ± 31[b]	264 ± 97[b]
Triglyceride	1,432 ± 750[d]	282 ± 120[c]	84 ± 514[b]
VLDL			
Cholesterol	251 ± 148[d]	74 ± 67[b]	216 ± 219[c]
Triglyceride	1,249 ± 681[c]	171 ± 119[b]	550 ± 360[b]
LDL			
Cholesterol	77 ± 55[b]	110 ± 34[c]	79 ± 30[b]
Triglyceride	70 ± 34	65 ± 18	71 ± 31
HDL			
Cholesterol	31 ± 7	35 ± 12	31 ± 11

[a] After 4-wk dietary periods starting with control values.
[b,c,d] Mean values in a row not followed by the same letter are significantly different ($p \leq 0.05$).
Source: Phillipson et al., 1985.

salmon oil was the finding that blood triglyceride concentrations were reduced by 63 and 80% in the Type IIb and Type V subjects, respectively, compared to concentrations originally observed on the control diet. The mechanism by which fish oils so dramatically lower blood triglyceride concentrations is unknown.

NUTRIENT BIOAVAILABILITY AS AFFECTED BY DIETARY COMPONENTS

Extrinsic (i.e., dietary) and intrinsic factors that affect nutrient bioavailability are listed in Table 10.12. Since both sets of factors can affect the absorption and utilization of a nutrient, it remains a difficult challenge to standardize these factors across a test group of human subjects and accurately measure nutrient bioavailability. The majority of research to date in humans has centered on the role of extrinsic factors in nutrient bioavailability. Studies with animal models help reduce or eliminate intrinsic variables, facilitating better evaluation of the roles of extrinsic factors (Fox et al., 1981). General and commonly used procedures to directly or indirectly assess nutrient bioavailability are indicated in Table 10.13 (Ja-

Table 10.12 Factors Affecting Nutrient Bioavailability

Extrinsic[a]	Intrinsic
Macronutrients	Age
Carbohydrates	Sex
Protein	Health status
Fat	Nutritional status
Micronutrients	Stress
Vitamins	
Minerals	
Other dietary components	
Tannins	
Dietary fiber	
Maillard products	
Phytic acid	
Ascorbic acid	

[a] Dietary in origin.

Table 10.13 Bioavailability Assays

	Ref.
In vitro solubility	Schricker et al. (1982)
	Clydesdale (1988)
Intestinal sections to measure transport	Jacobs et al. (1966)
	Van Campan (1973)
Human/Animal studies	
Balance	Douncey et al. (1977)
	Greger et al. (1978)
Biological index measurements	Alexander et al. (1983)
	Chao and Gordon (1983)
Relative biological values (in animals)	Steinke ad Hoplins (1978)
	Shah et al. (1979)
Isotopes	
Radionuclides	Aamodt et al. (1981)
	Spencer et al. (1978)
Stable isotopes	King et al. (1978)
	Carni et al. (1980)

cobs et al., 1966; Van Campen, 1973; Douncey et al., 1977; Greger et al., 1978; Steinke and Hoplins, 1978; Spencer et al., 1978; King et al., 1978; Shah et al., 1979; Carni et al., 1980; Aamodt et al., 1981; Schricker et al., 1982; Alexander et al., 1983; Chao and Gordon, 1983; Clydesdale, 1988). The bioavailability of the macronutrients, i.e., protein, carbohydrate, and lipid, has been assessed in humans or animals in studies employing growth, energy balance, digestibility, or a combination of these techniques (McGrandy et al., 1966; Roberts and Young, 1988; Joint FAO/WHO Expert Consultation). This literature is partially reviewed in the 10th edition of the *Recommended Dietary Allowances* (1989).

However, the preponderance of work dealing with nutrient bioavailability centers around vitamins and minerals (Gregory and Litherland, 1986; Ben-Amotz et al., 1989; House and Welch, 1989; Clifford et al., 1990). These determinations normally require availability of an index that is directly associated with the nutrient in question (see Fig. 10.2) and subject to quantification. Examples of biological or physiological functions associated with individual micronutrients and the indices used to measure nutrient status or bioavailability are reported in Table 10.14. The remaining portion of this chapter will deal with the elements iron, zinc, and copper.

An important synergistic nutrient interaction is the enhancing effect of animal protein on iron absorption, demonstrated by Cook and Monsen (1976). When subjects were fed meals containing beef, pork, chicken, or fish, the absorption of iron administered as Fe-59 was significantly enhanced compared to the absorption of Fe-59 administered with a meal containing ovalbumin as the control source of protein (Fig. 10.10). This observation has commonly been referred to as the "meat factor effect." When subjects were fed meals containing milk or eggs as the source of protein, Fe-59 absorption was similar to that observed in the control diet. It has been suggested that phosphorus associated with ovalbumin, milk, and egg protein binds iron and prevents its intestinal solubilization and/or absorption (Albright et al., 1984).

In experiments with the long-range goal of determining the mechanism for enhanced iron absorption by animal protein (Gordon and Godber, 1989), we demonstrated the same phenomenon in rats. It has been assumed that iron (II) is more biologically available than iron (III), and iron (II) salts of iron are commonly added to foods (Pla et al., 1973; Rees and Monsen, 1973). Therefore, in a recently completed experiment we examined how the oxidation state of iron in the stomach of the rat was affected by whether it was fed beef protein or milk protein (i.e., lactalbumin) (Kattelman et al., 1988).

Groups of rats were made anemic by feeding a low-iron diet (<6 μg

Table 10.14 Physiological or Biochemical Parameters Dependent on Micronutrients and Common Indices Used to Determine Nutrition Status or Nutrient Bioavailability in Animals and Humans

Micronutrient	Physiological or biochemical functions	Index
Vitamins		
Ascorbic acid	Reducing agent in hydroxylation reactions	Serum ascorbic acid
Vitamin E	Antioxidant/Cell membrane	Serum tocopherol
Vitamin A	Vision	Serum Vitamin A
Folic acid	Co-enzymes for one-carbon transfer reactions	Erythrocyte folate
Vitamin B_6	Aminotransferase reaction	Serum Pyridoxine
Minerals		
Ca	Bone growth/metabolism	Bone density
Mg	Neuromuscular transmission	Serum Mg
Fe	Hemoglobin for oxygen transport	Serum ferritin
Zn	Growth/RNA polymerase activity	Serum Zn
Cu	Decomposition of superoxide free radicals; collagen and elastin crosslinking	Serum Cu
Se	Removal of hydrogen peroxide and other free hydroperoxides	Erythrocyte glutathione peroxidase activity

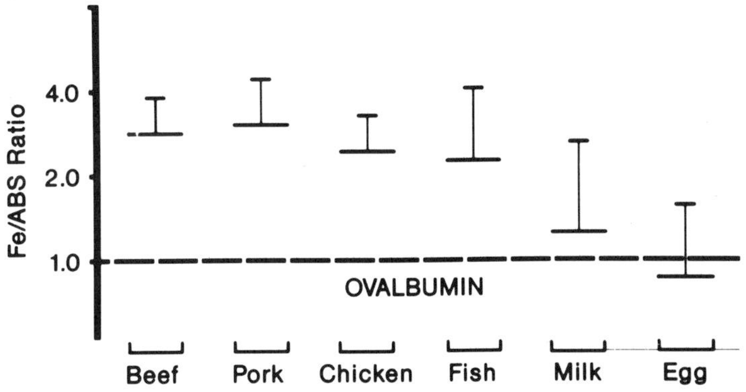

Fig. 10.10 The ratio of Fe-59 absorption among human subjects fed meals containing different sources of protein to the control meal containing ovalbumin. (*From Francis, 1985.*)

Fe/g) for 21 days. They were then fed diets containing either lactalbumin or beef, from which all the soluble protein and heme iron had been removed by extensive washing with distilled water. Each of the two protein diets was divided into two portions and supplemented with iron to contain 25 μg/g provided as ferrous sulfate [i.e., iron (II)] or ferric ammonium sulfate [i.e., iron (III)]. Diets were fed to the rats ad libitum for 10 days, after which the animals were killed. In addition to determining the final hemoglobin concentrations in these animals, the stomach contents were removed and analyzed for the oxidation state of the iron contained within. Only reduced, iron (II) ions will form a chromophor with the reagent 3-(2-pyridyl)-5,6-bis (4-phenylsulfonic acid)-1,2,4-triazine. The difference between the total amount of iron in the stomach and the amount determined as iron (II) is iron (III). The relative biological values (RBV) of the iron fed to these four groups of rats are reported in Table 10.15. The hemoglobin concentration observed in animals fed the diet containing lactalbumin and ferrous sulfate (25 μg Fe/g) was designated as the reference response value (i.e., relative biological value = 100%). The ratio comprised of the hemoglobin concentration of the test animals divided by that of the reference group multiplied by 100 is referred to as the RBV.

Two interesting and important results were observed. Both the iron (II) and iron (III) salts had the same ability to replenish hemoglobin in the anemic rats (Table 10.15). The washed beef protein diet significantly increased the RBV of both iron salts. Upon determining the oxidation

Table 10.15 Relative Biological Value (RBV) of Iron II and Iron III Determined in Iron-Deficient Rats Fed Diets Containing Lactalbumin or Washed Beef Protein

Dietary protein	Dietary iron oxidation state	RBV, pooled data (%) Oxidation state[a]	Protein[b]	Oxidation state[c]
Lactalbumin	Iron II	100[d]	100[d]	100
	Iron III	97[d]		(when fed iron II)
Washed beef	Iron II	130[e]	127[e]	103
	Iron III	121[e]		(when fed iron III)

[a] Mean value based on 6 animals, for each oxidation state of iron and each type of protein.
[b] Mean value based on 12 animals; pooled identical protein.
[c] Mean value based on 12 animals; pooled identical oxidation state.
[d,e] Mean values in a column not followed by the same letter are significantly different ($p \leq 0.05$).

state of iron in the stomach of these animals, no significant difference was observed between the two iron salts when consumed with the same protein (Table 10.16). However, the distribution between iron (II)/iron (III) ions was significantly different as affected by the two sources of protein. Animals consuming the diet with washed beef protein had a significantly higher iron (II)/iron (III) ratio compared to the rats fed the same amounts and sources of iron with lactalbumin as the protein source.

We hypothesize that animal protein catalyzes the conversion of nonheme dietary iron to predominantly the iron (II) oxidation state in the stomach. It is further hypothesized that the animal protein protectively binds iron (II) ions, ensuring that the iron remains soluble until absorbed into the intestinal mucosal cell (Fig. 10.11).

Ascorbic acid has also been reported to significantly improve the bioavailability of nonheme sources of iron (Gilloalg et al., 1983; Hallberg and Rosander, 1984). In addition to keeping iron in the reduced state, ascorbic acid may form a chelate with iron (II) ions, maintaining their solubility and improving absorption in the same way as suggested for animal protein.

Recent findings indicate that fructose may have an negative effect on copper utilization (Fields et al., 1983; Reiser et al., 1983). This observation is significant because more fructose is being added to foods, and the American diet may actually be low in copper (Holden et al., 1979). The sum of these suggestions is that a low copper status may be contributory to

Table 10.16 Distribution of Oxidation States of Iron (Iron II/Iron III) in Stomach Contents of Rats Fed Diets Containing Lactalbumin or Washed Beef Protein with Iron II or Iron III Iron Salts

Dietary protein	Dietary iron oxidation state	Iron II/Iron III, pooled data (%)		
		Oxidation state[a]	Protein[b]	Oxidation state[c]
Lactalbumin	Iron II	36/64[e]	42/58[e]	52/48
	Iron III	48/52[e]		(when fed ferrous)
Washed beef	Iron II	66/34[d]	70/30[d]	60/40
	Iron III	73/27[d]		(when fed ferric)

[a] Mean values based on 6 animals.
[b] Mean values based on 12 animals; pooled identical protein.
[c] Mean values based on 12 animals; pooled identical oxidation state.
[d,e] Mean values in a column not followed by the same letter are significantly different ($p \leq 0.05$).

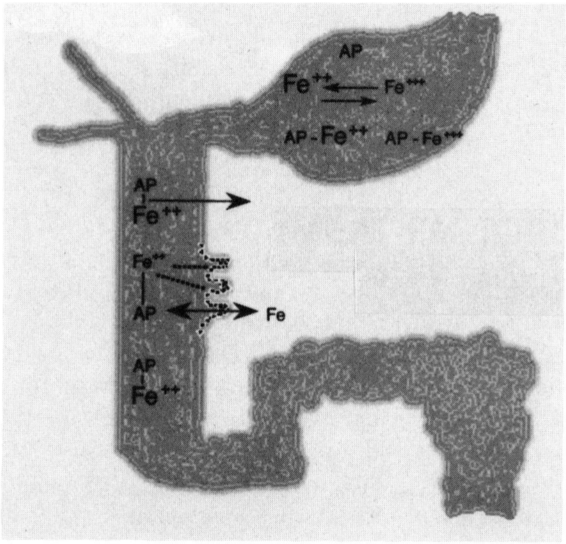

Fig. 10.11 Proposed mechanism by which beef protein enhances iron absorption, or "meat factor effect." Beef protein appears to promote the reduction of nonheme iron to its iron II oxidation state in the stomach. It is further hypothesized that beef protein and its degradation products bind with iron II ions, protecting their solubility and assuring subsequent absorption into intestinal mucosal cells.

coronary heart disease (Reiser et al., 1985). In studies with copper-deficient rats, it has been observed that male rats succumbed more rapidly to cardiac rupture when fed a copper-deficient diet with fructose compared to a diet containing glucose or starch. Copper is an essential component of the enzyme lysyl oxidase, and this enzyme is necessary for the synthesis of elastin, a structural protein in heart muscle. One suggestion presented to explain these findings is that fructose will bind copper, preventing its absorption. Although this is an interesting and still unexplained phenomena, it does not appear to have practical significance for human nutrition (O'Dell, 1990). Occurrence of these dramatic events in humans would require total copper deficiency, a state that has never been achieved or reported.

Phytic acid has long been known to have a negative influence on zinc bioavailability (O'Dell and Savage, 1960; Davis, 1974). In vitro phytic acid binds metals in the following order: Cu and Zn $>$ Co $>$ Mn $>$ Fe $>$ Ca 25 (Kratzer and Vahra, 1986). The discovery of zinc as an essential requirement for humans can be attributed to the deficiency of this nutrient observed among Iranians consuming bread with a high phytate concentration (Prasad et al., 1963; Prasad, 1983). For phytic acid to have significant effects on growth and development through impaired zinc nutrition, the diet must contain substantial amounts of foods high in phytic acid. A varied diet easily prevents any problem.

Using copper-depleted rats, we investigated the effects of zinc and phytic acid on copper bioavailability (Lee et al., 1988). Copper-deficient rats were initially divided into two groups of 18 animals. One group of rats had no phytic acid added to their diets, and the second group of animals were fed diets containing 1% phytic acid. Each of these groups was further divided into three groups and fed diets supplemented with 12, 30, or 230 μg Zn/g. All diets were further supplemented with copper to achieve 5 μg Cu/g. The animals were fed these diets for 3 days. Serum copper and zinc concentrations were determined, and these results are reported in Table 10.17. With diets containing normally adequate amounts of zinc (12 and 30 μg Zn/g), the inclusion of phytic acid in the diet significantly reduced serum zinc. The inhibitory effect of phytic acid on zinc bioavailability was only overcome by the addition of excess dietary zinc (230 μg Zn/g). Examining the effects of the same dietary treatments on serum copper concentrations, phytic acid had no effect on serum copper

Table 10.17 Effects of Dietary Zinc and Phytic Acid on Serum Copper and Zinc Concentrations in Copper-Deficient Rats[a]

Dietary phytate (%)	Dietary zinc (μg/g)	Serum concentration (μg/dl)	
		Zn	Cu
0	12	100[b]	43[b]
1	12	28[c]	49[b]
0	30	98[b]	32[c]
1	30	37[c]	46[b]
0	230	102[b]	22[d]
1	230	100[b]	23[d]

[a] Copper-deficient rats were fed diets containing zinc and phytic acid, as described, and 5 μg Cu/g for 3 days.
[b,c,d] Values in the same column not followed by the same superscript are significantly different ($p \leq 0.05$).

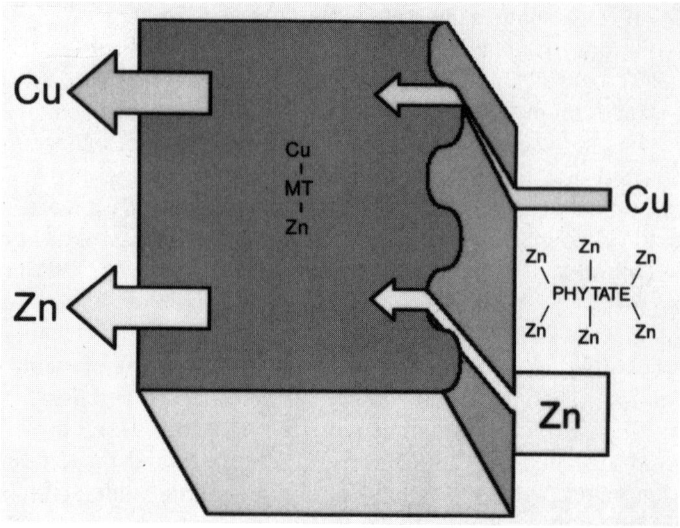

A

Fig. 10.12 Metallothionein (MT) is an intestinal mucosal cell protein that binds both zinc and copper and can regulate the transfer of both elements into the body (A). When phytic acid binds zinc in the intestinal lumen and in the absence of excess zinc, induction of MT synthesis in the mucosal cell does not occur and copper absorption is not hindered. In the absence of phytic acid in the lumen, or in the presence of excess zinc, metallothionein is induced within the intestinal mucosal cell, which subsequently binds copper preventing its transfer into the body (B).

when the dietary zinc concentration was 12 μg Zn/g. With dietary concentrations of 30 μg Zn/g, the feeding of 1% phytic acid helped to maintain serum copper at the concentration observed in the animals fed 12 μg Zn/g diet. Without phytic acid, serum copper concentrations were significantly lower when the dietary zinc concentration was 30 μg Zn/g. At the highest dietary zinc concentration (230 μg Zn/g), serum copper was significantly depressed, both with and without phytic acid, compared to the other groups. An explanation for the chemical and biochemical effects of phytic acid and zinc on serum copper is illustrated in Fig. 10.12.

With excess zinc in the diet, synthesis of the protein metallothionein is stimulated in the mucosal cell. This protein binds zinc and copper, possibly regulating their absorption and preventing excess amounts from reaching the systemic circulation. The result of having excess zinc is that copper transfer from the intestinal mucosal cell into the serum is reduced.

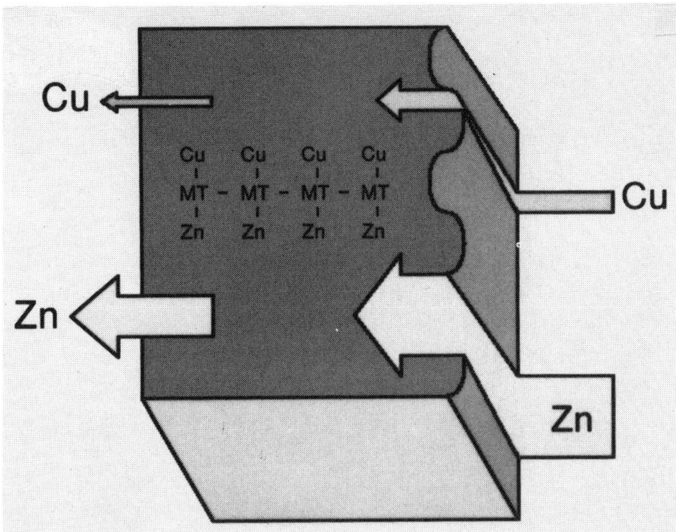

B

When phytic acid is part of the diet, it binds zinc in the intestinal lumen, preventing its absorption into the intestinal mucosal cell. Metallothionein synthesis does not increase in the intestinal mucosal cells, and copper is not retarded on its pathway into the serum.

Dietary fiber has the position of a food ingredient which enjoys the accolades of both the food science and nutritional communities for its functionality and positive contributions to human health (Staub et al., 1983; Gordon, 1989). Consistent recommendations are to increase dietary fiber intakes by 50–100% from our current intakes of 15–26 g/day (Federation of American Societies for Experimental Biology, 1987). The one concern regarding increased dietary fiber consumption is that it may bind minerals, preventing their absorption (James, 1980; IFT Expert Panel, 1989). The relationship between mineral absorption and dietary fiber intake is illustrated in Fig. 10.13. The suggestions that dietary fiber impairs mineral absorption are based primarily on in vitro binding studies (Eastwood and Mitchell, 1976; Thompson and Weber, 1981) and balance studies conducted in humans (Slavin and Marlett, 1980; Van Dokkum et al., 1982) and animals (Harmouth-Hoene and Schelenz, 1980; Bagheri and Gueguen, 1985). Stronger and more extensive studies are now available which indicate that dietary fiber does not interfere with mineral absorption and explain why this interference is unlikely to occur. These studies are discussed in a current review (Gordon, 1990).

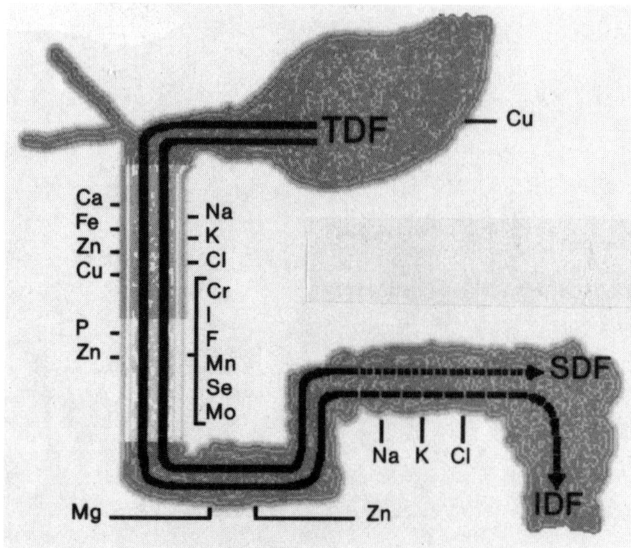

Fig. 10.13 Major areas of mineral absorption throughout the intestine. Both soluble and insoluble dietary fibers pass through the small intestine without being digested. There is a lack of convincing evidence that any dietary fibers adversely affect mineral absorption or retention in animals or humans. *(From Gordon, 1990.)*

IRON, ZINC, AND COPPER

Iron and copper are defined as transition elements because they have partially filled d-shell electron orbitals in any one of their common oxidation states (Cotton and Wilkinson, 1980). Iron in the iron (III) oxidation state has a $3d^5$ configuration, and the cupric (II) oxidation state has a $3d^9$ configuration. Zinc (II) has a $3d^{10}$ configuration, which theoretically prevents it being defined as a first transition series element, the two electrons

Fig. 10.15 Crystal field splitting energy levels for the ions of iron, zinc, and copper. Coordination numbers for each complex are indicated in parentheses. Electron distribution for each ion is not illustrated. Copper (I), iron (III), and zinc (II) can form tetrahedral coordination complexes. Iron (II), zinc (II), and iron (III) can form octahedral complexes. Copper (II) and zinc (II) can form tetragonal coordination complexes with a coordination number of 4 or a coordination number of 6 if they also coordinate with water or anions in addition to four other ligands. Copper (II) is most stable in a square planar complex with the ligands lying in the *x* and *y* axes relative to the central copper (I) ion.

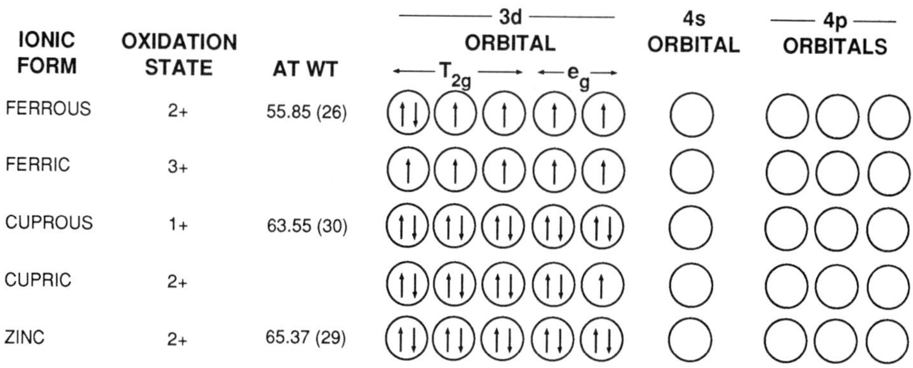

Fig. 10.14 The electronic structure for ions of iron, copper, and zinc. Each ion with its oxidation number (i.e., charge) is indicated along with its atomic weight and number of electrons (in parentheses) in the ground state. For all three elements, the outermost electrons occupy 3d orbitals, suggesting the possibility that all ionic forms of these elements could interact in biological system but as further influenced by stable energy levels for each ion and resulting orbital shape (see Fig. 10.15).

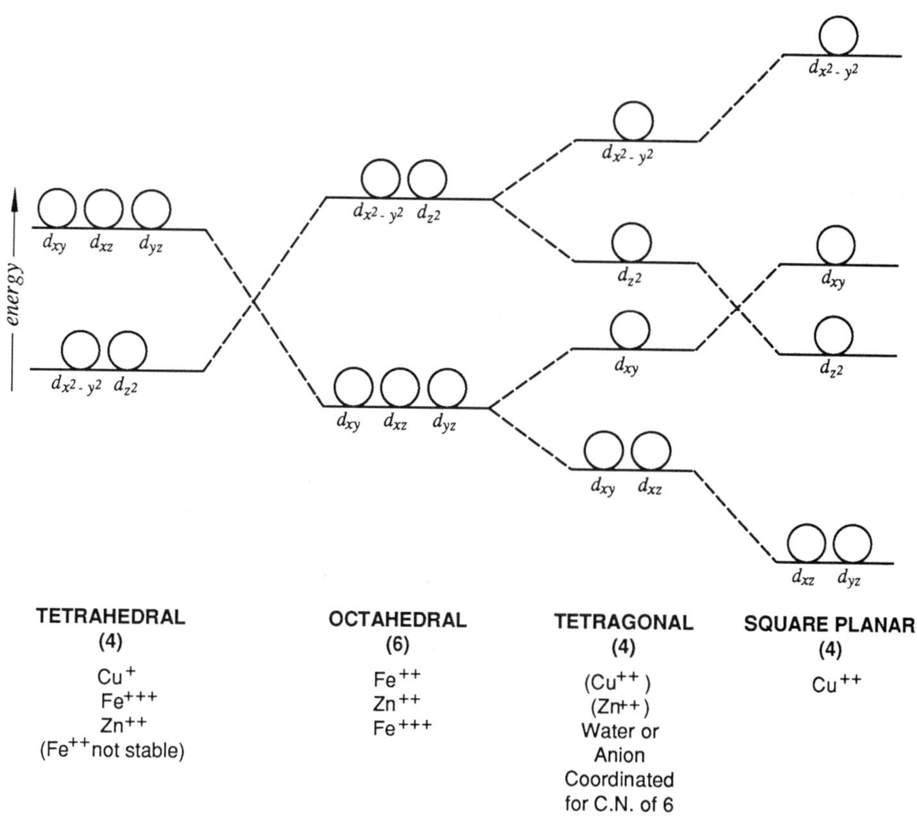

lost from its 4s shell help qualify zinc as such. Zinc displays many of the chemical and physical properties of the transition elements, specifically iron and copper, which are of a similar size. Based on these comparisons, zinc is considered a transition element along with iron and copper. The chemical similarities of these three elements among five oxidation states are illustrated in Figs. 10.14 and 10.15.

INTERACTIONS AMONG IRON, ZINC, AND COPPER

Hill and Matrone (1970) made the observation, "Those elements whose physical and chemical properties are similar will act antagonistically to each other biologically." The elements reviewed in their classic paper concerning potential biological interactions included Cu, Zn, Cd, Hg, Ag, Se, As, Sn, Te, V, and Cr. They provided an excellent explanation for how zinc will interact with copper, cadmium, and mercury because of similarities in their coordination chemistry. No reason is provided by the authors for not discussing iron in their paper. Limiting our discussion here to zinc and copper for the moment, both elements can attach to other molecules or ligands in a similar tetrahedral fashion. Copper (I) and copper (II) ions have four empty orbitals which will allow both oxidation states to form coordination complexes with four ligands (Fig. 10.14), but a tetrahedral complex is favored with copper (I) ions. Copper (I) ions favor a square planar tetrahedral complex with the copper (I) ion at the center of an x-y, two-dimensional, plane. Because the zinc (II) ion has the same electronic structure in its valance shell as the copper (I) ion, the potential for their biological interaction is chemically defined (Fig. 10.14). Theoretically, and based on the orbital arrangement of electrons, both iron (III) and iron (II) ions should be able to form tetrahedral complexes. However, when the orbital arrangement of electrons and their respective energy levels (Fig. 10.15) are considered, the chemical forms of copper, iron, and zinc most likely to form tetrahedral complexes are copper (I), iron (III), and zinc (II). With the similarity in electron orbitals (Fig. 10.14) and the energy levels for these orbitals (Fig. 10.15), the potential for these three elements to competitively interact in a biological system is chemically defined. How competition among these three elements may occur at the center of a tetrahedral complex is illustrated in Fig. 10.16.

These chemical properties of iron, zinc, and copper suggested to us that these three elements could interact with each other in a biological system, thereby mutually affecting their absorption and utilization (Gordon, 1985). Having now demonstrated multiple interactions among iron,

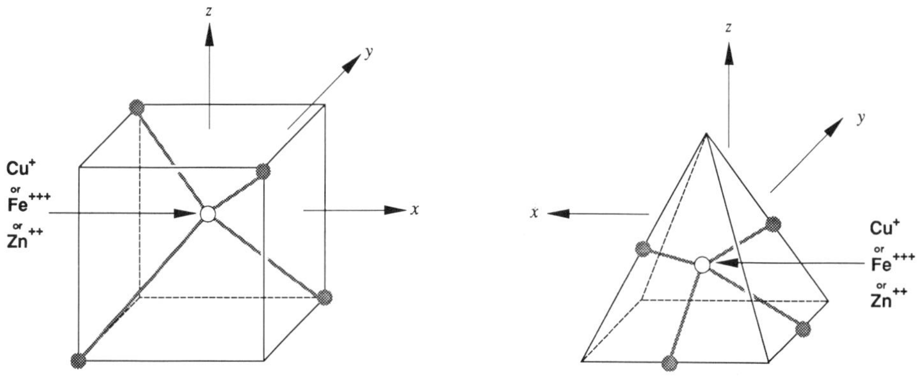

Fig. 10.16 A theoretical biological tetrahedral complex of which a central copper II, iron III, or zinc II can compete to bind with the four available ligands. Two illustrations are presented for the tetrahedral complex.

zinc, and copper, we feel they have significant implications in food science and nutrition.

To initially study iron, zinc, and copper interactions, a semipurified diet was formulated to strictly control the dietary concentration of each element (Gordon, 1989). Appropriate amounts of iron, zinc, and copper as their sulfate salts were incorporated in experimental diets prepared to give the following metal concentrations: 8.5, 10.5, 42.5, or 180 μg Zn/g; 2, 10, or 40 μg Cu/g; and 25, 90, or 270 μg Fe/g. Treatments were arranged in a $4 \times 3 \times 3$ factorial arrangement for statistical analysis with five animals per treatment group. The animals were allowed free access to feed and distilled water for 28 days. Upon terminating the experiment, hemoglobin, femur zinc, and liver copper concentrations were determined.

The effects of increasing dietary iron and zinc concentrations on hemoglobin concentrations at two dietary concentrations of copper are reported in Table 10.18. At each dietary copper concentration, there was a significant interaction between iron and zinc. These data have been plotted by surface response methodology (SAS, 1985) and are illustrated in Figs. 10.17 and 10.18 for dietary copper concentrations of 2 μg Cu/g and 10 μg Cu/g, respectively. With 2 μg Cu/g diet, increasing dietary zinc concentrations specifically caused a decrease in hemoglobin concentrations (Fig. 10.17), but with 10 μg Cu/g diet, increasing dietary zinc was first enhancing, and then inhibitory, to hemoglobin concentrations (Fig. 10.18). At both dietary copper concentrations, increasing dietary iron concentrations was first enhancing, and then inhibitory, to hemoglobin

Table 10.18 Effects of Dietary Iron, Zinc, and Copper on Hemoglobin Concentration in the Growing Rat*

	Dietary iron		
Dietary zinc (μg/g)	24 μg/g (g Hb,dl)	90 μg/g (g Hb/dl)	270 μg/g (g Hb/dl)
	Dietary Copper (2 μg/g)		
8.5	12.7^{A}	$13.5^{A,B}$	12.6^{A}
10.5	11.2^{A}_{a}	14.7^{A}_{b}	$11.0^{A,B}_{a}$
42.5	11.1^{A}	$12.1^{B,C}$	9.9^{B}
170.0	8.4^{B}_{a}	10.8^{C}_{b}	$10.2^{B}_{a,b}$
	Dietary Copper (10 μg/g)		
		Dietary Iron	
Dietary zinc (μg/g)	18 μg/g (g Hb/dl)	90 μg/g (g Hb/dl)	270 μg/g (g Hb/dl)
8.5	13.4^{A}	13.7	12.7^{B}
10.5	11.0^{B}_{a}	14.9_{b}	12.5^{B}_{a}
42.5	13.8^{A}	14.7	14.1^{A}
170.0	$11.9^{A,B}_{a}$	14.2_{b}	11.9^{B}_{a}

* Mean values of 5 animals per group. Mean values in a column not having the same superscript letter (capital) or mean values in a row not having the same subscript (lower case) are significantly different ($p \leq 0.05$).

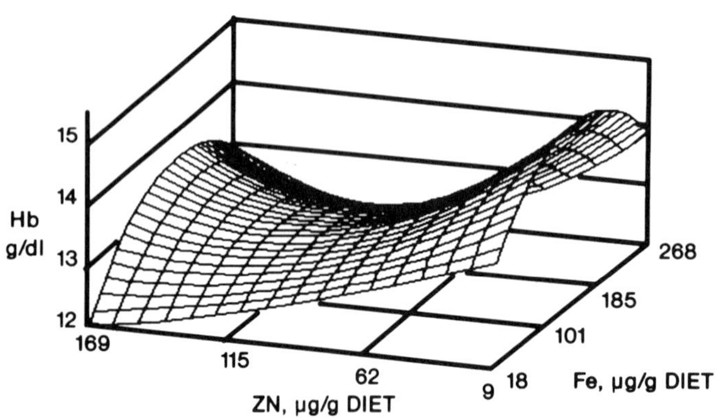

Fig. 10.17 Surface response curve illustrating the interactions between iron and zinc affecting hemoglobin concentration (g/dl) in the growing rat. Three dietary concentrations of iron (24, 90, and 270 μg/g) and four dietary concentrations of zinc (8.5, 10.5, 42.5, and 170 μg/g) were fed groups of rats ($n = 5$) in a $3 \times 4 \times 3$ factorial arrangement of treatments for 21 days. Only results for animals fed diets containing 2 μg Cu/g are shown. A significant two-way interaction is shown ($p \leq 0.01$).

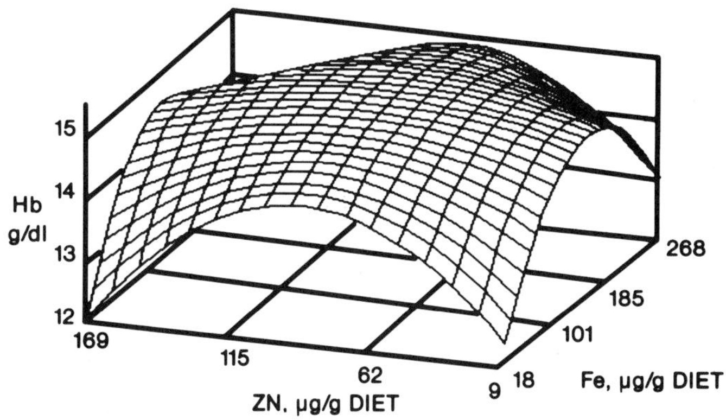

Fig. 10.18 Surface response curve illustrating the interaction between dietary iron and zinc affecting hemoglobin concentration (g/dl) in the growing rat. Only hemoglobin concentrations from rats fed diets containing 10 µg Cu/g are shown. See Fig. 10.17 for exact dietary iron and zinc concentrations. A significant two-way interaction is shown ($p \leq 0.01$). The difference between the surface response curves in Figs. 10.17 and 10.18 represents a significant three-way interaction among iron, zinc, and copper ($p \leq 0.01$) affecting hemoglobin concentrations.

Table 10.19 Effects of Dietary Iron, Zinc, and Copper on Femur Zinc and Liver Copper Concentrations in the Growing Rat

Mineral	µg/g	n	Femur Zn[a] (µg/g)	Liver Cu[a] (µg/g)
Iron	25	60	206.0[b]	14.1[b]
	90	60	195.7[c]	10.4[c]
	270	60	192.3[c]	7.9[d]
Zinc	8.5	45	170.1[d]	11.8[b]
	10.5	45	185.1[c]	12.6[b]
	42.5	45	211.4[c]	10.2[c]
	170.0	45	225.5[b]	8.6[d]
Copper	2	60	205.9[b]	5.1[c]
	10	60	194.7[b]	10.6[c]
	40	60	193.5[c]	16.7[b]

[a] Mean values reported on a dry weight basis represent pooled values for 12 groups of rats at each dietary iron and copper concentration and nine groups of rats at each dietary zinc concentration.
[b,c,d] Mean pooled values for each dietary concentration of iron, zinc, and copper in a column not followed by the same superscript are significantly different ($p \leq 0.05$). The authors recognize that significant interaction effects were observed in this study and that they are most important.

concentrations (Figs. 10.17 and 10.18). The data reported in Table 10.18 and illustrated in Figs. 10.17 and 10.18 represents significant two-way interactions. By comparing the difference in surface response curves in Figs. 10.17 and 10.18, a significant ($p \leq 0.01$) three-way interaction (i.e., multielement) among these three transition elements was observed.

The effects of varying dietary concentrations of iron, zinc, and copper on femur zinc and liver copper concentrations are reported in Table 10.19. Both femur zinc and liver copper concentrations are common indices used to assess the status or bioavailability of these two elements in rats (Momcilonic et al., 1975; Lo et al., 1980). Increasing dietary iron or copper depressed femur zinc. Increasing dietary concentrations of iron and zinc depressed liver copper concentrations. Significant interactions were observed, but the authors have taken the liberty of reporting pooled data in Table 10.19. Again, to help the reader visualize these interactions, the effects of increasing dietary concentrations of iron and zinc at a constant dietary copper concentration (10 μg Cu/g) on liver copper and femur zinc concentrations are illustrated in Figs. 10.19 and 10.20, respectively.

References in the scientific literature indicate these interactions do occur in humans (Festa et al., 1985; Hasckke et al., 1986), but more extensive experimentation will be required to indicate their significance to human nutrition and health. Nevertheless, our studies and the other references cited clearly demonstrate that the biological functioning of the nutrients are affected by their chemical properties.

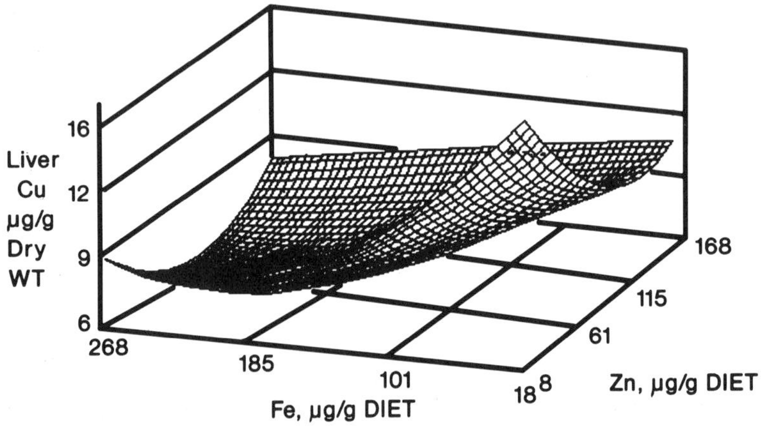

Fig. 10.19 Surface response curve illustrating the interaction between dietary iron and zinc affecting liver copper concentrations in the growing rat. See Fig. 10.17 for exact dietary Fe and Zn concentrations. A significant two-way interaction is shown ($p \leq 0.01$).

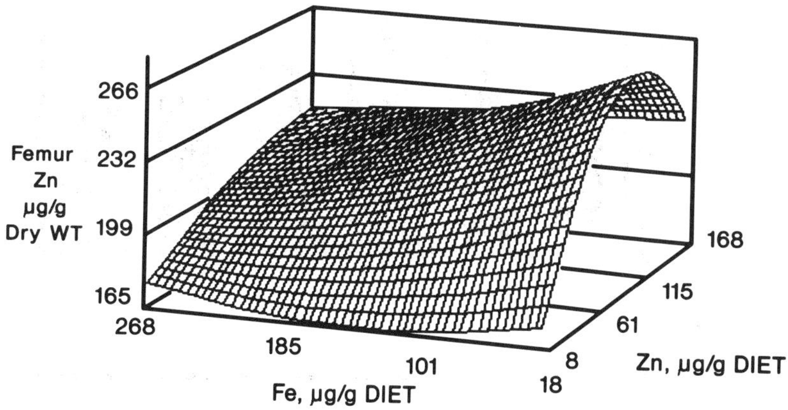

Fig. 10.20 Surface response curve illustrating the interaction between dietary iron and zinc affecting femur zinc concentrations in the growing rat. (See Fig. 10.17 for exact dietary Fe and Zn concentrations.) A significant two-way interaction is shown ($p \leq 0.01$).

CELL CULTURE

Interaction studies could be practically applied to optimize the dietary concentration of nutrients for maximum growth and maintenance. Using cells in culture (O'Dell, 1986; Tinker and Rucker, 1987), we have determined that the optimum growth requirement for one element is lower when used in combination with the other two elements (Lim and Gordon, 1990).

Using an intestinal mucosal cell line isolated from the duodenum of a rat (Negrel et al., 1983), cells were grown for a period of 5 days on Dulbecco's Minimum Essential Media (DMEM) containing 10% fetal bovine serum. The media was supplemented with iron, zinc, and copper as their sulfate salts. Cell growth was measured each day. Growth of these cells with increasing media concentrations of iron, zinc, or copper is reported in Fig. 10.21. The growth of this same cell line with the media supplemented with the optimum concentration for all three elements based on their interaction is shown in Fig. 10.22, and was significantly greater than cell growth on media without addition of these elements. Using a series of factorially arranged treatments (Lim and Gordon, 1990), it was determined that significantly better growth could be achieved by reducing the concentration of iron in the media from 2.5 to 1.25 µg Fe/ml and copper from 0.2 to 0.05 µg Cu/ml.

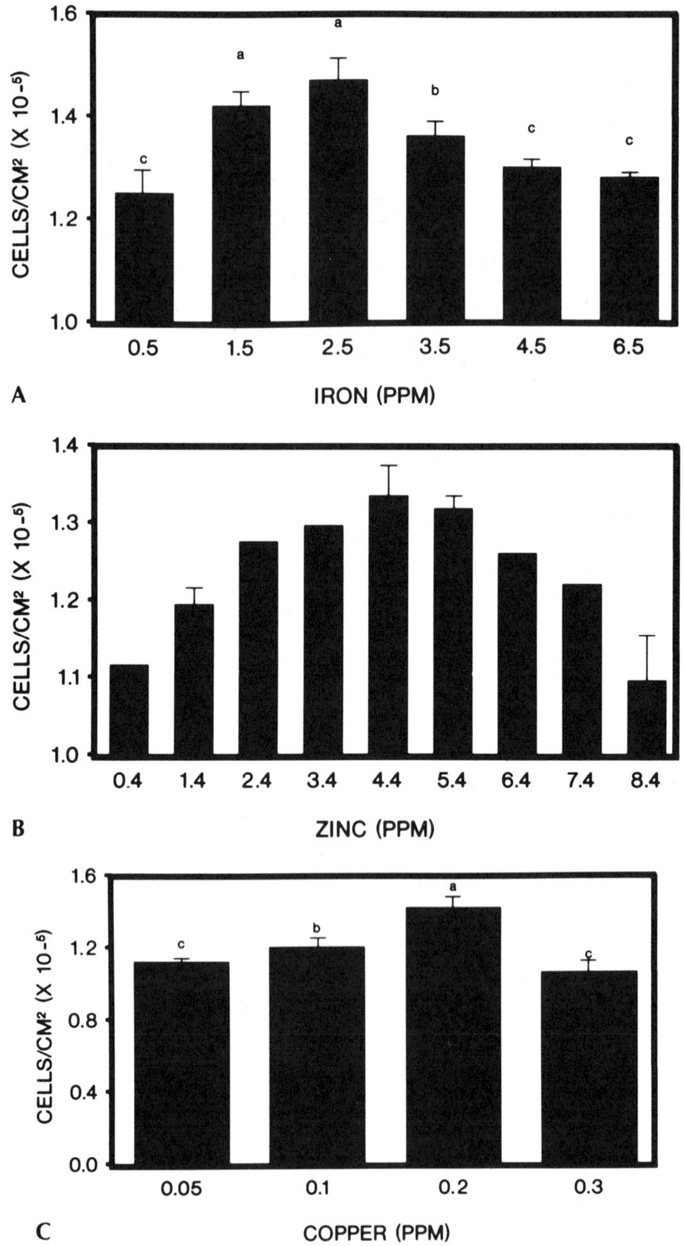

Fig. 10.21 Optimum iron (A), zinc (B), and copper (C) requirements for intestinal mucosal cells, IRD-98, grown on Dulbecco's Minimum Essential Media (DMEM) plus 10% fetal bovine serum after 4 days. Optimum mineral concentrations for IRD-98 intestinal mucosal cells were determined to be 2.5 μg Fe/ml, 4.4 μg Zn/ml, and 0.2 μg Cu/ml.

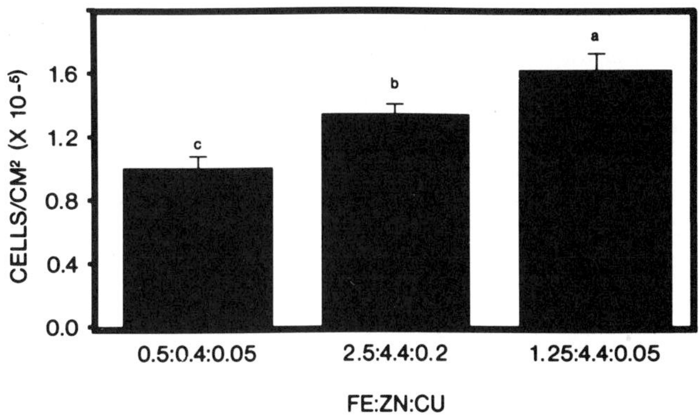

Fig. 10.22 Growth of intestinal mucosal cells, IRD-98, with different media concentrations of iron, zinc, and copper. The basal mineral concentrations in the growth media were determined to be 0.5 μg Fe/ml, 4.0 μg Zn/ml, and 0.05 μg Cu/ml. Cell growth on media containing optimum mineral concentrations for iron, zinc, and copper was significantly higher ($p \leq 0.01$) compared to growth of cells on basal media. Through factorial arrangement of media concentrations for iron, zinc, and copper, the highest cell growth ($p \leq 0.01$) was observed when media iron, zinc, and copper concentrations were 1.25, 4.4, and 0.05 μg/ml.

In addition to demonstrating these interactions, use of cells in culture is allowing us to more accurately determine the cellular and molecular mechanisms associated with these interactions.

RADIONUCLIDES

Radioactive forms of iron, copper, and zinc have played a vital role in helping us understand the metabolism of these elements and both the extrinsic and intrinsic factors that influence their bioavailabilities (Spencer et al., 1978; Aamodt et al., 1981).

Radioactive forms of these elements available for biological studies are listed in Table 10.20. The radionuclides of iron, zinc, and copper are produced by neutron capture reactions (Wang et al., 1965), and this is routinely accomplished at the Missouri University Research Reactor (MURR) (Columbia, MO). Iron-59 is produced from an enriched Fe-58 target. Zinc-65 is produced from Zn-64. Copper-67 is produced from enriched Zn-67 and is separated from the Zn-67 target by anion-exchange chromatography.

Table 10.20 Gamma Ray Energies and Abundance Associated with the Radionuclides Iron-59, Zinc-65, and Copper-67

Element	Half-life	Gamma ray energy, KeV[a]	Gamma ray abundance[b] (%)
Iron-59	44.5 days	142.65	1.02
		192.35	3.08
		1,099.25[c]	56.50
		1,291.60	43.20
Zinc-65	243.9 days	1,115.55[c]	50.70
Copper-67	61.92 hrs	91.27	7.00
		93.31	16.10
		184.58[c]	48.70

[a] Photons, KeV thousand electron volts.
[b] Indicates how fast different gamma rays are produced.
[c] Energy values used to detect and resolve iron, zinc, and copper radionuclides by high-resolution (Ge-Li) gamma ray spectroscopy.

Traditional techniques to measure gamma rays produced by radionuclides have relied on scintillation counting equipment with a sodium iodide (NaI) detector (Wang et al., 1965). This type of detector can register one gamma ray, or possibly two different rays, provided the energy difference between the gamma rays is a large enough (>100 KeV) to allow for peak resolution. High-resolution gamma ray spectroscopy utilizes a solid state germanium-lithium (Ge-Li) detector and can accurately resolve gamma ray peaks that differ by as little as a few KeV in energy (Fig. 10.23). The developmental hardware and software to use high resolution gamma ray spectroscopy for in vivo whole animal counting was developed at the MURR by Steven Morris and Kurt Zinn (1988).

To demonstrate the capabilities of this new technology and illustrate how the difference in the physical properties of these radionuclides can be used to determine mineral retention, the results of a simple experiment are reported (Gordon et al., 1990).

Rats were fed a standard diet containing 12 μg Zn/g, 25 μg Fe/g, and 6 μg Cu/g for 3 weeks. The animals were gavaged with a mixture of 1 μCi Zn-65, 1 μCi Fe-59, and 7 μCi Cu-67, and counted in a whole body counter with a Ge-Li detector (Fig. 10.23) for 7 days at 24-hr intervals. The simultaneous retention of these three isotopes is plotted in Fig. 10.24. Iron retention averaged 80%, zinc retention 60%, and copper retention 20% after 7 days. Although these values may appear high, they represent

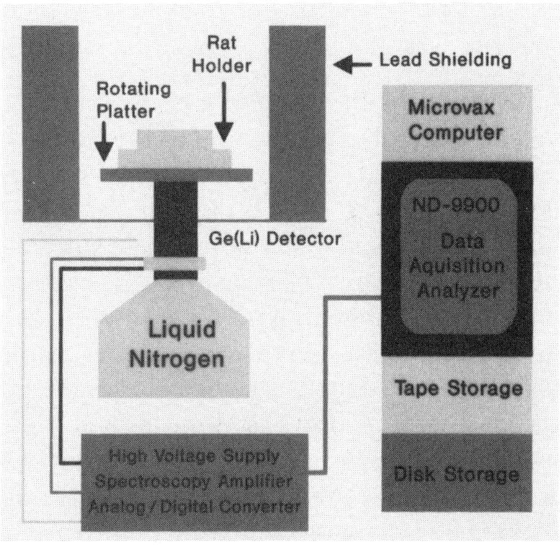

Fig. 10.23 In vivo whole body, small animal (i.e., rat) counter. Animals are restrained in a holder, which is secured to a 360° continuously rotating platter (20 rpm). This rotation ensures uniform geometry of the gamma rays for the Ge-Li high-resolution gamma ray detector. This detector with associated software instrumentation is capable of accurately resolving gamma ray peaks that differ by as little as a few KeV in energy. This allows for the simultaneous measurement of as many as 10–15 different gamma-emitting radionuclides.

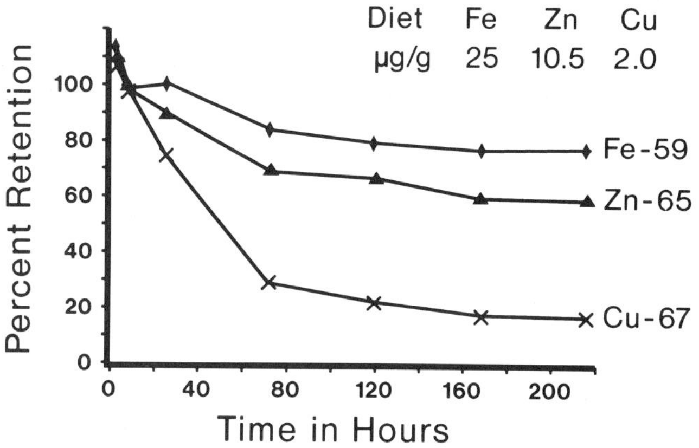

Fig. 10.24 Simultaneous measurements and retention of Fe-59, Zn-65, and Cu-67 in rats fed a diet with adequate amounts of iron, zinc, and copper. Male weanling rats fed a defined diet for 15 days were dosed by gavage with a mixture of the three radionuclides and counted in a high-resolution whole body counter for 7 days at 24-hr intervals. Each point on the graph represents percent radionuclide retention. All values were standardized to the amount of radioactivity determined in animals 5 hr after dosing with radionuclides.

the amount of absorption of the elements under the dietary conditions described. The reader must also be aware that the rat model absorbs iron and other elements at different ratios compared to humans. The percent retention of these radionuclides after 7 days in animals fed different dietary concentrations of iron, zinc, and copper is reported in Table 10.21. In addition to the data illustrated in Fig. 10.24, Table 10.21 contains retention values for animals fed diets with lower and higher concentration of iron, zinc, and copper. With increasing amounts of these minerals in the diet, their absorption is reduced. Use of these radionuclides and high-resolution gamma ray spectroscopy has allowed us to make an important finding regarding the interaction between iron and copper.

It is generally assumed that the intestine, in addition to being the site of nutrient digestion, may also play a vital role in the regulation of nutrient absorption (Charllon and Bothwell, 1983; Refsum and Schreiner, 1984). In previous experiments, we determined that dietary iron interferes with copper absorption. As dietary iron is increased, the amount of copper in the liver decreases. Based on these earlier observations, we assumed that the site of interaction was the intestine. Now having the ability to si-

Table 10.21 Retention of Iron-59, Copper-67, and Zinc-65 in Rats Fed Diets Containing Deficient, Adequate, and Excess Amounts of Iron, Copper, and Zinc[a]

	Dietary concentration (μg/g)	Whole body retention	(%)
Deficient			
Fe	3.2	Fe-59	92
Cu	0.7	Cu-67	30
Zn	10.5[b]	Zn-65	56
Adequate			
Fe	25.0	Fe-59	78
Cu	2.0	Cu-67	15
Zn	10.5	Zn-65	59
Excess			
Fe	250	Fe-59	19
Cu	40	Cu-67	5
Zn	170	Zn-65	13

[a] Animals were fed diets for 21 days, administered a mixture of the radionuclides, and counted in a whole body, high-resolution gamma counter for 7 days. Reported whole body retention values represent percent of administered dose retained after 7 days.

[b] Minimum dietary zinc concentration required to have adequate growth. A diet deficient in zinc (<8–10 μg Zn/g) would cause animals to stop eating and fail to thrive.

multaneously measure the radionuclides Fe-59 and Cu-67, we believe the adverse effect on copper by iron may actually occur in the liver (Pellett et al., 1991).

Six groups of rats, six animals per group, were arranged in a 3 × 2 factorial arrangement of treatments and fed diets containing 1 or 6 μg Cu/g diet and 2.5, 25, or 250 μg Fe/g diet. Animals were allowed free access to the diets and distilled water for 21 days and then administered 1 and 7 μCi of Fe-59 and Cu-67, respectively, by gavage. The animals were counted in a whole body, high-resolution gamma ray counter for 7 days.

Retention of Fe-59 and Cu-67 by these animals is illustrated in Figs. 10.25 and 10.26, respectively. Percentage Fe-59 retentions were highest in the animals fed an iron-deficient diet, 2.5 μg Fe/g, followed very closely by the group of animals fed a diet containing 25 μg Fe/g. In animals fed the diet with high dietary iron (250 μg Fe/g), iron retention was lowest and averaged 10%.

Based on our original hypothesis that iron interferes with copper absorption, we expected a decrease in copper retention as the amount of dietary iron was increased. Figure 10.26 shows that this does not occur.

In these same animals, we determined the concentration of circulating ceruloplasmin. Ceruloplasmin is considered to be a primary copper transport protein in the body. In this experiment and previous experiments in

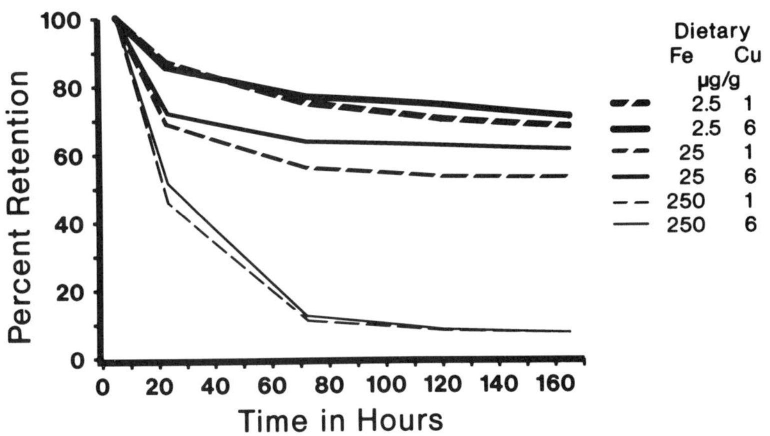

Fig. 10.25 Fe-59 retention in rats fed diets containing varying amounts of dietary iron and copper. Percent retention was significantly different ($p \leq 0.05$) between two groups of animals fed diets containing 250 μg Fe/g compared to four groups of animals fed diets with 2.5 or 25 μg Fe/g.

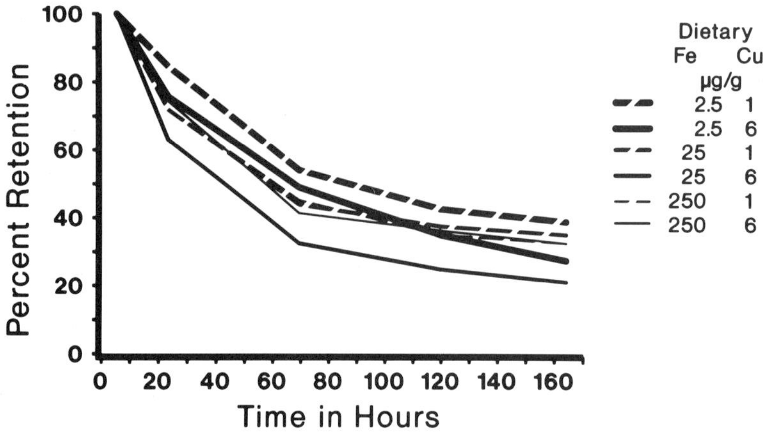

Fig. 10.26 Cu-67 retention in rats fed diets containing varying amounts of dietary iron and copper in a 3 (Fe) × 2 (Cu) factorial arrangement of treatments. Only retention of Cu-67 in animals fed diets containing 2.5 μg Fe/g and 1 μg Cu/g compared to animals fed diets containing 25 μg Fe/g and 6 μg Cu/g was significantly different ($p \leq 0.05$).

our laboratory, we have consistently observed a significant decrease in ceruloplasmin activity as dietary iron increased (Table 10.22).

With the available information, we now hypothesize that only a small inhibition of copper absorption occurs at the intestine caused by dietary iron. After absorption, iron and copper compete for storage and utilization

Table 10.22 Effects of Iron and Copper on Serum Copper Concentrations and Ceruloplasmin Activity in the Rat

Dietary Fe (μg/g)	Dietary Cu (μg/g)	n	Serum copper (μg/ml)	Ceruloplasmin activity (IU)
2.5	0.8	6	0.06 ± 0.06^a	0^a
2.5	5.7	6	1.20 ± 0.70^b	103 ± 21^b
25	0.8	6	0.07 ± 0.02^a	2 ± 2^a
25	5.7	6	1.30 ± 0.70^b	78 ± 17^c
250	0.8	6	0.09 ± 0.06^a	0^a
250	5.7	6	0.14 ± 0.12^a	11 ± 11^d

[a–d] Mean values ± S.D. in the same column not followed by the same superscript are significantly different ($p \leq 0.05$).

at metabolic sites in the liver and possibly other organs. The consequence of this competition is that excess iron interferes with the ability of the body to maintain normal ceruloplasmin homeostasis (Fig. 10.27).

The consequences of the negative effect of excess iron on copper nutrition in general, and on ceruloplasmin specifically, can be very serious to the host. Ceruloplasmin, in addition to being a copper transport protein, is an acute-phase protein elevated during inflammation, disease (e.g., cancer), and pregnancy (Cousins, 1985). It is theorized that ceruloplasmin and the copper it carries play vital roles, either positively or negatively, affecting these physiological conditions. If ceruloplasmin is important in these situations, excessive intake of iron may exacerbate sensitive conditions within the body.

We do not know the site or significance of every possible interaction among iron, zinc, and copper (Fig. 10.28) that can occur in the body, but the possibilities are numerous and present exciting challenges for us. The practical use of this information and the use of different physical forms

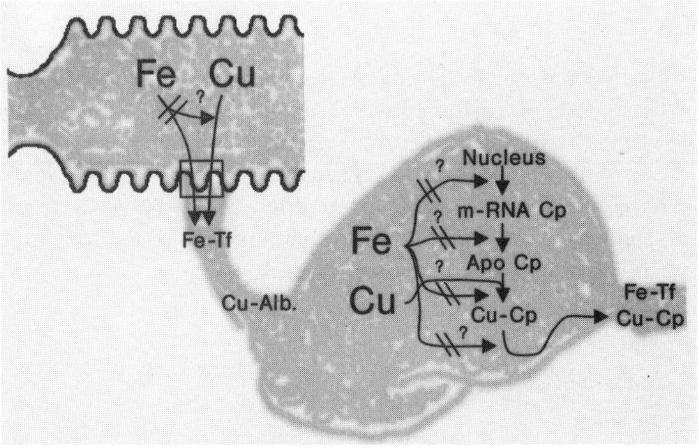

Fig. 10.27 The interaction between iron and copper does not appear to affect the absorption of copper into intestinal mucosal cells and subsequently into the blood. Iron and copper are transported from the intestine to the liver bound to transferrin (Fe-Tf) and albumin (Cu-Alb), respectively. In the liver, copper is transferred to ceruloplasmin (Cu-Cp), which is considered the major transport protein originating in the liver. Excess iron will interfere with the production of Cu-Cp, significantly lowering the concentration of this copper-containing protein in the blood. The exact molecular step at which excess iron interferes with the production of ceruloplasmin is unknown, but several possibilities are indicated. m-RNA = messenger RNA; Apo Cp = ceruloplasmin without copper; Cu-Cp = ceruloplasmin containing copper.

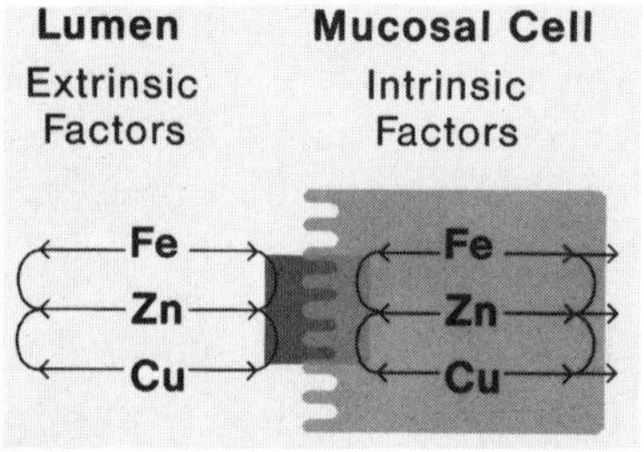

Fig. 10.28 Sites of potential interaction among the elements iron, zinc, and copper using the intestinal lumen and mucosal cell as examples. Within the lumen, extrinsic factors may affect the solubility and subsequent absorption of these nutrients. Although the sites and mechanism of absorption across the cell membrane (i.e., brush border in small intestine mucosal cells) are unknown, interactions among these elements are thought to occur at the cell membrane. Within the cell, mechanisms to control metal absorption by and transfer out of the cells may also be affected by imbalances in the concentration of these elements. The system can become more complex when interactions among transfer proteins in the blood and in different organs of the body are considered.

of these elements was part of our original reason for doing this research. The nutritionist is interested in the chemical and physical properties of nutrients to improve the nutritional quality. People should be advised against consuming excess iron as supplements, and the food industry must be cautious about the amount of iron added to foods. Also, the amount and ratios of all nutrients relative to each other should be considered when nutrient additions to foods are considered.

CONCLUSION

Nutrients are food ingredients. The physical and chemical properties of food ingredients affect their ability to be absorbed and utilized by the body. These same physical and chemical properties of nutrients result in biochemical and physiological changes within the body which relate to nutrition and health. Nutrients do not exist in isolation in foods or within the body. Their interaction also affects their absorption and metabolism, and again these interactions are directly influenced by the chemical and physical properties of nutrients.

Thus, chemical and physical forms and properties of food ingredients and nutrients and our understanding of them affect not only all aspects of food quality and acceptance, but also have far-reaching significance in the improvement of the nutritional and metabolic status of both humans and animals.

ACKNOWLEDGMENTS

The authors acknowledge the colleagues who have accomplished much of the work cited in this manuscript and with whom we have been associated. Much of this work has been supported through the University of Missouri Agricultural Experiment Station.

REFERENCES

Aamodt, R. L., Rumble, W. F., Johnston, G. S., Markley, E. J., and Henkin, R. I. 1981. Absorption of orally administered ^{65}Zn by normal human subjects. *Am. J. Clin. Nutr.* 34: 648–653.

Ahrens, E. H. Jr., Hirsch, J., Insull, W. Jr., Tsallis, T. T., Blomstrand, R., and Peterson, M. L. 1957. The influence of dietary fats on serum-lipid levels in man. *Lancet* 1: 943–953.

Albright, K. L., Gordon, D. T., and Cotterill, O. J. 1984. Release of iron from phosvitin by heat and food additives. *J. Food Sci.* 49: 78–82.

Alexander, A. R., Whanger, P. D., and Miller, L. T. 1983. Bioavailability to rats of selenium in various tuna and wheat products. *J. Nutr.* 113: 196–204.

Anderson, K. 1989. Ingredients for reduced-calorie foods. In *Foods For the 90's*, G. G. birch, G. Campbell-Platt, and M. G. Lindley (Ed.), pp. 149–168. Elsevier Applied Science, London.

Arciazewski, H. 1991. Fat functionality, reduction in baked foods. *INFORM* 2: 392–399.

Bagheri, S. and Gueguen, L. 1985. Effect of wheat bran and pectin on the absorption and retention of phosphorus, calcium, magnesium and zinc by the growing pig. *Reproduction Nutrition, Development* 25: 705–716.

Banks, W., Greenwood, C. T., and Muir, D. D. 1973. The structure of starch. In *Molecular Structure and Function of Food Carbohydrates*,

G. G. Birch and L. F. Green (Ed.), pp. 177–194. John Wiley & Sons, New York.

Bauernfeind, J. C. and Lachance, P. A., Ed. 1991. *Nutrient Additions to Food.* Food and Nutrition Press, Inc., Trumball, CT.

Beare-Rogers, J. L. 1983. Trans- and positional isomers of common fatty acids. *Adv. Nutr. Res.* 5: 171–200.

Ben-Amotz, A., Mokay, S., Edelstein, S., and Avron, M. 1989. Bioavailability of natural isomeric mixture as compared with synthetic all-trans-β-carotene in rats and chicks. *J.Nutr.* 119: 1013–1019.

Bjorn-Rasmussen, E., Hallberg, L., Isaksson, B., and Arvidsson, B. 1974. Food iron absorption in man: Applications of the two pool extrinsic tag method to measure heme and nonheme iron absorption from the whole diet. *J. Clin. Invest.* 53: 247–255.

Blackburn, N. A., Redfern, J. S., Jaris, M., Holgate, A. M., Hanning, I., Scarpello, J. H. B., Johnson, I. T., and Read, N. W. 1984. The mechanism of action of guar gum in improving glucose tolerance in man. *Clin. Sci.* 66: 329–336.

Boila, R. J., Salomons, M. D., Milligan, L. P., and Aherne, F. X. 1981. The effect of dietary propionic acid on cholesterol synthesis in swine. *Nutr. Rep. Intl.* 23: 1113–1121.

Bonanome, A. and Grundy, S. M. 1988. Effects of dietary stearic acid on plasma cholesterol and lipoprotein levels. *N. Engl. J. Med.* 318: 1244–1248.

British Nutrition Foundation's Task Force. 1990. *Report: Complex Carbohydrates in Foods.* The British Nutrition Foundation, Chapman and Hall, London.

Carni, J. J., James, W. D., Koirtyohann, S. R., and Morris, E. R. 1980. Stable traces iron-58 techniques for iron utilization studies. *Anal. Chem.* 52: 216–218.

Chao, L. S. and Gordon, D. T. 1983. Influence of fish on the bioavailability of plant iron in the anemic rat. *J. Nutr.* 113: 1643–1652.

Charllon, R. W. and Bothwell, T. H. 1983. Iron absorption. *Ann. Rev. Med.* 34: 55–68.

Cheftel, J. C., Cug, J-L., and Larient, D. 1985. Amino acids, peptides and proteins. In *Food Chemistry*, O. R. Fennema (Ed.), pp. 335–336. Marcel Dekker, Inc., New York.

Clifford, A. J., Jones, A. D., and Bills, N. D. 1990. Bioavailability of folates in selected foods incorporated into amino acid based diets fed to rats. *J. Nutr.* 120: 1640–1647.

Clydesdale, F. M. 1988. Mineral interactions in foods. In *Nutrient Inter-*

actions, C. E. Bodwell and J. W. Erdman, Jr. (Ed.), pp. 73–113. Marcel Dekker, Inc., New York.

Cook, J. D. and Monsen, E. R. 1976. Food iron absorption in human subjects. III. Comparison of the effect of animal proteins on nonheme iron absorption. *Am. J. Clin. Nutr.* 29: 859–867.

Consensus Development Conference on Diet and Exercise in Non-Insulin Dependent Diabetes Mellitus. 1987. *Diabetes Care* 10: 639–644.

Cotton, F. A. and Wilkinson, G. 1980. *Advanced Organic Chemistry*, 4th ed. John Wiley & Sons, New York.

Coulston, A., Greenfield, M., Kraemer, F., Tobey, T., and Reaven, G. 1980. Effect of source of dietary carbohydrate or plasma glucose and insulin responses to test meals in normal subjects. *Am. J. Clin. Nutr.* 33: 1279–1282.

Cousins, R. J. 1985. Absorption, transport and hepatic metabolism of copper and zinc: Special reference to metallothionein and ceruloplasmin. *Physiol. Rev.* 65: 238–309.

Crapo, P. A., Reaven, G., and Olefsky, J. 1976. Plasma glucose and insulin responses to orally administered simple and complex carbohydrates. *Diabetes* 25: 741–747.

Cummings, J. H., Englyst, H. N., and Wiggins, H. S. 1986. The role of carbohydrate in lower gut function. *Nutr. Rev.* 44: 50–54.

Cummings, J. H., Gibson, G. R., and Macfarlane, G. T. 1989. Quantitative estimates of fermentation in the hind gut of man. *Acta Vet. Scand.* (Suppl.) 86: 76–82.

Davis, N. T. 1974. Recent studies of antagonist interactions in the aetiology of trace element deficiency and excess. *Proc. Nutr. Soc.* 33: 293–298.

Dauncey, M. J., Shaw, J. C. L., and Urman, J. 1977. The absorption and retention of magnesium, zinc and copper by low birth weight infants fed pasteurized human breast milk. *Pediatr. Res.* 11: 991–997.

Dreher, M. 1987. *Handbook of Dietary Fiber*. Marcel Dekker, Inc., New York.

Duncan, K. H., Bacon, J. A., and Weinsier, R. L. 1983. The effects of high and low energy density diets on satiety, energy intake and eating time of obese and nonobese subjects. *Am. J. Clin. Nutr.* 37: 763–769.

Eastwood, M. A. and Mitchell, W. D. 1976. Physical properties of fiber: A biological evaluation. In *Fiber in Human Nutrition*, G. A. Spiller and R. J. Amen (Ed.), pp. 109–129. Plenum Press, New York.

Expert Panel on Population Strategies for Blood Cholesterol Reduction.

1990. *Report of the EPPSBCR*. National Cholesterol Education Program, NIH Publication No. 90-3046.

Federation of American Societies for Experimental Biology. 1987. *Physiological Effects and Health Consequences of Dietary Fiber*. Life Sciences Research Office, Center for Food Safety and Applied Nutrition, Food & Drug Administration, Department of Health and Human Services, Washington, DC.

Festa, M. D., Anderson, H. L., Dowdy, R. P., and Ellersieck, M. R. 1983. Effect of zinc intake on copper excretion and retention in men. *Am. J. Clin. Nutr.* 41: 285–292.

Fields, M., Ferretti, R. J., Smith, J. C., and Reiser, S. 1983. Effect of copper deficiency on metabolism and mortality in rats fed sucrose or starch diets. *J. Nutr.* 113: 1335–1345.

Fleming, S. E. and Arce, D. S. 1986. Volatile fatty acids: Their production, absorption, utilization, and roles in human health. *Clin. Gastroenterol.* 15: 787–814.

Fox, M. R. S., Jacobs, R. M., Jones, A. O. L., Fry, B. E. Jr., Rakowska, M., Hamilton, R. P., Harland, B. F., Stone, C. L., and Tao, S.-H. 1981. Animal models for assessing essential and toxic elements. *Cereal Chem.* 58: 6–15.

Francis, F. J. 1985. Pigments and other colorants. In *Food Chemistry*, O. R. Fennema (Ed.), pp. 550–557. Marcel Dekker, New York.

Furda, I. 1981. Simultaneous analysis of soluble and insoluble dietary fiber. In *The Analysis of Dietary Fiber in Food*, W. P. T. James and O. Theander (Ed.), pp. 163–172. Marcel Dekker, Inc., New York.

Gilloalg, M., Bothwell, T. H., Torrance, J. D., MacPhail, A. D., Derman, D. P., Bezwoda, W. R., Mills, W., Charlton, R. W., and Naget, F. 1983. The effects of organic acids, phytates and polyphenols on the absorption of iron from vegetables. *Br. J. Nutr.* 49: 331–337.

Glicksman, M. 1969. *Gum Technology in the Food Industry*. Academic Press, New York.

Glicksman, M. 1982. *Food Hydrocolloids, Vol. I, II, and III*. CRC Press, Inc., Boca Raton, FL.

Goldberg, I. and Williams, R., Ed. 1991. *Biotechnology and Food Ingredients*, pp. 153–191. Van Nostrand-Reinhold, New York.

Gordon, D. T. 1985. Interactions among Fe, Zn and Cu affecting hemoglobin status in the rat. *XIII International Congress of Nutrition*.

Gordon, D. T. 1989. Sites of the intestinal mucosal cell where the interactions among Fe, Zn and Cu occur affecting the bioavailability of these elements. In *Nutrient Availability: Chemical and Biological Aspects*,

D. A. T. Southgate, I. T. Johnson, and G. R. Fenwich (Ed.), pp. 147–149. AFRC Institute of Food Research, Norwich, U.K.

Gordon, D. T. 1990. Total dietary fiber and mineral absorption. In *Dietary Fiber*, D. Kritchevsky, C. Bonfield, and J. W. Anderson, (Ed.), pp. 105–127. Plenum, New York.

Gordon, D. T. and Godber, J. S. 1989. The enhancement of nonheme iron bioavailability by beef protein in the rat. *J. Nutr.* 119: 446–452.

Gordon, D. T., Zinn, K., Pellett, L., Kattelmann, K., Ratliff, V., and Markos, N. 1990. Interactions among Fe-59, Zn-65, and Cu-67 affecting their retention in the rat. *The FASEB J.* 4: A1158(5184).

Gordon, I. R. 1989. Fibre foods for the '90's. In *Foods for the '90's*. G. G. Birch, G. Campbell-Platt, and M. G. Lindley (Ed.), pp. 135–148. Elsevier Applied Science, London.

Greger, J. L., Abernathy, R. P., and Bennett, O. A. 1978. Zinc and nitrogen balance in adolescent females fed varying levels of zinc and soy protein. *Am. J. Clin. Nutr.* 31: 112–119.

Gregory, J. F., III and Litherland, S. A. 1986. Efficacy of the rat bioassay for the determination of biologically available vitamin B-6. *J. Nutr.* 116: 87–97.

Guyton, A. C. 1991. Digestion and absorption in the gastrointestinal tract. In *Textbook of Medical Physiology*, pp. 726–735. W. B. Saunders Co., Philadelphia.

Haber, G. A., Heaton, K. W., Murphy, D., and Burroughs, L. 1977. Depletion and disruption of dietary fiber. Effects on satiety, plasma glucose and serum insulin. *Lancet* 2: 679–683.

Hallberg, L. and Rossander, L. 1984. Improvement of iron nutrition in developing countries: Comparison of adding meat, soy protein, ascorbic acid, citric acid, and ferrous sulfate on iron absorption from a simple Latin American-type of meal. *Am. J. Clin. Nutr.* 45: 988–995.

Harmouth-Hoene, A. E. and Schelenz, R. 1980. Effects of dietary fiber on mineral absorption in growing rats. *J. Nutr.* 110: 1774–1784.

Hasckke, F., Ziegler, E. E., Edwards, B. B., and Fomon, S. J. 1986. Effect of iron fortification on infant formula on trace mineral absorption. *J. Fed. Gastroenterol. Nutr.* 5: 768–773.

Herman, R. W. 1974. Hydrolysis and absorption of carbohydrates, and adaptive response in the jejunum. In *Sugars in Nutrition*, H. L. Sipple and K. W. McNutt (Ed.), pp. 145–172. Academic Press, New York.

Hill, C. H. and Matrone, G. 1970. Chemical parameters in the study of in vivo and in vitro interactions of transition elements. *Fed. Proc.* 29: 1474–1481.

Holden, J. M., Wolf, W. F., and Mertz, W. 1979. Zinc and copper in self-selected diets. *J. Am. Diet. Assoc.* 75: 23–28.

House, W. A. and Welch, R. A. 1989. Bioavailability of an interactions between zinc and selenium in rats fed wheat grain intrinsically labeled with ^{65}Zn and ^{75}Se. *J. Nutr.* 119: 916–921.

IFT Expert Panel on Food Safety and Nutrition. 1989. Dietary fiber—a scientific status summary. *Food Technol.* 43: 133–139.

Illman, R. J., Topping, D. L., McIntosh, G. H., Trimble, R. P., Stares, G. B., Taylor, M. N., and Cheng, B-Q. 1988. Hypocholesterolemic effects of dietary propionate: Studies in whole animals and perfused rat liver. *Ann. Nutr. Metab.* 32: 97–107.

Jacobs, P., Bothwell, T. H., and Charlton, R. W. 1966. Intestinal iron transport: Studies using a loop of gut with an artificial circulation. *Am. J. Physiol.* 210: 694–700.

James, W. P. T. 1980. Dietary fiber and mineral absorption. In *Medical Aspects of Dietary Fiber*, G. A. Spiller and K. R. McPherson (Ed.), pp. 136–152. Plenum, New York.

Jandacek, R. J., Mattson, F. H., McNeely, S., Gallon, L., Vunker, R., and Glueck, J. 1980. Effects of sucrose polyesters on fecal steroid excretion by twenty-four normal men. *Am. J. Clin. Nutr.* 33: 251–259.

Jansen, G. R. 1981. Ingredients for reduced-calorie foods. In *Protein Quality in Humans: Assessment and in vitro estimation*, C. E. Bodwell, J. S. Adkins, and D. T. Hopkins (Ed.), pp. 118–147. AVI Publishing Co., Inc., Westport, CT.

Jenkins, D. J. A. and Jenkins, A. L. 1987. The glycemic index, fiber, and the dietary treatment of hypertriglyceridemia and diabetes. *J. Am. Coll. Nutr.* 6: 11–17.

Jenkins, D. J. A. and Wolever, T. M. S. 1981. Slow release carbohydrate and the treatment of diabetes. *Proc. Nutr. Soc.* 40: 227–235.

Jenkins, D. J. A., Wolever, T. M. S., Leeds, A. R., Gassull, M. A., Haisman, P., Dilawari, J., Goff, D. V., Metz, G. L., and Alberti, K. 1978. Importance of viscosity. *Br. Med. J.* L: 1392–1398.

Jenkins, D. J. A., Wolever, T. M. S., Taylor, R. H., Baker, H., Fielden, H., Baldwin, J. M., Bowling, A. C., Newman, H. C., Jenkins, A. L., and Goff, D. V. 1981. Glycemic index of foods: A physiological base for carbohydrate exchange. *Am. J. Clin. Nutr.* 34: 362–366.

Joint FAO/WHO Expert Consultation on Protein Quality Evaluation. 1990. Food and Agricultural Organization of the United Nations. Bethesda, Maryland.

Kattelman, K., Stringer, B., and Gordon, D. T. 1988. The oxidation state of iron in the alimentary tract of rats fed diets containing beef protein or casein. *Fed. Proc.* 2: A432.

Keys, A., Anderson, J. T., and Grande, F. 1965. Serum cholesterol response to change in the diet. IV. Particular saturated fatty acids in the diet. *Metabolism* 14: 776–787.

King, J. C., Raynolds, W. L., and Margen, S. 1978. Absorption of stable isotopes of iron, copper, and zinc during oral contraceptive use. *Am. J. Clin. Nutr.* 31: 1198–1203.

Kratzer, F. H. and Vahra, P. 1986. *Chelates in Nutrition.* CRC Press, Inc., Boca Raton, FL.

Krezawski, P. A., Nuttall, F. Q., Gannon, M. C., Billinton, G. J., and Parker, S. 1987. Insulin and glucose responses to various starch-containing foods in Type II diabetic subjects. *Diabetes Care* 10: 205–212.

Kruh, J. 1982. Effects of sodium butyrate, a new pharmacological agent, on cells in culture. *Mol. Cell Biochem.* 42: 65–82.

Kritchevsky, D. 1988. Effects of triglyceride structure on lipid metabolism. *Nutr. Rev.* 46: 177–181.

Lee, D., Schroeder, J. III, and Gordon, D. T. 1988. Enhancement of Cu bioavailability in the rat by phytic acid. *J. Nutr.* 118: 712–717.

Lim, S. and Gordon, D. T. 1990. Effects of Fe-Zn-Cu interaction on growth and metallothionein concentration in fetal rat small intestinal mucosal cells (IRD-98). *The FASEB J.* 4: 1158(5183).

Lineback, D. R. and Inglett, G. E., Ed. 1982. *Food Carbohydrates.* AVI Publishing Co., Inc., Westport, CT.

Lo, G. S., Settle, S. L., and Steinke, F. H. 1980. Bioavailability of copper in isolated soybean protein using the rat as an experimental model. *J. Nutr.* 114: 332–340.

Machlin, L. J. 1991. Vitatmin E. In *Handbook of Vitamins*, L. J. Machlin (Ed.), pp. 99–144. Marcel Dekker, Inc., New York.

Mattson, F. H. and Grundy, S. M. 1985. Comparison of effects of dietary saturated, monunsaturated, and polyunsaturated fatty acids on plasma lipids and lipoproteins in man. *J. Lipid Res.* 26: 194–202.

Mattson, F. H. and Volpenhein, R. A. 1972. Hydrolysis of fully esterified alcohols containing from one to eight hydroxy groups by the lipolytic enzymes of rat pancreatic juices. *J. Lipid Res.* 13: 325–328.

McConnell, A. A., Eastwood, M. A., and Mitchell, W. D. 1974. Physical characteristics of vegetables foodstuffs that could influence bowel function. *J. Sci. Food Agric.* 25: 1457–1461.

McGrandy, R. B., Barrows, C. H. Jr., Spanias, A., Meredith, A., Stone,

J. L., and Norris, A. H. 1966. Nutrient intakes and energy expenditures in men of different ages. *J. Gerontol.* 21: 581–587.

Mensink, R. F. and Katon, M. B. 1990. Effect of dietary trans fatty acids on high-density and low-density lipoprotein cholesterol levels in healthy subjects. *N. Engl. J. Med.* 323: 439–445.

Momcilonic, B., Belonje, B., Giroux, A., and Shoh, B. G. 1975. Total femur zinc as the parameter of choice for a zinc bioassay. *Nutr. Rep. Int.* 12: 197–203.

Monsen, E. R., Hallberg, L., Layrisse, M., Hegsted, D. M., Cook, J. D., Mertz, W., and Finch. 1978. Estimation of available dietary iron. *Am. J. Clin. Nutr.* 31: 134–141.

Negrel, R., Fampal, P., Nano, J., Covenel, C., and Ailhaud, G. 1983. Establishment and characterization of an epithelial intestinal cell line from rat fetus. *Experimental Cell Research* 143: 427–435.

Nutritional recommendations and principles for individuals with diabetes mellitus: 1986. 1987. *Diabetes Care* 10: 126–132.

O'Brien-Nabors, L. and Geraldi, R. C. 1986. *Alternative Sweeteners.* Marcel Dekker, New York.

O'Dell, B. L. 1984. Bioavailability of trace elements. *Nutr. Rev.* 42: 301–308.

O'Dell, B. L. 1986. Use of cells in culture to solve basic nutrition problems. *Nutr. Rev.* 44: 245–255.

O'Dell, B. L. 1990. Dietary carbohydrate source and copper bioavailability. *Nutr. Rev.* 48: 425–434.

O'Dell, B. L. and Savage, J. E. 1960. Effect of phytic acid on zinc availability. *Proc. Soc. Exp. Biol. Med.* 103: 304–306.

Pellett, L., Kattelmann, K., Zinn, K., Trokey, D., Forrester, I., and Gordon, D. T. 1991. The interaction between dietary Fe, Cu and stress on Cu-67 retention and serum ceruloplasmin (Cp) activity in rats. *The FASEB J.* 5:A1290(5335).

Perkins, E. G. and Visek, W. J., Ed. 1983. *Dietary Fats and Health.* American Oil Chemists' Society, Champaign, IL.

Phillipson, B. E., Rothrock, D. W., Connor, W. E., Harris, W. S., and Illingworth, D. R. 1985. Reduction of plasma lipids, lipoproteins, and apoproteins by dietary fish oils in patients with hypertriglyceridemia. *N. Engl. J. Med.* 312: 1210–1216.

Pla, G. W., Fritz, J. C., and Robinson, C. L. 1976. Relationship between the biological availability and solubility rate of reduced iron. *JAOAC* 59: 582–583.

Pla, G. W., Harrison, B. N., and Fritz, J. C. 1973. Comparison of chicks and rats as test animals for studying bioavailability of iron, with special reference to use of reduced iron in enriched bread. *JAOAC* 56:P 1369–1373.

Pomeranz, Y. 1985. *Functional Properties of Food Components*. Academic Press, Inc., New York.

Prasad, A. S. 1983. Clinical, biochemical and nutritional spectrum of zinc deficiency in human subjects: An update. *Nutr. Rev*. 14: 197–208.

Prasad, A. S., Miate, A., Farid, Z., Sandstead, H. H., and Schulert, A. R. 1963. Zinc metabolism in patients with the syndrome of iron deficiency anemia, hepatosplenomegaly. *J. Lab. Clin. Med*. 61: 531–549.

Prosky, L., Asp, N.-G., Schweizer, T. F., DeVries, J. W., and Furda, I. 1988. Determination of insoluble, soluble, and total dietary fiber in foods and food products. Interlaboratory study. *JAOAC*. 71: 1017–1028.

Recommended Dietary Allowances. 1989. National Academy Press, Washington, DC.

Rees, J. M. and Monsen, E. R. 1973. Absorption of fortification iron by the rat: Comparison of type and level of iron incorporated into mixed grain cereal. *Agr. Food Chem*. 21: 913–915.

Refsum, S. B. and Schreiner, B. B-I. 1984. Regulation of iron balance by absorption and excretion. *Am. J. Physiol*. 19: 867–874.

Reiser, S., Ferretti, R. J., Fields, M., and Smith, J. C. 1983. Role of dietary fructose in the enhancement of mortality and biochemical changes associated with copper deficiency in rats. *Am. J. Clin. Nutr*. 38: 214–222.

Reiser, S., Smith, J. C., Mertz, W., Holbrook, J. T., Scholfield, D. J., Powell, A. S., Canfield, W. K., and Canary, J. J. 1985. Indices of copper status in humans consuming a typical American diet containing either fructose or starch. *Am. J. Clin. Nutr*. 42: 242–251.

Roberts, S. B. and Young, V. R. 1988. Energy costs of fat and protein deposition in the human infant. *Am. J. Clin. Nutr*. 48: 951–955.

Rutenberg, M. W. and Solarek, D. 1984. Starch derivatives: Production and uses. In *Starch: Chemistry and Technology*, 2nd ed., R. L. Whistler, J. N. Bemiller, and E. F. Paschall, (Ed.), pp. 311–388. Academic Press, Orlando, FL.

Sakata, T., Hikusaka, K., Shiomura, Y., and Tomate, N. 1980. The stimulatory effect of butyrate on epithelial cells proliferation in the rumen of sheep and its mediation by insulin. Differences between *in vivo* and *in vitro* studies. In *Cell Proliferation in the Gastrointestinal Tract*, D.

R. Appleton, J. P. Sunter, and A. J. Watson (Ed.), pp. 271–276. Pitman, Tumbridge, U.K.

SAS. 1985. *SAS User's Guide: Basic Version, 5th ed.* SAS Institute, Inc., Cary, NC.

Schricker, B. R., Miller, D. D., and Van Campen, D. 1982. In vitro estimation of iron availability in meals containing soy products. *J. Nutr.* 1112: 1696–1705.

Schwartz, S. E., Levine, R. A., Singh, A., Scheidecker, T., and Track, N. S. 1982. Sustained pectin ingestion delays gastric emptying. *Gastroenterology* 83: 812–816.

Selvendran, R. R. 1984. The plant cell wall as a source of dietary fiber: Chemistry and structure. *Am. J. Clin. Nutr.* 39: 320–337.

Selvendran, R. R. 1985. Development in the chemistry and biochemistry of pectic and hemicellulosic polymers. *J. Cell Sci.* (Supl.) 2: 51–88.

Shah, B. C., Giroux, A., and Belonje, B. 1979. Bioavailability of zinc in infant cereals. *Nutr. Metab.* 23: 286–293.

Shen, J. L. and Jackson, R. H. 1991. Functional soy proteins designed as casein and egg white replacers. *Cereal Foods World* 36: 431–433.

Sipple, H. L. and McNutt, K. W., Ed. 1974. *Sugars in Nutrition.* Academic Press, New York.

Slavin, J. L. and Marlett, J. A. 1980. Influence of refined cellulose on human bowel function and calcium and magnesium balance. *Am. J. Clin. Nutr.* 33: 1932–1939.

Southgate, D. A. T. 1969. Determination of carbohydrates in foods. II. Unavailable carbohydrates. *J. Sci. Food Agr.* 20: 331–335.

Spencer, H., Kramer, L., Osis, D., and Norris, C. 1978. Effect of phosphorus on the absorption of calcium and on calcium balance in man. *J. Nutr.* 108: 447–451.

Staub, H. W., Mardones, B., and Shah, N. 1983. Modern dietary fibre product development and nutrient bioavailability. In *Dietary Fibre*, G. G. Birch and K. J. Parker (Ed.), pp. 37–60. Applied Science, London.

Steinke, F. H. and Hoplins, D. T. 1978. Biological availability to the rat of intrinsic and extrinsic iron with soybean protein isolates. *J. Nutr.* 108: 481–489.

Stephen, A. M. and Cummings, J. H. 1979. Water-holding by dietary fibre *in vitro* and its relationship to fecal output in man. *Gut* 20: 722–727.

Surgeon General's Report on Nutrition and Health. 1988. U. S. Government Printing Office, Washington, DC.

Swern, D., Ed. 1979. *Bailey's Industrial Oil and Fat Products.* John Wiley & Sons, New York.

Szczesniak, A. S. 1990. Texture: Is it still an overlooked food attribute? *Food Technol.* 44(9): 86–95.

Thompson, S. A. and Weber, C. W. 1981. Copper and zinc binding to dietary fiber sources: An ion exchange column method. *J. Food Sci.* 47: 125–131.

Tinker, D. T. and Rucker, R. 1987. Molecular biology applications in nutrition. *Nutr. Ref.* 45: 97–101.

Todd, P. A., Benfield, P., and Goa, K. L. 1990. Guar gum: A review of its pharmacological properties, and use as a dietary adjunct in hypercholesterolaemia. *Drugs* 39: 917–928.

Topping, D. L. 1991. Soluble fiber polysaccharides: Effect on plasma cholesterol and colonic fermentation. *Nutr. Rev.* 49: 195–203.

Trowell, H., Southgate, D. A. T., Wolever, T. M. S., Leeds, A. R., Gassull, M. A., and Jenkins, D. A. 1976. Dietary fibre redefined (letter). *Lancet* 1: 967.

Turnbull, A., Cleton, F., Finch, C. A., Thompson, L., and Martin, J. 1962. Iron absorption IV. The absorption of hemoglobin iron. *J. Clin. Invest.* 41: 1897–1907.

Van Campen, D. 1973. Enhancement of iron absorption from ligated segments of rat intestine by histidine, cysteine, and lysine: Effects of removing ionizing groups and of steroisomerism. *J. Nutr.* 103: 139–142.

Van Dokkum, A., Wesstra, A., and Schippeis, F. A. 1982. Physiological effects of fibre-rich types of bread. *Br. J. Nutr.* 47: 451–457.

Wang, C. H., Willis, D. L., and Loveland, W. D. 1965. Availability of radioisotope-labeled compound. In *Radiotracer Methodology in the Biological Environmental and Physical Sciences*, pp. 344–361. Prentice-Hall, Inc., Englewood Cliffs, NJ.

Wang, C. H., Willis, D. L., and Loveland, W. D. 1965. Gamma ray spectrometry using solid scintillation detectors. In *Radiotracer Methodology in the Biological Environmental and Physical Sciences*, pp. 142–167. Prentice-Hall, Inc., Englewood Cliffs, NJ.

Weiss, T. J. 1983. *Food Oils and Their Uses*. AVI Publishing Co., Inc., Westport, CT.

Wheby, M. S., Suttle, G. E., and Ford, K. T. III. 1970. Intestinal absorption of hemoglobin iron. *Gastroenterology* 58: 647–654.

Wilmshurst, P. and Crawley, J. C. W. 1980. The measurement of gastric transit time in obese subjects using ^{24}Na and the effects of energy content and guar gum on gastric emptying and satiety. *Br. J. Nutr.* 44: 1–9.

Wolf, M. J., Khoo, U., and Inglett, G. E. 1977. Partial digestibility of cooked amylomaize starch in humans and mice. *Die Stărke* 29: 401–405.

Wolin, M. J. 1981. Fermentation in the rumen and human large intestine. *Science* 213: 1463–1468.

Zinn, K. R. and Morris, J. S. 1988. High resolution gamma-ray spectroscopy as an *in vivo* tool for following the zinc-selenium interaction. In *Trace Elements in Man and Animals 6*, L. S. Hurley, C. L. Keen, B. Lonnerdal, and R. B. Rucker (Ed.), pp. 591–592. Plenum Press, New York.

11

The Transduction of Taste and Olfactory Stimuli

Joseph G. Brand

Monell Chemical Senses Center
Veterans Affairs Medical Center and
University of Pennsylvania
Philadelphia, Pennsylvania

Pankaj S. Shah

Monell Chemical Senses Center
Philadelphia, Pennsylvania

INTRODUCTION

Perception of the flavor of food is the result of complex integration of sensory inputs from stimuli affecting several sensory systems. These include the taste and olfactory senses as well as mechanoreceptive and thermoreceptive systems, and systems that respond to chemically induced pain. These sensory signals may be carried by one or more major neural systems from the receptor areas in the head and neck to the central nervous system (CNS). The twelve major nerves (numbered I through XII) that have sensory and motor functions in the head and neck region are referred to as cranial nerves. Many of the nontaste and nonolfactory sensations are carried by the trigeminal nerve (cranial nerve V). The functioning of these nontaste and nonolfactory sensory systems in oral and intranasal stimulation is the subject of a recent volume (Green et al., 1990).

Even though the appreciation of flavor is the result of multiple sensory inputs, the primary sensory inputs are from taste and olfaction. The sense of taste is initiated by specialized receptor cells located in the oral cavity. Their function is to recognize chemical stimuli and transduce this recognition event into a change in firing rate of a sensory nerve that forms close synaptic contact with the receptor cell. Receptor cells of olfaction have a similar purpose, although in the case of olfaction, the receptor cell is a bipolar neuron and the process of chemical recognition is transferred directly into a change in firing rate of this receptor neuron. This chapter describes the biochemical events that take place between the binding and recognition of a chemical stimulus and the alteration in firing rate of the sensory nerve, a process collectively known as transduction.

In humans, taste receptors are innervated (supplied with neural connections) by three cranial nerves. Papillae of the anterior two-thirds of the tongue are innervated by a branch of the VIIth cranial nerve called the chorda tympani branch. The posterior one-third, including the receptor cells of the vallate and foliate papillae, is innervated by cranial nerve IX. Cranial nerve X (the vagus) innervates receptors on the soft palate, glottis, and epiglottis. While the innervation patterns are in general the same across all mammals, there are species differences, and there is overlap among the innervated fields.

Taste cells are neuroepithelial structures with a life span of approximately 9–14 days (Beidler and Smallman, 1965; Farbman, 19880). They tend to respond to more than a single class of chemicals (e.g., salts, bitters, etc.) (Ozeki and Sato, 1972; Sugimoto and Teeter, 1991), and single fungiform papillae, containing even a single taste bud, are also responsive to multiple taste qualities (Harper et al., 1966). It would appear, therefore, that single taste cells contain receptor and transductive elements that allow them to recognize more than a single taste quality. The manner in which this information is coded by neurons and decoded by the brain is a matter of some debate (Scott, 1987).

Unlike taste receptor cells, which are innervated epithelial cells, the receptor cells for the sense of smell are bipolar neurons of the olfactory nerve, cranial nerve I. The receptor elements for olfactory stimuli are located on the ciliary processes that emanate from a terminal knob on olfactory receptor neurons. These neuronal receptors have a lifetime of approximately 30 days and are the only neuronal cell type that is in continuous renewal in the adult. Renewal and replacement of olfactory receptor cells (and taste receptor cells) ensures continued chemosensitivity in spite of environmental assaults to these receptor areas. The cilia of the olfactory neurons float in mucus and detect odors that are solubilized in the mucus sheet. The proximal end of the olfactory neuron makes synaptic contact with mitral/tufted cells in glomeruli of the olfactory bulb. Processing takes place within the bulb, before the signals are sent out of the bulb to the CNS (Scott and Harrison, 1987).

The manner in which olfactory stimuli are coded and interpreted is the subject of investigations at both the peripheral (receptor) and central levels. These studies have been dominated by the observation that the olfactory system is capable of recognizing thousands of odors. Coding and interpretation would, therefore, seem to be, of necessity, complex. The recent report of the cloning of a large family of what appears to be olfactory receptors (Buck and Axel, 1991) would lead one to speculate that much of the coding is performed at the peripheral level (Firestein, 1991). Yet, the CNS must still be performing a major role in interpretation and signal sorting, roles which are beginning to be defined (Kauer, 1987, 1991).

RECEPTOR BIOCHEMISTRY OF TASTE

Taste qualities are generally divided into four or five basic sensations. These are salty, sour, sweet, bitter, and umami. The latter quality has been described as an "amino acid" taste, typified by the stimulus monosodium glutamate (Kawamura and Kare, 1987; Yamaguchi, 1991) When studying the events of transduction at the cellular level in taste, one notes that these four or five sensory qualities are also distinguishable at a receptor level. Each classical quality apparently makes use of a unique transductive sequence and has associated with it a set of stimuli of varying but, in many cases, internally consistent structural attributes. It is this diversity of transductive sequences and structural or physical similarities of stimuli that support the existence of classical taste qualities (Kinnamon, 1988; Teeter and Cagan, 1990).

Salt Taste Transduction

Salty taste is associated with the unique taste of NaCl. While other alkali halides elicit varying amounts of saltiness, only NaCl and LiCl are described as primarily salty at concentrations above 0.1 M (Murphy et al., 1981). This fact and the observation that an acceptable tasting substitute for the stimulus, NaCl, has been difficult to find, led investigators to speculate on the nature of a receptor that could account for the specificity of the salt taste pathway. One transduction sequence that fulfills this specificity criterion is that involving a sodium-specific epithelial type ion channel (Shiffman et al., 1983; Heck et al., 1984). Ion channels are formed by specialized integral membrane proteins that allow the passage of ions based on various criteria including charge (+ or −) and size.

It is now generally accepted that salty taste transduction occurs via an epithelial ion channel (DeSimone et al., 1984, 1989; Brand et al., 1985; Avenet and Lindemann, 1989). This influx of Na into the cell depolarizes the cell membrane leading to alterations in voltage-sensitive ion channels that ultimately modulate intracellular calcium ion activity leading to neurotransmitter secretion. Sodium ions are removed from these cells via a Na-K ATPase pump (DeSimone et al., 1989). Evidence for this transduction sequence comes from studies of the transporting properties of the dorsal lingual epithelium (DeSimone et al., 1984) and neurophysiological studies evaluating the role of Na-channel inhibitors on salt-induced neural responses (Schiffman et al., 1983, 1990; Brand et al., 1985; DeSimone and Ferrell, 1985). In most animals studied, salt response is sensitive to the presence of the diuretic amiloride, and this sensitivity is Na- and Li-specific (but see McPheeters and Roper, 1985; Desor and Finn, 1989).

By measuring the firing rate of the chorda tympani nerve, it has been shown that amiloride will inhibit the neural response in rat due to NaCl and LiCl, but not that due to KCl, RbCl, or CsCl. The inhibition is apparently of a mixed kinetics type, suggesting the possibility that in taste transduction amiloride's antagonistic activities can be localized to multiple sites or multiple processes (Fig. 11.1). The failure of amiloride to completely eliminate responses due to NaCl may suggest a stimulatory action of Cl anion (Formaker and Hill, 1988) and/or that anions differentially modulate the diffusion of sodium to amiloride-insensitive or -inaccessible sites (Ye et al., 1991). The production and exploitation of a salt substitute or a salt taste enhancer will be made more efficient by taking this transduction mechanism into account.

Sour Taste Transduction

Sourness is related to the solution activity of the proton. While pH is often used as an index of sourness, it is obvious that pH alone cannot predict completely the perceived sour intensity of the solution nor its overall quality. The so-called "anion effect" is prominent and probably plays a role independent of the activity of the proton. Perception of sourness has been linked to various measures of acidity including titratable acidity (Beatty and Cragg, 1935), the quantity and character of the anion and the undissociated acid (Norris et al., 1984; Ganzevles and Kroeze, 1987), and, with organic weak acids, lipophilicity (Gardner, 1980).

The primary event in sour taste transduction appears to involve a pro-

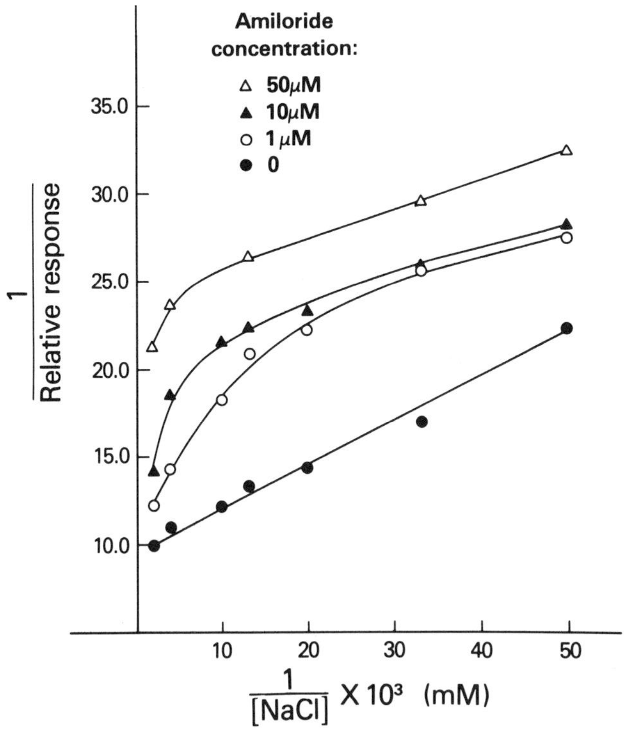

Fig. 11.1 Double reciprocal plot of the kinetics of the chorda tympani neural response to NaCl and three concentrations of amiloride in the rat. For each experiment where the preparation was stimulated during the presence of amiloride, the appropriate concentration of amiloride was continuously flowed over the tongue, while the stimulus (NaCl) was added as a 1-ml pulse on top of that flow. Response was normalized to a pulse of 500 mM NaCl. Between a series of stimulus presentations, the tongue was rinsed with water for 15 min. (*From Brand et al., 1985.*)

ton block of outward-going potassium channels in amphibians (Kinnamon and Roper, 1988), while in mammals, it may be an influx of protons through an amiloride-sensitive ion channel (Gilbertson et al., 1991). Such a block of K-channels will result in cellular depolarization provided a sufficient number of the K-channels are located at the apical end of the taste cell. Using a roving patch technique, Kinnamon et al. (1988) found that the apical end of taste cells contained a majority of the available proton specific K-channels. The role of the anion in the perception of sourness has not been rigorously studied at the receptor level. It would appear that it has multiple roles since its presence has been linked to effects on taste perception and salivary flow (Norris et al., 1984; Christensen et al., 1987). The latter events can alter intraoral pH (Christensen et al., 1987) and may come about due to stimulation of the trigeminal nerve.

Sweet Taste Transduction

Sweetness is generally associated with stimulation by naturally occurring sugars. Sweet-tasting compounds include selected members of such diverse chemical classes as carbohydrates, halogenated sugars, amino acids, aminoacyl sugars, peptides and proteins, terpenoids, *N*-sulfonyl amides, sulfamates, analines, polyketides, ureas, and heavy metal salts (e.g., lead acetate) (Walters et al., 1991). From a structure/activity viewpoint, many sweet-tasting compounds are found to have a proton donor group at a critical distance from a proton-accepting group—the original hypothesis of Shallenberger and Acree (1967). Subsequently, more inclusive theories were proposed, with some becoming quite complex (e.g., Walters et al., 1991). The intense search for sweeteners, with its possible huge market payoff, has provided receptor scientists with a number of useful molecular tools that are only just now being exploited.

The receptor and transductive events in sweet taste recognition have only recently begun to be characterized. The existence of inhibitors for sweet taste transduction and the specificity shown by these agents had long argued for sweet taste transduction being a receptor-mediated event. Yet, in spite of the isolation of a number of proteins that bind sweet-tasting compounds (Dastoli and Price, 1968; Shimazaki et al., 1986; Persaud et al., 1988), the actual isolation of sweet receptor proteins has been elusive; none of these proteins have been definitely identified as receptors per se. Yet, while the receptor(s) itself remains hypothetical, the sequence of events triggered by sweet stimulus binding to it has been characterized.

Sweet taste transduction involves production of the second messenger cyclic AMP (cAMP). A second messenger is a compound made inside the cell as a direct result of stimulation by another compound (the "first" messenger) to the outside of the cell. In the case of the taste receptor cell responding to a sweet-tasting stimulus, the cAMP produced intracellularly probably activates a cAMP-dependent protein kinase (Avenet et al., 1988; Kinnamon, 1988; Tonosaki and Funakoshi, 1988; Striem et al., 1989; Lancet and Ben-Arie, 1991). The phosphorylations stimulated by the kinase may lead to alterations in membrane ion channel activities that ultimately bring about changes in intracellular calcium ion activity, which triggers neurotransmitter release (Fig. 11.2a).

Striem et al. (1989) have demonstrated that the production of cAMP in lingual tissue of rat is G-protein–dependent. G-Proteins are a family of GTP-binding regulatory proteins that generally link a receptor protein with a second messenger generating system. In the rat, Striem et al. (1984) observed that the amount of cAMP produced from lingual tissue was approximately proportional to the sweetness intensity of the various stim-

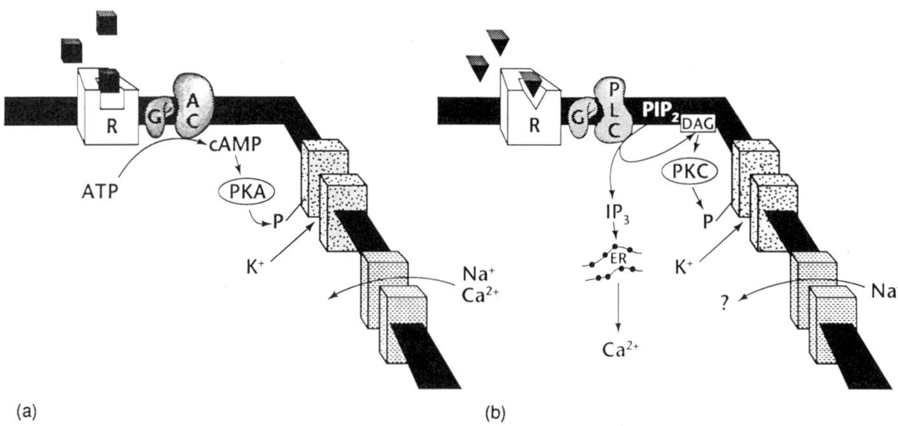

Fig. 11.2 Possible schemes for the transduction of sweet and bitter tastes. (a) Sweet taste is assumed to be mediated by one or more receptors (R) that recognize structural attributes of stimuli. This receptor, when binding the stimulus, activates a G-protein (G), which in turn activates adenylyl cyclase (AC) converting ATP to cAMP. The enhanced production of cAMP activates a cAMP-dependent protein kinase (PKA), which phosphorylate ion channels leading to a change in membrane potential. With a change in membrane potential, other voltage-sensitive ion channels may be activated or inhibited leading ultimately to release of neurotransmitter. (b) Bitter taste transduction may occur through several routes. One of these is via a receptor-mediated process. The receptor (R) recognizes the stimulus and activates a G-protein (G), which in turn activates a phospholipase C (PLC). This phospholipase promotes the production of two second messengers from phosphatidyl inositol-4,5-biphosphate (PIP_2). One of these, diacylglycerol (DAG), operates within the membrane to activate protein kinase C (PKC), which phosphorylates, among other proteins, ion channels. The other second messenger, inositol triphosphate (IP_3), is a cytosolic messenger that may release calcium ion from intracellular stores such as the endoplasmic reticulum (ER). Closure of an outward going K-channel results in a change in membrane potential, which may activate or inhibit other ion channels resulting ultimately in release of neurotransmitter.

uli used as agonists (agents which affect receptor systems) in the in vitro study. Likewise, divalent cations that reduce sweet taste perception also reduced cAMP production. In a definitive study, Striem et al. (1990) showed that methyl 4,6-dichloro-4,6-dideoxy-α-D-galactopyranoside (Jakinovich, 1983), an inhibitor of the neural response due to sweet tasting compounds, also inhibited the ability of those sweet-tasting compounds to generate cAMP in an in vitro assay.

The question of whether there are multiple numbers of sweet taste receptors is still open. It is known, for example, that all animals do not share responses to common chemical stimuli. Some animals, particularly carnivores, are insensitive to sweet-tasting carbohydrates (Beauchamp et al., 1977). Omnivores and herbivores, while generally sensitive to carbohydrates that humans find sweet, may not find them acceptable to the same degree, nor rank them in the same order of intensity vs. concentration (Kare, 1971). Among mammals, considerable divergence is found in the ability to taste artificial sweeteners. Hellekant, in a series of elegant studies, has provided evidence that within primates, one can discern a sensitivity pattern to sweeteners and sweet inhibitors that has remarkable evolutionary implications (Hellekant et al., 1981). For example, only Old World monkeys tend to share our sweet taste perceptual world, and this fact has implications for sensory testing of experimental sweeteners (see Naim et al., 1982). Hellekant has also amply demonstrated that sweet taste transduction in primates occurs on fibers relatively specific for this modality, and that these fibers carry information from both natural and artificial sweeteners (Hellekant and Ninomiya, 1991).

Bitter Taste

A large number of compounds with a great diversity of chemical structure taste bitter. While many of these compounds can be grouped into major structural classes—the quaternary amines, acetylated sugars, alkaloids, amino acids (primarily L-isomers), and certain inorganic salts—no chemical or physical features have been found that are common to all stimuli. This fact inevitably leads to two receptor hypotheses: that there are more than a single receptor for bitterness, or that there are no receptors at all, but rather that these stimuli achieve their taste effect by altering the lipid/protein environment of the plasma membrane of the taste cell to such a degree that intracellular ionic changes can take place via a change in membrane voltage (Kumazawa et al., 1986, 1988). The well-known genetic dimorphism on the ability to taste the thioureas (PTC) as well as studies on bitter taste in well-defined strains of mice argue for a multireceptor/transduction hypothesis but cannot at this time rule out the nonreceptor postulate.

Carbohydrates as bitter stimuli. The structure/taste relationships among carbohydrates have been reviewed by several investigators (see Lee, 1984; Belitz and Wieser, 1985; Maga, 1990). Among the free sugars, only β-D-mannose and gentiobiose have been reported to have bitter tastes. Interestingly, their α-isomers (i.e., α-D-mannose and isomaltose, respectively) are sweet tasting.

Among the carbohydrate derivatives, some interesting structure/taste correlations have been made. In glycosides, the lipophilicity of the aglycone has been directly correlated with bitterness, e.g., methyl glycosides of glucose, galactose, mannose, fructose, lyxose, and arabinose have been found to have varying degrees of sweetness, whereas the corresponding benzyl glycosides of all these compounds are very bitter (Birch and Lindley, 1973). The bitterness of glycosides increases gradually with increasing carbon chain length in the gluco-series, i.e., methyl glucoside is sweet, ethyl glucoside is bitter-sweet, the propyl derivative is bitter, the butyl is more bitter, etc.

Similarly, it was observed that converting the free hydroxyl groups to their ester derivatives also caused the sugars to taste bitter. All peracetylated mono- and disaccharides are bitter. A few di-, tri-, and tetrabenzoate derivatives of sucrose are also bitter. A similar trend was observed with deoxy sugars, i.e., the mono-deoxy derivatives of a series of mono- and disaccharides were found to be sweet, whereas their deoxy derivatives were always bitter (Birch and Lee, 1974).

For deoxy-halo derivatives of carbohydrates, substituting the hydroxyl groups with chloro-, bromo-, or iodo-moieties renders the monosaccharides primarily bitter. But in the disaccharide sucrose, an enhancement of sweetness is observed. With a few exceptions, the sweetness is directly proportional to the number of hydroxyl groups that are substituted by halogen atoms (e.g., 4,1′,6′-trichloro-sucrose, the new sweetener, sucralose). A similar trend was observed when we investigated some deoxy-halo derivatives of fructose. 6-Chloro fructose had been reported to be largely sweet with a trace of bitterness. When 1-chloro fructose was synthesized and examined, it was largely bitter with a trace of sweetness. Interestingly, 1,6-dichloro fructose is known to be bitter-sweet.

It appears the sweet-bitter dichotomy among carbohydrates occurs at a level of subtle changes at the hydroxyl groups. Several hypotheses for these sweet-bitter relationships have been proposed, but none of them can satisfactorily predict the taste qualities of all carbohydrates and their derivatives.

Amino acids as bitter taste stimuli. The structure/taste relationships of amino acids have been extensively investigated by several researchers (Solms et al., 1965; Schiffman and Engelhard, 1976; Maga, 1990). The paramount observation is that the short-chain neutral amino acids in both the L- and D-configurations are considered sweet (GLY, ALA, SER) or sweet/bitter (THR, VAL), while most all other L-isomers are bitter (L-PRO and L-GLU are exceptions). On the other hand, almost all D-isomeric forms are sweet. A similar trend is not observed in carbohydrate taste perception. In fact, the enantiomeric D- and L-sugars generally possess similar taste qualities (Shallenberger et al., 1969).

Between the two functionalities, namely, amino and carboxylate, the presence of an amino group is required for bitter taste perception, whereas modification of the carboxylate moiety indicates that it is not essential to bitterness. In some instances, this carboxylate modification has generated even more bitter compounds.

Increasing the steric bulk with lipophilic character in the side chain generally increases the bitterness of the amino acids, but the *t*-butyl group in the β position renders the compound tasteless. Similarly, substitution at the α-carbon makes the compound tasteless. Aromatic side chains generally increase the bitterness, e.g., L-phenylalanine and L-tryptophan are among the most bitter amino acids.

Generally, the intensity of bitter taste decreases with polar substituents in the side chain, e.g., hydroxy, amino, carboxylic, or carboxamide groups. Also, the distance of the polar group from α-amino moiety is inversely correlated with the influence of the polar functionality on bitterness.

Among azocycloalkane-2-carboxylic acids, it was found that increasing the size of the ring from 4 to 6 increased bitterness. When the alicyclic ring was present as a side chain a similar trend was observed up to a 9-carbon ring, after which the compounds became tasteless.

Peptides as bitter taste stimuli. Unlike amino acids, the peptides can taste bitter in both L- and D-series. Although L-phenylalanine methyl ester is bitter, its dipeptide ester with aspartic acid, namely, aspartame, is intensely sweet. Thus, the taste of a peptide cannot be easily correlated with that of its constituent amino acids. Several dipeptides have been investigated for their taste properties (Schiffman and Engelhard, 1976). Interestingly, no trend relating to bitterness and hydrophobicity was observed as in the case of amino acids and higher molecular weight peptides. The sequence of the constituent amino acids of a dipeptide was found to be an important factor in determining the taste. When seven pairs of dipeptides containing some residues in reverse orders were tested, none of them tasted the same. Most dipeptides were bitter in taste, although the constituent amino acids had sweet or other taste characteristics. Even the dipeptides containing identical amino acid residues did not taste the same as their constituent amino acids alone. For example, glycylglycine tastes bitter, whereas the amino acid glycine is known to be sweet (Schiffman and Engelhard, 1976). Thus, the relationships of bitterness or taste profiles and structures of dipeptides are not very clear.

The hydrophobicity of a peptide side chain has been correlated with increased bitterness in peptides. Also, it has been observed that proline-containing peptides that have a side chain of 5 carbon atoms or more generally impart a bitter taste. Several higher peptides containing proline residues have been shown to be bitter. Cyclopeptides were found to be more bitter than the corresponding open-chain peptides with the intensity of bitterness directly correlated with hydrophobicity. For the cyclopeptides, it was also observed that the taste was independent of the chirality of the peptides.

Transduction of bitter taste. The receptor and transductive events in bitter taste received little direct attention until the mid-1980s. At this point, some key observations were made. One reported that intracellular calcium ion activity was altered in intact taste cells with the addition of a bitter stimulus (Akabas et al., 1988). This increase in calcium resulted from a release of calcium from internal stores, a mechanism consistent with the prediction of formation of the second messenger inositol triphosphate (IP3) via a receptor-mediated process. Spielman and coworkers observed this production of IP3 in a congenic strain of mice sensitive to the bitter taste stimulus sucrose octaacetate (Spielman et al., 1991). This observation was supported by the work of Hwang et al. (1990), who demonstrated the presence of the IP3-generating system in rat taste tissue. Spielman and coworkers also reported a closure of outward going potassium channels in isolated mouse taste cells exposed to the bitter stimulus denatonium benzoate (Spielman et al., 1989). The picture that emerges, therefore, is one involving a receptor coupled to a G-protein that activates production of IP3 which releases calcium ion from internal stores. The calcium ion and/or the diacylglycerol produced could act to close outward going K-channels causing membrane depolarization (Fig. 11.2b).

It is important to recognize that the initial receptor recognition step is not absolutely required for transduction of some stimuli as long as they are capable of causing the release of calcium ion or closure of K-channels. Thus, quinine, a well-known hydrophobic bitter stimulus and K-channel blocker, could conceivably act directly on K-channels in the taste cell. Other hydrophilic bitter-tasting stimuli may presumably act via a receptor.

Other Tastes and Model Systems

Other taste primaries, such as metallic, soapy, etc., have been postulated from time to time, but their inclusion in a universal taste field has not received wide acceptance. Many of these "off-tastes" may be the result of olfactory stimulation or stimulation of the trigeminal nerve, and some are considered phantom tastes, often associated with neurological damage or pathology (e.g., Shafar and Glasg, 1965; El-Deiry and McCabe, 1990).

One primary taste not associated with the classic four, but perhaps equivalent to them, is that for the prototypical stimulus, monosodium glutamate (MSG) (Kawamura and Kare, 1987). This stimulus imparts a taste quality often called "umami," the Japanese word for delicious or savory. The taste of MSG has the interesting property of being enhanced by certain 5′-ribonucleotides, thereby forming a true taste synergism (Yamaguchi, 1987; 1991). That the umami taste is a primary taste quality is supported by work at the psychophysical level and the receptor level. Biochemical receptor-binding studies and recent results from reconstituted taste receptor membranes suggest the existence of a receptor site for glutamate and a mode of transduction involving an ion channel whose activity is controlled by stimuli that impart an umami taste, for example, the stimulus monosodium glutamate (Brand et al., 1991). These types of ion channels are said to be "gated" by the stimulus and are often considered to be receptor/ion channel complexes.

Earlier work had shown that glutamate was specifically bound to tissue from taste epithelium and that this binding was enhanced by some 5′-ribonucleotides (e.g., GMP) (Torii and Cagan, 1980). This biochemical observation was consonant with similar psychophysical observations on the synergistic effect. Recent studies from our laboratory have suggested that this receptor for glutamate may be a stimulus-gated ion channel (Brand et al., 1991). Plasma membranes from mouse circumvallate papillae of lingual epithelium were reconstituted into artificial lipid bilayers. Addition of glutamate to these otherwise silent bilayers induced a flow of current which was partially sodium selective (Fig. 11.3). Preliminary experiments suggest that this activity due to glutamate is also GMP-sensitive.

While there is much to be learned about taste transduction in mammalian systems, perhaps the best understood taste system in vertebrates is found in *Ictalurus punctatus*, the common channel catfish. This animal contains cutaneous taste receptors sensitive to amino acids (Caprio, 1975). Biochemically, three of these have been characterized. One, recognizing L-alanine and other short-chain neutral amino acids, is receptor-mediated

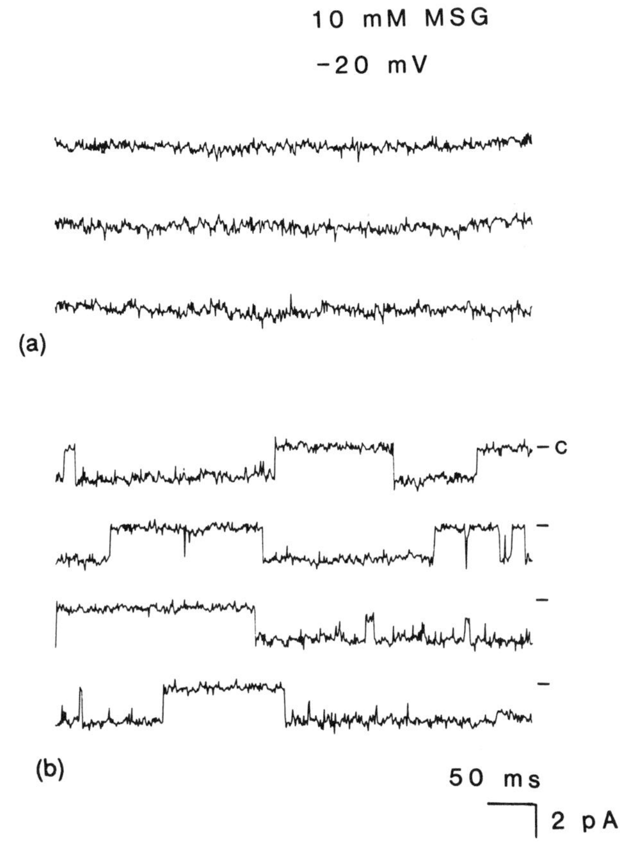

Fig. 11.3 Single-channel currents elicited at −20 mV in a phospholipid bilayer containing epithelial taste membranes from mouse lingual circumvallate papillae. Continuous current records before (a) and after (b) the addition of 10 mM monosodium glutamate to the bath. Inward currents (open) are displayed downward. The closed level is shown by the level −C. Slope conductance of this channel was 59 pS. *(From Brand et al., 1991)*

and G-protein regulated producing intracellular second messengers (Cagan, 1986; Brand et al., 1987; Kalinoski et al., 1989b). Another, sensitive to L-arginine (Kalinoski et al., 1989a), is a stimulus-gated ion channel (Teeter et al., 1990). A third, sensitive to L-proline, is also a stimulus-gated ion channel, but not the same molecular entity as that sensitive to L-arginine (Kumazawa et al., 1990). Thus, receptor-mediated production of second messengers and stimulus-gated ion channel-type receptors are common themes in taste transduction across many species (Kinnamon, 1988; Teeter and Cagan, 1990).

RECEPTOR EVENTS AND TRANSDUCTION IN OLFACTION

We are able to smell perhaps thousands of different odors. The way in which the peripheral receptor system is integrated with the CNS to permit discrimination of such a large number of volatile chemicals is the subject of intense investigation. This is a particularly important endeavor since it is the sense of smell that adds nuance and uniqueness to many flavors and foods. Without the olfactory component from food, eating becomes simply a chore. While the lack of olfactory ability is not life-threatening, it results in a life that is, as one anosmic had remarked, "missing a dimension."

The intense work of the past 5 years in olfaction has been responsible for the emergence of a coherent hypothesis on olfactory transduction. It is now apparent that the initial recognition of odors occurs on one or more of a large number of receptors. These are linked to one of at least three transduction mechanisms: either increased production of cAMP, increased production of IP3, or activation of a stimulus (odor) gated ion channel.

The existence of receptors that bind olfactory stimuli and that are unique to the olfactory epithelium had been postulated for many years. The receptors were biochemically characterized from a number of models, most notably from aquatic species (Rhein and Cagan, 1983; Bruch and Rulli, 1988; Novoselov et al., 1988). But the molecular nature of these remained elusive. Recently, Buck and Axel (1991) reported the cloning of nearly two dozen members of a large multigene family of putative olfactory receptors from rat. These clones encode integral membrane proteins of the seven transmembrane domain type. Subfamilies of these clones can be arranged that exhibit sequence divergence within the transmembrane domains suggesting that each subfamily may bind distinct structural classes of odor stimuli.

From biochemical studies showing odorant-stimulated production of second messengers, it was hypothesized that olfactory receptors interact with GTP-binding regulatory proteins (Pace et al., 1985; Huque and Bruch, 1986). At least one G-protein is unique to the olfactory system of the rat (Jones and Reed, 1989), while another may be unique to the olfactory system of catfish (Bruch and Kalinoski, 1987). These G-proteins modulate production of second messengers, at least two of which have been implicated in olfactory transduction. One of these second messengers, cAMP, is produced in response to a wide variety of odor stimuli (Pace et al., 1985; Sklar et al., 1986; Breer, 1990), while another, IP3, is produced in response primarily (although not exclusively) to odors that impart a putrid or otherwise unpleasant sensation (Boekhoff et al., 1990; Breer and Boekhoff, 1990). The messenger IP3 was originally discovered to be produced to the odor of pheromones in insects (Breer, 1990) and amino acid odorants in fish (Huque and Bruch, 1986).

Metabolic and ion channel events subsequent to the formation of cyclic AMP and IP3 indicate that the olfactory system contains ion channels in the ciliary membrane that are stimulated directly by cAMP or IP3, and that phosphorylation events and alterations in intracellular calcium ion activity are important regulatory and perhaps desensitization mechanisms in olfaction. Nakamura and Gold (1987) discovered a cAMP and cyclic GMP stimulated ion channel in olfactory cilia, while Restrepo et al. (1990) reported the existence of an IP3-stimulated ion channel in olfactory cilia. The former is a nonspecific cation channel, while the latter is primarily a calcium ion channel. Calcium entry into olfactory cells from the extracellular (mucosal) space is observed upon stimulation by some odors (Restrepo et al., 1990) (Fig. 11.4).

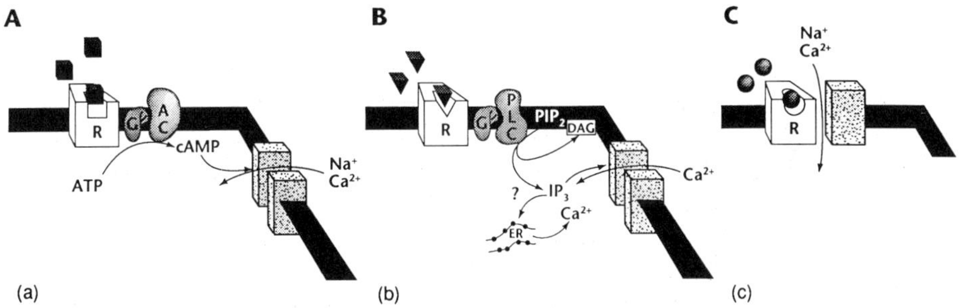

Fig. 11.4 Possible transduction sequences for olfaction. (a) Receptor-mediated production of the second messenger, cAMP. The receptor (R) recognizes a stimulus and activates a GTP-binding regulatory protein, G_{olf}. This G-protein (G) stimulates a type III adenylyl cyclase (AC) to produce cAMP. The messenger cAMP directly activates an ion channel that initiates a depolarization of the neuronal membrane. (b) Receptor-mediated production of the second messenger, IP_3. The receptor (R) recognizes a stimulus and activates a G-protein (G). The G-protein stimulates phospholipase C (PLC) to metabolize PIP_2 to IP_3 and DAG. The DAG presumably activates a protein kinase C. The IP_3 activates an ion channel directly and may also release Ca^{2+} from intracellular stores. Ion channel activation initiates depolarization of the neuronal membrane. (c) Stimulus-gated ion channel. The stimulus binds to the receptor portion of a receptor/ion channel complex. Binding to the receptor portion (R) causes the ion channel to open allowing influx of Na^+ and Ca^{2+} ion resulting in a depolarization of the neuronal membrane.

While the broad outlines of these two transduction sequences appear to be well described, the existence of other transduction sequences and modulatory influences on them are also quite likely. Stimulus (odor) gated ion channels have been identified through reconstitution studies (Labarca et al., 1988). Odorant binding proteins, observed in many species, may act to transport odors to the receptors or clear them from the receptor area (Pevesner et al., 1988; Pelosi and Tirindelli, 1989). Phosphorylation events have been detected (Heldman and Lancet, 1986) as have various enzymes normally associated with metabolism of xenobiotics, including olfactory-specific P450 enzymes and UDP glucoronosyl transferase (Nef et al., 1989; Lazard et al., 1991). The question of how or even whether these enzymes function in olfactory transduction is still open. They may act to detoxify volatile compounds and, in so doing, create an odorous substance. They may also alter existing odors, thereby changing the character of the odor. A recent observation indicates that the intensities of some—but not all—odors are diminished in some individuals if these odors are delivered in an oxygen-free inert atmosphere (Feigin et al., 1991).

CONCLUSIONS

Olfaction and taste play central roles in flavor detection and, together with mechano (Christensen, 1984), thermal, and pain sensory systems (Green et al., 1990), act in concert to permit a full appreciation of food flavor. The receptor mechanisms in taste and olfaction are distinguished by their use of a variety of transduction systems. In the taste system, each taste quality utilizes a particular transduction sequence. In the olfactory system, at least three transductive sequences are known, while a large family of putative receptor proteins has been cloned that presumably activate these transductive processes. Current studies in the chemical senses also emphasize the role that genetics plays in chemical communication and perception (Wysocki and Kare, 1991) and the interplay of nutrition and the chemical senses (Friedman and Kare, 1991). Continuing investigations of these processes will result in a more complete understanding of the coding of chemical stimuli and how the components of flavor are recognized and integrated for achieving a total impression of food quality.

ACKNOWLEDGMENTS

The authors are supported by NIH grants DC-00356, DC-00327, and the DC-00014-12 training grant from the National Institute on Deafness and

Other Communication Disorders and by a grant from the Veterans Affairs Department.

REFERENCES

Akabas, M. H., Dodd, J., and Al-Awquati, Q. 1988. A bitter substance induces a rise in intracellular calcium in a subpopulation of rat taste cells. *Science* 242: 1047–1050.

Avenet, P. and Lindemann, B. 1989. Chemoreception of salt taste. In *Chemical Senses Vol. 1: Receptor Events and Transduction in Taste and Olfaction*, J. G. Brand, J. H. Teeter, R. H. Cagan and M. R. Kare (Ed.), pp. 171–182. Marcel Dekker, Inc., New York.

Avenet, P., Hofmann, F., and Lindemann, B. 1988. Transduction in taste receptor cells requires cAMP-dependent protein kinase. *Nature* 331: 351–354.

Beatty, R. M. and Cragg, L. H. 1935. The sourness of acids. *J. Am. Chem. Soc.* 57: 2347–2351.

Beauchamp, G. K., Maller, O., and Rogers, J. G. 1977. Flavor preferences in cats. *J. Comp. Physiol. Psychol.* 91: 1118–1127.

Beidler, L. M. and Smallman, R. L. 1965. Renewal of cells within taste buds. *J. Cell Biol.* 27: 263–272.

Belitz, H.-D. and Wieser, H. 1985. Bitter compounds: Occurrence and structure/activity relationships. *Food Rev. Internat.* 1: 271–354.

Birch, G. G. and Lindley, M. G. 1973. Structural functions of taste in the sugar series: Effects of aglycones on the sensory properties of simple glycoside structures. *J. Food Sci.* 38: 665–667.

Birch, G. G. and Lee, C.-K. 1974. Structural functions of taste in the sugar series: Sensory properties of deoxy sugars. *J. Food Sci.* 39: 947–49.

Boekhoff, I., Tareilus, E., Stroutmann, J., and Breer, H. 1990. Rapid activation of alternative second messenger pathways in olfactory cilia from rats by different odorants. *EMBO J.* 9: 2453–2458.

Brand, J. G., Teeter, J. H., and Silver, W. L. 1985. Inhibition by amiloride of chorda tympani responses evoked by monovalent salts. *Brain Res.* 34: 207–14.

Brand, J. G., Bryant, B. P., Cagan, R. H., and Kalinoski, D. L. 1987. Biochemical studies of taste sensation. XIII. Enantiomeric specificity of alanine taste receptor sites of catfish, *Ictalurus punctatus*. *Brain Res.* 461: 119–128.

Brand, J. G., Teeter, J. H., Cagan, R. H., and Kare, M. R., Ed. 1989.

Chemical Senses Vol. I: Receptor Events and Transduction in Taste and Olfaction. Marcel Dekker, New York.

Brand, J. G., Teeter, J. H., Kumazawa T., Huque, T., and Bayley, D. L. 1991. Transduction mechanisms for the taste of amino acids. *Physiol. Behav*. 49: 899–904.

Breer, H. and Boekhoff, I. 1991. Odorants of the same odor class activate different second messenger pathways. *Chem. Senses* 16: 19–29.

Breer, H., Boekhoff, I., and Tareilus, E. 1990. Rapid kinetics of second messenger formation in olfactory transduction. *Nature* 345: 65–68.

Bruch, R. C. and Kalinoski, D. L. 1987. Interaction of GTP-binding regulatory proteins with chemosensory receptors. *J. Biol. Chem*. 262: 2401–2404.

Bruch, R. C. and Rulli, R. D. 1988. Ligand binding specificity of a neutral L-amino acid olfactory receptor. *Comp. Biochem. Physiol*. 91B: 535–540.

Bruch, R. C. and Teeter, J. H. 1989. Second messenger signalling mechanisms in olfaction. In *Chemical Senses Vol. 1: Receptor Events and Transduction in Taste and Olfaction*, J. G. Brand, J. H. Teeter, R. H. Cagan and M. R. Kare (Ed.), pp. 283–298, Marcel Dekker, Inc., New York.

Buck, L. and Axel, R. 1991. A novel multi-gene family may encode odorant receptors: A molecular basis for odor recognition. *Cell* 65: 175–187.

Cagan, R. H. 1986. Biochemical studies of taste sensation. XII. Specificity of binding of taste ligands to a sedimentable fraction from catfish taste tissue. *Comp. Biochem. Physiol*. 85A: 355–358.

Caprio, J. 1975. High sensitivity of catfish taste receptors to amino acids. *Comp. Biochem. Physiol*. 52A: 247–251.

Christensen, C. M. 1984. Food texture perception. *Adv. Food Res*. 29: 159–199.

Christensen, C. M., Brand, J. G., and Malamud, D. 1987. Salivary changes in solution pH: A source of individual differences in sour taste perception. *Physiol. Behav*. 40: 221–227.

Dastoli, F. R. and Price, S. 1968. A sweet-sensitive protein from bovine taste buds: Isolation and assay. *Science* 154: 905–907.

DeSimone, J. A. and Ferrell, F. 1985. Analysis of amiloride inhibition of chorda tympani taste response of rat to NaCl. *Am. J. Physiol*. 249: R52–R61.

DeSimone, J. A., Heck, G. L., Mierson, S., and DeSimone, S. K. 1984.

The active ion transport properties of canine lingual epithelium *in vitro*. Implications for gustatory transduction. *J. Gen. Physiol.* 83: 633–656.

DeSimone, J. A., Heck, G. L., Persaud, K. C., and Mierson, S. 1989. Stimulus-evoked transepithelial lingual currents and the gustatory neural response. In *Chemical Senses Vol. 1: Receptor Events and Transduction in Taste and Olfaction*, J. G. Brand, J. H. Teeter, R. H. Cagan and M. R. Kare (Ed.), pp. 13–34. Marcel Dekker, Inc., New York.

Desor, J. A. and Finn, J. 1989. Effects of amiloride on salt taste in humans. *Chem. Senses* 14: 793–803.

El-Deiry, A. and McCabe, B. F. 1990. Temporal lobe tumor manifested by localized dysgeusia. *Ann. Otol. Rhinol. Laryngol.* 99: 586–587.

Farbman, A. I. 1980. Renewal of taste bud cells in rat circumvallate papillae. *Cell Tissue Kinet.* 13: 349–357.

Feigin, A. M., Zinkevich, E. P., and Wysocki, C. J. 1991. Decrease in human olfactory sensitivity to some odors due to substitution of air in the nasal cavity by helium. *Chem. Senses* 16: 519.

Firestein, S. 1991. A noseful of odor receptors. *Trends in Neurosci.* 14: 270–272.

Formaker, B. K. and Hill, D. L. 1988. An analysis of residual NaCl taste response after amiloride. *Am. J. Physiol.* 255: R1002–R1007.

Friedman, M. I. and Kare, M. R. (Ed.) 1991. *Chemical Senses.* Vol. 4: *Appetite and Nutrition.* Marcel Dekker, New York.

Ganzevles, P. G. J. and Kroeze, J. H. A. 1987. The sour taste of acids. The hydrogen ion and the undissociated acid as sour agents. *Chem. Senses* 12: 563–576.

Gardner, R. J. 1980. Lipid solubility and the sourness of acids: Implications for models of the acid taste receptor. *Chem. Senses Flav.* 5: 185–194.

Gilbertson, T. A., Avenet, P., Kinnamon, S. C., and Roper, S. D. 1991. In situ recording from hamster fungiform taste cells: response to sour stimuli. *Soc. Neurosci. Abstr.* 17: 481–485.

Green, B. G., Mason, J. R., and Kare, M. R. (Ed.) 1990. *Chemical Senses.* Vol. 2: *Irritation.* Marcel Dekker, Inc., New York.

Harper, H. Wms., Jay, J. R., and Erickson, R. P. 1966. Chemically evoked sensations from single human taste papillae. *Physiol. Behav.* 1: 319–325.

Heck, G. L., Mierson, S., and DeSimone, J. A. 1984. Salt taste transduction occurs through an amiloride-sensitive sodium transport pathway. *Science* 223: 403–405.

Heldman, J. and Lancet, D. 1986. Cyclic AMP-dependent protein phosphorylation in chemosensory neurons: Identification of cyclic nucleotide-regulated phosphoproteins in olfactory cilia. *J. Neurochem.* 47: 1527–1533.

Hellekant, G. and Ninomiya, Y. 1991. On the taste of umami in chimpanzee. *Physiol. Behav.* 49: 927–934.

Hellekant, G., Glaser, D., Brouwer, J. N., and van der Wel, H. 1981. Gustatory responses in three prosimian and two simian primate species to six sweeteners and miraculin and their phylogenetic implications. *Chem. Senses* 6: 165–173.

Huque, T. and Bruch, R. C. 1986. Odorant- and guanine nucleotide-stimulated phosphoinositide turnover in olfactory cilia. *Biochem. Biophys. Res. Commun.* 137: 36–42.

Hwang, P. M., Verma, A., Bredt, D. S., and Snyder, S. H. 1990. Localization of phosphatidylinositol signalling components in rat taste cells: Role in bitter taste transduction. *Proc. Natl. Acad. Sci. USA* 87: 7395–7399.

Jakinovich, W. Jr. 1983. Methyl 4,6-dichloro-4,6-dideoxy-α-D-galactopyranoside: An inhibitor of sweet taste in gerbils. *Science* 219: 408–410.

Jones, D. T. and Reed, R. R. 1989. G_{olf}: An olfactory neuron specific G-protein involved in odorant signal transduction. *Science* 244: 790–795.

Kalinowski, D. L., Bryant, B. P., Shaulsky, G., Brand, J. G., and Harpaz, S. 1989a. Specific L-arginine taste receptor sites in the catfish, *Ictalurus punctatus*: Biochemical and neurophysiological characterization. *Brain Res.* 488: 163–173.

Kalinoski, D. L., Huque, T., LaMorte, V. J., and Brand, J. G. 1989b. Second messenger events in taste. In *Chemical Senses Vol. 1: Receptor Events and Transduction in Taste and Olfaction*, J. G. Brand, J. H. Teeter, R. H. Cagan and M. R. Kare (Ed.), pp. 85–101. Marcel Dekker, Inc., New York.

Kare, M. R. 1971. Comparative study of taste. In *Handbook of Sensory Physiology Vol. IV: Chemical Senses, Part 2, Taste*, L. M. Beidler (Ed.) pp. 278–292. Springer-Verlag, Heidelberg.

Kauer, J. S. 1987. Coding in the olfactory system. In *Neurobiology of Taste and Smell*, T. E. Finger and W. L. Silver (Ed.), pp. 205–231. John Wiley and Sons, New York.

Kauer, J. S. 1991. Contributions of topography and parallel processing to odor coding in the vertebrate olfactory pathway. *Trends Neurosci.* 14: 79–85.

Kawamura, Y. and Kare, M. R., Ed. 1987. *Umami: A Basic Taste*. Marcel Dekker, Inc., New York.

Kinnamon, S. C. 1988. Taste transduction: A diversity of mechanisms. *Trends Neurosci.* 11: 491–496.

Kinnamon, S. C. and Roper, S. D. 1988. Evidence for a role of voltage-sensitive apical K-channels in sour and salt taste transduction. *Chem. Senses* 13: 115–121.

Kinnamon, S. C., Dionne, V. E., and Beam, K. G. 1988. Apical localization of K-channels in taste cells provides the basis for taste transduction. *Proc. Natl. Acad. Sci. USA* 85: 7023–7027.

Kumazawa, T., Kashiwayanagi, M., and Kurihara, K. 1986. Contribution of electrostatic and hydrophogic interactions of bitter substances with taste receptor membrane to generation of receptor potentials. *Biochim. Biophys. Acta* 888: 62–69.

Kumazawa, T., Nomura, T., and Kurihara, K. 1988. Liposomes as a model for taste cells: Receptor sites for bitter substances including N—C=S substances and mechanism of membrane potential change. *Biochemistry* 27: 1239–1244.

Kumazawa, T., Teeter, J. H., and Brand, J. G. 1990. L-proline activates cation channels different from those activated by L-arginine in reconstituted catfish epithelial membranes. *Chem. Senses* 15: 603–604.

Labaraca, P., Simon, S. A., and Anholt, R. R. H. 1988. Activation by odorants of a multi-state cation channel from olfactory cilia. *Proc. Natl. Acad. Sci. USA* 85: 944–947.

Lancet, D. and Ben-Arie, N. 1991. Sweeet taste transduction. In *Sweeteners: Discovery, Molecular Design and Chemoreception*, D. E. Walters, F. T. Orthoefer, and G. E. DuBois (Ed.), pp. 226–236. American Chemical Society, Washington, DC.

Lazard, D., Zupko, K., Poria, Y., Nef, P., Lazarovits, J., Horn, S., Khen, M., and Lancet, D. 1991. Odorant signal termination by olfactory UDP glucuronosyl transferase. *Nature* 349: 790–793.

Lee, C.-K. 1984. The chemistry and biochemistry of the sweetness of sugars. *Adv. in Carbohydr. Chem. and Biochem.* 45: 199–351.

Maga, J. A. 1990. Compound structure versus bitter taste. In *Bitterness in Foods and Beverages*, R. L. Rouseff (Ed.), pp. 35–48. Elsevier, Amsterdam.

McPheeters, M. and Roper, S. 1985. Amiloride does not block taste transduction in the mudpuppy, *Necturus maculosus*. *Chem. Senses* 10: 341–352.

Murphy, C., Cardello, A. V., and Brand, J. G. 1981. Tastes of fifteen halide salts following water and NaCl: Anion and cation effects. *Physiol. Behav.* 26: 1083–1095.

Naim, M., Rogatka, H., Yamamoto, T., and Zehavi, U. 1982. Taste responses to neohesperidin dihydrochalcone in rats and baboon monkeys. *Physiol. Behav.* 28: 979–986.

Nakamura, T. and Gold, G. H. 1987. A cyclic nucleotide-gated conductance in olfactory cilia. *Nature* 325: 442–444.

Nef, P., Heldman, J., Lazard, D., Margalit, T., Jay, M., Hanucoglu, I., and Lancet, D. 1984. Olfactory-specific cytochrome P-450. cDNA cloning of a novel neuroepithelial enzyme possibly involved in chemoreception. *J. Biol. Chem.* 264: 6780–6785.

Norris, M. B., Noble, A. C., and Pangborn, R. M. 1984. Human saliva and taste responses to acids varying in anions, titratable acidity and pH. *Physiol. Behav.* 32: 237–244.

Novoselov, V. I., Krapivinskaya, L. D., and Fesenko, E. E. 1988. Amino acid binding glycoproteins from the olfactory epithelium of skate. *Chem. Senses* 13: 267–278.

Ozeki, M. and Sato, M. 1972. Responses of gustatory cells in the tongue of rat to stimuli representing four taste qualities. *Comp. Biochem. Physiol.* 41A: 391–407.

Pace, U., Hamski, E., Salomon, Y., and Lancet, D. 1985. Odorant-sensitive adenylate cyclase may mediate olfactory reception. *Nature* 316: 255–258.

Pelosi, P. and Tirindelli, R. 1989. Structure/activity studies and characterization of an odorant-binding protein. In *Chemical Senses* Vol. 1: *Receptor Events and Transduction in Taste and Olfaction*, J. G. Brand, J. H. Teeter, R. H. Cagan and M. R. Kare (Ed.), pp. 207–226. Marcel Dekker, Inc., New York.

Persaud, K. C., Chiavacci, L., and Pelosi, P. 1988. Binding proteins for sweet compounds from gustatory papillae of the cow, pig and rat. *Biochim. Biophys. Acta* 967: 65–75.

Pevsner, J., Sklar, P. B., Hwang, P. M., and Snyder, S. H. 1989. Odorant-binding protein. In *Chemical Senses* Vol. 1: *Receptor Events and Transduction in Taste and Olfaction*, J. G. Brand, J. H. Teeter, R. H. Cagan and M. R. Kare (Ed.), pp. 227–242. Marcel Dekker, Inc., New York.

Restrepo, D., Miyamoto, T., Bryant, B. P., and Teeter, J. H. 1990. Odor stimuli triggers influx of calcium into olfactory neurons of the channel catfish. *Science* 249: 1166–1168.

Rhein, L. D. and Cagan, R. H. 1983. Biochemical studies of olfaction:

Binding specificity of odorants to a cilia preparation from rainbow trout olfactory rosettes. *J. Neurochem.* 41: 569–577.

Rouseff, R. L., Ed. 1990. *Bitterness in Food and Beverages*. Elsevier, Amsterdam.

Schiffman, S. S. and Englehard, H. H. 1976. Taste of dipeptides. *Physiol. Behav.* 17: 523–535.

Schiffman, S. S., Lockhead, E., and Maes, F. W. 1983. Amiloride reduces the taste intensity of Na and Li salts and sweeteners. *Proc. Natl. Acad. Sci. USA* 80: 6136–6140.

Schiffman, S. S., Suggs, M. S., Cragoe, E. J., and Erickson, R. P. 1990. Inhibition of taste responses to Na salts by epithelial Na-channel blockers in gerbil. *Physiol. Behav.* 47: 455–459.

Shafar, J. and Glasg, M. D. 1965. Dysgeusia in the elderly. *Lancet* 1: 83–84.

Schallenberger, R. S. and Acree, T. E. 1967. Molecular theory of sweet taste. *Nature* 216: 480–482.

Shallenberger, R. S., Acree, T. E., and Lee, C. Y. 1969. Sweet taste of D- and L-sugars and amino acids and the steric nature of their chemoreceptor sites. *Nature* 221: 555–558.

Shimazaki, K., Sato, M., and Nakao, M. 1986. Photoaffinity labeling of thaumatin-binding protein in monkey circumvallate papillae. *Biochim. Biophys. Acta* 884: 291–298.

Scott, J. W. and Harrison, T. A. 1987. The olfactory bulb: Anatomy and physiology. In *Neurobiology of Taste and Smell*, T. E. Finger and W. L. Silver (Ed.), pp. 151–178. John Wiley and Sons, New York.

Scott, T. R. 1987. Coding in the gustatory system. In *Neurobiology of Taste and Smell*, T. E. Finger and W. L. Silver (Ed.), pp. 355–378. John Wiley and Sons, New York.

Sklar, P. B., Anholt, R. R. H., and Snyder, S. H. 1986. The odorant-sensitive adenylate cyclase of olfactory receptor cells. *J. Biol. Chem.* 261: 15538–15543.

Solms, J., Vuataz, L., and Egli, R. H. 1965. The taste of L- and D-amino acids. *Experientia* 21: 692–694.

Spielman, A. I., Mody, I., Brand, J. G., Whitney, G., MacDonald, J. F., and Salter, M. W. 1989. A method for isolating and patch clamping single mammalian taste receptor cells. *Brain Res.* 503: 326–329.

Spielman, A. I., Huque, T., Brand, J. G., and Whitney, G. 1991. The bitter taste of sucrose octaacetate is IP3-mediated. *Chem. Senses* 16: 409–410.

Striem, B. J., Pace, U., Zehavi, U., Naim, M., and Lancet, D. 1989. Sweet tastants stimulate adenylate cyclase coupled to GTP-binding protein in rat tongue membranes. *Biochem. J.* 15:121–126.

Striem, B. J., Yamamoto, T., Naim, M., Lancet, D., Jakinovich, W., and Zehavi, U. 1990. The sweet taste inhibitor methyl 4,6-dichloro-4,6,-dideoxy-α-D-galactopyranoside inhibits sucrose stimulation of chorda typani nerve and of the adenylate cyclase in anterior lingual membranes of rats. *Chem. Senses* 15: 529–536.

Sugimoto, K. and Teeter, J. H. 1991. Stimulus-induced currents in isolated receptor cells of the larval tiger salamander. *Chem. Senses* 16: 109–122.

Teeter, J. H. and Cagan, R. H. 1990. Mechanisms of taste transduction. In *Neural Mechanisms in Taste*, R. H. Cagan (Ed.), pp. 1–20. CRC Press, Inc., Boca Raton, FL.

Teeter, J. H., Brand, J. G., and Kumazawa, T. 1990. A stimulus-activated conductance in isolated taste epithelial membranes. *Biophys. J.* 58: 253–259.

Tonosaki, K. and Funakoshi, M. 1988. Cyclic nucleotides may mediate taste transduction. *Nature* 331: 354–356.

Torii, K. and Cagan, R. H. 1980. Biochemical studies of taste sensation. IX. Enhancement of L-[^{3}H]glutamate binding to bovine taste papillae by 5′-ribonucleotide. *Biochim. Biophys. Acta* 627: 313–323.

Walters, D. E., Orthoefer, F. T., and DuBois, G. E. 1991. *Sweeteners: Discovery, Molecular Design and Chemoreception*, American Chemical Society, Washington, DC.

Wysocki, C. J. and Kare, M. R. 1991. *Chemical Senses* Vol. 3: *Genetics of Perception and Communication*. Marcel Dekker, Inc., New York.

Yamaguchi, S. 1987. Fundamental properties of umami in human taste sensation. In *Umami: A Basic Taste*, Y. Kawamura and M. R. Kare (Ed.), pp. 41–73. Marcel Dekker, Inc., New York.

Yamaguchi, S. 1991. Basic properties of umami and effects on humans. *Physiol. Behav.* 49: 833–841.

Ye, Q., Heck, G. L., and DeSimone, J. A. 1991. The anion paradox in sodium taste reception: Resolution by voltage-clamp studies. *Science* 254: 724–726.

12

Disintegration and Segregation Kinetics of Dry Food Particulates

Micha Peleg

University of Massachusetts
Amherst, Massachusetts

INTRODUCTION

Size reduction of particulates is usually achieved through a deliberate operation. It can also occur incidentally during other processes. Examples of the latter are agglomerate disintegration during sieving and handling, attrition of grain during transport, and breakage of breakfast cereals. Many food particulates, such as instant coffee or breakfast cereals, are made of dry material and are, therefore, brittle and fragile. Their uncontrolled disintegration has undesirable effects such as nonuniform bulk density, segregation, and dust formation.

Most food engineering textbooks treat size reduction as a unit operation, and the process is described mainly in terms of a relationship between energy requirements and mean particle size. Notable examples are Rittinger's, Kick's, and Bond's laws (e.g., Loncin and Merson, 1979; Geankoplis, 1983). Such "laws," however, do not clearly address the process kinetics nor the specific breakage mechanisms at the particle level. They are also insensitive to the exact character of the particle size distribution and how it changes during the operation. Applications of theories that address these aspects of size-reduction processes have been developed mostly during and after the 1970s when computer modeling and data analysis became commonplace (Austin, 1971/2; Faulkner and Rimmer, 1983; McCabe et al., 1985). In this chapter the mathematical structure of such models and the physical principles on which they are based will be discussed. They will be demonstrated with recent examples especially developed for the behavior of food systems. The models' simpler predecessors, based on a more conventional kinetic approach, will also be discussed.

One of the most common features of uncontrolled disintegration, as already mentioned, is segregation of the fines that are produced during the process. Powder mixture segregation, also called demixing, is a well-understood physical phenomenon. Its detailed kinetics, however, have received much less attention than mixing, and have hardly been addressed at all in the food literature. In this chapter the principal mechanisms that regulate segregation will be briefly described, and indices to quantify the tendency of mixtures to segregate will be analyzed and compared.

DISINTEGRATION

Uni- and Bimodal Size Distributions

Size reduction processes can be analyzed in terms of events at the particle level or in terms of changes in the particle population. The latter is most commonly characterized by its size distribution (Herdan, 1960; Allen, 1981) and to a lesser extent by other features such as shape. Only very rarely is the size of a particle uniquely defined. Examples are spheres, cubes, or other particles with uniform and regular geometry. In most real particulate assemblies, the shape of the individual particle is not uniform, and therefore its characteristic size can be defined in more than one way. Furthermore, modern automated particle size analyses are based on different physical principles (e.g., optical counters, laser diffraction), and therefore their results need not always or fully agree. This problem is not only encountered in modern methods but can also affect the results of sieve analysis and even direct measurements with a microscope. Since the topics of this chapter are disintegration and segregation, this problem will not be further discussed. It will be assumed that a meaningful representative size can be determined, and all reference to size will be based on this assumption, irrespective of the method by which the size is actually measured.

The distributions most commonly used in powders classification after a size reduction process are the log-normal and Rosin-Rammler (Allen, 1981). The two are skewed to the right and, unless truncated, have an infinite range. Therefore, theoretically, they cannot give an accurate account of real particle populations that have a well-defined, finite size range, especially when their size distribution is skewed to the left. These problems can be eliminated by applying either the beta or a log beta distribution function (Peleg and Normand 1986, Popplewell et al., 1988; Barrett et al., 1991) or Johnson's S_B function (Yu and Standish, 1990). An advantage of the beta or log beta distribution is that it can be written in a form that makes the mode independent of the spread while providing the same fit as the log normal or Rosin-Rammler distribution. Particle populations are usually skewed to the right, but the emergence of size distributions that are skewed to left is not a theoretical possibility only. (Particle populations are called skewed to the left if most of the size range lies to the left of the mode, the size that occurs most frequently, and skewed to the right if most of the range lies to the right of the mode). Left-skewed populations may, in fact, be characteristic of the remnants of an almost uniform (or normally distributed) particle population during the intermediate stages of the disintegration process as shown in Fig. 12.1.

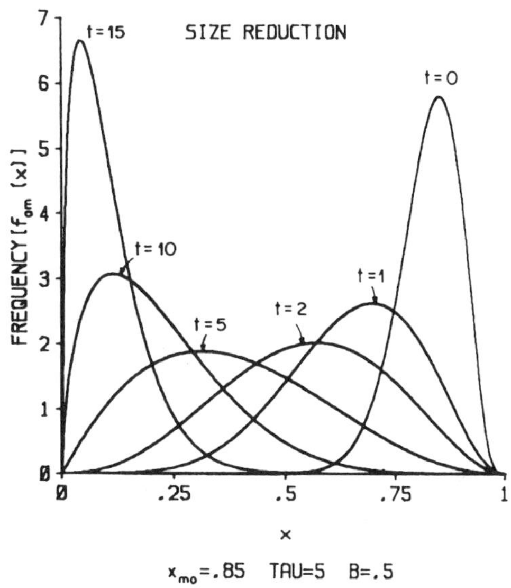

Fig. 12.1 Computer simulated normalized size distributions in a size reduction process using the beta function. Note the skewness reversal. *(From Peleg and Normand, 1986. Reproduced with permission of the American Institute of Chemical Engineers.)*

Only the size distribution of the *end* product is usually skewed to the right, and therefore skewness reversal must occur at some stage during the process.

In many disintegration processes the size distribution of the particle population is clearly bimodal (Coulson and Richardson, 1968; Malave-Lopez and Peleg, 1986a) as shown in Fig. 12.2 which depicts size distri-

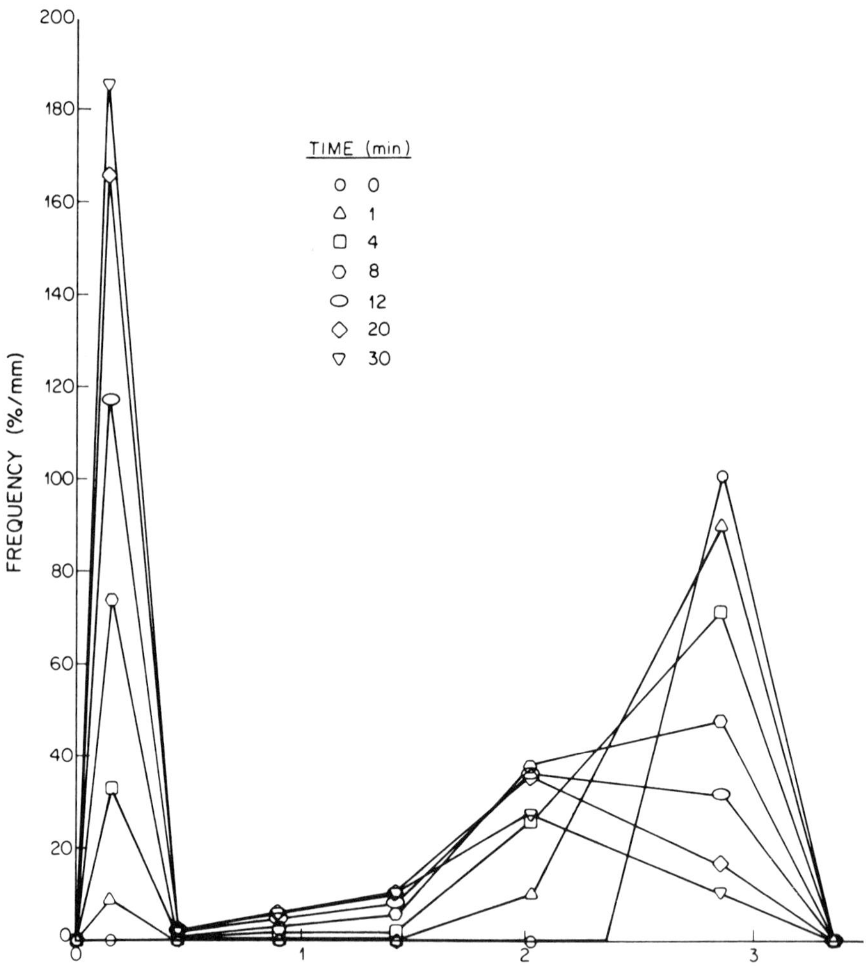

Fig. 12.2 Experimental size distributions of agglomerated instant coffee subjected to vertical vibrations. Note the bimodal character of the distributions. (*From Malave-Lopez and Peleg, 1986a. Reproduced with permission of the Institute of Food Technologists.*)

butions at various times during vibration-induced breakup of agglomerates. This feature is a manifestation of the existence of at least two breakage mechanisms (see below) and is inherent in systems where the disintegration product contains fines. Bimodal size distributions are described by (Everitt and Hand, 1981)

$$f(x) = Wf_1(x) + (1 - W)f_2(x) \tag{1}$$

where x is the size, $f_1(x)$ and $f_2(x)$ the frequency function of the fine and coarse functions, respectively, and W the weight fraction of the fines.

For some applications it is more convenient to express the relationship in terms of the normalized or dimensionless size (x), i.e.,

$$x = \frac{X - X_{min}}{X_{max} - X_{min}} \tag{2}$$

where the subscripts max and min refer to the upper and lower limits of the absolute size. If the size of the coarse and find fractions is beta distributed, then Eq. (1) can be written in the form (Popplewell et al., 1989)

$$f(x) = \frac{Wx^{a_1m_1}(1 - x)^{m_1}}{\int_0^1 x^{a_1m_1}(1 - x)^{m_1}dx} + \frac{(1 - W)\, x^{a_2m_2}(1 - x)^{m_2}}{\int_0^1 x^{a_2m_2}(1 - x)^{m_2}dx} \tag{3}$$

where a and m are constant and the subscripts 1 and 2 refer to the fine and coarse fraction, respectively. In this form of the beta distribution, each fraction's momentary mode x_{mi} (t) is independent of the spread (Peleg and Normand, 1986), i.e.,

$$x_m = \frac{a}{a + 1} \tag{4}$$

Consequently, when $a_i > 1$ ($x_m < 0.5$), the distribution (of each fraction) is skewed to the right; when $a_i < 1$, it is skewed to the left; and when $a_i = 1$ ($x_m = 0.5$), the distribution is symmetric.

The variance of the beta distribution of each fraction σ_i^2, which represents its spread, according to this model is given by

$$\sigma^2 = \frac{(am + 1)(m + 1)}{[(a + 1)m + 2]^2[(a + 1)\, m + 3]} \tag{5}$$

Even if m remains constant, the spread changes as a_i and consequently x_{mi} change.

Breakage Mechanisms

The direct cause of breakage of any kind is mechanical failure. The stress-strain conditions that lead to failure have been extensively discussed in the literature. Most failure criteria, based on a limiting stress, strain, or amount of absorbed energy are also applicable to particulates (Faulkner and Rimmer, 1983). Many particulates, however, have an irregular shape and, therefore, weak points. Further, they contain structural defects and imperfections that are always found, even in regular materials. Consequently, large particles, in which these are more likely to be found, are more susceptible to breakage than smaller particles. The result is the familiar observation that the energy required for size reduction increases dramatically as the particle size decreases. In fine powders, the creation of large new surfaces can also be a factor, but this is of no concern to us here.

Because each feed particle may have unique geometric features and mechanical history (e.g., direction and intensity of impact, orientation during friction with other particles or walls of equipment), it is more convenient to classify breakage in terms of product particles. Thus, the two major breakage mechanisms are shattering, which produces daughter particles of a size on the order of 2–5 times smaller than that of the parent particle, and erosion or abrasion, which produces mainly fines, about 10 or more times smaller than the original particles (Wei et al., 1977). A schematic view of the two mechanisms is shown in Fig. 12.3. (According to Menacho (1986), chipping, which results in daughter particles of intermediate size, can be considered a third mechanism.) In real systems the two mechanisms usually coexist, at least to some extent, and both types of daughter particles are produced simultaneously. The relative dominance of these two mechanisms usually changes as a result of the changes in the particle size distribution. Its mode of change is characteristic of the powder material and the mechanical conditions to which it is exposed (e.g., grinding, tapping, vibration).

The momentary state of a given system with respect to the erosion and shattering intensities can be quantified by an "erosion index." $I_E(t)$, which is defined as follows (Popplewell and Peleg, 1989b):

$$I_E = \frac{[W(t) - W_o][X_{mc}(0) - X_{mf}]}{(1 - W_o)[X_{mc}(0) - X_{mc}(t)]} \qquad (6)$$

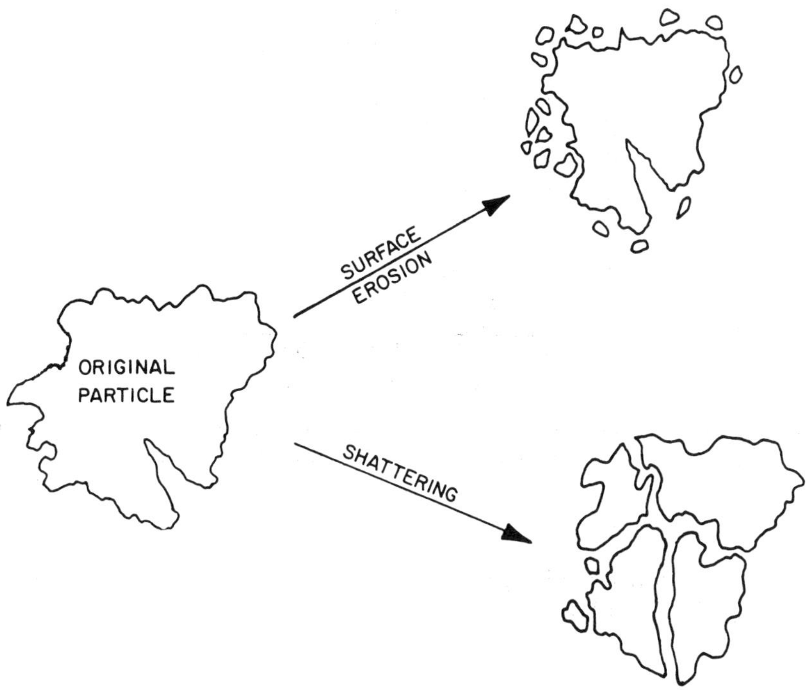

Fig. 12.3 Schematic view of the two principal mechanisms of particulates disintegration. (*From Popplewell et al., 1989. Reproduced with permission of the American Institute of Chemical Engineers.*)

where W(t) and W_o are the momentary and initial weight fraction of the fines, $X_{mc}(o)$ and $X_{mc}(t)$ the initial and momentary modes of the initial coarse fraction, and X_{mf} the mode of the fines. If $W_o = 0$ (no fines initially) and $X_{mf} \ll X_{mc}(0)$, Eq. (6) is reduced to

$$I_E = \frac{W(t)X_{mc}(0)}{X_{mc}(0) - X_{mc}(t)} \tag{7}$$

This dimensionless index basically represents the ratio between the amount of fines produced and the decrease in the mode of the original particle size. Thus, if absolutely no fines are produced, i.e., only shattering occurs, $I_E = 0$. The more prominent the erosion, the higher is the value of I_E, and when $I_E > 1$, erosion is the dominant mechanism. It is conceivable that in the progress of a disintegration process, the dominancy of shattering and erosion can be reversed. This was indeed ob-

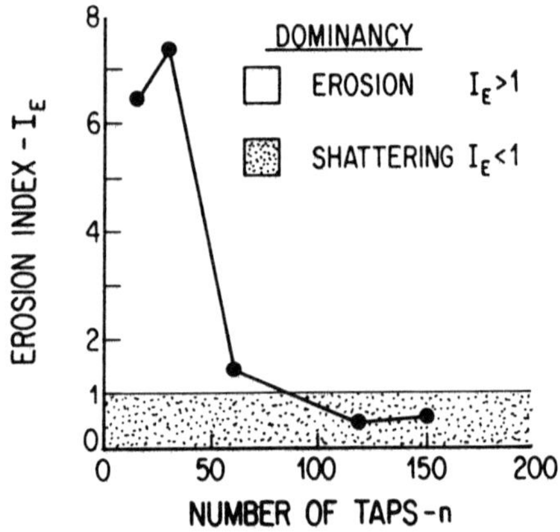

Fig. 12.4 Erosion index vs. number of tappings in the attrition of agglomerated instant coffee. Note the shift in dominancy from erosion to shattering as the particles become smaller.

served in the attrition of agglomerated instant coffee subjected to tapping, as shown in Fig. 12.4.

Rate Models

The kinetics of grinding processes has frequently been described in terms of a "first-order" model based on the disappearance rate of the original particles (Austin, 1971; Austin et al., 1981). This model yields a single term exponential decay equation of the kind

$$M(t) = M_o \exp\left(-\frac{t}{T}\right) \tag{8}$$

or a semi-logarithmic relationship in the form of

$$\frac{\ln M(t)}{M_o} = -\frac{t}{T} \tag{9}$$

where M(t) is the momentary weight of the fraction having the original particle size, M_o the total weight, and T a time characteristic of the system.

It has been realized (Fig. 12.5) that deviations from this relationship become significant when the particle's initial size is large or after long grinding times (Grandy and Fuerstenau, 1970; Gardner and Rogers, 1971;

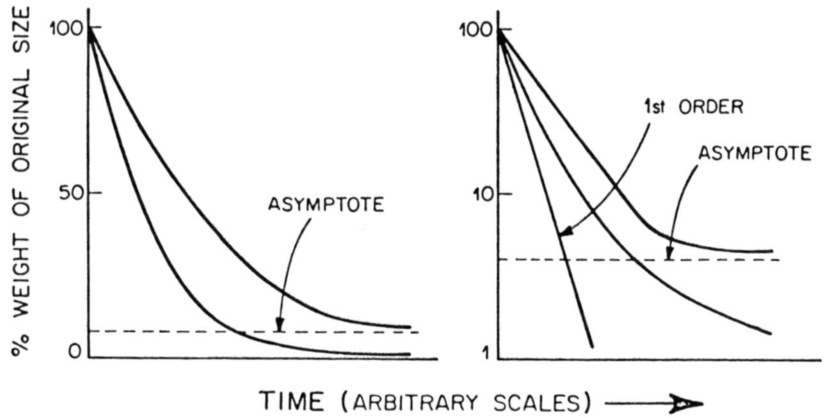

Fig. 12.5 Schematic views of an attrition curve. Left: linear scale; right: logarithmic scale. (*From Malave et al., 1988. Reproduced with permission of Taylor & Francis International Publishers.*)

Austin et al., 1981; Malave-Lopez et al., 1987). Malave-Lopez et al. (1987) tested three additional models to describe attrition data such as shown in Fig. 12.6: (1) a combination of a first and zero-order model,

$$\frac{M(t)}{M_o} = 1 - K\left[1 - \exp\left(-\frac{t}{T}\right)\right] \tag{10}$$

where T is the time characteristic and K a constant representing the weight fraction of disintegrated particles at $t \to \infty$; (2) a nonexponential model,

$$\frac{M_o - M(t)}{M_o} = \frac{t}{k_1 + k_2 t} \tag{11}$$

or in its linear form,

$$\frac{M_o}{M_o - M(t)}\, t = k_1 + k_2 t \tag{12}$$

where $1/k_2$ is the initial attrition rate and $1/k_2$ the asymptotic value of the disintegrated particles weight fraction; and (3) a two-term exponential model,

$$\frac{M(t)}{M_o} = a \exp\left(\frac{t}{T_1}\right) + (1 - a) \exp\left(\frac{t}{T_2}\right) \tag{13}$$

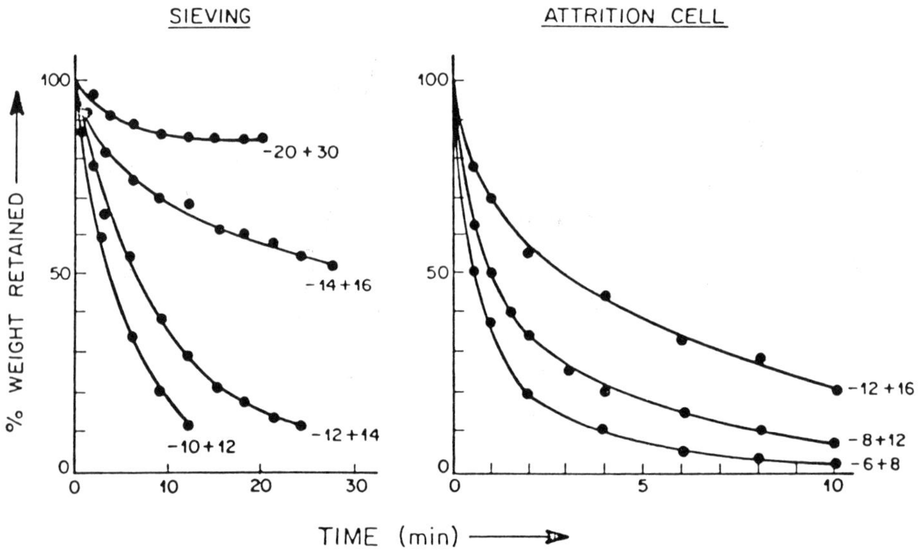

Fig. 12.6 Experimental attrition curves of instant coffee. Numbers in parentheses indicate the initial particle mesh size. (*From Malave et al., 1988. Reproduced with permission of Taylor & Francis International Publishers.*)

where a is a constant and T_1 and T_2 characteristic times. The applicability of these models was tested with breakage data for instant coffees (Figs. 12.7 and 12.8), where comparisons of their fit are also shown. For these coffees, the double exponential model was found to have the best fit. This may indicate that their size reduction is governed by at least two mechanisms which have distinct time characteristics. One is dominant in the disintegration of the large agglomerates in the absence of fines, and the other involves fines. Partial support for this conjecture can be found in the behavior of small agglomerates. Their breakage (Fig. 12.8) is not only characterized by low rate, as can be expected from mechanical strength considerations, but also by an almost perfect first-order kinetic. It should be stated, however, that statistical fit by itself does not establish the validity of a model, and it can be merely a reflection of the fact that most decay curves can be fitted by an exponential model having two or more terms.

Despite the fact that it contained only two experimentally determined constants as compared to the three of the double exponential model, the nonexponential model (Eq. 11, 12) was found to be almost as good as the double exponential model as judged by its statistical fit (Malave-Lopez

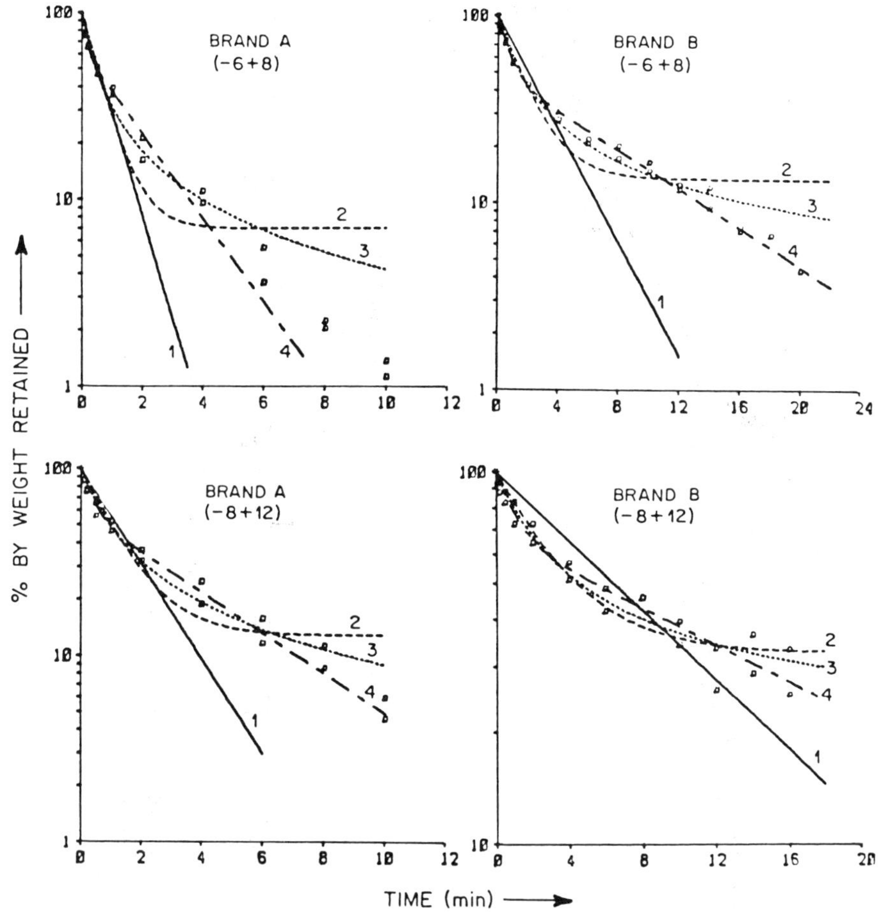

Fig. 12.7 The fit of various models to the attrition data of instant coffee. Models 1, 2, 3, and 4 are given in Eq. (8), (10), (11), and (13), respectively. Numbers in parentheses indicate the initial particle mesh size. (*From Malave et al. 1988. Reproduced with permission of Taylor & Francis International Publishers.*)

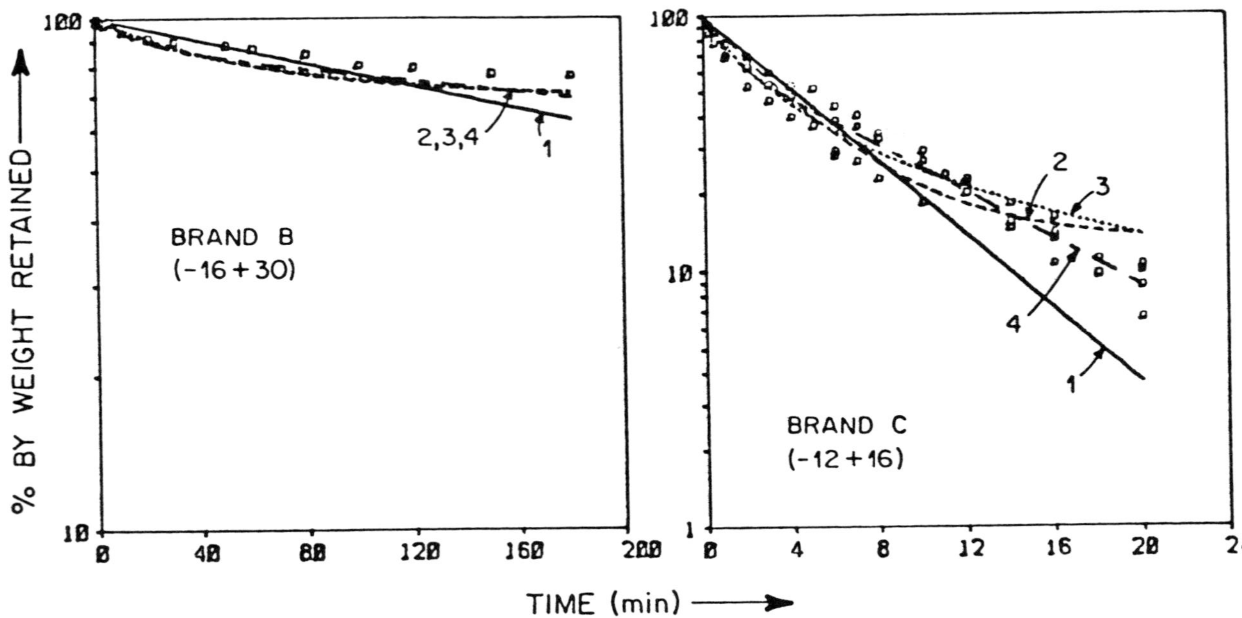

Fig. 12.8 Experimental attrition curve of a fine fraction of instant coffee and the fit of four models. Models 1, 2, 3 and 4 are given in Eq. (8), (10), (11) and (13), respectively. Note that the process appears as having a first order kinetics. Numbers in parentheses indicate the initial particle mesh size. (*From Malave et al. 1988. Reproduced with permission of Taylor & Francis International Publishers.*)

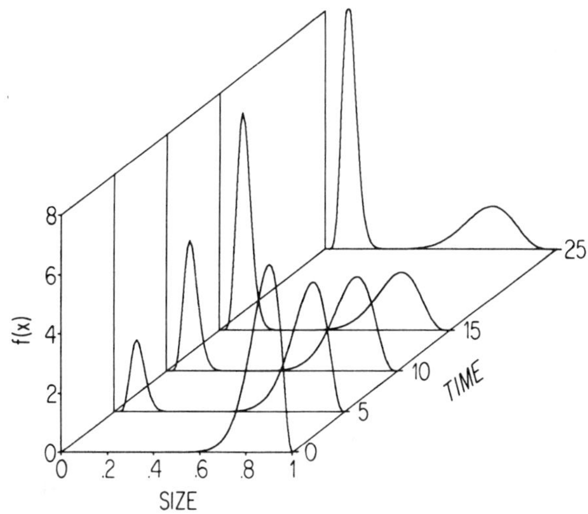

Fig. 12.9 Simulated particle size distributions in an erosion dominated disintegration process using Eq. (14) to (16). (*From Popplewell et al. 1989. Reproduced with permission of the American Institute of Chemical Engineers.*)

et al., 1987). This model facilitates comparisons between the behavior of different agglomerates or systems since it has only two parameters and these can be calculated by linear regression.

Simulation of Disintegration Processes with Empirical Models

Let us assume that a particle population having a fairly narrow size distribution is undergoing size reduction and that fines are produced in the process. Let us also assume that the mode and spread of the fines (group 1) are practically fixed and only their weight fraction increases with time. (This is not an unrealistic assumption as is evident from the behavior of agglomerated instant coffee shown in Fig. 12.2). Mathematically, this can be written as x_{ml} = const, σ_2^2 = const, and

$$W(t) = 1 - \exp\left(\frac{t}{T_{wl}}\right) \tag{14}$$

where T_{wl} is a characteristic time constant of the fines accumulation (Popplewell et al., 1989). Equation (14) implies that the rate of fines creation progressively decreases as the coarse fraction becomes smaller. It is obvious that as the original particles (group 2) disintegrate, their size decreases but their size spread increases. The increase, however, cannot continue indefinitely, and the spread will eventually decrease when their particle size approaches that of the fines. Mathematically, this can be expressed by (Popplewell et al., 1989)

$$x_{m2}(t) = x_{m1} + [x_{m2}(0) + x_{m1}] \exp\left(-\frac{t}{T_m}\right) \quad (15)$$

and (Peleg and Normand, 1986)

$$\sigma_2^2(t) = Bx_{m2}(1 - x_{m2}) \quad (16)$$

where $x_{m2}(0)$ is the dimensionless mode of the initial size, T_m a time constant representing the rate at which the original fraction mode decreases, and B a constant. (Built into these expressions is the assumption that the mode of the coarse particles, $x_{m2}(t)$, approaches that of the fines x_{m1} asymptotically.) If the assumptions on which Eqs. (14) to (16) are based are valid, $x_m(0)$, x_{m1}, and σ_1^2 are known or can be estimated then the whole disintegration process can be described by only three control parameters, namely T_w, T_m, and B. In fact, the difference between shattering and erosion-dominated processes will mainly be expressed by the

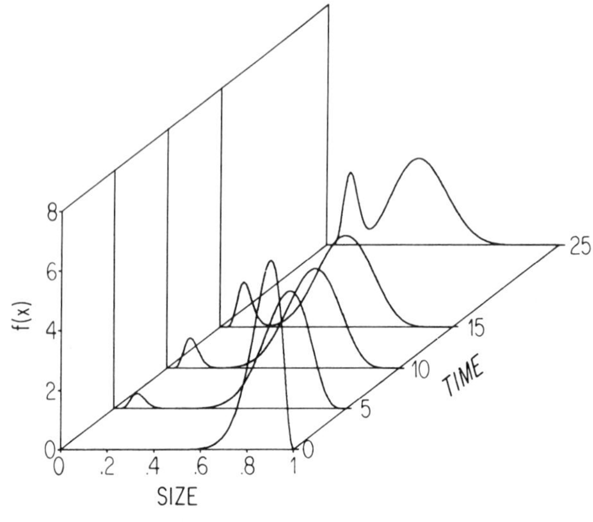

Fig. 12.10 Simulated particle size distributions in a shattering dominated disintegration process using Eq. (14) to (16). (*From Popplewell et al. 1989. Reproduced with permission of the American Institute of Chemical Engineers.*)

ratio between T_w and T_m. Thus, if $T_w \ll T_m$, the process is erosion dominated, and if $T_w \gg T_m$, shattering dominated. Examples of simulated curves using these equations are shown in Figs. 12.9 and 12.10. (The individual curves are mixed (double) beta distributions whose parameters a_1, a_2, m_1, and m_2 were calculated from the modes and variance of the fine and coarse fractions.)

Models Based on Breakage Probabilities and Mass Balance

A schematic view of the various possible steps in a particle disintegration process is given in Fig. 12.11. It shows that the daughter particles can fall into the various size slots directly or after successive breakage of intermediate product. It should also be borne in mind that what is presented as the original particle can itself be a product of a disintegrating particle of a larger size, but for our discussion we assign the subscript 1 to the largest particle size and n to the smallest. The division of different

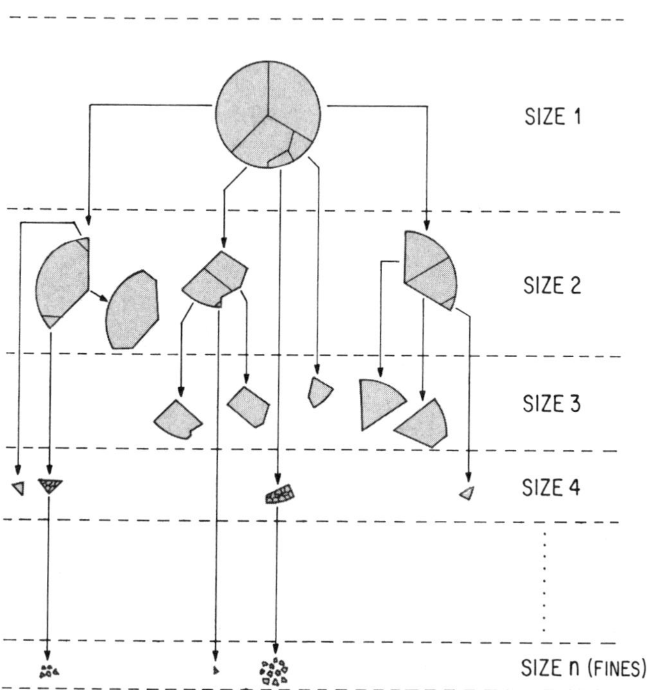

Fig. 12.11 Schematic view of various pathways in a particle size reduction.

size slots, as shown in the figure, ought to be done in light of two main considerations: that there will be enough slots of appropriate size to capture the system behavior, and that there will not be so many that the model is rendered unmanageable even with a computer. Now, treating the particle population as a whole we have to make two assumptions: that the probability of a particle of a given size disintegrating is a function of its size but is fixed in time, and that there is a set of specific probabilities that the daughter particles once formed will fall into the various size slots.

If we assume that the breakage in any size slot i except in the fines ($1 \le i < n$) is a first-order process, then we can write the following equations (Faulkner and Rimmer, 1983).
For the first size slot

$$\frac{dM_1}{dt} = -S_1M_1 \tag{17}$$

For slots i = 2 to n − 1

$$\frac{dM_i}{dt} = \sum_{j=1}^{i=1} S_j b_{ij} M_j - S_i M_i \tag{18}$$

For the fines slot n

$$\frac{dM_n}{dt} = \sum_{i=1}^{n-1} S_j b_{ij} M_j \tag{19}$$

where M_i is the mass fraction in size slot i, S_i or S_j the breakage rate constant characteristic of the particles occupying size slot i or j (units t^{-1}), and b_{ij} the probability that the disintegration products of particle in slot j will be of size i. This entails that

$$\sum_{i=j+1}^{n} b_{ij} = 1 \tag{20}$$

The first term on the right of Eq. (18) represents the particles that arrive in slot i as a result of disintegration of larger particles, and the second those lost as a result of disintegration and moving to slots of smaller size. The probabilities (b_{ij}) depend on the dominance of the breakage mechanism. Qualitatively they can be represented in a matrix of the kind shown in Fig. 12.12. The difference between the matrix of the erosion-dominated breakage and that found in older publications (e.g., Wei et al., 1977) is

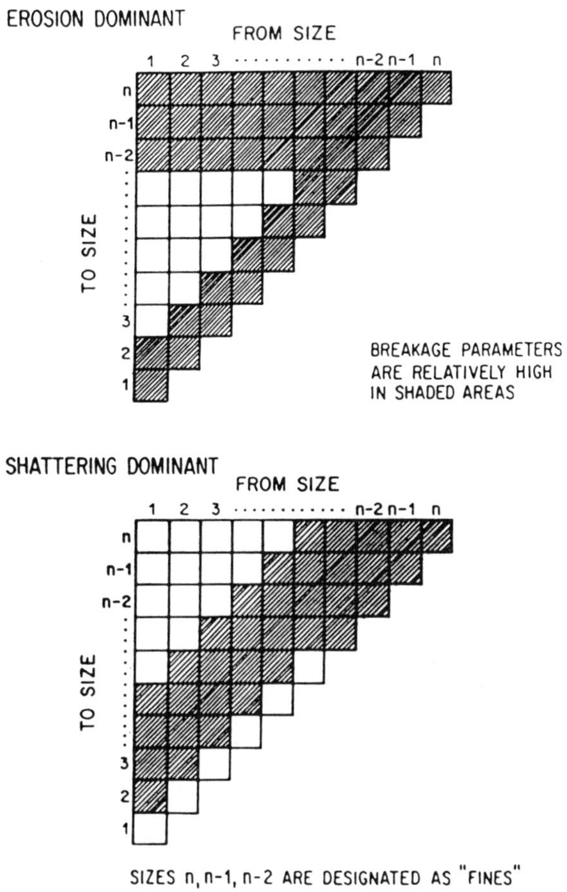

Fig. 12.12 Breakage matrices of an erosion and a shattering dominated size reduction. Shaded areas show sizes where daughter particles are likely to be. (*From Popplewell, 1989. Reproduced with permission.*)

that it allows for big particles to lose material to fines and still remain in their original size slot (Popplewell and Peleg, 1991a).

If the mathematical form of $b_i(x)$ and $S_i(x)$ is known or can be assumed, the size distribution after any desired time interval can be calculated by implementing mass balances over all the size slots simultaneously. The difference between shattering and erosion-dominated processes in such a case will be mainly expressed by the mathematical character of the expressions $b_{ij}(x)$; that is, whether the probability, of the formation of fines from the larger particles is high or low. It was recently shown that,

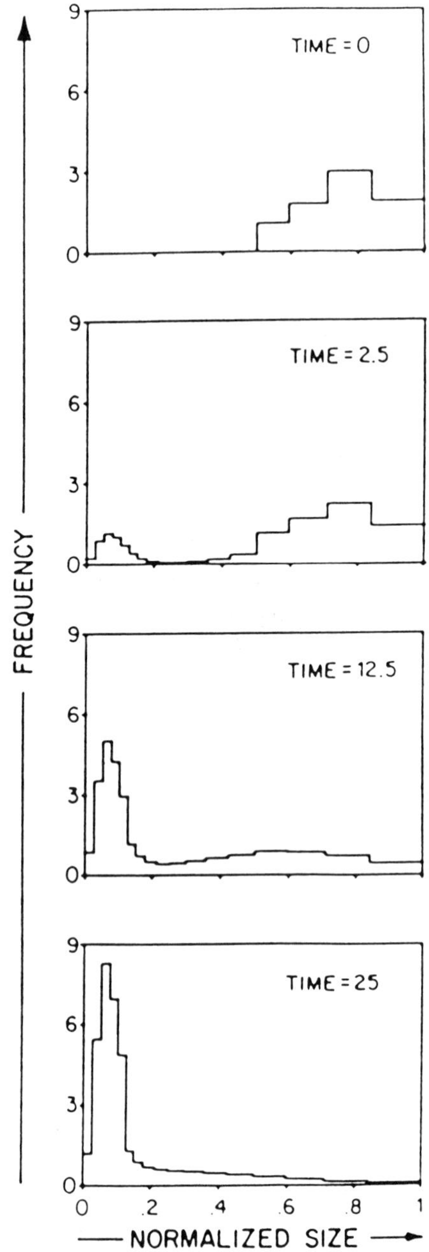

Fig. 12.13 Simulated sequence of particle size distributions during size reduction using a kinetic model with 17 size slots. (*From Popplewell, 1989. Reproduced with permission.*)

at least qualitatively, the mechanical attrition of instant coffees must be described by models in which the distribution of the probabilities, b(x), is bimodal in the large size slots and becomes unimodal in those of the small size. An example of a simulated attrition process based on these concepts and 17 size slots is given in Fig. 12.13.

SEGREGATION

Noninteractive ("Random") and Interactive ("Ordered") Mixtures

When particles of different species and/or size are mixed, they can form various types of arrays. Three types are schematically shown in Fig. 12.14. If the particles do not interact, each theoretically can occupy any unoccupied position in space. Therefore such mixtures have been called "random." When such a powder mixture is in motion, as in discharge, conveying, or vibration, the particles can change their relative position. This mobility is the main cause of segregation. In contrast, there are mixtures in which the fines are physically bound to the surface of the large particles (e.g., a food colorant adsorbed on crystalline sucrose or

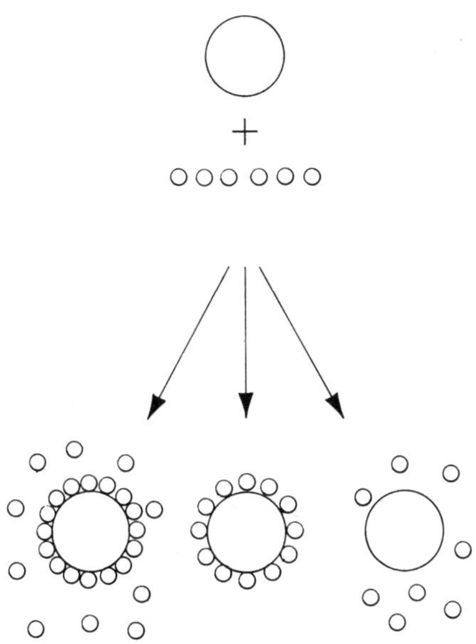

Fig. 12.14 Schematic view of non-interactive ("random"), interactive ("ordered") and noninteractive-interactive ("random-ordered") mixtures.

citric acid in a dry beverage mix). In such a case, the interaction results in a structure or "order." If the fines are not present in excess, their mobility is drastically curtailed, and they remain attached to their host particles even when the powder is in motion [Barbosa et al. (1985a)]. Consequently, such powders usually do not segregate. The situation is different when the fines are present in excess and the surface of large particles is saturated. In such a case, the excess fine particles form a noninteractive, or "random," mixture with the aggregates. Since their mobility remains intact, segregation can take place in the same manner as in a regular, noninteractive mixture.

Segregation is mainly the result of differences in particle size and, to a much lesser extent, of differences in shape or density (Williams, 1976). It takes place whenever the mixture is in motion, and it can occur even in mixers (Fan et al., 1990). The segregation pattern and intensity depend not only on the mixture type and composition, but also on the system geometry (Johanson, 1978). For example, the pattern can be different if the powder is discharged from a conveyor to a bin or subjected to vitration in a sealed container, and proper equipment design can reduce segregation of powder mixtures.

In several food systems, notably breakfast cereals and instant coffee, fines are not an ingredient, but the product of mechanical attrition. Once formed, however, the system becomes a noninteractive mixture and its segregation is regulated by the same factors that affect mixtures of particles with different composition.

Segregation Mechanisms

A schematic, and to a great extent exaggerated, view of the particles' positions in a powder mixture sample subjected to tapping is shown in Fig. 12.15. When the bed is lifted, the space below the large particles is filled with the smaller particles, and after the return the big particle finds itself in a slightly elevated position (see Fig. 12.15). After successive cy-

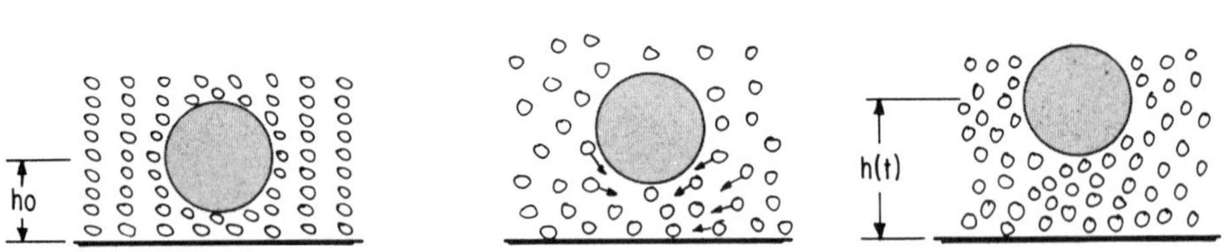

Fig. 12.15 Schematic view of the change in the relative position of the small and large particles during a mixture segregation.

cles, the big particle will appear to have climbed on the smaller particle to reach the top. The same is observed when the bed is subjected to vibration; the smaller particles always move to the bottom and the big ones to the top, even when their density is much higher than that of the smaller particles. The process has been simulated on a computer, albeit in two dimensions only, using the Monte Carlo method with the same result (Rosato et al., 1986).

When powders are discharged into bins, or from one conveyor to another, the system geometry is different and the rolling properties of the components can also be a factor. The pattern, i.e., where the segregated fines and coarse fractions are found, is to some extent predictable, and techniques to minimize segregation are available (Johanson, 1978).

In some cases, the smaller particles are so small that they form dust. Dusts, however, are treated as a separate class of particulates, and are not discussed in this work.

Segregation Indices

Segregation, or demixing, can be considered as the opposite of mixing, or mixing in reverse. In principle, therefore, any measure of the degree of mixedness can also serve as a measure of the degree of segregation. The problem of how to effectively assess the extent of mixing is an old one, and numerous mixing indices have been proposed as means of expressing it in quantitative terms (Weidenbaum, 1958; Orr, 1966; Schofield, 1976).

Most such indices are based on the deviation of the measured concentration of components in different locales from that predicted for a statistically uniform distribution in the system (e.g., a mixer or a container). Examples of such indices, I_M, are (Schofield, 1976)

$$I_M = \frac{\sigma_r}{S} \tag{21}$$

$$I_M = \frac{(\sigma_o - S)}{(\sigma_o - \sigma_r)} \tag{22}$$

$$I_M = \frac{(\sigma_o^2 - S^2)}{(\sigma_o^2 - \sigma_r^2)} \tag{23}$$

where σ_o is the theoretical standard deviation in an unmixed state, σ_r the theoretical standard deviation in a random mixture, and S the observed standard deviation between samples. Or, for a split cell of n rings (Olsen and Rippie, 1964)

$$I_M = \frac{\sum_{i=1}^{n} W_i(C_i - C_o)}{\sum_{i=1}^{n} W_i} \tag{24}$$

where W_i is the powder weight in ring i, C_i the concentration of an ingredient in that ring, and C_o its average concentration in the cell. The use of the index expressed by Eq. (24) in food was reported by Barbosa et al. (1985b), who measured the segregation intensity or sugar/starch mixtures.

A specific test and index for the segregation intensity of binary powder mixtures was developed by Williams et al. (Williams and Shield, 1967; Williams, 1976). They are based on tapping or vibrating a mixture sample in a split cylindrical cell and determining the weight of the coarse and fine fractions in the top and bottom halves of the cell after a given time or a set of time intervals. The segregation coefficient, C_s, was defined as

$$C_s = \frac{(W_{ct} - W_{cb})}{(W_{ct} + W_{cb})} \tag{25}$$

where W_{ct} and W_{cb} are the weight fractions of the coarse material in the top and bottom halves of the cell, respectively. In comparison with statistical mixing indices, William's segregation coefficient is simpler mathematically, and it can be determined experimentally more rapidly and with less effort. It was also found very effective in assessing, quantitatively, the effects of size differences and time on the segregation intensity of mixtures. Because of the way C_s is defined and determined, it is not sensitive in situations or systems where all the fines are concentrated in the bottom half of the cell. Such a situation can occur when the fines have a low overall concentration. An alternative segregation index, I_s, was recently proposed by Popplewell et al. (1989). It is defined by

$$I_s = \frac{(A - C_o)}{(1 - C_o)} \tag{26}$$

where A is the fines weight fraction at the very bottom of the cell and C_o the overall fines weight fraction in the cell. The magnitude of A can be determined experimentally from the fines concentration profile along the

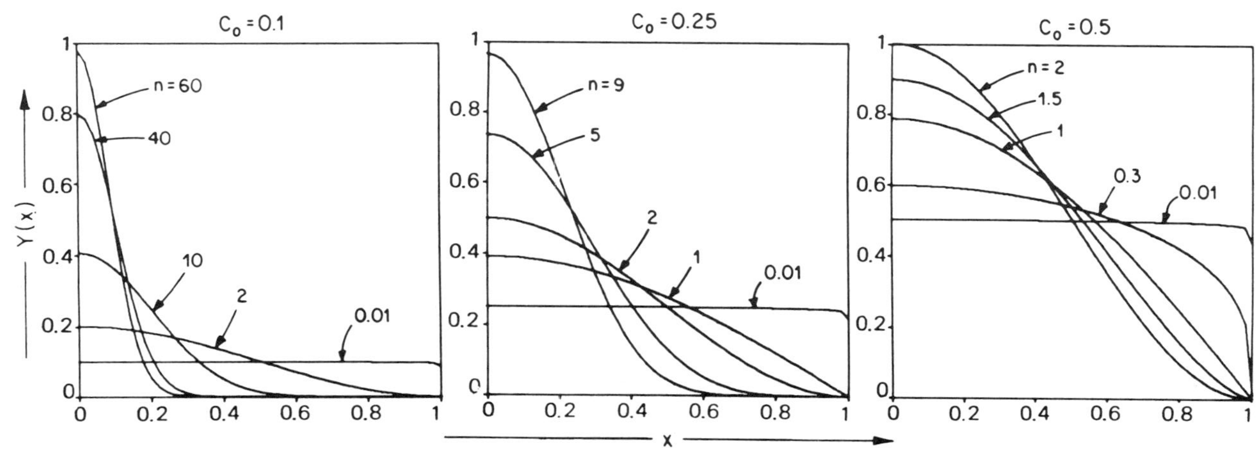

Fig. 12.16 Simulated fines concentration profiles at successive stages of shaking in a segregation cell. C_o is the fines overall weight fraction, x the normalized (dimensionless) cell height and Y(x) the fines weight fraction at a distance x from the bottom. The function used to generate the concentration profile is $Y(x) = A\cos^n(\pi x/2)$ where $A = C_o / \int_0^1 \cos^n(\pi x/2)dx$. A higher value of n represents a more advanced stage of segregation. (*From Popplewell et al. 1989. Reproduced with permission of Elsevier Sequoia S.A.*).

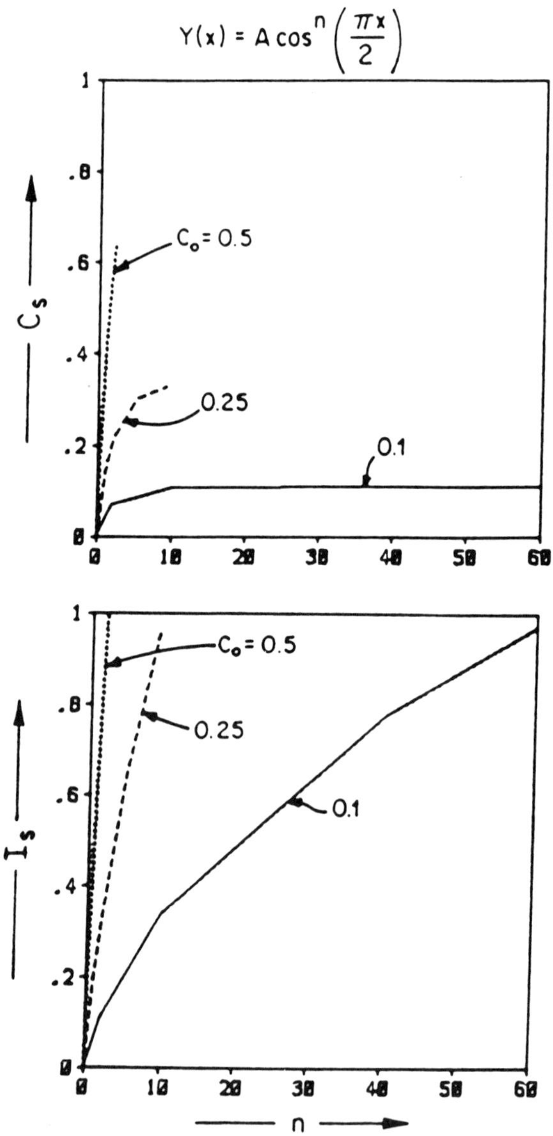

Fig. 12.17 The corresponding change in the magnitudes of C_s and I_s of the concentration profiles shown in Fig. 12.16.

cell or from the fines concentration in a narrow ring at the cell bottom. Like C_s, its theoretical range is from zero (no segregation at all) to one (complete segregation). It was shown through computer-generated fines concentration files, for different overall fines concentrations, that I_s is a more sensitive segregation index than C_s, especially in advanced stages when all or most of the fines have already reached the bottom half of the cell.

Since powders subjected to vibration or tapping may undergo compaction and since segregation can result in a density, changes along the cell, both indices, as well as the mixing indices based on statistical parameters, may give a somewhat distorted view of the exact state of the phenomenon in the cell. When compaction and density differences are taken into account, there are 24 distinct cases that ought to be treated and solved separately (Popplewell and Peleg, 1991b). Examples of cases include the compaction rate exceeding the segregation rate or vice versa as well as when the fines particles density is higher than that of the coarse particles or vice versa. It appears, however, that if density and volume reduction effects are not overwhelming, the advantage of the two segregation indices, C_s and I_s, achieved by their simplicity outweighs the importance of the errors introduced by not taking density changes into account. Simulated segregation profiles and the corresponding changes in C_s and I_s are shown in Figs. 12.16 and 12.17. It should be added that when real particulates are tested, the resultant scatter may be quite large and the profiles can be fitted by more than one model (Popplewell et al., 1989).

RELATION TO OTHER PHYSICOCHEMICAL PHENOMENA IN FOODS

Applicability of the Models to Other Food Systems

The previous parts of this chapter addressed the mechanical disintegration and segregation of food particulates at the bulk level. It was shown that concepts and models that are usually employed to quantify physicochemical phenomena governed by events at the molecular level can also be used to describe the behavior of particle assemblies. This is especially the case with respect to particle breakage. Size reduction can be treated as a process of first- or higher-order kinetics and treated by the same type of mathematical models that are used to describe the progress of chemical reactions. The bimodal size distributions and the existence of two major breakage mechanisms also have their equivalent in food chemistry. For example, enzymatic degradation of biopolymers can proceed by random cleavage of the molecule chain (''shattering''), removal of end units (''erosion''), or both. The resulting changes in the molecular weight distribution will follow a pattern very similar to that described in Figs. 12.9 and 12.10 (Popplewell et al., 1989).

Models based on population dynamics are a powerful tool to probe into the kinetics of processes that involve units whose number, size, and properties change with time, but because of their complexity they have not been frequently used in food research. The process of emulsification is a good example of where models of this kind are applicable and useful (Schubert and Ambruster, 1990). If in Fig. 12.11 the solid particles are replaced by oil droplets, the picture will not change qualitatively, and the same applies to Figs. 12.12 and 12.13. What will be different is the size and time scales and the corresponding breakage probability and rate parameters in Eqs. (18) to (20), which are regulated by different physicochemical phenomena. The same is true for emulsion coalescence where the same models are applicable but with the time direction reversed.

In principle, computer-simulated segregation patterns of the kind shown in Fig. 12.16 can also be used to describe the behavior of other food systems. Examples are creaming of unstable emulsions (with the x scale reversed) or sedimentation in cloudy beverages that contain fruit pulp. At least, theoretically, methods or indices of the type proposed for segregation intensity can also be used to quantify the severity of the separation problem in such systems. The spacial distribution pattern in such cases would more probably be determined on the basis of optical density rather than direct particle size measurements, but the concentration profiles themselves can be very similar.

Physicochemical Factors that Affect Breakage and Segregation

Under static conditions, that is, without mechanical energy input and motion, breakage and segregation do not occur. Thus, much of what has been described in this chapter is related to how these processes can be quantified once they have been induced by external factors, or how a given powder's tendency to disintegrate or segregate can be assessed. This was done by focusing on the mathematical structure of models or equations that can describe the behavior, not on the factors that affect the magnitude of their parameters or constants.

The kinetics of both disintegration and segregation processes, as previously mentioned, is determined by the mechanical history of the powder and by the particles physical properties. These include geometric features, such as size and shape, and mechanical properties, such as brittleness (the tendency to fail after small deformation) and strength (the ability to sustain stress). The latter are not only determined by the material of which the particle is made but also by its geometric structure (e.g., dry potato chips or corn flakes are fragile because they are thin and not because they are made of mechanically weak materials).

The properties of food particulates are strongly affected by moisture mainly through plasticization of their material or of their surface. The plasticization phenomenon and its association with glass transition are discussed in much detail in other chapters of this volume. Since plasticization means loss of brittleness, particles in such a state are less fragile. The mechanical properties of food particulates, however, e.g., extruded puffed particulates, need not be controlled exclusively through regulating the moisture contents. Plasticizing agents, glucose in starch, for example, can have an equivalent effect at low moisture levels (Ollett et al., 1991). Thus it is in principle possible to improve particulates' mechanical stability through controlled plasticization without compromising their biological or biochemical stability. The question how such a modification of the formulation will affect other quality attributes, e.g., taste or crispness, is outside the scope of this discussion. For more information see Katz and Labuza (1981) and Moreyra and Peleg (1981). In terms of the breakage models this will be manifested by smaller disintegration rate constants and a lesser tendency to produce fines.

Dissolution or plasticization of a particulate's surface as a result of moisture sorption or condensation, or melting, as in fat-containing powders, decreases surface roughness and considerably increases the powder cohesiveness by producing liquid bridges (Schubert, 1987). A smooth surface is less likely to provide sites for the initiation of fracture, another reason for the lesser tendency of "wet" particles to undergo brittle failure. More important, however, is the effect of interparticle cohesion on the powder bulk behavior. The active surface attracts the fines, thus producing "ordered mixtures" that do not segregate unless the fines are in great excess (Barbosa-Canovas et al., 1985a). Furthermore, since in cohesive powders the mobility of *all* the particles is greatly reduced, segregation in such a system is hardly a problem (Barbosa-Canovas et al., 1985b). But, as previously mentioned, the ability to control mechanical stability through moisture and texture modifications is limited by organoleptic and other considerations, such as the risk of caking (Peleg, 1983). It appears, therefore, that breakage and segregation problems are more effectively solved by control of the physical processes (e.g., maintaining a narrow size distribution to avoid segregation or devising a mixing order that will produce and stabilize ordered mixtures) and minimizing abuse by improved designs of the handling systems and/or the package.

ACKNOWLEDGMENTS

A significant part of this chapter is based on work supported by the Particulate and Multiphase Processes Program of the National Science Foundation (Grants CPE 8206765 and CBT 8520370). The author expresses his thanks to the National Science Foundation and the Massachusetts Agricultural Experiment Station for support, to Mr. R. J. Grant for his graphic aid, and to Professor H. G. Schwartzberg for valuable suggestions.

REFERENCES

Allen, T. 1981. *Particle Size Measurement,* 3rd ed. Chapman and Hall, London.

Austin, L. G. 1971/72. A review introduction to the mathematical description of grinding as a rate process. *Powder Technol.* 5: 1.

Austin, L. G., Bagga, P., and Celik, M. 1981. Breakage properties of some materials in a laboratory ball mill. *Powder Technol.* 28: 235.

Barbosa-Canovas, G. V., Rufner, R., and Peleg, M. 1985a. Micro-structure of selected binary food powder mixtures. *J. Food Sci.* 50: 473.

Barbosa-Canovas, G. V., Malave-Lopez, J., and Peleg, M. 1985b. Segregation in good powders. *Biotechnol. Prog.* 1: 140.

Barrett, A. M., Normand, M. D., and Peleg, M. 1991. A 'log-beta' vs the log-normal distribution for particle populations with a wide finite size range. *Powder Technol.* 66: 195.

Coulson, J. M. and Richardson, J. F. 1968. *Chemical Engineering.* Pergamon Press, Oxford.

Everitt, B. S. and Hand, D. J. 1981. *Finite Mixture Distributions.* Chapman and Hall, London.

Fan, L. T., Chen, Y. M., and Lai, F. S. 1990. Recent developments in solid mixing. *Powder Technol.* 61: 255.

Faulkner, B. P. and Rimmer, H. W. 1983. Size reduction. In *Encyclopedia of Chemical Technology,* 3rd ed., pp. 132–162. John Wiley and Sons, New York.

Gardner, R. P. and Rogers, R. S. 1975. A two component mechanistic approach for the comminution of material that exhibits heterogenous breakage characteristics. *Powder Technol.* 12: 247.

Geankoplis, C. J. 1983. *Transport Processes and Unit Operations.* Allyn and Bacon, Boston.

Grandy, G. A. and Fuerstenau, D. W. 1970. Simulation of non-linear grinding systems, Rodmill grinding, *AIME Trans.* 247: 348.

Harwood, C. F. 1977. Powder segregation due to vibration. *Powder Technol.* 16: 51.

Herdan, G. 1960. *Small Particle Statistics.* Butterworths, London.

Johanson, J. R. 1978. Particles segregation and what to do about it. *Chem. Eng.* 85: 183.

Katz, E. E. and Labuza, T. P. 1981. Effect of water activity on the sensory crispness and mechanical deformation of snack food products. *J. Food Sci.* 146: 403.

Lai, F., Hersey, J. and Staniforth, J. N. 1981. Segregation and mixing of fine particles in an ordered mixture. *Powder Technol.* 28: 17.

Loncin, M. and Merson, R. L. 1979. *Food Engineering—Principles and Selected Applications.* Academic Press, New York.

Malave-Lopez, J. and Peleg, M. 1986a. Patterns of size distribution changes during the attrition of instant coffees. *J. Food Sci.* 51: 691.

Malave-Lopez, J. and Peleg, M. 1986b. Mechanical attrition rate measurements in agglomerated instant coffees. *J. Food Sci.* 51: 687.

Malave-Lopez, J., Barbosa-Canovas, G. V., Peleg, M., and Zemelman, V. 1987. Kinetic models for the attrition of instant coffee. In *Particulate and Multiphase Processes,* T. Ariman and T. N. Vezruglu (Ed.), pp. 429–436. Hemisphere Pub., New York.

McCabe, W. L., Smith, J. C., and Harriott, P. 1985. *Unit Operations of Chemical Engineering.* McGraw-Hill, New York.

Menacho, M. 1986. Some solutions for the kinetics of combined fracture and abrasion breakage. *Powder Technol.* 49: 87.

Moreyra, R. and Peleg, M. 1981. Effect of equilibrium water activity on the bulk properties of selected food powders. *J. Food Sci.* 46: 1918.

Ollett, A. L., Parker, R., and Smith, A. C. 1991. Deformation and fracture behavior of wheat starch plasticized with glucose and water. *J. Mater. Sci.* 26: 351.

Orr, C. Jr. 1966. *Particulate Technology.* Macmillan, New York.

Olsen, J. L. and Rippie, E. G. 1964. Segregation kinetics of particulate solid systems. I. Influence of particle size and particle size distribution. *J. Pharm. Sci.* 53: 147.

Peleg, M. 1983. Physical properties of food powders. In *Physical Properties of Food,* M. Peleg and E. B. Bagley (Ed.), pp. 293–323. AVI, Westport, CT.

Peleg, M. and Normand, M. D. 1986. Simulation of size reduction and enlargement processes by a modified version of the beta distribution function. *AIChE J.* 32: 1928.

Popplewell, L. M. 1989. Description and modelling of disintegrating and segregating particle populations. Ph.D. thesis, University of Massachusetts, Amherst, MA.

Popplewell, L. M. and Peleg, M. 1989a. Theoretical comparison of two segregation indices for binary powder mixtures. *Powder Technol.* 58: 55.

Popplewell, L. M. and Peleg, M. 1989b. An erosion index to characterize fine production in size reduction processes. *Powder Technol.* 58: 145.

Popplewell, L. M. and Peleg, M. 1991a. A kinetic model for particulates disintegration processes that result in bimodal size distributions. *Powder Technol.* (in press).

Popplewell, L. M. and Peleg, M. 1991b. On the segregation of compressible binary mixtures subjected to tapping. *Powder Technol.* 67: 21.

Popplewell, L. M., Campanella, O. H. and Peleg, M. 1988. Description of normal, log-normal and Rosin-Rammler particle populations by a modified version of the beta distribution function. *Powder Technol.* 54: 119.

Popplewell, L. M., Campanella, O. H., and Peleg, M. 1989. Simulation of bimodal size distributions in aggregation and disintegration processes. *Chem. Eng. Prog.* 85: 56.

Rippie, E. G., Olsen, J. L., and Faiman, M. D. 1964. Segregation kinetics of particulate solid systems II. Particle density-size interaction and wall effect. *J. Pharm. Sci.* 52: 1360.

Rosato, A., Prinz, F., Standburg, K. J., and Swendsen, R. 1986. Monte Carlo simulation of particulate matter segregation. *Powder Technol.* 49: 59.

Schofield, C. 1976. The definition and assessment of mixture quality in mixtures of particulate solids. *Powder Technol.* 15: 1969.

Schubert, H. 1987. Food particle technology. Part I. Properties of particles and particulate food systems. *J. Food Eng.* 6: 1.

Schubert, H. and Armbruster, H. 1990. Principles of processing and stability of emulsions. In *Engineering and Food,* Vol. 1, W. E. L. Spiess and H. Schubert (Ed.), pp. 185–206. Elsevier Applied Science, London.

Wei, J., Lee, W., and Kramleck, F. J. 1977. Catalyst attrition and deactivation in fluid catalytic cracking systems. *Chem. Eng. Sci.* 32: 1211.

Weidenbaum, S. S. 1958. Mixing of solids. In *Advances in Chemical Engineering,* Vol. 2, T. B. Dreu and J. W. Woopes, (Ed.). Academic Press, New York.

Williams, J. C. 1976. The segregation of particulate materials. *Powder Technol.* 15: 245.

Williams, J. C. and Shields, G. 1967. The segregation of granules in a vibrated bed. *Powder Technol.* 1: 134.

Yip, C. W. and Hersey, J. A. 1977. Segregation in ordered powder mixtures. *Powder Technol.* 16: 149.

Yu, A. B. and Standish, N. 1990. A study of particle size distributions. *Powder Technol.* 62: 101.

13

Mechanistic Basis of Rheological Behavior of Foods

Edward B. Bagley

Agricultural Research Service, U.S. Department of Agriculture
Peoria, Illinois

INTRODUCTION

The general term "food" encompasses a wide range of materials, homogeneous and heterogeneous, single-component and multicomponent, single-phase and multiphase. The rheological behavior of foods (the way they deform and flow) reflects this compositional and structural complexity. To understand and use rheological information it is essential to have a mechanistic basis for the interpretation and correlation of the experimental data. Such interpretation of behavior in terms of theoretical mechanisms provides the guidelines needed to make sense of observations, to relate behavior to composition and structure, to predict and to modify properties, and to compare one experimental method with another.

The mention of firm names or trade products does not imply that they are endorsed or recommended by the U.S. Department of Agriculture over other firms or similar products not mentioned.

Since foodstuffs usually involve natural biopolymeric materials such as starch and protein, it is useful to be able to draw on the extensive experimental and theoretical information on synthetic polymers available from both academic and industrial sources. Often biopolymers are present in filled or composite systems, as, for example, in doughs where starch granules act as solid fillers dispersed in a gluten matrix. Extensive information on similar dispersions or suspensions is available in the chemical and mechanical engineering literature and elsewhere. More ordered composite systems would include laminated brittle foods, structures not unrelated to laminated composites widely used in a variety of structural applications.

Much remains to be done by the food scientist and engineer to fully exploit the knowledge developed in the polymer engineering and science areas in characterizing, modifying, and controlling food properties. Foodstuffs also provide their own special problems including susceptibility to biological and enzymatic activity, strong ionic and hydrogen-bonding interactions, combinations of hydrophobic and hydrophilic materials. Water is an ubiquitous component of most food systems, and this presents special problems. Finally, foodstuffs can change irreversibly with temperature, thus obviating the possibility of using certain high polymer techniques, such as time/temperature superposition methods. These problems offer food scientists opportunities for new scientific and technological developments.

RHEOLOGICAL MEASUREMENTS

Experimental methodology for measurement of rheological properties has been discussed extensively in the literature. Relatively recent references would include Dealy (1982), Prentice (1984), and Collyer and Clegg (1988). These works also include some theoretical background relevant to the measurements.

RHEOLOGICAL SYSTEMS

Polymer Solutions and Melts

Since foodstuffs will often include as a component high molecular weight polymeric (macromolecular) material, the properties and behavior of these macromolecules must be understood to appreciate their effects on foodstuff properties. The properties of these macromolecules will depend in the first instance on temperature. Figure 13.1 shows schematically

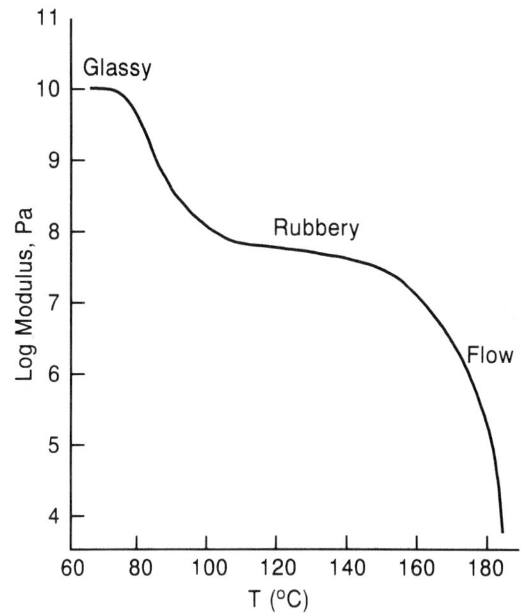

Fig. 13.1 Schematic representation of the modulus variation of polymeric materials with temperature.

the variation of the elastic modulus of a polymer with temperature, showing the change from a rigid brittle solid at low temperatures to rubbery material at intermediate temperatures. Finally, at high enough temperatures the polymer can be considered to be a melt, exhibiting flow. These transitions from brittle solid to rubber to melt are gradual, reflecting the changing free volume and increasing polymer segmental mobility with increasing temperatures. They do not represent clear-cut changes such as occur with a phase transition.

The melt flow region is of particular interest since the polymer viscosities measured in this region reveal considerable information on polymer size, structure, and melt organization. The viscosity will, in general, depend on shear rate, and typical viscosity/shear rate plots are shown schematically in Fig. 13.2. At low enough shear rates (curve A), the viscosity is independent of shear rate; this Newtonian limit is designated η_0. As the shear rate rises, the viscosity eventually starts to drop off with increasing shear rate, and for a certain shear rate region the behavior can be approximated by a power law relation, with A and C empirical constants:

$$\log(\eta) = A \log(\dot{\gamma}) + C \tag{1}$$

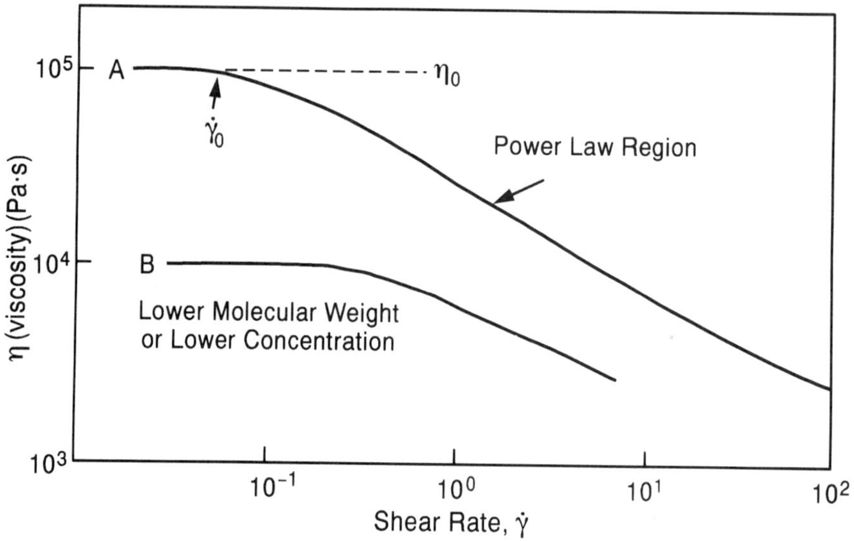

Fig. 13.2 Schematic representation of the variation of viscosity of polymer melts and solutions with shear rate.

At lower molecular weights similar curves are obtained, but are shifted to lower viscosities and show a Newtonian region which extends to higher shear rate levels. For polymers of low enough molecular weight, the viscosity is constant, that is, Newtonian, even to high shear rate levels. The actual details of the non-Newtonian portions of curves such as A of Fig. 13.2 are dependent not only on molecular weight but also on the details of the molecular weight distribution and the extent and type of chain branching (for polymers which are not linear). Further, the same general behavior is observed for polymer solutions. For polymer solutions the shift from curve A to curve B of Fig. 13.2 would reflect decreasing polymer concentration.

On Fig. 13.2, $\dot{\gamma}_0$ is the characteristic shear rate for the onset of non-Newtonian behavior. Often reduced viscosity plots are made, in which η/η_0 is plotted versus $\dot{\gamma}/\dot{\gamma}_0$. Such reduced plots, as noted by Graessley (1984) "depend only on molecular weight for polymer melts and concentrated solutions . . . data for narrow distribution samples of different polymer species, temperatures, concentrations, and chain lengths can be superimposed simply by shifts of scale along the viscosity and shear rate axes." This curve can be regarded as a reference base to compare with viscosity/shear rate plots of more complex polymer systems, for example, polymers of broader molecular weight distribution, chain branching, etc.

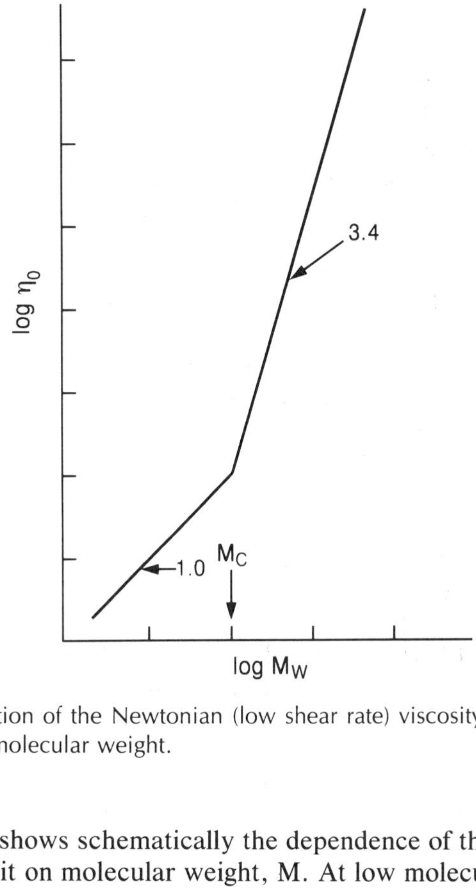

Fig. 13.3 Variation of the Newtonian (low shear rate) viscosity of polymers with weight average molecular weight.

Figure 13.3 shows schematically the dependence of the low shear rate Newtonian limit on molecular weight, M. At low molecular weights, the viscosity is proportional to the first power of molecular weight; above a critical value, Mc, the viscosity goes as the 3.4 power of M. This result is interpreted to mean that above a certain molecular weight polymer molecules become entangled and this effectively increases their molecular weight. The significance and implications of this concept for polymer flow and elastic behavior have been reviewed by Graessley (1974, 1984).

The relationships between the Newtonian viscosity levels, η_0, and molecular and compositional factors have long been of interest to polymer scientists and have been discussed in earlier textbooks (Middleman, 1968) as well as more recent works (Graessley, 1984). Thus, for a simple random coil molecule in solution or melt,

$$\eta_0 \propto \zeta_0 \left[\frac{So^2 Z \phi_1}{MV} \right]^2 \qquad (2)$$

where So^2 is the mean square radius of gyration of the unperturbed polymer, Z the number of monomer units in the polymer, V the specific volume, and ϕ_1 the volume fraction of the polymer in solution ($\phi_1 = 1$ for the melt). For M < Mc, a = 1; for M > Mc, a = 3.4. ζ_0 is a friction factor per monomer unit and has been estimated for a few polymers from monomeric diffusion coefficients (Graessley, 1984) or calculated from measured viscosities as described by Graessley (1968). Equation (2) provides a useful starting point for considering viscosity values. See also the discussion of Morris et al. (1981), where some of these concepts are applied to polysaccharide solutions, and the work of Launay et al. (1986).

Elastic moduli data also suggest the existence of critical molecular weights for entanglement. As discussed by Graessley (1984), the time-dependent relaxation modulus, G(t), when plotted as a function of time, goes from a high modulus level at very short times (10^{-10} sec) through a transition region where G(t) drops by 3 or 4 decades over times from 10^{-10} sec to 10^{-7} sec. (This is equivalent to changing temperature, as in Fig. 13.1). For low molecular weights, G(t) continues to drop steadily as the material flows. For higher molecular weights a plateau region exists, where G(t) is relatively constant over some decades of time (the rubbery region of Fig. 13.1). At long enough times, a terminal region is reached where the value of G(t) drops rapidly to low levels. The G(t) value at the plateau is independent of chain length, but the width of the plateau does depend on chain length. The existence of the plateau is attributed to chain entanglements.

The discussion above assumes the polymers take a random coil configuration. For some systems helix-coil transformations can occur, and under such circumstances, basic changes in type of rheological behavior will occur. Some polymers, xanthan gum, for instance, can behave like rigid rods, quite different in rheological detail from the random coil. The presence of charged groups in the polymer chair can have a profound affect on rheological behavior, as can the complexing that may occur in multicomponent systems (cf. Glass, 1989; Fanta and Christianson, 1991).

Finally, it must be understood that the brief discussion above of factors affecting polymer melt and solution behavior has not attempted to deal with some important, and relatively new, concepts which are proving valuable in furthering our understanding of polymer behavior. The interested reader will also find the review of scaling concepts in polymer physics (de Gennes, 1979) valuable. This reference includes discussion of solutions, melts, and gels as well as problems of phase separation. It should also be noted that in the last few years our understanding of polymer melt and solution flow has also been advanced by the development of the reptation model, where it is assumed that the polymer moves like a snake. This model of polymer flow and diffusion also deserves extensive discussion, but for this the reader is referred to the brief review of Larson (1988) and the references contained therein.

Dispersions/Suspensions

Many food systems consist of one phase dispersed or suspended in another, such as foams (air bubbles suspended in a liquid), emulsions (oil droplets in water), and doughs (starch granules in gluten). The suspended phase may consist of rigid particles or deformable particles. The rigid particle cases have been most extensively studied, and among the useful references for this field of work the review of concentrated suspensions by Gadala-Maria (1979) is valuable. While Gadala-Maria was particularly concerned with rigid particles, the work is of value to the food rheologist since it deals with concentrated systems of common occurrence in foods. Other relevant references would include Nielsen (1967) and Krieger (1983).

A mechanistic explanation as to how suspended particles will affect viscosity was given by Einstein. For dilute suspensions of rigid spheres the spheres increase the rate of energy dissipation during flow over that which would occur for the suspension solvent alone; the rigid spheres will increase viscosity. For dilute suspensions the ratio of the viscosity of the suspension, η_{sus}, to that of the pure solvent, η_0, is given by

$$\frac{\eta_{sus}}{\eta_0} = 1 + 2.5\phi \tag{3}$$

where ϕ is the volume fraction of the rigid spheres. In the limit as the volume fraction of spheres goes to zero, the intrinsic viscosity, $[\eta]$, is given as

$$[\eta] = \frac{\eta_{sus} - \eta_0}{\eta_0\phi} = 2.5 \tag{4}$$

and is equal to 2.5, independent of the size of the spheres. The equation is valid only to very small values of volume fraction, less than 0.02.

Many practical systems involve volume fractions of the order of 0.5, well beyond the limits under which Eq. (3) is valid. One approach to the treatment of such high loading levels was proposed by Mooney (1951), who used a mechanistic analysis to derive the relationship

$$\frac{\eta_{sus}}{\eta_0} = \exp\left[\frac{2.5\phi}{1 - k\phi}\right] \tag{5}$$

which gives the Einstein relation at low volume fraction and predicts infinite viscosity at a volume fraction of 1/k.

In considering the behavior of rigid spheres it must be recognized that when the volume fraction is high enough the spheres will assume a close packed configuration which precludes the possibility of flow. Thus k is 1.35 for close packing of uniform spheres (Nielsen, 1967) corresponding to a critical volume fraction of 0.77. Such a system might be regarded as having a yield point, since flow will not occur in any normal way at this value of ϕ.

Experimentally, for certain dispersions of rigid spheres, raising the volume fraction to levels above 0.5 leads to some very interesting phenomena. As the particle packing gets large, the behavior becomes dilatant, that is, the viscosity increases with increasing shear rate or rate of flow. Then flow becomes discontinuous. In a series of papers Hoffman provided a detailed mechanistic explanation for these effects (Hoffman, 1972, 1974, 1982). He showed that the cause of the flow instability was that "layers of spheres, packed in two-dimensional hexagonal arrangement at low shear rates, break up into less ordered arrays of spheres" (Hoffman, 1974). Hoffman presents a very useful and very detailed analysis and experimental data on a variety of systems including potato starch suspensions.

Mechanistic analyses such as given by Hoffman are important in considering a range of flow problems for food systems, including the behavior of food systems in extruders. Often, though data may look complex, relatively simple explanations can throw considerable light on experimental observations. For instance, what is the effect of a particle size distribution on the flow of suspensions? The effects can be very large indeed and have practical implications. Gadala-Maria (1979) shows the relative viscosity of a bimodal suspension, which goes through a minimum when plotted against the volume % of small spheres in the total solids. The particle size ratio also plays a part, and the observed effects, in which the viscosity of the bimodal distribution may be a decade or more below the viscosity of a monomodal distribution of the same total volume fraction, can be of considerable practical value.

Suspensions of nonspherical particles have been described in the literature both theoretically and experimentally. These include flakes, acicular particles, rods, discs, and so on. Further, the suspended particles can be deformable, rather than rigid, as in flexible threads. [The very interesting behavior of rods, discs, threads, and droplets in flow has been beautifully reviewed by Goldsmith and Mason (1967).] Starch systems are of special interest as starch granules suspended in water can be rigid or deformable depending on the temperature of the suspension. Thus if corn or wheat starch is heated in the gelatinization temperature range, the viscosity is initially very low (Christianson and Bagley, 1983). As cooking time increases, the starch granules start to swell and the volume fraction occupied by the particles increases because of this swelling; as the particles start to crowd each other, the viscosity rises (Fig. 13.4). If the particles are still largely rigid, the suspension will show dilatant behavior, that is, the viscosity can increase with increasing shear rate (Fig. 13.4). However, as the granule softens so that it is deformable in shear, the suspension becomes shear thinning, that is, the viscosity drops with increasing shear rate. Comparing the viscosities of swollen starch granule suspensions, measured at room temperature after cooking at temperatures of 60, 65, 70, or 75°C, all viscosity data can be superimposed using only the volume fraction of the swollen starch as the critical variable (Fig. 13.5) (Bagley and Christianson, 1982, 1983; Christianson and Bagley, 1983).

One interesting point concerns the effect of filler on the elastic prop-

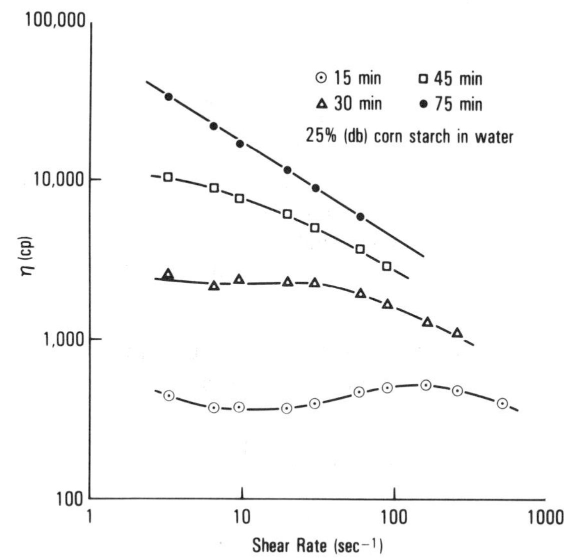

Fig. 13.4 Viscosity vs. shear rate for a 25% dispersion of cornstarch 60°C after cooking at 65°C for various times. After a short cook time (15 min, lowest curve), the starch granules have swollen enough to give a high viscosity but are still rigid enough to show dilatant behavior. As cook time increases, starch granule swelling increases (viscosity increases), but the dilatant behavior ceases, and at the longest cook time (75 min) the suspension of swollen, deformable granules shows typical shear thinning behavior. (*From Bagley and Christianson, 1983.*)

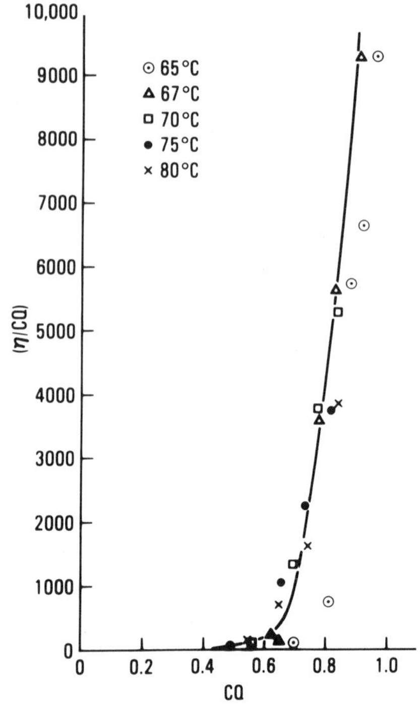

Fig. 13.5 Viscosity/Concentration data measured at 60°C for swollen starch suspensions prepared at different cook temperatures in the gelatinization range form a master curve, independent of cook temperature, when viscosity is given as a function of volume fraction (CQ). (*From Bagley and Christianson, 1982.*)

erties of a suspension. In general, similar equations hold for shear modulus and viscosity of dispersions (Nielsen, 1967), so that

$$\frac{\eta_{sus}}{\eta_0} = f(\phi) = \frac{G_{sus}}{G_0} \tag{6}$$

Since the ratio of viscosity to modulus gives a relaxation time, θ, then the ratio of these quantities for the filled and the unfilled system are the same, so that if θ is the relaxation time, then

$$\frac{\eta_{sus}}{G_{sus}} = \frac{\eta_0}{G_0} = \theta \tag{7}$$

Thus the addition of filler does not affect the relaxation time.

In special cases where certain molecular processes occur, such as crosslinking, the situation is not that simple. Starch/gluten mixtures were studied by Cavella et al. (1990), who comment that

> as the starch content increased, the dough elasticity decreased and at the same time viscous dissipation increased. This can be explained by considering that with increasing starch content, the volume concentration of proteins proportionally decreased and that, consequently, the network which they form after hydration will become, progressively, looser and therefore less elastic. By contrast, starch granules, which do not take part in the network formation enhance viscous dissipation during motion, which increased proportionally with increasing amounts of starch in the mixture.

Cavella et al. thus provided a mechanistic explanation for results that superficially did not meet expectation.

Application of the concepts of suspension rheology to polysaccharide and protein solutions has been discussed in considerably more detail by Morris et al. (1981), Launay et al. (1986), and Rha and Pradipasena (1986).

Crosslinked Systems

Many food systems can be classified as crosslinked systems. The crosslinks may be permanent covalent bonds, as they are in a vulcanized rubber, or the crosslinks may be temporary, generated either through molecular entanglements or through labile crosslinking bonds, which would include hydrogen bonds and ionic bonds. Crosslinked systems have been extensively studied, and one of the classic references describing the factors governing their behavior is Treloar (1975). There are, in addition, recent reviews of rubber elasticity (e.g., Heinrich et al., 1988).

For a material to exhibit rubberlike properties, Treloar enumerates three requirements that must be satisfied:

1. The presence of long-chain molecules with freely rotating links.
2. Weak secondary forces between the molecules.
3. Interlocking of the molecules at a few places to form a three-dimensional network.

The first of these conditions means that Brownian motion of the segments of the chain between crosslink points will ensure some average end-to-end distance between the crosslinks, which will depend on temperature, number of effective segments in the chain, and the length of a segment.

This first condition may be operative in a given system at sufficiently high temperature. At sufficiently low temperature, specifically below the glass transition temperature, the molecular motions of the segments become "frozen in," and the gel or rubber becomes a brittle, rigid solid. The influence of the glass transition on some food materials has been reviewed recently by Levine and Slade (1990).

The second criterion ensures that the molecules are exhibiting liquid-like behavior. This second condition can be violated in a given material at high enough deformations. Thus, when natural rubber is stretched, the molecular chains can become aligned and approach closely enough that crystallization can occur. This also serves to "freeze in" molecular motions, and the stretched, partially crystalline rubber becomes inelastic in the sense that further deformation of the rubber can only be achieved by large increases in applied stress, which soon reaches a level sufficient to fracture the polymer.

The third condition—that the chains be interlocked—ensures that irreversible deformation (flow) will not occur; the material will behave as a solid. A caveat is needed here, particularly for food systems, in which hydrogen bonding, ionic interactions, or other specific labile bonds may serve as crosslinking sites. These labile crosslinks can be of such a character that irreversible deformation can occur, so that the gel does, in fact, flow under the action of applied forces. This possibility must be kept in mind when examining stress/strain/time data for food systems.

In dealing with a specific gel or rubberlike material, the molecular factors affecting behavior must be kept in mind, as well as the fact that temperature is a critical parameter. Chain length and flexibility are significant factors. The presence of specific interactions must be considered. The extent of deformation of the sample in testing can change the response both quantitatively and qualitatively.

To a first approximation the response of a molecular network to applied forces or imposed strains is an entropic response—there is no change in the internal energy of the material on deformation. The work done is given by

$$W = -T\Delta S \tag{8}$$

which is given by

$$W = \frac{NkT}{2}[\lambda_1^2 + \lambda_2^2 + \lambda_3^2 - 3] \tag{9}$$

where λ_1, λ_2, λ_3 are the dimensions of the deformed sample whose initial dimensions were those of a unit cube. NkT = G, the shear modulus, and is related to the density and molecular weight between crosslinks, Mc, by

$$G = \frac{\rho RT}{Mc} \tag{10}$$

The modulus is proportional to the crosslink density; the higher the crosslink density, the more rigid the material becomes; for a given applied stress a smaller deformation occurs for a larger value of the modulus.

A more general approach to the development of relations between the stress and the strain in rubberlike materials (including viscoelastic liquids as well as solids) is to use the concepts of strain invariants, following Rivlin (see Treloar, 1975). For a general deformation of a unit cube to a solid of sides λ_1, λ_2, λ_3, the strain invariants are

$$I_1 = \lambda_1^2 + \lambda_2^2 + \lambda_3^2 \qquad (\text{Length})^2 \tag{11a}$$

$$I_2 = \lambda_1^2\lambda_2^2 + \lambda_2^2\lambda_3^2 + \lambda_3^2\lambda_1^2 \qquad (\text{Area})^2 \tag{11b}$$

$$I_3 = \lambda_1^2\lambda_2^2\lambda_3^2 \qquad (\text{Volume})^2 \tag{11c}$$

The significance of these strain invariants has been discussed by Kearsley (1989). Briefly, I_1 is an invariant of units a length square, I_2 refers to an area squared, and I_3 refers to a volume squared. If the volume of the material remains constant during a deformation, then $I_3 = 1$, and a particular strain energy function which experience has shown to be of special value is

$$W = C_1(I_1 - 3) + C_2(I_2 - 3) \tag{12}$$

The application of these concepts to the analysis of experimental results has been discussed extensively in the literature. Briefly, one advantage of having an expression for the strain energy function is that, for any given type of test deformation, simple shear, elongation, torsion, the relationship between stress and strain can be derived, and each type of experiment will yield the same values of the parameters. This approach has been used by Christianson et al. (1984) in examining a variety of gels systems using simple shear and torsion. Subsequently these shear and torsion results were compared with simple compression (Bagley et al. 1985). Using Eq. (12) (the Mooney-Rivlin relation, as it is sometimes called), values of the parameters ($C_1 + C_2$), which were in good agreement among themselves, were obtained from each experimental method.

In the simplest terms, the parameter C_1 is proportional to the crosslink density, while the interpretation of C_2 is more obscure. However, the basic relationship for the strain energy function serves as a starting point for the analysis of experimental rheological data. Further, the mechanistic treatment using the strain energy function brings into clear focus the mechanical interpretation of such data and outlines the directions to proceed if the data are more complex than the simple treatment outlined above. [See Treloar (1975, pp. 217–218) for a discussion of the need for care in drawing conclusions from agreement of results from one experimental procedure with the simple Mooney relationship.]

The advantage of having a mechanistic view of experimental data can be illustrated by two examples. The first is shown in Fig. 13.6, where two sets of uniaxial compression data for a gelatin gel are shown (Bagley et al., 1985). The data are plotted according to expectation from Eq. (12). The force data for the top curve of Fig. 13.6 were much higher than expected, and these high results are due to frictional effects occurring at the sample/platen interfaces. When this frictional effect is eliminated by using lubricated platens, the lower curve of Fig. 13.6 is obtained. This

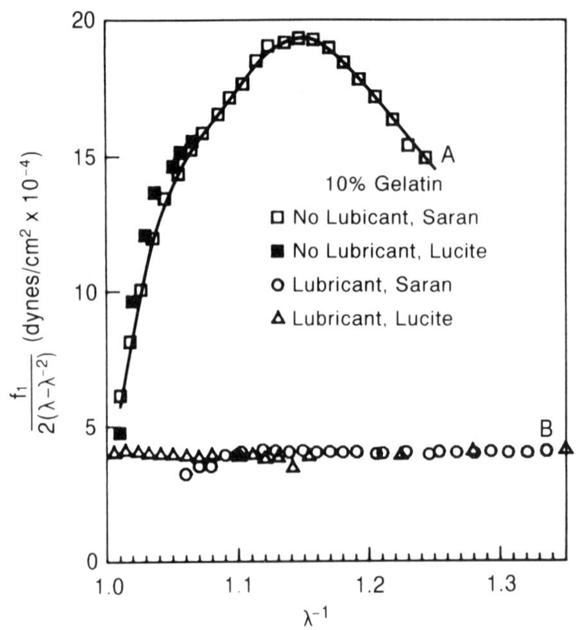

Fig. 13.6 Effect of lubrication on the force-deformation response of a 10% gelatin gel in uniaxial compression (*From Bagley et al., 1985.*)

shows that, in the absence of extraneous frictional effects, the gelatin is behaving in an ideal neo-Hookean way, with the slope of this lower curve being almost exactly zero (i.e., the C_2 term of Eq. (12) is zero). The three criteria mentioned above for ideal elastic behavior are undoubtedly met.

Other gel systems are not as ideal as gelatin, as is illustrated for starch gels in Figs. 13.7 and 13.8 (Christianson and Bagley, 1990). Figure 13.7 shows the raw data of stress versus a strain for starch gel. Two replicate cooks are shown with two replicate runs for each cook. These are shown to illustrate the reproducibility of the raw data and also to illustrate that the data in this form really do not tell us much about the reasons for the particular material behavior. Replotting according to Eq. (12) (Fig. 13.8) reveals an initial region at low deformations where the gels behave in a neo-Hookean fashion, with $C_2 = 0$. This permits the computation of an effective crosslink density from the intercept on the ordinate. Above a value of about 1.22, the material ceases to behave as a neo-Hookean material, due to the failure of one or more of the three conditions given above for ideal elastic behavior.

There is extensive data in the literature on behavior and properties of gels. Treloar devotes a chapter to the swelling phenomena which affect the properties of, for example, rubbers swollen with organic liquids. Hermansson (1988) reviews the gel structure of food biopolymers and shows dramatic photographs of structural factors, which clearly require careful

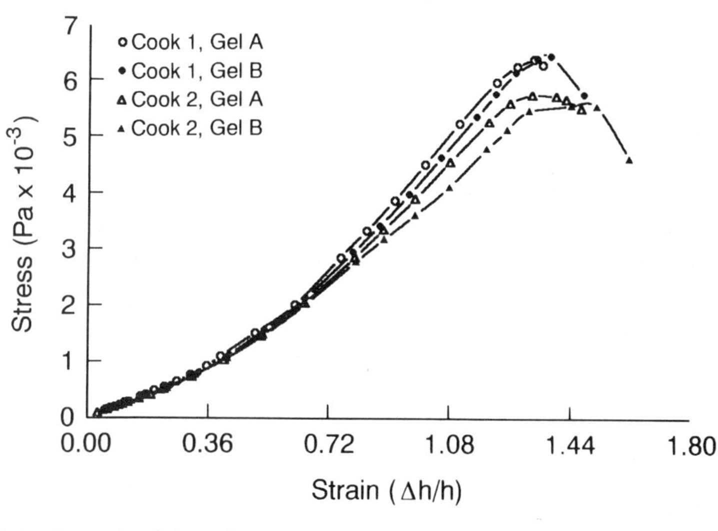

Fig. 13.7 Reproducibility of stress-strain curves for starch gels in lubricated uniaxial compression. (*From Christianson and Bagley, 1990.*)

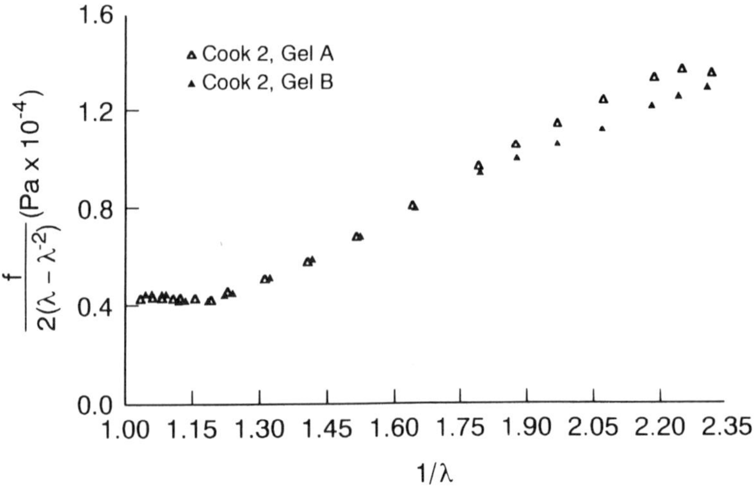

Fig. 13.8 Data of Fig. 13.7 replotted according to the Mooney-Rivlin relation, Eq. (12). At low deformation the gel behaves as a neo-Hookean material (zero slope).

consideration in applying the simple mechanistic concepts of gel behavior described above. Similarly, Brownsley and Morris (1988) discuss mixed and filled gels as models for foods. The effects of complexing between two polysaccharides and effects of fillers are discussed.

However, the critical point relevant to food systems was made by Hermansson, who notes, "In spite of all the work done we still know far too little about the microstructure of gels and the mechanism involved in gel formation, especially on the supramolecular level. Such knowledge is necessary in order to improve our understanding of relationships between the structure and other physical properties of food gels."

It is often convenient and conceptually useful to investigate model systems. Thus, Richardson et al. (1981) looked at the mechanical properties of filled gelatin gels. They used glass spheres, rods, and plates at various loadings and looked at the shear storage and loss moduli. Their analysis of the data started from the Einstein relation for the ratio of modulus of filled material divided by the modulus of the unfilled gel. Similarly, Ring and Stainsby (1982) dealt with filler reinforcement of starch, amylose, and gelatin gels using a variety of fillers. Information of this type, interpreting the properties of gels filled with inert fillers or fillers where something is known of the specific interactions, can be useful in providing base lines for the interpretation of the behavior of filled systems where complex interactions are operative.

Modeling for Data Correlation and Extrapolation

Mechanical models. Mechanical models have proven useful for describing the basic features of the observed rheological behavior of polymeric systems. There is an extensive older literature on the use of these models, for example Alfrey (1948) and more recently Nielsen (1974). By a judicious combination of springs (representing elastic effects described by moduli G_i) and dashpots (representing viscosities η_i), a wide range of behavior can be described. For example, a spring (modulus G_i) and a dashpot (η_i) in parallel will describe a time-dependent strain arising from applied stress governed by a relaxation time $\theta = \eta_i/G_i$. A number of these "Voigt" elements in series can be used to describe a material with a distribution of relaxation times. [The strain involved may be of several kinds: shear strain, torsional, compressional, etc. For large strain measurements a number of models are given in the literature. These have been summarized and discussed by Peleg (1984).]

For describing stress relaxation, a Maxwell element (a spring and dashpot in series) can be used with again the ratio of viscosity to modulus being a relaxation time.

For specific problems encountered in foods, particularly solid foods, which are rheologically nonlinear with response depending on deformation history, additional model elements can be used, as discussed by Peleg (1977a), and can be useful in treating fracture problems (Peleg, 1977b; Pollak and Peleg, 1980) and in computer-aided rheological characterization of solid foods (Miller et al., 1986).

The value of this model approach is that it provides a framework for correlation and comparison of data from measurements on complex materials. The weakness of mechanical models is the lack of direct coupling between the model elements and specific molecular or structural factors contributing to the material behavior. Model parameters are, as a general rule, to be considered as arbitrary curve fitting parameters, though none the less useful for that. For a more general approach to analysis of rheological data, a more mathematical approach via constitutive equations is needed.

Constitutive and semi-empirical models. Processing and quality control problems in the polymer industry over the years stimulated research into the rheological properties of polymeric systems, and this resulted in considerable progress in the understanding of these complex systems. In addition to a vast amount of experimental data, theoretical ideas were developed over the last 40 years, and the progress in developing constitutive relationships for polymer melts and solutions has been summarized by Larson (1988). Often, however, a given constitutive relation is complex, and approximate methods need to be used or semi-empirical equations developed.

A recent example of this latter approach is given in the work of Dickie and Kokini (1982), who used the Bird-Leider equation to model the flow behavior of a range of food materials including peanut butter, margarine, and dessert topping. Power law behavior is assumed and values of the power law parameters and time constants are obtained by fitting the data. In later work from the same laboratory, Dus and Kokini (1990) used the Bird-Carreau constitutive model to predict the nonlinear viscoelastic properties of a hard wheat flour dough. Five empirical constants including two time constants were extracted by them from the data. The time constants calculated differed by an order of magnitude, being 3800 sec and 50,000 sec. The long time constant would certainly imply some sort of cooperative process on a supermolecular (structural) scale for these doughs.

Other constitutive equations applied to food systems includes the K-BKZ integral model (often referred to in the literature as the BKZ model), used to interpret the stress relaxation data for doughs and explain the effects of testing rates (Bagley and Christianson, 1987). The Upper Convected Maxwell Model has been successfully applied to the correlation of biaxial extensional flow data (Bagley et al., 1988).

SUMMARY

Accumulation of rheological data on foodstuffs is easy. The critical issue is whether the data can be correlated and interpreted in molecular, structural, and processing terms, which will permit modification and control of properties to fit particular needs. There are opportunities for the food scientist to draw on extensive information available in the literature to develop mechanistic interpretations of the rheological data which will facilitate both the correlation and meaningful extrapolation of the data. Where the current knowledge base is inadequate, the food scientist has opportunities for original contributions which may well have applications beyond the realm of food properties alone.

REFERENCES

Alfrey, T. 1948. *Mechanical Behavior of High Polymers.* Interscience Publishers, Inc., New York.

Bagley, E. B. and Christianson, D. D. 1982. Swelling capacity of starch and the relationship to suspension viscosity—effect of cooking time, temperature and concentration. *J. Texture Studies* 13: 115–126.

Bagley, E. B. and Christianson, D. D. 1983. Yield stresses in cooked wheat starch dispersion. *Starch/Stärke* 35(3): S.81–86.

Bagley, E. B., Christianson, D. D., and Wolf, W. J. (1985). Frictional effects in compressional response in simple shear, torsion and lubricated uniaxial compression. *J. Rheol.* 29(1): 103–108.

Bagley, E. B. and Christianson, D. D. 1987. Stress relaxation of chemically leavened dough—data reduction using the BKZ elastic fluid theory. *J. Rheol.* 31(5): 405–413.

Bagley, E. B., Christianson, D. D., and Martindale, J. A. 1988. Uniaxial compression of a hard wheat flour dough: Data analysis using the upper convected Maxwell Model. *J. Texture Studies* 19: 289–305.

Browsley, G. J. and Morris, V. J. 1988. Mixed and filled gels—models for foods. In *Food Structure—Its Creation and Evaluation*, J. M. V. Blanshard and J. R. Mitchell (Ed.), pp. 7–23. Butterworths, London.

Cavella, S., Plazza, L., and Masi, P. 1990. Thermomechanical analysis of gluten-starch-water mixtures. Influence of testing conditions and composition. *Italian J. Food Sci.* 2(4): 235–248.

Christianson, D. D. and Bagley, E. B. 1983. Apparent viscosities of dispersions of swollen cornstarch granules. *Cereal Chem.* 60(2): 116–121.

Christianson, D. D., Navickis, L. L., Bagley, E. B., and Wolf, W. J. 1984. Rheological characterization of starch, starch-hydrocolloid and protein dispersions and gels in simple shear and torsion. In *Gums and Stabilizers for the Food Industry, 2*, G. O. Phillips, D. J. Wedlock, and P. A. Williams (Ed.), p. 123. Pergamon Press, New York.

Christianson, D. D. and Bagley, E. B. 1990. Behavior of corn, wheat, and rice starch gels in uniaxial compression. *Food Hydrocolloids* 4(4): 289–298.

Collyer, A. A. and Clegg, D. W. 1988. *Rheological Measurement*. Elsevier Applied Science Publishers, New York.

Dealy, John M. 1982. *Rheometers for Molten Plastics*. Van Nostrand Reinhold Co., New York.

De Gennes, Pierre-Gilles. 1979. *Scaling Concepts in Polymer Physics*. Cornell University Press, Ithaca, NY.

Dickie, A. M. and Kokini, J. L. 1982. Use of the Bird-Leider equation in food rheology. *J. Food Proc. Eng.* 5: 157–174.

Dus, S. J. and Kokini, J. L. 1990. Prediction of the nonlinear viscoelastic properties of a hard wheat flour dough using the Bird-Carreau Constitutive Model. *J. Rheol.* 34(7): 1069–1084.

Fanta, G. F. and Christianson, D. D. 1991. The influence of poly(ethylene co-acrylic acid) on the paste viscosity and gel rheology of cornstarch dispersion. *Cereal Chem.* 68(3): 300–304.

Gadala-Maria, F. A. 1979. The rheology of concentrated suspensions. Doctoral dissertation, Stanford University, University Microfilms International, Ann Arbor, MI.

Glass, J. E. (Ed.) 1989. *Polymers in Aqueous Media Advances in Chemistry Series.* American Chemical Society, Washington, DC.

Goldsmith, H. L. and Mason, S. G. 1967. The microrheology of dispersions. In *Rheology, Theory and Applications*, Vol 4. F. R. Eirich (Ed.), Academic Press, New York.

Graessley, W. W. 1974. The entanglement concept in polymer rheology. In *Advances in Polymer Science*, Vol. 16. Springer-Verlag, New York.

Graessley, W. W. 1984. Viscosity and flow in polymer melts and concentrated solutions. In *Physical Properties of Polymers*, American Chemical Society, Washington, DC.

Heinrich, G., Straube, E., and Helmis, G. 1988. Rubber elasticity of polymer networks: Theories. In *Advances in Polymer Science*, 85, Springer-Verlag, Berlin, Heidelberg.

Hermansson, A.-M. 1988. Gel structure of food biopolymers. In *Food Structure—Its Creation and Evaluation*, J. M. V. Blanshard and J. R. Mitchell (Ed.), pp. 25–40. Butterworths, London.

Hoffman, R. L. 1972. Discontinuous and dilatant viscosity behavior in concentrated suspension. I. Observation of a flow instability. *Trans. Soc. Rheology* 16(1):155–173.

Hoffman, R. L. 1974. Discontinuous and dilatant viscosity behavior in concentrated suspension II. Theory and experimental tests. *J. Colloid. Interfacial Sci.* 46(3): 491–506.

Hoffman, R. L. 1982. Discontinuous and dilatant viscosity behavior in concentrated suspensions III. Necessary conditions for their occurrence in viscometric flows. *Advances in Colloid and Interface Sci.* 17: 161–184.

Kearsley, E. A. 1989. Strain invariants expressed as average stretches. *J. Rheol.* 33(5): 757–760.

Krieger, I. M. 1983. Rheology of emulsions and dispersions. In *Physical Properties of Foods*, M. Peleg and E. B. Bagley (Ed.). AVI Publishing Co., Westport, CT.

Larson, R. G. 1988. *Constitutive Equations for Polymer Melts and Solutions.* Butterworths, Boston.

Launay, B., Doublier, J. L., and Cuvelier, G. 1986. Flow Properties of aqueous solutions and dispersions of polysaccharides. In *Functional*

Properties of Food Macromolecules, J. R. Mitchell and D. A. Ledward (Ed.). Elsevier Applied Science, Publishers, London & New York.

Levine, H. and Slade, L. 1990. Influence of the glassy and rubbery states on the thermal, mechanical and structural properties of doughs and baked products. In *Dough Rheology and Baked Product Texture*, H. Faridi and J. M. Faubion (Ed.). Van Nostrand Reinhold, New York.

Middleman, S. 1968. *The Flow of High Polymers*. Interscience Publishers, New York.

Miller, B. D. F., Peleg, M., Normand, M. D., Gonter, R. H., and Klein, E. 1986. A computer aided method for the rheological characterization and classification of solid food material. *J. Food Sci.* 51(1): 123–134.

Mooney, M. 1951. The viscosity of a concentrated suspension of spherical particles. *J. Colloid Sci.* 6: 162–170.

Morris, E. R., Cutler, A. N., Ross-Murphy, S. B., and Rees, D. A. 1981. Concentration and shear rate dependence of viscosity in random coil polysaccharide solutions. *Carbohydrate Polymers* 1: 5–21.

Nielsen, L. E. 1967. Mechanical properties of particulate filled systems. *J. Composite Materials* 1: 100–119.

Nielsen, L. E. 1974. *Mechanical Properties of Polymers and Composites*, Vol. I and II. Marcel Dekker, Inc., New York.

Peleg, M. 1977b. Contact and fracture elements as components of the rheological memory of solid foods. *J. Texture Studies* 8: 67–76.

Peleg, M. 1977a. Operational conditions and the stress-strain relationships of solid foods—Theoretical evaluation. (1977). *J. Texture Studies* 8: 283–295.

Peleg, M. 1984. A note on the various strain measures at large compressive deformations. *J. Texture Studies* 15: 317–326.

Pollak, N. and Peleg, M. 1980. Early indications of failure in large compressive deformation of solid foods. *J. Food Sci.* 45: 825–835.

Prentice, J. H. 1984. *Measurements in the Rheology of Foodstuffs*. Elsevier Applied Science Publishers, New York.

Rha, C. K. and Pradipasena, P. 1986. Viscosity of proteins. In *Functional Properties of Food Macromolecules*, J. R. Mitchell and D. A. Ledward (Ed.). Elsevier Applied Science Publishers, London, New York.

Richardson, R. K., Robinson, G., Ross-Murphy, S. B., and Todd, S. 1981. Mechanical spectroscopy of filled gelatin gels. *Polymer Bull.* 4: 541–546.

Ring, S. and Stainsby, G. 1982. Filler reinforcement of gels. *Prog. Fd. Nutr. Sci.* 6: 323–329.

Treloar, L. R. G. 1975. *The Physics of Rubber Elasticity*. Clarendon Press, Oxford, England.

14

The Kinetics of Nonenzymatic Browning

Theodore P. Labuza and Wendy M. Baisier

University of Minnesota
St. Paul, Minnesota

INTRODUCTION

Nonenzymatic browning (NEB) is one of the most prevalent chemical reactions that occurs in foods. Foods undergoing NEB usually exhibit a brown discoloration. The reaction also leads to off-flavors, off-colors, and textural changes and may cause loss of nutritional value. There are many different substrates and pathways for NEB. Oxidation of ascorbic acid in citrus fruits may cause loss of vitamin C with subsequent darkening of the fruit. Sugars can caramelize when exposed to excessively high temperatures. Maillard-type browning of reducing sugars and free amines in dehydrated potatoes results in off-flavors. In dry milk powder, similar Maillard browning results in decreased solubility and nutritional loss. The reaction is common for foods containing both reducing sugar and protein and enough moisture to act as a reaction phase.

MECHANISM OF THE MAILLARD REACTION

Despite its name, the reaction was actually discovered by an Englishman named Ling who, in 1908, first theorized that the color produced in the brewing process came from a reaction between sugars and protein (Ling and Malting, 1908). The man who gave the reaction its name and made it famous was a Frenchman named Louis Camille Maillard. In 1912 at the Sorbonne, he combined one part glycine with four parts glucose in water under physiological conditions. He noted that the solution turned progressively darker and that CO_2 was produced. He repeated the experiment with many different reducing sugars and amino acids. From the results, Maillard concluded that the reaction was relevant to many areas such as geology, plant pathology, and medicine. He made a very insightful prediction when he said that the reaction could have some therapeutic value for "better utilization of sugars" by diabetics (Monnier et al., 1989).

Hodge (1953) developed the scheme shown in Fig. 14.1 to illustrate the

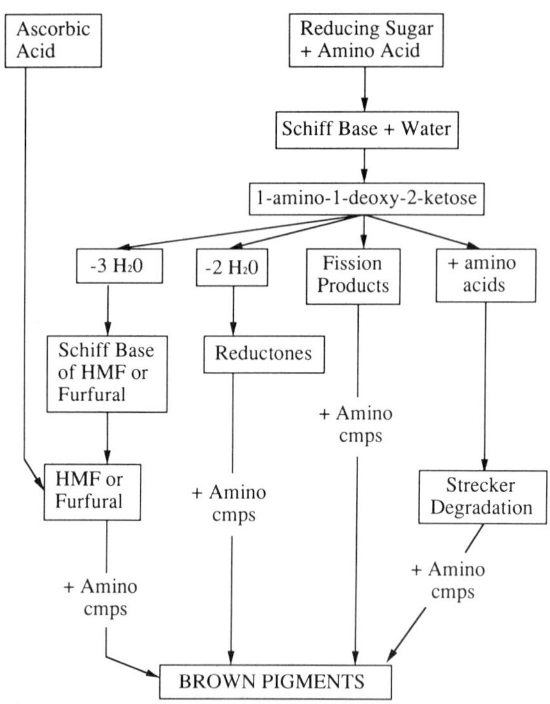

Fig. 14.1 Schematic of the Maillard reaction derived by Hodge (1953).

pathway of the reaction and categorized the reaction as follows:

Stage I Initial stage (colorless; no absorption in the near-UV)
- A. Sugar-amine condensation
- B. Amadori rearrangement

Stage II Intermediate stage (colorless or yellow, with strong absorption in the near-UV)
- C. Sugar dehydration
- D. Sugar fragmentation
- E. Amino acid degradation

Stage III Final stage (highly colored)
- F. Aldol condensation
- G. Aldehyde-amine polymerization; formation of heterocyclic nitrogen compounds

The first stage in the overall reaction is the combination of the carbonyl group of the reducing sugar with the free, uncharged amine group of the amino acid or protein with the loss of one mole of water. The trivalent nitrogen atom of the amine, which has an unshared electron pair, acts as a nucleophile towards the carbonyl group. Bases can catalyze the reaction by removing a proton from the nucleophile converting it from a weak to a strong nucleophile (O'Brian and Morrissey, 1989). The *N*-substituted glycosylamine that is formed then undergoes the irreversible Amadori rearrangement, in the case of aldoses, and the Heyn's rearrangement for ketoses. The Amadori rearrangement is the isomerization of the *N*-substituted aldosylamine to a 1-amino-1-deoxy-2-ketose, while the Heyns rearrangement changes a ketosylamine to a 2-amino-2-deoxyaldose (O'Brian and Morrissey, 1989). At room temperature, the formation of Amadori rearrangement products (ARPs) proceeds slowly, since a tautomeric shift to the open-chain form of the reducing sugar is required for the initial reaction to occur (see Fig. 14.6). The amount of acyclic form, for most monosaccharides and ARPs, is less than 1% of the total sugar in solution (Kaanane and Labuza, 1989; Yaylayan, 1990).

Hodge and Rist (1953) showed that the C-2 hydroxyl of an aldose is essential for producing a significant amount of browning, and it has been further shown that blocking the Amadori rearrangement step stops the formation of brown pigment entirely. Once the ARPs are formed, they are more stable than the glycosylamines in moist, acidic environments, although they are heat labile. Upon heating, the ARPs undergo dehydration and fission and yield colorless reductones as well as fluorescent substances, some of which may also be pigmented (Hodge, 1953). In 1975,

Namiki and Hayashi proposed a side reaction in which the initial condensation product may lead directly to browning while bypassing the Amadori rearrangement product. Nevertheless, the Amadori rearrangement product remains an important part of the accepted reaction pathway.

The second or intermediate stage involves the removal of the amino group from the sugar moiety followed by formation of more reactive compounds. Some of these fluorescent compounds and brown pigments may occur but at very low concentrations. The three degradation pathways that the 1-amino-1-deoxy-2-ketose (ARP) can take are dehydration of the sugar moiety, fragmentation of the sugar moiety, and Strecker degradation of the amino acid moiety (Hodge, 1953).

There are two types of sugar dehydration reactions, both of which depend on the pH of the system. Under acidic conditions, the nitrogen is protonated, and enolization involving the C-1 atom favors production of 1,2-eneaminol (Yaylayan, 1990). Dehydration and cyclization produce hydroxymethyl furfural (from hexoses) or furfural (from pentoses) with the possibility of amino acid regeneration. Formation of hydroxymethyl furfural in the glucose-glycine browning reaction under the conditions used by Maillard was demonstrated chromatographically by Chichester et al. (1952). Under basic conditions, enolization involving the C-3 atom favors the formation of a 2,3-enediol. In addition, the 1-amino-1-deoxy-2-ketose compound, a secondary amine, may react with a second mole of sugar via a second Maillard reaction to form a dikeoxyketose-amine (O'Brian and Morrissey, 1989). The sugar chain is dehydrated, with the loss of two molecules of water, and a reductone is formed (Hodge and Rist, 1952). Conjugated unsaturation is necessary for the stability of the reductone.

The second reaction that occurs with ARPs is fragmentation of the sugar moiety. The accepted mechanism for sugar fragmentation is dealdolization, the reverse of aldol condensation. Amines catalyze this reaction as they catalyze aldol condensation (Hodge, 1953). The products are aldols, amino-free polymers, and free amino compounds.

The third and last reaction that can occur with ARPs is Strecker degradation. This reaction takes place between a reductone and an α-amino acid and liberates carbon dioxide to form an aldehyde with one fewer carbon than the original amino acid. Maillard himself hypothesized that the carbon dioxide released came from the carboxyl group of the amino acid rather than from the sugar radical (Hodge, 1953). Stadtman et al. (1952) showed that under Maillard's conditions, over 80% of the carbon dioxide came from the carboxyl group of the amino acid. The other 10% came from the reaction of nonselectively labeled glucose with inactive glycine. Wolfrom et al. (1974) found that 90–100% of the carbon dioxide came from the amino acid carboxyl group. Production of carbon dioxide parallels aldehyde production and color formation. The aldehydes formed by Strecker degradation are a source of browning since they can condense with themselves, with sugar fragments, with furfurals, or with other dehydration products to form brown pigments. This pathway, however, is not the major color-producing reaction and is better known as the source of off-flavors associated with Maillard browning (Hodge, 1953).

The final stage of the reaction is where most of the color is produced. The chief reactions involved are thought to be aldol condensation, aldehyde-amine polymerization, and formation of heterocyclic nitrogen compounds such as pyrroles, imidazoles, pyridines, and pyrazines (Hodge, 1953). Polymerization of products from the second step and copolymerization with amino compounds yield the colored products (Warmbier et al., 1976). The chemistry at this point is very complex, but it is known that both water-soluble and water-insoluble pigments, called melanoidins, are produced (Reynolds, 1963). The melanoidins vary widely in molecular weight and contain several discrete chromophores (O'Brian and Morrissey, 1989). The empirical formula for melanoidins was thought to be $C_{67}H_{76}N_5O_{32}$, derived from the elementary analysis by Enders and Theis (1938). However, this only holds true for a glucose-glycine ratio of 1:1, since the molecular weight range very much depends on the amino acid-sugar ratio and on test conditions. For example, Rubinsztain et al. (1984) found that basic amino acids and sugars form melanoidins that have higher molecular weights than those formed by neutral or acidic amino acids. This indicates that cross-linking is probable. The same study revealed that the carboxyl acidity in the melanoidin structure can be correlated with the relative contribution of the sugar; the higher the sugar:amino acid ratio, the higher the carboxyl group concentration and the higher the concentration of total melanoidin pigment. The authors suggested that the melanoidin "backbone" is probably a sugar-moiety on which various nitrogen containing groups are grafted.

FACTORS INFLUENCING THE MAILLARD REACTION

Introduction

Scientists often study problems solely to gain knowledge about the world around us. With regards to the Maillard reaction, however, there are more practical motivations like food product quality, nutritional loss, or the advance of knowledge about the human aging process.

The formation of brown pigments and related browning aromas can be considered positive or negative, depending on the product. Examples of products that benefit from the reaction are roasted coffee, meat, baked bread, cereals, and processed maple syrup. On the negative side is formation of brown pigments in nonfat dried milk powder or dehydrated potatoes, leading to off-flavors and undesirable darkening of color. In dried milk, lactose reacts with the ϵ-amino group of lysine exposed from the casein micelle, resulting in severe reduction of the nutritional value of the product. Labuza and Saltmarch (1981), as well as many others, have studied this loss of protein quality for different foods.

The importance of nutritional loss is probably minimal to most people in the United States who enjoy a well-balanced diet. The main concern in this area is sterilized total parenteral (TPN) and enteral nutrition solutions used in hospital feeding. These solutions are administered to patients as their only source of nutrition, and thus the free amino acid content becomes very important (Massaro and Labuza, 1990). For such intravenous solutions, the reducing sugars and amino acids are kept in separate containers during both sterilization and storage. The system is cold sterilized by filtration, mixed together at the hospital pharmacy, and then used without a long period of storage. This is both costly and inconvenient for compromised patients, who cannot eat normally. Tube feeding solutions, however, are sterilized with everything mixed together and suffer serious browning, the implications of which are unknown.

Specific Factors

Several substrate-type and environmental factors influence the Maillard reaction—type of amino acid, pH, sugar type, solvent state, sugar:amine ratio, and temperature. A problem in trying to compare reported kinetic data for substrate loss, intermediate formation, and brown pigment formation from different investigators is the mode of reporting the results. Some workers provide data based on either sugar or amino acid concentration or just report brown pigment formation in terms of the optical density (OD) at different times. There is no consistency to reporting browning rates. Should it be reported as OD/mole/L where the moles can be based on amino acid or sugar concentration, or both; and should the length of the induction period be reported?

Type of Amino Acid

Lysine is often hypothesized as being the most reactive amino acid due to its two reactive amino groups (O'Brian and Morrissey, 1989). Despite this logical deduction, numerous studies have provided contradictory results. Wolfrom et al. (1974) studied the effect of amino acid type on the extent of browning in a 1:1 molar ratio of a mixture of D-glucose/amino acid heated at 65°C. Nine different amino acids present in orange juice as well as in other foods were compared. Results were presented as absorbance of a sample vs. time without any corrections for molar or weight concentrations. As can be seen in Fig. 14.2, the most intense and rapid color development was produced by L-arginine and 4-aminobutyric acid followed by glycine, alanine, serine, and L-proline. Aspartic acid, L-glutamic acid, and L-glutamine were found to behave similarly to glycine. Lysine was not included in this study. Unfortunately the pHs of the solutions were not stated or controlled and reports of the molar concentrations used were unclear. The unavailability of the precise molar concentrations used makes it difficult to compare to other studies or within the study itself. For example, does arginine actually form brown pigment faster on a molar basis? Making the assumption that the stated conditions for one part of the study were consistent with the others, browning rate constants were calculated on three different bases for each of the amino acids, as shown in Table 14.1. It can be seen that the order remains basically the same whether reported on an OD/day basis, an OD/day–mole amino acid, or a total amino acid and sugar basis. Since different amino acids may form different chromophores with different extinction coefficients and the pHs are sure to vary, this simple ranking may be incorrect.

In Fig. 14.2 it can also be seen that each amino acid shows a different

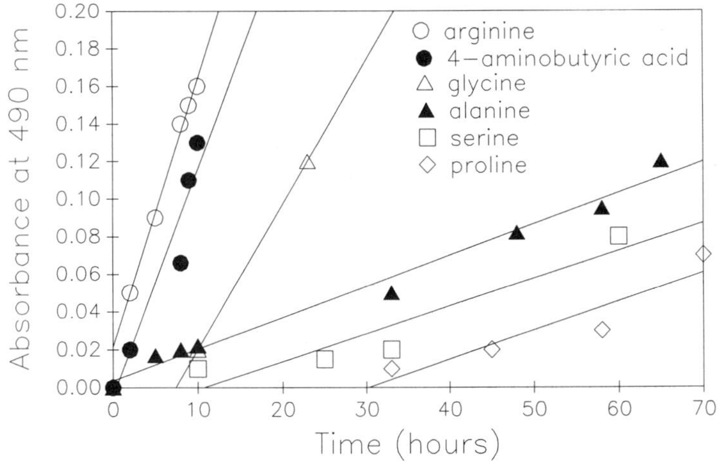

Fig. 14.2 Production of brown pigment by different amino acids at 65°C as a function of time. (*From the data of Wolfrom et al., 1974.*)

induction period. The rates in Table 14.1 are for the pseudo zero-order reaction segment after the induction period. Thus, serine has a seven times slower rate of browning than arginine, as well as a 25-hr induction period compared to no induction period for arginine. Since the different amino acids may produce different chromophores with different absorbance characteristics, as noted previously, one cannot say that the amino acid that browns the fastest also reacts (i.e., disappears) the fastest. Generally, little data is collected on amino acid disappearance kinetics.

Ashoor and Zent (1984) systematically studied the relative reactivity of amino acids and amides. The sugars and amines were combined in a 1:1 molar ratio of 0.005 M, each in a buffer at pH 9.0, and heated for 10 min at 121°C, after which the samples were evaluated spectrophotometrically for color at 420 nm. The authors classified the amino acids and amides into three groups according to relative reactivity in the Maillard reaction. The four amino acids classified as the most reactive were lysine, glycine, tryptophan, and tyrosine. In this study and many similar studies comparing amino acid reactivity, only one data point was taken and a high temperature was used. Since browning shows an induction period followed by a pseudo zero-order period, rate constants cannot be calculated, and the data cannot be compared to other studies.

Massaro and Labuza (1990) studied model total parenteral nutrition (TPN) solutions at pH 5.5–5.7 stored at 4°C or 30°C for 6 months. The authors found that lysine produced less color than tryptophan and cysteine on a molar basis (OD/mole amino acid-day) as shown in Table 14.2. The amino acid levels used were based on the level used in a finished commercial TPN formula with 1.1 M glucose. It can be seen that the browning rate of the lysine/tryptophan combination (OD/day) was close to the sum of each individually; however, with added cysteine, the rate of brown color formation was less than predicted by adding the rates of the three

Table 14.1 Browning Rate Constants for Different Amino Acids

Amino acid	OD/day (10^3)	OD/day–mole A[a] (10^3)	OD/day–mole A-mole R[a] (10^3)
Arginine	16.2	40.5	101.2
4-Amino-butyric acid	12.8	32.0	80.0
Glycine	6.7	16.8	41.9
Alanine	2.2	5.5	13.8
Serine	2.3	5.8	14.4
Proline	2.2	5.5	13.8

[a] A and R are initial molar concentrations of amino acid and glucose, respectivey (1:1 ratio).
Source: Wolfrom et al. 1947.

Table 14.2 Rate Constants for Brown Pigment Formation

Amino acid	Temp. (°C)	OD/day (10^5)	OD/day–mole A (10^3)	OD/day–mole A-mole R (10^3)
Lysine(L)	4	5.3	3.5	3.2
	30	12.5	8.2	7.5
Tryptophan(T)	4	14.8	42.4	38.5
	30	90.5	259.0	235.5
Cysteine(C)	4	12.1	21.3	19.4
	30	150.0	264.0	240.0
L and T	4	14.6	7.8	7.1
	30	105.0	56.5	51.4
L, T, and C	4	11.3	4.6	4.2
	30	132.0	54.2	49.3
Commercial	4	51.9	1.3	1.2
	30	750.0	19.1	17.4

Source: Massaro and Labuza, 1990.

amino acids together, suggesting inhibition by cysteine. The values reported for the browning rate (OD/day) in Table 14.2 are 100× smaller than those of Wolfrom et al. (1974) in Table 14.1, but when the values are equated by dividing by the initial substrate concentration, they are comparable, despite the 30°C temperature difference. This suggests that a more appropriate method of reporting browning data might be OD per initial mole amino acid and/or per initial mole reducing sugar.

Fry and Stegink (1982) studied two different TPN solutions; both containing 25% glucose and 4.25% amino acids, but one contained added electrolytes in millimolar amounts. The solutions were stored for 30 days at 4, 25, and 60°C. The amount of each amino acid incorporated into Maillard reactions products was quantitated by using C_{14}-labeled amino acids. pH was not measured, but the authors stated that histidine has enough buffering capacity to prevent any change in pH due to the addition of electrolytes. Massaro and Labuza (1990), however, found that in their TPN solution, which contained cysteine, pH dropped from 5.5 to 3.5 over a 60- to 70-day period, and then remained constant. Fry and Stegink (1982) concluded that proline and amino acids with hydrophobic side chains reacted more slowly than other amino acids. They found that tryptophan was the most reactive followed by serine and threonine. Somewhat like Ashoor and Zent (1984), who used only one point on a browning curve, Fry and Stegink (1982) used measurements at only three times and drew broad conclusions about the relatively reactivity of different amino acids.

Because of the differing lengths of lag periods and browning rates that follow, reported rates could be markedly in error.

An overview of the literature indicates a serious lack of data on the actual reactivity of different amino acids, especially with respect to storage of foods. Of even more concern is the reactivity of different proteins in foods. It has been assumed that the browning of proteins is related to the amount of ϵ-amino acid groups from lysine, other potential reacting nitrogen groups such as in histidine and tryptophan, and the number of end-terminal amino groups, which is molecular weight dependent. In the only study where the relative reactivity of different proteins was cursorily examined, Schnickels et al. (1976) reacted different proteins with glucose at a glucose-to-protein ratio of 1:3 in a model system at a reduced moisture content of 10–20 g water/100 g solids (a_w = 0.6–0.8) held at 35°C. The results reported in Table 14.3 show that there does not seem to be a relationship between the amount of available lysine and either the induction period or the browning rate, reported as OD/g solid-day. In the last column, their browning rate results are recalculated as OD/mg lysine-day, based on the measured lysine values. Except for fish protein, the rates are fairly close to a value of 1 OD/mg lysine-day. It is possible that in fish, about one half of the lysine is unavailable to react due to structural folding.

pH

pH has a very significant effect on the Maillard reaction. Generally it has been stated that the rate of browning (Ashoor and Zent, 1984) and

Table 14.3 Results of Browning and Lysine Loss during Storage at 35°C

Protein	Lysine (mg/g solid)	a_w	Induction time (days)	Browning rate (OD/g solid-day)	Browning rate (OD/mg lysine-day)
Wheat gluten + 10% lysine	33	0.72	7	35.0	1.1
Casein	14.2	0.70	13	10.2	0.7
Whey	8.6	0.78	20	8.5	1.0
Soy (TVP)	7.6	0.73	20	7.5	1.0
Fish conc.	12.8	0.68	8	5.9	0.5
Egg albumin	3.6	0.63	20	4.9	1.4

Source: Schnickels et al., 1976.

substrate loss (Dworschak and Orsi, 1977) increases with increasing pH, up to a pH of about 10, with little if any browning occurring below pH 6 (Ellis, 1959). Figure 14.3, calculated from the work of Ashoor and Zent (1984), demonstrates the effect of pH on the amount of brown pigment formed for three different amino acids. In this case, glucose (0.005 M) and amino acid (0.005 M) were heated at 121°C for 10 minutes. Since only one point was taken for each sample in this study, the conclusion that the pH maximum is between 9 and 10 may be inaccurate.

More recently Pilkova et al. (1990) studied the browning of the Heyn's rearrangement product of fructose and alanine. No information was given for the concentrations tested. The solutions, at pH 4–10, were heated at 110°C for 8 hours and the results for browning monitored spectrophotometrically at 520 nm. There was no induction period observed and the rates all followed zero order kinetics. Figure 14.4 shows the rate (OD/hr) vs. pH for this system plotted on a semilog plot and indicates a maximum rate at pH 7–8. No browning occurred at pH 4. Figure 14.4 shows a slope of about 1 on the left side of the rate maximum, as is typically found for many acid/base-catalyzed reactions, such as several of the intermediate steps in browning. For a pure base-catalyzed reaction, the log of the rate constant vs. pH will have a slope of 1 on either side of the pK_a for the amine group, which means a 10-fold increase or decrease in rate per pH unit change. In this case, however, the pK_a is near pH 10 for the amino acids, so this pattern is unusual and, in fact, the decrease above pH 8 is

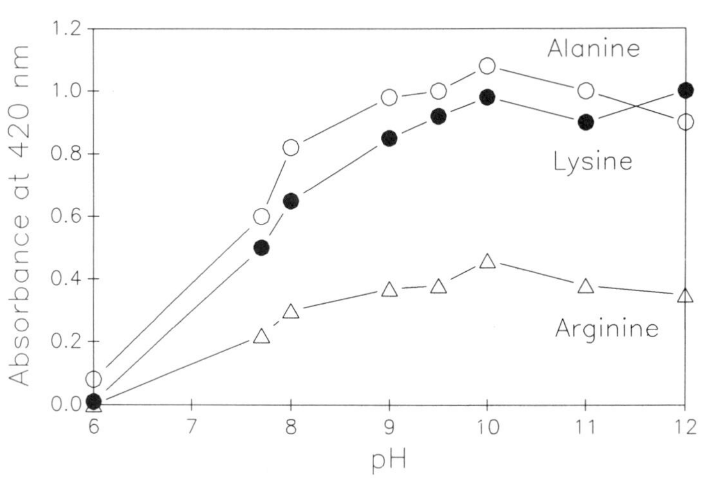

Fig. 14.3 Effect of pH on brown pigment formation at 121°C for 10 minutes. (*From Ashoor and Zent, 1984.*)

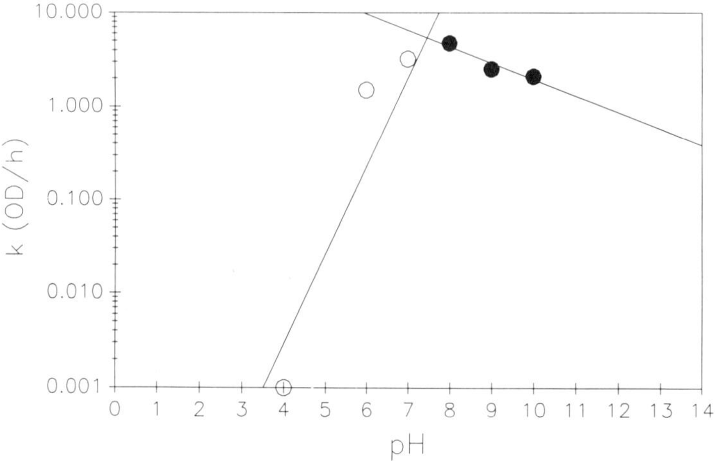

Fig. 14.4 The pH rate profile of browning. (*From Pilkova et al., 1990.*)

unexpected unless there are H^+ ion-catalyzed steps which now have become limiting. Bell and Labuza (1991a), in a study of aspartame loss, found a slope of 1 above the pK_a of 5.2, while a slope of -0.5 was found below this pH due to buffer anion reactions which were catalytic.

Dworschak and Orsi (1977) studied the effect of pH on the loss of two amino acids during Maillard browning with glucose at 130°C. Varying concentrations of tryptophan (0.01–0.07 M) and methionine (0.02–0.2 M) and glucose (0.01–1.0 M) in a 0.2 M phosphate buffer were used. The results, shown in Figure 14.5, follow a pattern similar to that of browning, as would be expected, but do not follow the expected increase in rate of 10 times for each pH unit decrease as did browning. The loss of both methionine and tryptophan were found to be first order, and there was a maximum in the rate around pH 9–10, close to the expected pK_as. As will be discussed subsequently, the initial step of the Maillard reaction is base-catalyzed, so that the reaction rate should increase with increasing pH, as it does. The drop in the reaction rate at pH values higher than 10 suggests again that there are some H^+ ion limitations or buffer effects. It should be noted that H^+ ions are also required to catalyze both Amadori and Heyns rearrangements (O'Brian and Morrissey, 1989). This trend of browning as a function of pH has been reported by other investigators (Schroeder et al., 1955; Underwood et al., 1959; Pomeranz et al., 1962; Wolfrom et al., 1974).

The pH dependence of the initial step of the Maillard reaction can be related to the amount of unprotonated form of the amino acid. The pK_a of the amino acid determines the concentration available in the unpro-

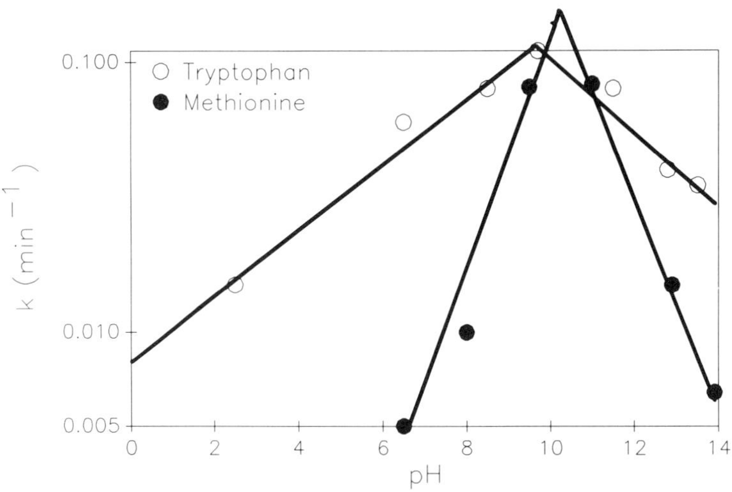

Fig. 14.5 Effect of pH on the degradation rate of tryptophan (○) and methionine (●) in the presence of glucose as a function of pH held at 130°C.

tonated form, which is the form that can react readily with aldose or ketose compounds. Equation (1) shows the equilibrium steps for the amine, H^+ ion interaction in solution.

$$RNH_3^+ \rightleftharpoons RNH_2 + H^+ \tag{1}$$

The equilibrium constant K_{eq} is

$$K_{eq} = \frac{[RNH_2][H^+]}{[RNH_3^+]} = \frac{k_b}{k_f} \tag{2}$$

The pK_a of an amino group is defined as the pH value where 50% of the amino group is protonated. The pK_a for the amine group in glycine is 9.6, while that for the ϵ-amino group of the lysine is 10.53 and 9.44 for the terminal amine of tryptophan. Table 14.4 shows the amount of glycine and lysine available in the unprotonated form as a function of pH. As can be seen, the higher the pH, the larger the percent of any of the amino acids that is in the unprotonated form and, therefore, more moles of amino acid can react with reducing sugars. Since this is a simple proton transfer reaction, as the amino group reacts, more of the unprotonated state forms very rapidly. Thus, the amount available decreases in direct proportion to the total amount reacted. Of critical importance is the fact that at pH 6 and below, which is representative of most foods, the amount of un-

Table 14.4 Percent Unprotonated Amino Group as a Function of pH

pH	Glycine	Lysine
1	2.5×10^{-7}	2.9×10^{-8}
2	2.5×10^{-6}	2.9×10^{-7}
3	2.5×10^{-5}	2.9×10^{-6}
4	2.5×10^{-4}	2.9×10^{-5}
5	2.5×10^{-3}	2.9×10^{-4}
6	0.025	2.9×10^{-3}
7	0.25	0.029
8	2.5	0.29
9	21	2.9
10	72	29
11	97	75

protonated amine is less than 0.02%. Thus the rate of browning will be small at typical pH values for foods, and the dependence of the rate on the total amine concentration will be significant. In addition, since some subsequent steps in the Maillard reaction are proton catalyzed and there are two competing mechanisms, one might expect to see a maximum as pH increases, as reported by Ashoor and Zent (1984), Dworschak and Orsi (1977), and Pilkova et al. (1990). The above studies, based on high temperature, short-time tests, suggested a pH lower limit of 4–6 for browning. However, Labuza and Saltmarch (1981) in a review of long-term storage studies of dry foods in which the initial pH before drying was less than 6, found many for which significant browning occurred during storage, e.g., dry whey powder. This suggests that the pH of these dry systems are lower than the initial value of the wet state or that the previous high temperature studies at low pH were too short (Bell and Labuza, 1991b).

Another factor causing pH rate dependence is related to the amount of acyclic form of the reducing sugars, the reactive form, as a function of pH. Cantor and Peniston (1940) showed that the amount of acyclic form increased with increasing pH for most sugars. For example, the amount of acyclic form increases about threefold for all of the sugars in the pH range 6.5–7.5 except for allose. In short, there is more reducing sugar available to react as the pH increases. Since the concentration of both reactive amine and reducing sugar becomes greater as pH increases, the overall rate of Schiff base formation should increase with increasing pH, but subsequent steps would show a decrease due to their proton dependence. The influence of pH on the brown pigment formation step is unknown, but brown pigment rate depends on all pH interactions in the total scheme.

Type of Sugar

For reducing sugars that participate in the Maillard reaction, the concentration of the acyclic or straight-chain form determines reactivity since it is only the acyclic form with free aldose or hexose group that can react. When a crystalline sugar is dissolved in water or other suitable solvent, it undergoes mutarotation, the equilibrium reaction between the α-pyranose, α-furanose, β-pyranose, β-furanose, and straight-chain forms as shown in Fig. 14.6. Several studies have attempted to quantitate the amount of straight-chain forms of different sugars at equilibrium. Cantor

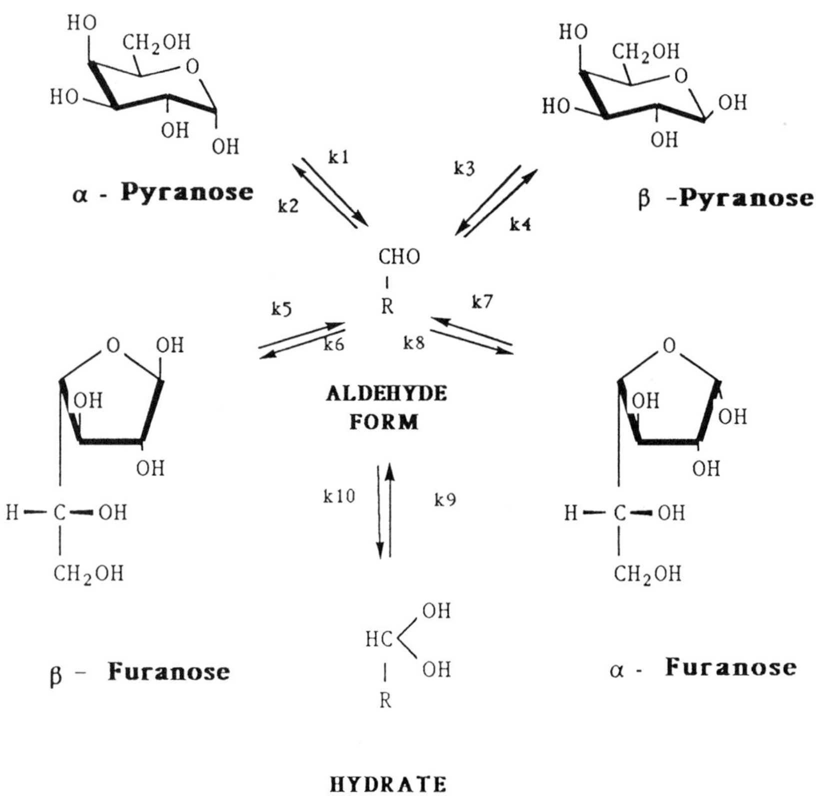

Fig. 14.6 Mutarotation of sugars in solution.

Table 14.5 Amount of Acyclic Form of Reducing Sugar (%) as a Function of pH at 25°C with an Initial Concentration of 0.1 M

	pH		
Sugar	6.5	7.0	7.5
D-Glucose	0.012	0.022	0.040
D-Galactose	0.070	0.085	0.140
D-Mannose	0.040	0.062	0.110
D-Xylose	0.100	0.130	0.360
D-Lyxose	0.150	0.180	0.360
L-Arabinose	0.130	0.220	0.400
D-Ribose	10.000	8.500	30.000
L-Allose	2.280	1.100	0.510

Source: Cantor and Peniston, 1940.

and Peniston (1940) used a crude polarographic technique to estimate the concentration of the acyclic form as a function of pH and sugar concentration as shown in Table 14.5. This method has many limitations which make the data suspect. If pH 6.5 is used for comparison, ribose has an equilibrium acyclic concentration of over 800 times larger than that of glucose. Generally, ribose browns faster than glucose.

Hayward and Angyal (1977) used circular dichroism to estimate concentrations of acyclic forms of sugars, as reported in Table 14.6. They assumed a molar extinction coefficient of unity for all of the sugars, which may not be correct. As can be seen by comparing the values in Tables 14.5 and 14.6, the two studies report absolute concentrations differing by

Table 14.6 Percentage of Acyclic Form of Reducing Sugars in Aqueous Solution at 20°C, pH 5.2–7.0

Sugar	% Acyclic form
D-Glucose	0.002
D-Mannose	0.005
D-Galactose	0.020
D-Xylose	0.020
D-Lyxose	0.030
D-Arabinose	0.030
D-Ribose	0.050
D-Fructose	0.700

Source: Hayward and Angyal, 1977.

a factor of 10 to more than 100, although the reactivity of the various sugars rank in approximately the same order. Delahay and Strassner (1952) found that polarographic wave heights, indicating the degree of free reducing form, were in the order arabinose > xylose > galactose > mannose > glucose, and also concluded that the bond between oxygen and hydrogen of the hydroxyl attached to the C-1 carbon is slightly acidic in aldoses. Therefore, the bond strength weakens as the pH increases, and the rate of ring-to-aldehyde transformation increases rapidly with the pH. As noted, the polarographic methods estimated greater concentrations of acyclic state than those estimated by circular dichroism.

In order to independently verify which technique gave more accurate results for the amount of acyclic state, Kaanane and Labuza (unpublished data) used Fourier Transform Infra-Red (FTIR) spectroscopy to estimate this amount. As a standard, they used the absorbance of the carbonyl stretch of butyraldehyde at 1770–1750 cm^{-1} in D_2O. The butyraldehyde standard was first used to determine the minimum concentration of carbonyl that could be detected by the method. The minimum was determined to be 1.6×10^{-5} mole/L. When a glucose solution (1.5 mole/L) was measured over pH 4–10, no peak was generated. Thus, the predicted acyclic concentration would be less than 1.6×10^{-5} mole/L, just under 0.001% of the total glucose concentration used. This is even less than the concentration reported by Hayward and Angyal (1977), but probably confirms that the polarographic data are in error. The fructose data at about pH 5 showed a small broad peak, as would be expected because of the effect of the OH^- groups on the absorbance peak. The calculated concentration of acyclic fructose was about 3×10^{-5} mole/L or 0.002%, which is still less than the other reported values.

In trying to relate the relative acyclic state concentrations to the browning rate, Burton and McWeeny (1963) reviewed the available literature. One of the studies discussed was a comparison of browning rate between glucose, mannose, and galactose performed by Burton et al. (1962). It was found that galactose browned faster than mannose, which browned faster than glucose, which follows the acyclic concentration pattern of Cantor and Peniston (1940). The same pattern was found by Lippich (1932) for mutarotation velocity. Burton and McWeeny (1963) concluded that the aldose/glycine reaction appeared to depend initially on the conformational stability of the aldose molecule as a whole. On this basis, pentoses are more reactive than hexoses, and browning is faster the higher the concentration of the acyclic state in solution.

Bunn and Higgins (1981) monitored the glycation rate of hemoglobin at 37°C with various monosaccharides, basically the first steps of the Maillard reaction. A concentration of 2 mM carboxyhemoglobin was incubated with 11 mM sugar (5.5:1 sugar:amine ratio) in Krebs-Ringer phosphate buffer at pH 7.3. Cyanoborohydride was present at 20 mM to reduce the Schiff base formed so that a more accurate and direct rate of glycosylation could be calculated without the error introduced because of the reverse reaction. The results for comparison of the different sugars are shown in Table 14.7. The rate of reaction with hemoglobin and various aldose-forming monosaccharides varied over a 100-fold range and correlated well with the estimated concentration of the sugar in the acyclic form using the Hayward and Angyal (1977) data. As seen in Table 14.7, glucose has the lowest rate of glycation of the aldoses with hemoglobin. The authors considered this to be a possible reason why glucose has emerged as the universal metabolic fuel through genetic selection. On the other hand, the rate of glycation of the ketoses as compared to glucose were much less than predicted based on the estimated acyclic state concentration. Since this was based on a direct measure of the glycated hemoglobin, it is not the same as the previously discussed problem of different absorbance of chromophores. Even with the ketoses, the ratio of the amount of acyclic state between sugars was not a good predictor of the rate of glycation.

Table 14.7 Rate Constants for Condensation of Monosaccharides (k) with Hemoglobin at 37°C

Sugar	$k(10^{-3}mM^{-1}hr^{-1})$	Ratio to glucose	% Acyclic form	Acyclic ratio to glucose
Aldoses				
D-Glucose	0.6	1	0.002	1
D-Mannose	3.6	6	0.005	2.5
Fucose	0.7	1.2	0.007	3.5
D-Allose	1.4	2.3	0.010	5
D-Galactose	2.8	4.7	0.020	10
D-Xylose	2.9	4.8	0.020	10
D-Talose	5.2	8.7	0.030	15
D-Ribose	10.0	16.7	0.050	25
D-Idose	55.0	91.7	0.200	100
Ketoses:				
L-Sorbose	1.0	1.7	0.200	100
D-Tagatose	0.3	2	0.600	300
D-Fructose	4.5	7.5	0.700	350
Xylulose	15.0	25	8.00	4000

Source: Bunn and Higgins, 1981.

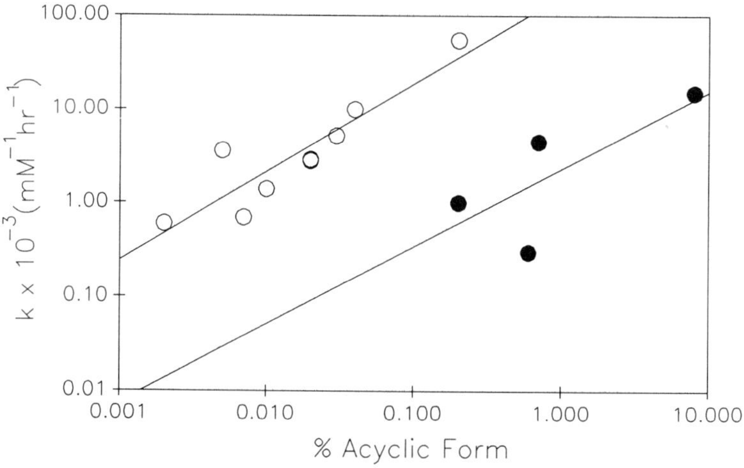

Fig. 14.7 Rate of glycation of hemoglobin at 37°C as a function of the percent acyclic form of reducing sugar in solution ○-aldose; ●-ketoses. (*From Kaanane and Labuza, 1989.*)

Kaanane and Labuza (1989) evaluated the data from this study and concluded that there was no clear relationship between the rate of glycation and the percent in the acyclic form if the aldoses and ketoses are considered together, but when considered separately, a log/log correlation, as shown in Fig. 14.7, was found. When the log of the rate constant is plotted vs. the log of the concentration of the acyclic state for each sugar class, a straight line was found, but as there is no theoretical basis for this, the log-log plot may only be fortuitous.

The effect of sugar structure was noted by Maillard himself. He found that the sugars brown fastest to slowest in the following order: D-xylose > L-arabinose > hexoses > disaccharides (Whistler and Daniel, 1985).

Solvent State

The formation of the acyclic form involves the transfer of three electrons and three protons from the solvent. Therefore, the nature of the solvent in terms of dielectric properties and viscosity affects the rate of transport of protons, thereby affecting mutarotation, the equilibration concentration of the acyclic form, and dependent chemical reactions. Kaanane and Labuza (1989) showed that in a study of the mutarotation kinetics of glucose and fructose in water vs. D_2O, fructose mutarotated

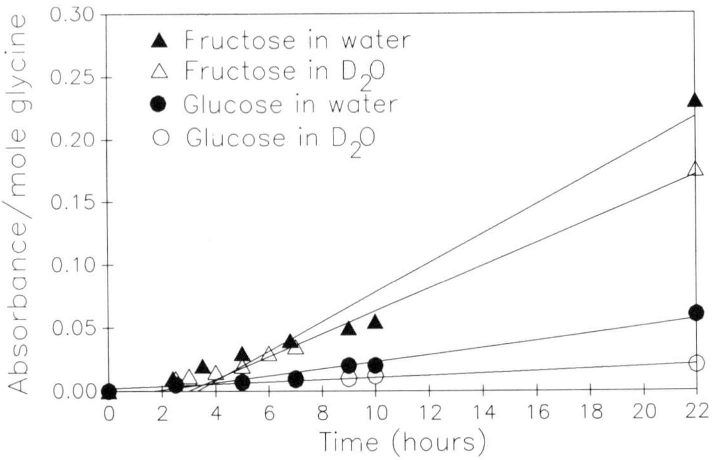

Fig. 14.8 Browning of glucose and fructose in water versus D_2O at 60°C. (*From Kaanane and Labuza, 1989.*)

about 10 times faster than glucose. Kaanane and Labuza (unpublished data) also measured the browning rates at 60°C for both glucose and fructose at the same concentration combined with glycine at a 1:5 molar ratio in both water and D_2O. The solutions, were prepared on a weight/volume of D_2O basis. As shown in Fig. 14.8, fructose browns much faster than glucose when reported on a OD/mole initial glycine basis. The fructose showed an induction period, while glucose did not enter a rapid rate phase over the short period of time studied. As noted, D_2O slowed both the mutarotation and browning process. Table 14.8 lists the calculated rates for both processes for all solutions tested. Although the temperatures differ, the relative rates should be the same. Glucose and fructose mu-

Table 14.8 Comparison of Mutarotation and Browning Rates between Glucose and Fructose in Water and D_2O

	Glucose		Fructose	
(min^{-1})	Mutarotation 24°C	Browning OD mole gly (hr^{-1}) 60°C	Mutarotation (min^{-1}) 24°C	Browning OD mole gly (hr^{-1}) 60°C
Water	0.0299	0.0039	0.2810	0.0110
D_2O	0.0068	0.0011	0.0596	0.0083

Source: Kaanane and Labuza, 1989.

tarotate 4.4–4.7 times faster in H_2O than in D_2O where proton transfer is theoretically slower by a factor of 2^2 or 4 times (Kaanane and Labuza, 1989). However the browning rate of glucose (0.25 M) with glycine (1.252 M) at 60°C in OD/mole glycine-hour is only three times slower in D_2O as compared to water. With fructose at the same concentration, D_2O slowed the browning rate by 1.3 times as compared to water. As noted earlier, the acyclic concentration of glucose is lower than that of fructose, thus proton and/or electron transport in the reaction media must have a strong influence on the relative browning rates of each. In addition, the type of chromophores that are formed and their absorbance may also differ by sugar type as well as amine type.

Labuza and Saltmarch (1981) reviewed the influence of the state of water, as characterized by its thermodynamic availability or water activity (a_w), on the rate of browning and substrate loss in reduced moisture systems.

With respect to the browning as a function of a_w, a min-max pattern is expected, as found by McWeeny (1973), and is shown in Fig. 14.9. Many studies have reported this, including Labuza et al. (1977) and Labuza and Saltmarch (1981). In general, the amount of browning after a given time or the rate of browning increases from the dry state to a maximum at a water activity of 0.5–0.8 and then decreases at higher water activities. Solute mobility is limited below the monolayer due to the higher binding energy and the glassy state that forms (Levine and Slade, 1989).

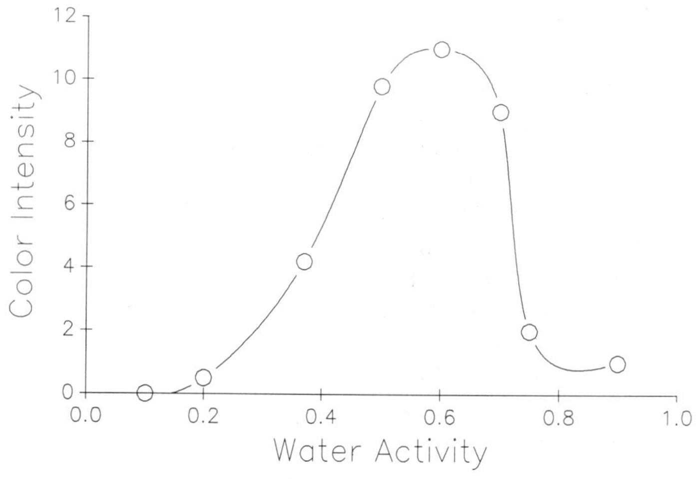

Fig. 14.9 The general influence of water activity on the extent of the Maillard reaction for a given dry food system.

When the water activity is increased, the solutes become more mobile as water is less bound, and the system enters a rubbery state. At low a_w, the water is tightly bound to surface polar sites by hydrogen bonding and is generally unavailable for reaction in solution (Rizvi and Peri, 1986). The critical point is related to the BET monolayer value, which usually occurs at water activities of 0.2–0.3 for most foods. At higher a_w, the solutes can dissolve and become mobile. The rate of browning decreases at higher water activities (0.7–0.8), as found by Eichner and Karel (1972). This is probably due to the increased dilution effect on solute concentration.

With respect to browning, water can retard the rate of the initial glycosylamine reaction by mass action since water is a product. On the other hand, water may enhance deamination reactions in the browning reaction sequence as observed by Reynolds (1963) for the production of furfural or hydroxymethylfurfural. For the overall reaction, as has been found by Eichner and Karel (1972) and Warmbier et al. (1976), both liquid and solid model systems containing glycerol had higher nonenzymatic browning rates than expected in the 0–0.5 a_w range, with maximas between 0.41 and 0.55, showing the effect of an available liquid reaction phase. Since glycerol is liquid at the temperatures studied and the sugars and amines can dissolve in it, there should be a higher rate at lower water concentrations.

Sugar:Amine Ratio

O'Brian and Morrissey (1989) stated that "it is generally agreed that an excess of reducing sugar over amino compound promotes the rate of Maillard browning," the explanation being that there are "mechanistic differences" in the destruction of the sugar compared to the amino acid. They, however, did not clarify what they meant by this. Lerici et al. (1990) stated in their introduction that the "velocity" of the Maillard reaction depends on the different characteristics of the sugars and amino acids, and on the different ratios between them. Of critical concern is that when one compares browning data between labs or within any one experiment, both the absolute concentrations as well as the ratio between amine and reducing sugar must be evaluated.

We can empirically state that browning rate is:

$$\textit{browning rate} = \frac{OD}{time} = k[\mathrm{R}][\mathrm{A}]$$

or, if k is constant,

$$k = \frac{\left(\frac{OD}{time}\right)}{[R][A]}$$

where R and A are reducing compound and amine, respectively. For example, if one states in a publication that the browning rate, in OD/time, increased by two and four times, respectively, as compared to a control in a study of a sugar:amine system with molar ratios of 1:0.5 for the control and 2:2 and 2:4 for the other two test conditions, did the researchers use twice as much sugar in the two test systems as compared to the control? If the sugar amount was kept constant, then the three systems are at ratios of 1:0.5, 1:1, and 1:2, respectively, using the control sugar level as 1. The ratio of the rates as OD/mole, based on added amine, is 1:2:4 as was found for the browning rate increase. This makes sense in that as you double the amine, you increase the rate of browning by two. If the ratios stated are based on absolute amounts of both sugar and amine, the ratios of rates are 1 × 0.5 to 2 × 2 to 2 × 4 or 1:8:16, which would suggest that, since browning increased only 1:2:4 times, above a certain level adding more reactant does not give a proportional increase in rate. If amine concentration was kept constant and the sugar concentration was varied, then the ratios are 2:1, 1:1, and 0.5:1, which would not make any sense as it would suggest that adding sugar decreases the rate. Researchers should always state the absolute values of both reducing sugar and amine to ensure correct interpretation. If there is a direct proportionality between rate and concentration, then dividing the observed OD/time by the concentration of the variable that was increased should be constant.

As noted here, the initial step of the formation of Schiff base depends on the concentrations of both sugar and amine. The actual development of the kinetics will be discussed subsequently. However, it is important to note that if one or the other substrate is at high enough concentration and the reaction period is short, the amount of brown pigment formed may not be affected by a change in those concentrations.

Dworschak and Orsi (1977) reported that for a 0.2 M phosphate buffer model system containing varying concentrations of glucose and either methionine or tryptophan, maximum browning occurred at about a 3:1 sugar to amine ratio. The concentration ranges used were 0.01–1.0 M glucose, 0.02–0.2 M methionine, and 0.01–0.07 M tryptophan. The samples were heated in closed ampules at 130°C for 2 hours. Figure 14.10

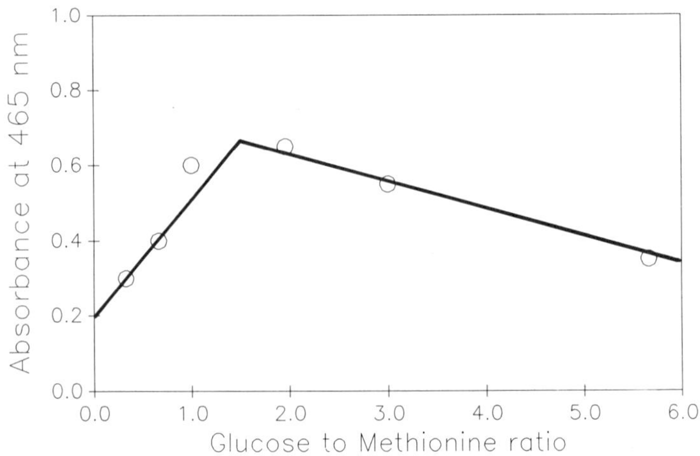

Fig. 14.10 Effect of changes in glucose and methionine concentration on the formation of brown color (heated at 130°C for 2 hr at pH 7).

shows their results for glucose and methionine. The sugar:amino acid ratios studied were 1:3, 1:4, 2:3, 1:1, 3:1, 3:2, and 4:1. Unfortunately both sugar and amine levels were varied to obtain these ratios. They concluded that for methionine the amount of brown pigment produced was the highest when glucose was in excess of methionine in the ratio of 2:3. No rate constants were presented because, as in other poorly designed studies, only one data point was taken for each system after the designated heating period. Once again, since only one point was taken, the rates reported can be very misleading due to differences in the length of the browning induction period. The right-hand side of the peak represents increasing ratios of sugar:amine, yet the apparent reaction rate decreases. The same argument applies to the left side of the graph, which has higher methionine:glucose ratios. The problem is that the absorbance is not adjusted for total concentration and for the various ratios, so any conclusion based on concentration ratios, as is theirs, is not justified. In fact, dividing the absolute browning by the sugar and amine concentrations produces no consistent pattern, therefore making interpretation difficult. The OD/concentration values, respectively, going from left to right are 33, 45, 60, 78, and 73.

Wolfrom et al. (1974), from a study of browning in a model orange juice system at 65°C containing various ratios of glucose to glycine, concluded that the browning rate increased with increasing glycine:glucose ratios in the range 0.1:1–5:1. The browning rates were calculated as OD/hr and were not adjusted for reactant concentration. Unfortunately, it is unclear what actual levels of either sugar or amine were tested. Thus specific conclusions cannot be made.

In contrast to these results, Warmbier et al. (1976), in a study of an intermediate moisture model system at a water activity of 0.52 (10–14 g H_2O/100 g solids) held at 45°C, concluded that the rate of browning increased as the molar ratio of glucose:lysine (present in casein) increased from 0.5 to 3.0, and then leveled off at a ratio of 5:1. This work is summarized in Fig. 14.11. In this study, they held the concentration of protein and thus reactive lysine constant, and varied the sugar concentration. The browning results were presented as OD/g solid/day, thus accounting for the total initial reactive solids (and thus both sugar and amino acid) on a weight basis. Dividing the browning rate by the glucose concentration would give the same pattern, since amine concentration is constant. In this case, even though we know the levels of the reactants, the actual influence of why the increased sugar results in a plateau is mechanistically unclear. Glucose and lysine loss were also monitored and individually followed pseudo first-order reaction rates.

Figure 14.12 demonstrates the effect of the glucose:lysine ratio on the loss rate constants for glucose and lysine. As previously noted, the browning rate was found to increase with increasing initial glucose:lysine ratio. The same was found for the rate loss constant for lysine loss. In contrast, the pseudo first-order rate loss constant for glucose decreased slightly as

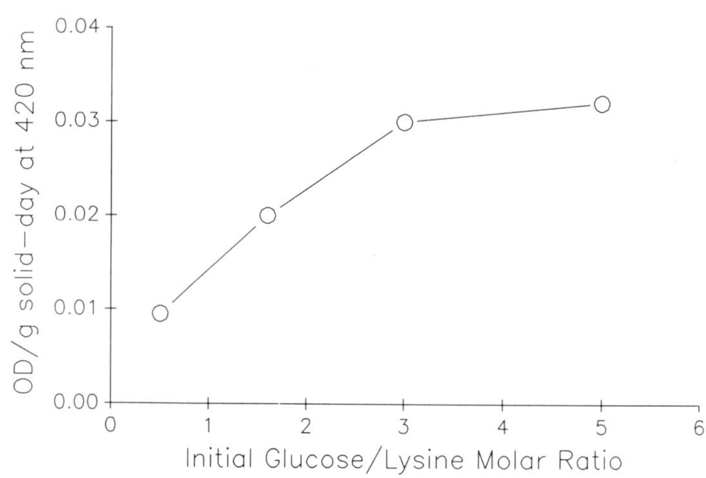

Fig. 14.11 Rate of brown pigment formation as a function of the initial glucose:lysine ratio in an intermediate moisture casein model system held at 45°C. (*From Warmbier et al., 1976.*)

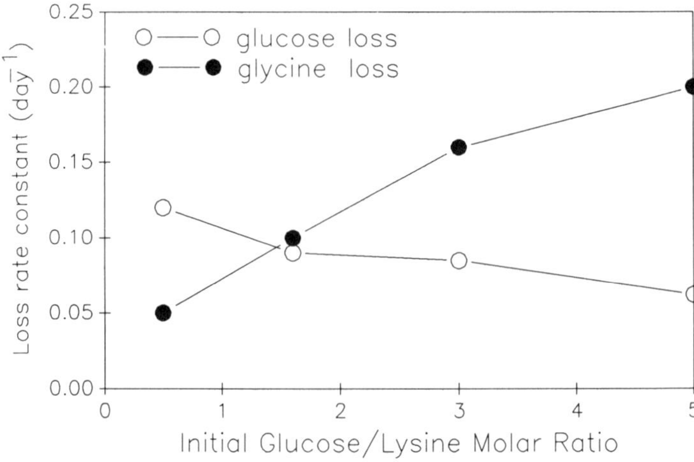

Fig. 14.12 Rate of glucose and glycine loss as a function of initial glucose: lysine ratio held at 45°C in an intermediate moisture casein model system. (*From Warmbier et al., 1976.*)

the initial glucose: lysine ratio increased. However, despite the reduced rate constant, the amount of glucose lost per unit time is greater as the glucose: lysine ratio increased. Since the overall rate of loss of glucose is equal to the rate constant times the initial concentration, it is not clear why the glucose loss rate constant would decrease as the amount of glucose increased. No statistical evaluations were made of the rate constants, so they may not be significantly different at the different ratios. Theoretically the rate constants for each should be factored by the concentration of the other reactant, based on the same mechanistic considerations presented for browning where:

$$\frac{-dG}{dt} = k_1[G][L] = \frac{-dL}{dt} \tag{3}$$

where G is glucose and L is lysine. If lysine is constant and glucose is increased, then

$$\frac{-dG}{dt} = [k_1L][G] = k_g[G] \tag{4}$$

and

$$\ln \frac{G}{G_0} = k_g t \tag{5}$$

Thus, the rate constant (k_g) for glucose should be the same at all glucose levels but, as was seen, is decreased slightly as the glucose level increased. On the other hand, for lysine,

$$\ln \frac{L}{L_0} = [k_1 G] = k_{lys}[G] \tag{6}$$

and thus the observed rate constant (k_{lys}) and the actual amount of loss should increase in proportion to the increased amount of glucose. It is obvious from Fig. 14.12, that the rate constant increase for a 10-fold increase in glucose was only fourfold. This suggests that perhaps there is some diffusion limitation or other limit on the reaction, although all were at the same a_w and about the same moisture content. Knowledge of the glass transition curve might help to clarify these results. It is also possible that the increased sugar level reduces the pH of the system and therefore reduces the reaction rate at higher concentrations (Bell and Labuza, 1991b). Finally, both the sugar and amine may be involved in other competing pathways, complicating the interpretation.

Lea and Hannan (1950) also observed a maximum in browning rate at a 3:1 ratio with a glucose/casein system held at 37°C and 70% relative humidity for 30 days. Six data points were taken, and the amino acid loss was represented as mg/g total nitrogen while the units for the browning rate were reported as OD/day. As in the Warmbier study, only the glucose was varied, with a 8:1 glucose-to-casein ratio being the maximum studied. Lea and Hannan (1949, 1950) did many experiments to determine the effect of numerous factors on the reaction. Their results are consistent with the factors discussed here, but an interesting point to note is that the plots of amino acid loss show an initial first-order loss which tapered off to a no-loss period after about 70% of the nitrogen disappeared, as measured by the Van Slyke method.

Most recently, Baisier and Labuza (1992), in order to better understand the above studies, did a thorough kinetic study to determine what effect, it any, the ratio of reducing sugar to amine had on the reaction rate constant for glucose loss, glycine loss, fluorescence formation, and brown pigment formation when stored in solution at 37°C. They used a liquid model system that consisted of 0.1 M phosphate buffer at pH 7, 1% sodium azide to inhibit microbial growth, and various ratios of glucose to glycine (0.125–8) covering the concentration range of 0.05–0.4 M (moles/L) for both reactants. They concluded that the absolute concentrations in addition to the molar ratios affect the rate constants, as would be expected by kinetic principles. The ratio alone cannot be used to make predictions about the amount of browning. Table 14.9 lists the rate constant for sam-

Table 14.9 Rate Constants for Samples with the Same Molar Ratio but Different Initial Absolute Concentrations

Sample (mole/mole)	Ratio Glu/Gly	$k_{\text{glu loss}}$ (day^{-1})	$k_{\text{gly loss}}$ (day^{-1})	k_{fluor} (F/day)	k_B (OD/day)	k_B OD/day[A][R]
0.1/0.1	1	2.20[a]	1.87[a]	6.17[a]	1.20[a]	120
0.2/0.2	1	3.82[b]	2.23[b]	38.46[b]	37.6[b]	950
0.3/0.3	1	7.49[c]	5.28[b]	52.09[c]	34.3[b]	381
0.4/0.4	1	7.68[c]	8.26[c]	100.79[d]	67.7[c]	425

[a–d] Values with the same superscript not different at $p < 0.05$.
Source: Baisier and Labuza, 1992.

ples that have the same molar ratio [1 M] but different absolute concentrations. Since the reaction is bimolecular, at least one molecule of each reactant must be present for the reaction to proceed. Therefore, it would be expected that the browning rate would increase when an additional amount of either substrate is added until the molar ratio is so high (or so low) that the substrate in lower concentration becomes limiting. As can be seen from Table 14.9, the molar ratio of reactants is not a good indicator of the effect on reaction rates. At a 1:1 molar ratio neither substrate is limiting, thus an increase in concentration causes the likelihood of the substrates meeting in solution to increase and in turn causes the rate constant to increase, but not in direct proportion to the amount over the whole concentration range. In fact, dividing the browning rate by the concentration gives values of 120, 950, 380, and 425, the meaning of which is hard to interpret, but suggests that the browning reaction rate is more complex than simply the absolute sugar and amine levels. For the absolute glycine or glucose rate constants, however, increasing the concentration of the other reactant gave approximately the requisite proportional increase in rate constant for loss, i.e., if glycine increased 4 times, the glucose loss rate constant increased 3.5 times. If glucose increased by a factor of 4, the glycine loss rate constant increased by 4.42 times. The rate of fluorescence increase, based on both sugar and amine concentration, also shows no recognizable pattern (610:961:578:625).

Temperature

As with other chemical reactions, the observed temperature dependence of the Maillard browning reaction has been successfully modeled using the Arrhenius concept in which the rate constant is exponentially related to the inverse absolute temperature for the simple reaction

$$A \rightarrow B \tag{7}$$

The rate loss of A (– dA/dt) or the rate of formation of B (dB/dt), assuming any concentration dependence, is

$$-\frac{dA}{dt} = \frac{dB}{dt} = k[A]^n \tag{8}$$

where k is the molecular rate constant and n is the order. It has been shown that the following is true for the temperature dependence of k:

$$k = k_0 e^{(-E_a/RT)} \tag{9}$$

where $-E_a/R$ is the slope of a plot of ln k versus 1/T, with T being in degrees Kelvin. E_a is the reaction activation energy, and R is the gas constant. For a multistep reaction, as in the case of the Maillard reaction, and for which the knowledge of the temperature/rate dependence of the individual intermediate steps is lacking, the measured E_a for browning is a complex function of the E_as for each step. As temperature changes, those steps with higher E_as will increase in rate more, and thus the pattern of intermediates and end products can change. For example, in lipid oxidation (Labuza, 1971), the overall rate constant for oxidation is represented in terms of the rate constants for the three steps of initiation (k_i), propagation (k_p), and termination (k_t), as follows:

$$\frac{dO_2}{dt} = \frac{k_p k_i^{1/2}}{(2k_t)^{1/2}} [ROOH]^{1/2}[RH] \tag{10}$$

thus,

$$k^* = \frac{k_i^{1/2} k_p}{(2k_t)^{1/2}} \tag{11}$$

and the overall activation energy is

$$E_a^* = \frac{E_{ai}}{2} + E_{ap} - \frac{E_{at}}{2} \tag{12}$$

A similar representation has not been made for the overall E_a for brown pigment formation.

In a review of the kinetics of the Maillard reaction, Labuza and Saltmarch (1981) found that for dry foods the E_a for browning varied with food type and water activity, with the E_as increasing as a_w decreased. The range of E_as for browning is shown in Table 14.10. The activation energies generally range from 16 to 30 kcal/mole.

Table 14.11 contains data for the temperature dependence of the lysine loss rate in the NEB reaction for various other foods. There is the same large range of activation energies. No such large data base exists for systems in which the loss of sugar and amine and the production of brown pigment were all simultaneously measured at several temperatures to get

Table 14.10 Activation Energy Parameters for the NEB Reaction in Different Foods

Food	a_w	Temperature	E_a (kcal/mole)	Reference
Dry cabbage	0.6	37	28	Mizrahi et al. (1970)
Dry cabbage	0.01	37	40	Mizrahi et al. (1970)
Dried potatoes	0.98	40–80	25	Hendel et al. (1955)
Dried potatoes	0.15	40–80	37	Ibid
Casein/glucose model system	0.52–0.60	25–45	33	Warmbier et al. (1976)
Glucose-glycine	1	60–100	16–22	Song and Chichester (1966)
Processed cheese	0.98	5–40	24	Thomas et al. (1977)
Apple juice	0.99	37–130	27	Hermann and Andrae (1963)
Figs	0.6–0.8	20–30	20	Copley and Van Arsdel (1964)
Raisins	0.6–0.8	20–30	24	Copley and Van Arsdel (1964)

Table 14.11 Activation Energy Parameters for Protein Quality Loss in NEB

Food	a_w	Temperature	E_a (kcal/mole)	Reference
Soybean meal	<0.1	100–120	30	Taira et al. (1966)
Dry cod	0.5	105–115	24	Miller (1956)
Bovine serum albumin	0.8	85–145	24	Carpenter et al. (1962)
NFDM	0.6	20–40	19	Ben Gara and Zimmerman (1972)
Soy-glucose model system	0.3–0.8	80–130	35	Thompson et al. (1976)
Casein/glucose model system	0.7	0–70	29	Lea and Hannan (1949)
Rice meal	>0.9	115–184	12.5	Tsao et al. (1978)
Pasta	0.5	40–90	13	Fabriani and Frantoni (1972)
Casein/glucose model system	0.41	25–45	23	Warmbier et al. (1976)

the activation energies and relative rates. Warmbier et al. (1976) did such a study for a casein/glucose model system and found an activation energy of 33 kcal/mole for browning, modeled as a pseudo zero-order reaction and 23 kcal/mole for lysine loss and 25 kcal/mole for glucose, both modeled as pseudo first-order reactions. The activation energies for sugar and amine loss could be expected to be similar, and the E_a for browning should be higher. Little data exists for the E_a of the intermediate reactions. A discussion with respect to fluorescent will be presented in the next section. Reineccius (1990) has postulated that the difference in amounts of Amadori reaction product formation at different temperatures is due to the differences in E_a for the different steps.

High temperatures (>100°C) are often used to reduce the time required for browning to occur, but could change the significance of various steps and therefore would not be representative of typical food storage conditions. As early as 1969, Spark spoke out against the use of the high temperatures (>100°C) that many researchers used in experiments dealing with the Maillard reaction. Frankel and Katchalsky (1937) suggested that there is a qualitative difference in the browning reaction at temperatures above 40°C. The effect of temperature on the increase in rate of reactant loss would follow the activation energy for the first step, while that for brown pigment formation would be different. This is simply illustrated by the work of Adrian (1982), who heated a glucose-glycine system for 1 hour at 90°C and for 9 hours at 70°C. At the end of the experiment the two systems had both lost the same amount of glycine, but the solution heated at the higher temperatures had twice the UV absorbance compared to the other.

Fluorescence (pH and Temperature Interactions)

The use of fluorescence to monitor Maillard browning began in the 1940s. Olcott and Dutton (1945) investigated the simultaneous development of color and fluorescence in the browning reaction. Friedman and Kline (1950) and Patton and Chism (1951) pointed out that the colored and fluorescing compounds were not identical and that the fluorogens may be precursors of the brown pigments.

White and Thistle (1943) and Pearce and Bryce (1947) studied the relationship between fluorescence and eating quality in dried egg powder. It was observed that fluorescence increased to a maximum and then decreased again, and Pearce (1944) discussed the development of fluorescence in various food products. Pearce (1950) studied the reaction in a solution of glycine and glucose or acetaldehyde with ammonia at pH 5.8 and temperatures from 27 to 93°C. He found that fluorescence also increased until it reached a maximum and then decreased, as found for the egg powder. The mathematics of this will be discussed later. He also found that decreasing the concentration of either reactant increased the amount of time required to reach a set fluorescence value, i.e., it reduced the reaction rate, although decreasing the concentration of glycine reduced the rate of fluorescence more than an equal reduction in the concentration of glucose. The effect of pH in the range 4–8 is shown in Fig. 14.13. There was no significant increase in the reaction rate in the pH range 4–6 at 100 and 120°F, while at 140°F, the rate increased significantly above a pH of 6, as found in other studies for browning above pH 6 at high temperature.

Above pH 6, the rate increase ($time^{-1}$) at 140°F per pH unit is about three times, less than the expected 10-fold increase found for browning, but similar to that found for amino acid loss, as shown previously. As expected, an increase in temperature reduced the amount of time required to reach the specified fluorescence value. Figure 14.14 from Pearce (1950)

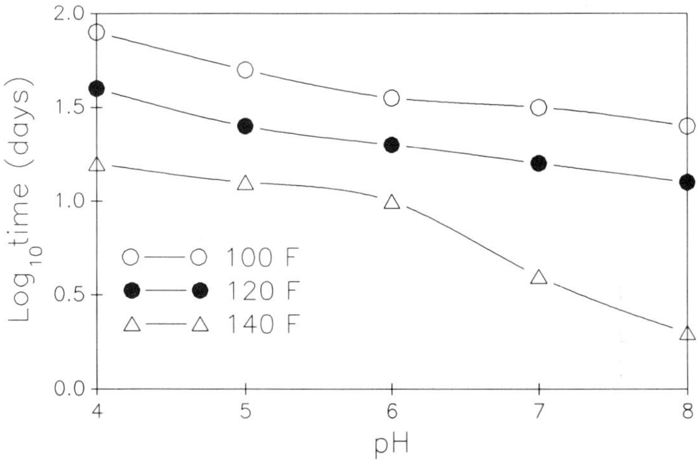

Fig. 14.13 The pH rate profile. (*From Pearce, 1950.*)

shows an Arrhenius-type plot of the time to reach a fluorescence value of 50 units for 1 M glucose with glycine at 0.1–1 M. These solutions were held at 60 to 93°C, and the pH was 5.8. As can be seen, straight lines were obtained at all molar ratios, with the time to reach 50 fluorescence units decreasing as the amount of amine increased. At all temperatures, an increase in the glycine ratio by a factor of 10 (0.1–1 M) decreased the time about 44-fold. He found the activation energy for all ratios was about 24 kcal/mole, quite similar to browning in solution. Although the author did not derive the kinetics of the reaction, it was apparent from the data that fluorescence formation proceeds to a maximum and then decreases in value. This observation has been corroborated by other early researchers (Kline and Stewart, 1948; Patton and Chism, 1951; Overby and Frost, 1952). Overby and Forst (1952) showed that fluorescing and colored compounds are not identical and that the peak of florescence precedes the increase in pigment formation, indicating that the fluorogens may be precursors of the brown pigments (Fig. 14.15).

Apparently, interest in fluorescence has declined over the years as exemplified by the fact that no recent studies of the Maillard reaction in food science have monitored fluorescent compound formation. Absorbance has become the quick and easy standard of monitoring the extent of the reaction. Only in the area of physiology has florescence continued to be used, primarily because absorbance is not an alternative in those systems due to the presence of naturally present interfering colors.

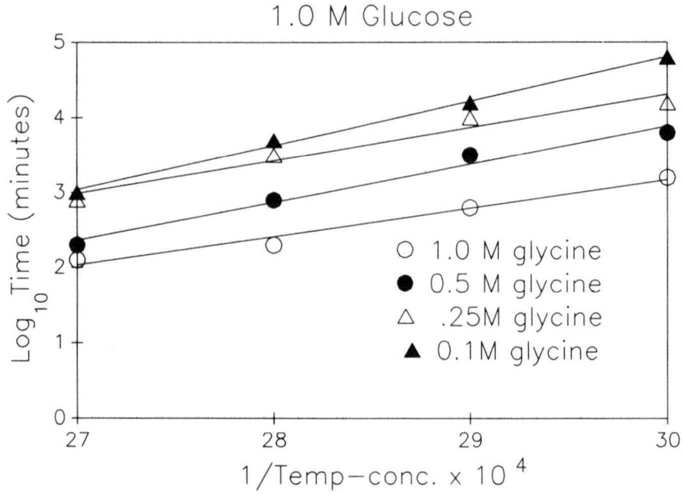

Fig. 14.14 Arrhenius plot of the time for a 1 M glucose solution to reach a fluorescence value of 50 when combined with varying amounts of glycine. (*From Pearce, 1950.*)

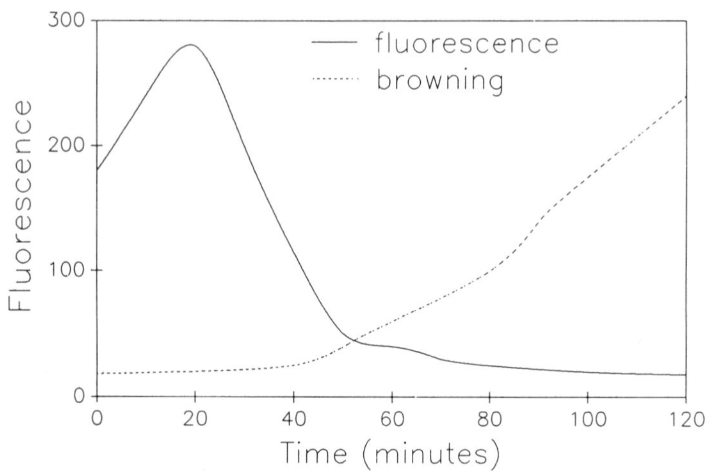

Fig. 14.15 Relation of fluorescence to color development in fibrin hydrolysate (5%) and glucose (5%) solution autoclaved at 115°C. (*From Overby and Frost, 1952.*)

KINETICS OF MAILLARD BROWNING

A simplified kinetic-based browning reaction mechanism in foods can be represented as:

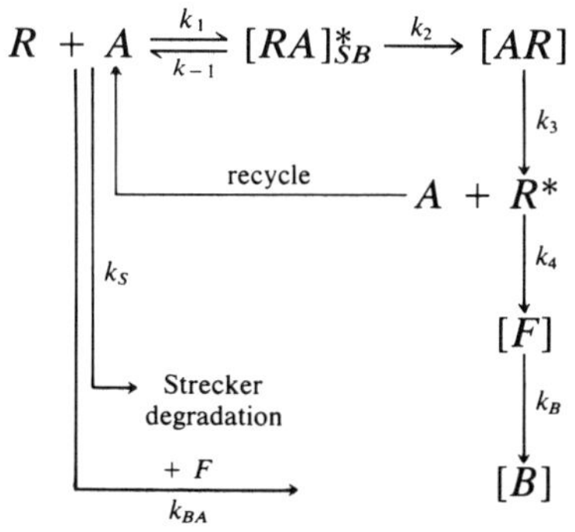

where R is the reducing sugar concentration, A is the reactive amine concentration, RA is the Schiff base, AR is the Amadori rearrangement product, and R* are the reactive reducing compound intermediates which may react and fluoresce, F are fluorescent compounds, and B is the brown pigment concentration. As noted earlier, the loss of sugar or amine in the early stages can be represented by

$$\frac{-d\mathrm{R}}{dt} = \frac{-d\mathrm{A}}{dt} = k_1[\mathrm{A}][\mathrm{R}] \tag{13}$$

which is an overall second-order reaction and assumes that the back reaction (k_{-1}) is negligible and that the amount of amine recycling, going into brown pigment or reacting via the Strecker degradation reaction, is small. Except for glycation studies, the food science literature has not used second-order kinetics, utilizing instead the assumption that either the amino acid or sugar is in excess.

As noted previously in Eqs. (5) and (6), if amine or reducing compound, respectively, are at high levels and not limiting, then the loss of the limiting component can be modeled as a pseudo first-order reaction in which a semilog plot of the fraction remaining is a straight line. Labuza (1984) has also shown this to be true even when neither component is limiting, if only data from the initial rate period (up to 30–40%) loss is used. If, indeed, the rate depends on both amine and reducing compound, and we assume k_2 to be very fast, then k_1 controls and the rate is overall second order and has the solution:

$$\frac{dx}{dt} = k_1[A_0 - AR][R_0 - AR] \tag{14}$$

where x represents either AR, $-A$, or $-R$. If the initial concentrations are the same ($A_0 = R_0 = C_0$) and the amount of amine regenerated as well as the back reaction (k_{-1}) is small, then

$$\frac{-dR}{dt} = \frac{-dA}{dt} = \frac{d[AR]}{dt} = k_1[C_0 - AR]^2 = k_1A^2 = k_1R^2 \tag{15}$$

Integrating for the sugar or the amine gives

$$\frac{1}{R} - \frac{1}{R_0} = k_1t = \frac{1}{A} - \frac{1}{A_0} \tag{16}$$

Thus a plot of 1/A or 1/R vs. time is a straignt line of slope k_1 ($mole^{-1}$ hr^{-1}).

If the concentrations are not the same initially, then the solution is

$$\frac{1}{[R_0 - A_0]} \ln \frac{A_0R}{AR_0} = k_1t \tag{17}$$

A plot of $\ln(A_0R/AR_0)$ or $\ln(R/A)$ vs. time gives a straight line with a slope $= k_1[R_0 - A_0]$. Thus, k_1 could be determined theoretically more accurately by using this second-order equation since this accounts for the loss of both reactants simultaneously. Figure 14.16 from the system of Baiser and Labuza (1991) previously described shows the second-order loss of glucose and glycine with glycose at 0.4 M. In all cases good straight lines were obtained to estimate k_1, as will be described subsequently.

Since the rate of formation of AR (dAR/dt) is equal to the rate of loss of either amine or reducing compound, then

$$\frac{d[AR]}{dt} = k_1[A][R] \tag{18}$$

which gives

$$[AR] = \frac{1}{A_0B_0} \ln \frac{AR - A_0}{AR - B_0} - \frac{1}{A_0B_0} \ln \frac{A_0}{B_0} = k_1t \tag{19}$$

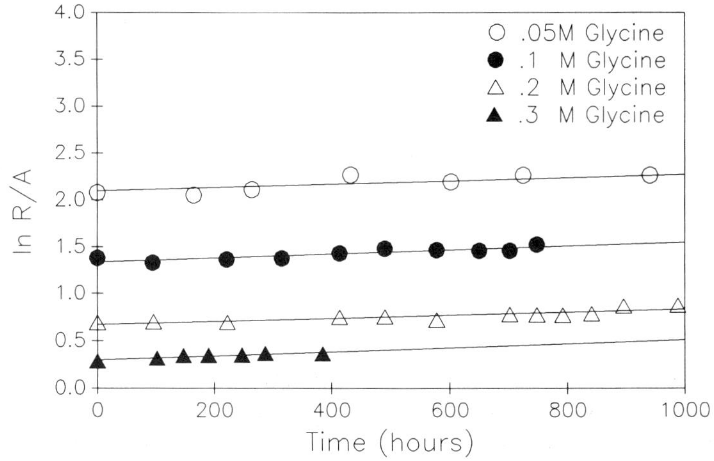

Fig. 14.16 Second-order loss plot for of glucose and glycine ($h^{-1}M^{-1}$) with an initial glucose concentration of 0.4 M at pH 7 and 37°C. (*From Baisier and Labuza, 1991.*)

This solution, if the task of measuring [AR] can be accomplished, also gives k_1 if $\ln(AR - A_0)/AR - B_0)$ is plotted versus time.

If either [A] or [R] are in excess, then for the sugar or amine loss the rate constant can be combined with the concentration of the compound in excess. Thus,

$$k_R = k_1[A_0] \tag{20}$$

and

$$k_A = k_1[R_0] \tag{21}$$

From this the rate of loss of sugar or amine is

$$\frac{-dR}{dt} = k_R[R] \tag{22}$$

and

$$\frac{-dA}{dt} = k_A[A] \tag{23}$$

If R = R_0 and A = A_0 at time zero and in each case the loss is small, Eqs. (22) and (23) can be integrated to obtain

$$\ln \frac{R}{R_0} = -k_R t \tag{24}$$

and

$$\ln \frac{A}{A_0} = -k_A t \tag{25}$$

Thus, a plot of the fraction of sugar or amine remaining vs. time gives a straight line of slope k_R or k_A, respectively. Figures 14.17 and 14.18 show first-order plots of glucose and glycine loss with slopes of k_R and k_A, respectively. Thus k_1 can be found by simply dividing by the initial concentration of the reactant not modeled, i.e.,

$$k_1 = \frac{k_R}{[A_0]} = \frac{k_A}{[R_0]} \tag{26}$$

As noted from these figures, the pseudo first-order assumption fits fairly well for up to 40% loss. Equations (24), (25), and (26) represent a pseudo

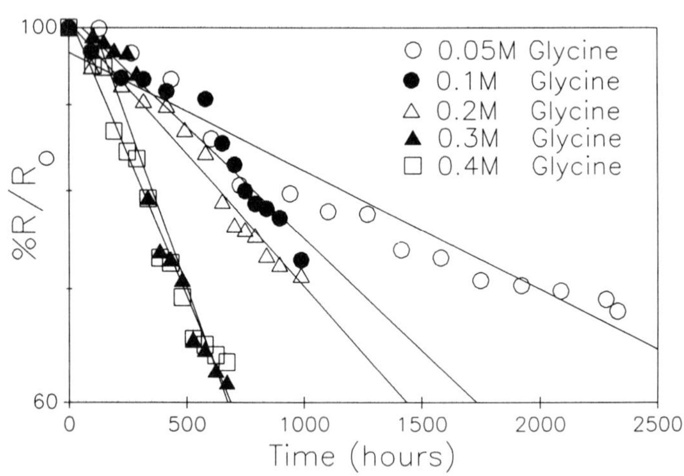

Fig. 14.17 First order loss of glucose (hr^{-1}) with an initial concentration of 0.4 M glucose and varying concentrations of glycine at pH 7 and 37°C. (*From Baisier and Labuza, 1992.*)

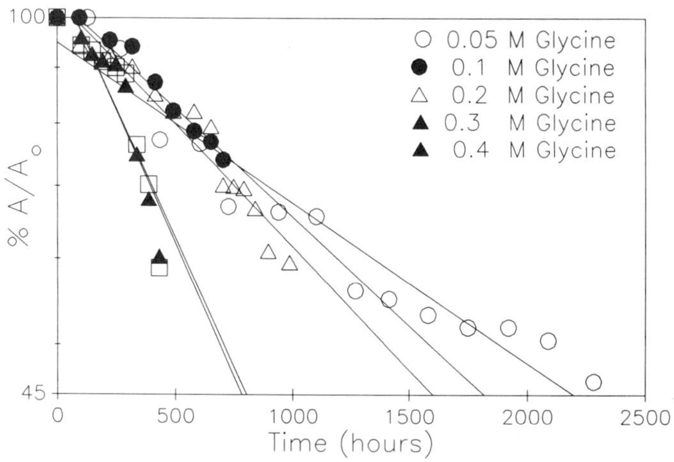

Fig. 14.18 First order loss of glycine (hr^{-1}) with an initial concentration of 0.4 M glucose and varying concentrations of glycine at pH 7 and 37°C. (*From Baisier and Labuza, 1992.*)

first-order loss and neglect the back reaction and the further reactions of the amine. Little to no data exist for the magnitude of the rate constant for these other reactions in food systems, but, as noted earlier, from Higgins and Bunn (1981), when glucose was present at 11 mM and hemoglobin was 2 mM, then $k_A = 0.0099\ hr^{-1}$ and $k_R = 0.002\ hr^{-1}$ and the pseudo first-order back reaction rate constant (k_{-g} or k_{-1}) was found to be $0.33\ hr^{-1}$. The value of k_1 thus was $9 \times 10^{-4}\ mM^{-1}hr^{-1}$ based on the amine loss and $10^{-3}\ mM^{-1}hr^{-1}$ based on glucose loss.

An important point to note from Eq. (20) is that the observed rate constant for glucose loss is only dependent on the initial glycine concentration and not the initial glucose concentration, while the observed rate constant for glycine loss is only dependent on the initial glucose concentration. From this we would expect that with an increase in the initial glucose concentration with the glycine concentration remaining constant, the rate constant for glucose loss would remain constant. On the other hand, if glycine concentration was increased with glucose remaining constant, the rate constant for glucose loss would increase. This, in fact, was found by Baisier and Labuza (1992) and helps to explain the confusion in the literature with respect to the sugar:amine ratio. Table 14.12 shows results for 0.1 M glycine in solution at pH 7.0 with increasing concentrations of glucose at 37°C. As the initial glucose concentration increases over the whole 0.04–0.5 M range, there was no significant change at the

Table 14.12 First-Order Rate Constant (hr^{-1}) for Glucose Loss in a System Containing 0.1 M Glycine with 95% Confidence Limits at 37°C and pH 7

Sample M/M	Ratio (Glu:Gly)	$k(hr^{-1} \times 10^4)$	R^2	Upper 95% limit (10^4)	Lower 95% limit (10^4)
0.04/0.1	0.4	1.25	(0.89)	1.65	0.85
0.06/0.1	0.6	1.32	(0.90)	1.71	0.93
0.1/0.1	1	1.14	(0.86)	1.52	0.76
0.2/0.1	2	1.17	(0.96)	1.39	0.95
0.3/0.1	3	1.16	(0.84)	1.73	0.58
0.5/0.1	5	0.99	(0.92)	1.52	0.46

Source: Baisier and Labuza, 1992.

95% confidence level in the rate constant using the kinetic criteria developed by Labuza (1984). Table 14.13 shows similar systems with 0.1 M glucose and varying concentrations of glycine. As the initial concentration of glycine increases, the rate constant increased in approximately the expected ratio of glycine concentration increase up to about a 0.5 molar ratio. Above that, the loss rate constant increase was less, suggesting

Table 14.13 First-Order Rate Constant (hr^{-1}) for Glucose Loss Containing 0.1 M Glucose with 95% Confidence Limits at 37°C at pH 7

Sample	Ratio (Glu:Gly)	$k(hr^{-1} \times 10^4)$ (R^2)	Upper limit (10^4)	Lower limit (10^4)	Expected increase	Actual increase
0.1/0.04	2.5	0.40^{a} (0.93)	0.59	0.20	1	1
0.1/0.06	1.7	0.66^{ab} (0.94)	0.90	0.43	1.5	1.65
0.1/0.1	1	1.00^{b} (0.92)	1.26	0.74	2.5	2.5
0.1/0.2	0.5	1.54^{b} (0.89)	1.98	1.10	5	3.85
0.1/0.3	0.33	2.52^{c} (0.97)	2.99	2.05	7.5	6.30
0.1/0.5	0.2	3.95^{c} (0.93)	5.46	2.44	12.5	9.88

[a–c] Values with the same superscript indicate no significant difference at $p < 0.05$.
Source: Baisier and Labuza, 1992.

some other reaction occurring such as regeneration of the amine through scission of the Amadori rearrangement product.

Jokinen et al. (1976) reported the loss of glucose in a soy protein/glucose system heated at 130°C as a polynomial curve fitting model, not taking into account the above kinetics. Their equation takes into account the effect of the system composition and appears as follows:

$$\frac{L}{L_0} = 0.581 + 0.466(\mathrm{pH}) - 0.0926(\mathrm{G}) - 0.0585(T) - 0.0068(t) + 0.0305(a_w)(\mathrm{S}) + 0.0331(\mathrm{S})(\mathrm{s}) \quad (27)$$

where L/L_0 is the fraction of available lysine, G is glucose content, T is temperature in degrees Celsius, t is time, S is sucrose content, and s is starch content. The fraction of lysine lost in this study never exceeded 50%. If everything is held constant except time, then one can show that

$$\frac{L}{L_0} = A + Bt \quad (28)$$

where A and B are constants. This is essentially a pseudo zero-order model and assumes an excess of sugar. Drawbacks of this model for use in food storage are the high temperatures (80–130°C) under which the data were collected and the fact that other sugars and proteins will behave differently, as was found by Schnickels et al. (1976). So it has no utility. In addition, most other researchers have reported the loss of reducing sugar and amine as a pseudo first-order reaction. Based on kinetics, however, a pseudo zero-order model is possible if the reaction extent is less than 50% and few data points over time are taken (Labuza, 1984). This model also neglects, as do most other studies, the fact that the amine is released from the Amadori compound during scission (k_3), and either reacts with more reducing sugar in the first step, enters the Strecker degradation reaction, or reacts with intermediates to form pigments which will change the observed rate constant.

Several researchers have observed what appears to be regeneration of the amino acid after significant loss. This is accounted for by the magnitude of the scission step, where AR is degraded (k_3). Thompson et al. (1976) and Wolf et al. (1977) in a study of a soy/glucose extruded mixture were the first to describe a recovery period for lysine. Since then, a study by Massaro and Labuza (1990) has corroborated that observation with different amino acids and glucose in a model TPN solution. Figure 14.19 shows a semi-log plot of the percent glycine remaining vs. time. It can be seen that the glycine loss initially follows a pseudo first-order loss

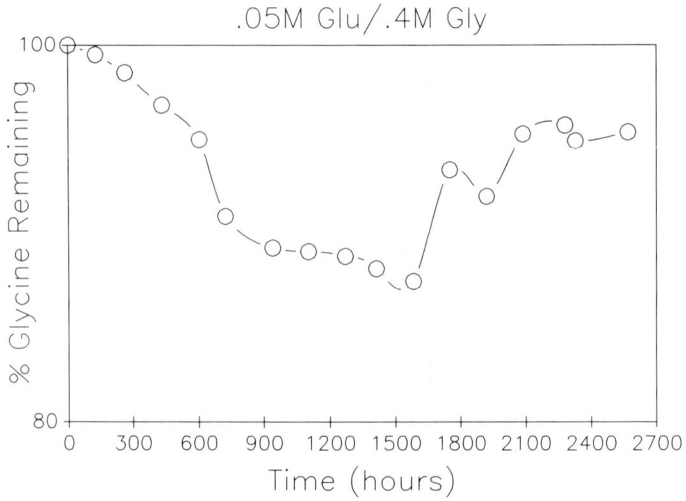

Fig. 14.19 First order of glycine showing the regeneration of glycine followed by the no loss period at pH 7 and 37°C. (*From Baisier and Labuza, 1992.*)

period until about 1800 hours, at which time the concentration increases dramatically and then levels off. In this study similar results were found for most of the systems with various initial concentrations of reactants.

The kinetics of this phenomena can be derived mathematically. From the model, the equation for change in concentration of amino acid (−dA/dt) as a function of time can be written as

$$\frac{-dA}{dt} = k_R[\mathrm{A}] - k_3[\mathrm{AR}] \tag{29}$$

where k_R is the observed rate constant for the formation of the Schiff base, assuming excess reducing compound, k_3 is the rate constant for the deamination of the Amadori rearrangement product, and [A] and [AR] are the concentrations of amino acid and Amadori rearrangement product, respectively, at time t. In this case one assumes that $k_R >>> k_{-1}$ so that the back reaction is negligible and that k_3 represents the overall backward formation rate of amine from [AR].

If we can assume that incorporation of the amino acid into the Strecker degradation reaction (k_s) and into pigment (k_{BA}) is small, then

$$A_0 - A = AR \tag{30}$$

Based on the prior work of Higgins and Bunn (1981), $k_{-1} \gg k_2$, therefore [RA] is small and can be neglected. In this case, A_0 is the initial concentration of amine and A is the concentration of amine at time t. Substituting Eq. (30) into Eq. (29), the equation for the model can be written as

$$\frac{-dA}{dt} = k_R A + k_3 A - k_3 A_0 = c_1 A - c_2 A_0 \tag{31}$$

where $c_1 = k_R + k_3$ and $c_2 = k_3$. Solving the equation, the final form can be written as

$$A = A_0 \left[\frac{c_2}{c_1} + \left(1 - \frac{c_2}{c_1}\right) e^{-c_1 t} \right] \tag{32}$$

Thus [A] at any time is a function of the initial concentration and the reaction rate constants k_R and k_3. The rate constant, k_R, is calculated as the initial rate constant, since in the early stages of amino acid loss via nonenzymatic browning, the reaction follows a pseudo first-order reaction relationship (Warmbier et al., 1976; Higgins and Bunn, 1981). After a certain time, based on the Bodenstein approximation, the change in A vs. time is negligible and

$$\frac{dA}{dt} = 0 = c_1 A_{eq} - c_2 A_0 \tag{33}$$

where A_{eq} = [A] during this steady-state period. Thus,

$$A_{eq} = \frac{k_3 A_0}{k_R + k_3} \tag{34}$$

From A_{eq}, the value of k_3 during the plateau can be calculated during the steady state phase by substitution of Eq. (34) into Eq. (33) and rearranging to get

$$k_3 = \frac{k_1 (A_{eq}/A_0)}{1 - (A/A_0)} = \frac{K_{eq} k_1}{1 - K_{eq}} \tag{35}$$

where $K_{eq} = A_{eq}/A_0$. If enough A vs. time data are taken, both at early and later stages, and A_0 is known, k_3 is thus easily calculated. Massaro and Labuza (1990) showed that Eq. (32) fit the data very well.

This model predicts an initial first-order decrease in free amino acid

has considered browning as a pseudo zero-order reaction, where the concentration followed by a steady-state phase in which the change in concentration of [A] is small. Figure 14.20, from Massaro and Labuza (1990, for a 45-to-1 glucose–to–total amino acid (lysine, tryptophan, and cysteine) ratio in a model TPN solution shows a plot of % lysine remaining vs. time. The system contained 1.1 M glucose, 0.0152 M lysine, 0.0035 M tryptophan, and 0.0057 M cysteine. As can be seen, the loss of lysine is initially rapid. Lysine is then regenerated and enters a no loss period after approximately 30% of the lysine had disappeared. Table 14.14 lists the $\%A_{eq}$ values for the three amino acids they studied derived from the plateau obtained.

Warmbier et al. (1976) studied browning in a low moisture system that contained microcrystalline cellulose, oil, casein, glycerol, glucose, and potassium sorbate. Glucose followed first-order kinetics as expected. Lysine loss was shown to be approximately first order at the start of the reaction, while the rate decreased as the reaction proceeded. The authors explained the deviation from strict first-order behavior by the fact that lysine can react to form brown pigment and can also participate in the Strecker degradation reaction, as has been hypothesized by Massaro and Labuza (1990), but assumed to be negligible.

The formation of the Amadori rearrangement product [AR] follows Eq.

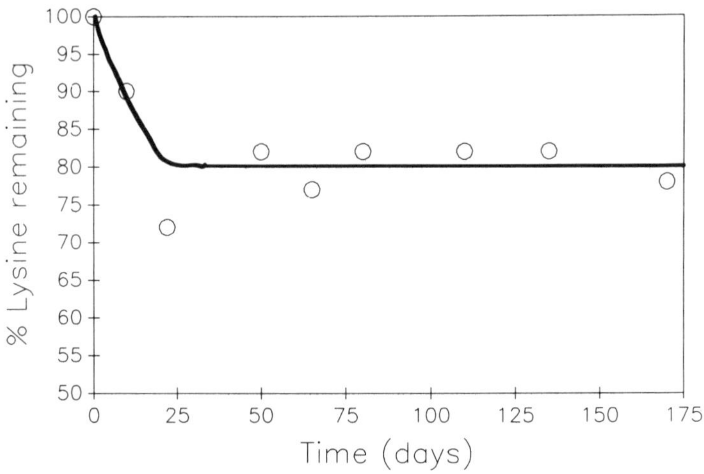

Fig. 14.20 Lysine loss of a glucose/lysine/tryptophan/cysteine system at pH 5.5 at 30°C. (*From Massaro and Labuza, 1990.*)

Table 14.14 % A_{eq} Values for TPN Solutions Containing the Specified Amino Acids

System	Temp.	$\%A_{eq}$ Lysine	$\%A_{eq}$ Tryptophan	$\%A_{eq}$ Cysteine
Lysine	4	77		
	30	71		
Tryptophan (T)	4		87	
	30		74	
Cysteine (C)	4			71
	30			73
L/T	4	71	68	
	30	58	74	
L/T/C	4	78	70	68
	30	79	64	70
Commercial	4	92	77	55
	30	78	56	59

Source: Massaro and Labuza, 1990.

(36) where we now account for both formation and loss and assume no back reaction to [RA]. Thus,

$$\frac{d[\mathrm{AR}]}{dt} = k_1[\mathrm{A}][\mathrm{R}] - k_3[\mathrm{AR}] \tag{36}$$

If the reducing compound is in excess, then the following is true:

$$\frac{d[\mathrm{AR}]}{dt} = k_R[\mathrm{A}] - k_3[\mathrm{AR}] \tag{37}$$

$$\frac{d[\mathrm{AR}]}{dt} = k_R \mathrm{A}_0 e^{-kt} - k_3[\mathrm{AR}] \tag{38}$$

Solving Eq. (38), a linear differential equation, the concentration of the rearrangement product with respect to time is

$$[\mathrm{AR}] = \frac{\mathrm{A}_0 k_R}{k_3 - k_R}\left(e^{-k_R t} - e^{-k_3 t}\right) \tag{39}$$

The formation of intermediates F, assuming similar reaction rate constants, can be expressed as

$$\frac{dF}{dt} = k_3[AR] - k_B[F] \tag{40}$$

The solution for F as a function of time will depend on the relative magnitudes of k_3 and k_B. Various nonlinear mathematical modeling procedures could be used to solve for these constants, if F and B could be determined as concentrations. Baisier and Labuza (1992), using nonlinear techniques and assuming $F_0 = 0$, solved for F. The solution is shown in Eq. (41).

$$F = k_3 k_A [R_0] \left\{ \frac{e^{-k_A t}}{(k_3 - k_A)(k_B - k_A)} + \frac{e^{-k_3 t}}{(k_A - k_3)(k_B - k_3)} + \frac{e^{-k_B t}}{(k_A - k_B)(k_3 - k_B)} \right\} \tag{41}$$

k_A is used when amine is in excess, and k_R is used in place of k_A when reducing compound is in excess. It should be expected that F would behave as expected for any intermediate; rise exponentially, slow down to reach a maximum and stay there for a period of time (the Bodenstein steady-state period) and then decrease, much like the hydroperoxide concentration in lipid oxidation, which was also modeled by Bodenstein steady-state kinetics (Labuza, 1971). Of critical importance is the fact that the fluorescent compounds have not been characterized. Thus the value of F cannot be put into a molar basis. If, in fact, the intermediate steps are many, and many different compounds form, then Eq. (41) cannot be reliably tested.

As discussed previously, White and Thistle (1943) and Pearce and Bryce (1947) studied the relationship between fluorescence and eating quality in dried egg powder. It was observed that fluorescence increased to a maximum and then decreased again, as would be expected for an intermediate. Pearce (1950) studied a solution of glucose and glycine or acetaldehyde with ammonia at pH 5.8 and temperatures from 27 to 93°C. He also found that fluorescence increased until it reached a maximum and then decreased again, as for the egg powder system, and that decreasing the concentration of either reactant increased the amount of time required to reach a set fluorescence value. These studies were done under accelerated conditions to evaluate the entire extent of the reaction in short periods of time. Such a dramatic increase and decrease of the fluorescence value would probably not be seen under more mild conditions. Figure

14.21 is a theoretical plot of the relationship between reactants and products in the Maillard reaction. It shows the initial reduction in concentration of reducing sugars and amines, the formation and degradation of intermediates such as the Schiff base, Amadori rearrangement product and fluorescent pigments, and the slow production of brown pigment.

From the kinetic scheme shown at the beginning of this section, the rate of formation of browning can be written

$$\frac{d\mathrm{B}}{dt} = k_B \mathrm{F} \tag{42}$$

Baisier and Labuza (1991), using the same nonlinear technique they used to solve for florescence, found a solution for B, the brown pigment, as

$$\mathrm{B} = [\mathrm{R}_0] + [\mathrm{B}_0] + [\mathrm{F}_0](1 - e^{-k_B t}) - k_3 k_A k_B [\mathrm{R}_0] \times \left\{ \frac{e^{-k_3 t}}{k_3(k_A - k_3)(k_B - k_3)} + \frac{e^{-k_A t}}{k_A(k_3 - k_A)(k_B - k_A)} + \frac{e^{-k_B t}}{k_B(k_3 - k_B)(k_A - k_B)} \right\} \tag{43}$$

As with fluorescence, the browning pigments have not been characterized well enough to test this equation adequately. Most of the literature

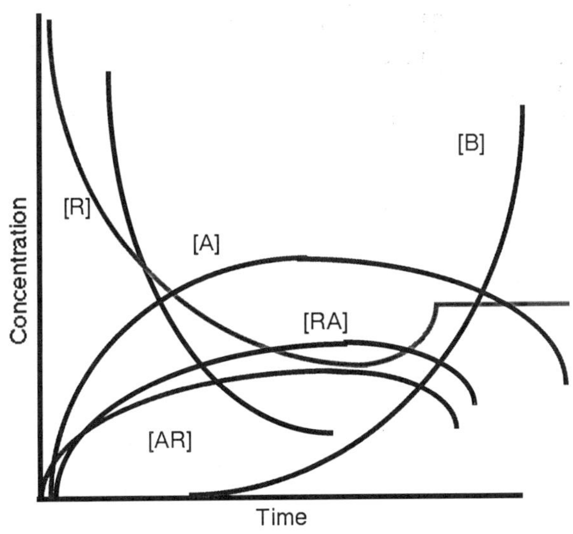

Fig. 14.21 Relationship between the reactants and products in the Maillard reaction as a function of time.

amount of browning (OD/g or OD/ml) is plotted vs. time and linear regression is used to get the rate constant (e.g, Warmbier et al., 1976; Labuza and Saltmarch, 1981). The units for browning are generally reported as either OD, OD/ml, or OD/g sample, none of which account for the influence of initial concentration of amine or sugar or their loss over time which, as we have shown, is very complex and not directly related to the concentrations used. In essence, using an assumed set of rate constants, calculated substrate losses, intermediate production, and pigment production produce the pattern shown in Fig. 14.21. The amount of browning should be negligible for a fairly long induction time or increase slowly since either k_B is small compared to the rest of the rate constants or the concentration of F is very small. A significant induction period has been found in many low temperature studies. This indicates that k_B is small. After F builds up, then B should begin to increase dramatically at a time-dependent rate. Both the induction and exponential rate periods could be modeled as pseudo zero order, depending on the quality of the data collected. Zero-order modeling is generally the practice. Importantly, at high temperature, the lag period should decrease or disappear altogether, and the increase in B may follow an exponential pattern over the whole time period.

Some published results do not fit to the initial zero-order assumption. For example, Haugaard et al. (1951) refluxed a mixture of glucose and glycine at 100°C and reported browning as the absorbance at 490 nm over a 5-hour time period. The author used three systems with varying concentrations of glucose and glycine as shown in Table 14.15. The authors used polynomial regression to develop the following model to predict amounts of browning versus time based on the concentrations of the substrates.

$$[B] = \frac{k_h[A]^2[R][t]^2}{2} \tag{44}$$

where [B] is the concentration of brown pigment, k_h is the reaction rate

Table 14.15 Browning Data Showing Sugar/Amine Effect

Solution	Glucose (M)	Glycine (M)	Slope(dB/dt) $\times 10^{-4}$	Slope/[Glu] $\times 10^4$
A	0.5	0.2	1.8	3.9
B	1.0	0.2	3.8	3.8
C	0.5	0.4	8.5	17.0

Source: Haugaard et al., 1951.

constant, [A] is the amino acid concentration, and [R] is the reducing sugar concentration at time t. Based on their model, the browning rate is the derivative of Eq. (44), assuming A and R go directly to B, i.e., ignoring the intermediate steps.

$$\frac{dB}{dt} = k_h[A]^2[R]t \tag{45}$$

This suggests that the rate increases as time increases and falls as the concentration of the two substrates decreases with time, which should be expected. The reaction is pseudo second order with respect to the amino acid concentration and first order with respect to the reducing sugar. Based on this, the rate of browning, over a narrow range, should double with a doubling of reducing sugar concentration, at the same amino acid concentration, as seen in Table 14.15 by simply dividing the slope of B vs. time by [R]. In addition, at a constant sugar concentration, when the amino acid concentration is doubled, the rate will go up by a factor of 4. Although there is no theoretical basis for this model, this equation implies that two moles of amino acid and one mole of reducing sugar react to form one mole of pigment and that the production of brown pigment is not linear with time, neither of which follows any of the prior derived kinetics. The system this equation was based on was heated at refluxing temperature and therefore would not represent the kinetics under normal storage conditions. Baisier and Labuza (1992) used their data to test the Haugaard model. The authors pointed out that since the initial reaction is bimolecular, predicting that a doubling of the reducing sugar concentration would cause a doubling of the browning rate would probably not be correct if all reactants went to pigment. The part of the Haugaard model that is unique is the prediction that doubling the amine concentration would cause a fourfold increase in the rate of brown pigment formation and that an increase in the rate of browning is linear with the sugar concentration. Table 14.16 from Baisier and Labuza (1992), shows the increase in the rate constant for brown pigment formation for samples with 0.4 M glucose and increasing concentrations of glycine at 30°C and pH 7, while Table 14.17 shows rate constants for samples with 0.4 M glycine and increasing concentrations of glucose. Although it appears that increases in glycine increase the observed zero-order rate, the increase is nowhere near a fourfold increase for every twofold increase in concentration and, in fact, seems to level off. Also, the browning rate increased more than linearly with increases in glucose, as seen in Table 14.17. The results of Haugaard et al. (1951) may have been fortuitous and a function of the conditions of the particular experiment.

Table 14.16 Zero-Order Rate Constant for Brown Pigment Formation as a Function of Increasing Glycine Concentration at Constant Glucose Level

Sample	k(OD/hr $\times$ 10^4)	Glycine ratio	Increase in k
0.4/0.05	2.97	1	—
0.4/0.1	46.34	2	15.6
0.4/0.2	48.91	4	16.5
0.4/0.3	62.30	6	21.0
0.4/0.4	67.72	8	22.8

Source: Baisier and Labuza, 1992.

A zero-order model for the kinetics of browning was used by Mizrahi et al. (1970) for cabbage, Warmbier et al. (1976) for a model casein/glucose system, and Waletzko and Labuza (1976) for a semimoist food, where

$$\frac{dB}{dt} = k_B = \text{rate}_{\text{browning}} \tag{46}$$

No induction period was assumed and the dependence of the rate of pigment formation on reactant concentration is accounted for in k_B, thus making k_B food composition specific. All of these studies were done between 20 and 45°C. Since, in many cases, the acceptable amount of browning is small, this is a fairly good assumption.

Song et al. (1966) and Song and Chichester (1966) developed a much more complicated model for the Maillard reaction based on the following presumed mechanism.

Table 14.17 Zero-Order Rate Constant for Brown Pigment Formation as a Function of Increasing Glucose Concentration at Constant Glycine Level

Sample	k(OD/hr $\times$ 10^4)	Glucose ratio	Increase in k
0.05/0.4	4.53	1	—
0.1/0.4	41.83	2	9.2
0.2/0.4	38.38	4	8.5
0.3/0.4	58.72	6	13.0
0.4/0.4	67.72	8	14.9

Source: Baisier and Labuza, 1992.

$$R + A \xrightarrow{k_1} F \tag{47}$$

$$F \xrightarrow{k_2} B_f \tag{48}$$

$$F + A \xrightarrow{k_3} B_m \tag{49}$$

R and A are reducing sugar and amine, respectively, B_f is colored by-product, and B_m stands for melanoidins. Many assumptions were made in this study, but were not adequately tested. They assumed that intermediates or by-products of side reactions contribute to melanoidin production and that α- and β-unsaturated enolic aldimines accumulate to a steady-state concentration and are the main precursors to melanoidins, following the pathway for F developed above. They concluded that the formation of hydroxymethyl furfural is not a main pathway to browning in the Maillard reaction. They assume that formation of intermediates is the rate-limiting step for the overall browning reaction, as we have done. According to McWeeny (1969), they did not prove that unsaturated aldimines are the key intermediates during and after the initial Maillard induction period.

The equation they produced, using our notation, is

$$\frac{dB}{dt} = \frac{k_B(k_{1+2+3+4})[R]}{(k_B[F])^{1/2}} + k_{BA}([A_0] - [B])([F] - [B]) \tag{50}$$

where R is the glucose concentration, F is the intermediate concentration, A is glycine concentration, B is total brown pigment concentration, and the ks represent specific rate constants.

CONCLUSIONS

From this review of the literature, it is clear that there are significant voids as well as conflicting results in the published literature on the kinetics of NEB. As mentioned previously, there are problems in comparing the reported kinetic data from different investigators for substrate loss, intermediate and brown pigment formation rates due to modes of reporting results. If a thorough job of understanding this reaction is to be accomplished, a standard method of reporting results of experiments is needed. Then, studies can be compared and cumulative conclusions reached. If this had been done in the past, we probably would not have the kinds of conflicting results discussed in the sections on the effect of amine type or sugar:amine ratio on browning rate. Several studies previously discussed used polynomial fits to describe the reaction. Since they have no theoretical basis, these do not help to explain the kinetics of the reaction and depend on specific experimental conditions. Conditions used in some of the studies reported in the literature are another area of concern. While increasing the temperature shortens the length of experiments, it does not provide valuable data about conditions generally encountered for food storage. Finally, until the fluorescent and brown pigments are fully characterized for molecular weight, theoretical equations based on the overall mechanism cannot be tested.

ACKNOWLEDGMENT

Published in the contribution series of the Minnesota Agricultural Experiment Station based on research conducted under Project No. 18-78.

REFERENCES

Adrian, J. 1982. In Handbook of Nutritive Value of Processed Food, Volume I: Food for Human Use, pp. 529–609. CRC Press, Inc. Boca Raton, FL.

Ashoor, S. H. and Zent, J. B. 1984. *J. Food Sci.* 49: 1206–1207.

Baisier, W. M. and Labuza, T. P. 1992. *Agr. Food Chem.* (in press)

Bell, L. and Labuza, T. P. 1991a. *J. Food Sci.* 56(1): 17–20.

Bell, L. and Labuza, T. P. 1991b. *Cryoletters* 128.

Ben-Gara, I. and Zimmerman, G. 1972. *J. Food Sci. Technol.* 9: 113.

Bunn, H. F. and Higgins, P. J. 1981. *Science* 213: 222–224.

Burton, H. S. and McWeeny, D. J. 1963. *Nature* 197: 266–268.

Burton, H. S., McWeeny, D. J., and Biltcliffe, D. O. 1962. *Chem. and Industr.* 219.

Cantor, S. M. and Peniston, Q. P. 1940. *JAOC* 62: 2113–2121.

Carpenter, K. J., Morgan, C. B., Lea, C. H., and Parr, L. J. 1962. *Brit. J. Nutr.* 16: 451.

Chichester, C. O., Stadtman, F. H., and Mackinney, G. 1952. *J. Am. Chem. Soc.* 74: 3418–20.

Copley, M. J. and Van Arsdel, W. B. 1964. *Food Dehydration* 2: 540.

Delahay, P. and Strassner, J. E. 1952. *J. Am. Chem. Soc.* 74: 893.

Dworschak, E. and Orsi, F. 1977. *Acta Aliment.* 6: 59.

Ellis, G. P. 1959. *The Maillard Reaction in Advances in Carbohydrate Chemistry*, M. L. Wolfrom, (Ed.), pp. 63–134. Academic Press, New York.

Eichner, K. and Karel, M. 1972. *J. Agr. Food Chem.* 20:218.

Enders, C. and Theis, K. 1938. *Brennst. Chem.* 19: 360–365, 402–407, 439–449.

Fabriani, G. and Frantoni, A. 1972. *J. Agri. Food Chem.* 20: 218.

Frankel, M. and Katchalsky, A. 1937. *Biochem J.* 31: 1595.

Frankel, M. and Katchalsky, A. 1938. *Biochem J.* 32: 1904.

Friedman, L. and Kline, O. L. 1950. *J. Biol. Chem.* 184: 599.

Fry, L. K. and Stegink, L. D. 1982. *J. Nutrition.* 112: 1631–1637.

Haugaard, G., Tumerman, L., and Silvestri, H. 1951. *J. Am. Chem. Soc.* 73: 4594.

Hayward, L. D. and Angyal, S. J. 1977. *Carbohydr. Res.* 53: 13–20.

Hendel, C. E., Silveira, V., and Harrington, W. O. 1955. *Food Technol.* 2: 433.

Hermann, J. and Andrae, W. 1963. *Nahrung* 7: 243.

Higgins, P. J. and Bunn, H. F. 1981. *J. Biol. Chem.* 256(10): 5204–5208.

Hodge, J. E. and Rist, C. E. 1952. *J. Am. Chem. Soc.* 74: 1494–1497.

Hodge, J. E. and Rist, C. E. 1953. *J. Am. Chem. Soc.* 75: 316.

Hodge, J. E. 1953. *J. Agric. Food Chem.* 1(15): 928–943.

Jokinen, J. E., Reineccius, G., and Thompson, D. R. 1976. *J. Food Sci.* 41: 816.

Kaanane, A. and Labuza, T. P. 1989. The Maillard Reaction in Foods. In *The Maillard Reaction in Aging, Diabetes and Nutrition*. J. Baynes (Ed.). A. R. Liss Press, Inc., New York.

Kline, R. W. and Stewert, G. F. 1948. *Ind. Eng. Chem.* 40: 919.

Labuza, T. P. 1971. *Crit. Rev. Food Tech.* 2: 335.

Labuza, T. P. 1984. *J. Chem. Ed.* 61(4): 348–388.

Labuza, T. P. and Saltmarch, M. 1981. *J. Food Sci.* 47: 92–96.

Labuza, T. P., Warren, K., and Warmbier, H. C. 1977. *Adv. Exp. Med. Biol.* 86B: 379.

Lea, C. H. and Hannan, R. S. 1949. *Biochim. Biophys. Acta* 3: 313–325.

Lea, C. H. and Hannan, R. S. 1950. *Biochim. Biophys. Acta* 4: 518–531.

Lerici, C. R., Barbanti, D., and Cherubin, S. 1990. *Lebensm.-Wiss. u.-Tech.*, 23: 289–294.

Levine, H. and Slade, L. 1989. *Agric. Food Chem.* 1: 315.

Ling, A. R. and Malting, J. 1908. *J. Inst. Brew.* 14: 494–521.

Lippich, F. 1932. *Biochem.* Z. 248: 280.

Massaro, S. A. and Labuza, T. P. 1990. *J. Food Sci.* 55(3): 821–826.

McWeeny, D. J. 1973. In *Molecular Structure and Function of Food Carbohydrates*. Birch G. and Green L. (Ed.), pp. 21–32. John Wiley and Sons, New York.

Miller, P. S. 1956. *J. Sci. Food Agric.* 7: 337.

Mizrahi, S., Labuza, T. P., and Karel, M. 1970. *J. Food Sci.* 35: 799.

Monnier, V. M., Sell, D. R., Satoshi, M., and Nagaraj, R. H. 1989. In *Ref. Finot.*, pp. 393–414.

Namiki, M. and Hayashi, T. (1975). *J. Agric. Food Chem.*, 23: 487–491.

O'Brian, J. and Morrissey, P. A. 1989. *Critical Reviews in Food Science and Nutrition*. 28(3): 211–248.

Olcott, H. S. and Dutton, H. G. 1945. *Ind. Eng. Chem.* 37: 1119.

Overby, L. R. and Frost, D. V. 1952. *J. Nutrition* 46: 539.

Patron, A. 1949. Ph.D thesis, University of Massachusetts, Amherst, MA.

Patton, A. P. and Chism, P. 1951. *Nature* 167: 406.

Pearce, J. A. 1944. *Can. J. Research* F, 22: 87.

Pearce, J. A. 1950. *Food Technol.* 4: 416–419.

Pearce, J. A. and Bryce, W. A. 1947. *Food Technol.* 1: 310.

Pilkova, L., Pokorny, J., and Davidek, J. 1990. *Die Nahrung*. 34(8): 759–764.

Pomeranz, Y., Johnson, J. A., and Schellenberger, J. A. 1962. *J. Food Sci.* 29: 350.

Reineccius, G. A. 1990. In *Ref. Finot.*, pp. 157–170.

Reynolds, T. M. 1963. *Adv. Food Res.* 12: 1. II. *Adv. Food Res.* 14: 168.

Rizvi, M. and Peri, C. 1986. *J. Food Technol.* 21: 199–208.

Rubinsztain, Y., Ioselis, P., Ikan, R., and Aizenshtat, Z. 1984. *Org. Geochem.* 6: 791–804.

Schnickels, R. A., Warmbier, H. C., and Labuza, T. P. 1976. *J. Agric. Food Chem.* 24(5): 901–903.

Schroeder, L. J., Iacobellis, M., and Smith, A. H. 1955. *J. Biol. Chem.* 212: 973–983.

Song, P-S,and Chichester, C. O. 1966. *J. Food Sci.* 31(6): 906–913.

Song, P-S, Chichester, C. O., and Stadtman, F. H. 1966. *J. Food Sci.* 31(6): 906–913.

Spark, A. A. 1969. *J. Sci. Fd. Agric.* 20: 308–316.

Stadtman, F. H., Chichester, C. O., and Mackinney, G. 1952. *J. Am. Chem. Soc.* 74: 3194–3196.

Taira, H., and Sukukai, Y. 1966. *Japan J. Nutr. Food* 18: 359.

Thomas, M., Turner, A. D., Abad, G., and Towner, J. M. 1977. *Milchwissenschaft* 32: 12–15.

Thompson, D. R., Wolf, J. C., and Reineccius, G. A. 1976. *Trans. ASAE* 19(5): 989–992.

Tsao, T. F., Frey, A. L., and Harper, J. M. 1978. *J. Food Sci.* 43: 1106–1108.

Underwood, J. C., Lento, H. G. Jr., and Willits, C. O. 1959. *Food Res.* 24: 181–184.

Waletzko, P. T. and Labuza, T. P. 1976. *J. Food Sci.* 41: 1338.

Warmbier, H. C., Schnickels, R. A., and Labuza, T. P. 1976. *J. Food Sci.* 41: 981–983.

Whistler, R. L. and Daniel, J. R. 1985. In *Food Chemistry*, 2nd ed. O. Fennema (Ed.)., pp. 69–137. Marcel Dekker, Inc., New York.

White, W. H. and Thistle, M. W. 1943. *Can. J. Research, D.* 21: 194.

Wolf, J. C., Thompson, D. R., and Reineccius, G. A. 1977. *J. Food Sci.* 42: 1540.

Wolfrom, M. L., Kashimura, N., and Horton, D. 1974. *J. Agr. Food Chem.* 792: 796.

Yaylayan, V. 1990. *Trends in Food Science & Technology* 1:(July): 20–22.

15

Kinetics of Lipid Oxidation

Marcus Karel

Rutgers University
New Brunswick, New Jersey

INTRODUCTION

Lipid Oxidation

Lipid oxidation is one of the two most important reactions in food systems. It is the cause of important deteriorative changes in food flavor, color, texture, and nutritional value. It may also result in formation of toxic and mutagenic agents. On the other hand, oxidation of lipids is also an important source of flavor compounds considered desirable. The above features are common also to the other complex of reactions in foods which shares the key importance in food technology: *nonenzymatic browning*. Furthermore, both reactions are linked by complex interactions. Browning reactions may be initiated, and final composition affected by products of lipid oxidation. Conversely lipid oxidation is known to be affected by products of nonenzymatic browning, some of which are antioxidants.

Nevertheless, it is possible to draw some important distinctions between the two sets of reactions. Browning occurs most readily in aqueous environments, is very temperature sensitive, and many of its steps are very pH sensitive. Lipid oxidation is less temperature dependent, occurs most readily in anhydrous systems, and is promoted by electromagnetic radiations. It has, therefore, the properties characteristic of *free-radical* reactions.

FREE RADICALS

Covalent bonds can break by homolytic or by heterolytic scission as shown in Eqs. (1) and (2).

$$\text{homolytic:} \quad A:B \rightarrow A\cdot \;+\; B\cdot \tag{1}$$

$$\text{heterolytic:} \quad A:B \rightarrow A:^{-} \;+\; B^{+} \quad \text{or } A^{+} \;+\; B:^{-} \tag{2}$$

Homolytic scission therefore yields free radicals (atoms or groups having an unpaired electron in the outer orbit), and heterolytic scission yields ions. Homolytic scission is more likely to occur in absence of solvent (especially water). Water, because of its dipolar character, promotes ionization:

$$OH_2 \cdots A{:}B \cdots OH_2 \rightarrow OH_2 \cdots A^{-} + B^{+} \cdots OH_2 \tag{3}$$

of course water itself is partially ionized:

$$H_2O \rightarrow OH^{-} + H^{+}$$

and can produce free radicals when irradiated with gamma rays or high energy electrons:

$$H_2O \rightsquigarrow OH\cdot + H\cdot \tag{4}$$

More persistent free radicals (with significant half-life in solvents as well as in gas phase) can occur when so-called "weak" bonds are broken. In food technology lipid oxidation provides the key example of such bonds in peroxides and hydroperoxides:

$$ROOR \; RO\cdot + RO\cdot \tag{5}$$

$$ROOH \rightarrow RO\cdot + OH\cdot \tag{6}$$

Other common free radicals in foods are those produced by thermally (roasting) or mechanically induced scission of C–C bonds and S–S bonds (milling, extrusion) and by hydrogen abstractions, notably by other free radicals:

$$\text{–SH} + R\cdot \rightarrow RH + \text{–S}\cdot \tag{7}$$

$$\begin{array}{c}\text{H}\\ \text{–C=C–C–C=C–}\\ \text{H H H H H}\end{array} + R\cdot \rightarrow RH + \begin{array}{c}\text{H}\\ \text{–C=C–}\underset{\cdot}{\text{C}}\text{–C=C–}\\ \text{H H}\quad\text{H H}\end{array} \tag{8}$$

The pentadiene group and the SH group are particularly abstraction sensitive. The radical formed by the pentadiene is stabilized by formation of conjugated double bonds:

$$\text{–}\underset{\cdot}{\text{C}}\text{–C=C–C=C–}$$

Notably, in lipid oxidation, peroxides formed by molecular oxygen addition to the free radical are conjugated. Thus 9,12-linoleic acid oxidizes to a 9-peroxide with double bonds in the 10,12 positions, or a 13-peroxide with double bonds in the 9,11 positions.

The key characteristic of free radicals is their lack of stability. When sufficient mobility exists (mobility may be reduced by trapping radicals in glasses), they recombine or attack other groups.

BASIC AUTOXIDATION SCHEME (BAS) OF THE LIPID OXIDATION CHAIN REACTION

The following scheme provides the commonly accepted basis for analyzing autoxidation kinetics. In the equations that follow, r refers to *rate* (e.g., concentration change per unit time), k is the appropriate rate constant (i.e., to obtain r, k must be multiplied by appropriate concentrations), RH is the oxidizing substrate, and NRP are nonradical products.

Step 1	Initiation	$\rightarrow R\cdot$	r_i;	(9)
Step 2	$R\cdot + O_2$	$\rightarrow RO_2\cdot$	k_2	(10)
Step 3	$RO_2\cdot + RH$	$\rightarrow ROOH + R\cdot$	k_3	(11)
	Termination			
Step 4	$R\cdot + R\cdot$	$\rightarrow$ NRP	k_4	(12)
Step 5	$RO_2\cdot + R\cdot$	$\rightarrow$ NRP	k_5	(13)
Step 6	$RO_2\cdot + RO_2\cdot$	$\rightarrow$ NRP	k_6	(14)

Steps 2 and 3 constitute the chain reaction propagation: the net effect is conversion of RH and of O_2 to ROOH and regeneration of a radical. Steps 4, 5, and 6 collectively give the termination rate r_t. In order to solve the above equations to give a simplified rate of oxidation, several simplifying assumptions are usually justified and widely used.

The key assumption that allows the development of rate expressions for free radical reactions is the Bodenstein or "steady-state" assumption:

$$\frac{d(\text{radical})}{dt} = 0 \tag{15}$$

Most persistent free radical reactions conform to this assumption, since if the concentration of radicals decreases with time, then the reaction "dies out," and an increase of radicals with time leads to explosions.

The second important assumption is that the three termination steps are related by

$$k_5 = \sqrt{k_4 k_6} \tag{16}$$

Utilization of the above assumptions allows us to derive an equation for overall rate of oxidation, which we may define as either

$$\bar{r} = \frac{d\,(\text{ROOH})}{dt} \tag{17}$$

or

$$\bar{r}' = \frac{d(\text{O}_2)}{dt} \tag{18}$$

If neither O_2 nor ROOH are involved in initiation or termination, of if their increase (decrease) in initiation or termination is negligible, then $\bar{r} = -\bar{r}'$. Such an assumption is obviously only an approximation. As an example of deviations, consider termination of 2 ROO· radicals, which often results in generation of O_2:

$$2\text{RO}_2\cdot \rightarrow \text{ROOOOR} \rightarrow \text{ROOR} + \text{O}_2 \tag{19}$$

Similarly ROOH is generated in Step 3, but is often consumed in initiation (as will be discussed below) or in ROOH decomposition leading to various (often volatile) products of autoxidation. The overall expression for $\bar{r}$ is shown in Fig. 15.1 (Reich and Stivala, 1969).

Assume $r_i = r_t \qquad \phi = k_5(k_4 k_6)^{-1/2}$

$$k_2(R\cdot)(O_2) = k_3(RO_2\cdot)(RH) = \bar{r}$$

Then:

$$\bar{r} = \frac{k_3(RH)r_i^{1/2}}{k_6^{1/2}\left(1 + \frac{2k_5k_3(RH)}{k_2k_6(O_2)} + \frac{k_3^2k_4(RH)^2}{k_2^2k_6(O_2)^2}\right)^{1/2}}$$

and if $\phi = 1$

Letting $B = k_3k_6^{-1/2} \qquad A = k_2k_4^{-1/2}$

$$\bar{r} = B(RH)r_i^{1/2}\{1 + A^{-1}B(RH)(O_2)^{-1}\}^{-1}$$

In this case a plot of $\bar{r}^{-1}$ vs. $(O_2)^{-1}\}^{-1}$ is linear.

Fig. 15.1 Basic autoxidation scheme (BAS): Simplifying assumptions allowing calculation of overall oxidation rate.

OXYGEN DEPENDENCE OF LIPID OXIDATION AND SIMPLIFIED KINETIC EXPRESSIONS FOR LOW AND HIGH OXYGEN CONCENTRATIONS

The termination of the oxidation reaction can occur by steps 4, 5, or 6. Which of these steps dominates depends on the oxidizing substrate (RH) and most significantly on oxygen pressure. The influence of substrate can be characterized by

$$\phi = \frac{k_5}{(k_4k_6)^{1/2}}$$

Figure 15.2 shows how the termination of radicals derived from ethyl linoleate depends on oxygen pressure, at a temperature of 25°C. It should be noted that since ϕ for a given substrate varies with temperature, the oxygen dependence of termination will also be temperature dependent. It is obvious that at "high" oxygen pressures termination occurs almost exclusively by step 6 and at "low" oxygen pressures by step 4. Under these conditions, using the steady-state assumption it can be shown that,

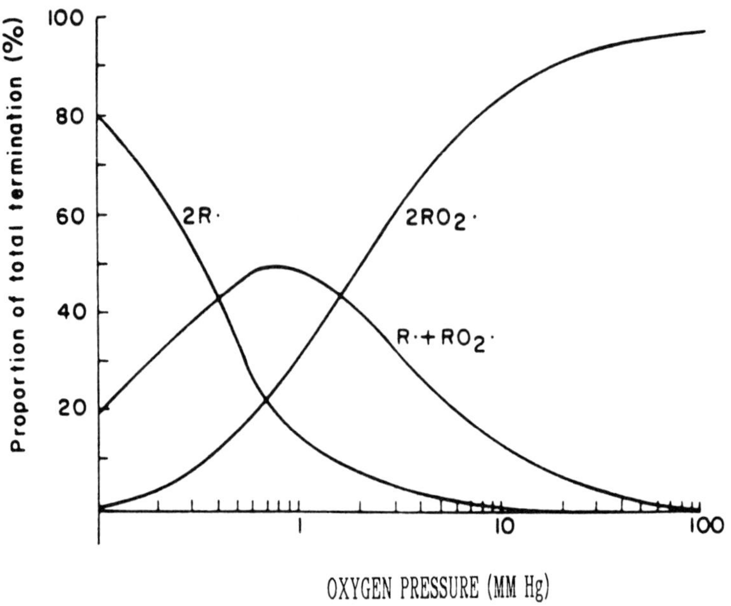

Fig. 15.2 Termination contributions of r_4, r_5, and r_6 for a substrate with O = 2.5 (*From Reich and Stivala, 1969.*)

for low oxygen pressure,

$$r_i = 2k_4 \, \mathrm{R}\cdot^2 \tag{20}$$

$$\mathrm{R}\cdot = \sqrt{r_i/2k_4} \tag{21}$$

$$r' = r_i^{1/2} \, (2k_4)^{-1/2} \, (\mathrm{RH}) \, (k_3) \tag{22}$$

(note rate independent of RH
for low O_2 pressure)

and for high oxygen pressure,

$$r_i = 2k_6 \, (\mathrm{RO_2}\cdot)^2 \tag{23}$$

$$\mathrm{RO_2}\cdot = r_i^{1/2} \, (2k_6)^{-1/2} \tag{24}$$

$$r = r_i^{1/2} \, (2k_6)^{-1/2} \, (\mathrm{RH}) \, (k_3) \tag{25}$$

(note rate independent of O_2
for high O_2 pressure)

For intermediate oxygen pressures, these equations may be modified to include a function representing oxygen dependence:

$$\bar{r} = k_3(\text{RH})\, r_1^{1/2}\, (2k_4)^{-1/2}\, f(\text{O}_2) \qquad (26)$$

$$f(\text{O}_2) = \frac{[\text{O}_2]}{b + [\text{O}_2]c} \qquad (27)$$

It should be noted that this function, noted on the basis of empirical data by Karel (1960), is in fact anticipated by the more complex expression shown in Fig. 15.1. The typical dependence of oxidation rate on oxygen pressure is shown in Fig. 15.3 for an avicel-lipid model system and a dry milk–lipid model system (Kacyn et al. 1983).

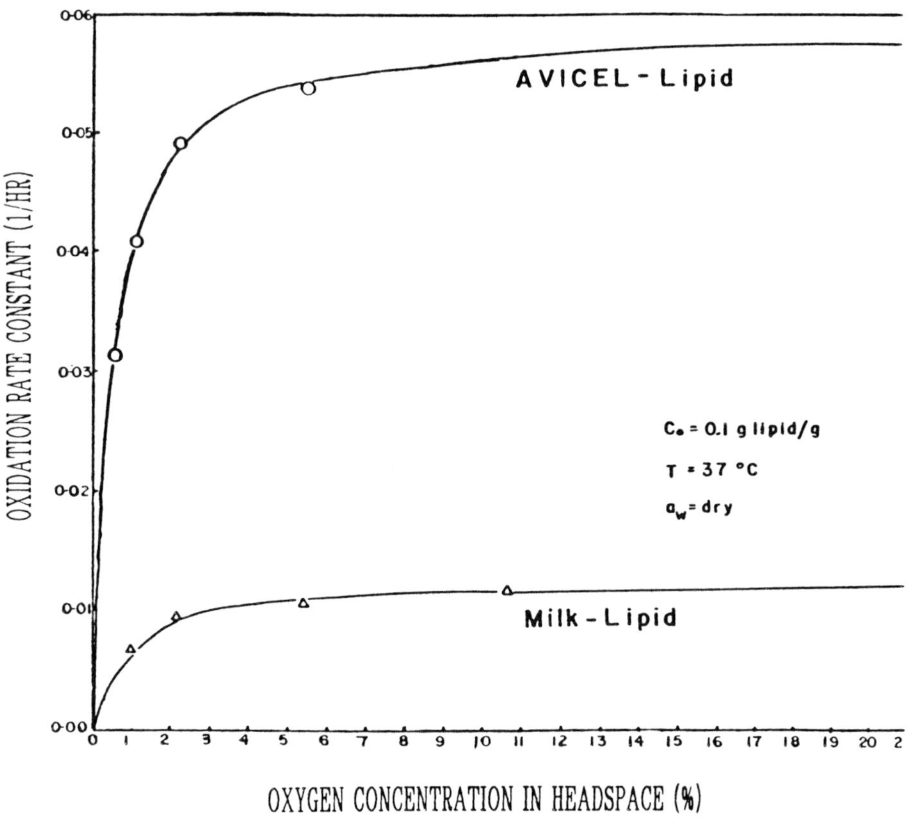

Fig 15.3 Dependence of oxidation rate on oxygen pressure (*From Kacyn et al., 1983.*)

EFFECT OF SUBSTRATE ON OXIDATION KINETICS

The major differences between substrates lie in rates of step 3 (hydrogen abstraction). It is well known that increasing the degree of unsaturation of fatty acids and of their esters and glycerides increases rate of oxidation. It is mechanistically evident that unsaturation, and especially the occurrence of one or more pentadiene groups, greatly increases ease of hydrogen abstraction and hence magnitude of k_3. Table 15.1 shows some values and activation energies for kinetic constants. Since k_3 depends strongly on both temperature and unsaturation, and k_2 is relatively independent, it is to be expected that k_2/k_3 ratio will change greatly as will $[O_2]$ dependence of the reaction as we change composition and/or temperature.

INITIATION

In our preceding discussion we have used the term r_i for the rate of initiation, without specifying the source or nature of radicals being generated. In lipid peroxidation there are three major routes for initiation. The first is initiation by electromagnetic radiation (e.g., light, gamma rays):

$$\mathrm{RH} \qquad \mathrm{R\cdot} + \mathrm{H\cdot} \qquad r_{ie} \tag{28}$$

The second is initiation by unimolecular or bimolecular hydroperoxide decomposition:

$$\mathrm{ROOH} \rightarrow \mathrm{RO\cdot} + \mathrm{\cdot OH} \qquad k_{i1} \tag{29}$$

$$2\ \mathrm{ROOH} \rightarrow \mathrm{ROO\cdot} + \mathrm{RO\cdot} + \mathrm{H_2O} \qquad k_{i2} \tag{30}$$

Table 15.1 Typical Magnitudes of Kinetic Constants in Oxidation of Hydrocarbons and Fatty Acids

Initiation	$k_{i,1} = 10^{-9}$ to 10^{-8} sec^{-1}	10–20 kcal/mole
	$k_{i,2} = 10^{-8}$ to 10^{-6} $(mole/L)^{-1}$ sec^{-1}	10–20 kcal/mole
Step 2	$k_2 = 10^6$ to 10^8 $(mole/L)^{-1}$ sec^{-1}	very low
Step 3	$k_3 = 1$ to 10^2 $(mole/L)^{-1}$ sec^{-1}	10–20 kcal/mole
Termination	$k_4 \approx k_5 \approx k_6 = 10^5$ to 10^8 $(mole/L)^{-1}$ sec^{-1}	very low

The third is by metal-catalyzed autoxidation:

$$\mathrm{ROOH} + \mathrm{Me}^{n+} \rightarrow \mathrm{RO\cdot} + \mathrm{Me}^{(n+1)+} + \mathrm{OH}^- \qquad (31)$$

$$\underline{\mathrm{ROOH} + \mathrm{Me}^{(n+1)+} \rightarrow \mathrm{ROO\cdot} + \mathrm{Me}^{n+} + \mathrm{H}^+} \qquad (32)$$

$$\text{Net:} \quad 2\,\mathrm{ROOH} + 2\mathrm{Me} \rightarrow 2\mathrm{Me} + \mathrm{RO\cdot} + \mathrm{ROO\cdot} + \mathrm{H_2O} \qquad k_m \quad (33)$$

It is evident that the net effect of the metal-catalyzed ROOH decomposition is the same as that of bimolecular decomposition; the rate constant k_m is, however, much larger than the autocatalytic constant k_{i2}. A special case of hydroperoxide decomposition is that of decomposition by metalloenzymes (e.g., lipoxydase). The enzyme-catalyzed rates are much higher than those in the case of metal catalysis.

The rate r_{ie} is proportional to flux of radiation and the other rates may be written

$$r_{i1} = k_{i1}(\mathrm{ROOH}) \qquad (34)$$

$$r_{i2} = k_{i2}(\mathrm{ROOH})^2 \qquad (35)$$

Initially, when ROOH is low, monomolecular decomposition predominates; as the ROOH concentration increases bimolecular decomposition takes over. A similar (simplified) equation can be written for metal-catalyzed oxidation:

$$r_{im} = k_{im}\,\mathrm{Me(ROOH)} \qquad (36)$$

This equation is simplified because the proportionality is often to a power of Me and ROOH which is not exactly equal to 1, reflecting a mixed autocatalytic and catalyzed decomposition.

By combining Eq. (22), (25), or (26) with Eq. (34), (35), or (36) we can obtain simplified equations for specific situations. Thus, if we are concerned with autocatalytic oxidation at high oxygen pressures (for most food lipids at room temperature this means above 4 or 5% oxygen at atmospheric pressure) we can combine Eqs. (35) and (25) to obtain

$$\bar{r} = \frac{d(\mathrm{ROOH})}{dt} = (k_{i2})^{1/2}(\mathrm{ROOH})k_3(\mathrm{RH})(2k_6)^{-1/2} = K_B(\mathrm{ROOH}) \qquad (37)$$

By combining $[(\mathrm{RH})\ (k_3)\ (k_{i2})^{1/2}(2k_6)^{-1/2}]$ into a single constant K_B, we assume that RH depletion is not significant. If this is not satisfactory, the equation may be written

$$\frac{d(\mathrm{ROOH})}{dt} = K'_B(\mathrm{ROOH})(\mathrm{RH}) = K'_B(\mathrm{ROOH})(\mathrm{RH}_0 - \mathrm{ROOH}) \quad (38)$$

Using the simpler form [Eq. (37)] we can integrate to obtain a very simple expression for extent of oxidation as a function of time:

$$\ln \frac{\mathrm{ROOH}}{\mathrm{ROOH}_0} = K'_B t \quad (39)$$

where $\mathrm{ROOH}_0 = \mathrm{ROOH}$ when $t = 0_2$ (Labuza, 1971).

Similar expressions can be derived for other forms of initiation. The conformance with Eq. (39) allows the plotting of extent of oxidation ROOH vs. time as a linear plot on a semilogarithmic scale for bimolecular initiation. $(\mathrm{ROOH})^{1/2}$ plotted against time will give a similar dependence for monomolecular initiation. Mixed initiation and metal-catalyzed initiation will give similar general trends but with dependence on ROOH and Me concentrations.

ANTIOXIDANTS

Antioxidants are compounds that break the chain reaction of oxidation by reacting with free radicals. The chemical nature of antioxidants is described in many texts and reviews. We shall refer to them generically as AH. The reactions affecting oxidation kinetics are (the radical can be either R· or ROO·):

$$\mathrm{R\cdot} + \mathrm{AH} \rightarrow \mathrm{RH} + \mathrm{A\cdot} \quad (40)$$

$$\mathrm{A\cdot} + \mathrm{R\cdot} \rightarrow \mathrm{AR} \quad (41)$$

The usual assumption in considering oxidation kinetics is to modify BAS by replacing reactions 4, 5, 6 with

$$\mathrm{AH} + \mathrm{R\cdot}\ (\text{or ROO}\cdot) \rightarrow \mathrm{NRP} \qquad k_A \quad (42)$$

It can be shown that under these simplifying conditions

$$\bar{r} = \frac{k_3\,(\mathrm{RH})\,r_i^{1/2}}{k_A\,(\mathrm{AH})} \quad (43)$$

An important feature of this assumption is that, since $r_i = r_t$ by the steady-state assumption,

$$-\frac{d(\mathrm{AH})}{dt} = r_i \tag{44}$$

Therefore the antioxidant is used up at a rate equal to the rate of initiation, and in the case of initiation being autocatalytic and bimolecular,

$$-\frac{d[\mathrm{AH}]}{dt} = k_{i2}\,(\mathrm{ROOH})^2 \tag{45}$$

This is an extremely important point since, as will be discussed later, it shows the futility of adding antioxidants to an already oxidized substrate. (The AH concentration will decrease in proportion to the *square* of the extent of preoxidation.) When initiation is metal catalyzed the loss of AH will also be directly proportional to Me^n where n is 1 or close to 1.0.

The loss of antioxidant will also result in a drastic increase in rate of oxidation when AH becomes depleted. Figure 15.4 shows schematically

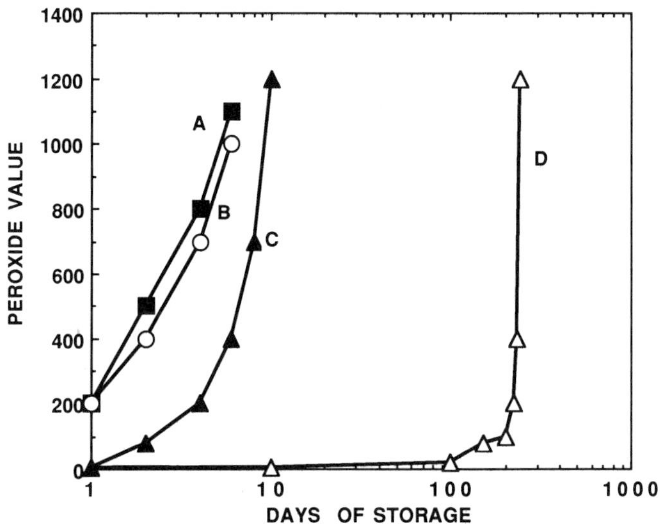

Fig. 15.4 Autoxidation of methyl linoleate in air at 24°C. (A) Irradiated with 4.65 Mrad, no antioxidant; (B) irradiated with 4.65 Mrad, 0.01% propylgallate; (C) not irradiated, no antioxidant; (D) not irradiated, 0.01% propylgallate. (*From Chapault and Mizuno, 1964.*)

the effect of initial degree of oxidation on the "lag phase" in oxidation observed empirically in pure systems as well as in foods and documented abundantly in literature.

OXIDATION IN COMPLEX SYSTEMS AND IN FOODS

The equation describing oxidation kinetics in pure systems can be applied to foods only with extreme caution. The degree of complexity that the kinetics can assume is at least vaguely evident when co-oxidation of as few as two compounds is considered. In Fig. 15.5 steps resulting from co-oxidation of compounds LH and RH are listed. Derivation of an ap-

BASIC AUTOXIDATION SCHEME FOR CO-OXIDATION OF 2 SUBSTRATES.

INITIATION

$ROOH \longrightarrow RO\cdot + OH$

$LOOH \longrightarrow LO\cdot + OH$

$ROOH + LOOH \longrightarrow ROO^{\cdot} + LO + HOH$

$\qquad\qquad\qquad\searrow LOO^{\cdot} + RO^{\cdot} + HOH$

$2\ ROOH \longrightarrow ROO^{\cdot} + RO^{\cdot} + HOH$

$2\ LOOH \longrightarrow LOO^{\cdot} + LO^{\cdot} + HOH$

PROPAGATION

$R + O_2 \longrightarrow ROO\cdot$

$L + O_2 \longrightarrow LOO\cdot$

$ROO\cdot + RH \longrightarrow ROOH + R$

$LOO^{\cdot} + LH \longrightarrow LOOH + L$

$LOO^{\cdot} + RH \longrightarrow LOOH + R$

$ROO^{\cdot} + LH \longrightarrow ROOH + L$

TERMINATION

$L\cdot + L\cdot \longrightarrow L{-}L$

$R\cdot + R\cdot \longrightarrow R{-}R$

$R\cdot + L\cdot \longrightarrow R{-}L$

$LOO^{\cdot} + LOO^{\cdot} \longrightarrow LOOOOL$

$ROO^{\cdot} + ROO^{\cdot} \longrightarrow ROOOOR$

$LOO^{\cdot} + ROO^{\cdot} \longrightarrow LOOOOR$

$ROO^{\cdot} + R^{\cdot} \longrightarrow ROOR$

$ROO^{\cdot} + L^{\cdot} \longrightarrow ROOL$

$LOO^{\cdot} + R^{\cdot} \longrightarrow LOOR$

$LOO^{\cdot} + L^{\cdot} \longrightarrow LOOL$

Fig. 15.5 Steps in the BAS scheme for co-oxidation of two substrates (RH and LH).

propriate BAS scheme is certainly manageable in this case if the relevant constants are known. In foods, however, we deal with many oxidizable substrates, some of which are lipid soluble, others (e.g., phospholipids) of which are either water soluble or concentrate at the oil/water interface.

Furthermore, foods contain metals, other prooxidants, various antioxidants, and "inert" compounds which can bind lipids, complex metals, and scavenge oxygen directly. Finally, as indicated previously, lipid oxidation products can participate in nonenzymatic browning, and reaction products of nonenzymatic browning can have pro- or, more typically, antioxidant properties. Nevertheless, the kinetic features described previously apply remarkably well to many food storage and processing problems. In particular, oxygen dependence, dependence on initial degree of oxidation and temperature dependence can be reasonably well correlated by the above equations in foods containing substantial unsaturated lipid contents, such as fried foods, frying oils, animal products, and oil-seeds. Prediction of oxidative stability is, however, much more difficult in low moisture foods (because of the rather complex effect of water) and in emulsions and in amorphous (glassy) foods (because of heterogeneities in oil distribution in the system.

EFFECTS OF WATER ON KINETICS OF LIPID OXIDATION

Water activity is a major factor affecting the rate of lipid oxidation in foods. It has long been recognized that very low water contents in fat-containing foods are conducive to rapid oxidation, and the literature documenting this phenomenon is voluminous (Karel and Yong, 1981).

Several hypotheses have been advanced to explain water's retarding effect on lipid oxidation. One factor that may be important is production of antioxidants through nonenzymatic browning. The effectiveness of products of nonenzymatic browning was demonstrated early by Griffith and Johnson (1957) and was one of the mechanisms hypothesized by Karel (1960). Several investigators in Japan and Europe have confirmed recently that intermediates in the complex set of recations we term nonenzymatic browning are effective antioxidants. Since high water activity promotes browning, this may be one of the explanations for the observed effects.

Studies on purified model systems containing no components capable of forming antioxidants through browning, however, also show retardation by increasing water content, and the explanation for these effects if probably based on the effect of water on initiation and termination steps of the chain reaction.

In purified systems, water interferes with the normal bimolecular decomposition of hydroperoxides by hydrogen bonding with amphipolar hydroperoxides formed at the lipid-water interface. In the presence of trace metals added to model systems of lyophilized emulsions, humidification at moderate levels retards oxidation because of hydration of metal ions. The reduction in rate depends on the type and hydration state of the metal salt added as well as on water content. The monomolecular rate period is primarily affected by this mechanism, although bimolecular rates are also decreased (Karel et al., 1967).

Relatively less is known about water activity's effect on termination, but it is to be expected that water would increase mobility and hence promote recombination of free radicals.

Water may also be expected to influence free-radical interactions between oxidizing lipids and other food components by influencing the concentrations of initiating radicals present, the degree of contact and mobility of reactants, and the relative importance of radical transfer recombination reactions. In particular, reactions between peroxidizing lipids and proteins are strongly affected by water activity. Increasing water activity promotes the formation of Schiff base type of crosslinks between proteins, with bifunctional fragments of oxidized lipids incorporated as crosslinking agents. On the other hand, at low water activities, free-radical crosslinks predominate, and proteins crosslink without the incorporation of lipid fragments (Funes and Karel, 1981).

Retardation of peroxidation by increasing water content is reversed at high water activity. As water contents are increased above monolayer coverage in foods and in model systems, resistance to diffusion of solutes decreases and solubilization becomes significant. In systems containing chelating agents and antioxidants, high water contents allow solubilization of chelating agents that sequester metals, as well as solubilization of antioxidants. This effect lowers the oxidation rate. However, at high water activities, water may accelerate oxidation by solubilizing catalysts or by inducing swelling macromolecules such as proteins to expose additional catalytic sites.

In systems in which there are both surface lipids and lipids encapsulated within a carbohydrate polysaccharide or protein matrix, the surface lipids oxidize readily when exposed to air (Gejl-Hansen and Flink, 1977; Shimada et al., 1991). The encapsulated lipids, however, do not oxidize until the structure of the encapsulated matrix is modified or destroyed by absorption of water. Maintenance of this structure is of considerable importance in protecting unsaturated lipids from oxidation in dry preparation. An example of this effect is shown in Fig. 15.6 (Kacyn et al., 1983).

Specific effects of water on overall lipid oxidation kinetics have been investigated by several researchers. Quast and Karel (1972) reported that

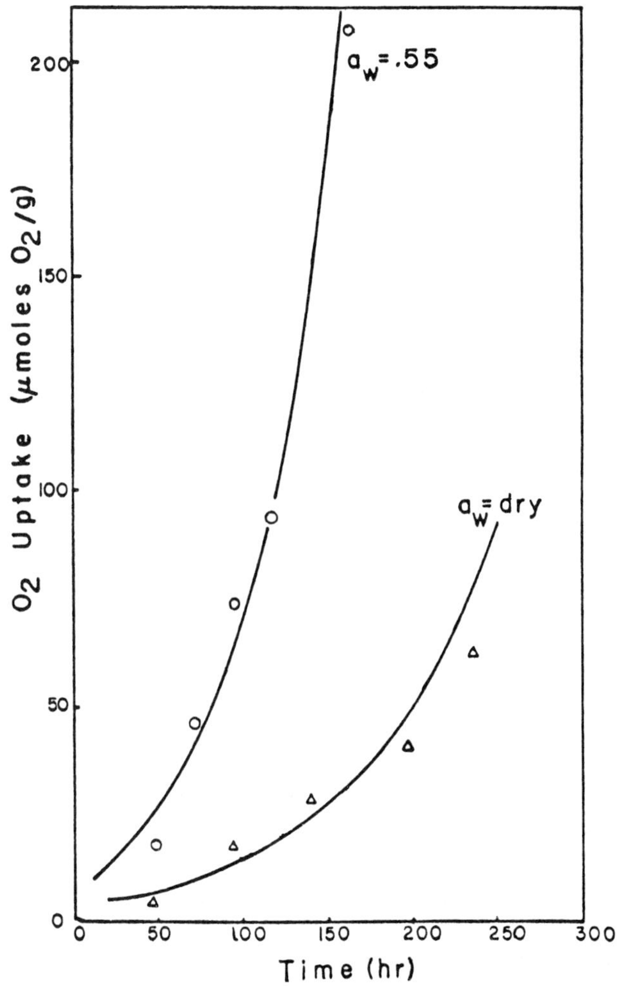

Fig. 15.6 Effect of water-induced structure collapse on oxidation of lipid in a skim-milk matrix. Comparison of dry and humidified milk systems. (*From Kacyn et al., 1983.*)

oxidation of lipids in potato chips in the range of a_w from 0 to 0.4 was proportional to $a_w^{-1/2}$. Haralampu and Karel (1983) reported that rate of oxidation of beta-carotene at low moisture contents was given by the equation:

$$\bar{r} = b_1 a_w^{-1} + b_2$$

where b_1 and b_2 were constants. The equation applied in a a_w range of 0 to 0.7.

Control of Oxidation in Foods

The practical steps one can take based on kinetic considerations are listed in Table 15.2 (Karel, 1984a, b). It should be obvious that the minimizing of the oxygen concentration would require effective removal of oxygen from within the dry food as well as from headspace. We have stressed the need to minimize free radical initiation. It should also be pointed out that exposure to light and other radiations will catalyze initiations and should therefore be avoided. Another group of initiators often neglected is due to volatile decomposition products of oxidation. El-Magoli et al. (1980) have shown that oxidation of foods could be accelerated not only by direct incorporation of preoxidized lipids, but even by exposure to volatile breakdown products from oxidizing lipids. Therefore, even the presence in the same container of an already oxidized food will tend to speed up the oxidation of an unoxidized food. In choosing antioxidants it is also important to add them in a manner that assures good contact with the lipids they are meant to protect, and also to add them before oxidation of the food has progressed very far.

Another important consideration is the exclusion from the dehydrated food of oxidizing agents during the dehydration process. Work by Tsai and Hudson (1985) has confirmed that cholesterol oxidation products, including potentially dangerous cholesterol α-oxides, can be detected in commercial dry yolk, and that the presence of these oxides is strongly correlated with the use of direct gas-fired driers. It is strongly suspected that the oxidation of cholesterol is linked to the presence of nitrogen oxides.

In a direct gas-fired system, liquid egg was dried by the flue gas, which contained the combustion products of natural gas with an empirical chem-

Table 15.2 Principles of Control of Oxidation in Dry Foods which Follow from Kinetic Principles

Minimize oxygen concentration
Minimize rate of initiation by
  Minimizing initial oxidation level or contamination with oxidized components,
  Minimize pro-oxidant (metal) level,
  Maintain water activity at optimal level.
Choose effective antioxidants and assure their addition to sensitive lipids *before* significant level of oxidation is reached.

ical formula $C_1H_{3.8}$. Complete combustion of natural gas produces water, carbon dioxide, and minor amounts of carbon monoxide (less than 10 ppm), nitrogen dioxide, and nitric oxide. The last two compounds are usually referred to collectively as nitrogen oxides or NO_x. Concentrations of NO_x in flue gas vary from 80–160 ppm in home hot water boilers and commerical cooking ovens to several hundred parts per million in large utility boilers. Nitrogen oxides are known stable free radicals. They may serve as initiators of cholesterol autoxidation (Smith, 1981).

ACKNOWLEDGMENT

This is paper D-10209-3-91 of the New Jersey Agricultural Station, and their support is greatly appreciated.

REFERENCES

Chipault, J. R. and Mizuno, G. R. 1964. *J.A.O.C.S.* 41: 468.

El-Magoli, S., Karel, M., and Yong, S. 1980. *J. Food Biochem.* 3: 111.

Funes, J. and Karel, M. 1981. *Lipids* 16: 347.

Geil-Hansen, F. and Flink, J. M. 1977. *J. Food Sci.* 42: 1049.

Griffith, T. and Johnson, J. 1957. *Cereal Chem.* 34: 159.

Haralampu, S. G., Saguy, I. and Karel, M. 1986. *Mathematical Modelling* 7: 1–13.

Kacyn, L. J., Saguy, I., and Karel, M. 1983. *J. Food Process. Preserv.* 7: 193.

Karel, M. 1960. Effects of water and oxygen on chemical reactions. Ph.D. thesis, M.I.T., Cambridge, MA.

Karel, M. 1984a. *J. Chem. Ed.* 61: 335–339.

Karel, M. 1984b. Activities Report R&D Associates 36(1): 50–55.

Karel, M., Labuza, T. P., and Maloney, J. F. 1967. *Cryobiology* 3: 1288.

Karel, M. and Yong, S. 1981. *Water Activity: Influences on Food Quality.* Academic Press, New York.

Labuza, T. P. 1971. *Food Technol.* 2: 355.

Nonhebel, D. C., Tedder, J. M., and Walton, J. C. 1979. *Radicals.* Cambridge University Press, Cambridge, England.

Porter, N. A., Weber, B. A., Weenen, H., and Khan, J. A. 1980. *J. Amer. Chem. Soc.* 102: 5597.

Quast, D. and Karel, M. 1971. *J. Food Technol.* 6: 95.

Reich, L. and Stivala, S. S. 1969. *Autoxidation of Hydrocarbons and Polyolefins*. Marcel Dekker, New York.

Shimada, Y., Roos, Y., and Karel, M. 1988. *J. Agr. Food Chem.* 39: 637.

Smith, L. L. 1981. *Cholesterol Autoxidation*. Plenum Press, New York.

Tsai, L. S. and Hudson, C. H. *J. Food Sci.* 50: 229.

16

Physical Consequences of Thermal Reactions in Food Protein Systems

David W. Stanley and Rickey Y. Yada

University of Guelph
Guelph, Ontario, Canada

INTRODUCTION

This chapter will deal with the physical properties of certain reactions catalyzed by heat that occur in the protein fraction of foods. It is becoming apparent that in order to fully understand the complex behavior of food proteins it is necessary to have an appreciation of their structure. Thus, this subject will also be addressed. Much greater information is now available concerning the structure of proteins since the equipment and expertise necessary for its gathering have become more accessible to food scientists, but, as will be seen, this subject continues to challenge researchers.

The major reactions to be considered will be those that occur during relatively mild heating and that result in the formation of structures, i.e., denaturation, crosslinking, and aggregation. Food scientists now realize that the simplistic view of the mechanisms of these reactions that have been frequently advanced by protein chemists fail to hold for the complex mixture of components found in food. Certain physical approaches such as thermodynamics and kinetics can be applied to these situations, however, and can yield useful information. One of the more important characteristics of a protein in dictating functional characteristics is its thermal stability. This appears to result from certain structural parameters that can be altered through chemical modification, as will be mentioned subsequently. It is the theme of this chapter that, even though a thorough understanding of the reactions occurring to proteins during heating is not available at present, the application of basic physical techniques can yield information useful to the food scientist.

MECHANISMS OF THERMAL REACTIONS IN PROTEINS

The most common processing step imposed on food proteins is thermal treatment of varying severity. This triggers a variety of reactions, but under atmospheric conditions denaturation, crosslinking and aggregation dominate. More severe heat treatments can generate many deleterious consequences and lead to the formation of complex compounds, some of which have potential nutritional consequences.

Definition

Proteins in food systems undergo many reactions as a consequence of processing and storage. Paramount among these, however, is denaturation, a term widely used by food scientists and protein chemists, but, perhaps, employed differently by both groups. For food researchers, denaturation is of great importance since it has such a large influence on functionality. Proteins are important structure builders in food systems and, based on the dogma that structure dictates function, it would be expected that altered conformational states would differ in functional properties. A definition of this reaction that satisfies both basic and applied researchers is not easy to achieve, but, for the purposes of this chapter, the following will be adopted: Protein denaturation is a conformational change in native protein structure of sufficient consequence to produce an observable difference in functional properties.

Denaturation is extremely difficult to define, mainly because of its imprecise nature and nonfood origin. Although many definitions have been advanced, perhaps the most accepted currently by chemists is that proposed by Tanford: "(I)t is simply a major change from the original native structure, without alteration of the amino acid sequence, i.e., without severance of any of the primary chemical bonds which join one amino acid to another'' (Tanford, 1968). This has often been condensed to a single word—unfolding. This chemical definition generally fails in two significant ways: it lacks precision (is protein denaturation a ''simple'' reaction? what is a ''major'' change? how is ''native'' structure defined?) and it is not suited to food systems. Contrast the conditions employed in chemical experiments with those used in food processing: the simple, one-step, thermally activated, native ↔ denatured chemical model for a single species of small, monomeric globular protein in which high dilution rates are used, pHs and ionic strengths are regulated to minimize protein-protein interactions and a uniform, slow heating rate is employed, versus food processing in which a heterogenous group of often large, oligomeric proteins in intimate contact with fats, simple and complex carbohydrates, etc., and existing in a relatively concentrated state at pHs and ionic strengths that promote protein-protein interactions is exposed to a frequently uneven heating regime that can result in partial or severe damage and initiate many side reactions. Given these vastly differing conditions, it would be remarkable if comparable mechanisms were to dominate.

The functional definition chosen here for denaturation is quite similar to that given by the distinguished biochemist Hans Neurath as cited by Colvin (1964), ''any nonproteolytic modification of the unique structure of a native protein giving rise to definite changes in chemical, physical or biological properties.'' In this same paper Colvin makes a plea for the abolition of the term ''denaturation'' since it is such a ''broad, fuzzy canopy.'' Indeed, food scientists, perhaps even physical scientists, must get only slight satisfaction from seeing this term in such common scientific use. While there is merit in this suggestion, since denaturation is defined in so many different ways by various groups of researchers, the chances of this seem low, if only because of its convenience and its seeming ability to convey an extraordinarily difficult aggregate process in one word. On a further point, it will be noted that many definitions stipulate that denaturation excludes the breaking of covalent bonds. More strictly, one may suppose that what is meant is that the intramolecular peptide bonds joining amino acids are not ruptured. However, as will be seen, covalent bonds are, in fact, often broken during heating, specifically disulfide linkages, which can then reform intra- or intermolecularly and produce important changes in structure and function. For example, Kauzmann (1959), whose definition of denaturation is similar to that of Neurath, specifically includes changes in configuration that result from severance of colvalent bonds because of disulfide crosslinking. The practical importance of such reactions are apparent from work such as that of Noh and Richardson (1989), who have demonstrated the ability of heat to catalyze disulfide exchanges that affect the properties of milk during further processing.

Several other points regarding denaturation should be addressed before proceeding. The first concerns the "native" state of proteins before any reaction begins. It seems unlikely that all protein molecules exhibit exactly the same conformation in situ and, if even if they did it, would be, first, challenging to substantiate because of the changes that can occur during measurement (Joly, 1965), and second, inconsequential to the food scientist, because "native" state depends upon the complete prior history of the tissue. Next, the aspect of reversibility should be considered. Again, chemical work has established that, under certain specific laboratory conditions, restoration of the original conformation and concomitant function is possible, but in food-processing operations, especially structure formation, this can often be excluded. On the other hand, recovery of certain enzyme activities after such heat treatments as blanching and pasteurization does occur and is of importance (Whitaker, 1977; Schwimmer, 1981). This points out the significant conceptual difference between loss of enzyme activity due to conformation changes and denaturation. Tanford (1970) suggests that although many chemical studies of denaturation have been carried out under conditions of thermodynamic reversibility, this probably only reflects the preference of the researcher and does not imply that the process is usually reversible for most proteins. Uncontrolled denaturation at high temperatures almost invariably leads to aggregation and nonreversibility, indeed it is often the aim of food scientists to control this reaction so as to produce desirable structures as will be discussed in the section on food protein gels. Yet it is possible to "uncook" some food proteins; Lumry and Biltonen (1969) cite how this may be done through dissociation of the aggregate in strong reagents followed by dilution and dialysis under controlled conditions.

Lehninger (1973) has estimated that there are a total of 10^{10}–10^{12} different kinds of protein molecules existing in the spectrum of living species. Obviously, biochemists have only examined a handful of these in any depth. Thus, it seems a bit pretentious to assume that all of the large number of food proteins must fit into the mold generated from studying 0.000001% of the population. One point of agreement among all researchers is that applying sufficient perturbation such as heat to a protein results in altered structure and associated changes in function. This area of structure-function relationships is of primary importance to food scientists and will be considered next.

Relationship of Structure to Function in Food Proteins

Structural studies of food proteins are increasing as a result of the awareness that functional properties are a direct consequence of conformation. It has proven useful when discussing the structure of proteins to use the convenient terms of primary, secondary, tertiary, and quaternary levels of organization; this nomenclature as devised by Linderstrøm-Lang and Bernal (Lapanje, 1978) recognizes the hierarchical nature of the protein complex. This is a vast area of study, but a competent overview of protein structure and its measurement has been provided recently by Bell and Bell (1988). It is generally accepted that denaturation affects the secondary, tertiary, and quaternary levels of organization. In order to understand how this event proceeds it is necessary to review the mechanism of conformation creation. The conformation of a protein molecule is thought to be dictated first by the genetic code that determines the primary sequence and then by parameters that produce the transition from a linear polypeptide to a three-dimensional folded structure. Primary sequence is important since it determines the type of secondary structures formed. These include conformations such as several α-type helices, several β-type structures (pleated sheets and bends or turns), and the often invoked random coil. How these secondary structures organize into a three-dimensional form constitutes the tertiary structure of a protein.

The mechanisms of de novo protein folding and refolding after denaturation are not well known. A recent review by Fischer and Schmid (1990) suggests that large proteins are less likely to refold after denaturation than small proteins because of the presence of "folding domains" in large proteins which can act independently during folding. On the other hand, Kamata et al. (1979) has obtained some evidence suggesting that the large soy 11S storage protein refolds after the application of heat or treatment with various denaturants. Heat denaturation can result in the presence of some residual secondary structures in proteins; Tanford (1968) cites experiments where approximately 25% of the native structure is retained, although this has been attributed to increased ordering of water around apolar groups (Privalov and Gill, 1988). The structured region is thought to be mainly apolar in nature, and either a portion of the original conformation or a new region formed after unfolding. This may influence the refolding reaction. The rate of protein folding is quite variable—folding can take anywhere between milliseconds and hours, with smaller proteins generally folding faster than large ones. It would seem that a protein can follow different pathways during this reaction, perhaps determined by protein-related factors such as the conformation of proline residues, the ability to reform disulfide bonds and the propensity of intermediates with exposed apolar surfaces to aggregate. External agents such as temperature, pH, concentration, ionic strength, etc. are probably also of importance. An important point to remember is that many proteins demonstrate a variety of stable and metastable forms that differ in their exact conformation. For those interested in the area of protein folding, the works of Lumry and Biltonen (1969), Anfinsen (1973), Anfinsen and Scheraga (1975), Whitaker (1977), Rashin (1984), Fischer and Schmid (1990), and Myers (1990) are recommended.

Germane to the understanding of the protein structure hierarchy and function is an understanding of protein domains. For the longest time, the concept of one structural gene = one polypeptide chain = one functional unit was the accepted dogma of protein biochemists (Benzer, 1957). However, in the past 20 years there has been growing evidence that a domain and not the polypeptide chain should be considered the basic protein unit (Edel et al., 1957; Kirschner and Bisswanger, 1976). Domains may be either structural or functional. Initial studies regarding domains involved limited proteolysis which identified subregions of the polypeptide chain that could often be excised without loss of properties such as ligand-binding or enzyme activity. This observation led to the proposition of functional domains. With the refinement of protein structure analysis, especially three-dimensional analysis, it then became evident that functional domains consisted of one or more structural domains, and that these structural domains were geometrically separate entities with molecular weights of approximately 20,000 daltons. It is now generally accepted that all globular proteins can be subdivided into such subregions and that most functional domains with molecular weights above 20,000 presumably possess more than one structural domain. The "active site" of a functional domain is, therefore, usually located at the interface of two structural domains. An example of this concept is given in Fig. 16.1 for the enzyme glutathione reductase. The domain theory is of importance when considering denaturation since structural disruption appears to proceed in sev-

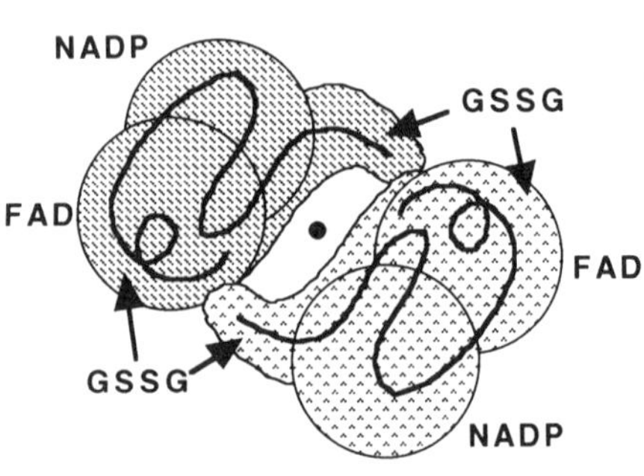

Fig. 16.1 Depiction of protein domains. Glutathione reductase is a dimeric enzyme consisting of two subunits, each with three domains: one binds FAD, one binds NADP, and the third forms the interface. The general course of the two chains is shown by the dark line; the substrate (GSSG) binds between the subunits (arrows). (*Adapted from Schulz et al., 1978.*)

eral discrete stages, each of which corresponds to an "all-or-none" breakdown of these cooperative blocks (Privalov and Gill, 1988).

Further evidence for the importance of structural domains as basic units was obtained by comparing the three-dimensional structure of various proteins. The same structural domain identified by its characteristic chain fold can occur in different proteins. For example, the immunoglobulin domain has been identified in a number of proteins including superoxide dismutase, human leukocyte antigen, and immunoglobulin G (Schulz and Schirmer, 1979). Based on these findings, i.e., subdivision of globular proteins into structural domains of equal size and the occurrence of a given structural domain in different proteins, it is now believed that the majority of proteins are constructed in a modular fashion with structural domains acting as the components. With this in mind, the original concept of one structural gene = one polypeptide chain = one functional unit should be refined to one polypeptide chain = one or more functional domains; one functional domain = one or more structural domains (Schulz and Schirmer, 1979).

In the present work emphasis will be placed on the quaternary structure of proteins. The reasons for this include: (1) many food proteins are large and oligomeric; and (2) the influence of heat on these structures and subsequent structure formation is of major importance in food systems. That many of the large proteins, i.e., those over 50 kd, consist of more than one polypeptide chain, termed subunits, was first discovered by Svedberg using ultracentrifugal methods, and the term quarternary structure to describe this oligomeric system was proposed by Bernal in 1958 (Bell and Bell, 1988). We shall use the term subunit to denote a protein chain possessing an internal covalent and noncovalent structure that may join with other subunits through noncovalent forces or disulfide bonds to form an oligomeric macromolecule. A sampling of oligomeric proteins of interest to food scientists is given in Table 16.1. The number of subunits per protein has been reported to range from two to as high as 24 for apoferritin (Klotz et al., 1975; Bell and Bell, 1988), although the majority of proteins exhibiting a quaternary structure consist of two or four identical subunits. Previously, it was mentioned that the nature of the in situ "native state" cannot be determined with any precision; this is now compounded by the equilibrium established between the monomers and the assembled protein. This step can be concentration dependent or determined by the availability of a cofactor (Bell and Bell, 1988), leading to significant amounts of protein structures and substructures in various states of fabrication. The biological significance of noncovalently linked large proteins has been pointed out by Klotz et al. (1975) to include the following advantages: (1) less genetic material is required to code for identical subunits than for a

Table 16.1 Some Examples of Oligomeric Food Proteins

Protein	Molecular weight (d)	Subunits
Lactoglobulin	35,000	2
Hemoglobin	64,500	4
Avidin	68,300	4
Lipoxygenase	108,000	2
Tyrosinase	128,000	4
Lactate dehydrogenase	140,000	4
7S Soy protein	200,000	9
Invertase	210,000	4
Catalase	232,000	4
Collagen	300,000	3
11S Soy protein	350,000	12
Legumin	360,000	6
Myosin	475,000	6

Source: Adapted from Klotz et al., 1975; Cheftel et al., 1985; Chambers et al., 1990.

single great protein; (2) the chances for a genetic error are reduced since even if there is a mistake in some of the subunits, a proportion of the assemblies will still be functional; (3) the dynamic equilibrium set up between the subunits and the final structure allows another layer of controlling metabolic activities.

It is important to examine how the subunits of these proteins are held together, since the strength of these bonds and how they react to perturbations determines their structural conformation and functional properties. Of the noncovalent interactions possible, including hydrogen bonding, dipolar interactions, ionic interactions, and hydrophobic interactions, the latter seem most important in food proteins such as muscle contractile proteins, casein micelles, and plant storage proteins (Murray et al., 1981; Hultin, 1985; Swaisgood, 1985; Damodaran, 1989). Hydrophobic interactions are not classified as true bonding but result from apolar side chains associating to avoid contact with polar solvents of which water is the primary example. By orienting themselves toward the interior of the protein, where a lower dielectric constant is found, hydrophobic interactions are fostered. The number of apolar groups, their degree of hydrophobicity, and the location in the chain determine the overall contribution of hydrophobic interactions to protein structure.

It is usual for about half of the apolar residues to be buried within the protein, and of the polar residues about a quarter can be found in the interior of the molecule where they participate in the coordination of metals or form part of an enzymic active site (Chothia, 1984). Protein hydrophobicity can be calculated or determined experimentally; the work of Nakai and Li-Chan (1988) covers the application of these methods to food materials and will be brought up subsequently. Some hydrophobicity values for proteins are given in Table 16.2. It may be seen that many food proteins and especially some plant storage proteins are characterized by high hydrophobicity values. Up to 70% of these proteins may be apolar and 30–40% totally apolar. It has been estimated that proteins containing more than 28% apolar amino acids cannot conceal them all within the molecule, leaving some exposed at the surface (Kinsella, 1982). How does this influence the quaternary structure of these molecules? Murray et al. (1981) discusses this through the concept of "hydrophobic patches" in which protein molecules exhibiting high hydrophobicity cannot sufficiently "hide" them in the interior of the protein, thus resulting in areas

Table 16.2 Some Examples of Hydrophobicity Values of Food Proteins

Protein	HΦ (cal/res)
Gliadin	1300
Bovine Serum Albumin	1120-1000
Rapeseed 12S	1050
α-Lactalbumin	1050
β-Lactoglobulin	1050
Oat Isolate	1040
Fababean 140 kd	1023
Soy Micelle	1000
Actin	1000
Apo-ferritin	1000
Elastin	990
Conalbumin	980
Ovalbumin	980
Fababean 340 kd	924
Lysozyme	890
Collagen	880
Myosin	880
Ribonuclease A	870
Soy Isolate	822
Casein	725
Sunflower isolate	597
Whey protein	387
Gluten	349
Pea isolate	277

Source: Adapted from Damodaran, 1989; Myers, 1990.

of apolar side chains on the surface. The tendency of these regions to interact through self-association and thereby exclude high dielectric constant solvent is great, particularly at close distances, and leads to relatively stable and strong bonding. It is to be expected that when these proteins do disassociate, the uncovered apolar areas would increase the subunits effective hydrophobicity and, under certain conditions, lead to aggregation. An attempt to represent this interaction may be seen in Fig. 16.2. It is widely held that hydrophobic interactions play a major role in stabilizing proteins, but some scepticism exists; Privalov and Gill (1988) contend that van der Waals interactions between apolar groups are of more importance in this function. The influence of heat on the structure and function of food proteins will now be examined.

Influence of Heat on Protein Structure

The first influence of increasing temperature on native proteins is to promote structural alterations. At which temperature and to what extent these changes occur are determined by the parameter of thermal stability. The thermal stability of a particular protein probably results from a large number of small stabilizing structural factors; Mozhaev and Martinek (1984) identify a number of these, including amino acid composition, compact packing/protein-protein contacts, binding of metals and other prosthetic groups, as well as intramolecular interactions and linkages. It appears that the overall strategy is to add rigidity to the structure through the introduction of more noncovalent bonds (Zale and Klibanov, 1986).

While it is widely held that hydrophobic interactions are among the major contributors to stability, and a correlation between hydrophobicity and thermostability has been reported for many thermophilic proteins,

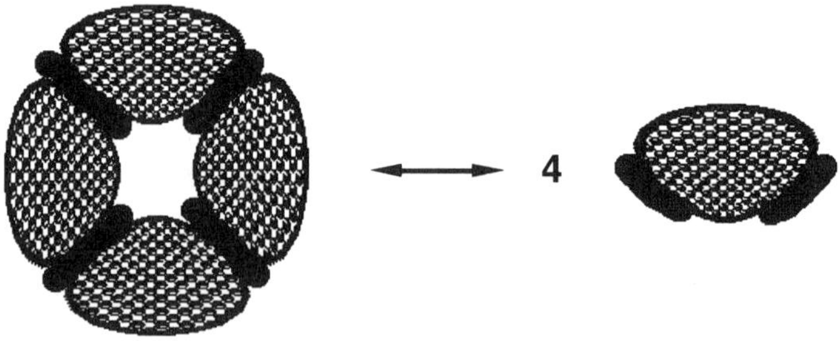

Fig. 16.2 Oligomeric protein consisting of four subunits held together by hydrophobic interactions undergoing dissociation to yield subunits with increased hydrophobicity.

other views exist (see Nakai and Li-Chan, 1988). Since hydrophobic interactions are the only stabilizing forces that increase with temperature, at least up to 60–90°C (Damodaran, 1989; Myers, 1990), it has been proposed that thermal denaturation occurs when the stabilizing effect of increasing hydrophobic interactions are finally overcome by the deterioration of other stabilizing interactions (Brandts, 1964). The influence of heating on the hydrophobic bond is diagrammed in Fig. 16.3. Specifics of heat effects will now be examined; as well, a recent review has covered this subject with regards to food proteins (Kilara and Sharkasi, 1986).

Conformational reactions. Heating of globular proteins in dilute solutions under controlled conditions will produce effects ranging from minor, reversible conformational modifications to major, irreversible processes. These changes have received much attention in biochemical laboratories, but the role of heat in disturbing specific bonding forces in food proteins has not been studied as thoroughly. Mild heat treatment, up to, say, 70–80°C, has a major effect on secondary, tertiary and quaternary structure of most proteins. Brandts (1964) has examined the relative significance of two of the major contributors to these structures, hydrogen bonding and hydrophobic interactions, and also conformation entropy to the stabilization of chymotrypsinogen during heating. Figure 16.4 shows that hydrogen bonding remains constant throughout the process while hydrophobic interactions increase. At higher temperatures, however, these forces become overwhelmed by the destabilizing force of conformational

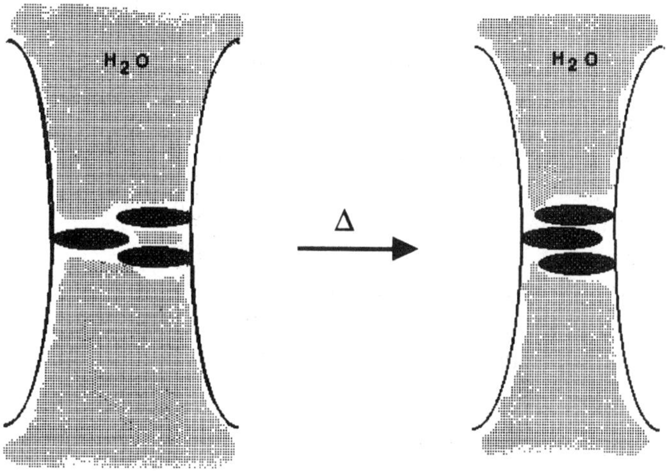

Fig. 16.3 Heat (Δ) enhancement of hydrophobic interactions. (*Adapted from Lumry and Biltonen, 1969.*)

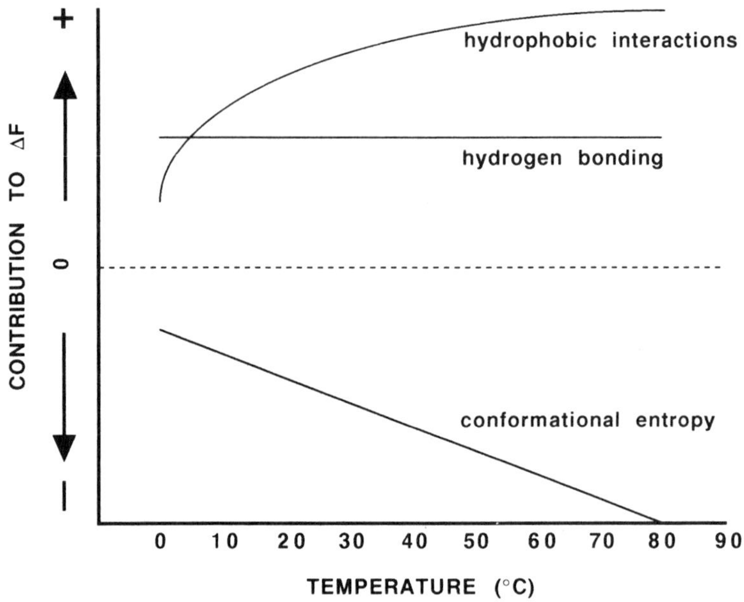

Fig. 16.4 Influence of heat on contributions of hydrogen bonding, hydrophobic interactions, and conformational entropy to the stability of chymotrypsinogen. (*Adapted from Brandts, 1964.*)

entropy. This latter term is related to a molecular description of the unfolding process; while long segments of unfolded chain attached at only one end make a large contribution, the presence of disulfide bonds in the unfolded portion restricts the freedom of the chain and reduces this parameter (Lumry and Biltonen, 1969). Thus, it would be expected that the denaturation temperature would be influenced by the degree of hydrophobicity, and this is the case when the conformation and primary structure of the protein allows the apolar amino acid residues to interact and cooperate. While a more complete thermodynamic consideration of this reaction will be undertaken later in this chapter, for now it is important to notice that denaturation involves a transformation from an ordered to a disordered state, with a concomitant overall increase in entropy.

Protein denaturation may be characterized as initially a cooperative process that in its latter stages becomes a cascade (Feeney, 1980). This cooperativity is reflected in large entropic terms. It has been proposed that the special requirements of apolar side chains to self-associate impart additional cooperativity (Brandts, 1969). Myers (1990) provides data on a group of food proteins showing that the observed enthalpy of reaction is related to hydrophobicity, denaturation temperature, and the nature of aggregation produced by heat. Although protein denaturation is often considered an "all-or-nothing" phenomenon, it seems unlikely that all portions of the protein react simultaneously, thus giving rise to the concept of protein domains, perhaps higher in cooperative apolar residues, that are more thermally stable. Zale and Klibanov (1986) report that most of the higher structures in ribonuclease were not apparent at 90–100°C, although it will be remembered that other researchers have found significant contents of native structure in heat-treated proteins (Tanford, 1968).

Changes in conformation are induced by heating. Jacks et al. (1975), working with heated peanut storage protein, demonstrated that the degree of randomness increased during a thermal treatment, while α-helical content decreased; the study of Deshpande and Damodaran (1989) on the bean storage protein phaseolin provided similar results (Table 16.3). Overall, then, the effect of heat on proteins is to disrupt many of the bonds involved in the higher orders of structure. These steps have often been collectively termed unfolding, but in food-processing situations this may be inappropriate. Some recent work on thermostability of enzymes may add to our knowledge in this area. Klibanov and associates have studied two enzymes, lysozyme and ribonuclease, to determine mechanisms of their inactivation resulting from heat (Ahern and Klibanov, 1985; Zale and Klibanov, 1986). In both cases, the processes found to be responsible for the irreversible loss of enzyme activity during heating at 90–100°C included, at pH 6 and 8, a combination of disulfide interchange (a reaction that can lead to cleavage of disulfide bonds linking polypeptide chains), β-elimination of cystine residues, and deamidation of Asn and/or Gln residues, while at pH 4 the latter step unites with hydrolysis of peptide

Table 16.3 Conformational Parameters (%) of Heated Proteins

Protein	Treatment	α-Helix	β-Sheet	Random
Arachin	Unheated	14.6	27.0	58.4
	150°C	9.6	29.4	61.0
	170°C	5.0	22.8	72.2
	190°C	4.3	19.7	76.0
Phaseolin	Unheated	10.5	50.5	27.5
	5 min @ 99°C	7.0	45.0	36.0
	10 min @ 99°C	6.5	46.0	36.0
	15 min @ 99°C	7.0	46.5	35.5
	30 min @ 99°C	7.5	45.0	35.5

Source: Adapted from Jacks et al., 1975; Deshpande and Damodaran, 1989.

bonds. In the case of ribonuclease it was possible to partition the rate constant into individual contributions by these reactions (Table 16.4). While it is to be borne in mind that these results pertain to enzyme inactivation, they do indicate that some reactions occur during the heating of proteins ≤100°C that involve disruptions in primary structure and that pH has a major influence on the mechanisms involved.

A further effect of heating on proteins, and one that not only can lead to conformational changes but is also of great interest to nutritionists (Finot, 1982), involves the destruction of amino acids. Because of the nutritional implications of this reaction, many food systems have been studied. For example, results have been cited by Neucere and Cherry (1982) for peas and by Fukushima (1980) for defatted soybean flour that show losses of lysine, cystine, arginine, tryptophan, histidine, and serine, among others. The degree of loss is related to the severity of the heat treatment and the nature of the original material. Damage to these components undoubtedly induces conformational alterations through such mechanisms as exposure of charged groups that alter the isoelectric point. Higher temperatures can cause other reactions in the constituent amino acids of proteins including racemization and the formation of various cyclic derivatives; a common consequence of these events is protein-protein reactions via covalent bonds (Cheftel et al., 1985).

Protein-protein reactions. As the conformation of proteins is altered under the influence of heat, protein-protein interactions become more prevalent. These can involve covalent or noncovalent bonding. Covalently bonded proteins may be considered as crosslinked. Friedman (1977) defined crosslinking as a "durable combination of (usually large) distinct elements at specific places to create a new entity that has different properties as a result of the union." With respect to heated proteins, crosslinks can form among protein units or between proteins and nonprotein components. These bondings are not unknown in native proteins; both collagen (Stimler and Tanzer, 1977; Mechanic, 1977; Ledward, 1986) and

Table 16.4 Rate Constants of Irreversible Inactivation of Lysozyme at pH 6 and 90 or 100°C

Process	Rate constant (hr^{-1}) at: 90°C	100°C
Directly measured overall process	0.56	4.1
Peptide bond hydrolysis	0	0
Deamidation of Asn/Gln residues	0.15	4.1
Disulfide interchange	0.31	0
Destruction of cystine residues	0.054	0

Source: Adapted from Ahern and Klibanov, 1985; Zale and Klibanov, 1986.

elastin (Anwar et al., 1977) derive some of their mechanical strength from crosslinks.

With large oligomeric proteins such as found in plant storage tissue, one of the first effects of heat is dissociation into subunits (Catsimpoolas and Meyer, 1970; Stanley, 1989). It might be expected that this, in conjunction with the exposure of previously buried apolar groups during unfolding (Fig. 16.2), would increase the surface hydrophobicity of the protein. Table 16.5 provides some hydrophobicity data for heated proteins

Table 16.5 Hydrophobicity, Solubility, and Emulsifying Activity of Native and Heated Proteins

Protein	Treatment	Surface HΦ	Solubility index (%)	Emulsifying act.(m^2/g)
Bovine serum albumin	Native	325	100	148
	Heated	304	27	140
β-Lactoglobulin	Native	426	100	96
	Heated	192	6	51
Soy isolate	Native	95	26	42
	100°C, 0.25 min	97	26	51
	100°C, 0.5 min	131	24	56
	100°C, 1.0 min	150	14	48
	100°C, 2.0 min	144	4	58
Ovalbumin	Native	6	100	24
	80°C, 1.5 min	38	91	34
	80°C, 2.0 min	87	76	35
	80°C, 2.5 min	95	71	45
	80°C, 3.0 min	142	49	49
Pea isolate	Native	66	43	61
	80°C, 1 min	77	35	66
	80°C, 2 min	104	30	35
	80°C, 4 min	100	25	37
	80°C, 7 min	121	20	25
Phaseolin	Native	51	—	—
	99°C, 5 min	361	—	—
	99°C, 10 min	348	—	—
	99°C, 15 min	441	—	—
	99°C, 30 min	446	—	—
Rapeseed isolate	Native	65	28	60
	100°C, 0.5 min	75	16	63
	100°C, 1.0 min	77	11	49
	100°C, 1.5 min	86	4	41
	100°C, 2.0 min	78	3	40

Source: Adapted from Voutsinas et al., 1983; Deshpande and Damodaran, 1989.

and it may be seen that large increases occur with most, but not all, food proteins. The subunits liberated during heating can then crosslink by various mechanisms. Examples of these reactions are given in Fig. 16.5. Disulfide bond formation and intermolecular peptide bonds (isopeptides) are crosslinking mechanisms frequently investigated in food systems. Isopeptides can result from the reaction of the ϵ-amino group of lysine with the carboxyl group of aspartic or glutamic acids or the amide groups of glutamine or asparagine; crosslinks may also originate from the degradation of cystine, of which lysinoalanine is the most well known (Hurrell and Carpenter, 1977; Neucere and Cherry, 1982). Although their formation is favored by higher temperatures and an alkaline environment, it seems likely that they can occur, if to a lesser degree, during typical food-processing conditions.

Little agreement exists in the literature about which reaction will occur in a particular protein under a given set of conditions. One major factor that is often discounted in such systems is the protein/water ratio. This parameter would seem to dictate the nature of the protein-protein interactions occurring, much as the starch/water ratio determines the events during heating of starch-containing systems (Biliaderis, 1983). Differential scanning calorimetry has been used to advantage for these investigations; Table 16.6 provides a summary of available data showing the marked dependence of thermal properties on hydration effects. That water content

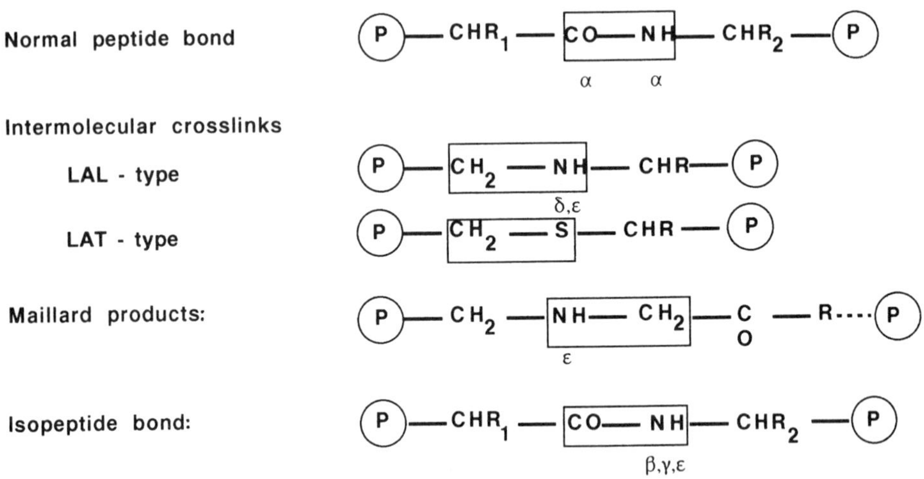

Fig. 16.5 Bond types formed during heat-catalyzed crosslinking of proteins. LAL = lysinoalanine type; LAT = lanthionine type; circled P = protein molecule; Greek letters show carboxyl or amino groups. (*Adapted from Stanley, 1989.*)

Table 16.6 Dependence of Protein Thermal Properties on Water Content

Protein	Water content	T_d (°C)	ΔH
Myoglobin	2.3	122	0.4 (mcal/mg)
	9.5	89	1.6
	15.6	82	2.7
	20.6	79	3.6
	35.2	75	4.7
Chymotrypsinogen A	10	99	170 (kJ/mol)
	20	80	290
	30	72	400
	40	70	520
	50	67	560
Soy flour	10	115	—
	20	109	—
	30	102	—
	40	97	—
	50	89	—

Source: Adapted from Hägerdal and Martens, 1976; Fujita and Noda, 1981; Sheard et al., 1986.

can influence reaction mechanisms in protein systems is shown from the work of Sheard et al. (1986) in which heat treatment of soy protein at high water contents led to aggregation by hydrophobic interactions while decreasing water levels changed this mechanism to favor covalent disulfide bonding. In general, it would appear that lowered water contents and higher processing temperatures promote covalent bonding forces at the expense of noncovalent ones.

Intermolecular peptide bonds, having a higher energy of formation, might be expected to occur under more severe conditions, such as those found during thermal extrusion of proteins, and evidence for this has been presented (Noguchi, 1989; Stanley, 1989). On the other hand, it would seem that in some cases this type of covalent bonding can be produced in a milder environment. Thus, Otterburn et al. (1977) provide evidence for some isopeptides in lactalbumin heated at 85°C and higher levels in egg albumin, zein, and casein treated at 115°C. It is worth remembering that significant structural, functional, and textural changes can be produced by relatively few chemical alterations (Aguilera and Stanley, 1990).

If, in addition to heat, alkaline conditions are employed during the treatment, admittedly not an overly common food processing option, another type of protein-protein interaction should be considered. Lysinoalanine (LAL) and similar compounds may be produced with the concomitant formation of covalent crosslinks. This subject has been addressed in papers by, among others, Sternberg and Kim (1977), Whitaker (1980), Freidman (1982), and Schwass and Finley (1984). While heat and elevated pH undoubtedly accelerate this reaction, can it occur under more conventional conditions? The presence of LAL has been reported in a wide variety of home prepared foods, commercial products and ingredients (Sternberg and Kim, 1977), and these authors also report the results of a laboratory experiment in which casein was heated under different conditions (Fig. 16.6). It may be seen that considerable quantities of LAL are formed even at neutral and acidic pH and 100°C. Again, protein concentration was an important variable in the reaction and, during the linear portion of the process, LAL production was exponentially related to casein concentration and, therefore, water availability.

The above discussion should not diminish the role of noncovalent bonding during the heating of proteins. Increased exposure of apolar amino acid groups and the heat stability of hydrophobic interactions make this type of bonding particularly important. These bonding forces often lead to structure formation and will be discussed further later.

In summary, covalent protein-protein interactions during heating proceed primarily by way of disulfide and isopeptide bond formation and crosslinks resulting from LAL production. Of course, other factors not discussed here such as pH and ionic strength can have a significant effect on these reactions, but now protein-nonprotein reactions will be explored.

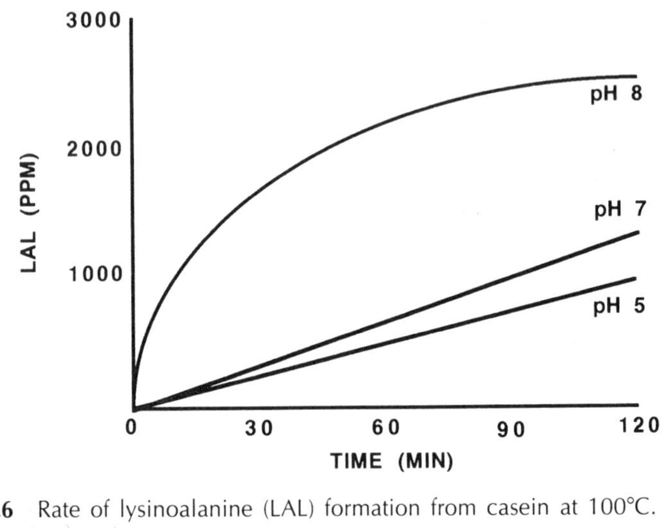

Fig. 16.6 Rate of lysinoalanine (LAL) formation from casein at 100°C. (*Adapted from Sternberg and Kim, 1977.*)

Protein-nonprotein reactions. Denaturation reactions in food rarely occur in the absence of other constituents. Lipids and carbohydrates are usually the most plentiful, and proteins, under the influence of heat, can covalently interact with both of these. In general, deteriorative reactions in proteins take place more quickly when these other elements are present (Satterlee and Chang, 1982). Perhaps the most conspicuous process in this category is the Maillard reaction. Because of the importance of nonenzymatic browning in many different environments and because it involves numerous food components, this series of reactions has been studied extensively. Recent reviews have been published by Feeney et al. (1975), Tanaka et al. (1977), Feeney and Whitaker (1982), Waller and Feather (1983), and O'Brien and Morrissey (1989). Maillard reactions proceed following the interaction of ϵ-amino groups from proteins and the aldehyde group of reducing sugars, which can occur as a result of mild heating (Hurrell and Carpenter, 1977). Feeney (1980) gives the example of liquid egg white where, as a result of a relatively high concentration of glucose and an alkaline pH of above 9, a reaction can occur at room temperature. Further heat treatment can produce inter- and intramolecular covalent crosslinkages and associated nutritional deterioration (Finot, 1982). The concomitant decrease in protein digestibility has been attributed to the formation of crosslinks between amino acid residues that reduce availability of the protein to enzyme attack (Satterlee and Chang, 1982).

The Maillard reaction is known to be influenced greatly by the water activity of the environment (Labuza, 1980a; Rockland and Nishi, 1980) and typically is maximized at intermediate a_w levels. As an example, Labuza and Saltmarsh (1981) examined the kinetics of browning in whey powder. Table 16.7 shows that the rate constant of the disappearance of lysine is positively influenced by reaction temperature and peaks at an a_w of around 0.4–0.5.

Interactions of proteins with reducing sugars can have important side

Table 16.7 Rate Constant ($days^{-1} \times 10^3$) for Lysine Loss from Hygroscopic Whey Powder Stored at Different Temperatures and Water Activities

a_w	25°C	35°C	45°C
0.33	1.2	6.9	18.5
0.44	2.7	13.9	30.0
0.65	1.8	9.6	15.7

Source: Adapted from Labuza and Saltmarsh, 1981.

effects on the properties of the resulting products. Kato et al. (1981) studied the influence of the Maillard reaction on ovalbumin by reacting this protein with glucose at 50°C and 65% RH. Loss of solubility was highly correlated with decreasing α-helix content over the storage period (Table 16.8); evidence of denaturation and crosslinking was observed during the reaction. Also, the carbohydrate moiety was thought to stabilize the Maillard product against the effect of heat, leading to the generalization that glycoproteins are more resistant to thermal stress. As will be seen, carbohydrate moieties can also contribute to thermal stability.

Proteins can react with a variety of other food constituents in reactions that are accelerated by heat. Lipoproteins, a major component of biological membranes, undergo lipid oxidation readily with the products of this process including free radicals and aldehydes such as malonaldehyde, and then enter into covalent crosslinking with proteins. These steps can have adverse effects on food quality and have been implicated in deteriorative reactions ranging from the PSE defect in pork, to textural toughening in frozen fish, to a loss of vigor in stored seeds, to senescence in fruits and vegetables (Stanley, 1991). Other compounds such as flavors and pigments can also interact with proteins. Arai (1980) has reviewed this subject, and in some cases, such as the interaction of soy protein with 1-hexanal, the reaction is promoted by heat denaturation.

Measuring the Effect of Thermal Changes in Proteins

In order to detect heat-induced changes in proteins, the food scientist can measure conformation directly or indirectly, or focus on the results of

Table 16.8 Averaged Changes in the Protein Fractions of an Ovalbumin/Glucose Mixture during Storage at 50°C and 65% RH

Storage time (days)	Solubility (%)	α-Helix content (%)
0	50	100
2	25	82
4	25	75
6	20	71
10	6	64
15	3	45

Source: Adapted from Kato et al., 1981.

changes such as various functional properties. There are many methods available for these purposes, but space restrictions limit this discussion to a few examples. Lapanje (1978) and Kilara and Sharkasi (1986) provide a more comprehensive listing.

Direct Structural Measurements

The four organizational levels of protein structure previously identified include: (1) the primary structure as determined by the amino acid sequence; (2) secondary structure referring to the stable conformation of the polypeptide chain that can exist in forms such as helices, β-sheets, and β-turns; (3) the polypeptide chain folds into a final three-dimensional shape that describes its tertiary structure; (4) for proteins consisting of more than one polypeptide chain, quaternary structure is their organization into a joint structure. The conformation of a protein molecule is determined by the application of the genetic code and parameters causing the transition from a linear polypeptide to a three-dimensional folded structure. Recently, research has established that certain proteins (e.g., protein disulfide isomerase, peptidyl-protyl-transisomerase) are involved in cellular proteins folding events (reviewed by Fischer and Schmid, 1990). The primary sequence is of importance since it determines the conformation formed. The final native structure may be thought of as being determined by "the totality of interatomic interactions" inherent in the molecule (Anfinsen, 1973). Many experimental approaches have been developed to examine the conformation of proteins. Several of these will now be surveyed.

General methods. X-ray diffraction crystallography of single crystals is the most powerful direct approach to determining protein structure since it can divulge the location and bonding of all atoms in the polypeptide chain except hydrogen. The complexity of data interpretation, equipment cost, requirement for protein crystallization, and the assumption that crystalline structure remains unchanged upon solvation in water often makes this method inaccessible and of little use to the food scientist. It is possible to predict partially secondary structure of proteins from their amino acid sequences, and this is known for some food proteins (Damodaran, 1989). Detailed three-dimensional structures can also be obtained by nuclear magnetic resonance (NMR) methods. A different application of this technology was used recently by Lambelet et al. (1989) to study protein denaturation in whey proteins. Low resolution NMR instrumentation was applied to generate proton transverse magnetization decay curves from which relaxation times were calculated. The parameter measured in these experiments was a decrease in mobility of the water fraction bound to the protein as a result of heating. This method is said to respond to protein unfolding but not other thermal processes such as gelation. Comparison of the NMR procedure to the standard chemical method involving nitrogen determination of filtrates resulted in consistently lower values for the former measurement, indicating that NMR analysis is influenced more by unfolding than aggregation. This approach has definite benefits in the study of thermal events in proteins and will likely become more widely used.

Information on the secondary and/or tertiary structure of proteins can be obtained by physical techniques, such as infrared spectroscopy, optical rotatory dispersion (ORD), and circular dichroism (CD), through the use of spin probes and by fluorescence spectroscopy. The latter technique has recently been shown to be applicable to studying thermal denaturation of oat proteins (Ma and Harwalkar, 1988a). Chemical approaches to evaluating protein conformation include chemical modification, susceptibility to proteolysis and the use of antibodies. Protein denaturation involves the entire spectrum of conformational transitions including those occurring in the peptide chain and in the side chains so that the method selected should reflect both these changes.

Raman spectroscopy. Raman spectroscopy is a technique that, because of recent advances in instrumentation combining a laser as a powerful light source with rapid computing procedures, offers some advantages in studying the structure of proteins. It is a light scattering procedure based on the inelastic scattering of visible light that gives information on the vibrational motion of atoms in molecules, somewhat analogous to infrared spectroscopy, but complementary to it, since water is an acceptable solvent. Protein samples can be used directly for analysis since dilution or addition of reagents is not required. Perhaps most important for the consideration of denaturation is that the spectrum reflects contributions from both the peptide backbone and side chains. Also, the technique is particularly sensitive to α-helix content. The application of this method to proteins has been reviewed by Frushour and Koenig (1975), Painter (1984), Peticolas (1984), Mendelsohn (1986), and Tu (1986). These authors should be consulted for theoretical discussions of the Raman effect.

The Raman spectrum of a protein is the sum of the bands representing the peptide backbone and those from the side chain residues. Denaturation involves unfolding of the backbone and changes in the side chain microenvironment. The results of Raman spectroscopy are suited to detecting these structural transitions from an ordered to a disordered state. Thermal denaturation has been studied by this method for several proteins (see

Frushour and Koenig, 1975; Painter, 1984; Tu, 1986). Thermal denaturation of proteins can lead to cleavage of disulfide bonds, a reduction in the content of ordered chain conformations with the appearance of the random coil structure, and the interaction of previously buried aromatic side chains with solvent molecules. Complete unfolding, however, seems much more likely to occur with the conditions of strong denaturants employed in basic biochemical investigations rather than in food-processing surroundings where acid and heat predominate (Hermansson, 1978). It is of interest to note that, using this sensitive technique, several investigators have come to the conclusion that the thermal denaturation of globular proteins is best represented by a gradual stepwise unfolding rather than a sharp transition between only two states (Frushour and Koenig, 1975; Chen and Lord, 1976).

More recently, Hirotsuka and Nakai (1991) have examined molecular interactions in soy proteins as part of a larger study investigating the formation of lecithin-protein complexes. Raman spectra of a 10% 7S-enriched fraction heated for 30 minutes at 80°C showed conformational changes in the peptide backbone without extensive exposure of hydrophobic side chains; in fact, tyrosine groups were buried upon heating. Emulsifying activity increased, however, with this treatment. This behavior was anomalous since there was a general positive correlation between emulsifying activity and surface hydophobicity. Previous work by this group on a commercial soy isolate (Voutsinas et al., 1983) described increasing hydrophobicity with time of heating at 80°C. On the other hand, Li-Chan and Nakai (1989 and unpublished data) heated lysozyme solutions and observed subsequent increases in the exposure of aliphatic and aromatic side chain groups that correlated well with increased hydrophobicity. The difference in these two proteins could be a result of the complicated quaternary structure of the 7S molecule, in which subunits are bonded by hydrophobic interactions. Other studies on the 7S soy globulin have shown that heating does not cause formation of stable dissociation products, even at low ionic strength; aggregation occurs on heating and is enhanced by the presence of salt (Hermansson, 1978).

A particular application of Raman spectroscopy concerns studies of the crosslinking mechanisms involved in gel formation. Clark and Lee-Tuffnell (1986) describe the usefulness of this procedure in characterizing the role of β-sheet formation during heat setting of protein gels. This method has the advantage being relatively unaffected by sample turbidity, and has established the appearance of β-sheet in gels. Raman and IR are suited for such studies, and more recently CD methodology has been applied to protein gelling. Figure 16.7 provides some data gathered in this way that allow a calculation of the proportions of various secondary struc-

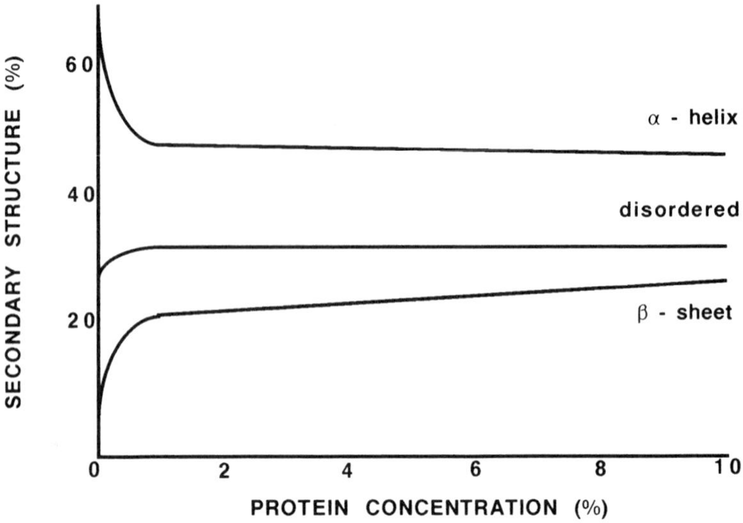

Fig. 16.7 Proportions of α-helix, β-sheet, and disordered conformations calculated from circular dichroism spectra data as a function of bovine serum albumin concentration for heat-set (75°C) aggregation. (*Adapted from Clark and Lee-Tuffnell, 1986.*)

tures, but at this point only solutions of aggregates made at subgelling concentrations and clear gels can be studied.

Indirect Structural Measurements

It is often easier to measure the effect of heat on a complex group of constituents such as is found in food by indirect methods that avoid the complications of confusing interactions by focusing on the whole sample. In this section differential scanning calorimetry and measurements of hydrophobicity will be considered.

Differential scanning calorimetry. Differential scanning calorimetry (DSC) measures conformational changes in heated proteins secondarily by evaluating the energy necessary to disrupt the native structure. The absorption of heat, as applied through a controlled temperature program, is recorded as an overall endothermic peak, the area of which is directly proportional to the net thermodynamic enthalpic change. Parameters obtainable from the thermal record include the overall enthalpy change (ΔH), and from this an estimate of the thermal energy required to denature the protein, and the temperature of denaturation (T_d). This method has been used widely in assessing thermal changes to food proteins since it yields dynamic data as a function of heating temperature and has the ability to utilize high concentrations of heterogeneous samples, such as is found in commercial preparations.

As will be described subsequently, both thermodynamic and kinetic data can be obtained from the thermograms, although it must be remembered that what appears to be single peak is in fact made up of many endothermic and exothermic contributions. For example, the overall endotherm associated with thermally induced protein denaturation results from a mixture of reactions such as the endothermic disruption of hydrogen bonds, the exothermic breaking of hydrophobic interactions, and other events occurring during heating (Arntfield et al., 1990a). Of particular importance is the differentiation between denaturation and aggregation, since the physical state of the sample is not a factor, a problem that often precludes the use of optical instruments where light scattering by insoluble particles produces artifacts. Aggregation is generally an exothermic reaction that lowers the overall enthalpy value (Myers, 1990), but the energies involved are quite small. However, Bacon et al. (1989) have reported that by adjusting the experimental conditions of pH, ionic strength, and protein concentration it was possible for them to separate the denaturation endotherm from the aggregation exotherm of the DSC thermal curve. Zarins and Marshall (1990) produced a similar effect by the addition of various alcohols. Several reviews have covered the application of DSC to foods and their protein components (Wright, 1982; Biliaderis, 1983; Lund, 1983; Arntfield et al., 1990a; Ma, 1990). Steim (1965), Jackson and Brandts (1970), Privalov and Khechinashvili (1974), Biltonen and Freire (1978), Privalov (1979a), Sturtevant (1980), Sandu and Lund (1985, 1986), and Myers (1990) take a more basic approach, while Ma et al. (1990) cover instrumentation and techniques.

Food scientists working with biological tissues usually encounter mixed proteins in the concentration range of 5–20%. Protein chemists, on the other hand, study isolated proteins, usually in aqueous solutions at low concentrations and at pH values far away from the isoelectric point so as to discourage protein interactions and aggregation. For this reason, scanning microcalorimeters have been developed that operate in the area of 1% protein (Jackson and Brandts, 1970; Privalov, 1979b, 1982; Danilenko et al., 1985a,b; Privalov and Gill, 1988). This, of course, brings up the question of the influence of concentration on DSC data and, therefore, the applicability of results obtained from one system to the other. Several studies have looked at the influence of concentration on thermal parameters of proteins (Hägerdal and Martens, 1976; Fujita and Noda, 1981;

Sheard et al., 1986). These studies show that water concentration plays an important role in DSC results. Thus, it seems unlikely that conclusions based on dilute protein solutions examined by microcalorimetry are directly transferable to much of the work done on food proteins using higher concentrations. The use of concentrated protein solutions maximizes protein-protein interactions and results in calorimetric data with a marked dependence on water content. There is some evidence that protein molecules in high concentrations do not unfold totally, since the protein-protein interactions overwhelm the ability of the protein to compete for water hydrogen bonds.

At higher protein concentrations additional thermal events also occur. The thermogram represents the sum total of all endothermic and exothermic occurrences. Thus, the overall endotherm seen during denaturation is a combination of endothermic reactions such as the contribution of hydrogen bonds and exothermic reactions including the ordering of water around exposed apolar groups resulting from the breaking of hydrophobic interactions and also exothermic protein aggregation (Privalov and Khechinashvili, 1974; Arntfield and Murray, 1981; Biliaderis, 1983).

Another reason that DSC has been accepted for protein studies is its ability to predict or correlate with functional properties of importance in food systems. For example, Hermansson (1979) and Arntfield and Murray (1981) have demonstrated the usefulness of the DSC record as an indicator of plant protein denaturation. This type of information is of commercial importance and can be used, for instance, to establish quality control standards for either incoming ingredients or finished products. Of particular interest is the use of the DSC procedure to study gel formation in thermally treated protein systems. This topic will be addressed subsequently in this review.

Hydrophobicity. The measurement of protein hydrophobicity is not a trivial undertaking. Advanced techniques including fluorescence quenching, hydrophobic probes, and hydrophobic chromatography are necessary to obtain reliable data, although it is possible to get an approximation through applying rating factors to the results of amino acid analysis (Nakai and Li-Chan, 1988). A recent review provides a comparison of the methods used currently to measure protein hydrophobicity and gives two new approaches for quantifying the degree of side chain exposure—protein magnetic resonance and laser Raman spectroscopy (Nakai et al., 1991). All apolar residues do not exhibit the same degree of hydrophobicity; Myers (1990) divides them into three groups with different temperatures of maximum strength. The applicability of data from these procedures to functional properties has been shown repeatedly; in particular Catsimpoolas and Meyer (1970) and Voutsinas et al. (1983) have related hydrophobicity measurements to gelation phenomena in plant proteins. Several different types of hydrophobicity measurements are possible, yielding different parameters. It would seem that surface hydrophobicity, theoretically the hydrophobicity of the protein surface, gives the best correlations with functional properties of food proteins (Damodaran, 1989).

Protein unfolding by thermal means, implying an increase in hydrophobicity and a decrease in tertiary and quaternary structure, has a major effect on various functional properties, but there seems to be no consistent pattern among different species. Table 16.5 gives some examples of these types of data from Voutsinas et al. (1983). While all proteins became insoluble to various degrees as a result of the heating treatment, the response of hydrophobicity and emulsification capacity was variable, yet related. Thus, Kato and Nakai (1980) reported a significant correlation between surface hydrophobicity and emulsification activity in various proteins, although another structural factor, such as chain flexibility, is probably also involved in this functional property. Subsequent efforts by Matsudomi et al. (1985) in which only the 11S fraction (see below) of soy protein was used showed a linear correlation of emulsion stability with surface hydrophobicity. This was not the case, however, for the 7S fraction alone (Hirotsuka et al., 1984).

In general, though, the data of Voutsinas et al. (1983) indicate that heated oligomeric proteins increased in hydrophobicity and those proteins that did not either were small, single-chain molecules with significant disulfide bond involvement assayed under extremely low pH (≤ 4) conditions, i.e., bovine serum albumin and β-lactoglobulin, or a protein where the exposed apolar groups are thought to interact to produce an apparent decline in hydrophobicity, i.e., whey proteins, or a protein whose native state is a random coil, i.e., casein. Thus, it appears that Nakai and Li-Chan (1988) may be correct when they state " the molecular basis for these processes are not understood well enough to predict the behavior of different food proteins." On the other hand, Dr. Nakai's laboratory has pioneered studies aimed at explaining these phenomena, and it does appear that the concept of hydrophobicity is useful for this purpose. The basic tenet of this view is that as proteins unfold or denature, steric hindrance is reduced and hydrophobic side chains that were previously buried become exposed and able to produce a structuring effect on nearby water molecules, significantly altering protein surface properties.

Measurements of Function

Often the food scientist's major concern is not a fundamental consideration of such thermal events as denaturation or crosslinking but, rather, imparting desirable functional properties such as texture to a proteinaceous, multicomponented fabricated food. Without sophisticated instrumentation to either directly or indirectly measure these protein parameters, functional properties become the focus of attention. In many cases, these criteria do reflect the influence of either intentional or unintentional heating on the protein ingredients. Many functional properties exist, and all are of interest to some segment of the food industry. The present commentary will be limited mainly to those of importance in the formation of heat-generated structured gels from plant proteins.

Protein gels. For those not familiar with the literature in the area of protein structures, the terminology can be confusing. In Table 16.9 the nomenclature employed by Hermansson (1978, 1979) is given, to which this discussion will adhere when possible. Essentially, structure formation depends on the balance between protein-protein interactions and protein-solvent interactions. Of particular interest are the structures formed during gelation. Food gels have been in production for many years; structurally, gels are continuous, three-dimensional networks of crosslinked polymeric molecules (Oakenfull, 1984) in which asymmetric particles join together at only a relatively few points and the solvent, usually water, is entrapped in large quantities. Thus, in common protein gels these particles form the discontinuous solid phase and water is the continuous liquid phase.

Gel formation is of increasing interest to those investigating structural

Table 16.9 Nomenclature of Protein Structure Formation

Chain—Solvent Mechanisms
Dissociation—separation of intact subunits
Denaturation—major conformational changes without alteration of amino acid sequence
Solubilization—dissolution, often defined operationally
Chain—Chain Mechanisms
Precipitation—separation from solution, often defined operationally
Association—changes at the molecular level characterized by weak bonds at specific binding sites, e.g., monomer, dimer, subunit equilibrium
Aggregation—general collective term for nonspecific protein–protein interactions including flocculation, coagulation and gelation
Flocculation–a colloidal aggregation due to random protein–protein interactions not involving major conformational changes
Coagulation—random aggregation involving major conformational changes
Gelation—nonrandom aggregation involving major conformation changes

Source: Adapted from Hermansson, 1978, 1979.

interactions in food systems, since it affords an opportunity to produce under controlled conditions a structure amenable to rheological measurement, and also a food material of commercial interest, mainly because it can hold in matrix form a large amount of water as well as providing a conduit for nutrients, micronutrients, flavors, and colors. Further, gels offer the possibility of substituting one protein for another in food systems (Beveridge et al., 1984). To obtain gels from globular proteins it is necessary to use concentrations about one order of magnitude higher than that required for gelation of polysaccharides or gelatin (Paulson and Tung, 1989). The formation and properties of food protein gels have been the subject of several reviews (Clark and Lee-Tuffnell,1986; Ledward, 1986; Edwards et al., 1987; Oakenfull, 1987; Hamann et al., 1990; Fligner and Mangino, 1991), a major symposium (IFT, 1984) and a recent book (Harris, 1990), (see also Chapter 7).

One of the earlier studies on thermally induced protein gels that still retains significant applicability is the work of Tombs (1970, 1974). Solutions of bovine serum albumin and peanut arachin were heated and the resulting gels or aggregates observed. Measurement techniques included determining molecular weight of aggregates by osmometry, gel formation by gel time, and gel structure by electron microscopy. As might be expected, initial protein concentration had a decisive influence on the structures formed. Increasing concentrations led to more rapid aggregation and the formation of structures with more molecules per unit. Gelation did not occur until the degree of aggregation was well advanced. Chemical studies showed that the major structural association occurring during heating albumin was the formation of intermolecular disulfides. This protein is unusual in containing 17 available disulfide pairs that become more reactive as the temperature increases, but the protein does not unfold to a great degree, probably due to remaining intrachain S-S bonds. Other work on this protein has shown that heat does not induce increased hydrophobicity, as often occurs with other species, although solubility decreases markedly (Voutsinas et al., 1983).

Results for the large plant storage protein were widely different. Here, the first effect of increasing temperature was to induce the subunits to dissociate and aggregation occurred only after this step. This had been predicted by Armstrong et al. (1979) among others, and has subsequently been confirmed for other plant proteins exhibiting quaternary structure. For example, Deshpande and Damodaran (1989) heated solutions of phaseolin, the oligomeric major storage protein of common beans, and observed disruption of quaternary and tertiary structure; Bacon et al. (1989) thermally denatured the vicilin protein of pea and found this led to the formation of soluble aggregates, presumably the result of interaction among heat disassociated subunits; Ma and Harwalkar (1987) studied an oligomeric oat globulin and found thermal treatments promoted dissociation of subunits. Other researchers, using nonthermal denaturing methods, have reported similar results (Prakash, 1986; Grinberg et al., 1989; Chambers et al., 1990). Thus, it would seem usual that dissociation (if applicable), perhaps denaturation (see subsequent discussion), and then association is the normal course of events in thermal protein gel formation.

The nature of the protein interactions in thermal gels is varied, and may involve all those that were active in the native molecules. In oligomeric proteins it would be expected that hydrophobic bonding, of great importance in subunit association, would predominate. No gel will form, however, below a certain critical protein concentration. In soy systems this value is around 7–8% (Catsimpoolas and Meyer, 1970; Schmidt and Morris, 1984); this point will be discussed in more detail subsequently. Also, different structures can result from different concentrations. Clark and Lee-Tuffnell (1986) cite results from a heat-set bovine serum albumin system presented in the form of a phase diagram (Fig. 16.8). Two types of gels were produced, a rigid, brittle but less elastic opaque gel with heterogeneous pore sizes, favored at lower pH, and a weaker but more elastic and homogeneous clear gel, favored at higher pH. Hermansson has also had success in producing two different gels from a given protein, in this case bovine myosin, by manipulating ionic strength (Hermansson et al., 1986). These two states are termed aggregated and ordered by Hermansson (1986a), who considers this behavior to be the result of an equilibrium between attractive and repulsive forces that can be disturbed

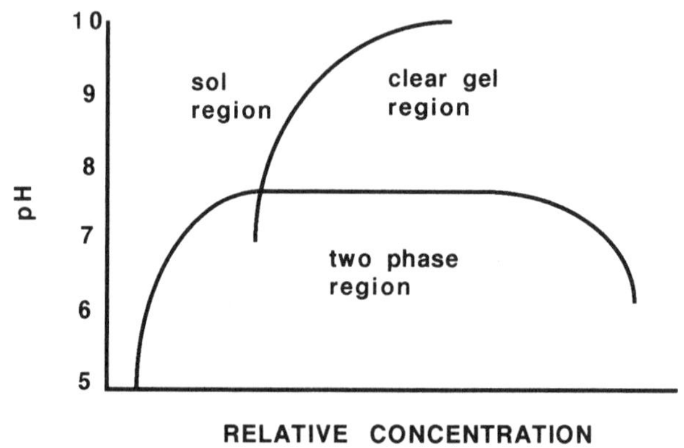

Fig. 16.8 Apparent phase behavior in a bovine serum albumin gel as a function of protein concentration and pH. (*Adapted from Clark and Lee-Toffnell, 1986.*)

by small changes in such variables as temperature, ionic strength, concentration, and pH. Aggregated gels are regarded as composed of proteins in different stages of phase separation. Thus, in a situation similar to that occurring during thermal extrusion of proteins, parameters under the control of the experimenter can be varied to produce gels of widely different functional properties. Familiar examples of aggregated protein food gels include yogurt and other milk protein gels, blood plasma gels, soy protein products such as tofu, and meat protein gels (Hermansson, 1986a).

Another way of viewing the two types of gel networks formed by proteins is given by Damodaran (1989), who believes that the factor determining gel category is the ratio of net charge to average (as opposed to surface) hydrophobicity. Since it has been found that altering pH or ionic strength will cause changes in the type of gel produced, this idea would seem to have merit.

Water holding. In considering thermal protein gels it is important not to overlook their major component—water. One of the most important functional properties of gels is their water-holding capacity. Hermansson (1986b) has reviewed this subject as applied to aggregated and ordered protein gels. As aggregation proceeds in a gel system, locally stronger protein-protein interactions will increase and phase separation grows. This leads to a loss in water-holding capacity since bigger pores or capillaries are formed between the aggregates and the water can be more easily pressed out. Water holding seems to be dictated by pore size with a more open structure giving rise to lower water holding than a dense network structure. Heating above the gelation temperature can produce phase separation and reduced water-holding capacity in some protein gels as a result of a more coarse structure with larger aggregates and pores, while increasing salt concentration and elevating pHs on the alkaline side of the isoelectric point improves these parameters (Fig. 16.9). These data are for whey proteins but in the case of soy 11S protein possessing a complex quaternary structure, increased heating led to a transition from an aggregated disordered gel to an ordered structure consisting of strands. This is thought to occur because of the thermal requirement, greater in the presence of salt, for dissociation and subsequent reassociation of subunits (Hermansson, 1986b).

Another point to take into account is the nature of the water being bound by the gel, since interactions with the protein can reduce its mobility. Through the use of techniques such as NMR, calorimetry, and Raman spectroscopy, it has been concluded that "bound" water is only a small percentage of the total water found in food gels and that it has little or no effect on gel properties (Busk, 1984; Hermansson, 1986b). The

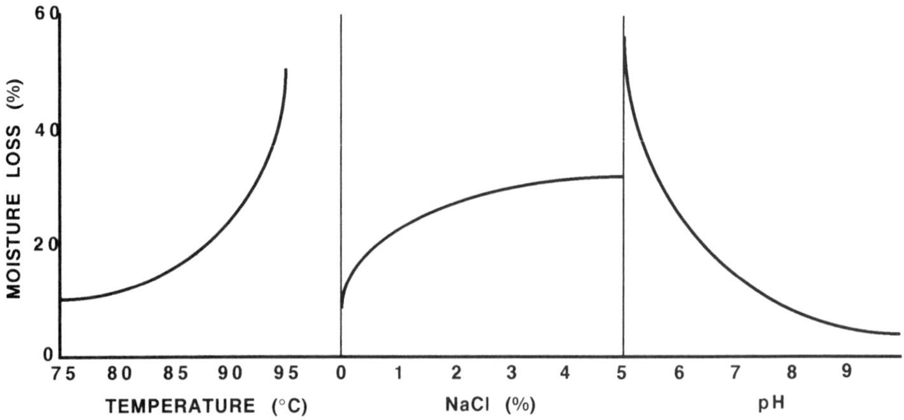

Fig. 16.9 Moisture loss from 12% whey protein gels as a function of temperature, NaCl concentration, and pH. (*Adapted from Hermansson, 1986b.*)

latter author provides a review of methodologies that can be applied to gels in order to obtain estimates of water-holding properties. Many of these are easy to apply and can be performed in most food laboratories.

Rheological testing. The gelation of proteins is a rheological occurrence so that in order to monitor the process it is usual to analyze gels with texture-measuring equipment. Protein gels, however, do not make ideal rheological materials and care must be taken in measuring their properties. Rheological evaluation of gels presents some distinct problems, the chief of which is their viscoelastic behavior. Whereas food scientists are often called upon to deal with solid and liquid foods, gels exhibit both these characteristics. Rheological theory dictates that the properties of viscoelastic materials must be ascertained using time-dependent tests, i.e., the textural attributes of viscoelastic foods vary with time under deformational forces. Tests employed include the creep curve (strain vs. time at constant stress) and the relaxation curve (stress vs. time at constant strain). A creep compliance curve relates changes in the ratio of strain to shear stress with time. Its shape represents the stretching and breaking of bonds between the basic structural elements of a material. Kamata et al. (1988) and Kamata and Kinsella (1989) have used creep behavior to study protein gels; a typical example is shown in Fig. 16.10.

Static tests, however, are still widely practiced in the food industry, and of these the most common is that of gel strength. In this method, the rigidity of gels is measured as the initial slope of a force-deformation curve made by a compression test using a universal testing machine. This is an

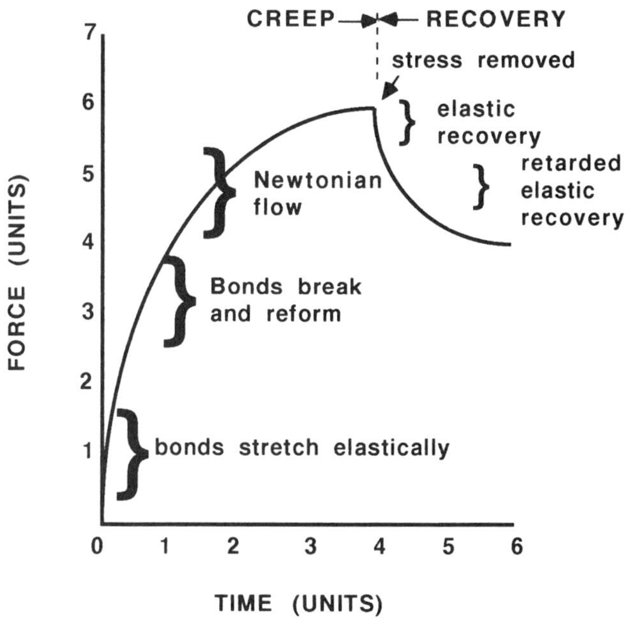

Fig. 16.10 Typical creep compliance curve. (*Adapted from Aguilera and Stanley, 1990.*)

example of a "single point" measurement, and it has often proved inadequate, the reason for this being the nature of the material being examined. The compression test mentioned above is suitable for solid foods—those characterized as possessing definite volume and shape and able to withstand small deformations while regaining their original shape. On the other hand, liquid foods are described rheologically by viscosity measurements designed to evaluate resistance to flow. Gels must be viewed as hybrids between solids and liquids.

The best methods currently available for characterizing viscoelastic foods are through dynamic testing in which sinusoidally oscillating stress or strain is applied to the sample. This is also referred to as small amplitude sinusoidal testing, and it has proved capable of characterizing the gelation process. For an ideal elastic solid the stress will be in phase with the strain, and for an ideal liquid there will be a 90° phase difference between the stress and strain. On the other hand, if the material is viscoelastic then the phase angle will be between 0° and 90°. The results obtained from dynamic testing experiments can quantitate the degree of viscoelasticity possessed by the gel, even though the cost of the equipment is high. Dynamic shear measurements are made using small strains so that the rheological properties obtained are as close as possible to those of the undisturbed sample, thus avoiding artifacts associated with sample perturbation during testing.

It is often possible to form the gel in situ between the plates of the rheometer. The results of such dynamic experiments, obtained by comparing the sinusoidal fluctuations of stress and strain, are often expressed using the parameters G′, G″, and δ. G′ and G″ are stress/strain ratios; in G′, the storage modulus, the ratio is based on the in-phase stress component; in G″, the loss modulus, it is based on the component that is 90° out of phase. Another way of viewing this is that G′ is a measure of energy stored due to elastic deformation during each test cycle and G″ measures the energy dissipated as heat due to viscous flow with the sample. The loss angle (δ) is given by the ratio of G″ to G′ and represents the phase difference separating the sinusoidally varying shear stress and strain signals. A perfectly elastic material would possess a δ of 0, whereas for a perfectly viscous material it would be 90°. It would be expected that as a gel forms and the sample becomes more elastic that G′ values will increase and δ would decrease (Beveridge et al., 1984, 1985). Reviews of methods to measure rheological changes in heat-set protein gels have been presented by Tung (1978), Mitchell (1980), Clark and Lee-Tuffnell (1986), Hamann (1987), Hamann et al. (1990), and Hamann (1991).

Microstructural analysis. Thus far, we have seen how methods including DSC, hydrophobicity measurement and rheological procedures are useful in gel characterization. One more technique to add to this list is microstructural analysis. With this approach it is possible to visualize gel structure and the influence upon it of processing variables. Light microscopy, scanning electron microscopy, and transmission electron microscopy have all been used to advantage in examining gels (Davis and Gordon, 1984), but scanning electron microscopy has proven especially useful. Using this technique Hermansson et al. (1986) and Hermansson (1988) have examined heated myosin protein gels and found evidence for two quite different gel structures, depending upon the conditions employed in the manufacturing process. The structure of fine gels requires the resolution of transmission electron microscopy, while aggregated gels can be viewed with scanning electron or, at lower magnifications, light microscopy.

In all of the microscopic methods, however, the high degree of hydration makes artifacts a real possibility. One way to overcome this problem is through the use of cryogenic sample preparation for scanning electron microscopy that leaves the water in situ, but to lower the temperature to a point where the vapor pressure is reduced and the escape of water and other volatiles is negligible. In this technique samples are frozen in liquid nitrogen at −210°C in order to freeze as fully as possible all the available water. Following processes to remove ice and coating of the sample to eliminate charging, the prepared sample is examined on a cold stage maintained at about −180°C. All the preparative steps can be completed within a few minutes of freezing. This method has been reviewed by Aguilera and Stanley (1990) and examples of gels treated in this way are shown in Fig. 16.11.

Application to a food system. With the methodologies discussed thus far at their disposal, it should be possible for food scientists to obtain a better understanding of thermally induced gel formation in plant proteins. Soy

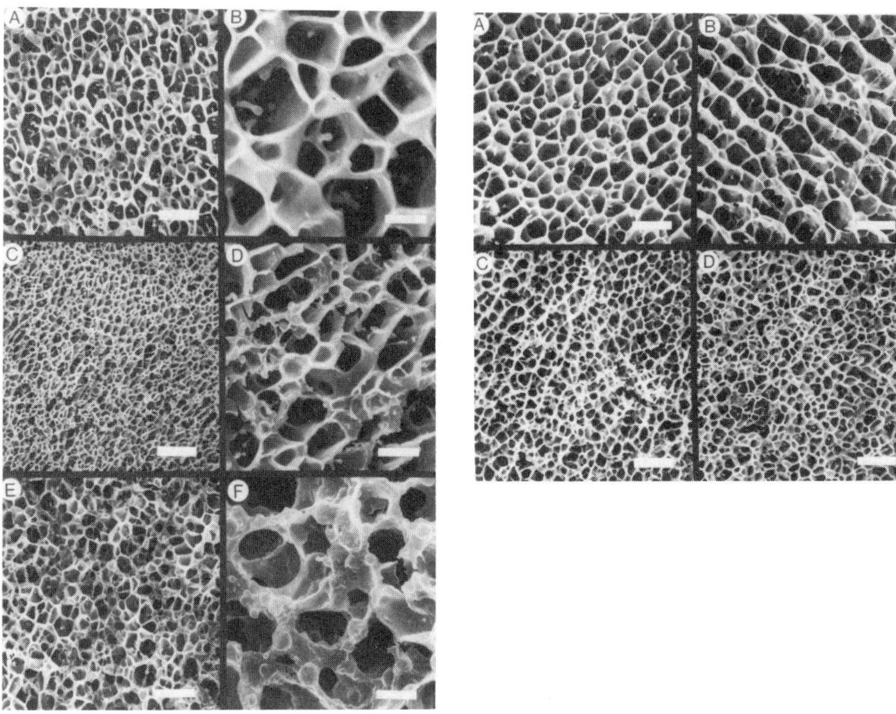

Fig. 16.11 (Right) Effect of total solids (TS) and pH on structure of a hydrocolloid-based milk gel as determined by cryo-SEM (bar = 30 μm). (A) 5.5% TS, pH 6.5; (B) 5.5% TS, pH 7.0; (C) 10.5% TS, pH 6.5; (D) 10.5% TS, pH 7.0 (Left) Effect of fat on gel structure (TS 10.5%, pH 7.0) as determined by cryo-SEM (bar A,C,E = 30 μm; bar B,D,F = 6 μm). A,B, 0% fat; C,D, 2% fat; E,F, 4% fat. (*Courtesy of Dr. S. Y. Xu.*)

proteins can form stable and firm gels at temperatures between 90° and 100°C, depending on concentration. These proteins have achieved some success as functional ingredients in foods, and considerable research effort has gone into studying the influence of heating on their functional properties with the aim of improving texture. They offer a major challenge to the food scientist because of their complex nature. Denaturation and other heating effects have been known for some time to influence the functional properties of these molecules (Wu and Inglett, 1974).

Soy Proteins. In order to understand how thermotropic gelation occurs in plant proteins it is necessary first to go into the structural makeup of these intricate molecules, as exemplified by the soy system. Soy proteins are currently used in many foods. The storage proteins of soybeans occur mainly in protein bodies and consist of four fractions, based on sedimentation behavior, ranging from 2 to 15S. The two major species are the 7S or β-conglycinin fraction and the 11S or glycinin fraction. The former has a MW of around 200 kd and is composed of at least six combinations of trimer subunits not joined by disulfide bonds while the latter is characterized as being 350 kd and containing from three to eight acidic and basic subunits that are packed into hexagonal structures, some of which are connected by disulfide bonds. The 11S protein is reported to contain 21 disulfide linkages, one function of which is to render the protein more heat resistant to dissociation. Both these species have only a small α-helix content and are primarily constituted of β-sheet and disorder regions. These components can undergo complex association-dissociation interactions, the final products of which depends upon concentration, pH, ionic strength, temperature and calcium concentration (Hermansson, 1978, 1979, 1986a; Koshiyama et al., 1980–81; Schmidt and Morris, 1984; Utsumi et al., 1984; Danilenko et al., 1985a; Clark and Lee-Tuffnell, 1986; Oakenfull, 1987; Ledward and Mitchell, 1988).

Effect of Heat on Soy Proteins. The first step in producing a thermotropic gel is to heat a dispersion of the protein. One effect of heat on the soy system is to produce conformational changes involving destruction of the quaternary structure. Koshiyama et al. (1980–81) demonstrated an increase in hydrophobicity during heating of isolated 11S soy protein that, at an ionic strength of 0.1, appeared related to dissociation of subunits; Voutsinas et al. (1983) also observed increasing hydrophobicity during heating of commercial soy preparations. Further work in this area showed that heating mixtures of 7S and 11S proteins led first to dissociation and then specific interaction among the subunits promoted by hydrophobic interactions and disulfide bonding (German et al., 1982; Utsumi et al.,

1984). Utsumi and Kinsella (1985a) reported that the soluble macromolecular complexes formed from heating soy isolate are composed mainly of basic subunits of 11S protein associated with β subunits of 7S protein via electrostatic interactions. Association of basic subunits through disulfide bonds was also observed. Matsudomi et al. (1985) studied the influence of ionic strength on surface hydrophobicity in 11S soy protein during heating. It was found that at 90°C an increment in ionic strength from 0.01 to 0.5 resulted in a 50% increase in hydrophobicity. Further work with this protein using calorimetry showed that denaturation temperature increased concomitantly with ionic strength (Danilenko et al., 1985a). Isolated 11S protein registered a thermal transition temperature of 80°C while that of 7S protein was lower at 67°C (Babajimopoulos et al., 1983). German et al. (1982) reported 92°C and 77°C; this difference between the two proteins was attributed to the large number of disulfide bonds in the former molecule. Similar results were obtained using a soy isolate, 84.5°C and 72°C, respectively (Damodaran, 1988).

It is an interesting academic point to consider whether the heating protocols employed with soy proteins produce denaturation in the classical sense, i.e., unfolding and a loss of structure. German et al. (1982) cite structural studies on heated soy proteins where, under similar conditions, secondary structure is not altered, and conclude that while heating dissociates the subunits, they are not unfolded. Thus, even though many studies have shown an increase in hydrophobicity as a result of heating, it could be that this is a result of exposure of the apolar patches mentioned previously becoming uncovered. Also, if the associative forces become sufficiently powerful, aggregation could result without the prerequisite of unfolding. Ma and Harwalkar (1987) have come to this conclusion as a result of their studies with oat storage proteins, but it must be noted that oat proteins are characterized by both a high hydrophobicity and denaturation temperature (Myers, 1990).

The studies discussed in this section suggest that the influence of heat on soy proteins is to promote subunit dissociation and then aggregation in which hydrophobic bonding plays a strong role but where the exact nature of the interaction is dictated by factors such as pH, ionic strength, and concentration. In order for a suitable gel to form there must be a balance between the associative and disassociative forces among the protein molecules. Disulfide bonding is thought to be important in gel formation as well, since the addition of reducing agents has been reported to inhibit this reaction (Schmidt and Morris, 1984). Mori et al. (1986) and Shimada and Cheftel (1988) have also presented evidence for disulfide bonding in soy protein gels. Damodaran (1989) dissents from this view, pointing out the possibility that rather than form structure directly, the role of disulfide bonds may be to effectively increase polypeptide chain length.

Is heating necessary for the formation of soy gels? Beveridge et al. (1985) and Hermansson (1986a) have shown that gels can be formed from heated commercial soy preparations which were previously denatured to the point of exhibiting no DSC endotherm. Preheating of soy protein dispersions was reported by Shiga and Nakamura (1987) to improve the rheological and water-holding properties of the resultant gels, but Ledward and Mitchell (1988) found exactly the opposite; in their work, preheating caused a reduction in gel strength. On the other hand, soy proteins can be overheated; Catsimpoolas and Meyer (1970) found that a temperature of 125°C resulted in chemical degradation and aggregates that did not form a gel during cooling, while Shimada and Cheftel (1988) suggested that the reason soy gels soften above 130°C is due to disulfide bond breakdown.

Soy gels are classified as thermosetting in that they are formed by heating and cannot be remelted without destroying the primary structure (Rector et al., 1989). These gels can be remelted to the soluble aggregate state under certain conditions, however, and this step is a function of the protein concentration (Catsimpoolas and Meyer, 1970). Of course, gels that are heated will also be cooled and Damodaran (1988) has provided evidence that the cooling step is important in structure formation of 11S protein. CD spectra showed that upon cooling 11S dispersions in the presence of 0.5 M NaCl the thermally unfolded protein partially refolds into an ordered structure containing more α-helix than β-sheet conformation. These results build on earlier work that suggested this protein partially refolds after a heat treatment or the application of denaturants (Kamata et al., 1979). Paulson and Tung (1989) found rapeseed gels increased in firmness during cooling and hypothesized that this reflects reforming of hydrogen bonds that had been weakened by the prior heating treatment.

Effect of Protein Concentration. The influence of heat on soy proteins is mediated by the concentrations used. Mori et al. (1981) exposed 11S protein to heat and found that whether the heated protein undergoes disaggregation or gel formation is controlled by concentration; low levels of protein favored disaggregation while high levels led to gel formation. The effects of heat and ionic strength on dissociation-association reactions in soybean protein fractions was studied through electrophoresis by Iwabuchi and Yamauchi (1984) who reported heated 7S proteins will dissociate and then recombine into insoluble aggregates at high concentration and ionic strength while reducing these conditions led to the proteins remaining dissociated. These results were later substantiated by Nakamura et al. (1986). Thus, soy proteins exhibit a gel point or a protein concentration below which a gel will not form. Both Bikbov et al. (1985) and Nakamura et al. (1986) agree that the gel point for 11S proteins is below that for 7S proteins, although the numerical value will depend upon the experimental conditions and the definition used for gelation. The former group published hydrophobicity values for the subunits of 11S and 7S soy proteins and found that they correlated with gelling ability, suggesting that hydrophobicity can be used as a measure of gelling ability, at least for related proteins.

The mechanical strength of the gel also depends upon concentration, since it is related to the number of crosslinks formed per protein chain. Damodaran (1989) provides data showing a linear relationship between gel strength and (protein concentration)^2 for both soy isolate and 7S protein. This author also gives a plot of (gel strength)^2 versus protein molecular weight in which it appears that, at a given protein concentration, the physical properties of gels are influenced by molecular size.

Effect of Ionic Strength and pH. Many researchers have investigated the influence of these two parameters on soy gel properties. A summary of this work is useful in understanding the basics of gel formation. The general effect of monovalent salts is to stabilize the quaternary structure of soy proteins by promoting protein-protein interactions while alkaline or acidic pHs lead to dissociation into subunits through protein-solvent interactions and charge repulsions. Salts support aggregation since they reduce the diffuse part of the electric double layer. These effects are not as straightforward as they might appear, however. Hermansson (1979) showed that aggregation was increased by NaCl concentrations up to 0.2 M, but at higher levels it was suppressed. Below 0.1 M, dissociation into subunits still occurred, yet aggregation was inhibited, probably due to the electric double layer and repulsion forces in the absence of mobile counterions. It would be predicted that the stabilizing effect of salts would increase the gelation temperature of soy proteins, and this has been demonstrated (Hermansson, 1986a).

Optimum gelation occurs when attractive and repulsive forces are in balance; this implies a narrow range of conditions for this event, and it would be expected that a strong interaction occurs between pH and ionic strength. Also, they could influence other characteristics of the system. For example, Arntfield et al. (1990b) demonstrated an effect of salts on hydrophobic interactions. This group, among others (e.g., German et al., 1982; Harwalkar and Ma, 1987; Sathe et al., 1989) have taken results that salts influence thermal stability in relation to their position in the lyotrophic series as evidence for involvement of hydrophobic interactions in protein thermal stability. However, Chambers et al. (1990) point out that the mechanism by which these salts produce dissociation of subunits is not yet fully understood and counsel caution in the interpretation of such data.

7S, 11S, and Soy Protein Gels. The two major storage proteins of soy are, as has been seen, of quite different compositions and properties. Gels made from them would also be expected to possess dissimilar structures and characteristics. Summarizing the discussions by Mori et al. (1981), Babajimopoulos et al. (1983), Utsumi and Kinsella (1985a,b), Bikbov et al. (1985), Nakamura et al. (1986), and Hermansson (1986a) on the nature of 7S, 11S and soy gels, it would appear that 11S gels form via disulfide bonds and various noncovalent bonding, while 7S gels do not form disulfide bonds. 7S gels are harder, perhaps due to their higher density, but 11S gels are more elastic, probably as a result of the disulfides. As previously mentioned, Hermansson (1986a) classifies protein gels into two types—gels formed by "random" aggregation and gels formed by association of molecules into strands in a more ordered way. More highly ordered gels are characterized by greater clarity and elasticity. On the basis of clarity 7S gels would seem to be ordered and 11S gels are aggregated, but this classification is arbitrary. Gels made from complete soy proteins exhibit properties in between the 7 and 11S molecules. This observation might not seem remarkable, but considering that various 7 and 11S subunits interact preferentially, it does not necessarily follow that the resulting gels should be intermediate in their properties.

Mechanisms of Soy Gel Formation. The mechanisms involved in gel formation in heated protein dispersions have received much attention, beginning with Ferry (1948), who suggested a two-step mechanism; an initiation step involving unfolding or dissociation of the proteins is followed by an aggregation and association step that, under appropriate thermodynamic conditions, leads to gel formation. When it became known that aggregation and association are not necessarily the same, i.e., aggregation can yield soluble macroaggregates or an insoluble coagulum depending on the conditions used, this additional step was added. This proposal has been shown to apply to both 7S and 11S gels (Mori et al., 1981; Nakamura et al., 1986). As a result of data demonstrating the importance of the cooling step to structure formation, Damodaran (1988) established a more elaborate scheme in which a partially refolded intermediate is in equilibrium with the heat denatured and gel states. The extent of gel network formed is a function of the relative magnitude of the equilibrium constants that would dictate the strength and elasticity of the gel. These mechanisms are summarized in Fig. 16.12.

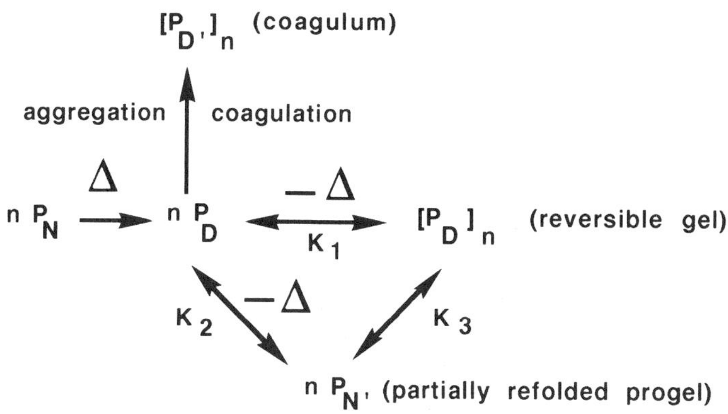

Fig. 16.12 A proposed mechanism for thermal gelation of globular proteins. n number of native protein (P_N) molecules are heated (Δ) to form n number of denatured (P_D) protein molecules which, upon cooling (−Δ) form a reversible gel $[P_D]_n$. $P_{N'}$ is the protein in a partially refolded state formed during cooling. K_1, K_2, and K_3 are equilibrium constants. If the protein contains a high level of apolar amino acid residues it may form an irreversible coagulum-type gel $[P_{D'}]$. (*Adapted from Damodaran, 1988.*)

From the considerable research that has been done with soy proteins, there emerges a concept of gel formation dependent on, first, heating of an aqueous dispersion to promote disassociation of the subunits and possibly their unfolding, resulting in exposure of previously buried apolar side chains and reactive sulfhydryl groups. As temperature rises hydrophobic interactions increase while hydrogen bonding decreases. Attractive and repulsive forces compete in this dynamic system, but at some point protein-protein interactions become dominant and aggregation occurs. It must be emphasized that large conformational changes are not necessary for this process; Lumry and Biltonen (1969) point out that small displacements may produce more significant alterations in protein structure than large ones. During cooling hydrogen bonds can reform, adding to gel stability and perhaps restoring some of the original secondary structure. Characteristics of the final gel, some of which, such as clarity, gel strength, gel point, and elasticity, can be of great importance to food scientists, are dictated by these forces. As pointed out by the work of Hermansson (1978) and also by Arntfield et al. (1989), if denaturation/unfolding proceeds quickly, so that the molecules can orient themselves prior to aggregation, a fine gel network, characterized by higher clarity and elasticity, will result. If attractive forces are stronger, chain association will be more random and a coarser, more opaque gel will form.

Further increasing the attractive forces results in the production of a coagulum. Of the experimental parameters that control the attractive/repulsive balance, several, including protein concentration, heating and cooling rates, pH, ionic strength, chemical modification of proteins, gelling additives and protein mixtures, are under the control of the manufacturer.

It should be appreciated by this point how complex the art and science of gel formation can be. While many studies have focused on individual parts of the problem one creative approach that seems productive is to attempt the simultaneous measurement of several gel parameters. Arntfield et al. (1989) has used this approach to advantage while studying the gelling characteristics of vicilin, the major storage protein of faba beans. Figure 16.13 depicts changes in protein conformation (hydrophobicity), DSC response and structure development (G′) during gel formation at a heating rate of 2°C/min. An endotherm typical of denaturation occurs at about 80°C; following this, a steady increase in hydrophobicity is seen and then, at about 95°C, G′ values start to rise, indicative of gelation.

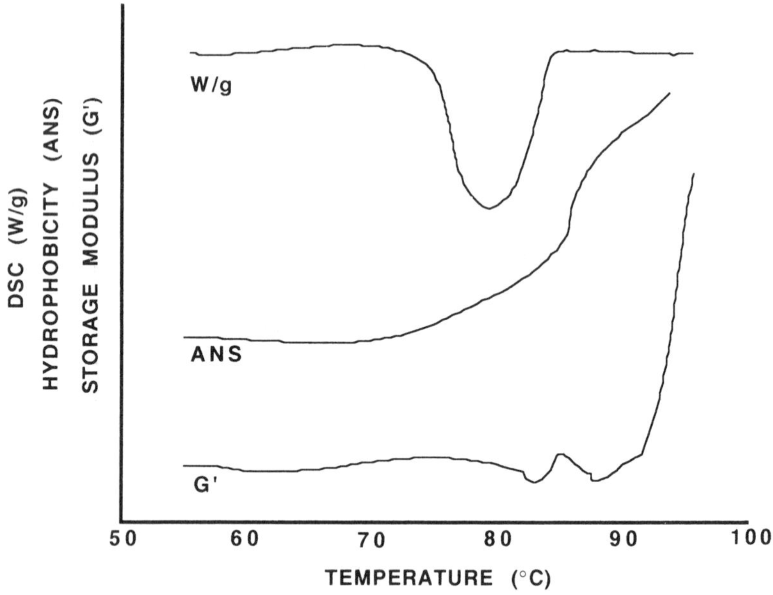

Fig. 16.13 Changes in protein conformation, as monitored by hydrophobicity (ANS) and DSC (W/g), and structure development, as monitored by storage modulus (G′), in pH 7, 10% fababean vicilin heated at 2°C/min. (*Adapted from Arntfield et al., 1989.*)

These data substantiate a gelling mechanism in which unfolding leads to increased hydrophobic intermolecular attractive forces, culminating in gelation. In this work the effects of pH, salt, and SDS were also investigated (Table 16.10). Here it may be seen that these factors have a significant influence on final gel quality and that many of the measurements seem interrelated. Both acidic and alkaline conditions lowered enthalpy values. A weak elastic network was obtained at acidic pHs while a stronger but less elastic network resulted from alkaline conditions. The concentration of NaCl showed a significant influence on denaturation temperature and storage modulus, while these latter two parameters were inversely proportional. Enthalpy values were not greatly altered, suggesting no major conformational changes, but, as evidenced by increasing denaturation temperatures, a higher stability accompanied salt addition. Increasing SDS concentration was accompanied by significant decreases in enthalpy and denaturation temperature which indicated considerable conformational alterations. Correspondingly lower storage modulus results reflected an inability to form viable structures; in this case this event

Table 16.10 Effect of pH, NaCl Concentration, and SDS on Thermal Properties and Gel Characteristics of 10% Vicilin

Treatment	ΔH (J/g)	T_d(°C)	G′(Pa)
pH 3	0	—	176
pH 4	5.0	67.8	485
pH 5	14.4	80.3	265
pH 6	11.1	82.0	251
pH 7	13.1	79.1	532
pH 8	14.1	80.3	838
pH 9	7.4	77.0	779
0.1 M NaCl	15.9	78.1	1902
0.2 M NaCl	14.8	82.0	1012
0.3 M NaCl	16.4	83.1	587
0.4 M NaCl	21.4	84.8	502
0.5 M NaCl	17.1	84.2	514
2.5 mM SDS	12.3	78.2	1131
5.0 mM SDS	11.4	78.2	1387
7.5 mM SDS	12.0	78.0	562
10 mM SDS	13.9	77.4	754
25 mM SDS	8.1	77.9	140
50 mM SDS	6.2	76.7	0.4

Source: Adapted from Arntfield et al., 1989.

occurred between 10 and 25 mM SDS. By obtaining data of this sort it becomes possible to examine relationships among variables in terms of their contribution to structure formation.

In this section it has been shown that approaches to measuring functional properties related to the structure of heated proteins vary considerably in sophistication. Often in food studies it is thought necessary to sacrifice some degree of accuracy for the utility of obtaining large quantities of data that are familiar to the industry, such as gel strength. On the other hand, it has been seen that the application of fundamental physical tests to food systems such as gels can reveal much about their structure and point the way to some potentially quite useful products.

THERMODYNAMICS AND KINETICS OF PROTEIN REACTIONS

Thus far in this chapter the goal has been to present an overview of heat-catalyzed reactions occurring in proteins from the vantage point of the food scientist. As one of the stated goals of this symposium is to point out how useful physical chemical techniques can be to those researchers studying food, it now becomes important to focus on two of the most powerful tools of this discipline, thermodynamics and kinetics, and how they can be used to advantage in these studies. Other authors in this volume (see Chapters 1, 14, 15) have dealt with the principles of thermodynamics and kinetics, allowing the present discussion to center on their application to proteins. Thermodynamics as a field of study is concerned with heat changes in a system undergoing chemical reaction and how energy is redistributed among the various species during this reaction. It can be used to predict the feasibility and extent of chemical reactions but not the rates and mechanisms of these reactions, the latter being the purview of the branch of physical chemistry known as kinetics. Kinetic studies are useful to food scientists since they provide information that cannot be gained solely from thermodynamic data. On the other hand, kinetic studies alone often fail to answer completely the questions asked by those investigating complicated food protein systems.

The presence of a quaternary structure brings into question some of the assumptions commonly made during protein denaturation studies. It may help to understand these concepts by looking at some basic physical chemistry involved in protein denaturation. As a model for simple, reversible denaturation of a single-chain protein, the most commonly employed thermodynamic reaction is a two-state concept. This results from observations of "all-or-none" behavior for the physical properties of proteins that change totally over a short transition range. For example, Rha and Pradipasena (1986) have considered protein viscosity and found a strong temperature dependence as the molecule uncoils. Studies of proteins in solution have shown that the native state is characterized by compactness, where the structure is maintained mainly by hydrogen bonds and hydrophobic interactions. Water interacts with polar surface groups, and apolar groups are concentrated toward the hydrophobic interior of the molecule. According to this view, chemical denaturation is defined as a rearrangement of conformation resulting from alterations in secondary and tertiary structure or, in one word, unfolding. The thermodynamic and kinetic aspects of this reaction are as follows:

$$N \text{ (native)} \leftrightarrow D \text{ (denatured)} \tag{1}$$

$$K = \frac{[D]}{[N]} \tag{2}$$

$$\Delta F = \Delta H - T\Delta S \tag{3}$$

$$\Delta F = -RT \ln K \tag{4}$$

$$\ln K = \frac{-\Delta H}{RT} + \frac{\Delta S}{R} \tag{5}$$

$$C_p = \frac{d(\Delta H)}{dT} \tag{6}$$

where these symbols stand for the standard thermodynamic constants. Note that from the assumption of reversible denaturation, Eq. (1), an equilibrium constant can be obtained where this value is the ratio of the two concentrations, Eq. (2), and from the Gibbs free energy equation, Eq. (3), the various thermodynamic constants can be calculated through the use of Eq. (4) and Eq. (5), although the validity of this approach is questionable. Myers (1990) estimates that the accuracy of such calculations is only $\pm 30\%$. Also, Eq. (1) implies a two-state behavior of proteins during denaturation. It is possible that this may occur under certain conditions (Tanford, 1970; Privalov, 1979a), but, as pointed out by Lumry and Biltonen (1969) the most important feature of this model is that it allows simple thermodynamic analyses of conformational transitions. Brandts (1969), who has used this representation to advantage in very significant work on protein denaturation, states that, theoretically, any reaction can be treated as a two-state transition by arbitrarily dividing the sequence into two parts. In any case, it is difficult to substantiate the two-state mechanism, and evidence exists which suggests that this is an oversimplification for at least some proteins (e.g., Privalov, 1982; Privalov and Gill, 1988).

It would be useful in this discussion if these thermodynamic principles could be related to protein structure. An overall view of the literature supports the concept that, based on Eq. (3), where ΔH is the change in enthalpy and ΔS is the change in entropy, increases in conformational entropy during denaturation are offset by negative entropy changes resulting from the forcing of apolar side chains into a polar solvent (see Fig. 16.4). While the free energy decreases as temperature increases, this is not a linear function. Considering the 0–100°C range, a maxima is reached (interestingly close to the optimum physiological temperature) before the decline begins. Privalov and Gill (1988) equate this temperature of maximum thermodynamic stability to the point where the entropies of the native and denatured states are equal.

The difference in free energy (ΔF) between the native and denatured states may be considered as the "work of disruption" of the native structure and is a measure of its stability (Privalov and Gill, 1988). These authors regard this value to be valid for a small globular protein, but larger proteins with multiple domains require a set of values, specifying the stability of each region and their interaction. Entropy is often used as an index for folding/unfolding of the protein, while enthalpy is expressed in terms of heat as joules or calories. How enthalpy and entropy change in relation to ΔF during a theoretical two-state protein denaturation reaction is shown in Fig. 16.14. It may be seen that while both enthalpy and entropy progress through large minima and maxima, ΔF is relatively unchanged. This is the concept of enthalpy-entropy compensation, which one group has termed "the fundamental thermodynamic feature of protein conformation change" (Lumry and Biltonen, 1969). Enthalpy-entropy compensation is a broad and far-reaching concept based on cooperative molecule/water interaction and ordering, with important implications to many reactions occurring in food. Specifically, it is thought that upon unfolding, exposure of apolar groups causes an ordering of adjacent water molecules via hydrogen-bonded clathrates with an associated change in heat capacity (C_p, see subsequent discussion).

This general subject was reviewed by Labuza (1980b) and the papers of Brandts (1969) and Lumry and Biltonen (1969) are recommended with respect to this effect in protein denaturation. From this work it becomes obvious that a major factor in determining the thermal stability of all proteins is water. Two basic types of interaction are thought possible; the first involves the disruption of intramolecular hydrogen bonds during unfolding to form new hydrogen bonds with solvent water, and the second is the exposure of apolar side chains during unfolding that then order

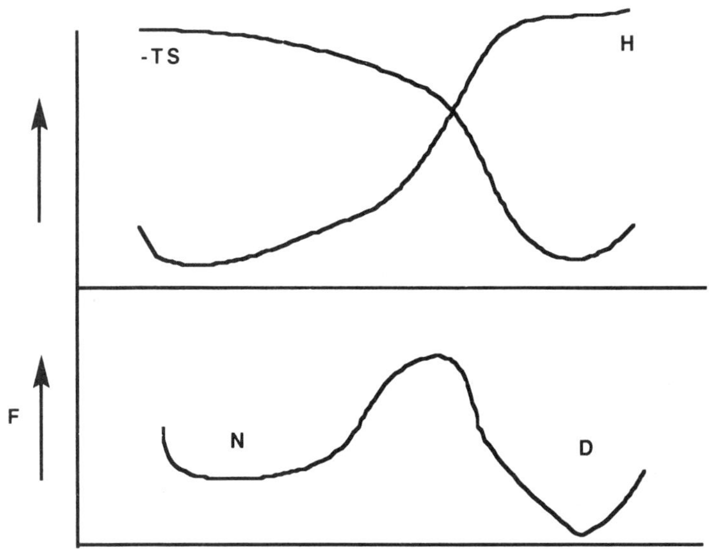

Fig. 16.14 Enthalpy (H), entropy (S), and free energy (F) changes during the unfolding of chymotrypsinogen. (*Adapted from Lumry and Biltonen, 1969.*)

solvent water. The importance of these reactions can be appreciated when the work of Privalov is considered. This author approximated that at least 80% of the temperature dependence of denaturation enthalpy could be explained by the ordering of water around exposed apolar groups (Privalov, 1979a; Fujita and Noda, 1981). Even though it was later considered that this value may be an overestimate (see Privalov, 1982), protein-water interactions are of primary significance in thermal stability (Tanford, 1970, 1980), and a correlation between hydrophobicity and thermostability has been found for many thermophilic proteins (Bigelow, 1967; Ikai, 1980; Merkler et al., 1981; Mozhaev and Martinek, 1984). Thus, it follows that concentration effects should be carefully considered in experimental work. This reemphasizes an important difference in the way protein chemists and food scientists work with and, hence, view proteins. Also, this type of finely balanced behavior is reminiscent of what was observed in gel formation of protein dispersions and may reflect the same overall mechanism.

In the case of protein denaturation, although there are large changes in enthalpy and entropy, the net ΔF is small, as shown in Table 16.11. Studies of this type have been made possible by a revised Eq. (3) in which

Table 16.11 Calculation of ΔF for Unfolding of Chymotrypsin at pH 3.0

Source	Contribution (kcal)
ΔH (H_2 bonds + hindered rotation)	184
ΔF (abnormal ionization + repulsion)	−9
$-T\Delta S$ (conformational entropy)	−364
ΔH (hydrophobic bonds)	−123
$-T\Delta S$ (hydrophobic bonds)	319
Net =	7

Source: Adapted from Lumry and Biltonen, 1969.

the most important factors contributing to the change in free energy have been identified (Brandts, 1964, 1969):

$$\begin{aligned}\Delta F = {} & \sum \Delta H \text{ (peptide } H_2 \text{ bonds)} \\ & + \sum \Delta F_{sc} \text{ (hydrophobic side chains)} \\ & - T \sum \Delta S \text{ (conformational entropy)}\end{aligned} \tag{7}$$

The first two factors in Eq. (7) tend to stabilize proteins, while the last is destabilizing. This sets up a balance of forces within the molecule between these two components, and the net result of this balance is determined by temperature. This was modeled in Fig. 16.4, where it will be appreciated that as temperature increases there comes a point at which the destabilizing effect of conformational entropy overcomes the stabilizing effect of hydrophobic bonding (which itself begins to decrease at temperatures above those shown in the figure). This is the temperature of denaturation, at which ΔF tends to 0 and K tends to 1.

It has been shown that these generalized equations, based on the assumption of a two-state transition, can be applied to many small, compact globular proteins (Brandts, 1964, 1969; Tanford, 1968, 1970; Privalov and Khechinashvili, 1974; Privalov, 1979a). However, this approach has been challenged for the larger, subunited proteins often encountered by food scientists. Privalov (1982) points out that there are no assurances that the assumption of a two-state transition holds even for globular proteins showing a sharp denaturation transition by different techniques. From a consideration of C_p effects, it can be determined that for many proteins their different domains unfold as independent units. It follows from this that the stability of the native protein is decided by the parameters of the least stable subunit. Also, it is tempting to equate the folding domains mentioned previously with the domains that appear during unfolding.

These thermodynamic considerations have been useful in exploring the mechanisms operating during denaturation under the simplified conditions given previously. The question remains if they are as helpful in food-processing situations. Realistically, the idea of ΔF proceeding orderly from positive to negative with a spontaneous conformational change at exactly 0 becomes difficult in a food system and suggests the possibility of kinetic rather than thermodynamic stability.

Differential scanning calorimetry has emerged as the technique of choice for acquiring thermodynamic and kinetic data of thermal transitions in proteins. The works of Borchardt and Daniels (1957), Steim (1965), Privalov and Khechinashvili (1974), Biltonen and Freire (1978), Sturtevant (1980), Wright (1982), Biliaderis (1983), Lund (1983), Sandu and Lund (1985, 1986), Arntfield et al. (1990a), and Myers (1990) may be consulted for general principles and specific applications. The popularity of this method rests on the ability to gather rapidly a large amount of accurate data that can be converted to both thermodynamic and kinetic parameters. With respect to protein denaturation, Privalov (1979a) and Privalov and Gill (1988) showed that a significant increase in heat capacity (C_p, the quantity of heat needed to raise the temperature a unit amount) occurred as a result of the conformational changes in proteins during denaturation. DSC instrumentation is designed to measure the excess of heat capacity of a sample over that of a reference material as a function of temperature. The excess enthalpy is calculated by integration of the curve of excess heat capacity versus temperature [Eq. (6), Kirchhoff's equation]. Studies of protein denaturation by DSC have shown that a substantial increase in heat capacity occurs during unfolding.

In Findlay et al. (1986) and Findlay and Barbut (1990) derivations of some of the thermodynamic considerations important to protein denaturation are presented as are ways to collect these data from the thermal curve. Biliaderis (1983) and Myers (1990) summarize some of the important conclusions from theoretical experimentation on single-chained proteins. In a simple DSC analysis of proteins, the two parameters obtained are T_d, the denaturation temperature, and ΔH, the denaturation enthalpy. While the former is an indication of the thermostability of the sample, the latter, theoretically the amount of heat energy needed to produce the conformational change, is more difficult to interpret (Myers, 1990). Although it might be expected that higher ΔH values would imply a greater thermodynamic stability, the choice of operating parameters such as heating rate, solvent, and concentration, as well as impurities in the sample make conclusions difficult under experimental conditions used for food samples. The presence of two or more materials capable of a thermal transition, such as several species of protein, or one oligomeric protein, further complicates interpretation of DSC data since it must be borne in mind that the thermal curve is a composite. Also, at the high protein concentrations and heating rates used to simulate real processing conditions, the assumption of reversibility no longer holds since aggregation results, which is an exothermic process (Biliaderis,1983). Table 16.12 provides a gathering of thermodynamic data for denaturation of some food proteins.

How may thermodynamic data gathered by DSC be useful to the food scientist interested in heating effects on proteins? Arntfield et al. (1990a) point out that such information can be employed to measure conformational changes in proteins occurring during processing or storage. Thus, native proteins yield a measurable enthalpy while loss of conformation or denaturation would decrease this parameter. Additionally, the DSC cell could be used to simulate the effects of certain processes on food proteins. One area where these results would be expected to be of particular relevance is in determining the degree of prior heat treatment. This directly influences certain functional properties and the value of the protein to a food processor. Thus, enthalpy measurements could be used in

Table 16.12 Thermodynamic Data for Protein Denaturation

Protein	Denaturing conditions	T_d	ΔF	ΔH	ΔS
Ribonuclease	pH 3.2, 30°C	—	12.9[a]	222[a]	690[b]
Chymotrypsinogen	pH 3.0, 25°C	—	30.5	163	439
Myoglobin	pH 9.0, 25°C	—	56.8	176	397
Egg albumin	Thermal	65.0[c]	—	132[d]	316[e]
Hemoglobin	Thermal	60.5	—	76	153
Peroxidase	Thermal	70.1	—	185	466
Amylase	Thermal	60.0	—	42	52.3
Oat storage protein	Thermal	113	—	7.2[f]	—
Bovine serum albumin	Thermal	62	—	5.2	—
Rapeseed storage protein	Thermal	85	—	5.0	—
Soy storage protein	Thermal	98	—	5.0	—

[a] kJ/mol.
[b] J/mol·°K.
[c] °C.
[d] kcal/mol.
[e] cal/mol·°K.
[f] cal/g.
Source: Adapted from Labuza, 1980b; Cheftel et al., 1985; Myers, 1990.

setting specifications for incoming proteinaceous ingredients (Arntfield and Murray, 1981). For such a "denaturation index" to be beneficial there should be a predictable relationship between enthalpy and the processing parameter (heating treatment, time of storage, etc.). Data available in the literature for preheating (Armstrong et al., 1979; Aboagye and Stanley, 1985; Ma and Harwalkar, 1988b) and storage (Arntfield et al., 1990a) of unfractionated plant proteins fail to show a linear relationship between these parameters and residual enthalpy. These results do not rule out the use of this index for quality control purposes, but they do suggest caution in its interpretation. This effect could be due to changes in the relative amounts of the various species as denaturation progresses (Myers, 1990) or more complex reactions occurring differentially within these oligomeric proteins (Aboagye and Stanley, 1985). Also, as an additional example of anomalous calorimetric behavior in proteins, Ma and Harwalkar (1984) reported that the prolamins and glutelins fractions of oat proteins, while exhibiting some reasonable functional properties in the native state, had no thermal response in the DSC cell.

One further example of the application of thermodynamic data obtained by DSC to solving problems associated with plant proteins will be given. Storage of the dried bean *Phaseolus vulgaris* under tropical conditions of temperature and humidity will cause the seeds to remain hard during cooking. This so-called "hard-to-cook" defect adversely affects consumption patterns and nutritional value (Stanley and Aguilera, 1985). In order to understand the mechanism of this failure to soften, it becomes important to know how beans react to heating. Previous work has indicated legumes soften during cooking primarily through thermally induced changes to pectic substances in the middle lamella in which water plays a major role (Sefa-Dedeh et al., 1978; see also Rao and Lund, 1986). However, often factors such as protein denaturation and starch gelatinization may also be important in this process since they also depend upon water availability. To investigate these possibilities, DSC thermal curves of flour prepared from beans stored under different conditions were compared using a high sample/water ratio (Table 16.13). Enthalpy of starch gelatinization increased as a result of the adverse storage conditions, indicating that higher energies are required for this reaction under conditions of limited water availability, while gelatinization temperature was unaffected. On the other hand, protein enthalpies decreased during storage; prior research (Hohlberg and Stanley, 1987a) has established the presence of active proteases in beans, while protein chromatography and amino acid analysis indicated a tendency for high molecular weight storage proteins to undergo proteolysis as a result of tropical storage, which may explain the DSC data.

Table 16.13 Temperatures and Enthalpies of Protein Denaturation and Starch Gelatinization Bean Flours

Storage conditions	Cooked hardness (N)	Protein denaturation		Starch gelatinization	
		T_d(°C)	ΔH (J/g)	T_g(°C)	ΔH (J/g)
Tropical	530	109	1.47	91	4.94
Temperate	194	108	1.60	90	3.14

Source: Adapted from Garcia-Vela and Stanley, 1989.

Additional data were collected as a function of heating time. Examination of the thermal records and chemical analysis showed that these events could be described by apparent first-order kinetics (Table 16.14). Beans stored under tropical conditions thermally soften much more slowly than those stored under temperate conditions; softening of beans stored under temperate conditions does not seem restricted by pectin breakdown, which occurs much more rapidly than softening. Softening of beans stored under tropical conditions may be restricted by pectin breakdown since this rate constant is markedly lowered by the hardening phenomenon when compared to other events. Since water is necessary for the hydrolysis of pectin, this could reflect a failure of water to penetrate the cell wall, perhaps being held back by hydrophobic ligninlike polymers at the surface. Starch gelatinization occurs rapidly in both samples and, thus, would not seem to play a role in softening.

From these two examples and the many others available in the literature (e.g., Arntfield et al., 1990a; Ma, 1990) it is clear that data such as these are extremely useful to the food scientist. They can serve as a quality control tool to evaluate functional properties and help in providing an understanding of chemical and physical events occurring during pro-

Table 16.14 Apparent First-Order Rate Constants for Thermal Softening, Protein Denaturation, and Starch Gelatinization in Cooked Beans

Storage conditions	Apparent first-order rate constant (min^{-1})				
	Thermal softening	Protein denaturation	Starch gelatinization	Seed coat softening	Pectin breakdown
Tropical	0.0039	0.0443	0.307	0.0101	0.040
Temperate	0.0137	0.0652	0.307	0.0216	0.100

Source: Adapted from Garcia-Vela and Stanley, 1989; Plhak et al., 1989; Stanley et al., 1989; del Valle and Jackman, unpublished data.

cessing. Care must be taken, however, in applying thermodynamic data to food systems. Kilara and Sharkasi (1986) cite a relevant example of applying such parameters obtained for a pure protein to an impure food system.

Kinetic parameters of protein denaturation can also be determined by DSC using, in fact, the same data that were obtained from thermodynamic studies. In this case emphasis is placed on the dynamics of the reaction as opposed to its energetics.

$$\pm \frac{dA}{dt} = kA^n \tag{8}$$

$$k = Ze^{-E_a/RT} \tag{9}$$

The basic equation for rate of change of a species (A) with time (t) is given in Eq. (8) where k is the rate constant ($time^{-1}$), dependent upon temperature, and n is the apparent reaction order. The temperature dependence of the rate constant is determined by the Arrhenius equation, Eq. (9), which states that k is an exponential function of absolute temperature (T), where Z is the preexponential or Arrhenius frequency factor, R is the gas constant, and E_a is the activation energy in J mol^{-1} (Labuza and Kamman, 1983). Several different approaches are available to determine these constants, the two most frequently encountered being those of Borchardt and Daniels (1957) and the peak temperature dynamic method (see Sandu and Lund, 1985, 1986). Table 16.15 gives some rep-

Table 16.15 Apparent First-Order Rate Constants for Various Reactions in Food Proteins

System	Conditions	k
Protein denaturation	Thermal	
Egg albumin		2.5×10^{-4} sec^{-1}
Hemoglobin		4.3
Peroxidase		5.1
Amylase		8.0
Maillard browning—model	Glucose/lysine, 45°C, pH 5, a_w.90	3.0×10^{-3} hr^{-1}
Maillard browning—whey	45°C, a_w 0.65	2.1×10^{-2} $days^{-1}$
LAL—casein	100°C, pH7	2.1×10^{-2} min^{-1}
LAL—soy protein	40°C, pH 12	2.3×10^{-2} min^{-1}
Crosslinks—muscle	121°C, Asp-Lys	4.3×10^{-2} hr^{-1}

Source: Adapted from Hurrell and Carpenter, 1977; Sternberg and Kim, 1977; Labuza, 1980b; Labuza and Saltmarsh, 1981; Satterlee and Chang; 1982; Petriella et al., 1985.

resentative kinetic data for various reactions in food proteins and the wide variation in time units reemphasizes the heterogenity of these events on a temporal scale.

Another example provides kinetic constants determined for in situ muscle myosin and for isolated bean and oat globulins (Table 16.16). Heating rate had no significant effect on the constants, although other data (not shown) indicated a general tendency for denaturation temperature to increase with heating rate. For both proteins it was possible to represent denaturation as a single-step reaction with an apparent reaction order above 1; high ($R^2 > 0.99$) correlations obtained from the Arrhenius plots are an indication of the goodness of fit of the kinetic model to the thermal scans. Although thermal denaturation is often referred to as an apparent first-order reaction (Myers, 1990), the larger reaction order seen here may be a consequence of the unsuitability of the two-state model because of intermediate reactions, dissociation of these polymeric proteins prior to denaturation as observed by Deshpande and Damodaran (1989) or aggregation.

Brown and Yada (1991) investigated the effect of endogenous carbohydrate on protein denaturation and thermal stability in an aspartic glycoenzyme from *Mucor miehei*. Results from plots of isothermal denaturation indicated that both *M. miehei* proteinase and a carbohydrate-reduced *M. miehei* enzyme (37% of the carbohydrate was removed through periodate oxidation) demonstrated first-order, two-state behavior. Subsequent measurements of activation energies for denaturation of the proteinases indicated that deglycosylation of *M. miehei* proteinase brought about a substantial destabilization (Table 16.17). Similar trends were noted for equilibrium studies in the presence of guanidine hydrochloride, where calculation of apparent thermodynamic parameters indicated that a reduction in the carbohydrate of *M. miehei* proteinase resulted in a decrease in stability. Data such as these are useful in relating

Table 16.16 Kinetic Parameters for Thermal Transitions in Food Proteins Determined by DSC at a Heating Rate of 10°C/Min

Protein	Conditions	T_d (°C)	E_a (kJ/mol)	n	log Z (min^{-1})
in situ Beef myosin	Aged 4 d, 2.4 μm	57	824	1.35	131
Bean 7S protein	Isolated	110	1050	2.58	144
Oat globulin protein	Isolated	114	505	2.50	69

Source: Adapted from Findlay et al., 1986; Hohlberg and Stanley, 1987b; Ma and Harwalkar, 1988b.

Table 16.17 Kinetic and Apparent Thermodynamic Parameters for the Thermal Denaturation of *M. miehei* Proteinases and Deglycosylated Proteinases

		Proteinase	
Parameter	Method	*M. miehei*	*M. miehei* (deglycosylated)
E_a (kJ/mol)	Enzyme activity	458 ± 41	339 ± 30
	UV spectroscopy	437 ± 12	370 ± 19
ΔF (kJ/mol)	Enzyme activity	55.3 ± 0.8	28.6 ± 1.0
	UV spectroscopy	41.6 ± 0.6	—

Source: Adapted from Brown and Yada, 1991.

important functional properties such as thermal stability to compositional effects, in this case carbohydrate content. They also have practical significance, since they suggest a new route for the thermal destabilization of this enzyme that would then allow it to be used in cheesemaking.

Another overall application of kinetics is to gain insight into the structural features of the reaction in question and to use this information to determine the chemical mechanism(s) involved. This is an extremely difficult task, yet some advances have been made by biochemical investigators under controlled laboratory conditions, although applications to food systems are generally lacking. The works of Tanford (1961) and Cleland (1977) may be consulted, while Lapanje (1978) covers the kinetics of protein denaturation from a mechanistic viewpoint.

The general importance of thermodynamics and kinetics in the study of food proteins can hardly be disputed. Often these parameters can be derived from the same data. Caution must be exercised, however, in the application of these values, if only because the assumptions of reversibility and two-state denaturation are rarely met under the conditions and concentrations used in food systems.

CONCLUSIONS

This chapter has attempted to collate and integrate information obtained from both protein biochemists and food protein chemists as it relates to the physical properties of thermal reactions in food protein systems, especially protein denaturation. Vital to the optimal use and prediction of functionality in food-related proteins is a better understanding of the nature of the relationships between structure and function. Whereas substantial knowledge has been gained towards this goal from protein biochemists, the information obtained must be applied with some caution. For the protein biochemist, proteins are studied in isolation and under conditions where protein-solvent interactions are encouraged. This is in direct contrast to the food scientist, who is often faced with situations in which the protein component is only one of several ingredients in a complex food system where a multitude of interactions may occur. Many food proteins are oligomeric and concentrations in foods are such that protein-protein interactions are often promoted. These conditions have made it difficult for the food scientist to provide an accurate mechanistic explanation of the effects of heat on food protein functionality. Despite these difficulties, concepts and principles of protein structure, protein domains, hydrophobicity, thermodynamics, and kinetics of protein unfolding, first put forth by protein biochemists, are being adopted by food scientists to help explain and understand such functional properties as gelation, thereby avoiding the empiricism that has existed in the past. In addition, an increase in refined techniques to quantitatively measure both structure and/or functional parameters has contributed to this effort. The ability to accurately predict functionality based on structural parameters, once thought to be unachievable, is now within sight.

ACKNOWLEDGMENTS

Original research described in this chapter was supported by the Natural Sciences and Engineering Research Council, the Ontario Ministry of Agriculture and Food and The International Development Research Centre.

REFERENCES

Aboagye, Y. and Stanley, D. W. 1985. *Can Inst. Food Sci. Technol. J.* 18: 12.

Aguilera, J. M. and Stanley, D. W. 1990. *Microstructural Principles of Food Processing and Engineering* Elsevier Applied Science, London.

Ahern, T. J. and Klibanov, A. M. 1985. *Science* 228: 1280.

Anfinsen, C. B. 1973. *Science* 181: 223.

Anfinsen, C. B. and Scheraga, H. A. 1975. *Adv. Prot. Chem.* 29: 205.

Anwar, R. A., Gerber, G. E., and Baig, K. M. 1977. In *Protein Cross-linking. Nutritional and Medical Consequences,* M. Friedman (Ed.), p. 709. Plenum Press, New York.

Arai, S. 1980. In *Chemical Deterioration of Proteins* J. R. Whitaker and M. Fujimaki (Ed.), p. 195. American Chemical Society, Washington, DC.

Armstrong, D. L., Maurice, T. J., and Stanley, D. W. 1979. In *Functionality and Protein Structure,* A. Pour-El (Ed.), p. 147. American Chemical Society, Washington, DC.

Arntfield, S. D. and Murray, E. D. 1981. *Can. Inst. Food Sci. Technol. J.* 14: 289.

Arntfield, S. D., Ismond, M. A. H., and Murray, E. D. 1990a. In *Thermal Analysis of Foods,* V. R. Harwalkar and C.-Y. Ma (Ed.), p. 51. Elsevier Applied Science, London.

Arntfield, S. D., Murray, E. D., and Ismond, M. A. H. 1990b. *J. Agric. Food Chem.* 38: 1335.

Arntfield, S. D., Murray, E. D., Ismond, M. A. H., and Bernatsky, A. M. 1989. *J. Food Sci.* 54: 1624.

Babajimopoulos, M., Damodaran, S., Rizvi, S. S. H., and Kinsella, J. E. 1983. *J. Agric. Food Chem.* 31: 1270.

Bacon, J. R., Noel, T. R., and Wright, D. J. 1989. *J. Sci. Food Agric.* 49: 335.

Bell, J. E. and Bell, E. T. 1988. *Proteins and Enzymes.* Prentice-Hall, Inc., Englewood Cliffs, NJ.

Benzer, S. 1957. In *The Chemical Basis of Heredity,* W. D. McElroy and B. Glass (Ed.), pp. 70–93. Johns Hopkins University Press, Baltimore, MD.

Beveridge, T., Arntfield, S. D., and Murray, E. D. 1985. *Can. Inst. Food Sci. Technol. J.* 18: 189.

Beveridge, T., Jones, L., and Tung, M. A. 1984. *J. Agric. Food Chem.* 32: 307.

Bigelow, C. C. 1967. *J. Theoret. Biol.* 16: 187.

Bikbov, T., Grinberg, V., Danilenko, A., Chaika, T., Vaintraub, I., and Tolstoguzov, V. 1985. *J. Agric. Food Chem.* 33: 912.

Biliaderis, C. G. 1983. *Food Chem.* 10: 239.

Biltonen, R. L. and Freire, E. 1978. *CRC Crit. Rev. Biochem.* 5: 85.

Borchardt, H. J. and Daniels, F. 1957. *J. Am. Chem. Soc.* 79: 41.

Brandts, J. F. 1964. *J. Am. Chem. Soc.* 86: 4302.

Brandts, J. F. 1969. In *Structure and Stability of Biological Macromol-*

ecules, (Ed.) S. N. Timasheff and G. D. Fasman, p. 213. Marcel Dekker, Inc., New York.

Brown, E. D. and Yada, R. Y. 1991. *Biochem. Biophys. Acta* 1076: 406.

Buck, G. C. 1984. *Food Technol.* 38(5): 59.

Catsimpoolas, N. and Meyer, E. W. 1970. *Cereal Chem.* 47: 559.

Chambers, S. J., Carr, H. J., and Lambert, N. 1990. *Biochim. Biophys. Acta* 1037: 66.

Cheftel, J. C., Cuq, J.-L., and Lorient, D. 1985. In *Food Chemistry,* 2nd ed., O. R. Fennema (Ed.), p. 245. Marcel Dekker, Inc., New York.

Chen, M. C. and Lord, R. C. 1976. *Biochemistry* 15: 1889.

Chothia, C. 1984. *Ann. Rev. Biochem.* 53: 537.

Clark, A. H. and Lee-Tuffnell, C. D. 1986. In *Functional Properties of Food Macromolecules,* J. R. Mitchell and D. A. Ledward (Ed.), p. 203. Elsevier Applied Science Publishers, London.

Cleland, W. W. 1977. *Adv. Enzymol.* 45: 273.

Colvin, J. R. 1964. In *Proteins and Their Reactions,* H. W. Schultz and A. F. Anglemier (Ed.), p. 69. AVI Publishing Co., Westport, CT.

Damodaran, S. 1988. *J. Agric. Food Chem.* 36: 262.

Damodaran, S. 1989. In *Food Proteins,* J. E. Kinsella and W. G. Soucie (Ed.), p. 21. American Oil Chemists' Society, Champaign, IL.

Danilenko, A. N., Grozav, E. K., Bikbov, T. M., Grinberg, V. Y., and Tolstoguzov, V. B. 1985a. *Int. J. Biol. Macromol.* 7: 109.

Danilenko, A. N., Rogova, E. I., Bikbov, T. M., Grinberg, V. Y., and Tolstoguzov, V. B. 1985b. *Int. J. Peptide Protein Res.* 26: 5.

Davis, E. A. and Gordon, J. 1984. *Food Technol.* 38(5): 99.

Deshpande, S. S. and Damodaran, S. 1989. *Biochim. Biophys. Acta* 998: 179.

Edel, G. M., Cunningham, B. A., Gall, W. E., Gottlieb, P. D., Rutishauser, U., and Waxdal, M. J. 1969. *Proc. Natl. Acad. Sci. USA* 63: 78.

Edwards, S. F., Lillford, P. J., and Blanshard, J. M. V. 1987. In *Food Structure and Behaviour,* J. M. V. Blanshard and P. Lillford (Ed.), p. 1. Academic Press, London.

Feeney, R. E. 1980. In *Chemical Deterioration of Proteins,* J. R. Whitaker and M. Fujimaki (Ed.), p. 1. American Chemical Society, Washington, DC.

Feeney, R. E. and Whitaker, J. R. 1982. In *Food Protein Deterioration. Mechanisms and Functionality,* J. P. Cherry (Ed.), p. 201. American Chemical Society, Washington, DC.

Feeney, R. E., Blankenhorn, G., and Dixon, H. B. F. 1975. *Adv. Prot. Chem.* 29: 135.

Ferry, J. D. 1948. *Adv. Protein Chem.* 4: 1.

Findlay, C. J. and Barbut, S. 1990. In *Thermal Analysis of Foods,* V. R. Harwalkar and C.-Y. Ma (Ed.), p. 92. Elsevier Applied Science, London.

Findlay, C. J., Parkin, K. L., and Stanley, D. W. 1986. *J. Food Biochem.* 10: 1.

Finot, P. A. 1982. In *Modification of Proteins. Food, Nutritional and Pharmacological Aspects,* R. E. Feeney and J. R. Whitaker (Ed.), p. 91. American Chemical Society, Washington, DC.

Fischer, G. and Schmid, F. X. 1990. *Biochemistry* 29: 2205.

Fligner, K. L. and Mangino, M. E. 1991. In *Interactions of Food Proteins,* N. Parris and R. Barford (Ed.), p. 1. American Chemical Society, Washington, DC.

Friedman, M. 1977. In *Protein Crosslinking. Nutritional and Medical Consequences,* M. Friedman (Ed.), p. v. Plenum Press, New York.

Friedman, M. 1982. In *Food Protein Deterioration. Mechanisms and Functionality,* J. P. Cherry (Ed.), p. 231. American Chemical Society, Washington, DC.

Frushour, B. G. and Koenig, J. L. 1975. In *Advances in Infrared and Raman Spectroscopy,* Vol. 1, R. J. H. Clark and R. E. Hester (Ed.), p. 35. Heyden and Son, Ltd., London.

Fujita, Y. and Noda, Y. 1981. *Int. J. Peptide Protein Res.* 18: 12.

Fukushima, D. 1980. In *Chemical Deterioration of Proteins,* J. R. Whitaker and M. Fujimaki (Ed.), p. 237. American Chemical Society, Washington, DC.

Garcia-Vela, L. A. and Stanley, D. W. 1989. *J. Food Sci.* 54: 1284.

German, B., Damodaran, S., and Kinsella, J. E. 1982. *J. Agric. Food Chem.* 30: 807.

Grinberg, V. Y., Danilenko, A. N., Burova, T. V., and Tolstoguzov, V. B. 1989. *J. Sci. Food Agric.* 49: 235.

Hamann, D. D. 1987. *Food Technol.* 41(3): 100.

Hamann, D. D. 1991. In *Interactions of Food Proteins,* N. Parris and R. Barford (Ed.), p. 212. American Chemical Society, Washington, DC.

Hamann, D. D., Purkayastha, S., and Lanier, T. C. 1990. In *Thermal Analysis of Foods,* V. R. Harwalkar and C.-Y. Ma (Ed.), p. 306. Elsevier Applied Science, London.

Harris, P. (Ed.). 1990. *Food Gels,* p. 478. Elsevier Applied Science, London.

Harwalkar, V. R. and Ma, C.-Y. 1987. *J. Food Sci.* 52: 394.

Hägerdal, B. and Martens, H. 1976. *J. Food Sci.* 41: 933.

Hermansson, A.-M. 1978. *J. Texture Studies* 9: 33.

Hermansson, A.-M. 1979. In *Functionality and Protein Structure,* A. Pour-El (Ed.), p. 81. American Chemical Society, Washington, DC.

Hermansson, A.-M. 1986a. *JAOCS* 63: 658.

Hermansson, A.-M. 1986b. In *Functional Properties of Food Macromolecules,* J. R. Mitchell and D. A. Ledward (Ed.), p. 273. Elsevier Applied Science Publishers, London.

Hermansson, A.-M. 1988. In *Food Structure—Its Creation and Evaluation,* J. M. V. Blanshard and J. R. Mitchell (Ed.), p. 25. Butterworths, London.

Hermansson, A.-M., Harbitz, O., and Langton, M. 1986. *J. Sci. Food Agric.* 37: 69.

Hirotsuka, M. and Nakai, S. 1991. *J. Agric. Food Chem.* in press.

Hirotsuka, M., Taniguchi, H., Narita, H., and Kito, M. 1984. *J. Food Sci.* 49: 1105.

Hohlberg, A. I. and Stanley, D. W. 1987a. *J. Agric. Food Chem.* 35: 571.

Hohlberg, A. I. and Stanley, D. W. 1987b. *J. Food Process. Preserv.* 11: 31.

Hultin, H. O. 1985. In *Food Chemistry,* 2nd ed., O. R. Fennema (Ed.), p. 725. Marcel Dekker, Inc., New York.

Hurrell, R. F. and Carpenter, K. J. 1977. In *Protein Crosslinking. Nutritional and Medical Consequences,* M. Friedman (Ed.), p. 225. Plenum Press, New York.

IFT. 1984. *Gelation in food protein systems. Food Technol.* 38(5): 59–109.

Ikai, A. 1980. *J. Biochem.* 88: 1895.

Iwabuchi, S. and Yamauchi, F. 1984. *J. Food Sci.* 49: 1289.

Jacks, T. J., Neucere, N. J., and McCall, E. R. 1975. *Int. J. Peptide Protein Res.* 7: 153.

Jackson, W. M. and Brandts, J. F. 1970. *Biochemistry* 9: 2294.

Joly, M. 1965. *A Physico-chemical Approach to the Denaturation of Proteins.* Academic Press, London.

Kamata, Y. and Kinsella, J. E. 1989. *J. Food Sci.* 54: 170.

Kamata, Y., Okubo, K., and Shibasaki, K. 1979. *Agric. Biol. Chem.* 43: 1219.

Kamata, Y., Rector, D., and Kinsella, J. E. 1988. *J. Food Sci.* 53: 589.

Kato, A. and Nakai, S. 1980. *Biochim. Biophys. Acta* 624: 13.

Kato, Y., Watanabe, K., and Sato, Y. 1981. *J. Food Sci.* 46: 1835.

Kauzmann, W. 1959. *Adv. Prot. Chem.* 14: 1.

Kilara, A. and Sharkasi, T. Y. 1986. *CRC Crit. Rev. Food Sci. Technol.* 23: 323.

Kirschner, K. and Bisswanger, H. 1976. *Ann. Rev. Biochem.* 45: 143.

Kinsella, J. E. 1982. In *Food Protein Deterioration. Mechanisms and Functionality,* J. P. Cherry (Ed.), p. 301. American Chemical Society, Washington, DC.

Klotz, I. M., Darnall, D. W., and Langerman, N. R. 1975. In *The Proteins,* 3rd ed., Vol. 1. H. Neurath and R. L. Hill (Ed.), p. 293. Academic Press, New York.

Koshiyama, I., Hamano, M., and Fukushima, D. 1980–81. *Food Chem.* 6: 309.

Labuza, T. P. 1980a. *Food Technol.* 34(4): 36.

Labuza, T. P. and Kamman, J. F. 1983. In *Computer-Aided Techniques in Food Technology,* I. Saguy (Ed.), p. 71. Marcel Dekker, Inc., New York.

Labuza, T. P. and Saltmarsh, M. 1981. *J. Food Sci.* 47: 92.

Labuza, T. P. 1980b. *Food Technol.* 34(2): 67.

Lambelet, P., Berrocal, R., and Ducret, F. 1989. *J. Dairy Res.* 56: 211.

Lapanje, S. 1978. *Physicochemical Aspects of Protein Denaturation,* John Wiley & Sons, New York.

Ledward, D. A. 1986. In *Functional Properties of Food Macromolecules,* J. R. Mitchell and D. A. Ledward (Ed.), p. 171. Elsevier Applied Science Publishers, London.

Ledward, D. A. and Mitchell, J. R. 1988. In *Food Structure—Its Creation and Evaluation,* J. M. V. Blanshard and J. R. Mitchell (Ed.), p. 219. Butterworths, London.

Lehninger, A. L. 1973. *Short Course in Biochemistry,* p. 3. Worth Publishers, Inc., New York.

Li-Chan, E. and Nakai, S. 1989. In *Food Proteins,* J. E. Kinsella and W. G. Soucie (Ed.), p. 232. American Oil Chemists' Society, Champaign, IL.

Lumry, R. and Biltonen, R. 1969. In *Structure and Stability of Biological Macromolecules,* S. N. Timasheff and G. D. Fasman (Ed.), p. 65. Marcel Dekker, Inc., New York.

Lund, D. B. 1983. In *Physical Properties of Foods,* M. Peleg and E. B. Bagley (Ed.), p. 125. AVI Publishing Co, Inc., Westport, CT.

Ma, C.-Y. and Harwalkar, V. R. 1984. *J. Agric. Food Chem.* 32: 144.

Ma, C.-Y. and Harwalkar, V. R. 1987. *Cereal Chem.* 64: 212.

Ma, C.-Y. and Harwalkar, V. R. 1988a. *J. Agric. Food Chem.* 36: 155.

Ma, C.-Y. and Harwalkar, V. R. 1988b. *J. Food Sci.* 53: 531.

Ma, C.-Y., Harwalkar, V. R., and Maurice, T. J. 1990. In *Thermal Analysis of Foods,* V. R. Harwalkar and C.-Y. Ma (Ed.), p. 1. Elsevier Applied Science, London.

Ma, C.-Y. (1990). In *Thermal Analysis of Foods,* V. R. Harwalkar and C.-Y. Ma (Ed.), p. 149. Elsevier Applied Science, London.

Matsudomi, N., Mori, H., Kato, A., and Kobayashi, K. 1985. *Agric. Biol. Chem.* 49: 915.

Mechanic, G. L. 1977. In *Protein Crosslinking. Nutritional and Medical Consequences,* M. Friedman (Ed.), p. 699. Plenum Press, New York.

Mendelsohn, R. 1986. In *Spectroscopy in the Biomedical Sciences,* R. M. Gendreaum (Ed.), p. 151. CRC Press, Inc., Boca Raton, FL.

Merkler, D. J., Farrington, G. K., and Wedler, F. C. 1981. *Int. J. Peptide Protein Res.* 18: 430.

Mitchell, J. R. 1980. *J. Texture Studies* 11: 315.

Mori, T., Nakamura, T., and Utsumi, S. 1981. *J. Food Sci.* 47: 26.

Mori, T., Nakamura, T., and Utsumi, S. 1986. *J. Agric. Food Chem.* 34: 33.

Mozhaev, V. V. and Martinek, K. 1984. *Enz. Microb. Technol.* 6: 50.

Murray, E. D., Myers, C. D., Barker, L. D., and Maurice, T. J. 1981. In *Utilization of Protein Resources,* D. W. Stanley, E. D. Murray, and D. H. Lees (Ed.), p. 158. Food and Nutrition Press, Inc., Westport, CT.

Myers, C. D. (1990). In *Thermal Analysis of Foods,* V. R. Harwalkar and C.-Y. Ma (Ed.), p. 16. Elsevier Applied Science, London.

Nakai, S. and Li-Chan, E. 1988. *Hydrophobic Interactions in Food Systems.* CRC Press, Inc., Boca Raton, FL.

Nakai, S., Li-Chan, E., Hirotsuka, M., Vazquez, M. C., and Arteaga, G. 1991. In *Interactions of Food Proteins,* N. Parris and R. Barford (Ed.), p. 42. American Chemical Society, Washington, DC.

Nakamura, T., Utsumi, S., and Mori, T. 1986. *Agric. Biol. Chem.* 50: 1287.

Neucere, J. N. and Cherry, J. P. 1982. In *Food Protein Deterioration. Mechanisms and Functionality,* J. P. Cherry (Ed.), p. 135. American Chemical Society, Washington, D.C.

Noguchi, A. 1989. In *Extrusion Cooking,* C. Mercier, P. Linko, and J.

M. Harper (Ed.), p. 353. American Association of Cereal Chemists, Inc., St. Paul, MN.

Noh, B. S. and Richardson, T. 1989. In *Food Proteins,* J. E. Kinsella and W. G. Soucie (Ed.), p. 267. The American Oil Chemists' Society, Champaign, IL.

O'Brien, J. and Morrissey, P. A. 1989. *CRC Crit. Rev. Food Sci. Nutr.* 28: 211.

Oakenfull, D. 1984. *J. Food Sci.* 49: 1103.

Oakenfull, D. 1987. *CRC Crit. Rev. Food Sci. Nutr.* 26: 1.

Otterburn, M., Healy, M., and Sinclair, W. 1977. In *Protein Crosslinking. Nutritional and Medical Consequences,* M. Friedman (Ed.), p. 239. Plenum Press, New York.

Painter, P. C. 1984. In *Food Analysis. Principles and Techniques,* Vol. 2, *Physicochemical Techniques,* D. W. Gruenwedel and J. R. Whitaker (Ed.), p. 511. Marcel Dekker, Inc., New York.

Paulson, A. T. and Tung, M. A. 1989. *J. Agric. Food Chem.* 37: 319.

Peticolas, W. L. 1984. In *Spectroscopy of Biological Molecules,* C. Sandorfy and T. Theophanides (Ed.), p. 153. D. Reidel Publishing Co., New York.

Petriella, C., Resnik, S. L., Lozano, R. D., and Chifire, J. 1985. *J. Food Sci.* 50: 622.

Plhak, L. C., Caldwell, K. B., and Stanley, D. W. 1989. *J. Food Sci.* 54: 326.

Prakash, V. 1986. *Int. J. Peptide Protein Res.* 28: 192.

Privalov, P. L. 1979a. *NETSU* 6(2): 56.

Privalov, P. L. 1979b. *Adv. Prot. Chem.* 33: 167.

Privalov, P. L. 1982. *Adv. Prot. Chem.* 35: 1.

Privalov, P. L. and Khechinashvili, N. N. 1974. *J. Mol. Biol.* 86: 665.

Privalov, P. L. and Gill, S. J. 1988. *Adv. Prot. Chem.* 39: 191.

Rao, M. A. and Lund, D. B. 1986. *J. Food Process. Preserv.* 10: 311.

Rashin, A. A. 1984. *Biopolymers* 23: 1605.

Rector, D. J., Kella, N. K., and Kinsella, J. E. 1989. *J. Texture Studies* 20: 457.

Rha, C. and Pradipasena, P. 1986. In *Functional Properties of Food Macromolecules,* J. R. Mitchell, and D. A. Ledward (Ed.), p. 79. Elsevier Applied Science Publishers, London.

Rockland, L. B. and Nishi, S. K. 1980. *Food Technol.* 34(4): 42.

Sandu, C. and Lund, D. 1985. *Thermochim. Acta* 88: 453.

Sandu, C. and Lund, D. 1986. In *Food Engineering and Process Applications,* M. LeMaguer and P. Jelen (Ed.), p. 301. Elsevier Applied Science Publishers, London.

Sathe, S. K., Mason, A. C. and Weaver, C. M. 1989. *J. Food Sci.* 54: 319.

Satterlee, L. D. and Chang, K.-C. 1982. In *Food Protein Deterioration. Mechanisms and Functionality,* J. P. Cherry (Ed.), p. 415. American Chemical Society, Washington, DC.

Schmidt, R. H. and Morris, H. A. 1984. *Food Technol.* 38(5): 85.

Schulz, G. E. and Schirmer, R. H. 1979. *Principles of Protein Structure.* Springer-Verlag, New York.

Schulz, G. E., Schirmer, R. H., Sachsenheimer, W., and Pai, E. F. 1978. *Nature* 273: 120.

Schwass, D. E. and Finley, J. W. 1984. *J. Agric. Food Chem.* 32: 1377.

Schwimmer, S. 1981. *Source Book of Food Enzymology.* AVI Publishing Co., Westport, CT.

Sefa-Dedeh, S., Stanley, D. W., and Voisey, P. W. 1978. *J. Food Sci.* 43: 1832.

Sheard, P. R., Fellows, A., Ledward, D. A., and Mitchell, J. R. 1986. *J. Food Technol.* 21: 55.

Shiga, K. and Nakamura, Y. 1987. *J. Food Sci.* 52: 681.

Shimada, K. and Cheftel, J. C. 1988. *J. Agric. Food Chem.* 36: 147.

Stanley, D. W. 1989. In *Extrusion Cooking,* C. Mercier, P. Linko, and J. M. Harper (Ed.), p. 325. American Association of Cereal Chemists, Inc., St. Paul, MN.

Stanley, D. W. 1991. *CRC Crit. Rev. Food Sci. Technol.* 30: 487.

Stanley, D. W. and Aguilera, J. M. 1985. *J. Food Biochem.* 9: 277.

Stanley, D. W., Wu, X. and Plhak, L. C. 1989. *J. Texture Studies* 20: 419.

Steim, J. M. 1965. *Arch. Biochem. Biophys.* 112: 599.

Sternberg, M. and Kim, C. Y. 1977. In *Protein Crosslinking. Nutritional and Medical Consequences,* M. Friedman (Ed.), p. 73. Plenum Press, New York.

Stimler, N. P. and Tanzer, M. L. 1977. In *Protein Crosslinking. Nutritional and Medical Consequences,* M. Friedman (Ed.), p. 675. Plenum Press, New York, NY.

Sturtevant, J. M. 1980. In *Bioenergetics and Thermodynamics: Model Systems,* A. Braibanti (Ed.), p. 391. Reidel Publishing Co., New York.

Swaisgood, H. E. 1985. In *Food Chemistry,* 2nd ed., O. R. Fennema (Ed.), p. 791. Marcel Dekker, Inc., New York.

Tanaka, M., Kimiagar, M., Lee, T.-C., and Chichester, C. O. 1977. In *Protein Crosslinking. Nutritional and Medical Consequences,* M. Friedman (Ed.), p. 321. Plenum Press, New York.

Tanford, C. 1961. *Physical Chemistry of Macromolecules.* John Wiley & Sons, Inc., New York.

Tanford, C. 1968. *Adv. Protein Chem.* 23: 121.

Tanford, C. 1970. *Adv. Protein Chem.* 24: 1.

Tanford, C. 1980. *The Hydrophobic Effect. Formation of Micelles and Biological Membranes,* 2nd ed. John Wiley & Sons, Inc., New York.

Tombs, M. P. 1970. In *Proteins as Human Food,* R. A. Lawrie (Ed.), p. 126. The AVI Publishing Co., Inc. Westport, CT.

Tombs, M. P. 1974. *Faraday Discuss. Chem. Soc.* 57: 158.

Tu, A. T. 1986. In *Spectroscopy of Biological Systems,* R. J. H. Clark and R. E. Hester (Ed.), p. 47. John Wiley & Sons, Ltd., New York.

Tung, M. A. 1978. *J. Texture Studies* 9: 3.

Utsumi, S. and Kinsella, J. E. 1985a. *J. Food Sci.* 50: 1278.

Utsumi, S. and Kinsella, J. E. 1985b. *J. Agric. Food Chem.* 33: 297.

Utsumi, S., Damodaran, S., and Kinsella, J. E. 1984. *J. Agric. Food Chem.* 32: 1406.

Voutsinas, L. P., Cheung, E., and Nakai, S. 1983. *J. Food Sci.* 48: 26.

Waller, G. R. and Feather, M. S. (Ed.) 1983. *The Maillard Reaction in Foods and Nutrition.* American Chemical Society, Washington, DC.

Whitaker, J. R. 1977. In *Food Proteins,* J. R. Whitaker and S. R. Tannenbaum (Ed.), p. 14. AVI Publishing Co., Westport, CT.

Whitaker, J. R. 1980. In *Chemical Deterioration of Proteins,* J. R. Whitaker and M. Fujimaki (Ed.), p. 145. American Chemical Society, Washington, DC.

Wright, D. J. 1982. In *Developments in Food Proteins*—1, B. J. F. Hudson (Ed.), p. 61. Applied Science Publishers Inc., Englewood, NJ.

Wu, Y. V. and Inglett, G. E. 1974. *J. Food Sci.* 39: 218.

Zale, S. E. and Klibanov, A.M. 1986. *Biochemistry* 25: 5432.

Zarins, Z. M. and Marshall, W. E. 1990. *Cereal Chem.* 67: 35.

Index